World Geomorphological Landscapes

Series editor

Piotr Migoń, Wroclaw, Poland

More information about this series at http://www.springer.com/series/10852

Alain Demoulin
Editor

Landscapes and Landforms of Belgium and Luxembourg

Springer

Editor
Alain Demoulin
Department of Physical Geography
and Quaternary
University of Liège
Liège
Belgium

ISSN 2213-2090 ISSN 2213-2104 (electronic)
World Geomorphological Landscapes
ISBN 978-3-319-58237-5 ISBN 978-3-319-58239-9 (eBook)
DOI 10.1007/978-3-319-58239-9

Library of Congress Control Number: 2017939306

Printed on acid-free paper

This Springer imprint is published by Springer Nature
The registered company is Springer International Publishing AG
The registered company address is: Gewerbestrasse 11, 6330 Cham, Switzerland

Series Editor Preface

Landforms and landscapes vary enormously across the Earth, from high mountains to endless plains. At a smaller scale, nature often surprises us creating shapes which look improbable. Many physical landscapes are so immensely beautiful that they received the highest possible recognition—they hold the status of World Heritage Sites. Apart from often being immensely scenic, landscapes tell stories which not uncommonly can be traced back in time for tens of million years and include unique geological events such as meteorite impacts. In addition, many landscapes owe their appearance and harmony not solely to the natural forces. For centuries, and even millennia, they have been shaped by humans who have modified hillslopes, river courses, and coastlines, and erected structures which often blend with the natural landforms to form inseparable entities.

These landscapes are studied by geomorphology—'the science of scenery'—a part of Earth Sciences that focuses on landforms, their assemblages, surface and subsurface processes that moulded them in the past and that change them today. To show the importance of geomorphology in understanding the landscape, and to present the beauty and diversity of the geomorphological sceneries across the world, we have launched a book series *World Geomorphological Landscapes*. It aims to be a scientific library of monographs that present and explain physical landscapes, focusing on both representative and uniquely spectacular examples. Each book will contain details on geomorphology of a particular country or a geographically coherent region. This volume presents the geomorphology of Belgium and Luxembourg—two adjacent European countries which may seem too small, monotonous and inconspicuous to afford a separate volume in the series. The authors show us that this is not the case. Quite to the contrary, we have an opportunity to learn that subdued lowland and upland terrains may have fascinating histories that go millions of years back in time. Belgium and Luxembourg host classic examples of incised ancient plateaus, deep weathering, picturesque sandstone formations, inherited periglacial landforms, spectacular karst and recent faulted escarpments due to intraplate tectonics. There is a lot to read about!

The World Geomorphological Landscapes series is produced under the scientific patronage of the International Association of Geomorphologists (IAG)—a society that brings together geomorphologists from all around the world. The IAG was established in 1989 and is an independent scientific association affiliated with the International Geographical Union (IGU) and the International Union of Geological Sciences (IUGS). Among its main aims are to promote geomorphology and to foster dissemination of geomorphological knowledge. I believe that this lavishly illustrated series, which keeps to the scientific rigour, is the most appropriate means to fulfil these aims and to serve the geoscientific community. To this end, my great thanks go to Prof. Alain Demoulin for adding this book to his busy agenda, successfully coordinating the large team of authors, and delivering such an exciting illustrated story to read and admire. I also acknowledge the excellent work of all individual authors who accepted to share their expert knowledge of their countries with the global geomorphological community. On a more personal note, I was once privileged to visit some geomorphic landscapes of Belgium under the expert guidance of the late Prof. Albert Pissart. He was one of the great supporters of the IAG when the Association matured and would surely be pleased that his country is now so nicely presented within the IAG-endorsed book series.

Wroclaw, Poland Piotr Migoń

Contents

Contributors

Cecile Baeteman Quaternary Environments and Humans, Royal Belgian Institute of Natural Sciences, Brussels, Belgium

François Barbier Global Operations and Components at Flextronics International Ltd, Singapore, Singapore

Koen Beerten Engineered and Geosystems Analysis, Belgian Nuclear Research Centre SCK•CEN, Mol, Belgium

Dominique Bonjean Scladina Cave Archaeological Centre, Sclayn-Andenne, Belgium

Frédéric Boulvain Pétrologie Sédimentaire, Université de Liège, Liège, Belgium

Nils Broothaerts Department Earth and Environmental Sciences, Division of Geography and Tourism, KU Leuven, Leuven–Heverlee, Belgium

Thierry Camelbeeck Section of Seismology—Gravimetry, Royal Observatory of Belgium, Brussels, Belgium

Jozef Deckers Division of Soil and Water Management Department of Earth and Environmental Sciences, KU Leuven, Leuven–Heverlee, Belgium

Augustin Dekoninck Department of Geology, University of Namur, Namur, Belgium

Alain Demoulin Department of Physical Geography and Quaternary, University of Liège, Liège, Belgium

Olivier Dewitte Department of Earth Sciences, Royal Museum for Central Africa, Tervuren, Belgium

Kévin Di Modica Scladina Cave Archaeological Centre, Sclayn-Andenne, Belgium

Roland Dreesen Geological Survey of Belgium, Royal Belgian Institute of Natural Sciences, Brussels, Belgium

Christian Dupuis Géologie Fondamentale et Appliquée, University of Mons, Mons, Belgium

Paul De Smedt Heverlee, Belgium

Michiel Dusar Geological Survey of Belgium, Royal Belgian Institute of Natural Sciences, Brussels, Belgium

Camille Ek Université de Liège, Liège, Belgium

Michel Erpicum Department of Geography, University of Liège, Liège, Belgium

Vincent Hallet Department of Geology, University of Namur, Namur, Belgium

Eric Hallot ISSeP, Liège, Belgium

Irénée Heyse Department of Geography, University of Ghent, Ghent, Belgium

Geoffrey Houbrechts Department of Physical Geography and Quaternary, University of Liège, Liège, Belgium

Rik Houthuys Geoconsultant, Halle, Belgium

Jos Janssen Werkgroep Geologie, LIKONA (Limburgse Koepel Voor Natuurstudie), Provinciaal Natuurcentrum, Genk, Belgium

Etienne Juvigné University of Liège, Liège, Belgium

Birgit Kausch Service Géologique du Luxembourg/Geological Survey of Luxembourg, Bertrange, Luxembourg

Robert Maquil Retired Diekirch, Luxembourg

Tom Maris Department of Biology, University of Antwerp, Wilrijk, Belgium

Johan Matthijs VITO, Mol, Belgium

Patrick Meire Department of Biology, University of Antwerp, Wilrijk, Belgium

Bastiaan Notebaert Department Earth and Environmental Sciences, Division of Geography and Tourism, KU Leuven, Leuven–Heverlee, Belgium; FWO Fund for Scientific Research, Flanders, Belgium

Myriem Nouri Department of Geography, University of Liège, Liège, Belgium

André Ozer University of Liège, Liège, Belgium

Alexandre Peeters Department of Physical Geography and Quaternary, University of Liège, Liège, Belgium

François Petit Department of Physical Geography and Quaternary, University of Liège, Liège, Belgium

Stéphane Pirson Service public de Wallonie, Jambes, Belgium

Jean Poesen Department of Earth and Environmental Sciences, KU Leuven, Leuven-Heverlee, Belgium

Yves Quinif Géologie fondamentale et appliquée, University of Mons, Mons, Belgium

Gilles Rixhon Institute of Geography, University of Cologne, Cologne, Germany

Joël Rodet UMR 6143 CNRS, Laboratory of Geology, University of Rouen, Mont Saint Aignan, France

Gilles Ruffet Géosciences Rennes, Rennes Cedex, France

Lennert Schepers Department of Biology, University of Antwerp, Wilrijk, Belgium

Stijn Temmerman Department of Biology, University of Antwerp, Wilrijk, Belgium

Michel Toussaint Retired Ouffet, Belgium

Jean Van Campenhout Department of Physical Geography and Quaternary, University of Liège, Liège, Belgium

Miet Van Den Eeckhaut Arcadis Belgium, Brussels, Belgium

Dany Van Uytven Werkgroep Geologie, LIKONA (Limburgse Koepel Voor Natuurstudie), Provinciaal Natuurcentrum, Genk, Belgium

Jef Vandenberghe Department of Earth Sciences, VU Amsterdam, Amsterdam, The Netherlands

Noël Vandenberghe Department Earth and Environmental Sciences, Division of Geology, Leuven, KU Leuven, Leuven-Heverlee, Belgium

Kris Vanneste Section of Seismology—Gravimetry, Royal Observatory of Belgium, Brussels, Belgium

Tom Vanwalleghem Hidrología e Hidráulica Agrícola, Univeristy of Cordoba, Córdoba, Spain

Koen Verbeeck Section of Seismology—Gravimetry, Royal Observatory of Belgium, Brussels, Belgium

Michèle Verhaert Department of Geology, University of Namur, Namur, Belgium

Gert Verstraeten Department Earth and Environmental Sciences, Division of Geography and Tourism, KU Leuven, Leuven–Heverlee, Belgium

Luc Willems Pétrologie Sédimentaire, University of Liège, Liège, Belgium

Johan Yans Department of Geology, University of Namur, Namur, Belgium

Morphogenic Setting and Diversity of Processes and Landforms: The Geomorphological Regions of Belgium

1

Alain Demoulin

Abstract

Following a few words about the historical context of geomorphological research in Belgium and Luxembourg, the main geomorphic regions of the two countries are presented. The dominant controls exerted on morphogenic processes by lithology and elevation (i.e., tectonic background) lead to distinguish northern Belgium, corresponding to the low-elevation Cenozoic Belgian basin where sedimentation generally prevails over erosion, and the Ardennian Paleozoic massif of southern Belgium and northern Luxembourg, uplifted and strongly incised and eroded in the Plio-Quaternary. In between are the transitional plateaus of Middle Belgium. South of the Ardenne, Belgian Lorraine and Luxembourgian Gutland pertain to the homoclinal landscapes of the Paris basin. Finally, a brief overview is presented of the processes and landforms treated in the successive chapters.

Keywords

Geomorphological regions of Belgium and Luxembourg • Morphogenic processes • Geomorphology in Belgium

1.1 Introduction

At the moment of writing an introduction for this book about landscapes and landforms of Belgium and Luxembourg, I gathered ideas that I attempted to formulate as originally as possible, in order to attractively introduce the two countries. However, as I was hesitating about the best way to begin with my subject, I decided to have first a look at introductory texts in the already published books of the series. This led me to discover that, beyond large countries richly provided with magnificent landscapes, most of the smaller territories systematically claimed the same wide diversity of landscapes that I had imagined to make a hallmark of Belgium and Luxembourg's geomorphology. And it is true indeed that, whatever their size, countries more anonymous geomorphologically also contain a variety of landscapes and landforms inherited from the equitably distributed fourth dimension, namely time, which allowed contrasted climatic contexts and tectonic events to succeed each other and leave specific imprints along their long geomorphic history. Perhaps is the value of these landscapes not linked to their grandeur but other qualities such as rarity or excellent preservation of landforms, representativeness of important, though not spectacular, processes, or a very well-documented context of evolution may make them equally or even more interesting. The latter fact, i.e., a wealth of data of all kinds (geomorphology, geology, climate, vegetation, human interference) and extended studies, characterizes particularly Belgium where first valuable geological observations were made by naturalists as early as the second half of the eighteenth century (de Limbourg 1770, 1774) and first synthetic views on the physical geography of the country were published by Houzeau already in 1854.

Besides unveiling the most scenic landscapes of Belgium and Luxembourg, the main aim of the community of Belgian geomorphologists, grouped into the Belgian Association of Geomorphologists (BAG) and seconded by many colleague geologists, in writing this volume has been to produce a

A. Demoulin (✉)
Department of Physical Geography and Quaternary, University of Liège, Sart Tilman, B11, 4000 Liège, Belgium
e-mail: ademoulin@ulg.ac.be

A. Demoulin (ed.), *Landscapes and Landforms of Belgium and Luxembourg*, World Geomorphological Landscapes, DOI 10.1007/978-3-319-58239-9_1

reference book about geomorphic processes and the resulting landscapes and landforms in the two countries. Indeed, beyond texts that were devoted, mainly or partly, to physical geography, with geomorphological analysis limited at best to a few considerations about fluvial processes (Omalius d'Halloy 1828; Houzeau 1854; Fourmarier 1934), the single existing book dealing explicitly with the geomorphology of Belgium (Pissart 1976) is long outdated, this all the more as the last four decades saw the numbers of geomorphic studies, data sets, new results and ideas increase exponentially and cover all facets of the geomorphological research. This book has been built thus as the most exhaustive possible work presenting the latest geomorphological understanding of the countries' landscapes, not only to help all geoscientists and interested people travelling in Belgium and Luxembourg apprehend the visited landscapes but also to become a benchmark that hopefully will long be used in future research.

1.2 Geomorphic Regions in Belgium and Luxembourg

With respective areas of 30 528 and 2586 km^2, Belgium and Luxembourg together represent a tiny region, i.e., ~0.02% of the earth's land surface or ~0.3% of the surface of Europe. They encompass an accordingly limited range of elevations, from 0 m asl along the Belgian part of the southern North Sea's coastline to 694 m at Botrange and 693 m at Weisser Stein, the two highest summits of NE Ardenne, in eastern Belgium (Fig. 1.1). Beyond the small Ijzer catchment of western Belgium and coastal lowlands drained by artificial ditches directly into the North Sea, most of the country is distributed more or less equally between the Scheldt catchment in the north and the Meuse catchment in the south. The only exception is the Sauer catchment which, as a sub-tributary of the Rhine, covers the SE-sloping margin of SE Ardenne and most of the territory of Luxembourg.

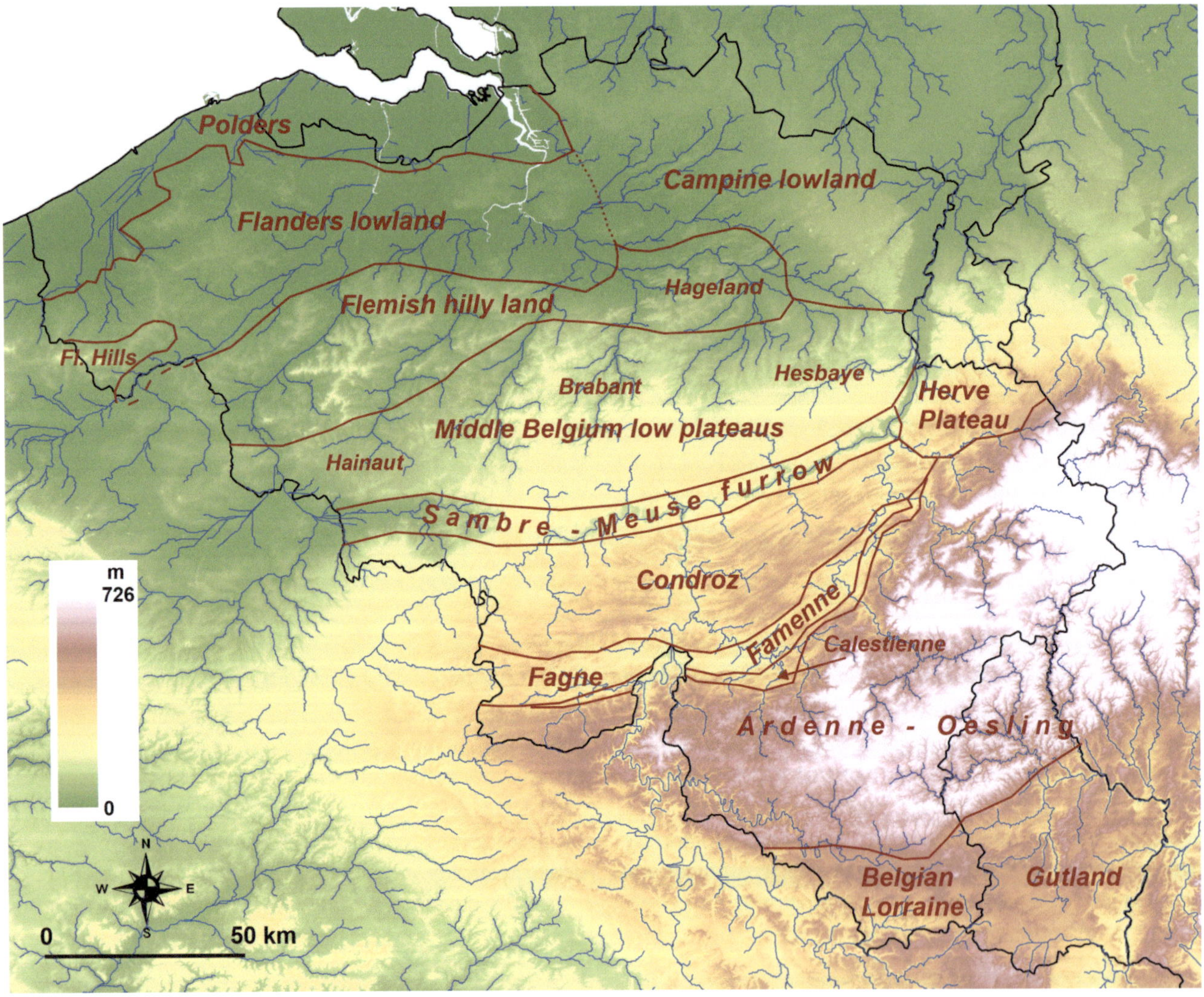

Fig. 1.1 The main geomorphic regions of Belgium and Luxembourg, delineated over the background SRTM 1″ DEM

Although some morphological maps of Belgium distinguished up to 73 geomorphic units (De Moor and Pissart 1992), a reasonable first-order division into ten main regions or so, whose differences in landscape are determined chiefly by elevation and geological constraints, provides a more meaningful basis for highlighting the contrasted geomorphic settings that compose the Belgian and Luxembourgian territories. A succession of essentially E-W trending regions may be followed roughly from NNW to SSE (Fig. 1.1). Starting from the shoreline in the NW, the first region (Polders) corresponds to the flat tract of reclaimed coastal plain and Scheldt tidal flats lying mostly below 7 m asl (or TAW, for Tweede Algemene Waterpassing, i.e., second general levelling). It goes south- and eastward into the lowlands of Low Belgium, whose elevation mostly remains between 10 and 50 m asl. The geology of these lowlands is mainly Cenozoic clays and sands buried under extended coversands of Late Pleistocene age. One distinguishes the lowlands of Flanders in the west, corresponding to the area of the Flemish valley, Middle to Late Pleistocene equivalent of the present middle and lower Scheldt basin, and the Campine (or Kempen in Dutch) in the east, an area rising slightly eastward to an up to ~85-m-high tilted low plateau underlain by Middle Pleistocene Meuse gravels. Between these proper lowlands and the low plateaus of Middle Belgium is a transition zone with elevations between 50 and 100 m asl. Beyond local names, this transitional area is best described as the Flemish hilly land, where, from the eponymous Flemish hills in the west to the Hageland in the east, the landscape is characterized by generally elongate, ferricrete-capped hills dominating by a few tens of metres a network of wide valley floors. Moreover, this is the zone where the coversands of the north progressively give way to finer aeolian deposits heralding the typical loess of Middle Belgium.

The next geomorphic unit corresponds to the low plateaus of Middle Belgium (Fig. 1.1). Coincident with the up to 25-m-thick loess belt that crosses Belgium from west to east, straddling also the major water divide between the Scheldt and Meuse basins, these plateaus display a gently undulating, weakly incised topography rising progressively from ~100 m asl in the north to 220 m asl in the south. Laterally, the western Hainaut plateau hardly exceeds 100 m in elevation, lying thus significantly lower than the central Brabant and eastern Hesbaye (Haspengouw in Dutch) plateaus. The varying deposits of the Meso-Cenozoic Belgian basin directly underlying the loess cover determine to some extent the character of each plateau landscape. In the east for example, while the Tertiary sands and clays that extend beneath the loess cover of northern Hesbaye allowed many broad valleys to develop within a generally humid landscape of rich pastures and orchards, the Cretaceous chalks on which the loess rests in the southern part of the plateau imposed a distinct stamp of dryness to a monotonous landscape whose uniformity is barely interrupted by a network of dry valleys largely concealed by thick loess. In the south, the low plateaus of Middle Belgium are abruptly interrupted by the Sambre-Meuse axis, a ~130-km-long, 80- to 150-m-deep, ENE-trending straight furrow carved by the Meuse and Sambre rivers along a shear zone probably active during the Miocene (Demoulin 1993). Along most of its length, this axis follows the Upper Carboniferous Namur syncline, extending parallel to the Variscan front and continued westward by the Haine basin toward Denain, in France. Intense mining in the coal basins of the Namur syncline during the nineteenth and twentieth centuries contributed anthropogenic landforms such as huge slag heaps to the Sambre-Meuse axis landscape. Based on geomorphic considerations, Colbeaux et al. (1977) defined the North-Artois shear zone extending from Liège to the Dover Strait and corresponding also more or less to the Variscan front line. The present low-level activity of this zone seems supported by instrumental seismicity data (e.g., Camelbeeck et al. 2007).

The Sambre-Meuse axis represents a major discontinuity in the geomorphology of Belgium, separating the fairly uniform topography of Low and Middle Belgium in the north from the deeply incised higher landscapes of High Belgium in the south. The bulk of High Belgium corresponds to the folded Paleozoic Ardenne massif, i.e., the western continuation of the Rhenish shield (see Chap. 2). Its subdivision in geomorphic regions is mainly a product of the long-term geomorphological evolution of the massif, which created distinct erosional levels (see Chap. 5), and the varying lithology of the basement. Climbing the southern hillslopes of the Meuse and Sambre valleys, one reaches the first massive ridge of the Condroz. The Condroz region extends as an erosional landscape whose envelope surface rises from ~240 m asl close to the Sambre-Meuse axis to ~350 m in its southern part. Following the ENE-trending Variscan grain, its structurally controlled topography displays alternating ridges cut in sandstones that outcrop along anticline axes and wide valleys carved in limestones of the intervening synclines, forming a typical Appalachian-type landscape with levelled ridge tops attesting an ancient, now incised erosion surface. As all geomorphic regions of High Belgium, the general elevation of the Condroz progressively decreases westward, where it mostly stays below 300 m asl west of the Upper Meuse valley (Fig. 1.1). East of Liège, the Herve Plateau, or *Pays de Herve*, is a hybrid area juxtaposing characteristics of the Hesbaye and the Condroz. Indeed, a frame made of the same Condrusian ridge-and-valley topography cut in the Paleozoic basement supports there a broad arcuate backbone ridge composed of extended remnants of the Cretaceous cover that also underlies southern Hesbaye, west of Liège.

South of the Condroz, a narrow, up to 10 km-wide tract of lower lying land has been carved as periglacial pediments in Upper Frasnian and Lower Famennian shales highly prone to frost-shattering (Fig. 1.1). At 200–220 m elevation, this E-W depression, called Famenne or Fagne east and west of the Meuse, respectively, lies 100–150 m below the Condrusian topography. Fairly uniform in the east, its surface is more irregular west of the Meuse, where Frasnian bioherms form scattered hills. To the south, the Fagne-Famenne rapidly gives way to a 200- to 300-m-high slope leading to the high Ardenne. However, this slope is interrupted by a narrow bench cut into Middle Devonian limestones at 260–300 m asl, the Calestienne, which hosts many caves developed through the action of acidic waters coming down from the Ardenne (see Chap. 8). South of the Calestienne, the main part of the slope, locally reaching ~200 m in height but rapidly diminishing west of the Meuse, corresponds to an erosional scarp between stepped Paleogene erosion surfaces.

In the heart of High Belgium but also in northern Luxembourg where it bears the name of Oesling, the higher topography of the Ardenne *s.s.*, corresponding to the Lower Devonian core of the massif and its Cambrian inliers, may again be approximated as an erosional plateau landscape, this time at elevations that decrease from 600 to 700 m asl in the east to 450–500 m in the south and 350–400 m west of the Meuse (Fig. 1.1). This plateau physiognomy inherited from etchplanation processes that continued the Mesozoic levelling of the massif in Paleogene times has been partly obliterated, mainly along the massif's margins, by the recent downcutting of deep valleys in response to regional Plio-Quaternary uplift. The geomorphic region of the Ardenne thus consists of central plateau landscapes moderately incised and covered by meagre pastures, and steeper marginal areas interrupted by deep valleys of every size and mainly supporting forests. Finally, south of the Ardenne-Oesling, Belgian Lorraine and Luxembourgian Gutland together make a last homogeneous geomorphic region sharply contrasting with it. Belonging to the Paris basin and its NE extension into the Luxembourg syncline, this region is separated from Ardenne-Oesling by another marked erosional scarp and displays a classical cuesta landscape cut into alternating harder and weaker rocks of the Lower and Middle Mesozoic cover outcropping in this part of the basin.

1.3 Geomorphological Themes

This book attempts at providing the most exhaustive possible overview of the geomorphology of Belgium and Luxembourg. As there is no particular logic between the various geomorphic settings presented in the chapters, we conceived it as a tour of the two countries, starting in Luxembourg, then travelling northward and anticlockwise across Belgium before closing the circle and ending our geomorphological journey in southern Belgium and Luxembourg. However, neither all geomorphic regions nor all geomorphic-specific contexts and mechanisms could be treated. For instance, we judged that, though very typical, the Appalachian topography of the Condroz did neither possess the originality nor offer the recent findings required to make a valuable contribution to the book. Likewise, we gave up the idea of describing geomorphic processes of more local meaning, such as the periglacial carving of the Famenne depression, or dealing with phenomena already tackled in several other chapters, like the development of Neogene cryptokarsts. We briefly review hereafter the regions, landforms, and topics dealt with in the book (Fig. 1.2).

Highlighting the contrast between the Paleozoic paleogeographies and tectonic history of the Ardenne basement in the south and the mostly depositional Meso-Cenozoic evolution of the Belgian part of the Anglo-Belgian basin in the north, Chap. 2 sets the scene and provides the geological background necessary for a sound understanding of the geomorphic processes and evolutionary models presented in the next chapters. By contrast, Chap. 3, devoted to the climate of Belgium and Luxembourg, discusses only the range of observed temperatures and precipitations in the frame of the present atmospheric circulation over Belgium and Luxembourg as a snapshot example of their spatial gradients across the study area. While it helps to appreciate the climatic context of modern geomorphic processes in the two countries, it does not deal with the evolution of past regional climates, which will be evoked when required in the course of specific studies.

Starting our tour in Luxembourg, Chap. 4 presents and discusses the very specific landscapes produced by the Plio-Quaternary geomorphic evolution affecting the thick Lower Liassic sandstones outcropping in Gutland (Fig. 1.2). This topic has awakened sufficient geomorphological interest worldwide to persuade authors to dedicate books and journal's special issues to it (e.g., Young et al. 2009; Migon and Viles 2015) and the Luxemburgian case is a very representative example of sandstone landscapes and landforms in the mid-latitudes. Travelling north to the Ardenne-Oesling and its stepped erosion surfaces, the next chapter (Chap. 5) refers to a key area in the history of long-term geomorphology and erosion cycles, Davis (1899) having already commented on this area at the end of the nineteenth century. However, recent advances in dating of the saprolites associated with old surfaces have brought strong support to the updated understanding of the latter's preserved remnants, whereas the accumulated geomorphic and weathering data

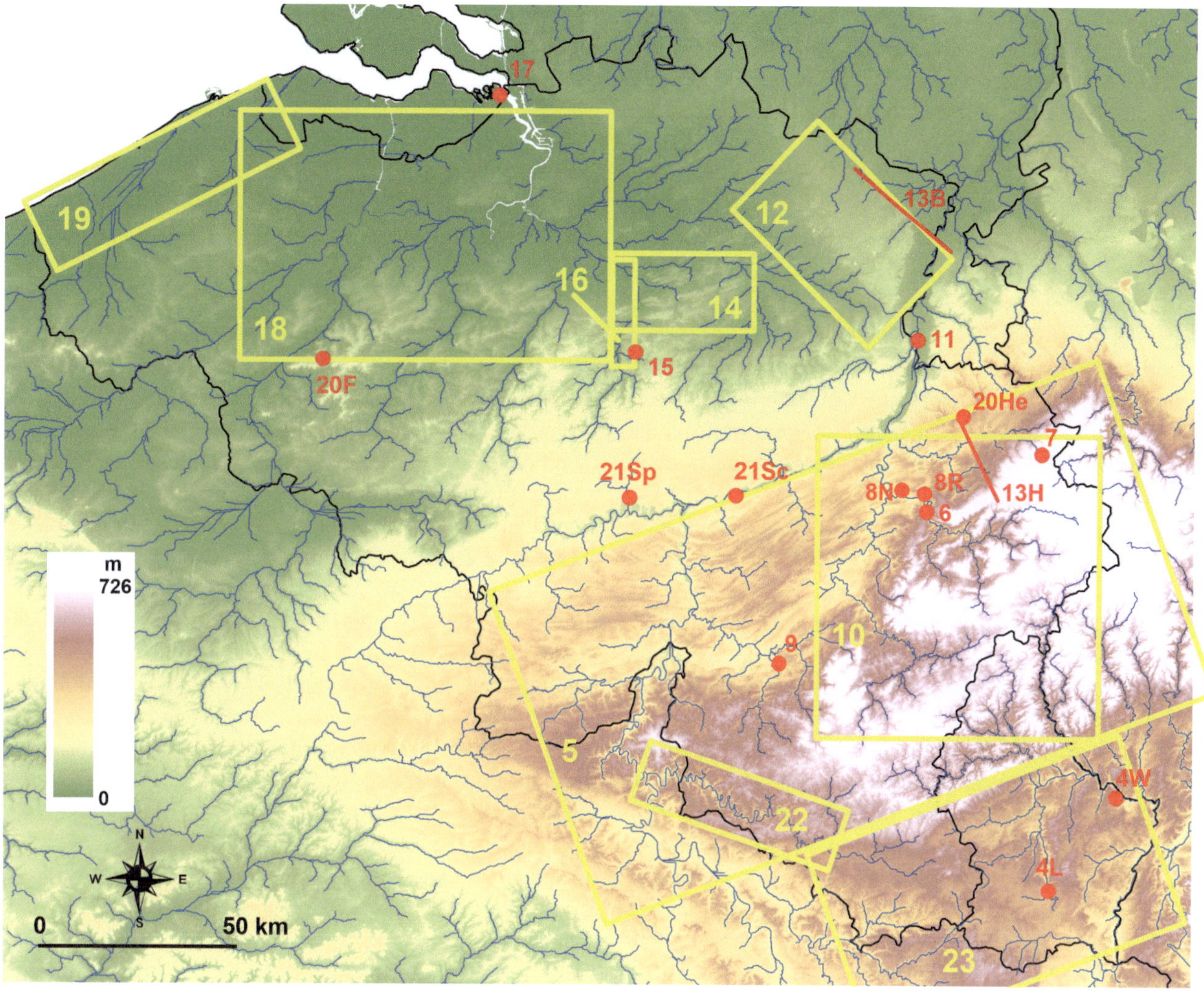

Fig. 1.2 Location of the regions (*yellow rectangles*) and sites (*red dots or lines*) described in this book. *Numbers* refer to the book chapters. *B* Bree scarp. *F* Flemish Ardennes. *H* Hockai fault zone. *He* Herve Plateau. *L* Luxembourg. *N* Noû Bleû. *R* Remouchamps. *Sc* Scladina Cave. *Sp* Spy Cave. *W* Wolfschlucht

sets make the Ardenne-Oesling a good example where the geomorphic evidence clearly contradicts the common interpretation of recently acquired fission track and cosmogenic radionuclide denudation data.

Chapters 6 and 7 consider more local but highly meaningful phenomena attesting the conditions of the Quaternary geomorphic evolution of NE Ardenne. A remarkable scenic section of the Amblève incised valley is presented first, focused on the analysis of a boulder-bed reach with blocks up to several m^3 in volume descended from the local hillslopes. An intriguing, and still not definitively solved, issue is raised of the processes by which many such big boulders were transported over 90 km before being included in terrace deposits of the Lower Meuse. The second of these two chapters describes world-famous features among a diversity of periglacial landforms and deposits inherited from the cold conditions of the Late Quaternary in the Ardenne, namely ramparted depressions attesting the development of lithalsas in NE Ardenne during the Younger Dryas. Dense fields of such periglacial ramparted depressions are especially well preserved in the Hautes Fagnes plateau (Fig. 1.2), which is known as the nicest and best documented place in the world to study relict lithalsas.

The twin Chaps. 8 and 9 deal with karstic phenomena developed in a narrow zone of Middle Devonian limestone, which borders to the north the Lower Devonian Ardenne. There, acidic waters flowing down from the siliceous Ardennian basement have given rise to the development of several extended, and touristically renowned, cave systems. Chapter 8, which presents the famous Remouchamps cave and the newly discovered nearby Noû Bleû karstic system, emphasizes mainly the relationships between the surface and underground evolution of karstic areas and describes many superb speleothem types found in pristine state in the Noû Bleû cave, inaccessible to the public. Chapter 9 is devoted to the karstic system of Han-sur-Lesse, located more to the west (Fig. 1.2). It stresses hydrogeological aspects of the karstic evolution that led to the subterranean cutting off of a meander of the surface Lesse River, and also takes the opportunity of field evidence in the Han cave to expose the concept of two-stage karstification through ghost-rock weathering of limestone massifs. To finish with the geomorphology of the Ardenne region, Chap. 10 brings together recently published new data about, and interpretations of, the Quaternary incision of the Ardennian drainage network in response to combined climatic and tectonic signals. Cosmogenic radionuclide dating of a Middle Pleistocene terrace level revealed its diachronic character, confirmed by the analysis of a set of tectonic knickpoints whose upstream migration is still going on in the Ardennian basin of the Meuse. These observations have led on to a reinterpretation of the classical scheme of climatically triggered Quaternary terrace staircases in the valleys of many European rivers, taking better account of the specific response of drainage systems to tectonic perturbations. In passing, the origin of the course of the Meuse is also briefly reconsidered.

Looking at geomorphic phenomena north of the Ardenne, Chap. 11 comes back to karstic landscapes, but in the particular context of Cretaceous chalk massifs. The described example is a mixture of interacting surface and underground karstic forms well exposed by a dense network of underground quarries within the Montagne Saint-Pierre, a well-known chalk massif along the Belgian-Dutch border where the first *Mosasaurus* fossils were uncovered in the nineteenth century (Van Marum 1790) (Fig. 1.2). Moving further north, Chap. 12 is devoted to the Campine plateau of NE Belgium and proposes a synthetic overview of various geomorphic problems related to the evolution of this area. Beyond offering interesting perspectives on the implications of differential vertical motion between the Roer Valley Graben and its Campine shoulder, the Meuse Quaternary evolution, and landforms related to the accumulation and reworking of Weichselian aeolian sands, this chapter introduces the issue of landscape and landform preservation and valorization in densely populated areas. The next chapter (Chap. 13) is also partly devoted to the Campine area and, more widely, to the morpho- and seismotectonics of assumed and confirmed active seismogenic faults in eastern Belgium. Tectonic landforms and seismic hazard in eastern Belgium are essentially related to the activity of the major Feldbiss fault system bounding to the south the actively subsiding Roer Valley Graben, part of the Lower Rhine segment of the Cenozoic European Rift System, but also to related faults such as the Hockai fault zone of NE Ardenne, whose rupture in 1692 caused one of the most violent historical earthquakes in Europe north of the Alps. The paleoseismological results obtained along the Feldbiss fault during the 1990s and the 2000s were pioneering in evidencing the potential hazard of strong earthquakes in assumed stable intraplate areas (e.g., Camelbeeck and Meghraoui 1996).

Moving west, Chap. 14 proposes a brand new hypothesis about the origin of the elongate hills characterizing the Hageland landscape, in central Middle Belgium (Fig. 1.2). Doing so, it highlights how geology and long-term geomorphology are intricately related and how careful joint geological and geomorphological investigations can shed convincing new light on old problems. A few kilometres south of the Hageland, Chap. 15 comes back to more recent geomorphic processes having induced barely known landforms in the heavily anthropogenically reworked landscapes of the loess belt of Middle Belgium. Mapping and analysing currently largely inactive gullies in the Meerdaal Forest, south of Leuven, and closed depressions under forest and in open cropland, the authors demonstrate how gullies that developed in a phase of land clearing and agricultural activities during the Roman period or earlier were only preserved, as a rather unique occurrence, in areas that remained continuously forested during the last 2 ky. They also point to the human origin of closed depressions as ancient quarries, stressing the human–environment interactions in the (pre)historical landscape evolution of Middle Belgium. Chapter 16 deals with another aspect of the same area, namely the complex evolution of floodplains in the Late Glacial and Holocene as exemplified by the Dijle catchment. It underlines how climatically triggered natural processes and later interference of human action successively caused the incision of large meanders into Weichselian braided river deposits, the cut-off and abandonment of these meanders, peat accumulation, deforestation-induced renewed floodplain sedimentation, and the development of a smaller-amplitude meander belt, often artificially cut-off in the last centuries.

With Chap. 17, we move to western Belgium and, first, the Scheldt estuary (Fig. 1.2). With a full salinity gradient along a length of ~ 160 km, this typical lowland river estuary is one of the best preserved among those debouching in the North Sea. This study discusses the present intertidal areas morphodynamics along the single-channel Belgian part of the estuary, focusing on the extended Saeftinghe brackish marsh (~ 30 km^2). After a review of the functions of intertidal areas and the processes that shape them, it describes how recreation of tidal flats by the Flemish Sigmaplan is used, notably as flood control areas, to remedy the changes in estuary dynamics caused by previous human works. Chapter 18 also deals with the Scheldt basin, the Quaternary, and especially Late Pleistocene evolution of which it reconstructs. Called the Flemish Valley, the Late Pleistocene drainage system of the Middle and Lower Scheldt encompasses a large part of Low Belgium where detailed investigations of complex time-varying sedimentation patterns allowed for reconstructing the history of a remarkable set of nested terraces, likely initiated by a catastrophic dam breaching in the southern North Sea and paced by glacio-eustatic variations. As for Chap. 19, it presents an overview of the Holocene evolution of the coastal plain of Belgium, with detailed histories of two protected areas, namely De Moeren at the western end of the plain and the Zwin region at the other end.

Chapter 20 focuses on particular geomorphic processes and landforms, namely those related to landsliding, rather than to a specific region of Belgium or Luxembourg. Indeed, though still almost unknown in Belgium 20 years ago, ancient landslide scars have been mapped, and present landsliding activity noticed, in many regions of Middle Belgium since then. Two case studies are presented, one from the Flemish Ardennes, a hilly area of the Middle Scheldt basin in western Belgium, and the other one from the Herve Plateau in eastern Belgium. Despite different geological settings, both areas show similar lithological frames, with alternating subhorizontal sands (liquefiable in the Herve Plateau) and weak clays. They also show the same contrast between large deep-seated ancient (Holocene) landslides and small present-day reactivation of these existing landslides. While the present reactivation episodes are climatically triggered, the ancient landslides might be of seismic origin in the two regions. Chapter 21 is no more linked to a particular geomorphic region as it deals with all paleontological and archaeological remains attesting the presence and activities of Neandertals in Belgium. While the Spy and Scladina caves are described as emblematic examples of caves where Neandertal fossils have been found in Wallonia (Fig. 1.2), open-air sites are shown to be also highly instructive with respect to the understanding of territorial exploitation by Neandertals. While all cave sites are by necessity located in Wallonia, a small fraction of the open-air findings have been reported from Flanders.

Chapters 22 and 23 end our tour of Belgium by exploring two last geomorphic landscapes in SE Belgium and southern Luxembourg. Chapter 22 describes the remarkable elongate incised meanders of the Semois valley in southern Ardenne, several of which have been classified as exceptional geoheritage of Wallonia, and discusses their genetic link with the relative orientation between meander loops and the regional slaty cleavage. Chapter 23 presents the cuesta landscape of the NE rim of the Paris basin in Belgian Lorraine and Luxembourgian Gutland, not only providing a careful description of the morphology of the successive cuestas but also paying attention to weathering and duricrusting phenomena and briefly stressing remaining issues about the early evolution of the drainage network.

Finally, it seemed meaningful to dedicate a specific chapter to the geomorphosite concept, showing how it fundamentally intermingles with the notions of geosite and geodiversity and is also to some extent related to those of bio- and cultural diversity. This chapter (Chap. 24) proposes a brief review of Belgian sites classified as exceptional and displaying a specific geomorphic component. It highlights the different approaches by the public and the authorities of landscape and landform conservation in Wallonia (southern Belgium), where really scenic and fairly well-preserved landscapes occur in sparsely populated rural and forested areas, and in Flanders (northern Belgium), whose monotonous natural landscapes are strongly obliterated by a very dense land occupation and exploitation by man, and thus are a priori much less appealing.

References

Camelbeeck T, Meghraoui M (1996) Large earthquakes in northern Europe more likely than once thought. EOS Trans 77:405–409 American Geophysical Union

Camelbeeck T, Vanneste K, Alexandre P, Verbeeck K, Petermans T, Rosset P, Everaerts M, Warnant R, Van Camp M (2007) Relevance of active faulting and seismicity studies to assessments of long-term earthquake activity and maximum magnitude in intraplate northwest Europe, between the Lower Rhine Embayment and the North Sea. In: Stein S, Mazzotti S (eds) Continental intraplate earthquakes: science, hazard, and policy issues. Geological Society of America special paper 425, pp 193–224

Colbeaux JP, Beugnies A, Dupuis C, Robaszynski F, Sommé J (1977) Tectonique de blocs dans le sud de la Belgique et le nord de la France. Ann Soc Géol Nord 97:191–222

Davis WM (1899) The Peneplain. Am Geolo 23:207–239

de Limbourg R (1770) Mémoire sur l'histoire naturelle d'une partie des Pays-Belgique. Mém Acad Imp R Sci B-Lett 1(1777):195–219 (Bruxelles)

de Limbourg R (1774) Mémoire pour servir à l'histoire naturelle des fossiles des Pays-Bas. Mém Acad Imp R Sci B-Lett 1(1777):369–417 (Bruxelles)

Demoulin A (1993) L'origine de l'axe Sabre-Meuse. Ann Soc Géol Belg 116:29–41

De Moor G, Pissart A (1992) Les formes du relief. In: Denis J (ed) Géographie de la Belgique, Crédit Communal, Bruxelles, pp. 130–216

Fourmarier P (1934). Vue d'ensemble de la géologie de la Belgique. Ses enseignements dans le domaine de la géologie générale. Vaillant-Carmanne, Liège, p 198 (Ann Soc Géol Belg Mém)

Houzeau JC (1854) Essai d'une géographie physique de la Belgique, au point de vue de l'histoire et de la description du globe. Hayez, Bruxelles, p 331

Migon P, Viles H (eds) (2015) Sandstone geomorphology. Landscape formation, field mapping, research methods. Zeitschr für Geomorphologie 59(Suppl. 1):268

Omalius d'Halloy JBJ (1828) Mémoires pour servir à la description géologique des Pays-Bas, de la France, et de quelques contrées voisines. D. Gerard, Namur, p 307

Pissart A (ed) (1976) Géomorphologie de la Belgique. Laboratoires de Géologie et Géographie Physique, University of Liège, 224 p

Van Marum M (1790) Beschrijving der beenderen van den kop van eenen visch, gevonden in den St Pietersberg bij Maastricht, en geplaatst in Teylers Museum. Verhandelingen Teylers Tweede Genootschap, pp 383–389

Young R, Wray R, Young A (2009) Sandstone landforms. Cambridge University Press, UK, p 314

An Introduction to the Geology of Belgium and Luxembourg

2

Frédéric Boulvain and Noël Vandenberghe

Abstract

Belgium and the Grand-Duchy of Luxembourg show surprising geological diversity over their small combined area of 33,114 km^2. Almost all types of sedimentary rocks crop out and are generally preserved along well-described and easily accessible sections or in quarries. Several sections are known worldwide and are visited for stratigraphic or sedimentological purposes. Magmatic rocks are not abundant and metamorphic rocks are restricted to slates. The stratigraphic scale ranges from the Cambrian to the Quaternary, which translates to a half billion years of Earth history. This chapter provides a comprehensive overview of the different stratigraphic units, starting from the oldest and ending with the youngest. Modern stratigraphic schemes highlight formations' geometries and interrelations. Some of the most remarkable units are further detailed. The two orogenic phases that shaped the Lower Paleozoic inliers and the Devonian-Carboniferous faulted and folded belt, i.e. the Caledonian and Variscan orogeny, are also addressed.

Keywords

Caledonian inliers • Variscan fold-and-thrust belt • Brabant massif • Ardenne allochthon • Mesozoic sedimentation in Belgium and Luxembourg • Cenozoic Belgian basin

2.1 Introduction

Despite the limited size of the Belgian and Luxembourgian territories, the cumulative thickness of their geological formations is thought to reach 18 km. Stratigraphically, these formations range from the Lower Cambrian to the Quaternary, with only minor hiatuses. More than a half billion years of Earth history is thus exposed.

Books and papers on the geology of Belgium and Luxembourg are too numerous to be cited here. Still worth being mentioned are the "Prodrome d'une description géologique de la Belgique" (Fourmarier 1954) and the "Manuel de la Géologie du Luxembourg" (Lucius 1952). More recently, the regional geological guidebooks 'Ardenne Luxembourg' (Waterlot et al. 1973) and 'Belgique' (Robaszynski and Dupuis 1983) must be cited. Boulvain and Pingot (2015) have proposed a synthesis of the geology of Wallonia (southern Belgium).

The subsurface of Belgium and Luxembourg is mostly characterized by the presence of sedimentary rocks. Magmatic rocks are subordinate and metamorphism has never reached more than the epizone (green slate facies). The sedimentation periods and subsequent deformation events shaped the Belgian and Luxembourgian substrates into four major sedimentary-structural units, namely (1) the Lower Paleozoic inliers, (2) the Devonian-Carboniferous faulted and folded belt including the former, (3) the homoclinal Triassic-Jurassic series and (4) the subhorizontal

F. Boulvain (✉)
Pétrologie Sédimentaire, Université de Liège,
Sart Tilman, B20, 4000 Liège, Belgium
e-mail: fboulvain@ulg.ac.be

N. Vandenberghe
Department Earth and Environmental Sciences,
Division of Geology, Leuven, KU Leuven,
Celestijnenlaan 200E, 3001 Leuven-Heverlee, Belgium
e-mail: noel.vandenberghe@kuleuven.be

A. Demoulin (ed.), *Landscapes and Landforms of Belgium and Luxembourg*,
World Geomorphological Landscapes, DOI 10.1007/978-3-319-58239-9_2

Cretaceous (4a) Cenozoic (4b) cover (Figs. 2.1 and 2.2). The faulted and folded belt includes three tectonic units, the Brabant parauthochton (BP), the Haine-Sambre-Meuse thrust sheets (HSM) and the Ardenne allochthon (AA), separated by the Midi-Eifel thrust fault (F). The Ardenne allochthon is structured in major anticlines and synclines, from north to south, the Dinant Syncline and its eastern equivalent in the Vesdre-Aachen area, the Ardenne Anticline, the Neufchâteau-Wiltz-Eifel Syncline and the Givonne Anticline.

The following provides a description of the different sedimentary units, from the oldest to the youngest. The two orogenic phases that shaped the Lower Paleozoic inliers and the Devonian-Carboniferous faulted and folded belt will also be briefly addressed.

2.2 The Caledonian Cycle: The Lower Paleozoic Inliers

The sediments that constitute the Lower Paleozoic inliers were deposited during the Caledonian sedimentary-tectonic cycle. In the beginning, Belgium and Luxembourg belonged to the Avalonia microplate, part of the Gondwana supercontinent situated around the South Pole (Fig. 2.3). At the Cambrian/Ordovician transition, Avalonia separated from Gondwana and started to drift away. During the Middle and Late Ordovician (Sandbian—Hirnantian), a first continental collision between Avalonia and the Baltica microplate was responsible for the Ardenne phase of the Caledonian orogeny and, throughout the end of the Silurian, the merged Avalonia-Baltica microplates collided with Laurentia

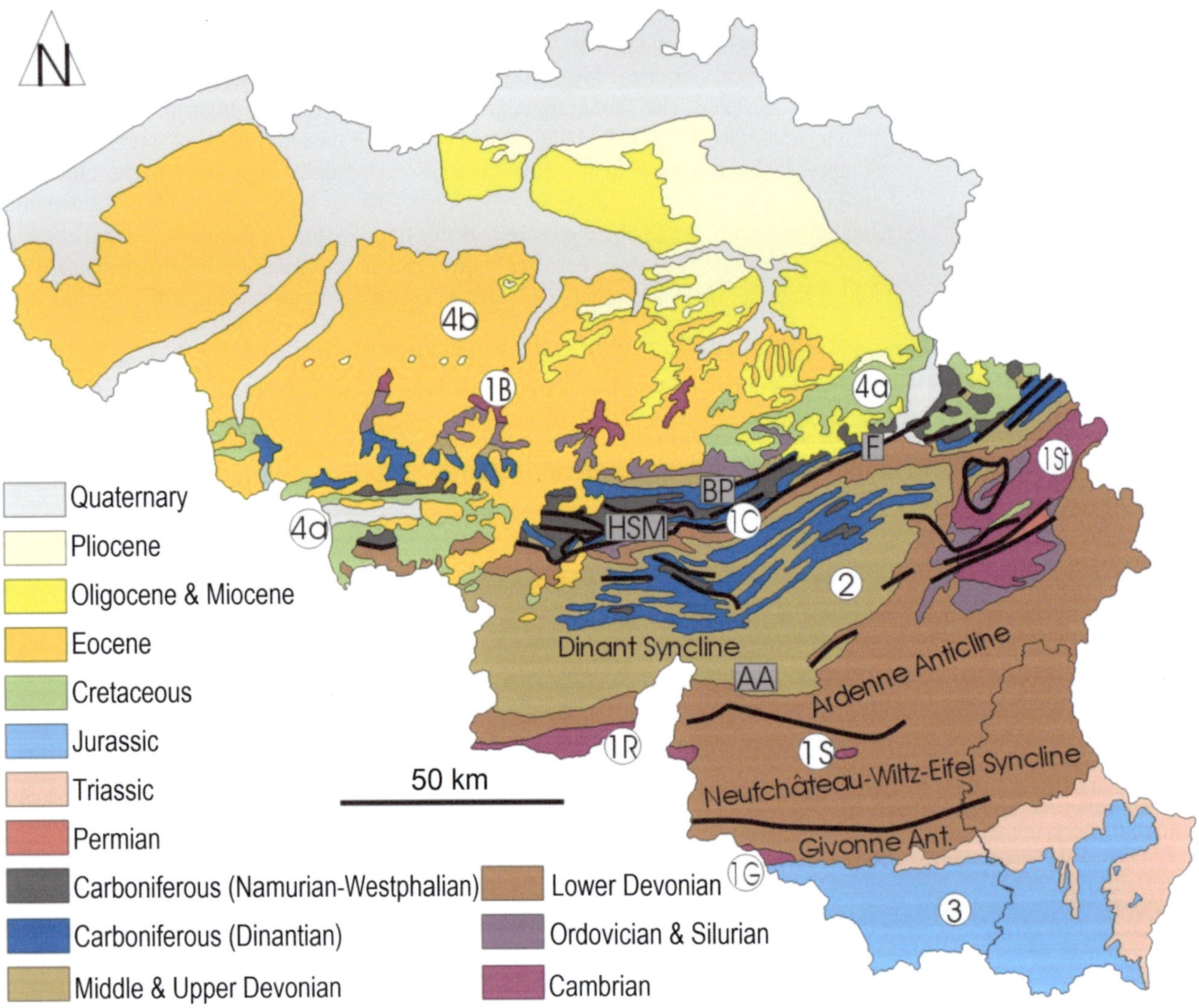

Fig. 2.1 Schematic geological map of Belgium and Luxembourg. *BP* Brabant parautochthon. *AA* Ardenne allochthon. *F* Midi-Eifel Fault. *HSM* Haine-Sambre-Meuse thrust sheets. *1B* Brabant Lower Paleozoic (LP) inlier. *1C* Condroz LP inlier. *1St* Stavelot LP inlier. *1R* Rocroi LP inlier. *1S* Serpont LP inlier. *1G* Givonne LP inlier. *2* Devonian-Carboniferous faulted and folded belt. *3* Homoclinal Triassic-Jurassic series. *4a* Cretaceous subhorizontal cover. *4b* Cenozoic subhorizontal cover. *Thick black lines* denote thrust faults

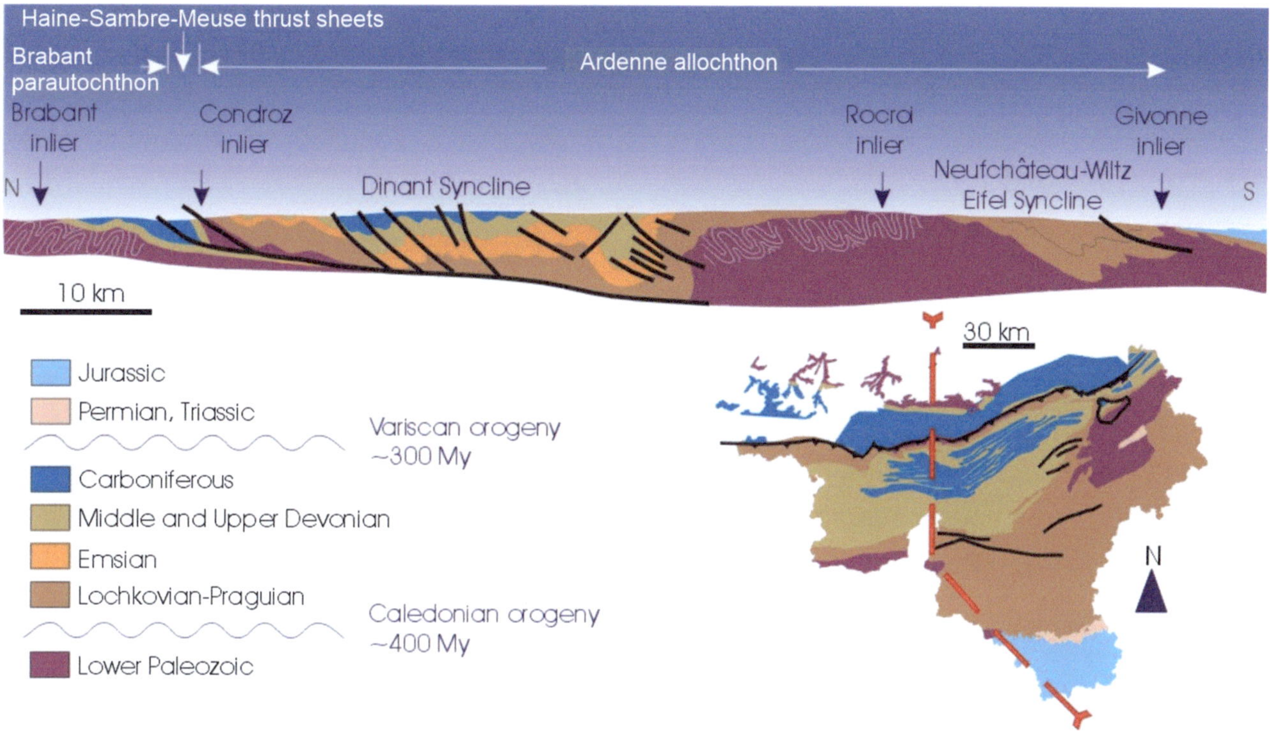

Fig. 2.2 Geological cross-section of South Belgium (Wallonia). *Thick black lines* are for Variscan thrust faults

(another supercontinent, located further north), giving rise to the Brabant phase of the Caledonian orogeny.

Six Paleozoic inliers dating back to this cycle crop out in Belgium. From north to south these are the Brabant, Condroz, Stavelot, Serpont, Rocroi, and Givonne massifs (Fig. 2.4). Sedimentation in the Condroz and Brabant inliers is so similar that Herbosch and Verniers (2014) have proposed a common sedimentary basin. The other inliers are part of the Ardenne basin, located further south.

According to geochemical studies, the Brabant Lower Paleozoic formations were deposited on a crystalline basement dated back 1800 Ma, topped by volcanic material (André 1991). This hypothesis was recently confirmed by the age of detrital zircons from the old basement cropping out in Hunsrück (Wartenstein Gneiss, Germany), related to the Panafrican orogeny (Linnemann et al. 2012).

The Lower Paleozoic sedimentation is organized in three supersequences, starting with littoral or platform sandstones and ending with deep hemipelagic or turbiditic deposits. These large-scale sequences (10–50 My) recorded major paleogeographic changes due to plate movements (Cocks and Torsvik 2002). The first supersequence spreads from the Lower Cambrian to the Lower Ordovician and corresponds to the separation of Avalonia from Gondwana, associated with the opening of the new marine domain of the Rheic Ocean. The second supersequence ends during the Late Ordovician and is likely related with the Ardenne Caledonian orogenic phase. The third supersequence is only observed in the Brabant and Condroz inliers (Table 2.1) because it developed during the Late Ordovician and Silurian, when the Ardenne was undergoing the Caledonian orogeny.

The first supersequence (Lower Cambrian-Lower Ordovician), observed in all but the Condroz inliers, is described hereafter as a representative example (Fig. 2.5, Table 2.1). Within the Ardenne massifs, the quartzitic sandstones of the Deville Group characterize the Lower Cambrian. These light-coloured, massive sandstones are believed to have been deposited on a shallow detrital platform. The Middle and Upper Cambrian correspond to the Revin Group, which are dark-coloured formations dominated by slates. Turbidites are widespread and the youngest Cambrian formation even contains black shale, suggesting a deep and anoxic sedimentation basin. The supersequence ends in the Stavelot massif with the lower part of the Salm Group, which exposes alternating green sandstone and slate strata corresponding to distal and Bouma-type turbidites (Lamens 1985).

A significant magmatic episode occurred during the Ordovician-Silurian. This episode is well documented in the Brabant massif, with lava flows, volcano-sedimentary deposits, and volcanic pipes such as the 2-km-wide Quenast pipe made of a quartz-rich microdiorite. This event is now interpreted as intracontinental calcalcaline magmatism

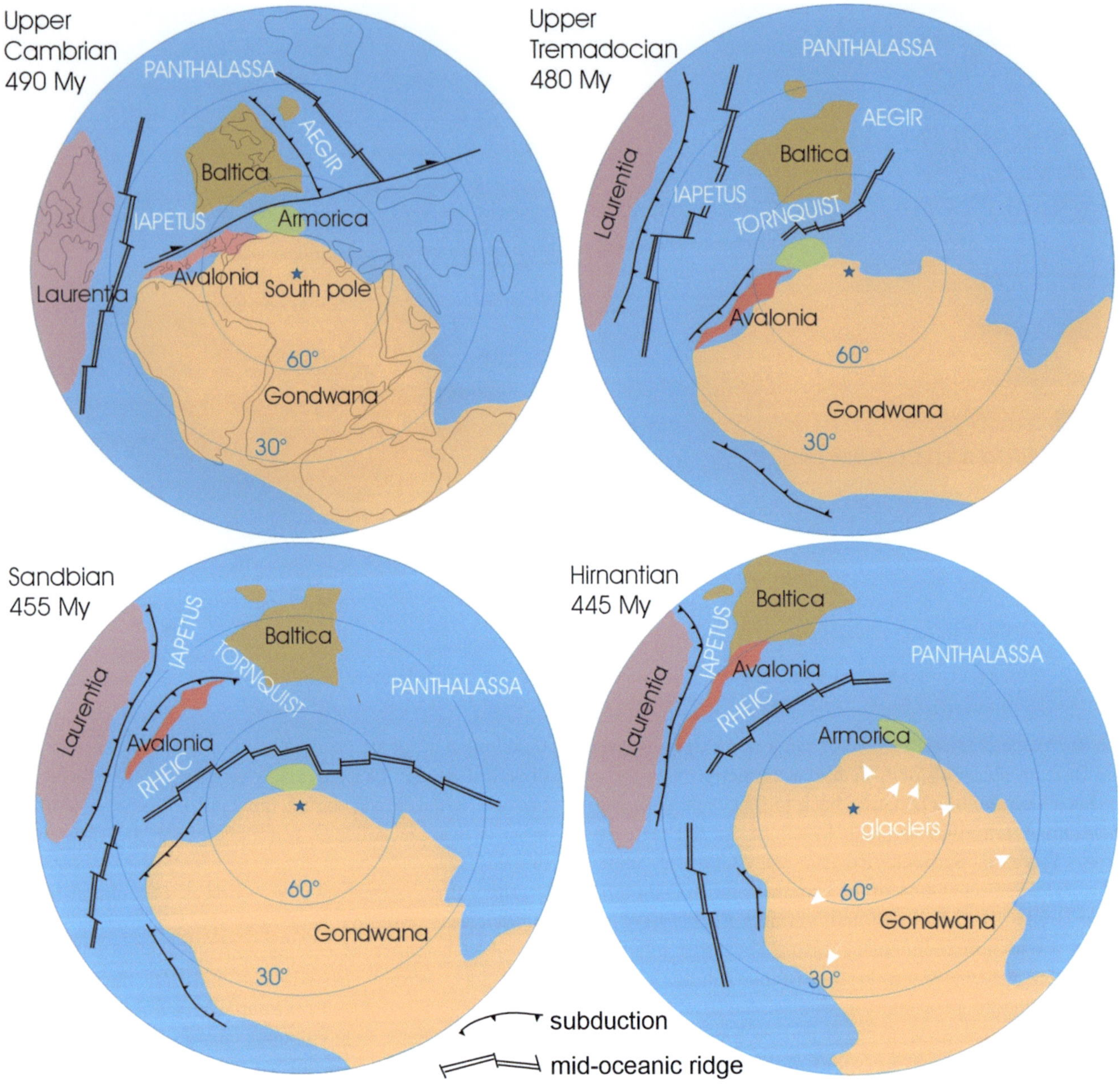

Fig. 2.3 Paleogeography of the Lower Paleozoic. *White arrows* indicate ice movements. Simplified after Cocks and Torsvik (2002)

related to the Avalonia-Baltica docking phase (Linnemann et al. 2012). It probably ended with the main compressive phase of the Caledonian orogeny.

As a consequence of the diachronism of the Caledonian orogeny mentioned above, the first post-Caledonian sediments are also diachronic, attributed to the lowermost Lochkovian in the Ardenne and to the Givetian around the Brabant massif. Another major difference is that the Variscan orogeny affected the Ardenne and Condroz inliers but not the Brabant massif, located north of the Variscan deformation front. As a result, the Ardenne and Condroz inliers were affected by two orogenic phases while the Brabant massif underwent only the Caledonian orogeny.

In the Ardenne inliers, the Caledonian orogeny is largely characterized by thrust sheets and tight north-verging folds. The presence of many slumps often obscures the structural interpretation (Meilliez and Lacquement 2006). The Condroz inlier shows south-verging folds associated with a north-dipping schistosity. With regard to the Brabant massif, Sintubin and Everaerts (2002) and Debacker (2012) have proposed a detachment mechanism between the Cambrian core and the Ordovician-Silurian border (see Fig. 2.4).

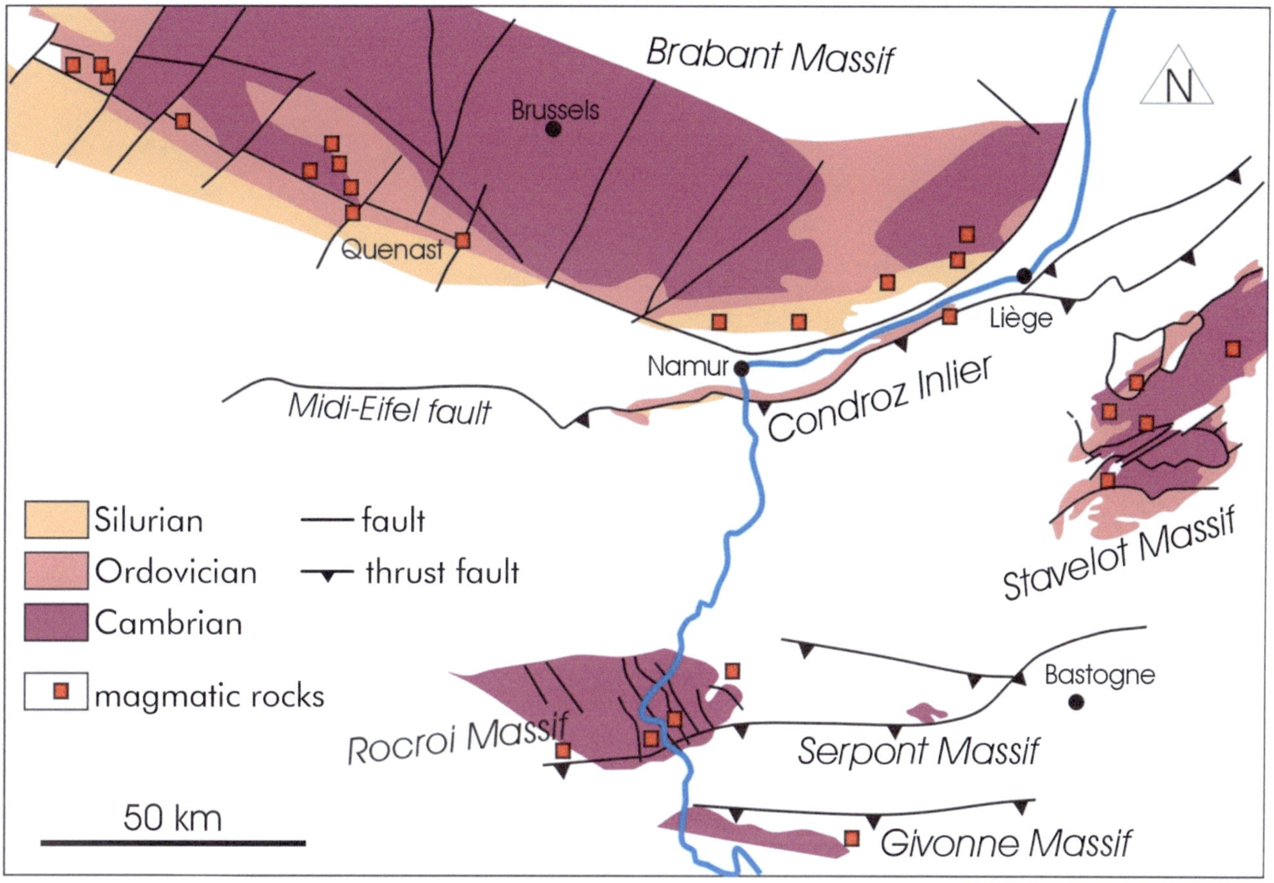

Fig. 2.4 Simplified geological map of the Belgian Lower Paleozoic inliers (below the Meso-Cenozoic sedimentary cover, compare with Fig. 2.1) (modified after De Vos et al. 1993)

Table 2.1 Age and thickness of formations in the different Lower Paleozoic inliers

Inliers	Stratigraphy	Thickness
Brabant	Lower Cambrian-Upper Silurian	>13 km
Condroz	Middle Ordovician-Upper Silurian	>1.5 km
Stavelot	Lower Cambrian-Middle Ordovician	>3 km
Serpont	Cambrian	>0.8 km
Rocroi	Lower Cambrian-Middle Ordovician?	>2.5 km
Givonne	Cambrian	>1.5 km

Indeed, the core displays steeply plunging folds associated with a subvertical schistosity whereas the borders show moderately plunging folds with a north-dipping schistosity (Debacker et al. 2005). The centripetal character of the deformation is markedly diachronic (by ~30 My) throughout the Brabant massif (Debacker et al. 2005).

2.3 The Variscan Cycle

The Variscan sedimentary-tectonic cycle took place in Belgium and Luxembourg during the Late Paleozoic. It began in the Early Devonian with renewed detrital sedimentation on the shores of the Old Red Sandstone Continent and ended with the Variscan orogeny in the Late Carboniferous. We review hereafter the successive series that make this sedimentary cycle, ending in an orogenic period responsible for the formation of the Pangea supercontinent.

2.3.1 The Lower Devonian Detrital Formations

The Lower Devonian crops out in large areas of the Ardenne Anticline, continued into the Luxembourgian Eisleck, and in the Neufchâteau-Wiltz-Eifel Syncline. Sediments mainly consist of sandstones, siltstones, slates, and shales. The

Fig. 2.5 **a** Quartzitic sandstones from the Deville Group, Rocher des Quatre Fils Aymon. **b** Black slate from the Revin Group, Saint-Nicolas. Rocroi inlier

regional seascape followed the east-west trending southern coast of the Old Red Continent, bordered by the northern Rheic Ocean (Fig. 2.6). In this passive margin setting, sediment thickness increases rapidly southward. The Lower Devonian totals 1.3 km along the northern border of the Dinant Syncline, 3.1 km along its southern border and 4.5 km in the Neufchâteau-Wiltz-Eifel Syncline (Fig. 2.7). There is no Lower Devonian sedimentation north of the Midi-Eifel Fault (Fig. 2.1). The sediment supply originated from the Old Red Sandstone Continent.

The first Lower Devonian sediments deposited on the Caledonian basement are conglomerates interpreted as continental alluvial fans (Meilliez 2006) (Fig. 2.8a). They rapidly pass upwards to versicoloured shales and siltstones including sandstone lenses (Fig. 2.8b), in patterns typical of alluvial plain and channel systems. The first marine sediments are littoral and platform sandstones/quartzites or shales/slates (Goemaere and Dejonghe 2005). They are younger along the northern border of the Dinant Syncline (Pragian) than along its southern border (Lochkovian), reflecting the progression of the Lower Devonian marine transgression. The transgression peak was reached during the Pragian, with external platform shales and slates in the south (La Roche Formation) and fluvio-littoral sandstones in the north (Acoz Formation). The Emsian shows a marked regressive trend, with fluvio-littoral environments prograding southward at the expense of the marine facies. The most spectacular unit is the deltaic Burnot Formation, which includes several hundred metres of red conglomerates, sandstones, and siltstones (Corteel and De Paepe 2003).

2.3.2 The Middle Devonian Mixed Carbonate-Detrital Formations

The Middle Devonian formations crop out along the borders of the Dinant Syncline and its eastern equivalents. They are also present in the Haine-Sambre-Meuse thrust sheets and the Brabant parautochthon.

The Middle Devonian was marked by a more drastic transgressive regime. The rising sea level was responsible for an extension of the ocean north of the future Midi-Eifel Fault, up to the Brabant parautochthon. Simultaneously, the Lower Devonian detrital facies gave way to argillaceous limestone and to first carbonated platforms (Fig. 2.6b). The Eifelian marks the transition between the old detrital and the new carbonate world. Facies are still mixed and carbonate platforms, laterally restricted, were still surrounded by shale. However, at the onset of the Givetian, a huge carbonate platform was established over southern Belgium. The contemporaneous coast was located near the Brabant massif. This spectacular development of carbonates was probably related to a warmer climate in an area that was then situated between the Equator and the Southern Tropic, combined with a dramatic decrease in detrital supply coming from the Old Red Sandstone Continent.

Along the southern border of the Dinant Syncline, the well-developed Givetian platform shows 450 m of limestone including fore-reef, reef, and lagoon environments (Boulvain et al. 2009) (Figs. 2.9 and 2.10). This thickness decreases northerly to $\sim$100 m of typical littoral carbonates.

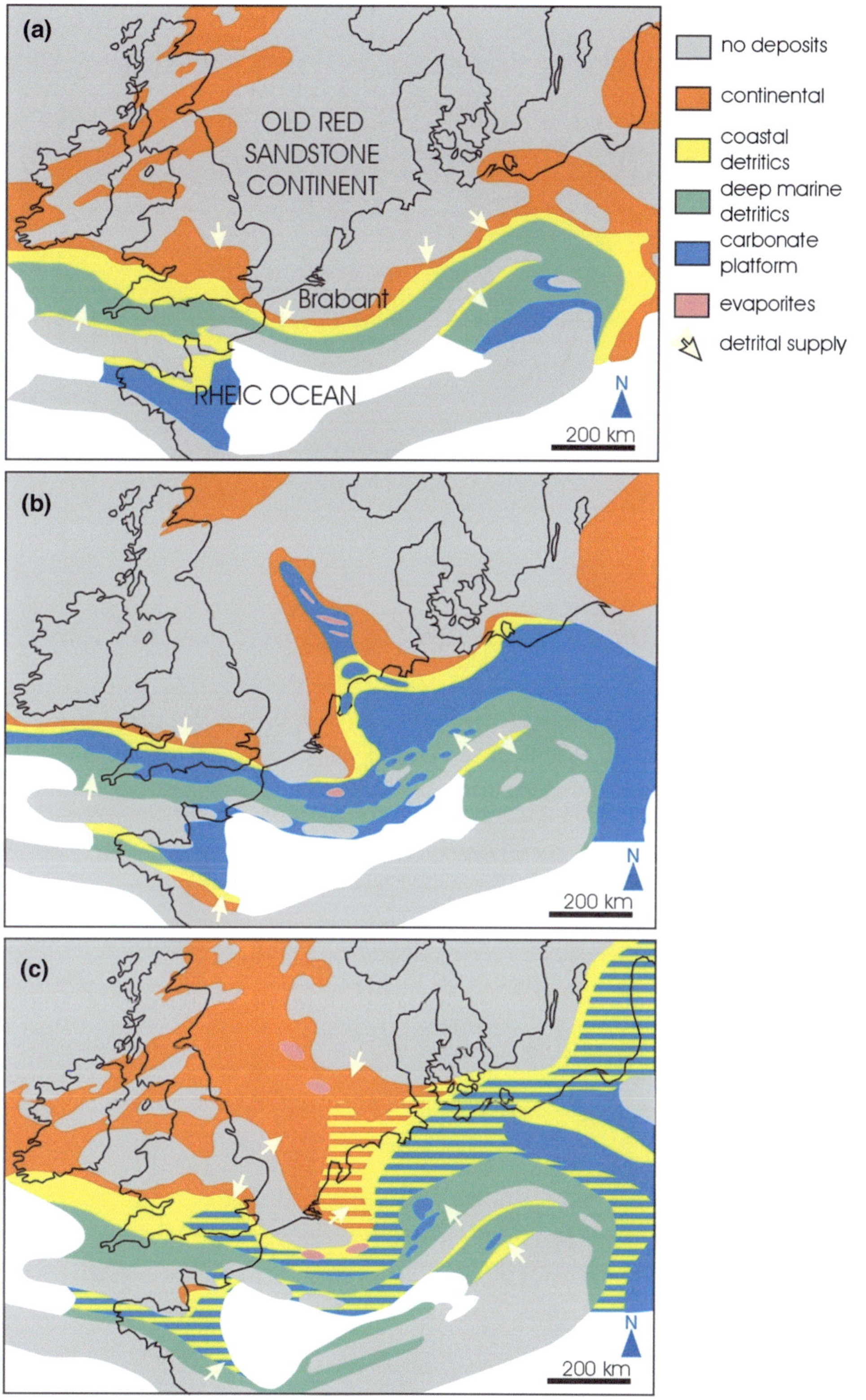

Fig. 2.6 Schematic paleogeographical maps of northwestern Europe during the Devonian. **a** Lower Devonian, **b** middle Devonian and **c** upper Devonian (simplified after Ziegler 1982)

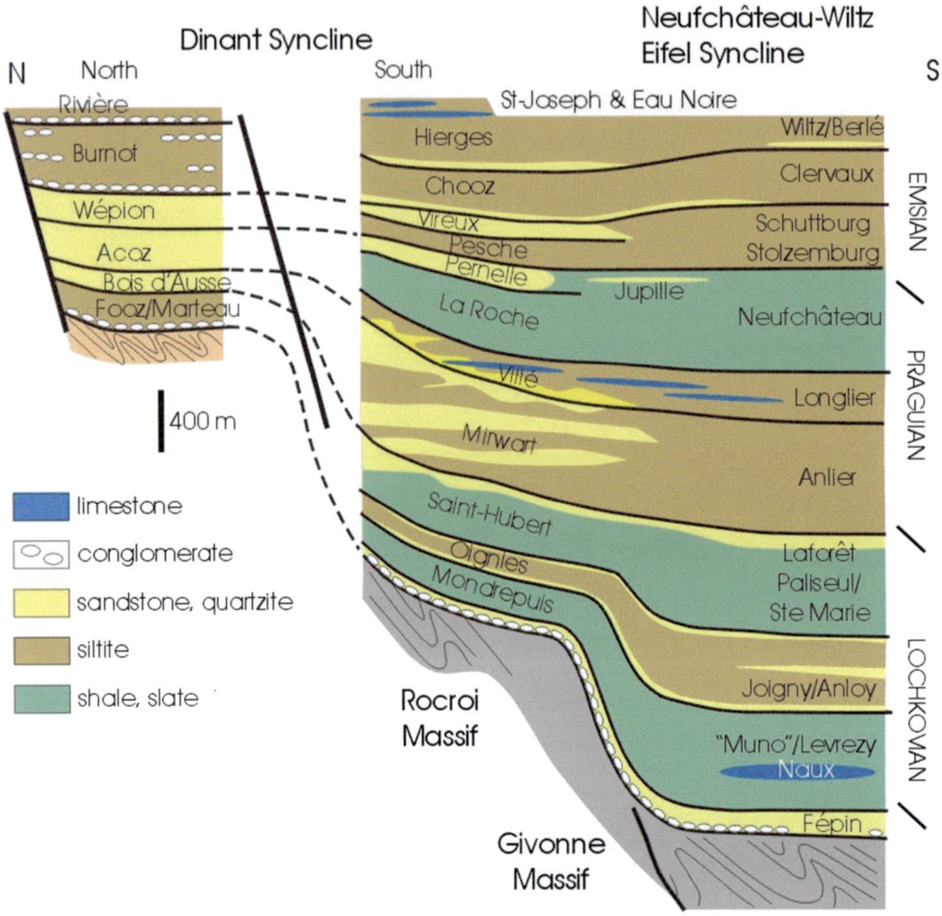

Fig. 2.7 Synthetic north-south transect through the Dinant and Neufchâteau-Wiltz-Eifel synclines before Variscan tectonism, showing the Lower Devonian formations. Lateral thickness variations are attributed to syn-sedimentary normal faults

Fig. 2.8 Lower Devonian sediments. **a** Conglomerate, Ninglinspo and **b** versicoloured shale and siltstone, Helle valley. Marteau Formation, Lochkovian

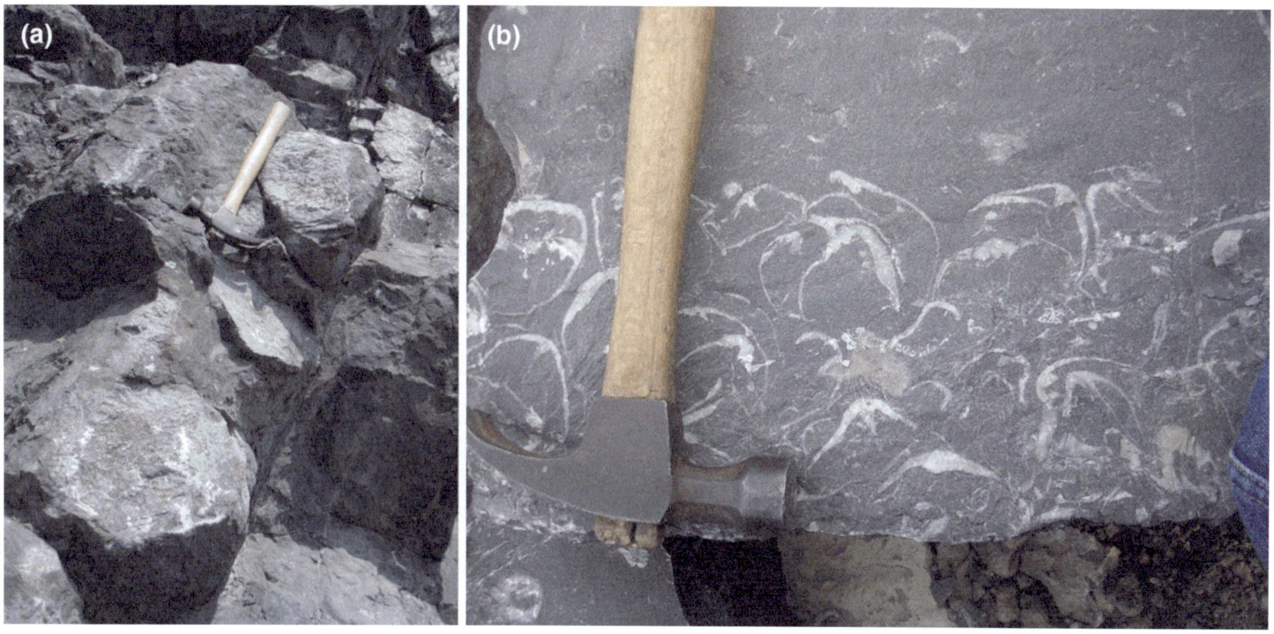

Fig. 2.9 Givetian limestone. **a** Coral colonies, Resteigne and **b** coquina bed (storm deposit), Couvin

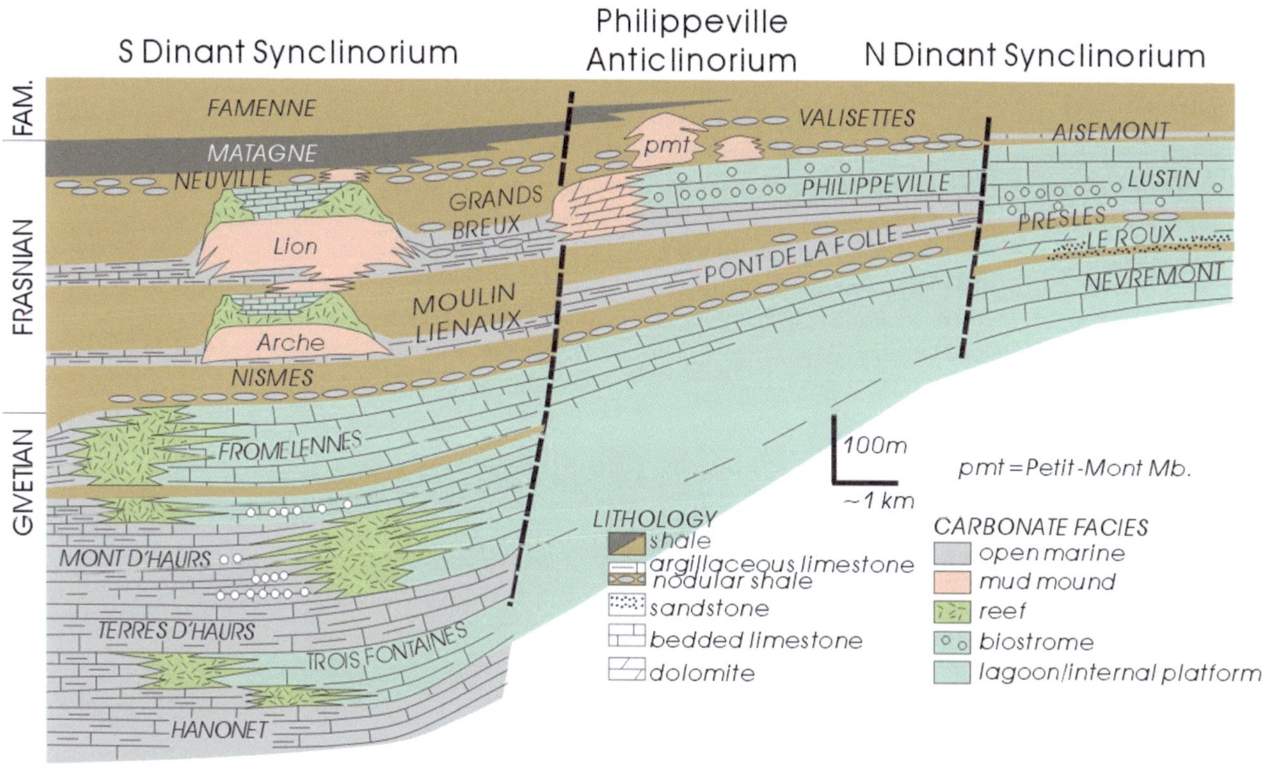

Fig. 2.10 Synthetic north-south transect through the Dinant Syncline before Variscan tectonism showing the Givetian and Frasnian formations

2.3.3 The Upper Devonian Mixed Carbonate-Detrital Formations

During the Frasnian, a transgressive phase brought the coastline farther north, perhaps flooding the entire Brabant massif. A shale unit, several tens of metres thick (Nismes Formation), concealed the entire drowned Givetian platform. After this episode, a new carbonate platform developed, shifted northward relative to the Givetian one (Fig. 2.10). The southern border of the Dinant Syncline shows three stratigraphic levels bearing Frasnian carbonate mounds (Boulvain 2007). These are, in stratigraphic order, the Arche, the Lion and the Petit-Mont members (Fig. 2.10). In the Philippeville Anticline, only the upper level contains mounds (Petit-Mont Member), the other carbonate mound levels being replaced laterally by bedded limestone, with local back-reef character. At the northern border of the Dinant Syncline and in the Brabant parautochton, the entire Frasnian consists of bedded limestone and argillaceous strata (Da Silva and Boulvain 2004).

The Frasnian Petit-Mont carbonate mounds of Belgium are probably the earliest studied among Palaeozoic carbonate mounds worldwide. This remarkable interest shown by generations of geologists derives from the outcrop number and quality, with 69 known carbonate mounds, the majority of which were actively quarried for 'marble' (this word being not used here in a strict petrographic meaning). Consequently, several hundred square metres of sawn sections are accessible for studies (Fig. 2.11). Embedded in shale and nodular shale, the Petit-Mont mounds are 30–80 m thick and 100–150 m in diameter. The initial carbonate mound facies consists of red limestone with sponges, becoming progressively enriched in crinoids and corals, then in stromatopores (calcified sponges) and cyanobacteria. The red pigment was produced by microaerophilic iron bacteria.

After the drowning of the Frasnian carbonate platform and its burial under transgressive shale, the Famennian Stage marked a seascape's complete renewal (Fig. 2.6c). A clear regressive trend brought the coastline back to the south of the Brabant massif and carbonates were replaced by detrital sediments (Thorez et al. 2006). Clastic sedimentation began with the Famenne shale, deposited below the storm wave base, followed by the Esneux Formation, rhythmically alternating shale and sandstone in the storm wave zone, and, finally, after a minor carbonate episode, the Montfort and Evieux formations, consisting in littoral and fluvio-littoral sandstones. These so-called Condroz sandstones are still used in many public and private buildings.

2.3.4 The Dinantian Carbonates

The Carboniferous in Belgium is traditionally subdivided into three series, the Dinantian, the Namurian, and the Westphalian. Carbonates, detrital sediments, and mixed detrital sediments and coal dominate these series, respectively (Fig. 2.12). During the Carboniferous, the Rheic Ocean, located near the Equator, was closing, as evidenced by the forthcoming collision of Gondwana and Laurussia (i.e., the Old Red Sandstone Continent). This major tectonic event gave rise to the Variscan mountain belt and to the formation of the supercontinent Pangea.

Following the fluvio-littoral environment of the Upper Famennian, the Dinantian sedimentation marks a return to pure marine conditions. A dramatic decrease of detrital supply favoured the resumption of a carbonate factory, and a

Fig. 2.11 Frasnian carbonate mounds (Petit-Mont Member). **a** The Beauchâteau mound is 30 m high and is in nearly *horizontal position*. Philippeville Anticline and **b** *red* "marble" with sponges

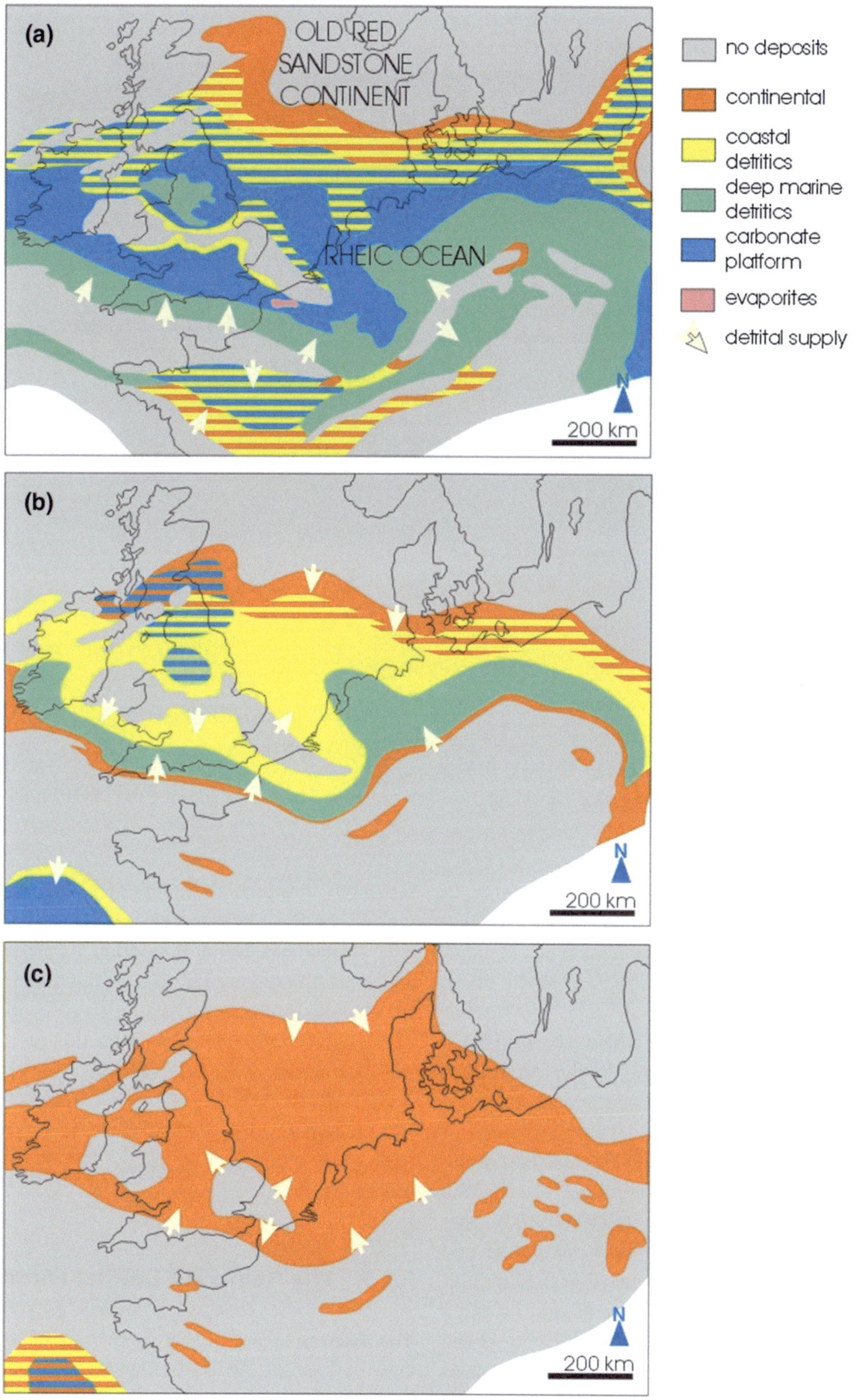

Fig. 2.12 Schematic paleogeographical maps of northwestern Europe during the Carboniferous. **a** Dinantian, **b** Namurian and **c** Westphalian. (simplified after Ziegler 1982)

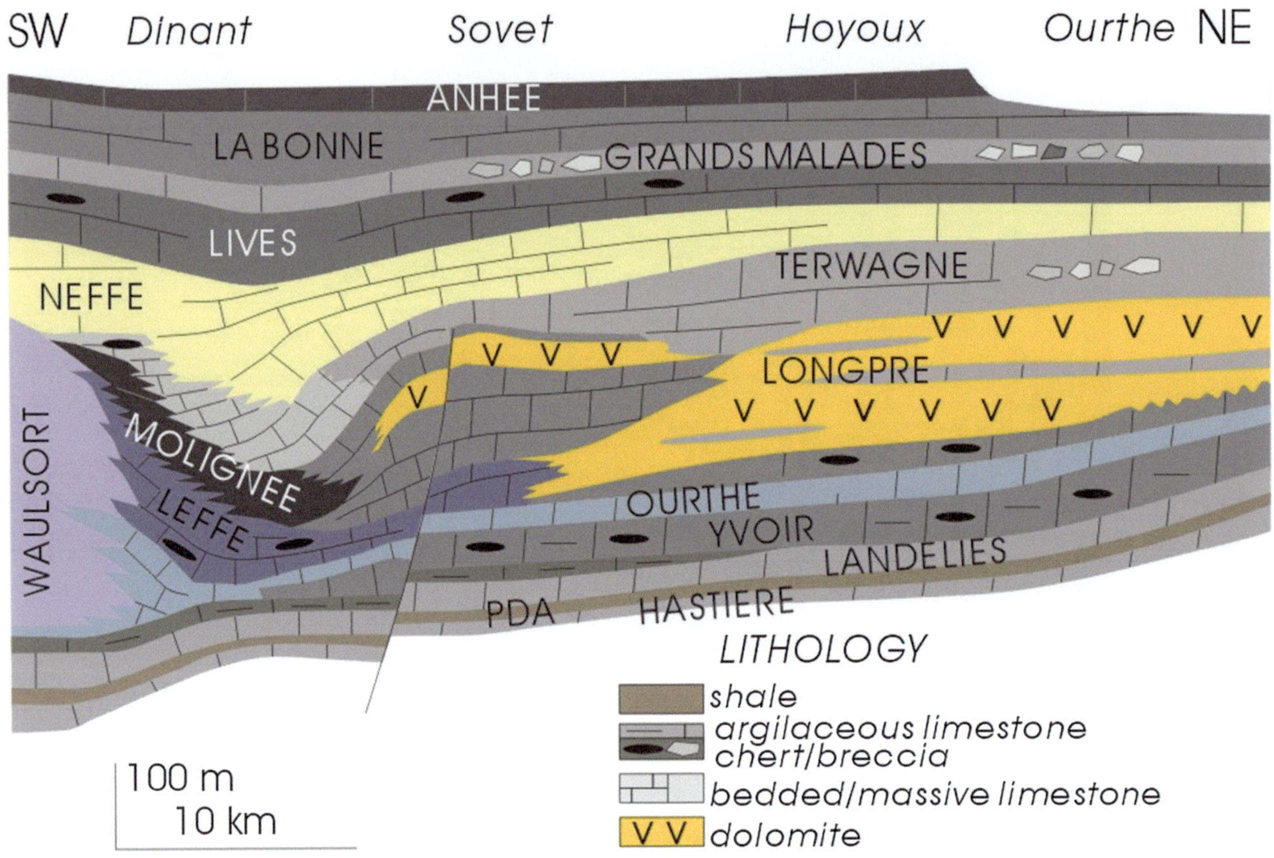

Fig. 2.13 Synthetic north-south transect through the Dinant Syncline before Variscan tectonism showing the Dinantian formations. *PDA* Pont d'Arcole. See text for explanation

large carbonate platform subsequently developed south of the Brabant massif (Fig. 2.12a). This platform was divided into several sedimentary areas with different types of limestones in variable accommodation spaces (Hance et al. 2001). The maximum thickness of the Dinantian reaches 2.5 km in the Hainaut area. Several Dinantian formations from the Dinant-Condroz areas are briefly described, because of their economic significance or international status (Fig. 2.13).

The first Dinantian formations, up to the Ourthe Formation, show only moderate lateral variation. The Ourthe Formation is extensively quarried for its crinoid-rich limestone, misleadingly known as 'Petit granit', as are several equivalent units from the Hainaut area (Fig. 2.14a). Following this formation, a significant differentiation occurred between a shallow, locally dolomitic platform (Condroz and Namur areas) and a more subsiding area where increased accommodation space allowed for carbonate-mound building (Fig. 2.13). These mounds, called Waulsortian reefs, are nearly 400 m high and are rich in bryozoans and sponges (Lees et al. 1985). They are surrounded by chert-rich flank sediments (Fig. 2.14b). The final filling of the Dinant trough included restricted fine-grained black limestone (the world-famous 'black marble' of the Molignée Formation). The subsequent Neffe Formation, consisting of bioclastic shoals, is a very pure light grey limestone intensely quarried for chemical purposes. Dissolution of evaporite beds in the overlying Grands Malades Formation was responsible for the formation of the Grande Brèche, which is a decametres-thick collapse breccia. Finally, the Anhée Formation registered a gradual decrease in carbonate production, progressively replaced by fine-grained detrital sediments.

2.3.5 The Namurian Detrital Formations

The Namurian series, corresponding to the Serpukhovian and the base of the Bashkirian, is essentially characterized by several hundred metres of black shale with subordinate marine limestone (Nyhuis et al. 2014), followed by shale and coarse-grained sandstones or conglomerates. Small beds of coal announce the future development of the great equatorial coal forest (Fig. 2.12b).

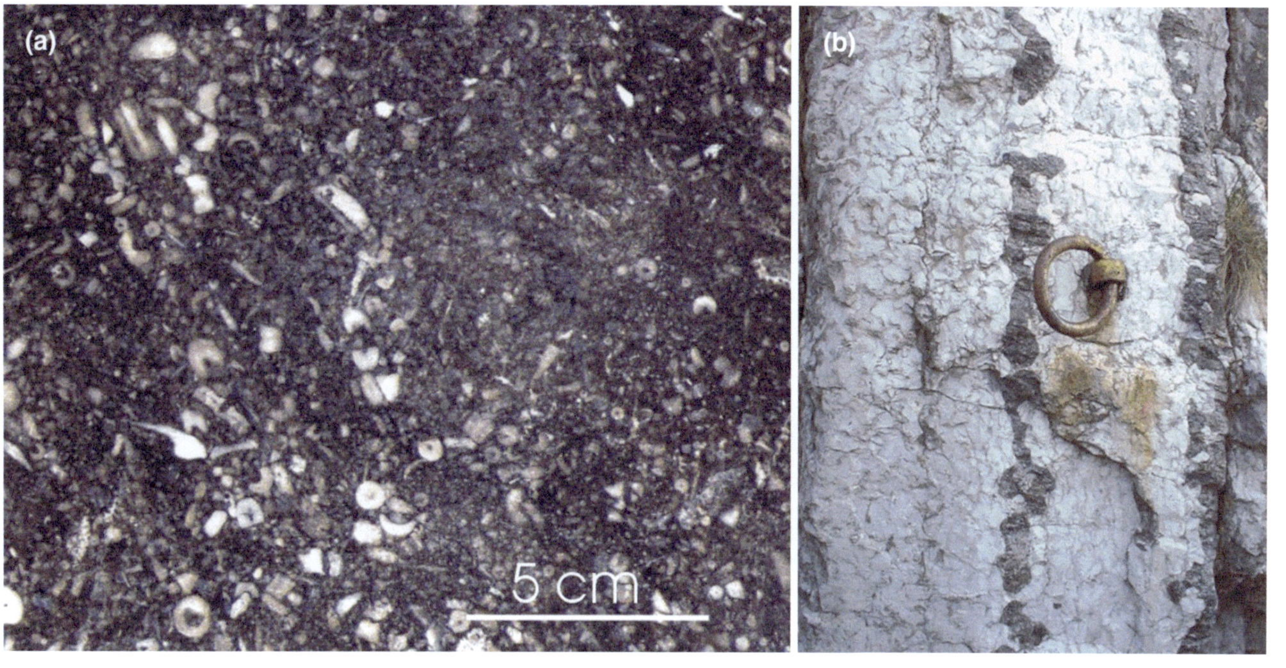

Fig. 2.14 Dinantian carbonates. **a** Polished crinoid-rich limestone (*Petit-granit*) from the Ourthe Formation and **b** chert-rich flank sediments of the Waulsortian buildups. Rocher Bayard, Dinant

2.3.6 The Westphalian Coal Measures

The Westphalian (Bashkirian-Moscovian) includes the bulk of the coal measures in Belgium. The uplift of the Variscan mountain belt led to a general retreat of the seas, giving way to extended lagoon and marsh environments (Fig. 2.12c). Huge amounts of detrital sediments coming from the Variscan mountains mixed with plant remains and accumulated in subsiding areas, now locally forming over 2-km-thick formations. Even the Brabant massif was covered by kilometres of sediment. The Westphalian sedimentation shows a characteristic cyclicity, each cycle starting with a sandstone bed, followed by a coal seam and topped by shale. The sandstone corresponds to coastal and fluvial sediments affected by soil formation and topped by organic matter from plants of the equatorial forest accumulated in a reducing swamp environment, itself overlain by floodplain and lacustrine shale. The ocean was confined to the Netherlands and made only episodic incursions, depositing marine sediments whose fossils are stratigraphically very useful.

2.3.7 The Variscan Orogeny

The remains of the Variscan mountain belt crop out from Spain to Bohemia, going through the Vosges, Massif Central, Ardenne, and Cornwall. Variscan tectonics is responsible for the general structure of Belgium, with the major Midi-Eifel thrust fault separating the Brabant parautochthon in the north from the Ardenne allochthon in the south. The Ardenne allochthon itself is structured in large-scale anticlines and synclines that extend eastward into Luxembourg (Figs. 2.1 and 2.2). The currently accepted model involves closing of the Rheic Ocean during the Late Carboniferous and continental collision between Laurussia and Gondwana (Matte 1986). Belgium and Luxembourg are part of the Rheno-hercynian fold-and-thrust belt (Ardenne allochthon) and of the Variscan foreland (Brabant parautochthon), which show the following characteristics in our region (see Fig. 2.1):

- The southern limb of the Ardenne Anticline and the Neufchâteau-Wiltz-Eifel Syncline are affected by large-scale thrust faults (Schavemaker et al. 2012). North-verging overturned folds are predominant with an axial planar cleavage.
- The Caledonian massifs (Stavelot, Rocroi, Serpont, Givonne) are polycyclic domains that were affected by both the Caledonian and Variscan orogenies. The respective influence of the two deformations is still a matter of discussion. In the Rocroi massif, for example, the similarity of major folds between the Lower Paleozoic and the Lochkovian cover, together with a cleavage affecting both units (Fig. 2.15), argues in favour of the prevalence of the Variscan structuration.
- The Dinant Syncline's southern border is characterized by north-verging overturned folds, whose inverted limbs are frequently affected by subhorizontal inverse faults. The cleavage dips south.

Fig. 2.15 The Lower Devonian Fépin conglomerate forms an unconformity over the Cambrian Revin Group, Fépin. Note the throughgoing Variscan cleavage

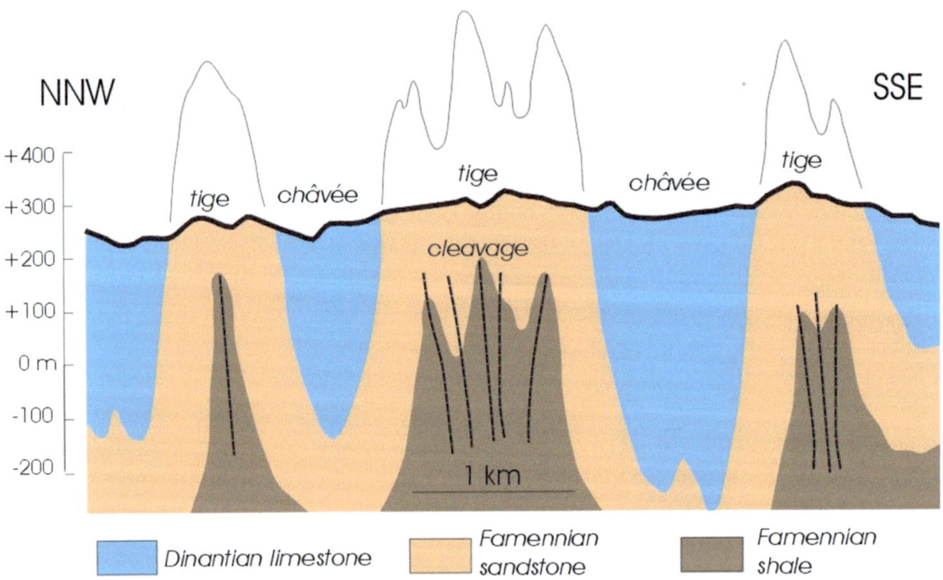

Fig. 2.16 Schematic cross-section in the Condroz area (Achêne) showing the relation between geological structure, lithology, and topography. 'Tige' and 'châvée' are the local names given to ridges and troughs, respectively. Height exaggeration 4×

- In the Dinant Syncline, folds are generally upright with a planar axial cleavage. The presence of folded sandstone and limestone formations is responsible for the development of the well-known condruzian morphology, with elongated hills following the axis of hard sandstone anticlines and troughs being cut in the soluble limestone that outcrops in the synclines (Fig. 2.16).
- North of the Midi-Eifel thrust fault, several complex tectonic units form the 'Haine-Sambre-Meuse thrust sheets' area as a consequence of the thrusting of the Ardenne allochthon onto the Brabant parauthochton (Belanger et al. 2012) (Fig. 2.17). Beds are vertical or overturned and longitudinal faults are common.
- North of the Haine-Sambre-Meuse thrust sheets, folds, and faults are weakly developed and the beds of the Middle Devonian-Carboniferous Brabant parautochthon cover the Brabant massif with a dip of 10°–20° to the south. Theoretically, the Variscan front is located north of the last deformational structure observed in the Brabant parautochthon.

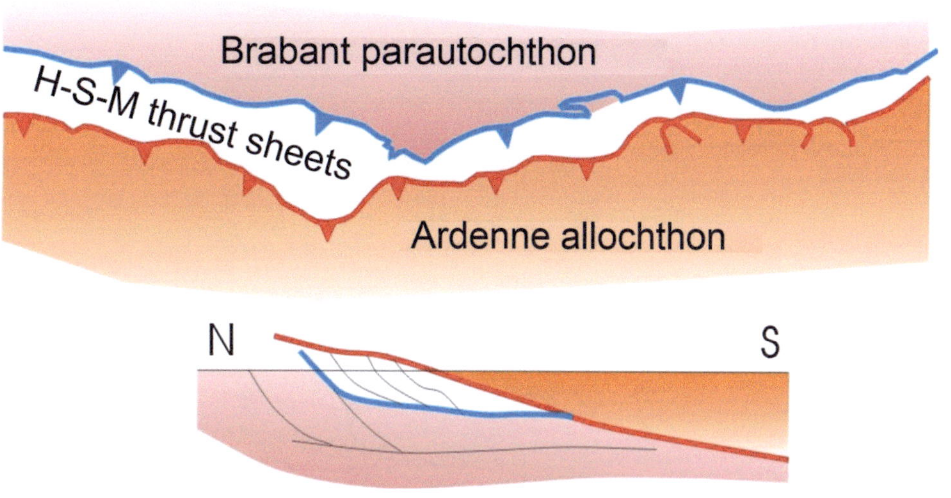

Fig. 2.17 Tectonic relations between the Haine-Sambre-Meuse (H-S-M) thrust sheets and the Ardenne allochthon, in map view and cross-section

A well-developed, though low-grade, metamorphic phase (anchizone-epizone, around 400 °C) is recorded in the Ardenne (Fielitz and Mansy 1999). Geographically, this narrow, ~120-km-long metamorphic area extends from SW Ardenne to N Luxembourg. The Variscan metamorphism was responsible for the transformation of shale into fine slate, used locally for roof tiles. The age of the Variscan metamorphism is pre-orogenic, probably related to burial.

2.4 The Post-orogenic Times

Since the Variscan orogeny, no tectonic phase has affected the Belgian and Luxembourgian regions other than through far-field epeirogenic effects and limited brittle deformation (Vandycke 2002; Havron et al. 2007; Demoulin and Hallot 2009). However, limited subsidence has made further sedimentation possible. Homoclinal or horizontal unfolded formations have been locally accumulating from the Permian to the present.

During the Permian, the Pangea supercontinent moved into an arid climate, due to generalized continental conditions induced by its huge surface. Permian formations are uncommon in Belgium and Luxembourg, cropping out only in the Malmédy graben of NE Ardenne (alluvial fan conglomerate) and determined in the deep subsurface in the Campine basin (Dusar et al. 2001).

2.5 The Homoclinal Triassic-Jurassic Series

The geodynamic context of the Mesozoic series is that of the break-up of Pangea. Epicontinental seas, connected with either the Paris basin or Germany and the Netherlands, episodically covered our countries. Climate was still warm due to low latitude, and the erosion of the Variscan mountains brought large amounts of detrital material into the seas. It was not until the Middle Jurassic (Bajocian) that a carbonate factory started up again (Fig. 2.18a, b).

The Triassic-Jurassic series are confined to the Belgian Lorraine in SE Belgium and Guttland in Luxembourg. Together with a very low dip angle to the south, alternating weak and hard strata have been responsible for the development of cuestas (Lucius 1952; Maubeuge 1954; see also Chap. 23). The earliest formation covering the Ardenne-Eisleck basement corresponds to alluvial systems (clay and gravels of the Habay Formation in Belgium, red sandstones from the Buntsandstein in Luxembourg, Fig. 2.19). Their age is gradually younger westward because the Triassic coastlines progressed from east to west (Fig. 2.19). The first marine influence is recorded by the Muschelkalk formations in Luxembourg, with sandstone and marls with evaporites and encrinites. In Belgium, the first marine unit is the Attert Formation with clay and dolomitic marls including evaporitic pseudomorphs. The Mortinsart Formation ends the Triassic marine transgressive cycle with littoral sandstone and marls, topped by alluvial clay.

The Jurassic marine transgression was more important and deposited several tens of metres of alternating calcareous marls and sand or sandstones. The prominent Luxembourg Formation consists of alternating littoral sand and sandstone reaching nearly 100 m in thickness (Van den Bril and Swennen 2009) (Figs. 2.19 and 2.20). The upper part of this diachronic formation, still actively exploited for building stone, grades laterally eastwards into the marls of the Arlon Formation. Pliensbachian and Toarcian formations then correspond to fine-grained dark marls, locally rich in organic matter, indicating quiet sedimentation conditions on an anoxic sea floor. In Luxembourg, the top of the Toarcian is sandier with ironstone beds (the '*Minette*'), which fully

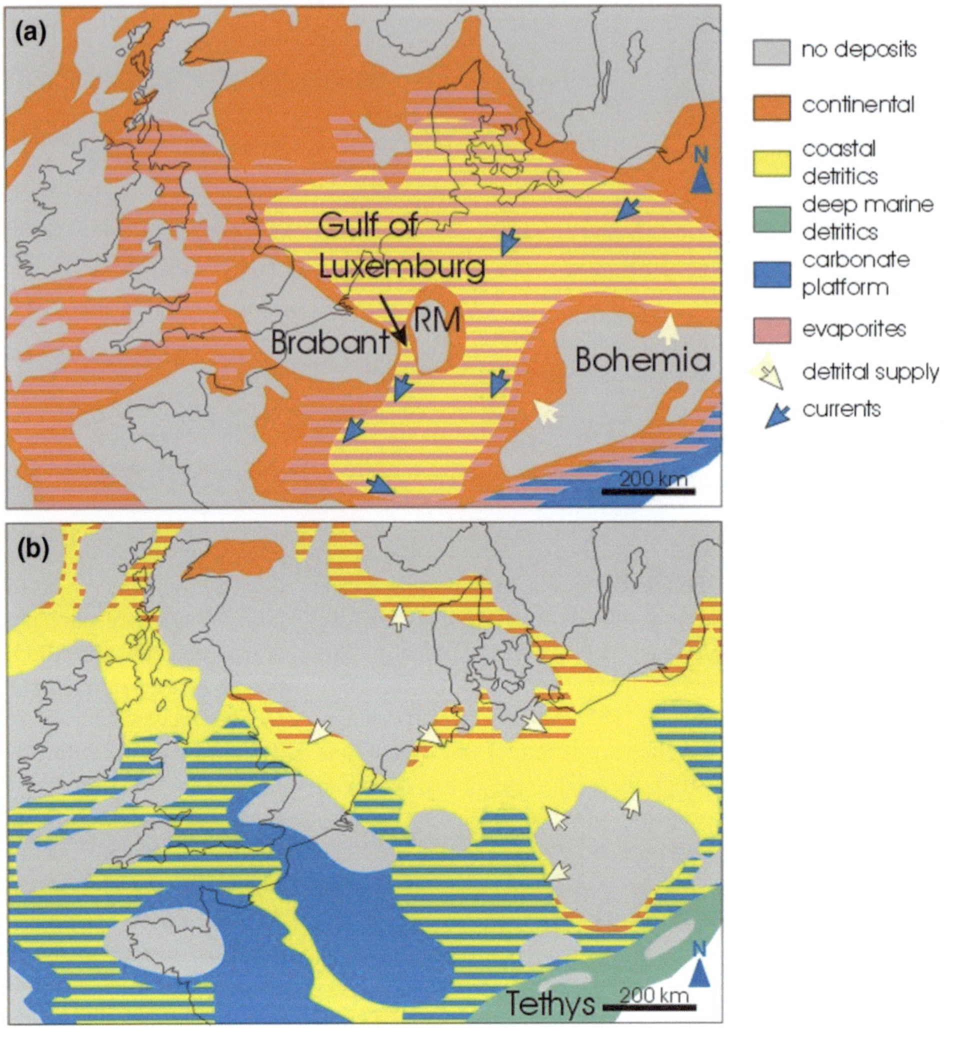

Fig. 2.18 Schematic paleogeographical maps of northwestern Europe during the Triassic and the Jurassic. **a** Keuper and **b** Sinemurian-Aalenian. *RM* Rhenish massif (simplified after Ziegler 1982)

developed through the Aalenian (Bintz and Storoni 2009). In Belgium, the contemporaneous beds were largely eroded during an episode of emersion. The end of the Jurassic sedimentary record occurred during the Bajocian in both countries. This stage is characterized by the development of a carbonate platform. In Luxembourg, the Audun-le-Tiche Limestone includes spectacular 20-m-thick, 200-m-wide coral reefs.

2.6 The Cretaceous Cover

During the Cretaceous, Pangea was fully broken up. The newly formed Atlantic Ocean started to influence our regions, and Belgium and Luxembourg reached latitudes between 40° and 60°N. The Late Cretaceous was remarkable for its very high sea levels, turning Europe into an archipelago (Fig. 2.21). In Belgium, Cretaceous units crop out in the Mons basin and in the Tournai, Hesbaye, and Herve regions (Fig. 2.1). Though largely dominated by chalk that accumulated from the Coniacian (Mons basin) or the Campanian (Hesbaye and Herve areas) to the Maastrichtian, other types of sediments are also represented.

As an example, the Mons series is briefly described (Fig. 2.22). During the Early Cretaceous, continental facies corresponded to Wealdian alluvial gravels, sands, clay, and lignite. In the north of the basin, Wealdian sediments filled sinkholes where the famous iguanodons of Bernissart were notably trapped (Bultynck 1989). Then, the Albian Sea, coming from the Paris basin, flooded the Mons gulf and deposited littoral gravels, glauconitic sand and sponge-rich marls. The Cenomanian and Turonian stages are characterized by glauconitic marls and silicified limestone (e.g., Baele 2003). From the Coniacian onwards, the detrital supply vanished and the chalk sea was installed in the Mons region. Some Maastrichtian chalk units are sufficiently rich in phosphate debris to justify underground mining. In the Hesbaye-Herve region, the chalk sea was established only in the Campanian, probably in parallel with the complete flooding of the Brabant massif (Felder 1975).

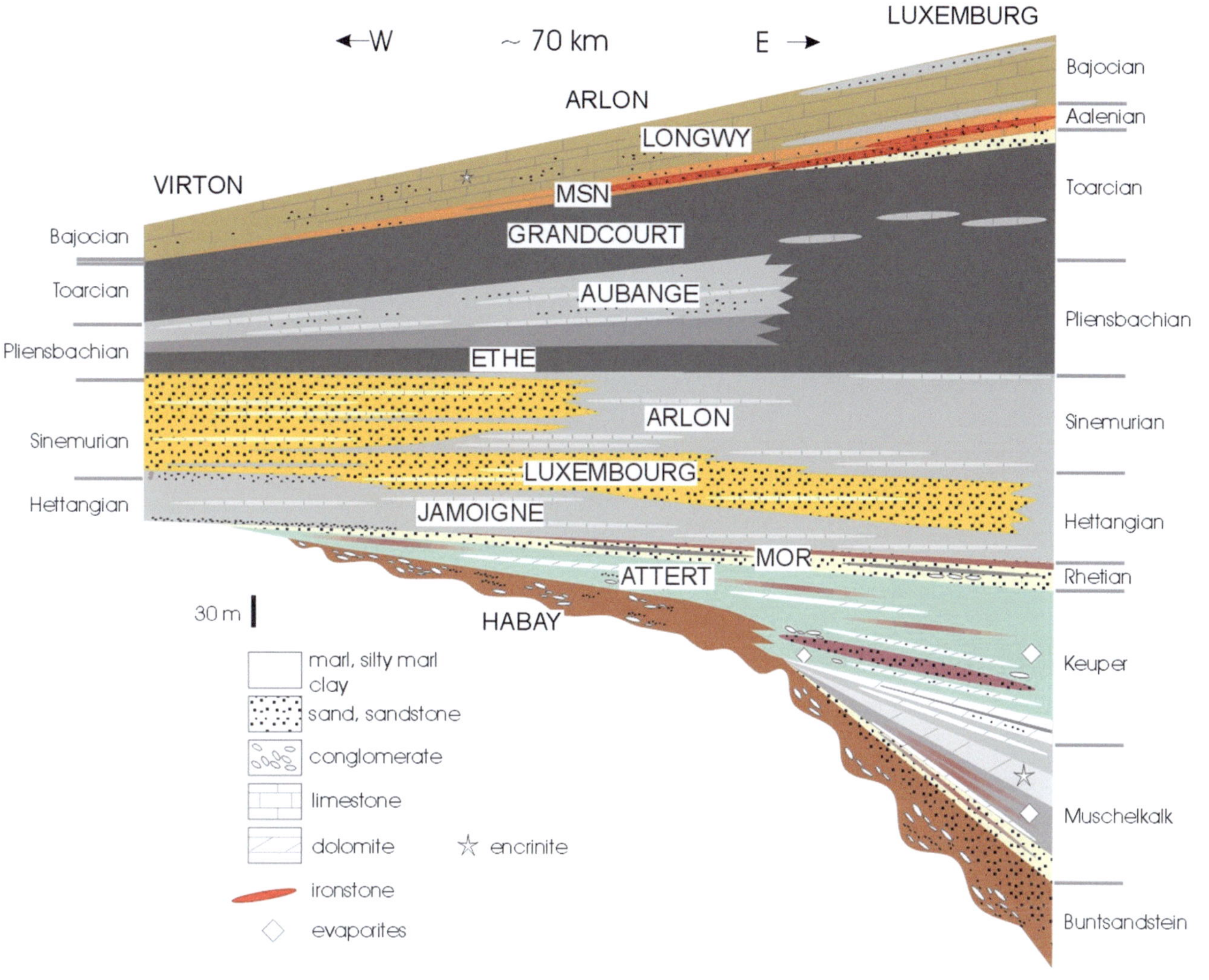

Fig. 2.19 Synthetic east-west transect through the Triassic-Jurassic cover of Belgian Lorraine and Luxembourgian Guttland. *MOR* Mortinsart Formation. *MSN* Mont-Saint-Martin Formation (modified after Boulvain et al. 2001)

2.7 The Cenozoic

Cenozoic sedimentary deposits cover most of the area north of the Sambre-Meuse drainage line and only smaller outliers occur over the Condroz and Ardenne (Fig. 2.1).

The Cenozoic strata in Belgium (Borremans 2015) have formed at the southern rim of the North Sea basin. At the turn of the Meso- to Cenozoic, the North Sea was a shallow dish-shaped area subsiding due to thermal cooling over a failed Mid-Mesozoic rift. At the beginning of the Cenozoic, the southern North Sea extended into the Paris basin (Fig. 2.23) and Alpine tectonic forces caused vertical uplift pulses in the area resulting in hiatuses in the Danian chalk deposition.

During the Selandian (61.6–59.2 Ma), chalk was eroded at a large scale along the margins of the North Sea basin, and redeposited. In Belgium, the resulting Gelinden marls contain subtropical Thetyan-type tree leaves (Steurbaut 1998) (Fig. 2.24).

At the Paleocene/Eocene transition (56 Ma), the North Atlantic Ocean opened further, preceded by a broad thermal uplift that started already in the Middle Paleocene. The earliest indirect indication of related volcanic activity in the Franco-Belgian basin is the middle Thanetian 'tuffeau', a calcareous fine sandstone in the Hannut Formation (Fig. 2.24), which owes its light weight to the voids left by the numerous, now dissolved sponge spiculae, the abundance of which is associated with the excess ash in the sea water at that time. Thin bentonite beds and dispersed glass shards appear only locally in the Belgian basin at the very beginning of the Eocene. At the same time, the dissolved silica from the Middle Thanetian beds allowed for groundwater silcrete formation, notably silicifying the overlying Hoegaarden swamp cypress forest (Fairon-Demaret et al. 2003) (Fig. 2.25). This very early Eocene was a turbulent time with marked climatic changes and vertical uplift of the Bray-Artois and Brabant blocks, resulting in a variety of

Fig. 2.20 View of the interdigitation of the Luxembourg and Arlon formations in the Tontelange quarry, 5 km north of Arlon, southeast Belgium

continental and coastal deposits overlying a major erosive unconformity and grouped in the Tienen Formation (Fig. 2.24) with a duration of only ~0.5 million year. At Dormaal, the basal Tienen Formation deposits have yielded the earliest modern mammal fauna in Europe.

Drifting away from the spreading North Atlantic led to a thermal subsidence in the North Sea basin, which was collecting clay-dominated sediments during most of the Ypresian (56–47.8 Ma) (Steurbaut 2006). Fine sand deposits only formed in the coastal realm. The clay mineralogy in these deposits is mostly smectite, interpreted as the weathering product on land or in the sea of basalts and basaltic pyroclastics. Vertical trends in the clay-to-silt ratio of these Ieper Group sediments (Fig. 2.24) are related to global sea-level variations.

The final Ypresian and the transition to Lutetian time are characterized by the reappearance of dominant sandy deposits, erosive bases of successive sequences and hiatuses. All these features reflect uplift of the Artois-Brabant area, resulting in the separation of the Paris basin and the southern North Sea from the Lutetian onwards. The most prominent, decametres-deep erosive contact occurs at the base of the Brussels sand Formation in Brabant, with an associated erosive channel located offshore Oostende. These approximately north- to northeast-trending channels are supposed to represent the last connection between the Paris and Belgian basins, cut during a global low sea level (Vandenberghe et al. 2004). During the Lutetian, Bartonian, and Priabonian, (47.8–33.9 Ma), a few transgressive pulses, the most extensive of which caused deposition of the Asse-Ursel clay members (Fig. 2.24), did not cover more than northwestern Belgium, while the area to the east remained uplifted above sea level. This major tectonic rearrangement was a response to the Pyrenean deformation pulse at the end of the Eocene. Since the Late Ypresian, also the Hainaut area has remained above sea level and the Brabant block became drowned only sporadically, while subsidence of the Campine Basin, restarted in the latest Cretaceous, continued uninterruptedly (Fig. 2.24).

A renewed transgression over eastern Belgium and even over the Ardenne occurred in the earliest Oligocene and deposited fine sands in a shallow sea. The Sint-Huibrechts-Hern sand Formation (Fig. 2.24) of the Tongeren Group is part of this transgression, as are the sands preserved in large karstic depressions in the Condroz. Since the Early Eocene, the climate was gradually cooling but a sharper cooling occurred worldwide at the end of the short-lived earliest Oligocene marine incursion (De Man et al. 2004). At the same time, in parallel with the sea withdrawal, a short period of widespread soil formation occurred in the newly emerged landscape. In the swampy coastal plain of the subsequent major Rupelian transgression lived a rich tetrapod fauna among which were the

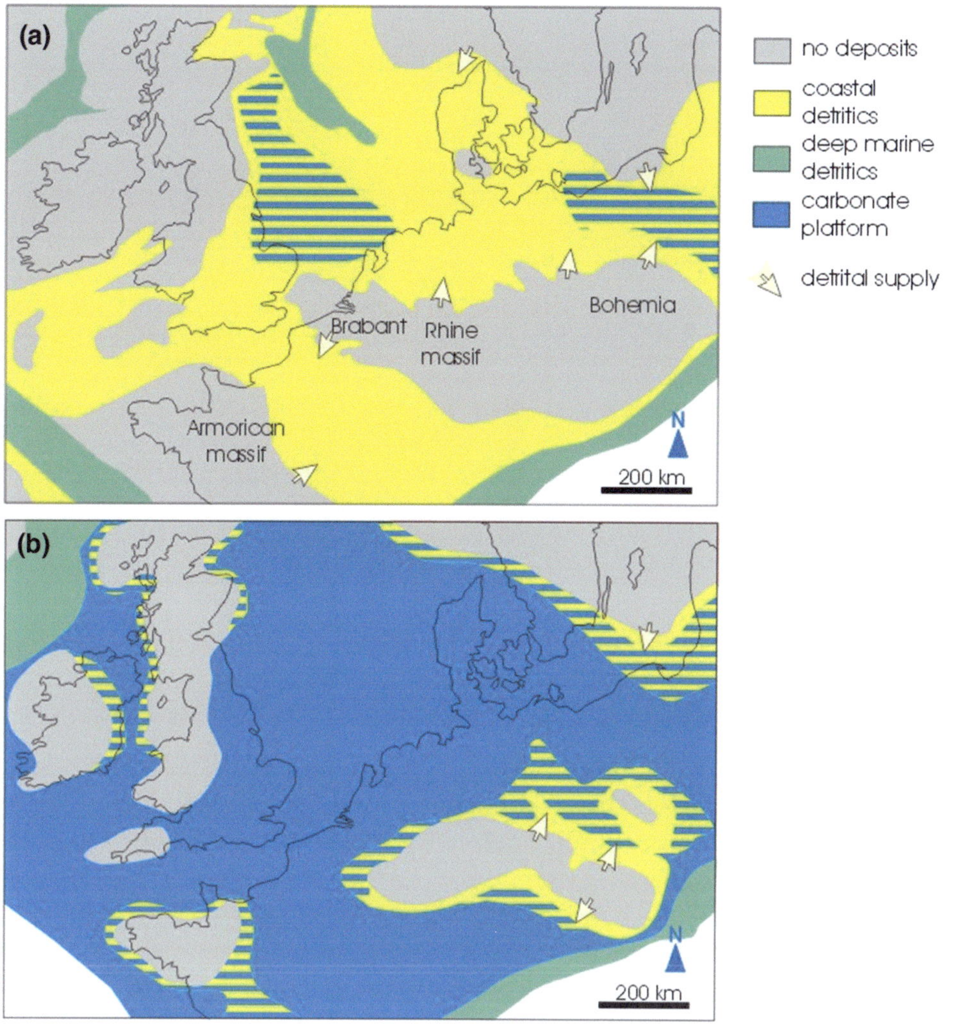

Fig. 2.21 Schematic paleogeographical maps of northwestern Europe during the Cretaceous. **a** Lower Cretaceous and **b** upper Cretaceous (simplified after Ziegler 1982)

first mammals of Asian origin resulting from the faunal turnover known as the Grande Coupure.

During the Rupelian (33.9–28.1 Ma), a large part of north Belgium subsided and became inundated, and the Boom clay Formation (Fig. 2.24) was deposited (Fig. 2.26). Since the Pyrenean tectonic rearrangement in the area, the clay mineralogy had changed to dominantly illitic with associated kaolinite. The Boom Clay displays a banded lithology that reflects water depth changes at a pace controlled mainly by obliquity oscillations and explained by the waxing and waning of a major ice cap that had developed on Antarctica since the abrupt cooling in the very early Rupelian (Vandenberghe et al. 2014a).

By the end of the Rupelian, differential vertical tectonics affected again north Belgium, resulting in uplift of the Antwerp area and erosion of a considerable thickness of previously deposited Boom Clay while in Limburg, in the east, the Upper Rupelian deposits turned from deeper water clay into fine sands of the Eigenbilzen Formation. Such shallowed waters resulted from the broad regional uplift preceding the subsidence of the Roer Valley Graben (RVG) that affected northeast Limburg during the Chattian (28.1–23 Ma). Mainly sandy Upper Oligocene deposits are preserved almost exclusively inside the graben, up to a few hundred metres thick in the very northeast. The northwest-southeast trending RVG boundary faults in eastern Belgium are reactivations of earlier faults that have their origin in the deeper Paleozoic faults. The main boundary fault east of which thick Chattian deposits are preserved is known as the Mol-Rauw Fault (see Fig. 12.1 of Chap. 12).

Outside the graben to the west, sedimentation resumed after a long hiatus spanning most of the Chattian and the Aquitanian. Resulting from a combination of global low sea level and uplift induced by Alpine tectonics (known as the Savian pulse), this hiatus is widespread in the North Sea basin. The latest Aquitanian and Burdigalian deposits in

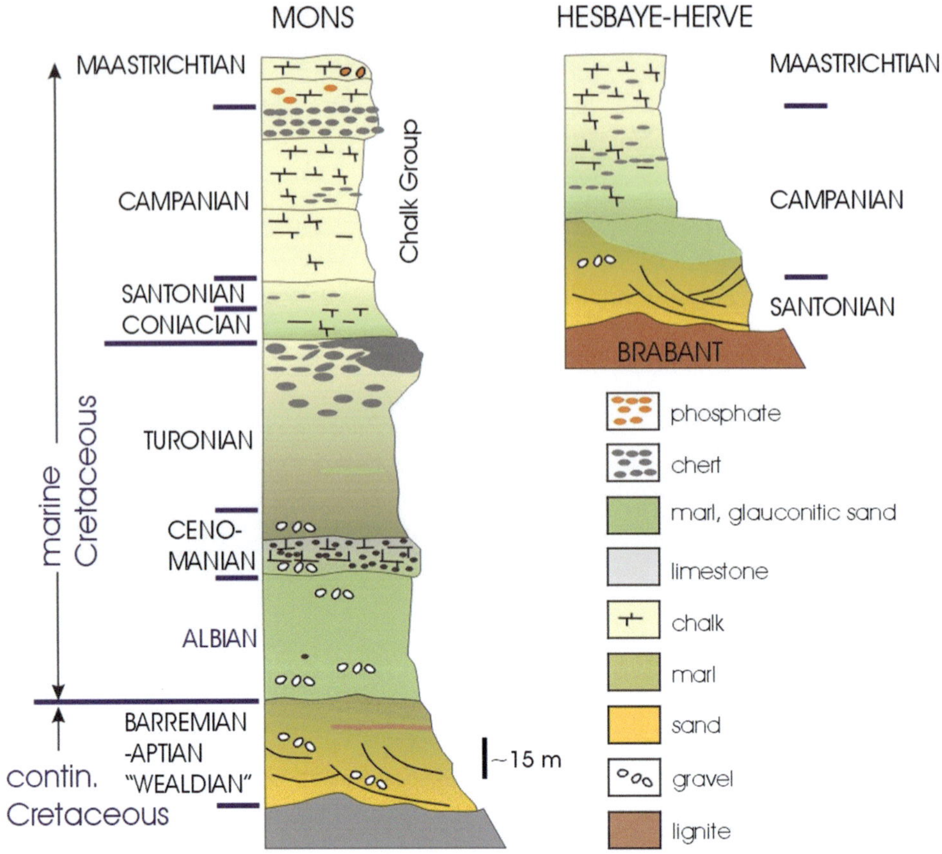

Fig. 2.22 Schematic logs of the Cretaceous cover in the Mons basin and the Hesbaye-Herve region, showing the time lag between the two chalk series

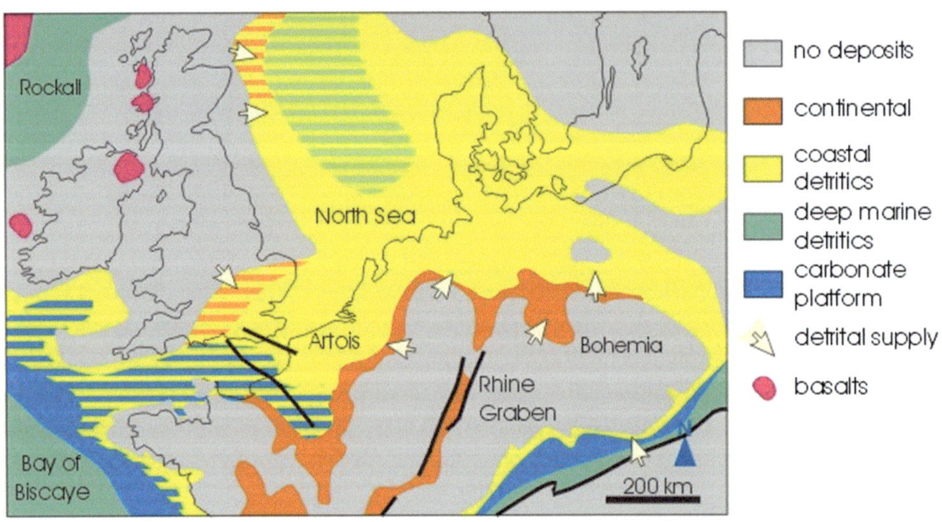

Fig. 2.23 Schematic paleogeographical map of northwestern Europe during the Paleogene (simplified after Ziegler 1982)

north Belgium are shallow-water, glauconite-rich sands of the Berchem and Bolderberg formations (Louwye and Laga 2008) (Fig. 2.24). Several sedimentation pulses can be recognized, possibly continuing into the Middle Miocene. During the Middle Miocene (16.0–11.6 Ma), a quartz-rich continental sand developed in NE Belgium, including the pure quartz Opgrimbie sand facies (Fig. 2.27) and lignite tentatively correlated with the Frimmersdorf seam of the main Lower Rhine brown coal area. Also during the Middle Miocene, important vertical tectonic activity rearranged the Rhine river system and created the major Mid-Miocene Unconformity in the North Sea. In north Belgium, an erosional hiatus occurred at that time between the Berchem-Bolderberg formations and the Upper Miocene deposits of the Diest and Kasterlee formations, representing the remains of an end Serravallian fluvial system draining

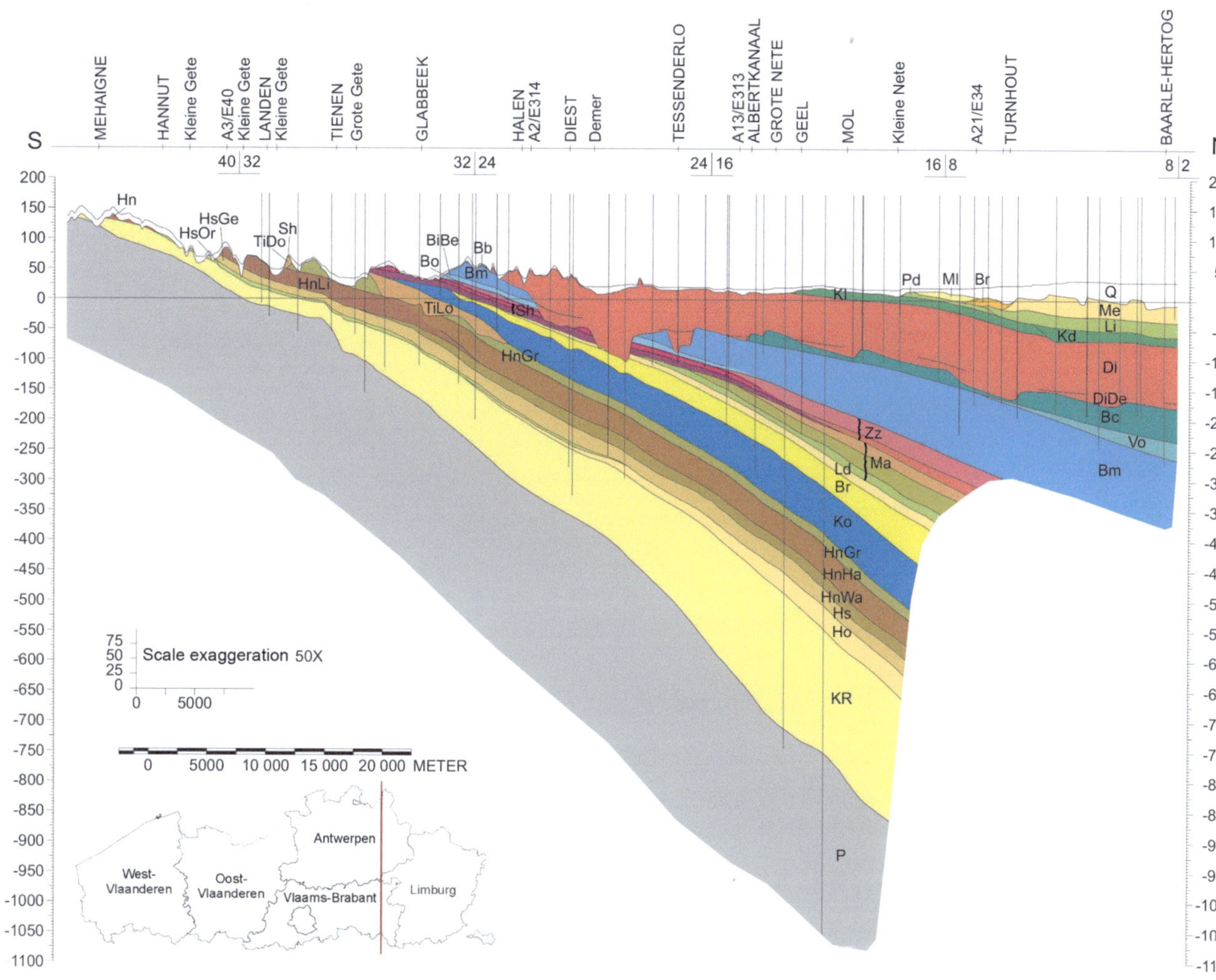

Fig. 2.24 Synthetic geological cross-section between Hannut in the south and Baarle Hertog in the north (location is shown by the *red* line in the *inset*). *P* Paleozoic. *KR* Cretaceous. *Ho* Houthem Formation (Fm) (Danian). *Hs* Heers Fm (Selandian), with HsO Orp Member (Mbr) and HsGe Gelinden Mbr. HnHa Halen Mbr, HnWa Waterschei Mbr, and HnGr Grandglise Mbr of the Hannut Fm (Thanetian). TiDo Dormaal Mbr, and TiLo Loksbergen Mbr of the Tienen Fm. *Ko* Kortrijk Fm of the Ieper Group (Ypresian). *Br* Brussels Fm. *Ld* Lede Fm. *Ma* Maldegem Fm, including the Asse and Ursel Mbrs. *Zz* Zelzate Fm. *Sh* Sint-Huibrechts-Hern *Fm* BiBe Berg Mbr of the Bilzen Fm. *Bm* Boom Fm (Rupelian). *Vo* Voort Fm (Chattian). *Bc* Berchem Fm (Lower to Middle Miocene). *DiDe* Dessel Mbr of the Di Diest Fm (Tortonian to Messinian). *Kd—Kl* Kattendijk and Kasterlee Fms (Messinian). *Li* Lillo Fm (Pliocene). *Me* Merksplas Fm (Pliocene). *Pd* Poederlee Fm (Pliocene). *Ml* Mol Fm (Pliocene). *Br* Brasschaat Fm (Quaternary). *Q* Quaternary (modified after D.O.V. (http://www.dov.vlaanderen.be) and Borremans 2015). *Vertical lines* locate boreholes. *Numbers* refer to geological map sheet numbering

towards the RVG depression. The present geometry of this fluvial system was reshaped during its subsequent transgressive inundation. The Diest Formation deposits in this northeast-trending valley system in the Hageland probably represent a slightly earlier sedimentation pulse than the westward prograding Campine Diest Formation extending from the Rhine delta mouth in the east to the Antwerp area in the west (Vandenberghe et al. 2014b). The classical identification of the Diest Formation in the iron-sandstone-capped hills of southwest Flanders and northern France is debatable (see Chap. 14).

During the Late Miocene and Pliocene (5.3–2.6 Ma), the influence of the prograding Rhine delta became gradually more visible in the sediments of north Belgium. The marine glauconitic fine sands of the Kasterlee Formation in Campine and the suite of Pliocene glauconitic coastal marine deposits in the Antwerp area became progressively replaced by the westward shifting quartz-rich estuarine sands of the Mol Formation, which is part of the Kiezeloolite Formation occurring typically in the RVG in the east. Also the oldest gravels of the Meuse, known as the *Traînée mosane*, are considered Pliocene in age.

Fig. 2.25 **a** Loose white quartz sand of the earliest Eocene Tienen Formation showing a sedimentary silcrete layer of regional extension in central to east Middle Belgium. **b** Top of the same formation including tree stumps (*arrows*) of the silicified Hoegaarden swamp cypress forest in the white sand. The overlying *darker* sand is the glauconitic cross bedded Mid-Eocene Brussels Formation

The Quaternary deposits in Belgium are relatively thin. Coastal plain deposits occur only in the Early Pleistocene formations in north Belgium and in the later Pleistocene series along the present Belgian coast (see Chap. 19). Other Quaternary layers are mainly fluviatile sediments and aeolian loam and coversands, displaying periglacial structures acquired during glacial periods (see Chap. 18). The interplay between tectonic uplift and river erosion and deposition during glacial–i-nterglacial cycles has resulted in the common occurrence of river terraces (see Chap. 10). The Campine Plateau consists of Middle Pleistocene Rhine and Meuse deposits bordered to the west by the Mol-Rauw Fault. It now stands in relief because of deeper erosion of the more erodible Pliocene sand outcropping to the west of the plateau (see Chap. 12). Holocene deposits show significant thickness only along the present coast and in the river valleys (see Chaps. 16 and 18).

Fig. 2.26 The Rupelian Boom clay Formation in the Argex clay pit at Kruibeke-Burcht, near Antwerp. The layering in the clay represents obliquity-driven cycles of varying water depth. The *upper darker clays* contain more land-derived organic particles than the paler lower clays. *Thin white horizons* are septaria layers, which have given the name 'Septarien Ton' to this clay in Germany. The *inset* shows a vertical and a horizontal sections across a septaria

Fig. 2.27 Mid-Miocene Opgrimbie quartz sand facies in the Sigrano exploitation pit near Heerlen, The Netherlands, ~15 km east of the Dutch-Belgian border at Maastricht. Shallow-water sedimentary current structures are cross-cut by subvertical fractures caused by the activity of a nearby fault of the Roer Valley Graben. The *top* cover consists of Pleistocene loam and gravel

2.8 Conclusions

Belgium and Luxembourg are countries where stones have always played a significant role. The history of people and of their underground is intertwined. The immemorial mining of diverse resources, from chert through coal and ironstone to limestone has shaped the landscapes and societies. Sedimentary rocks largely characterize the Belgian and Luxembourgian substrate. The alternation of sedimentation periods and deformation events has shaped the underground into several major sedimentary-structural units, from the Lower Paleozoic inliers through the Devonian-Carboniferous faulted and folded belt, and the homoclinal Triassic-Jurassic series to the subhorizontal Cretaceous and Cenozoic covers. The Variscan fold-and-thrust belt includes two major tectonic units, namely the Brabant parauthochton and the Ardenne allochthon, separated by the Midi-Eifel thrust fault and the Haine-Sambre-Meuse thrust sheets. The Ardenne allochthon is itself formed into major anticlines and synclines.

Acknowledgements F. Boulvain is grateful to all those who shared their remarks and observations when visiting outcrops in Belgium and Luxembourg. Special thanks to J-L. Pingot, A. Herbosch, A. Delmer, M. Hennebert, E. Juvigné, S. Dechamps, J. Thorez, M. Coen-Aubert and J-M. Marion. Robin Weatherl is acknowledged for linguistic help.

References

André L (1991) The concealed crystalline basement in Belgium and the "Brabantia" microplate concept: constraints from the Caledonian magmatic and sedimentary rocks. Ann Soc Géol Belgique 114:117–139

Baele JM (2003) Identification of post-Variscan supergene evolution in marine cherts and residual silicified deposits from Belgium. Géol France 1:39–42

Belanger I, Delaby S, Delcambre B, Ghysel P, Hennebert M, Laloux M, Marion JM, Mottequin B, Pingot JL (2012) Redéfinition des unités structurales du front varisque utilisées dans le cadre de la nouvelle carte géologique de Wallonie (Belgique). Geol Belg 15:169–175

Bintz J, Storoni A (2009) Les Minerais de Fer Luxembourgeois. Schortgen, Esch-sur-Alzette, p 24

Borremans M (ed) (2015) Geologie van Vlaanderen. Academia Press, 491p. ISBN 978 90 382 2433 6

Boulvain F (2007) Frasnian carbonate mounds from Belgium: sedimentology and palaeoceanography. In: Álvaro JJ, Aretz M, Boulvain F, Munnecke A, Vachard D, Vennin E (eds) Palaeozoic Reefs and Bioaccumulations: Climatic and Evolutionary Controls. Geol Soc London, Spec Publ 275:125–142

Boulvain F, Belanger I, Delsate D, Dosquet D, Ghysel P, Godefroit P, Laloux M, Roche M, Teerlynck H, Thorez J (2001) New lithostratigraphical, sedimentological, mineralogical and palaeontological data on the Mesozoic of Belgian Lorraine: a progress report. Geol Belg 3:3–33

Boulvain F, Mabille C, Poulain G, Da Silva AC (2009) Towards a palaeogeographical and sequential framework for the Givetian of Belgium. Geol Belg 12:161–178

Boulvain F, Pingot JL (2015) Genèse du sous-sol de la Wallonie. Acad roy Belgique, Bruxelles, p 208

Bultynck P (1989) Bernissart et les iguanodons. Inst roy Sci nat Belgique, 115 p

Cocks LRM, Torsvik TH (2002) Earth geography from 500 to 400 million years ago: a faunal and palaeomagnetic review. J Geol Soc London 159:631–644

Corteel C, De Paepe P (2003) Boron metasomatism in the Brabant Massif (Belgium): geochemical and petrographical evidence of Devonian tourmalinite pebbles. Geol en Mijnbouw 82:197–208

Da Silva AC, Boulvain F (2004) From palaeosols to carbonate mounds: facies and environments of the Middle Frasnian platform in Belgium. Geol Q 48:253–266

Debacker TN (2012) Folds and cleavage/fold relationships in the Brabant Massif, southeastern Anglo-Brabant Deformation Belt. Geol Belg 15:81–95

Debacker TN, Dewaele S, Sintubin M, Verniers J, Muchez P, Boven A (2005) Timing and duration of the progressive deformation of the Brabant Massif, Belgium. Geol Belg 8:20–34

De Man E, Ivany L, Vandenberhe N (2004) Stable oxygen isotope record of the Eocene-Oligocene transition in the southern North Sea Basin: positioning the Oi-1 event. Netherlands J Geosci 83(3):193–197

Demoulin A, Hallot E (2009) Shape and amount of the Quaternary uplift of the western Rhenish shield and the Ardennes (western Europe). Tectonophysics 474:696–708

De Vos W, Verniers J, Herbosch A, Vanguestaine M (1993) A new geological map of the Brabant Massif, Belgium. Geol Mag 130:605–611

Dusar M, Langenaeker V, Wouters L (2001) Permian-Triassic-Jurassic lithostratigraphic units in the Campine basin and the Roer Valley Graben (NE Belgium). Geol Belg 4:107–112

Fairon-Demaret M, Steurbaut E, Damblon F, Dupuis C, Smith T, Gerrienne P (2003) The in situ *Glyptostroboxylon* forest of Hoegaarden (Belgium) at the Initial Eocene Thermal Maximum (55 Ma). Rev Paleobotany Palynol 126:103–129

Felder WM (1975) Lithostratigrafie van het Boven-Krijt en het Dano-Montien in Zuid-Limburg en het aangrenzende gebied in Zagwijn. Rijks Geol Dienst 63–72

Fielitz W, Mansy JL (1999) Pre- and synorogenic burial metamorphism in the Ardenne and neighbouring areas (Rhenohercynian zone, central European Variscides). Tectonophysics 309:227–256

Fourmarier P (ed) (1954) Prodrome d'une description géologique de la Belgique. Soc géol de Belgique, Liège, p 826

Goemaere E, Dejonghe L (2005) Paleoenvironmental reconstruction of the Mirwart Formation (Pragian) in the Lambert quarry (Flamierge, Ardenne, Belgium). Geologica Belgica 8:37–52

Hance L, Poty E, Devuyst FX (2001) Stratigraphie séquentielle du Dinantien type (Belgique) et corrélation avec le nord de la France (Boulonnais, Avesnois). Bull Soc Géol France 172:411–426

Havron C, Vandycke S, Quinif Y (2007) Interactivité entre tectonique méso-cénozoïque et dynamique karstique au sein des calcaires dévoniens de la région de Han-sur-Lesse (Ardennes, Belgique). Geol Belg 10:93–108

Herbosch A, Verniers J (2014) Stratigraphy of the Lower Palaeozoic of the Brabant Massif, Belgium. Part II: the Middle Ordovician to lowest Silurian of the Rebecq Group. Geol Belg 16:115–136

Lamens J (1985) Transition from turbidite to shallow-water sedimentation in the Lower Salmian (Tremadocian, Lower Ordovician) of the Stavelot Massif, Belgium. Sedim Geol 44:121–142

Lees A, Hallet V, Hibo D (1985) Facies variation in Waulsortian buildups. Part I. A model from Belgium. Geol J 20:138–153

Linnemann U, Herbosch A, Liégeois JP, Pin C, Gärtner A, Hofmann M (2012) The Cambrian to Devonian odyssey of the Brabant Massif within Avalonia: a review with the new zircon ages, geochemistry,

Sm-Nd isotopes, stratigraphy and palaeogeography. Earth Sci Rev 112:126–154

Louwye S, Laga P (2008) Dinoflagellate cyst stratigraphy and palaeoenvironment of the marginal marine Middle and Upper Miocene of the eastern Campine area, northern Belgium (southern North Sea Basin). Geol J 43:75–94

Lucius M (1952) Manuel de la Géologie du Luxembourg, vue d'ensemble sur l'aire de sédimentation luxembourgeoise. Imprimerie de la Cour V. Buck, 406 p

Matte P (1986) La chaîne varisque parmi les chaînes paléozoïques péri-atlantiques, modèle d'évolution et position des grands blocs continentaux au Permo-Carbonifère. Bull Soc Géol France 8(II):9–24

Maubeuge PL (1954) Le Trias et le Jurassique du sud-est de la Belgique. In: Fourmarier P (ed) Prodrome d'une description géologique de la Belgique. Soc géol de Belgique, Liège, pp 385–416

Meilliez F (2006) La discordance éodévonienne de l'Ardenne: caractérisation stratigraphique et paléo-environnementale de la Formation de Fépin et ses conséquences. Géol France 1–2:29–33

Meilliez F, Lacquement F (2006) La discordance éodévonienne de l'Ardenne: structure du site de Fépin et conséquences sur les interprétations géodynamiques de l'Ardenne. Géol France 1–2:73–77

Nyhuis C, Rippen D, Denayer J (2014) Facies characterization of organic-rich mudstones from the Chokier Formation (lower Namurian), south Belgium. Geol Belg 127:311–322

Robaszynski F, Dupuis C (1983) Belgique. Guides géologiques régionaux, Masson, Paris, p 204

Schavemaker Y, De Bresser JHP, Van Baelen H, Sintubin M (2012) Geometry and kinematics of the low-grade metamorphic "Herbeumont shear zone" in the High-Ardenne slate belt (Belgium). Geol Belg 15:126–136

Sintubin M, Everaerts M (2002) A compressional wedge model for the Lower Palaeozoic Anglo-Brabant Belt (Belgium) based on potential field data. Geol Soc London, Spec Publ 201:327–343

Steurbaut E (1998) High-resolution holostratigraphy of Middle Paleocene to Early Eocene strata in Belgium and adjacent areas. Palaeontographica, A 247(5–6):91–156

Steurbaut E (2006) Ypresian. In Dejonghe L (ed) Current status of chronostratigraphic units named from Belgium and adjacent areas. Geologica Belgica 9(1–2):73–93

Thorez J, Dreesen R, Streel M (2006) Famennian. In: Dejonghe L (ed) Current status of chronostratigraphic units named from Belgium and adjacent area. Geol Belg 9:27–45

Vandenberghe N, Van Simaeys S, Steurbaut E, Jagt J, Felder P (2004) Stratigraphic architecture of the Upper Cretaceous and Cenozoic along the southern border of the North Sea Basin in Belgium. Neth J Geosci 83(3):155–171

Vandenberghe N, De Craen M, Wouters L (2014a) The Boom Clay geology. From sedimentation to present-day occurrence. A review. Memoirs Geol Surv Belgium 60:76

Vandenberghe N, Harris W, Wampler M, Houthuys R, Louwye S, Adriaens R, Vos K, Lanckacker T, Matthijs J, Deckers J, Verhaegen J, Laga P, Westerhoff W, Munsterman D (2014b) The implications of K-Ar dating in the Diest Sand Formation on the paleogeography of the Upper Miocene in Belgium. Geol Belg 17 (2):161–174

Van den Bril K, Swennen R (2009) Sedimentological control on carbonate cementation in the Luxembourg Sandstone Formation. Geol Belg 12:3–23

Vandycke S (2002) Paleostress records in Cretaceous formations in NW Europe: extensional and strike-slip events in relationships with Cretaceous-Tertiary inversion tectonics. Tectonophysics 357:119–136

Waterlot G, Beugnies A, Bintz J (1973) Ardenne, Luxembourg. Guides géologiques régionaux, Masson, Paris, p 206

Ziegler PA (1982) Geological atlas of the Western and Central Europe. Shell Int Petrol Maatsch, 130 p

The Climate of Belgium and Luxembourg

3

Michel Erpicum, Myriem Nouri and Alain Demoulin

Abstract

The present climate of Belgium and Luxembourg is shown to be oceanic warm-temperate, benefitting from the warming effect of the North Atlantic Drift. Mean annual air temperatures are around 10 °C and vary spatially mainly as a function of elevation. Annual temperature amplitudes are in the 13–17 °C range. Annual rainfall depths vary from ~700 mm in western Belgium to 1300–1400 mm in the wettest areas of NE and SW Ardenne. Belgium and Luxembourg are located in the zone of seasonal shift of the north polar front and the associated mid-latitude, or polar front jet stream.

Keywords

Climate • Temperature • Precipitation • Atmospheric circulation

3.1 Introduction

Located between 49° 30′ and 51° 30′ N, and 2° 30′ and 6° 30′ E, Belgium and the Grand-Duchy of Luxembourg (GDL) are entirely situated in the 0–700 m elevation range, with a 66-km-long coast where Western Belgium meets the southern North Sea and maximum elevations in eastern Belgium (694 m asl) and northern Luxembourg (560 m asl). They are dominantly crossed by SW moist air masses (Fig. 3.1) coming from the Atlantic Ocean and having first passed across NW France or the British Isles. Such dominant winds in this mid-latitude region determine a *Cfb* climate according to the updated Köppen-Geiger classification (Peel et al. 2007), i.e., a warm-temperate climate without dry season (oceanic type), with warmest month's mean temperature <22 °C and mean temperature of the four warmest months >10 °C. Except in easternmost Belgium and parts of GDL, where January's mean temperature is between 0 and −0.5 °C, the coldest month's mean temperature is >0 °C in the two countries. Owing to the presence of the North Atlantic Drift, which continues the Gulf Stream along the coasts of Western Europe and, especially during winter, warms the air masses coming from the west, the mean annual temperature of 9.7 °C recorded at Brussels over the period 1833–2015 (Statbel 2016) is fairly high for latitudes around 50° (Alexandre et al. 1992). Nevertheless, the dominant westerlies occasionally give way to advection of continental air masses coming from the east. Originating in high pressures centered in Scandinavia or Eastern Europe, these masses may cause prolonged episodes, sometimes several weeks long, of very dry and hot (during summer) or very cold (during winter) conditions over Belgium and GDL. These recurrent westward general circulations explain why only 700–800 mm annual rainfall is observed in most of Low and Central Belgium, which is rather unimpressive with regard to the oceanic character of the climate.

M. Erpicum (✉) · M. Nouri
Department of Geography, University of Liège, Sart Tilman, B11, 4000 Liège, Belgium
e-mail: michel.erpicum@ulg.ac.be

M. Nouri
e-mail: myriem.nouri@doct.ulg.ac.be

A. Demoulin
Department of Physical Geography and Quaternary, University of Liège, Sart Tilman, B11, 4000 Liège, Belgium
e-mail: ademoulin@ulg.ac.be

A. Demoulin (ed.), *Landscapes and Landforms of Belgium and Luxembourg*, World Geomorphological Landscapes, DOI 10.1007/978-3-319-58239-9_3

Fig. 3.1 *Rows* of inclined poplars along a canal in Flanders as seen by the Flemish painter Théo van Rysselberghe (oil on canvas, 1894), showing the effect of the dominant westerlies

3.2 Main Climatic Features

3.2.1 Temperature

The −0.6 °C/100 m vertical gradient in air temperature is the main factor responsible for altitude-dependent spatial variations of temperature across Belgium and GDL. However, throughout the year, the proximity of the sea also contributes to milder temperatures in NW Belgium in the case of air advection from the west. Overall, the isotherms are markedly parallel to the contour lines, with mean annual air temperatures between 10.5 and 11 °C in the lowlands of NW Belgium, around 10 °C in the low-elevated plateaus (100–200 m asl) of Central Belgium, and below 9 °C in the core of the Ardenne-Oesling (or Eislek, in Luxembourgian dialect) uplifted area (>450 m asl). However, this pattern of elevation-dependent temperatures is locally altered by

- the nature of the ground and its cover, such as the predominance of sandy soils producing, for example, above-average daily and annual temperature amplitudes in Campine
- the foehn effect, which warms the rain shadow side of Ardennian summit ridges whose windward flank has intercepted heavy rainfall and dried the downward winds
- site effects related to the topography, which induce for instance lower minimum temperatures in the Ardennian valley bottoms than on the plateau under clear sky, justifying the plateau setting of most Ardennian villages (Fig. 3.2)
- site effects related to densely urbanized areas.

The amplitude of mean monthly temperature variations over the year is fairly uniform throughout Belgium and GDL, from ~13 °C at Oostende, North Sea coast, through ~15 °C at Bastogne, central Ardenne, and Leopoldsburg, Campine, to ~17 °C in the slightly more continental setting of Luxembourg City. As for annual mean temperatures, they display a 0.8 °C interannual variability (1σ) overprinting a long-term trend since 1833 that fairly replicates the global warming over the same period. Going farther back in time, historical archives attest that the twelfth century (i.e., the end of the medieval climatic optimum, when vineyards were common in Belgium) was marked by warm springs and dry

Fig. 3.2 Thick frost in the Amblève valley *bottom* near Aywaille on December 3, 1989 (see location on Fig. 3.4). The temperature inversion is marked by the *horizontal upper* limit of the frost

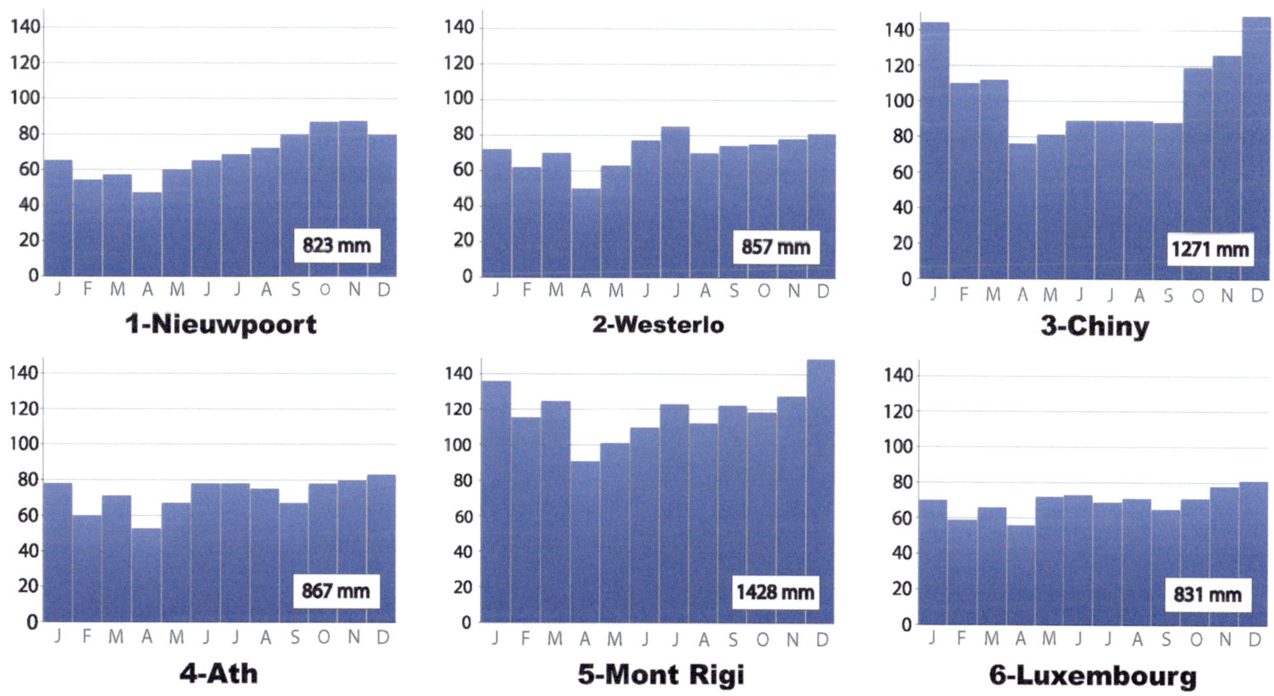

Fig. 3.3 Monthly precipitation histograms for six selected stations (see location on Fig. 3.4), averaged over the 1981–2010 period (modified after IRM 2016b)

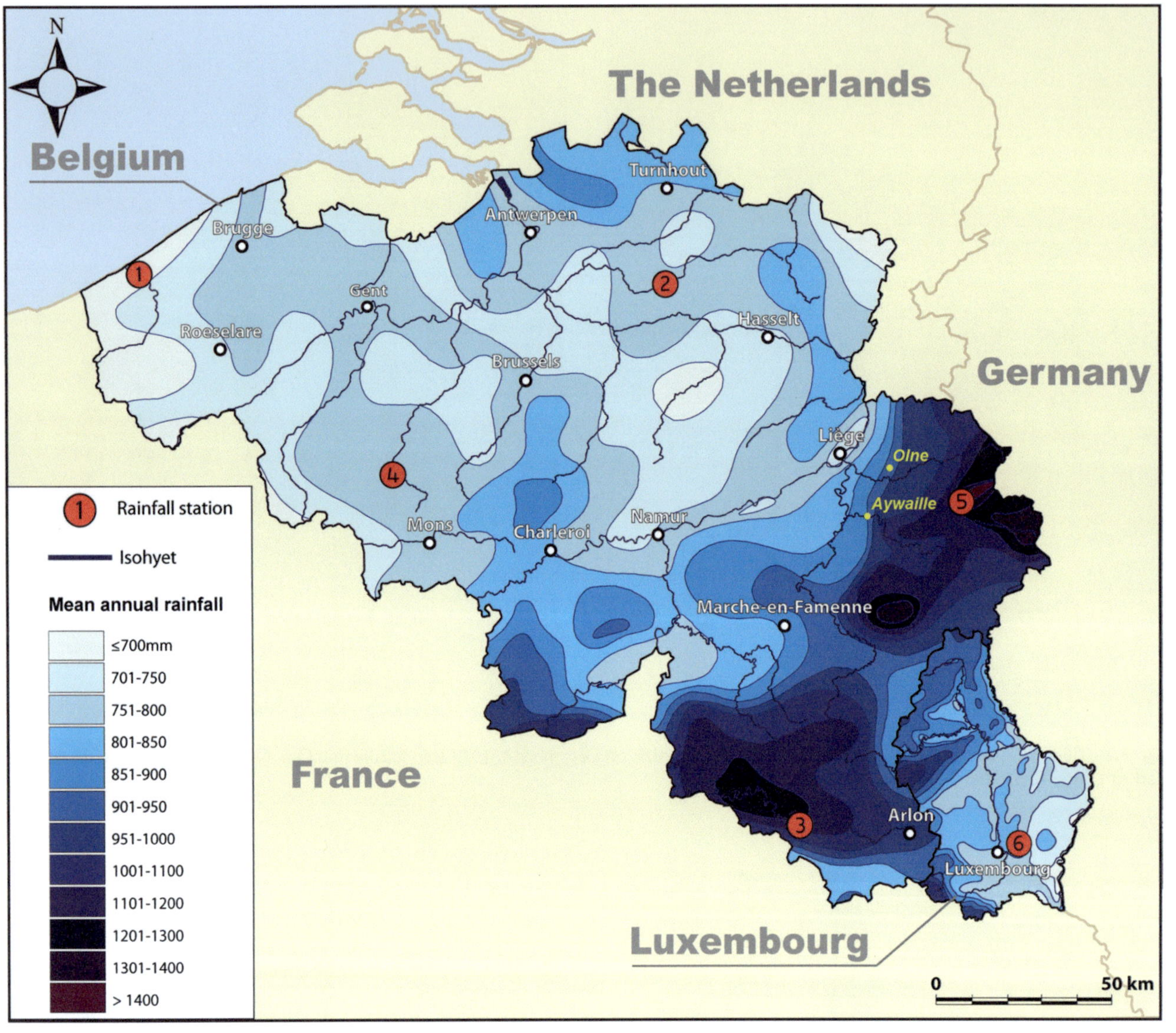

Fig. 3.4 Mean annual precipitations, averaged over the 1961–1990 and 1981–2010 periods for Belgium and Luxembourg, respectively

summers, in sharp contrast with the highly humid springs and summers and cold winters that caused the great famines of the fourteenth century (Alexandre 1987) and possible landsliding (see Chap. 20). After temporary warming during the fifth century, the Little Ice Age was marked in Belgium and GDL by very cold winters and wet, cold springs from 1550 to 1850.

3.2.2 Precipitation

Annual rainfall depths are fairly equally distributed over the year, with a poorly marked seasonal minimum in early spring (April) (Fig. 3.3). Annual rainfall range from $\sim$700 mm in westernmost Belgium, in downwind areas of east-central Belgium, and in SE GDL (Moselle valley) to 1300–1400 mm in the Hautes Fagnes plateau of NE Ardenne and in the lower Semois area of SW Ardenne (Pfister et al. 2005; IRM 2016a) (Fig. 3.4). At Brussels, mean annual precipitations amount to 805 ± 119 mm, indicating non-negligible interannual variability and showing a minimum of 406 mm in 1921 (Statbel 2016). The oceanic character of the Belgian and GDL climate is underlined by the high percentages of rainy days, with daily precipitation exceeding 1 mm for more than 33% of the days in any season. The residual analysis of the linear regression between yearly precipitation amount and altitude (Alexandre et al. 1992) supports the correlation but displays residual patterns that suggest either its nonlinear character and/or an orographic effect (Alexandre et al. 1998). Although precipitation regimes may significantly vary across Belgium and GDL (Fig. 3.3), a weak monthly rainfall minimum is observed

Fig. 3.5 0.35-m-thick snow cover at Olne, 220 m asl, southern Herve Plateau, on December 26, 2010 (see location on Fig. 3.4)

during the winter–spring transition (February–April) in most weather stations (Alexandre et al. 1992). Geomorphologically effective precipitations take the form of either periods of prolonged rainfall causing widespread flooding (e.g., summer 1980: 243 mm in Uccle from June 21 to July 20; winters 1993–94 and 1994–95) or storms and intense convective rainfall events liable to provoke flash floods in small catchments (e.g., 168 mm/24 h at Brasschaat, Campine, September 15, 1998).

Although snow is here a minor climatic component that shows high interannual variability, a continuous snow cover may occasionally persist for more than one month over the High Ardenne plateau above 500 m. Snowfalls are more frequent in January–February, generally accumulating less than 0.3 m of snow (Fig. 3.5). Since 1950, a >1-m-thick snow cover was observed only three times in the Hautes Fagnes plateau, namely in February 1952 and 1953 and in the beginning of March 1988. Between 1983 and 2014, the average yearly number of days with a snow cover varied from 12 to 16 at 100 m asl (Brussels area, where snow hardly reaches the ground during the mildest winters) to 22 at 300 m asl. On the Ardennian summits, this number significantly increases from west to east (40 days in Saint-Hubert at 550 m asl, 60 days at the top of the Baraque de Fraiture at 650 m asl, 75 days in the Hautes Fagnes area between 650 and 700 m asl). Noteworthily, rapid melting of the snow cover combined with a heavy rainfall event during a sudden return to milder conditions is responsible for ~50% of the floods recorded in the Ardenne valleys.

3.2.3 Other Climatic Features

A main characteristic of Belgian and GDL climates is the high numbers of overcast days (60–75% from November to March) and days with cumuliform clouds (80% on average from April to October). December mean sunshine durations are consequently as low as 45, 35, and 45 h at Oostende (Belgian coast), Botrange (NE Ardenne) and Luxembourg airport, respectively. These same locations record July

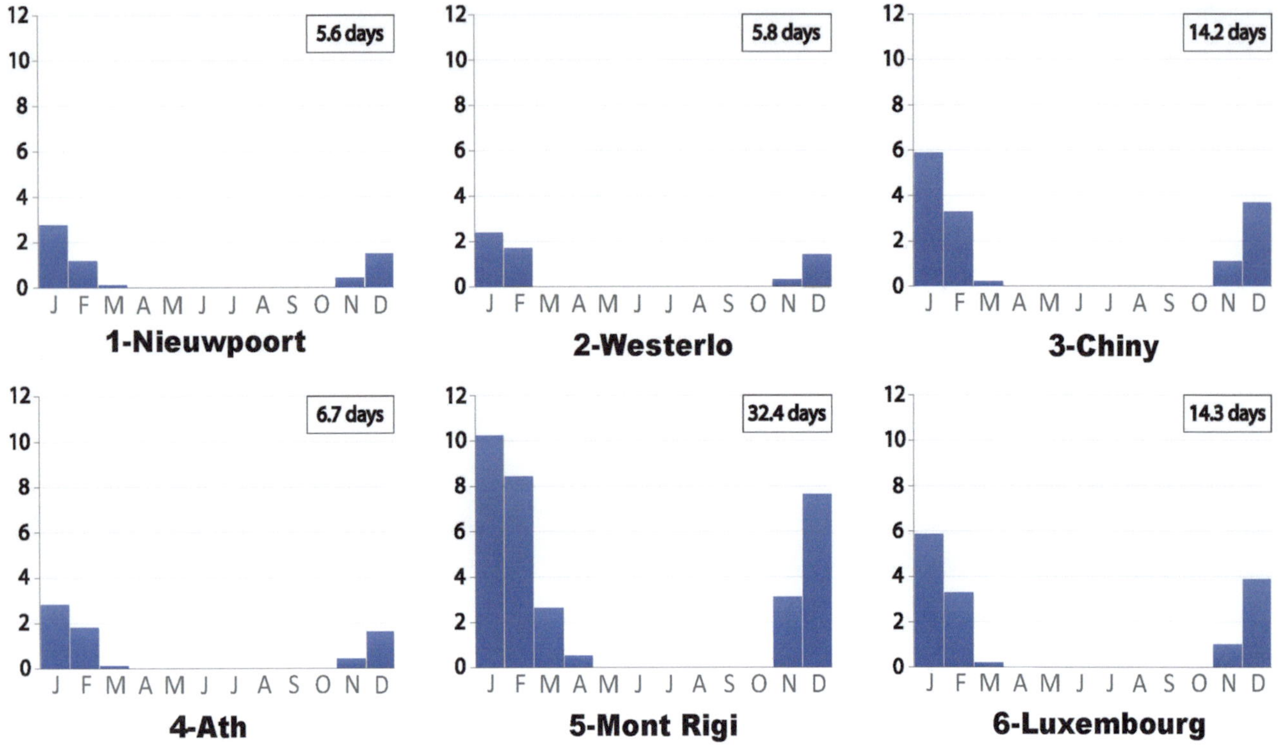

Fig. 3.6 Monthly numbers of days of continuous frost (maximum temperature <0 °C) for six selected stations (see location on Fig. 3.4), averaged over the 1981–2010 period (modified after IRM 2016b)

means of 225, 190, and 250 h and annual means of 1730, 1400, and 1730 h over the period 1981–2010, i.e., about 20–25% of the potential sunshine duration in winter and 40–50% in summer.

Whereas snow is not so frequent in Belgium and GDL, days with night frost are much more common, from ~50 per year near the coast to >120 in the areas of the Ardennian plateau above ~580 m asl. Likewise, yearly numbers of days with continuous frost (maximum temperature <0 °C) increase from 6 in the coastal area to >30 in High Ardenne above 580 m (Fig. 3.6). Severe cold conditions occasionally cause temperatures to drop between −20 and −30 °C, especially in the incised Ardennian valleys prone to high daily amplitudes of temperature.

3.3 Atmospheric Circulation

The latitudinal position of Belgium and GDL locates them in the zone of seasonal shift of the polar front and the associated mid-latitude, or polar front jet stream. The polar front is characterized by the convergence of cold polar and warm subtropical air masses, inducing a strong temperature gradient that generates the barotropic eddy-driven jet streams and embedded storm tracks, bringing moist air from the ocean over Europe. While a stronger Azores High pushes the polar front northward in summertime, allowing warm subtropical air to invade Belgium and GDL and moving storm tracks further north, its winter weakening combined with a deepening of the Icelandic Low has the opposite effect, bringing the polar front and its attendant low pressures, cold air and rainfall-laden westerlies over our region. In addition to this seasonal shift of the polar front, weather interannual variability in Belgium and GDL is controlled by the varying strength of the Azores High and Icelandic Low, imposing varying pressure gradients over the North Atlantic that are known as the North Atlantic Oscillation (NAO). A large pressure gradient between strong high and low centers (positive NAO index) reduces the polar front jet undulations and strengthens it in a more northern position, notably causing mild and wet winters in Belgium and GDL such as that of 2006–07, but also storms. Conversely, weakened Icelandic Low and Azores High determine a negative NAO. Westerlies are displaced southward; they slow and start to undulate, allowing frequent southward intrusions of polar air. Belgian winters tend then to be colder and drier and summers are cool and humid, such as in 1980.

References

Alexandre P (1987) Le Climat de l'Europe au Moyen Age. Contribution à l'histoire des variations climatiques de 1000 à 1425, d'après

les sources narratives de l'Europe occidentale. Ecole des Hautes Etudes en Sciences Sociales, Paris, p 808
Alexandre J, Erpicum M, Vernemmen C (1992) Le climat. In: Denis J (ed) Géographie de la Belgique. Crédit Communal de Belgique, Brussels, pp 87–128
Alexandre J, Erpicum M, Mabille G, Cornet Y (1998) Précipitations atmosphériques et altitude. Prélude à une cartographie des montants annuels et mensuels en Belgique. Publications de l'Association Internationale de Climatologie 11:219–226
IRM (2016a) http://www.meteo.be/meteo/view/fr/16788784-Atlas+Climatique.html
IRM (2016b) https://www.meteo.be/meteo/view/fr/27484519-Climat+dans+votre+commune.html
Peel M, Finlayson B, McMahon T (2007) Updated world map of the Köppen-Geiger climate classification. Hydrol Earth Syst Sci Dis 4:439–473
Pfister L, Drogue G, Poirier C, Hoffmann L (2005) Evolution du climat et répercussions sur le fonctionnement des hydrosystèmes au Grand-Duché de Luxembourg au cours des 150 dernières années. Ferrantia 43:85–100
Statbel (2016) http://statbel.fgov.be/fr/statistiques/chiffres/environnement/climat/

4 Landscapes and Landforms of the Luxembourg Sandstone, Grand-Duchy of Luxembourg

Birgit Kausch and Robert Maquil

Abstract

The Liassic Luxembourg Sandstone outcrops on ~ 350 km^2, edged by a cuesta escarpment, in Gutland, the southern part of Luxembourg. It is a homogeneous up to 100-metres-thick unit in which rivers of all sizes have incised steep cliffs separating extended plateaus. The Luxembourg Sandstone landscapes display typical geomorphological features at various scales, mainly shaped by gravitational and fluvial processes that varied in intensity with climatic and hydrological conditions. The cliffs have developed by landsliding in the marly lower part of the slopes and rockfall in the overlying sandstone. Old and recent landslides have affected many slopes. Structures induced or enhanced by weathering are observed everywhere on the sandstone free faces. This chapter describes the landscapes of the so-called Luxembourg's Little Switzerland, with its impressive rock formations of, e.g. the Wolfsschlucht, and sandstone landforms reshaped by fortification works along the Alzette and Petruss valleys in Luxembourg City.

Keywords

Sandstone landforms • Complex slope evolution • Gutland • Sandstone weathering

4.1 Introduction

The Luxembourg Sandstone (unit li2 of the geological map of Luxembourg), which outcrops over ~ 350 km^2, covers large parts of southern Luxembourg, also called Gutland (Fig. 4.1). This unit varies in thickness from a few to ~ 100 m and displays a regular joint pattern. Rivers have cut through the sandstone, forming impressive cliffs and bizarre landforms sculptured in the rock by erosion (Fig. 4.2). Rockfalls and landslides have played an important part in the landscape formation. Typical weathering forms involving dissolution and calcite crystallization are also observed on exposed surfaces. Rock overhangs, caves and open joints have been used as shelters in ancient times, while rocky promontories and plateaus were preferred settlement areas, with many castles still perched on them. Natural resources like water and building stones have been exploited by man since his early days, as attested by archaeological findings going back to the Palaeolithic (Le Brun-Ricalens and Valotteau 2005). Numerous artistic and touristic descriptions and sketches of sandstone landscapes in Luxembourg exist, but Lucius (1907, 1952) was among the firsts to give a scientific account of their forms and evolution (Fig. 4.3).

The Lower Liassic sandstone forms with the underlying Triassic and overlying younger Jurassic sediments the Gulf of Luxembourg, a NE extension of the Paris Basin into the Rhenish Shield (Fig. 4.1). The sediments were deposited by seas transgressing progressively on a pre-Triassic peneplain cut into the Paleozoic rocks of the Variscan Ardenne and Eislek. The sandstone facies is diachronic and passes westward to younger formations (Bintz et al. 1973; Muller 1980).

B. Kausch (✉)
Service Géologique du Luxembourg/Geological Survey of Luxembourg, 23, rue du Chemin de Fer, 8057 Bertrange, Luxembourg
e-mail: b.kausch@web.de

R. Maquil
91, rue Clairefontaine, 9220 Diekirch, Luxembourg
e-mail: robert.maquil@pt.lu

A. Demoulin (ed.), *Landscapes and Landforms of Belgium and Luxembourg*, World Geomorphological Landscapes, DOI 10.1007/978-3-319-58239-9_4

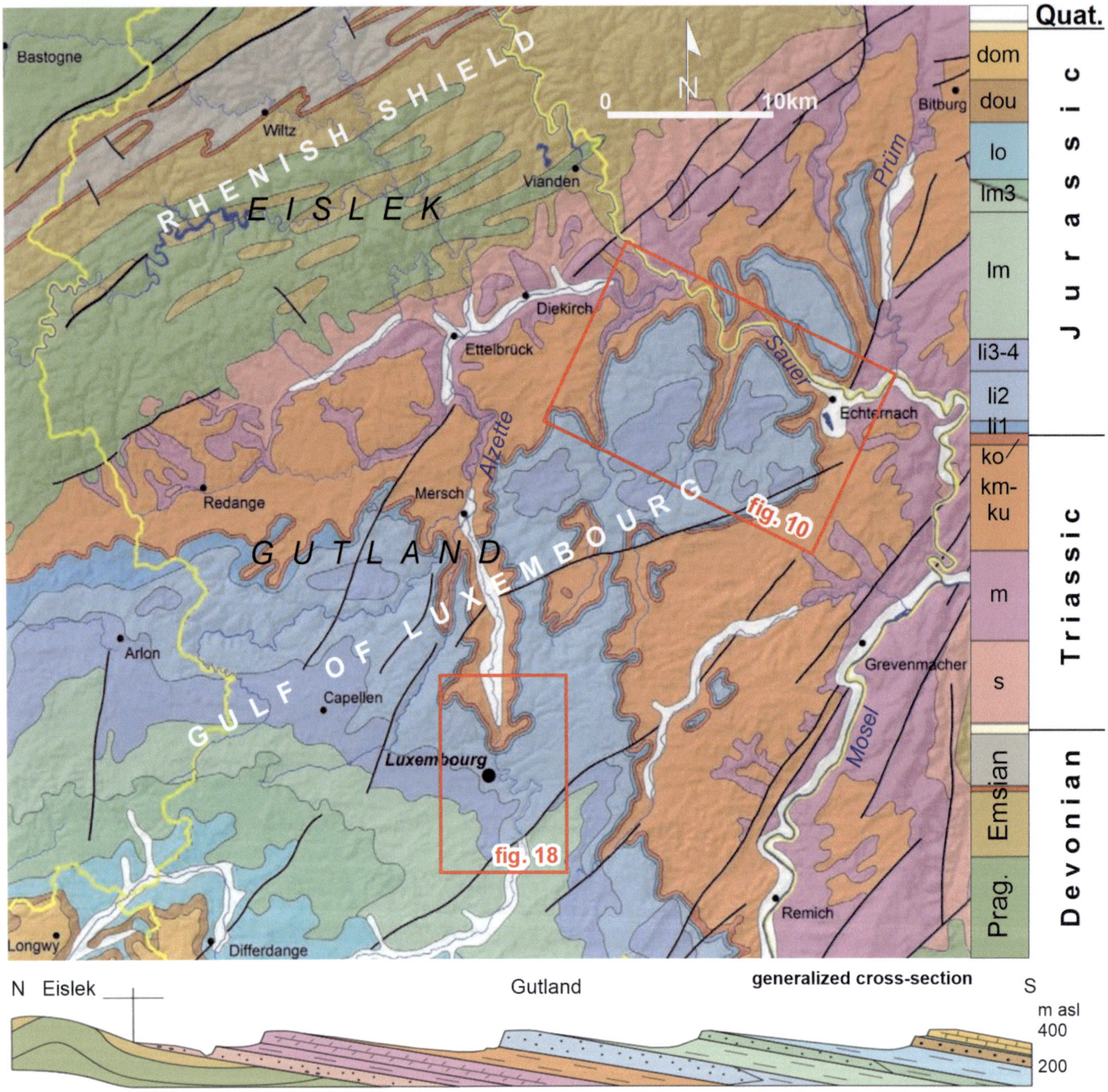

Fig. 4.1 Simplified geological map and generalized cross-section of Luxembourg. li2 (*light blue*) Luxembourg Sandstone outcrop area. The *red rectangles* delimit the areas described in the text, namely Luxembourg's Little Switzerland (Fig. 4.10) and Luxembourg City area (Fig. 4.18). © SGL 2015

The Mesozoic cover has an overall thickness of ~1000 m. It was slightly folded into a wide synclinal structure with numerous minor undulations plunging gently to the southwest. The Luxembourg Sandstone nicely exhibits such fold structures in the study area. Lithologically, the alternation of harder and softer strata is remarkably expressed in the very typical cuesta landscape of Gutland and Lorraine (the Belgian equivalent of Gutland, see Chap. 23) (Fig. 4.1).

The Luxembourg Sandstone forms the Gutland's main cuesta, which is uninterruptedly followed from Germany, through Luxembourg and Belgium, to France. Isolated buttes in front of the cuesta frequently correspond to minor synclinal undulations (Fig. 4.1). The plateau is more intensely dissected in the northeastern outcrop area, in Germany, where the thickness of the sandstone gets more and more reduced due to the SW-plunging direction of the strata and to Tertiary erosion, causing discontinuities in the cuesta relief (Hövel et al. 2015). The Sauer River crosses the sandstone area epigenetically, separating the Luxembourgian plateaus of the Mullerthal and their double cuesta to the southwest

Fig. 4.2 **a** Central part of the *Wolfsschlucht* gorge, opposing a subvertical escarpment on the uphill side of the gorge (*right*) and the ~30° inclined face of a slab tilted to the valley (*left*). In the background, needle monolith isolated from the escarpment and weakly tilted toward the gorge. **b** Sandstone cliff of the Bock promontory overlooking the Alzette valley in Luxembourg City. Remnants of the fortress with a defence gallery dominate the cliff. Together with the bridge, they formed the outer defence system of the fortress

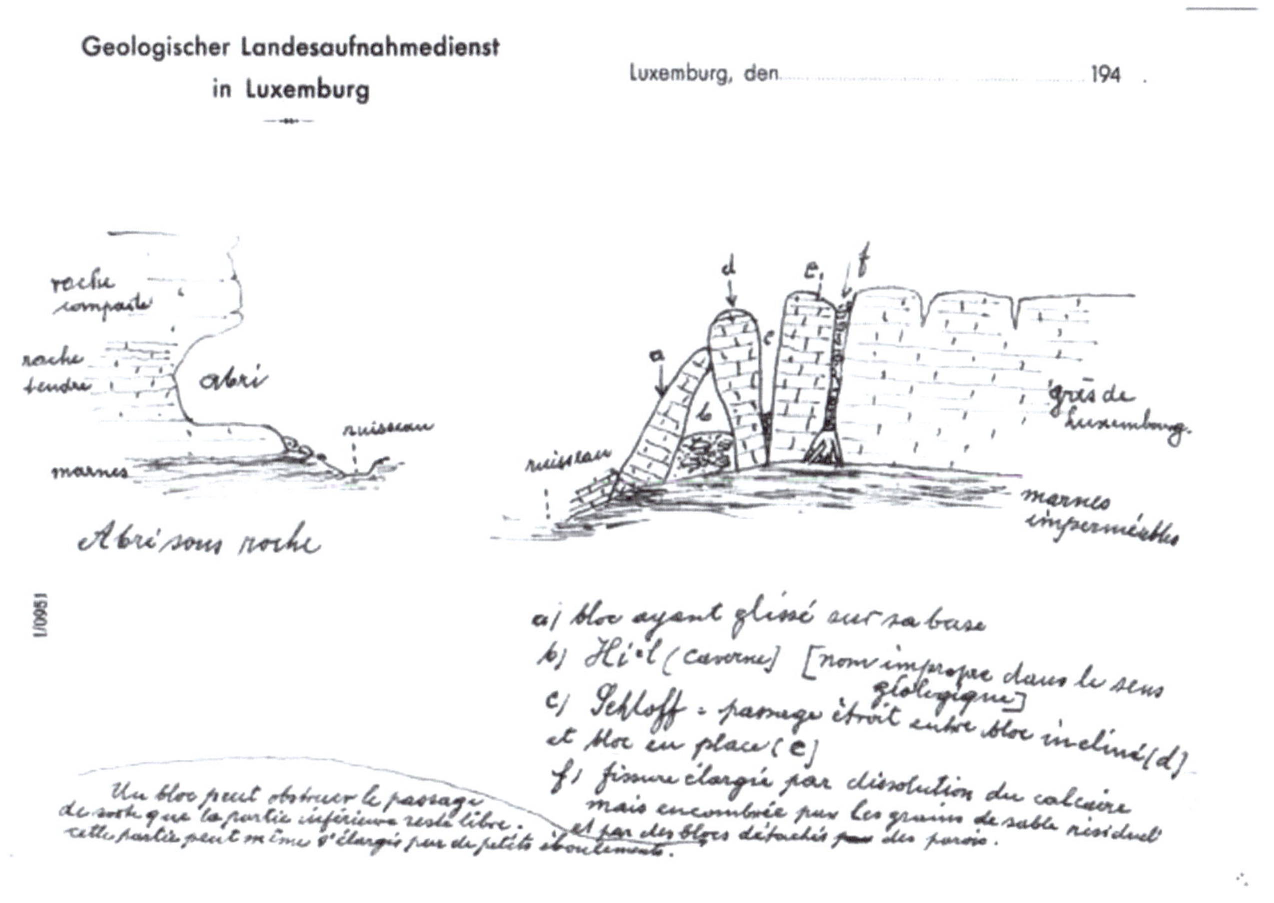

Fig. 4.3 Early sketch featuring sandstone landforms in Luxembourg's Little Switzerland, drawn by Lucius (1952), founder of the Geological Survey of Luxembourg

from the dissected plateaus around Ferschweiler, isolated by the Prüm and smaller creeks to the northeast. The N-flowing obsequent Alzette, a tributary of the Sauer, incised a deep valley into the sandstone and the underlying Triassic rocks at and downstream of Luxembourg City, with sandstone cliffs dominating gentler slopes developed in Keuper marls.

While the cuesta scarps are thought to be fairly stable Late Tertiary to Early Quaternary features (Busche et al. 2005; see also Chap. 23), the evolution of the valley side cliffs is more recent, probably still active during the last glacial period and, to some extent, even the Holocene. Through their control on mass movement efficiency, climatic conditions, groundwater flow and river transport determined the speed of cliff evolution. This evolution is illustrated hereafter in two picturesque and touristic areas of the Luxembourg Sandstone area, namely Luxembourg's Little Switzerland and the site of Luxembourg City. Little Switzerland, a part of the Mullerthal region (Fig. 4.1), has a long touristic vocation due to its bizarre landforms, such as the escarpments of the Wolfsschlucht (wolf's gorge, Fig. 4.2a), formed by a succession of large mass movement events. The City of Luxembourg (Fig. 4.1) has been built on top of incised sandstone plateaus above the Alzette and its small tributary, the Petruss. Man exploited this special topographic position, strongly fortifying the place over the centuries, shaping the landscape and exposing the up to 50-m-high cliffs.

4.2 Geological and Geomorphological Background

Rock properties and mineral composition are paramount in the understanding of weathering and erosion processes. Mineralogy is a primary control on the nature of the weathering products, while geotechnical properties help predict the rock's reaction to physical and chemical weathering and their susceptibility to mass movements. Mineralogically, the bulk of the Lower Liassic Luxembourg Sandstone is made of quartz grains and variable proportions of carbonate ($CaCO_3$) matrix (Fig. 4.4). Quartz grains are well sorted in the fine to medium sand fractions. Clay minerals are only present in small amounts, forming with carbonates and small percentages of quartz up to decimetre-thick marly interlayers. Accessory minerals are essentially iron oxides and hydroxides. In confined settings and under the permanent water table, pyrite is stable and the rocks are gray in colour.

Two contrasted sandstone types are equally represented in the formation, namely sandstone and calcareous sandstone (Fig. 4.5). The yellowish sandstone contains less than 10% carbonate matrix. It has a very high porosity (up to 30%) and permeability (up to 10^{-2} m/s) yielding an excellent filtration capacity and a high-quality aquifer. With up to 60% calcitic matrix, the lighter coloured calcareous sandstone is much denser, having almost zero porosity and a very small permeability. Uniaxial compression strength is high but variable, possibly in relation with the degree of latent cracking state. Crack formation affects only the calcareous sandstones. The compression strength of the porous sandstones is low.

The Luxembourg Sandstone as a whole is cut through by a nearly vertical, sub-orthogonal network of primary joints with a metre- to decametre-wide spacing. These joints define large blocks or slabs and influence strongly the layout of the drainage system. Joints and fissures are mostly closed on the plateaus but may be widely opened by dissolution in lower lying zones of water infiltration or by unloading along the plateau edges, guided by the evolution and opening of the valleys (Fig. 4.6). A secondary fracture system is associated with the faults. These large-scale fracture and joint patterns are independent of the networks of small joints developed in the calcareous sandstones and often opened by dissolution.

Figure 4.7 shows generalized cross-sections through the slopes formed by the Luxembourg Sandstone and the underlying Triassic strata. The lithological profile is characterized by alternating steep slopes on hard rocks (sandstone, limestone, dolomite) and gentler slopes on soft rocks (marl, claystone). Thick packages of hard rock form cliffs and escarpments, whereas smaller ones support minor steps in the landscape. Each hard rock layer shows a high but thickness-dependent degree of fracturation, thus ensuring a fracture permeability that allows and guides waterflow even in low-porosity rocks. One notes that the Alzette valley north of Luxembourg City and the Sauer valley in the Mullerthal area are developed in similar stratigraphical and lithological positions, with valley sides consisting of Luxembourg Sandstone cliffs dominating gentler low slopes on Lower Liassic and Upper Triassic marls and, in the case of the Sauer valley, a bottom cliff cut in Middle Triassic dolomites (Fig. 4.1).

As sketched in the geomorphological profile (Fig. 4.7b), slope profiles evolve by physical and chemical weathering and soil and rock creep. Fracturation and rock and block fall produce screes. Dissolution of carbonates converts marly rocks to clay, and sandstones to sand. The regolith thickness varies from very thin on, e.g. the Steinmergelkeuper (Middle Keuper, alternation of dolomitic marl and dolomite) to several metres on marly and gypsiferous material. Depending on the geomechanical properties, slope angles vary from $\sim 25°$ on marls to 45° for screes and 60° for altered cliffs. In relation with the succession of harder and softer layers, creep accumulates colluvial wedges at the foot of each steepened reach of the slopes. In the landscapes of the Luxembourg Sandstone, hillslopes appear frequently disturbed by landsliding, especially where the underlying highly sensitive Rhaetian (Upper Keuper) claystones are in contact with groundwater.

While the sandy facies gradually goes into the underlying Marls of Elvange (li1 in Fig. 4.1) at the base of the Luxembourg Sandstone, the passage to the overlying Marls and Limestones of Strassen (li3) is more abrupt and often marked by an erosion surface. Fed by direct infiltration from the top

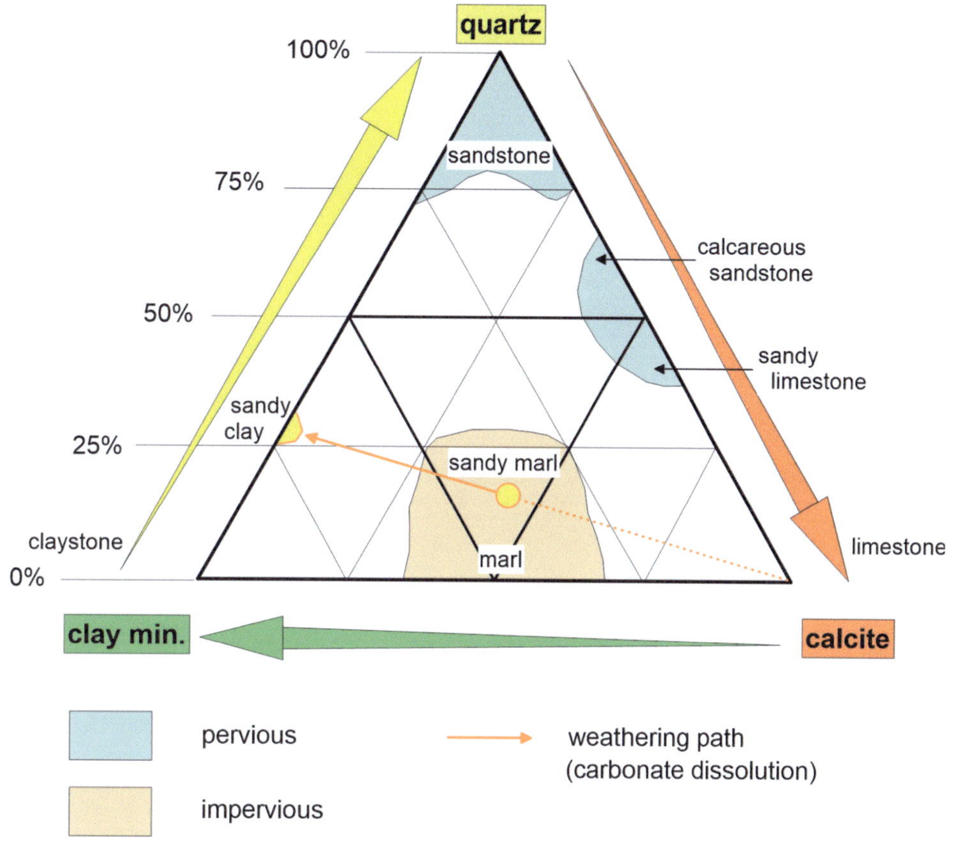

Fig. 4.4 Mineralogy of li2 (Lower Liassic) sandstones (siliceous and calcareous) and marls. The li2 unit consists of ~45% sandstones, ~45% calcareous sandstones, ~2% marls and ~6% sandy marls. Secondary minerals are mainly iron oxides and hydroxides, and pyrite. The *thin orange arrow* describes the weathering trajectory from *bluish* sandy marl (*yellow dot*, see also Fig. 4.8) to *yellowish* loose rock

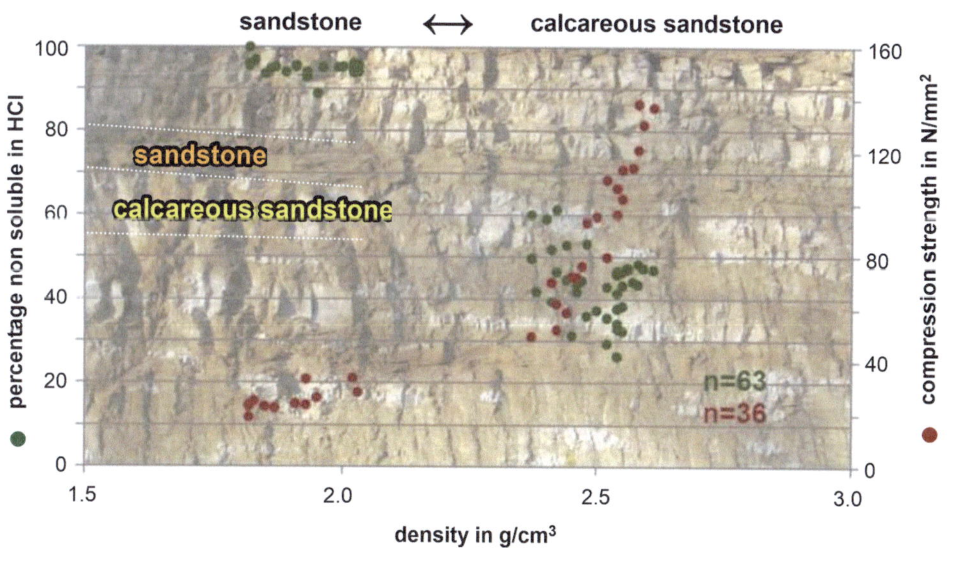

Fig. 4.5 Geotechnical properties of the sandstones and calcareous sandstones. Owing to their high porosity, the sandstones are much lighter and less resistant to compression than the calcareous sandstones

of the plateaus, groundwater circulating through either fractures or matrix porosity accumulates at the base of the reservoir layers and flows in dip direction to outlets, drains or to deeper layers. Outflow of water is either diffuse or concentrated as springs. Many large sources mark out the base of the sandstone, one of the largest, with a discharge of >3000 m^3/day, being located close to the hamlet of Mullerthal. Small perched groundwater bodies may temporarily form on top of the marly interlayers (Fig. 4.8).

The Mesozoic sediments of the Gulf of Luxembourg have been uplifted and tilted, and underwent regional erosional levelling and river incision during the Cenozoic. The

Fig. 4.6 View of the sandstone cliffs in the White Ernz valley at Larochette, with Larochette castle in the *far right*. Numerous large open joints perpendicular to the valley axis cut through the massif, the cliff surface following the second main joint direction

ancestor of the modern drainage system evolved on an erosion surface, still preserved on the top of the cuestas, that Lucius (1948), Louis (1953), Demoulin (1995) and Löhnertz (1994, 2003), Löhnertz et al. (2011) date to Tertiary times while Le Roux and Harmand (2014) argue for a Cretaceous age. As evidenced by the courses, independent of the dip direction of the strata, of the Moselle and its (sub)tributaries Sauer and Alzette, this river system evolved epigenetically. Fluvial incision into the surface probably began sometime at the end of the Neogene but especially accelerated at $\sim$0.8 Ma, when an active mantle plume under the Eifel region is assumed to have caused strong and fast regional uplift (Meyer and Stets 1998; Garcia-Castellanos et al. 2000; van Balen et al. 2000). The cuesta landscape started then to form. There is consensus that the position of the main cuesta escarpments, at current altitudes of $\sim$400 m, has changed only slightly since the early times of their formation (Tricart 1949; Busche et al. 2005; Liedtke et al. 2010). Narrow valleys with steep cliffs and flat valley bottoms developed in the first stages of incision into the sandstone, as long as rivers were powerful enough to remove the material delivered by the hillslopes. In case of insufficient energy, the coarse material of rockfalls accumulated on the valley floor as debris cones, stabilizing the cliffs and preserving V-shaped transverse profiles of the valleys. However, as soon as the underlying marls and claystones were exposed by further incision, slide processes tended to widen the valley floors. Following the hypothesis of Louis (1953) and Löhnertz (1994, 2003) for rivers in the nearby S Eifel, the observation that the valley of the smaller Alzette is broader than that of the larger Sauer might perhaps indicate that a pre-Quaternary (Paleogene?) Alzette valley had been incised, then filled up and, finally, re-excavated in more recent times.

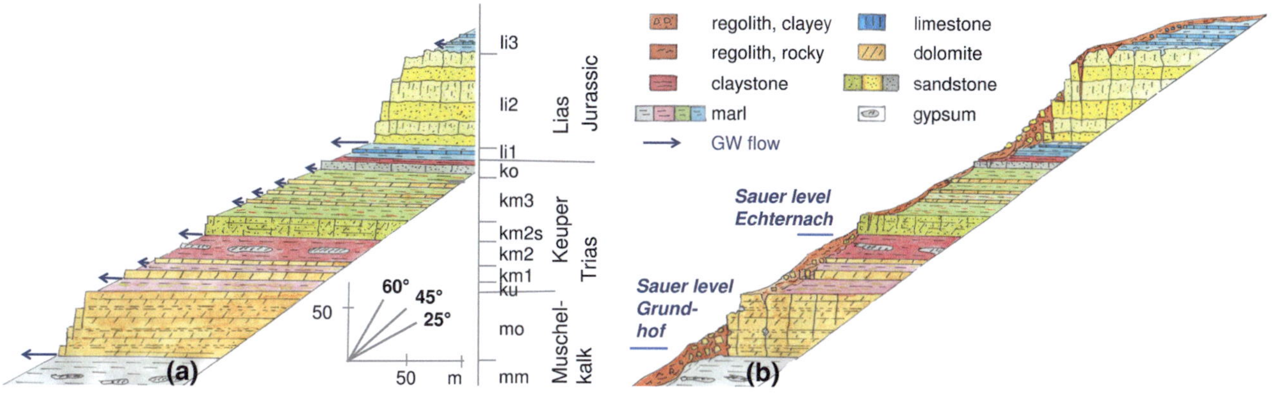

Fig. 4.7 Schematic cross-sections through the Luxembourg Sandstone and the underlying Triassic rocks. **a** Theoretical "lithological" slope profile and **b** actual geomorphological profile, including regolith and debris cover. The slope is shaped chiefly by stone and rockfall from the cliffs and creep over the gentler slope sections. In the Sauer valley (Little Switzerland), the Muschelkalk dolomites form a cliff at the base of the slopes while the sandstone escarpment makes the edge of the plateau. Together, they constitute a so-called double cuesta. The Alzette valley north of Luxembourg exposes only the upper third of this profile. *GW* groundwater

4.3 Luxembourg's Little Switzerland and the Wolfsschlucht (Wolf's Gorge)

Luxembourg's Little Switzerland (LLS) is part of the larger Mullerthal region (Fig. 4.1). Due to their majestic cliffs, bizarre rock formations and impressive weathering structures, its sandstone landscapes are touristically promoted since the last quarter of the 19th century. First hiking trails were installed as early as 1879 toward the famous cascade of the Schiessentümpel, near the hamlet of Mullerthal (Fig. 4.10), or 1881 from Echternach to the impressive gorges of the Wolfsschlucht (Figs. 4.2, 4.9 and 4.10) (Massard 2012).

Geologically, the LLS is located close to the northern limit of the Luxembourg Sandstone outcrop area and its bordering cuesta. The frontslopes of a doubled cuesta makes here the southern side of the Sauer valley (Figs. 4.10 and 4.11). The N-flowing Black Ernz, White Ernz and Lauterborn creek incise the sandstone plateau locally covered by residual Marls and Limestones of Strassen (li3 in Fig. 4.1). The network of smaller creeks mostly traces the primary joint system of the sandstone. The attitude of the 60- to 90-m-thick sandstone is determined by the SW-plunging Weilerbach syncline and its smaller undulations interrupted by a few mainly N-trending faults (Fig. 4.10). The underlying Marls of Elvange (li1) and Triassic strata are exposed in the valley of the Sauer and in the lower course of the Black Ernz, with a well-expressed small cliff in Middle Triassic dolomites near Grundhof (Fig. 4.10). About 100 m of soft marls separate this cliff from the proper frontslope of the sandstone cuesta. Springs mark the base of the sandstone body where the strata locally dip toward the escarpment.

Figure 4.11 shows an excerpt of the geomorphological map of Luxembourg by Désiré-Marchand (1985) near Echternach. The split frontslope exposed along the southern side of the Sauer valley features here the cuesta of the Luxembourg Sandstone as a double cuesta (not drawn as such on Fig. 4.11). North of the Sauer, the Ferschweiler plateau around Ernzen is also limited to the east by an important escarpment. East of the White Ernz and south of the Thull cut off meander (Figs. 4.10 and 4.11), the top of the cuesta culminates at altitudes of 400–410 m as a preserved remnant of the Tertiary topography. Residual soft Marls and Limestones of Strassen on the dipslope of the cuesta attest that this erosion surface levelled indifferently rocks of contrasted resistance. The weaker rocks were then eroded chiefly during the Quaternary.

4.3.1 Cliff and Gorge Formation

The Wolfsschlucht is a system of escarpments and gorges that extends parallel to the valley side of the Sauer and breaks apart the upper part of the frontslope of the cuesta north of the Erelchen plateau (Fig. 4.12). The gorges are up to 200 m in length, in the mean ~20 m wide and similarly deep. The most spectacular eastern part is 50 m long and up to 40 m deep. The uphill side of the system is characterized by almost vertical cliffs following the primary joint direction, while the downslope gorge walls are formed by large sandstone slabs inclined by ~30° (Lucius 1907) (Figs. 4.2 and 4.9). Blocks and slabs are also limited laterally by major joint surfaces approximately perpendicular to the primary one.

Fig. 4.8 Fresh outcrop of the Luxembourg Sandstone at Reuland, mostly made of strongly fractured calcareous sandstones underlain by an impervious *bluish layer* of marl locating a perched water table in the sandstones. On top of the outcrop, a thin regolith smoothens the slope

While the upper escarpments are discontinuous and scattered with debris cones, the lower parts of the slopes to the Sauer are generally steep, irregular, and interrupted by benches, a morphology that unequivocally betrays a slope evolution by rotational and translational landsliding affecting the whole hillslope and escarpment. Field evidence and the overall topography on maps and DEMs allow identification of three large interconnected landslides (Fig. 4.12). These landslides are ~300 m in width on average. Scarp size and drillings through more than 30 m of alternately sandy and marly regolith in the slipped bodies point to very deep-seated surface of rupture. They probably were successively active for a long period of time, with last movements possibly in historical times as suggested by a major episode of slope sliding that forced a ~40 m northward displacement of the Sauer Holocene channel in the central zone (B, Fig. 4.12). In LLS in general and the plateaus north of Echternach in particular, most large movements are assumed to have taken place mainly during the last Glacial through destabilization of the escarpment foot by active solifluction in the marls and clays (Römer 2002; Reinheimer et al. 2010). The rockfall that induced the Irrel waterfall in the Prüm valley, ~5 km north of Echternach (Fig. 4.11), has been dated to 15,000–13,000 years BP (Hill, in Schröder-Lanz 1984; Dittrich et al. 1997).

The landslide A of Fig. 4.12 is characterized by many large springs (with a total discharge rate of about 600 m^3/day) and diffuse outlets emerging at various altitudes in the middle of the slope, which might indicate that this slide remains the most prone to reactivation. In the upper part of landslide zone A, spectacular escarpments with numerous tilted blocks are observed (Fig. 4.9). However, the larger slide B is responsible for the most impressive central part of the Wolfsschlucht (Fig. 4.2), which opened as a result of gliding, tilting, toppling and backtilting movements of rock slabs.

To the SE of landslide C (Fig. 4.12), talus slopes with alternating scree cones with elements up to several m^3 in size

Fig. 4.9 Another view of the Wolfsschlucht's western part. The plateau is on the *left*, limited by the subvertical escarpment, the sandstone slabs in the *right* are tilted downslope. The interlayering of thicker calcareous sandstone and thinner sandstone is enhanced by systematic differential erosion

and stony regolith embedding isolated blocks bury the base of the sandstone escarpment. Locally, blockfalls may be directly related to the toppling and disintegration of sandstone monoliths. Large monoliths may also move longer distances downslope by falling, bouncing and rolling, as exemplified by the 100 m^3 sandstone block lying in the Sauer River at Weilerbach, ~2 km to the NW of the Wolfsschlucht. There, chaotic slopes scattered with many blocks attest recent escarpment collapse.

Along the Erelchen plateau escarpment, the contact between the vertical sandstone face and the scree cones locally occurs through an up to 10-m-wide bench, which is key to the understanding of the mechanism of whole slope evolution (Fig. 4.13). Rockfall-induced accumulation of blocks at the base of the sandstone cliff loads the upper part of the slope cut in the underlying Elvange marls and Rhaetian clays highly prone to landsliding. The combination of such loading with episodes of increased infiltration through the Luxembourg Sandstone induces high hydraulic pressures in the underlying limestones and Rhaetian sandstones, which in turn cause multiple rotational landsliding in the marls and clays up to the base of the sandstone escarpment, inducing flattening of the upper part of the slipped body and preparing further sandstone slab toppling (Fig. 4.13). As a result, slightly displaced slabs leave 1- to 2-m-wide open joints with parallel walls, such as the *Devil's Crevice* (*"Brèche du Diable"*, Fig. 4.12) or the *Siewenschleff* at Berdorf (Fig. 4.14). Thick calcareous tufa accumulated in the now dry *Devil's Crevice*, originating in water percolation through the Marls and Limestones of Strassen preserved locally on top of the plateau and infiltration in the formerly not wide open joint.

While many slabs are tilted toward the valley by up to 30° (Lucius 1907), others are backtilted, creating triangular caves like the Raiberhiel (Fig. 4.14b). In the Wolfsschlucht, depending on the position of their respective centre of gravity with respect to the circular rupture surface that propagated from the downslope marls and clays below the base of the sandstone escarpment, many tilted slabs lean against backtilted blocks (Fig. 4.13). This occurs when the curved shear surface ends up in a vertical joint of the sandstone, causing rotation and up to 45° backtilt of the outward slab and destabilizing and tilting the next one.

Downslope, multiple rotational landsliding occasionally evolved into translational movements that reached the foot of the valley side, as attested by the 500-m-long reach of the Sauer that was displaced by ~40 m toward the inner side of the broad channel meander just upstream of Echternach

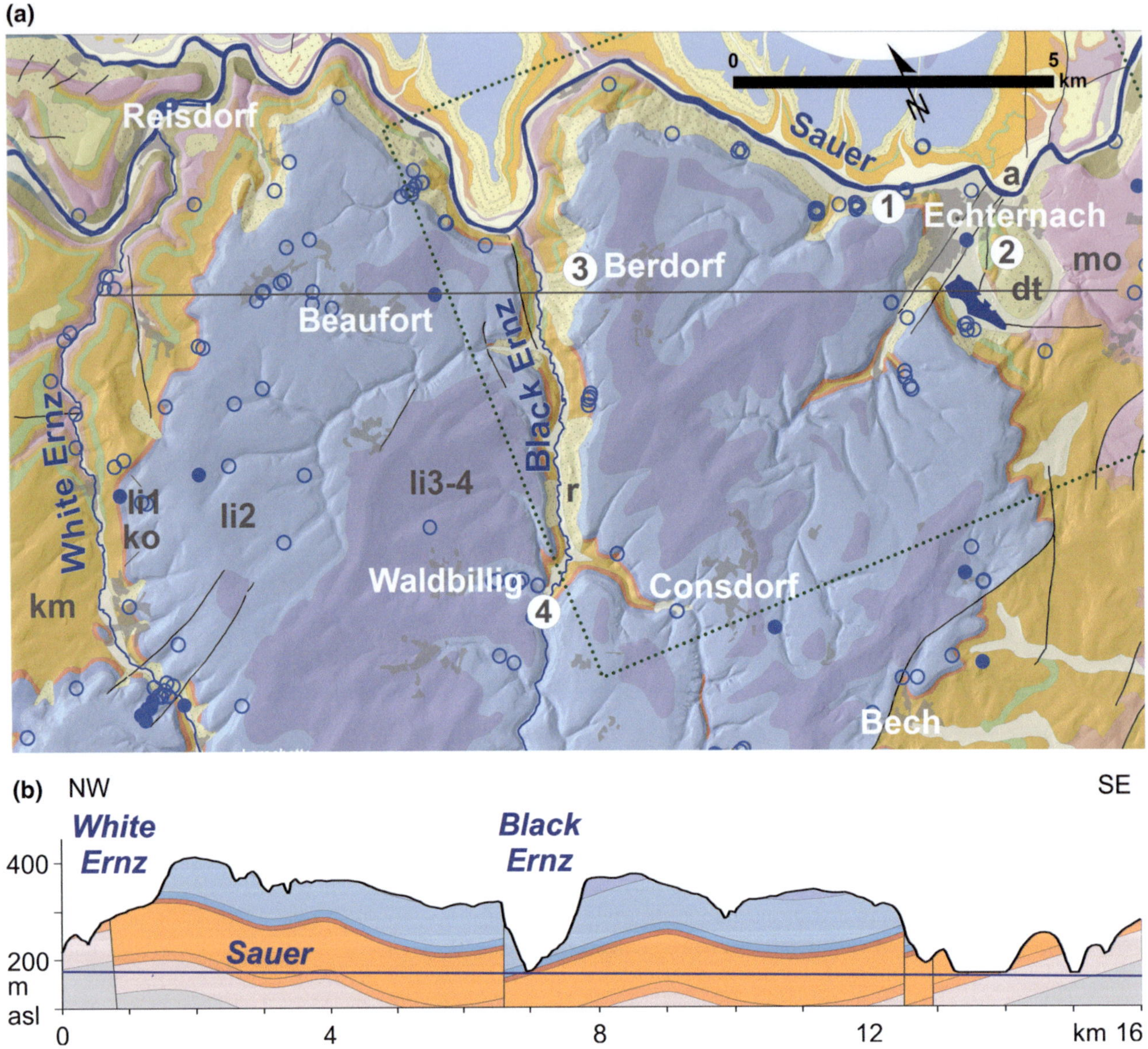

Fig. 4.10 **a** Geology, draped on the DEM of the Mullerthal region (see location in Fig. 4.1). Below the Luxembourg Sandstone (*li2*) that forms the main plateau are Lower Liassic marls and limestones (*li1*) and Keuper (*ko*, *km*) and Muschelkalk (*mo*) rocks. The sandstone is in turn locally covered by marls and limestones (*li3-4*). Fluvial terraces of the Sauer are noted 'dt'. *1* Wolfsschlucht, *2* cut-off meander of the Sauer at Thull, *3* Raiberhiel and Siewenschleff (Fig. 4.14), *4* Schiessentümpel cascade (Fig. 4.15). *Open and filled blue circles* sources and wells. The *green dotted rectangle* locates the geomorphological map excerpt shown in Fig. 4.11. **b** Schematic cross-section along the *black line* crossing the northern part of the sandstone plateau in **a**. The long profile of the eastward flowing Sauer is projected onto the cross-section. Map colours conform to the international colour code. © SGL, ACT

(Figs. 4.12 and 4.13). In case of compound landslides, the usual hard-to-answer question is whether the system evolved progressively or retrogressively (i.e. down- or uphill). Here, several observations point to the latter, namely (1) the relative freshness of many rock fragments in the scree cone with respect to more altered material in regoliths elsewhere in LLS and (2) the overall freshness and length of the escarpments, which both point to the young age of the escarpments, and (3) the increased incision and lateral erosion of the Sauer after the Thull meander was cut off at the Tardiglacial-Holocene transition (Coûteaux 1970), which probably induced the destabilization of the foot of the slope, whereas 6-m-thick alluvial deposits stabilize it now.

Sandstone cliffs and their rim of scree cones and stony regolith are also found in all small narrow valleys, which have however not yet incised the underlying marls and clays and consequently do not display the lower lying gentle slopes. Though still active currently, e.g. through dissolution

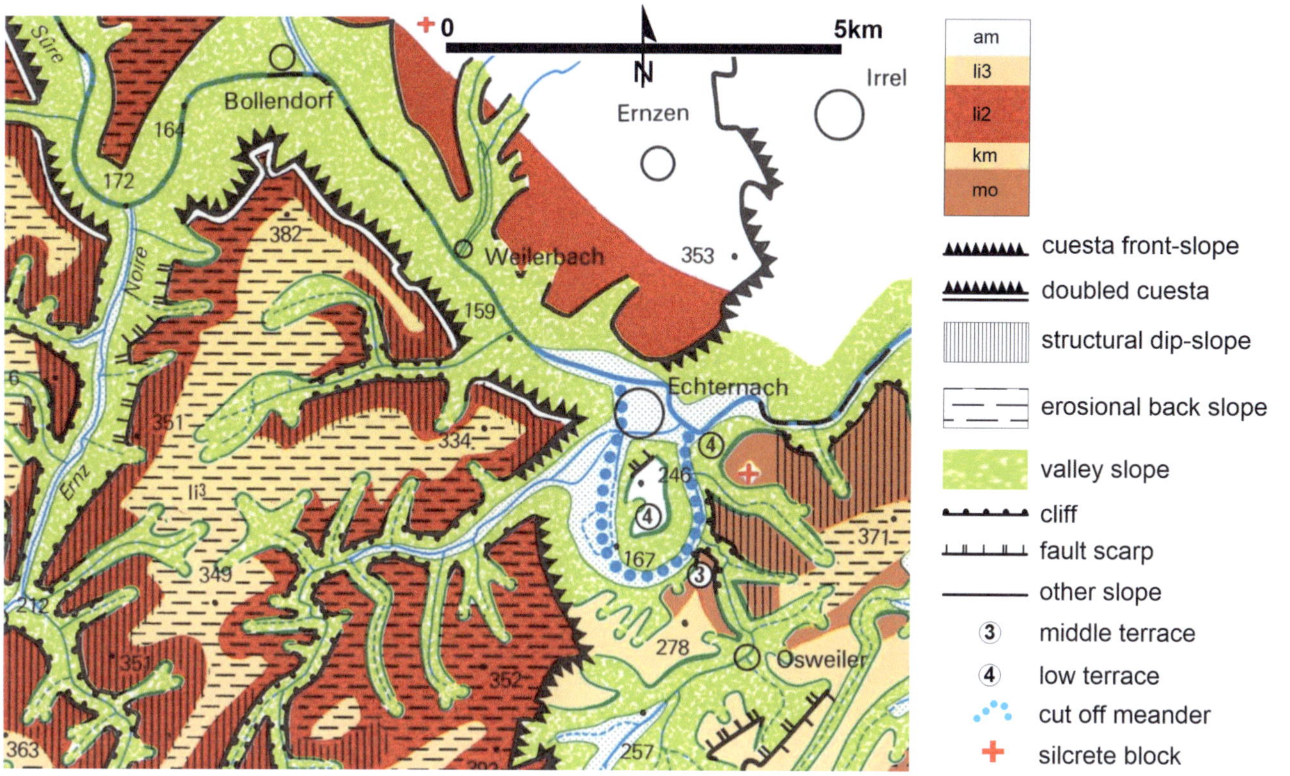

Fig. 4.11 Excerpt of the geomorphological map (1:100,000) showing the importance of the structural component in the landscape and geomorphology of the Echternach region (Désiré-Marchand 1985). *mo* upper Muschelkalk. *km* middle Keuper. *li2* Luxembourg sandstone, Lower Lias. *li3* Strassen marls and limestones, Lower Lias. *am* modern alluvium

of the carbonate matrix, weathering processes affected every sandstone outcrop especially during the cold phases of the Pleistocene, opening joints and liberating blocks by the combined action of freeze-thaw cycles and vegetation roots. Currently, most of the material fallen at the cliff foot cannot be removed by the creeks, which are able to transport only sand and gravel. Many rivers (e.g. Black Ernz, Alzette, Lauterborn creek) but not all (e.g. White Ernz) display prominent knickpoints in their long profile at, or upstream of, the point where they cross the contact between the Luxembourg Sandstone and the underlying weaker rocks. In the Black Ernz, the *Schiessentümpel* cascade marks the base of such a knickpoint near the hamlet of Mullerthal (Figs. 4.10 and 4.15). While the knickpoint is cut into the sandstone, the channel bottom downstream of it exposes the basal marls, causing many large springs to flow out at the channel's level. Upstream of Mullerthal, the Black Ernz valley shows a strongly asymmetric transverse profile, which might to some extent be related to the slight synclinal undulation cut obliquely by the river in this area (Fig. 4.10 b). The river flows along the western valley side, where it removed all debris delivered by the hillslope except the largest sandstone blocks, which clutter the river bed and form the picturesque cascade.

4.3.2 Overview of Weathering Structures in the Sandstone

Many weathering structures and microforms have developed in the body and on the free faces of the Luxembourg Sandstone, partly in relation with valley incision. They also depend on the contrasted carbonate content between the yellow sandstone and the lighter coloured fractured calcareous sandstone. The sandstones show grain size variations that induce changes in porosity at the layer scale. As for the calcareous sandstones, despite their very low porosity, they may develop high fracture permeabilities. They are also characterized by the greatest structural variability, alternating continuous layers of variable thickness and lenses or rounded masses aligned or scattered within non-calcareous sandstones. Chemical weathering mainly occurs through carbonate dissolution by water circulating in the outcropping sandstones and at the base of the sandy regolith. The carbonate dissolution front, undulating in function of the joint density, generally lies a couple of metres beneath the regolith. Secondary calcite precipitation occurs within the weathering structures or as tufa deposits in open joints and in creek channels. Pyrite oxidation occurs in non-dissolved gray sandstones above the water table,

(a)

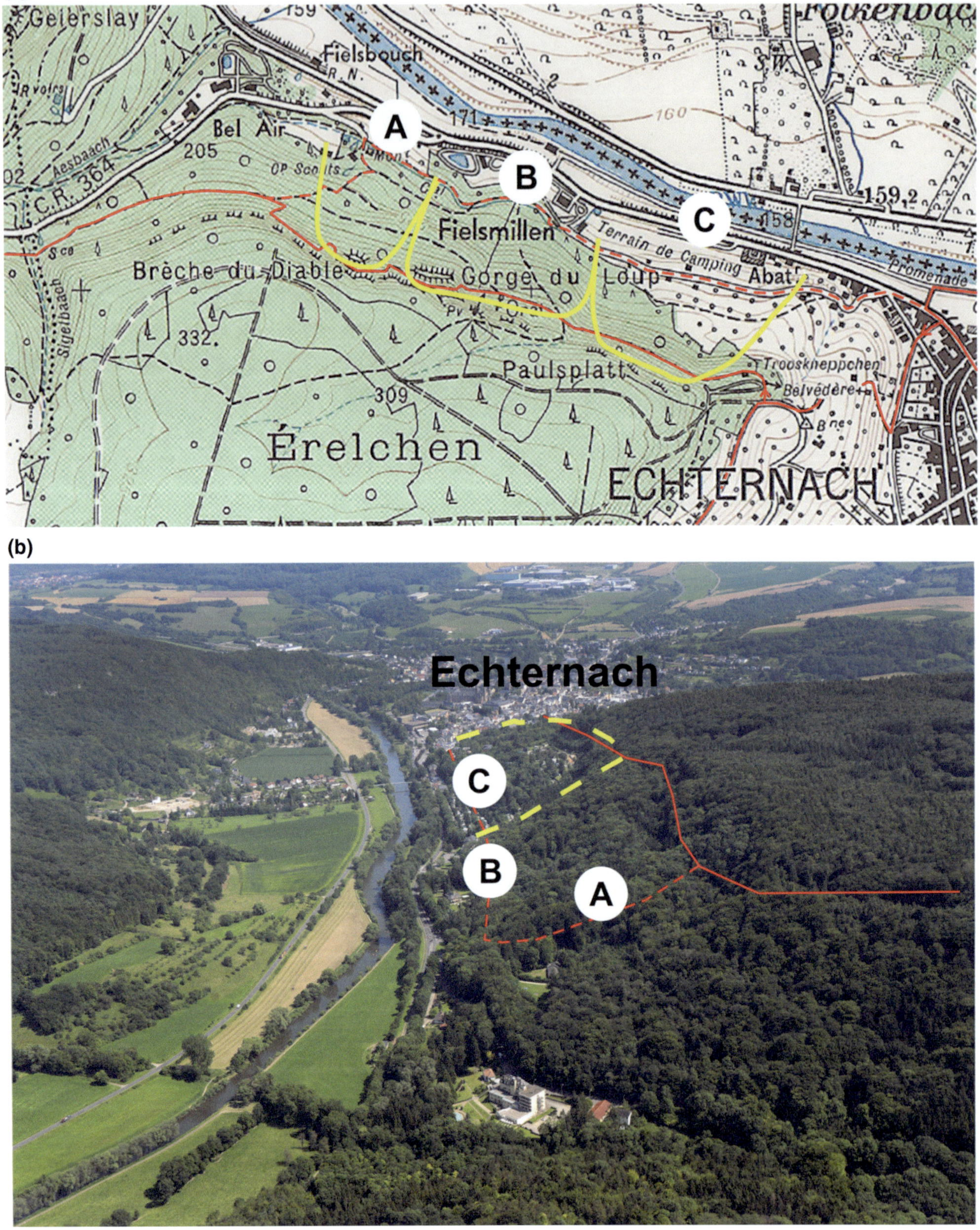

(b)

Fig. 4.12 **a** Historical topographic map (1:20,000, 1979) of the Erelchen Plateau, south of the Sauer valley at Echternach, locating the Wolfsschlucht ("*Gorge du Loup*", along the *red-coloured* hiking trail) at the top of the cuesta escarpment. *Yellow lines* delimit three large landslide scars (*A*, *B*, and *C*) © ACT. **b** Aerial photograph of the same area, taken from the WNW. One notes how landslide *B* displacement pushed the Sauer channel northward © Rol Schleich

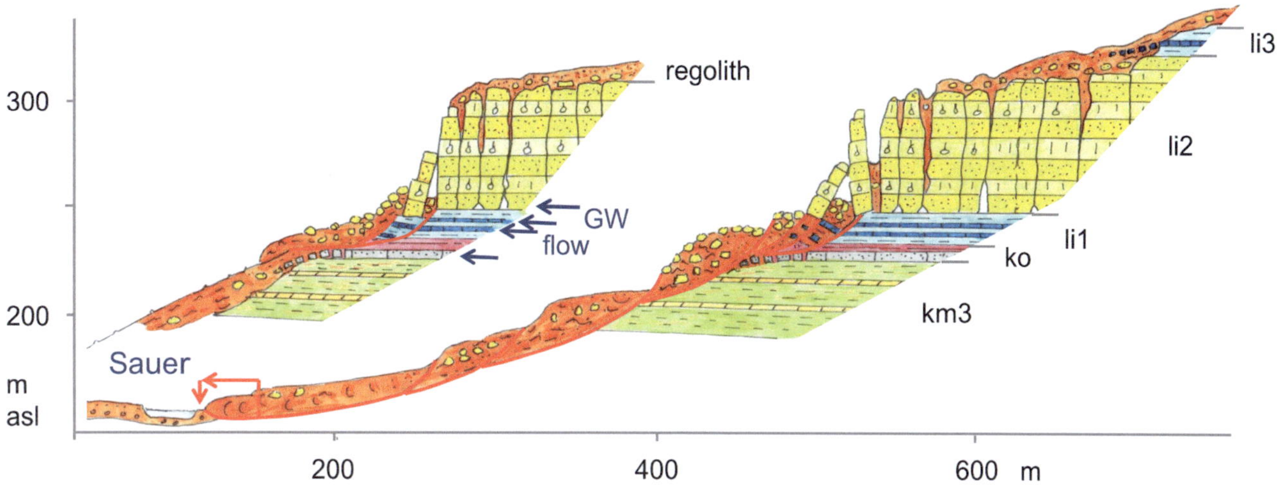

Fig. 4.13 Cross-sections showing the evolution of slope failure and gorge opening at the height of landslide *B* and Wolfsschlucht (see Fig. 4.12). Successive individual slides assumed to have evolved retrogressively (see text) are represented. At the base of the slope, the Sauer channel was displaced by ~40 m. At the *top*, tilt or backtilt motions of destabilized individual sandstone slabs along the cuesta edge are determined by the location of their respective centre of gravity with respect to the underlying shear surface

Fig. 4.14 **a** *Siewenschleff* (seven narrow passages). Horizontal displacements of several decimetres to 2 m are observed in the two primary joint directions. **b** *Raiberhiel* (robbers' cave). Narrow cave formed by backtilting above a rotational slide motion. Inclined stratification on the backtilted slab evidences the rotation angle. Both photographs taken near Berdorf (see Fig. 4.10, point 3)

inducing carbonate dissolution and the formation of secondary gypsum, which is subsequently leached by percolating waters.

4.3.2.1 Large Open Fissures

Large open irregularly walled fissures of the primary network are seen everywhere in the sandstone outcrop area (Fig. 4.6). Runoff concentrating in the topographical depressions of the plateau surface infiltrated the fissures and opened them progressively by dissolution of the carbonate matrix (up to more than 40% of total volume) and removal of the sandy residue by lateral flow toward the valleys. This mechanical emptying of the fissures was probably most efficient during the Pleistocene periods of active river incision, in general warm/cold transitions (e.g. Vandenberghe 2008; Demoulin et al. 2012), which allowed for sufficient hydraulic gradients. Nowadays, these deep open joints are sealed by a weathering mantle, which collapses sometimes during rainy periods.

4.3.2.2 Dissolution in Calcareous Beds

Dissolution along the fissures of the calcareous sandstones is very common, leaving all kinds of figures from wide open fissures giving the appearance of a set of teeth to the layer (Fig. 4.16b, c) to broadly spaced pillars in the final steps of a layer's dissolution and, locally (e.g. in the upper part of the

Fig. 4.15 *Schiessentümpel* cascade in the Black Ernz valley near the hamlet of Mullerthal (see Fig. 4.10, point 4). Large sandstone blocks collapsed from the cliffs crowning the valley sides cannot be removed by the current river and form the cascade

Wolfsschlucht escarpment), to keyhole-like voids indicating groundwater flow under pressure. The latter imply a groundwater level above the observation level, thus suggesting formation at a very early stage of valley incision and scarp formation. Similar structures found at different altitudes all over the sandstone area might document the incision history of the creeks into the sandstone (Adamovič et al. 2015).

4.3.2.3 Elliptical Cavities in the Sandstone

When calcareous sandstone nodules are embedded in porous sandstones, percolating water concentrates at the outer surface of the impervious nodules and dissolves the calcitic matrix of their outer rim, isolating progressively the nodules from the surrounding rock. If liberation is complete, the nodule may fall out, leaving a cavity in the sandstone (Fig. 4.16d). This very common process mostly observed on exposed joint walls is active even today in zones of diffuse infiltration of water close to exposed surfaces.

4.3.2.4 Honeycomb Weathering

Honeycomb structures (Fig. 4.16e) are also frequently found. Surfaces showing narrow ridges ringing shallow depressions are common, while zones of almost spherical cm-size hollows are less frequent. It seems that shallow depressions are often precursors of the spherical hollows. Honeycomb, absent in the Wolfsschlucht area, is particularly abundant on the escarpments near Nommern, NE of Mersch (Fig. 4.1). Occurrences are limited to the porous sandstone, where honeycombs follow the layering and associated grain size variations. Their formation remains unclear, possibly involving the formation of a crust by evaporation of outward-diffusing water and calcite recrystallization, then crust perforation and disintegration of the subcrustal sandstone in regularly spaced hollows and, finally, re-induration of the hollowed surface. The process could be repeated cyclically, as described by Robinson (2007). Salt action is thought to take often an active part in honeycomb formation (Robinson 2007). In the case of the Luxembourg Sandstone, salt may come from pyrite oxidation in fresh gray sandstones liberating sulfate ions or from residual gypsum leaching. Löhr (2012) dates honeycomb formation by archaeological evidence to early Postglacial times, though evidence from modern periglacial environments (Weise 1983) suggests it might also date back to colder periods.

4.3.2.5 Differential Weathering and Erosion

Differential weathering and erosion often nicely enhance sedimentary structures like cross bedding and layering (Fig. 4.16a). Less resistant sandstone layers are in depression while the hard calcareous sandstones stand out. Beside mechanical erosion by wind and rain, chemical dissolution, aided by the presence of moss, acts mainly where humidity is sufficient in the more porous sandstones.

4.3.2.6 Crust Formation

Crusts are quite common in the City of Luxembourg but rare in the Mullerthal region. Thin, often black crusts are produced by crystallization of various atmospheric salts incorporating dust particles. They are mostly limited to porous sandstones, the surface of which they protect. Once the crust is removed, the freshly exposed rock surface is very prone to abrasion of sand particles loosened by the dissolution of the carbonate matrix.

4.3.2.7 Rock Overhangs in Creek Valleys

Large regular rock overhangs with rounded concave forms are observed in many creek valleys at different heights above the valley bottom (Fig. 4.16f). They are believed to have developed essentially by lateral fluvial erosion and abrasion of the soft sandstone by the bed load during high discharge events. At the same time, the calcareous sandstone was broken up and plucked along small fissures, locally creating small cascades. However, other overhangs might have formed as a result of either freeze-thaw cycles or weathering enhanced by the soil moisture (Cílek and Žák 2007; Robinson 2007; Römer 2002).

4.4 Luxembourg City in Its Natural Landscape

The City of Luxembourg (Fig. 4.1) has been founded around 963 AD on the rocky Bock promontory of the plateau dominating the river Alzette and its small tributary, the Petruss. From the 12th century onward, it was progressively fortified to one of the most important fortresses of Europe,

Fig. 4.16 Typical weathering structures covering the sandstone rock faces in Luxembourg's Little Switzerland. **a** Differential weathering and erosion between interlayered resistant calcareous and weaker porous sandstones, **b** and **c** fissures opened by dissolution in the calcareous sandstone layers, **d** holes created by surficial dissolution and liberation of calcareous sandstone nodules out of porous sandstone layers, **e** honeycomb weathering structures and **f** rock overhang

called "Gibraltar of the North" (Fig. 4.17). Spanish and even older construction works were largely expanded by the Marquis de Vauban under the French occupation in the late 1680s. He employed up to 15,000 people, exploited many quarries and, besides overground works, constructed about 4 km of underground galleries. Older works were

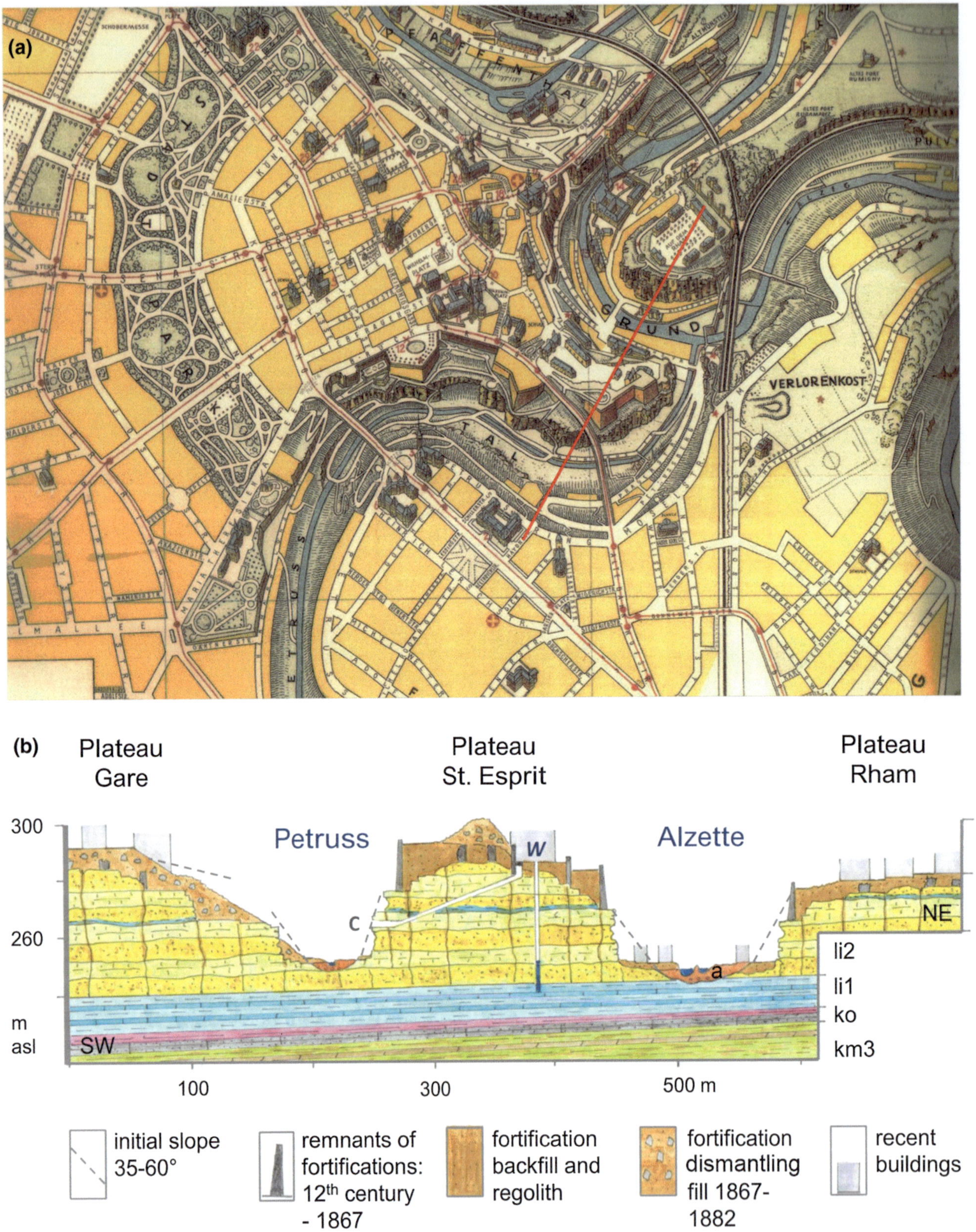

Fig. 4.17 **a** Excerpt of an historical city map of Luxembourg City (Gronzka 1942) by R. Gronzka, showing the Alzette and Petruss valleys and the cleaned-up cliffs surmounted by the fortress walls. **b** Schematic geological cross-section through the Alzette and Petruss valley and the plateaus of Luxembourg City (*bold red line* in **a**) showing the importance of both surface and underground defence works. *w* fortress well; *c* casemates: 17 km of underground galleries were spared of 23 km before dismantling; *a* alluvium; *li2* sandstones and subordinate marls (see Figs. 4.4 and 4.5); *li1* limestone and marl; *ko* clay-and sandstone, *km3* dolomitic marl and dolomite. *Dashed lines* reconstruct initial slopes, showing the importance of valley remodelling. Remnants of the fortification walls, buried under 19th century filling along the plateau SW of the Petruss, are in *dark gray*

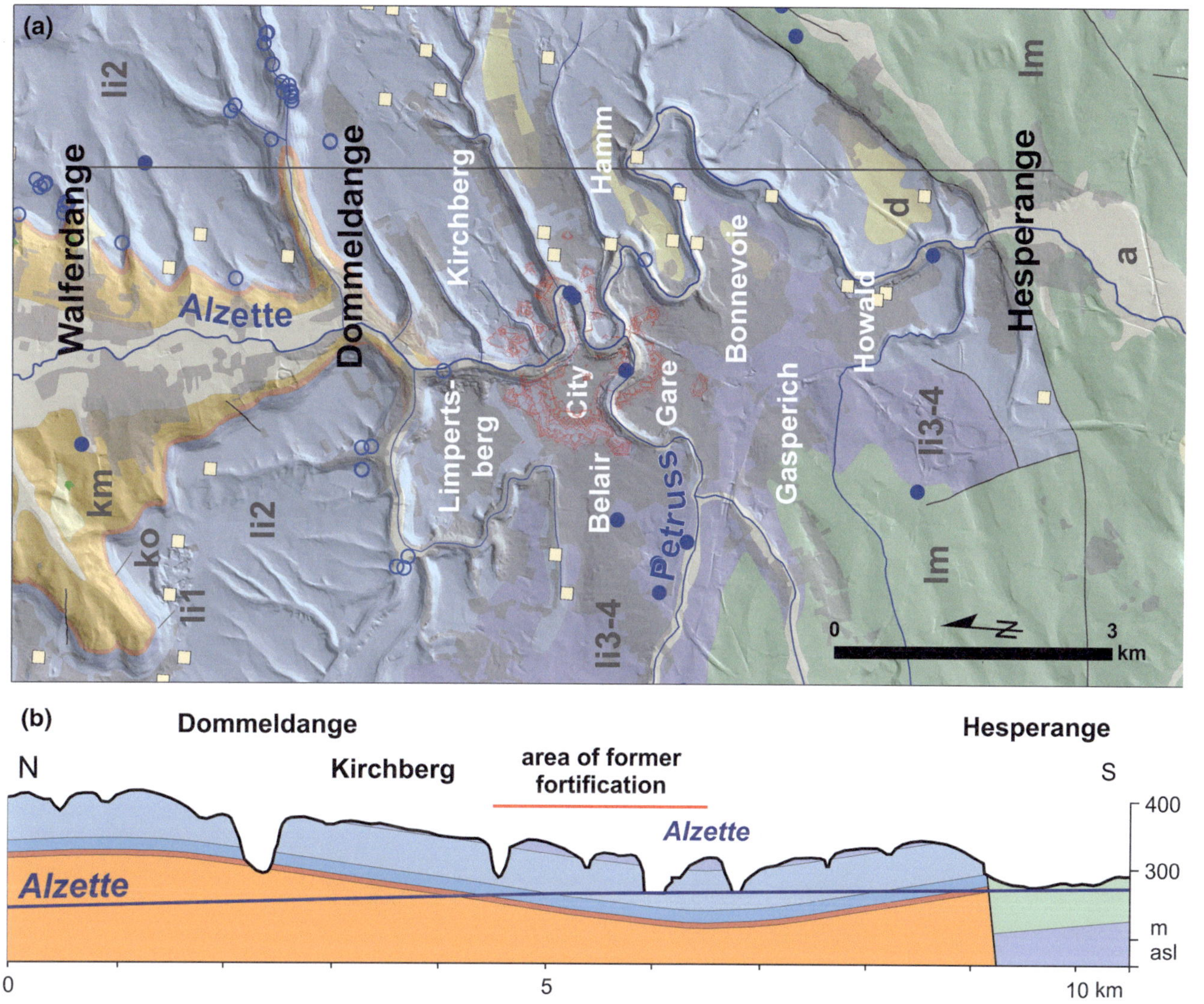

Fig. 4.18 **a** Geology, draped on the DEM of the Luxembourg city area (see location in Fig. 4.1). See Fig. 4.10 for the stratigraphic legend. Sandy marls and limestones (*lm*) are in *green*. Scattered undifferentiated Cenozoic deposits and weathering products (*d*) are found on the plateaus. *Open* and *filled blue circles* sources and wells. *Light yellow squares* main quarries. Compare the extent of modern Luxembourg (*gray areas*) with the area enclosed in the fortifications (*thin red lines*) © SGL, ACT. **b** Schematic cross-section along the *black line* in **a** showing the undulated structure of the Mesozoic cover, cut by an E-striking fault near Hesperange. The long profile of the northward flowing Alzette is projected onto the cross-section

maintained and completed during later occupations. While important fortifications had to be demolished in 1867 following a clause of the Treaty of London, interest arose in keeping witness remnants, notably for touristic reasons. From the late 19th century, the city expanded in different phases on the plateaus and the floor of the deeply incised valleys. The remnants of old quarters and fortifications are registered in the UNESCO World Heritage List since 1994.

The City of Luxembourg extends mainly on the undulated SW-plunging strata of the Luxembourg Sandstone (Fig. 4.18b), covered only to the west and southwest by progressively more continuous and thicker marls (li3-4 and lm in Fig. 4.1). By contrast, Lower Liassic marls and Keuper sediments underlying the sandstone are observed north of Luxembourg City, in the valleys of the Alzette and its larger tributaries. The orientation of the small valleys is largely determined by the primary joint system of the sandstone outcrop area, which is limited by a fault in the south.

The difference in rock resistance to weathering and erosion is particularly visible in the incision of the Alzette valley. The valley floor is ≥ 1 km wide north and south of the sandstone outcrop area whereas it is <0.3 km where the river flows on the sandstone. South of Hesperange (Fig. 4.18a), a very broad valley with flat hillslopes is cut in thick friable Middle Liassic marls. North of Luxembourg City, the sandstone cliffs dominating slopes similar to those

Fig. 4.19 **a**. WSW-looking view from Fetschenhof on the fortifications of Luxembourg City, drawn around 1839 by J.M. W. Turner. Watercolor, gouache and ink on blue paper. The central Bock promontory leads to the Plateau of the city and, in the *left background*, the Plateau Saint Esprit. *Left middle ground* the Rham Plateau, separated from the Plateau of the city by the Alzette valley. *Right middle ground* the Kirchberg Plateau (Courtesy and collection of the National Museum of History and Art, Luxembourg). **b**. Lithography illustrating the dismantling of the Luxembourg fortifications in the second half of the 19th century, by P. Blanchard

described in the Sauer valley at Echternach (Fig. 4.7) limit the width of the valley to ~2 km. These cliffs are almost continuous down to the Mersch area, where they go into the escarpments of the WSW- to SW-striking sandstone cuesta (Fig. 4.1). Topographical map analysis and geotechnical surveys evidenced numerous large old landslides on the marly slopes, supposed to have been active at the end of the last Glacial but now stabilized by Holocene fluvial accumulation at the foot of the slopes. Landsliding occurred along shear surfaces in the Rhaetian claystones, in a context of steeper slopes and higher and much more variable groundwater level and hydraulic pressures than today. Currently, only small, though numerous, landslides are triggered by human activity on these slopes.

At Luxembourg City, the Alzette and Petruss have dissected the sandstone body into several wide plateaus (Fig. 4.18). The Alzette valley width between cliffs varies from ~50 m at the entrance of the river in the sandstones, north of Hesperange, to ~300 m. The Petruss, a small tributary whose valley is however of similar width, joins the

Alzette at an altitude of 245 m, ~50 m below the historic centre of the town located on the top of the plateau. One may still observe how the fortress was made more impregnable by building up to 12 m high walls above the high subvertical cliffs along the two valleys (Fig. 4.17). The size and extent of the fortress walls are spectacularly depicted on numerous watercolor paintings by the English Romantic landscape painter J.M.W. Turner (Fig. 4.19a), as is also the importance of the dismantling activities of the fortress in the 19th century by a lithography of P. Blanchard (Fig. 4.19b). The box valley (*Kastental*) shape of the valleys, with steep vertical cliffs and flat valley bottom (Fig. 4.17b), has been sculptured by man, who removed the regolith at the foot of the cliffs to straighten the steep valley walls. Cliff shaping always followed the sandstone's primary joint system in order to satisfy the military aim of avoiding dead angles. The valley floor of the Alzette was widened not only for military reasons but also to allow construction of the medieval craftspeople quarter of Grund. An old 50-m-deep well in the central plateau of Saint Esprit reached the groundwater table and provided the fortress with drinkable water in case of siege. The quasi-absence, except locally in the upper part of the cliffs, of typical weathering features similar to those described in LLS underlines the extent of the cleaning up by the fortress builders. Now, moss and lichen extensively cover favourably orientated rock faces mainly of the porous sandstone layers. More generally, first signs of differential erosion already reappear and encrusting is common. In contrary to Vauban's time, when no vegetation was tolerated on the cliffs and walls, bushes and small trees are now allowed to grow on the cliffs for environmental reasons, causing however more frequent stone or rock fall than formerly. Continuous cliff cleaning up, stability inspection and, occasionally, stabilization works are thus necessary to ensure long-term protection.

4.5 Conclusion

The picturesque landscapes of the Mullerthal region with their long touristic vocation in Luxembourg's Little Switzerland have preserved their natural character and nicely expose all geomorphological landforms typical of sandstone areas. The valleys incised in the plateau of the Lower Liassic Luxembourg Sandstone display an all the more large variety of rock sculpturing features as this Formation alternates beds of porous sandstone and calcareous sandstone. The region is accordingly characterized by a high diversity of weathering structures and erosion forms at all scales, which participated in, and therefore allow the reconstruction of, its Quaternary landscape evolution dominated by valley incision and gravitational mass movements. These easily accessible landscapes in the heart of West Europe are the quintessence of sandstone geomorphology. As such, they have high scientific and educational values that deserve to be preserved and valorized. This has recently come in addition to ecological and biodiversity considerations to decide the creation of the natural park Mellerdall. We also seized the opportunity of the Luxembourg City site to highlight how man always took advantage of the terrain characteristics to develop and protect his oldest settlements but also how he concurrently became a major factor of landform evolution.

References

Adamovič J, Mikuláš R, Navrátil T (2015) Spherical and ellipsoidal cavities in European sandstones: a product of sinking carbonate dissolution front. Zeitschrift für Geomorphologie 59 (Suppl. 1):123–149

Bintz J, Hary A, Muller A (1973) Luxembourg. In: Pomerol C (ed) Guides géologiques régionaux: Ardenne Luxembourg, pp 135–192

Busche D, Kempf J, Stengel I (2005) Landschaftsformen der Erde. Bildatlas der Geomorphologie. Wissenschaftliche Buchgesellschaft, Darmstadt

Cílek V, Žák K (2007) Late Glacial and Holocene sedimentation under sandstone rock shelters of Northern Bohemia (Czech Republic). In: Härtel H, Cílek V, Herben T, Jackson A, Williams R (eds) Sandstone landscapes. Academia, Prague, pp 133–138

Coûteaux M (1970) Etude palynologique des dépôts quaternaires de la Vallée de la Sûre à Echternach et à Berdorf, et de la Moselle à Mertert. Institut Grand-Ducal de Luxembourg, Section des sciences naturelles, physiques et mathématiques, Archives 34:297–336

Demoulin A (1995) Les surfaces d'érosion méso-cénozoïques en Ardenne-Eifel. Bulletin de la Société géologique de France 166:573–585

Demoulin A, Beckers A, Rixhon G, Braucher R, Bourlès D, Siame L (2012) Valley downcutting in the Ardennes (W Europe): Interplay between tectonically triggered regressive erosion and climatic cyclicity. Neth J Geosci 91:79–90

Désiré-Marchand J (1985) Notice de la carte géomorphologique du Grand-Duché de Luxembourg. Publications du Service Géologique du Luxembourg, Bulletin 13/1985

Dittrich D, Bartels L, Vogel K (1997) Neue Ergebnisse zur Geologie des Ferschweiler Plateaus und des Heiderückens (Zentrale Trier-Bitburger Mulde)—Tektonik, Keuper-und Lias-Stratigraphie sowie ein Beitrag zur genetischen Deutung des Naturdenkmals « Irreler Wasserfälle ». Mainz Geowiss Mitt 26:55–98

Garcia-Castellanos D, Cloetingh S, van Balen R (2000) Modelling the Middle Pleistocene uplift in the Ardennes/Rhenish Massif: thermo-mechanical weakening under the Eifel? Glob Planet Change 27:39–52

Gronzka R (1942) Plan der Festungs- und Rosenstadt Luxemburg, Neuauflage. Städtisches Verkehrsamt Luxemburg, Luxemburg

Hövel H, Alberti M, Dikau R (2015) Qualitative Auswertung von Laserscandaten zur Erfassung von Massenbewegungsflächen am südlichen Ferschweiler Plateau (Südeifel). Mainz Geowiss Mitt 43:185–208

Le Brun-Ricalens F, Valotteau F (2005) Patrimoine archéologique et Grès de Luxembourg: un potentiel exceptionnel méconnu. In: Ries C, Krippel Y (eds) Sandstone landscapes in Europe. Past, present and future, Ferrantia 44, pp 77–81

Le Roux J, Harmand D (2014) Les enseignements du relief de côtes dans l'est du Bassin Parisien. Bull Inf Géol Bassin Paris 51(4): 20–35

Liedtke H, Deshaies M, Gamez P, Harmand D, Preusser H (2010) Die Oberflächenformen in der Grenzregion Saarland-Lothringen-Luxemburg. Les formes de relief dans la région frontalière Sarre-Lorraine-Luxembourg. Forschungen zur deutschen Landeskunde, vol 259, Leipzig

Löhnertz W (1994) Grundzüge der morphologischen Entwicklung der südlichen Eifel im ältesten Tertiär. Mainz naturwiss Archiv Beih 16:17–38

Löhnertz W (2003) Eocene paleovalleys in the Eifel: mapping, geology, dating and implications for the reconstruction of the paleosurfaces and vertical movements of the lithosphere at the edges of the Rhenish shield. Géol Fr 2003(1):57–62

Löhnertz W, Lutz H, Kaulfuß U (2011) Eifel–Mosel–Hunsrück. In: Deutsche Stratigraphische Kommission (ed) Stratigraphie von Deutschland IX, Tertiär, Teil 1. Schriftenreihe der Deutschen Gesellschaft für Geowissenschaften, vol 75, pp 376–415

Löhr H (2012) Umweltgeschichte—Zwischen Natur-und Kulturgeschichte. In: Clemens L, Felten F, Schnettger M (eds) Kreuz, Rad, Löwe. Rheinland-Pfalz, ein Land und seine Geschichte, Ph. von Zabern Verlag, Mainz, pp 179–209

Louis H (1953) Über die ältere Formenentwicklung im Rheinischen Schiefergebirge, insbesondere im Moselgebiet. Münchner Geographische Hefte 2, München, p 97

Lucius M (1907) Einige landschaftlichen Eigentümlichkeiten der Umgebung Echternachs geologisch betrachtet. Bull de la Société des Naturalistes luxembourgeois 17:159–167

Lucius M (1948) Entstehung und Entwicklung des Luxemburger Flusssystems. Bulletin de la Société des Naturalistes luxembourgeois 53:17–48

Lucius M (1952) Correspondance inédite du 29 octobre 1952 adressée au Dr. E. Schneider. Archives MNHA no d'inventaire 2002–13, original at the SGL

Massard J (2012) Die Geschichte des Echternacher Verschönerungsvereins. http://massard.info/pdf/verschoenechternach.pdf, 1 Nov 2015

Meyer W, Stets J (1998) Junge Tektonik im Rheinischen Schiefergebirge und ihre Quantifizierung. Z dtsch geologischen Ges 149: 359–379

Muller A (1980) Luxembourg. In: Géologie des Pays européens: France, Belgique, Luxembourg. Publications du 26e congrès géologique international, Paris, pp 577–594

Reinheimer L, Krauter E, Wehinger A, Alberti M (2010) Untersuchungen zur Hanginstabilität am Ferschweiler Plateau (Südeifel). Mainz Geowiss Mitt 38:73–98

Robinson D (2007) Geomorphology of the inland sandstone cliffs of Southeast England. In: Härtel H, Cílek V, Herben T, Jackson A, Williams R (eds) Sandstone landscapes. Academia, Prague, pp 133–138

Römer W (2002) Aspekte der Schichtstufenentwicklung am Ferschweiler Plateau (Südeifel) und ihre Simulation im theoretischen Modell. Petermanns Geogr Mitt 146:40–47

Schroeder-Lanz H (1984) Morphologische Probleme der Südwesteifel (Einzugsgebiet Kyll und Prüm) mit Beiträgen von H. Weiler und J. Hill. In: Jätzold R (ed) Der Trierer Raum und seine Nachbargebiete, Trierer Geographische Studien 6, pp 37–64

Tricart J (1949) La partie orientale du Bassin de Paris. Etude morphologique. SEDES, Paris, p 474

Van Balen R, Houtgast R, van der Wateren F, Vandenberghe J, Bogaart P (2000) Sediment budget and tectonic evolution of the Meuse catchment in the Ardennes and the Roer Valley Rift System. Glob Planet Change 27:113–129

Vandenberghe J (2008) The fluvial cycle at cold-warm-cold transitions in lowland regions: A refinement of theory. Geomorphology 98:275–284

Weise O (1983) Das Periglazial. Geomorphologie und Klima in gletscherfreien, kalten Regionen. Gebr. Borntraeger, Berlin, p 199

Erosion Surfaces in the Ardenne–Oesling and Their Associated Kaolinic Weathering Mantle

5

Alain Demoulin, François Barbier, Augustin Dekoninck, Michèle Verhaert, Gilles Ruffet, Christian Dupuis and Johan Yans

Abstract

This chapter deals with the long-term geomorphology of the Paleozoic Ardenne–Oesling massif of S Belgium and the landforms that currently attest the very long persistence and high resistance to erosion of landscape elements created as far back in time as the Lower Cretaceous in a region of predominantly continental regime and constant low to moderate elevation since the Permian. We first present recent results regarding the antiquity of kaolinic weathering mantles preserved in four sites on or close to the plateau surfaces characteristic of the Ardennian landscape, showing that they record at least three main weathering phases, namely in the Early Cretaceous (extending in the early Late Cretaceous), the Late Paleocene, and the Early Miocene. Then, having provided a working definition of the erosion surface and listed criteria helping recognize them in the landscape, we focus on the Hautes Fagnes massif, the highest NE part of the Ardenne–Oesling. We describe the geometric and sedimentary evidence that allow one to recognize a succession of four erosion surfaces formed during the Cretaceous and the Cenozoic and reconstruct the morphogenetic evolution of the area that responded to the variable interplay of the marine transgression-regression, climatic, and tectonic controls. Finally, we present the few estimates of long-term denudation rates available for the Ardenne massif, fission track and cosmogenic nuclide studies both suggesting Cenozoic denudation rates in the order of a few 10 m/Myr, and comment on their compatibility with field observations.

Keywords

Long-term geomorphology • Kaolinic weathering • Saprolite • Erosion surface • Ardenne • Oesling

A. Demoulin (✉)
Department of Physical Geography and Quaternary, University of Liège, Sart Tilman, B11, 4000 Liège, Belgium
e-mail: ademoulin@ulg.ac.be

A. Dekoninck · M. Verhaert · J. Yans
Department of Geology, University of Namur, rue de Bruxelles 61, 5000 Namur, Belgium
e-mail: augustin.deconinck@unamur.be

M. Verhaert
e-mail: michele-alexandra.verhaert@unamur.be

J. Yans
e-mail: johan.yans@unamur.be

G. Ruffet
Géosciences Rennes, UMR 6118, University of Rennes1, Campus de Beaulieu, 35042 Rennes Cedex, France
e-mail: gilles.ruffet@univ-rennes1.fr

C. Dupuis
Géologie Fondamentale et Appliquée, University of Mons, rue de Houdain, 9, 7000 Mons, Belgium
e-mail: christian.dupuis@umons.ac.be

F. Barbier
Global Operations and Components at Flextronics International Ltd, Singapore, Singapore
e-mail: barbier.francois@gmx.com

A. Demoulin (ed.), *Landscapes and Landforms of Belgium and Luxembourg*, World Geomorphological Landscapes, DOI 10.1007/978-3-319-58239-9_5

5.1 Introduction

It has long been recognized that erosion surfaces are a main component of the landscape in the southern part of Belgium and the northern Grand-Duchy of Luxembourg, where they impose the levelled horizons and extended flat areas characterizing the Paleozoic Ardenne–Oesling Plateau (Oesling, or Éislek in Luxembourgian, is the name given to the part of the Rhenish massif situated in the Grand-Duchy of Luxembourg). Bearing witness to the Mesozoic and Cenozoic geomorphological evolution of the region, they constitute the long-term frame in which incision of the Plio-Quaternary drainage network has not yet so much nibbled that one could not identify this pre-Quaternary heritage.

The Ardenne massif is famous for his well-preserved erosion surfaces, the so-called peneplains that W. Davis commented on already in the end of the nineteenth century (Davis 1896, 1899) after having visited the area with A. de Lapparent during one of his European travels. Already in that time went the discussion on the age of the surfaces and on the uplift history of the massif revealed by their arrangement. This testing by Davis of his normal cycle of erosion in the Ardennian landscape was possible because, ten years earlier, Gosselet (1888) had provided a very modern overview of the nature and age of the post-Paleozoic sediments sparsely preserved on the surfaces and had discussed the tectonic deformations implied by their distribution.

However, although many studies dealt with the Ardennian erosion surfaces in the first half of the twentieth-century, building on the Davis' theory or sometimes adapting it, like Baulig (1926) with his multifaceted Ardennian platform, it was a long time before one realised that these surfaces had been formed under climatic conditions totally different than those currently prevailing. Being the product of a long evolution under tropical conditions, they essentially respond to the etchplanation model of Wayland (1933), possibly flavoured with features of the pedimentation model of King (1947), much more than to the normal erosion model of Davis (1899). Gullentops (1954) was probably the first to show real interest in the remnants of the weathering mantle often accompanying the old surfaces and to take advantage of them in his reconstruction of the long-term geomorphic history of the massif. Recently, qualitative observations gave way to quantitative data about the age of the Ardennian kaolinic saprolites, which bring additional support to the chronology of such reconstructions (Yans 2003; Thiry et al. 2006).

In this chapter, the geomorphic history of the Ardenne–Oesling massif will be reexplored through the analysis of four archetypal sites of the Ardenne. We first describe the deep weathering profile exposed at Transinne, in the last quarry active in kaolin exploitation in west-central Ardenne, and present age data. Then, moving to the eastern part of the Ardenne (Plateau des Tailles area, Bihain site), we provide a new Ar–Ar dating of the weathering. The Hautes Fagnes Plateau, a recently uplifted area that has preserved several well-developed erosion surfaces of Cretaceous and Paleogene age will also be described because it offers a condensed view of various modes of their evolution. Finally, new dating of supergene minerals from the northern part of the Ardenne (Beez and Morialmé sites, Entre-Sambre-et-Meuse) will be discussed against the background of the Ardenne evolution.

5.2 Weathering Mantles

5.2.1 Geographical and Geological Setting

Located in the southern part of Belgium, the Ardenne region geologically corresponds to the Ardenne allochthon, which is part of the Rheno-hercynian fold-and-thrust belt. Belanger et al. (2012) recently distinguished the following structural units, from north to south across the Paleozoic of Belgium: the Brabant Massif and the base of its Devonian cover unaffected by the Variscan orogeny, the Brabant para-utochthon, the Haine-Sambre-Meuse overturned thrust sheets (HSM-OTS) and the Ardenne allochthon where the Transinne and Bihain sites are located (Fig. 5.1). The third presented site, Morialmé, is situated in the Entre-Sambre-et-Meuse area, which is the northwestern part of the Ardenne allochthon, whereas the fourth one, Beez, is located in the HSM-OTS.

Thick saprolites are widespread in Belgium (e.g., Quesnel et al. 2002; Yans 2003; Dejonghe and Boni 2004; Dusar and Lagrou 2007). Yans (2003) highlighted at least ten areas that experienced period(s) of deep chemical weathering in Belgium. Figure 5.2 provides an overview of those affected by weathering period(s) well constrained in time in and around the Ardenne, but additional areas with unconstrained timing of weathering may also be listed (e.g., Welkenraedt, Hautes Fagnes, Rocroi, northern part of the Mons Basin) (Yans 2013). The weathering led to the formation of saprolites up to 80- and 65 m-thick in the Brabant Massif and the Haute-Lesse area, respectively.

North of the Ardenne, the long phase of weathering and kaolinitisation, that affected the Paleozoic basement of the Brabant Massif to depths reaching a preserved maximum of 80 m started in the Early Cretaceous and continued up until the Cenomanian along the southern margin of the massif and the Santonian or later along its northern margin, depending on the local time of flooding and onset of subsequent marine sedimentation (Dusar and Lagrou 2007). The saprolites are widely preserved, with thickness varying strongly even at the local scale (Mees and Stoops 1999). In the Belgian

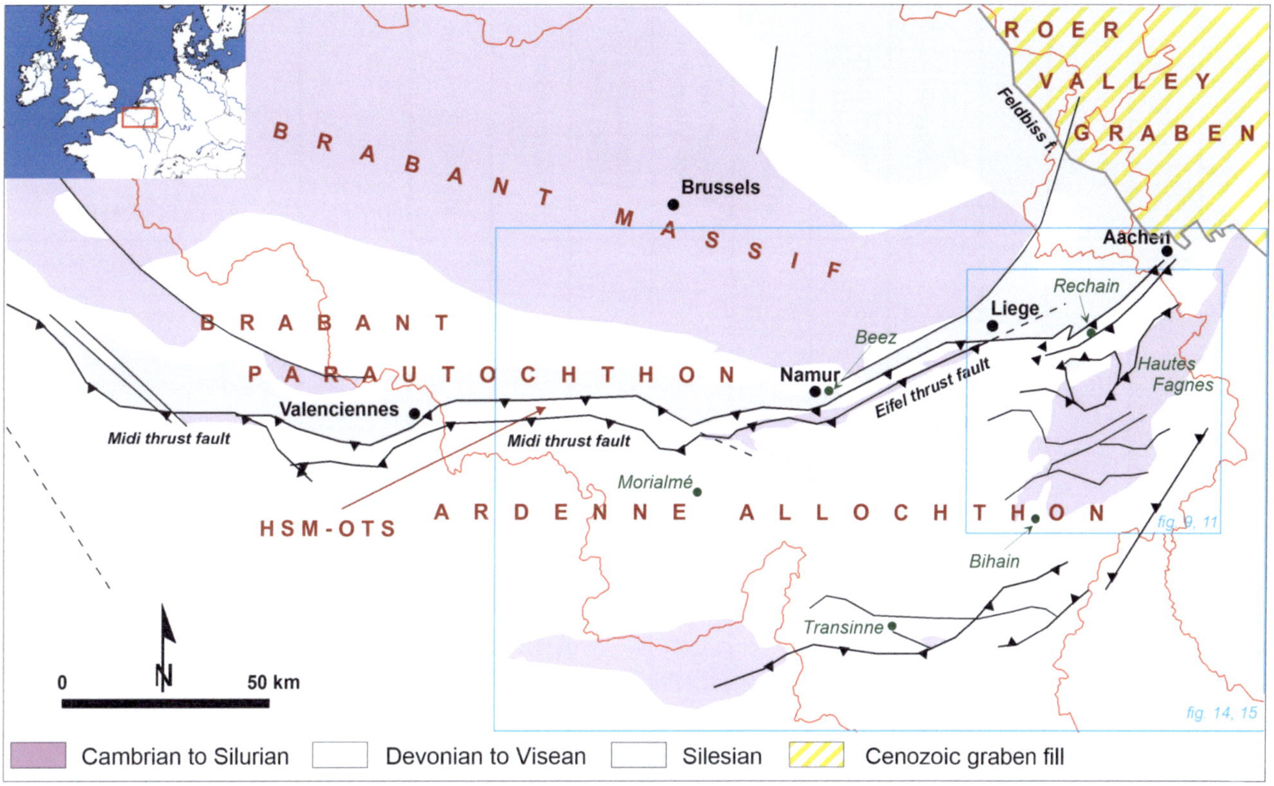

Fig. 5.1 Main structural units of the Paleozoic in Belgium and Luxembourg (stripped of their Meso-Cenozoic cover in Middle and Low Belgium). Four units are distinguished within the Paleozoic basement (Brabant Massif, Brabant parautochthon, HSM-OTS: Haine-Sambre-Meuse overthrust sheets, and Ardenne allochthon). The Variscan front is marked by the Midi and Eifel thrust faults, from west to east. In the NE, up to 1.5-km-thick Cenozoic deposits were accumulated in the Roer Valley Graben. *Green* dots and names locate the sites mentioned in this chapter

Lorraine, south of the Ardenne, the Lower Devonian sediments were similarly weathered before being covered by Permo-Triassic sediments (Boulvain et al. 2000).

West of the Ardenne, in the eastern Mons Basin, weathering of glauconiferous levels of the Grandglise Member (Hannut Formation, Late Thanetian) was likely responsible for the formation of the so-called Landenian quartzites and Sparnacian silcretes (e.g., Dupuis 1979; Quesnel et al. 2002), possibly in relation with volcanic ashfalls (Baele et al. 2016), before they were buried beneath the Ypresian sediment cover.

5.2.2 Characterization and Dating of Key Ardennian Saprolites

Residual saprolites are especially widespread in the little dissected plateau areas of central Ardenne. While occurrences, more limited in extent and depth, are also present outside the Early Devonian core of the massif, it is remarkable that the summit ridges of the Condroz area, at elevations between 250 and 400 m north of central Ardenne, are totally devoid of more than very local spots of in situ weathering (Gullentops 1954; Alexandre and Thorez 1995). Although the Ardennian saprolites were developed by ferrallitic processes under climates warmer than present, the ferruginous upper part of the profiles has mostly been removed by later erosion of the uplifted plateau and is only preserved in scattered places around the massif (e.g., at the Borne-de-Fer, French–Luxembourgian border; Théveniaut et al. 2007), where uninterrupted Mesozoic and Cenozoic tectonic stability strongly limited denudation. Elsewhere, only the saprolite sensu Nahon (1991), i.e., the bulk of the deeply weathered bedrock below the lateritic soil, is (partly) preserved. Depending on bedrock lithology, the neoformed minerals are essentially made of kaolinite, often poorly crystallized and mainly derived from the chlorite component of the Ardennian shales (Alexandre and Thorez 1995), and iron oxides and hydroxides. Quartz is the dominant residual mineral over sandstone and quartzitic bedrock. The saprolite retains the sedimentary and metamorphic structures of the parent rock, as exemplified by the Transinne outcrop, in west-central Ardenne, where stratification is clearly preserved in the weathered material (Fig. 5.3). The distribution of weathering depths may be highly unequal at the local scale over the Ardennian erosion surfaces because erosion

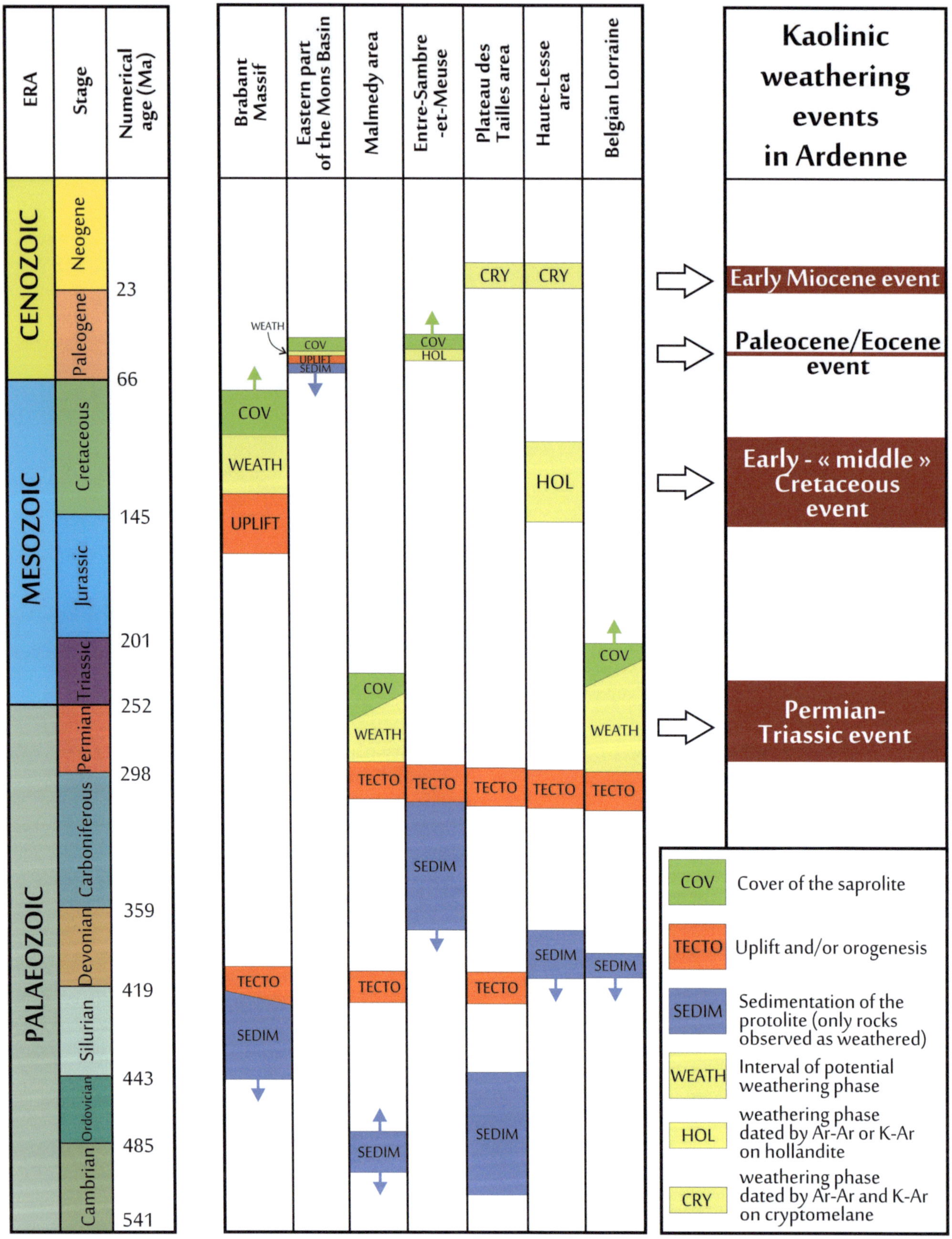

Fig. 5.2 Chronology of sedimentation (only rocks affected by deep weathering are represented here), geodynamic activity—essentially uplift, weathering, and burial, defining the history of the saprolites recognized in Belgium (and Luxembourg)

led to preservation of only the roots of the original weathering mantles and the weathering front is very irregular, mainly in relation with the variable degree of bedrock fracturing and the fracture geometry.

In order to improve the understanding and, especially, the chronology of the long-term geomorphic evolution of the Ardenne–Oesling, its erosion surface landscape and the associated kaolinic weathering mantles, we focused on four sites sampling saprolites in central Ardenne (Transinne in the Haute-Lesse area and Bihain in the Plateau des Tailles), the Entre-Sambre-et-Meuse (Morialmé), and the Haine-Sambre-Meuse overturned thrust sheets (Beez, a karst-controlled site near Namur). The dating of weathering periods is obviously very helpful for reconstructing the chronology of paleosurface development because, by definition, the erosion surface itself leaves few datable traces of its evolution. Moreover, it contributes to highlight the interplay between weathering and erosion processes, including the estimation of denudation rates and sediment budgets in the adjacent basins (e.g., De Putter et al. 2015).

Transinne

The Paleozoic basement of the Ardenne and adjacent areas experienced intense weathering during successive post-Variscan periods. These weathering phases generated kaolinic profiles that reach a depth of 65 m below the levelled plateau surface cut by the Transinne quarry at 445 m asl (Fig. 5.4). By contrast, outcrops in the nearby valley of the Haute Lesse expose fresh, though densely jointed bedrock at elevations of ~330 m. Discovered by Jerôme in 1907, the kaolin has been mined since 1922 in this area and is now used for the cement and tile industries. In Transinne, the mineralogy of the saprolite developed over Late Lochkovian shales and sandstones mainly consists of neoformed kaolinite, iron, and manganese oxides in various

Fig. 5.3 View of the kaolinized Lochkovian shales and sandstones at Transinne (see location in Fig. 5.1) showing the preservation of the rock structures by the saprolite

proportions, with residual illite and quartz (Dupuis et al. 1996; Yans 2003).

Weathered mantles may contain neoformed cryptomelane, a potassium-bearing Mn oxide with tunnel structure, and hollandite, its barium-bearing equivalent that may also include potassium cations. Belonging to the hollandite supergroup (Biagioni et al. 2013), both are suitable for radiometric dating using the K–Ar system. The $^{39}Ar–^{40}Ar$ and $^{40}K–^{40}Ar$ methods have been widely used on Mn oxides to date weathering periods around the world (e.g., Vasconcelos et al. 1995), including in Western Europe (e.g., Hautmann and Lippolt 2000; Dill et al. 2010). Dating hollandites and cryptomelanes of the Transinne section by these methods, Yans (2003) showed that the saprolite in Transinne is the result of a polyphase weathering. K–Ar ages of hollandites in the upper part of the profile range between 120 and 135 Ma (Early Cretaceous), while Ar–Ar and K–Ar results consistently point to ages around 21 Ma (Early Miocene) for cryptomelanes of the basal part of the profile. Additional K–Ar ages on hollandites provided by Thiry et al. (2006) date the intermediate part of the profile to ~88–94 Ma (early Late Cretaceous). Moreover, paleomagnetic dating on iron (hydr)oxides sampled in the upper part of the Transinne profile yielded ages around 110–120 Ma, fully supporting the K–Ar ages obtained in that part of the profile (Yans 2003; Thiry et al. 2006). This vertical succession of ages is consistent with the long-term development of the weathering mantle by downward progression of the weathering front whenever the climatic and tectonic conditions were favourable to deep chemical weathering. The very old age of the upper part of the preserved profile underlines that Cenozoic denudation was almost zero in interior parts of the Ardennian erosion surfaces, and the ages obtained close to the weathering front suggest that the Early Miocene was the last period favourable to resumption of kaolinization in the Ardenne. Finally, Pb–Pb dating of secondary U-bearing phosphates emplaced *per descensum* in fractures located just below the weathering front and probably closed since the time of phosphate precipitation might indicate active weathering as early as the Late Permian-Early Triassic,

Fig. 5.4 The Transinne quarry cuts into Lochkovian shales and sandstones outcropping on the Selandian erosion surface of western and central Ardenne, here at 445 m asl, exposing a ~65-m-thick mantle of deeply kaolinized rocks (© Google Earth 2015)

further confirming the antiquity of parts of the Ardennian landscape (Yans and Dupuis 2005). However, this dating requires further investigation because lead is very mobile during weathering.

Bihain

Saprolite thicknesses up to 50 m have also been recorded over phyllites and quartzites of the Cambrian, Revin Group, of the Stavelot massif, for instance in the Plateau des Tailles area (Cosan 1969) (Fig. 5.5). At Bihain, a hamlet established at 580–600 m asl on a remnant of an Early Paleocene erosion surface (see below), the saprolite formed on the phyllites of the Ordovician Ottré Formation. Here too, weathering led to the neoformation of kaolinite, iron oxi-hydroxides and iron–manganese oxides (Fig. 5.6); Fransolet (1979) already identified the presence of cryptomelane in the lower part of the saprolite of this area. Recent $^{39}Ar-^{40}Ar$ dating of homogeneous cryptomelane and of a mixture of cryptomelane-lithiophorite suggests an Early Miocene age for the lower part of the weathered profile at Bihain (Fig. 5.7), consistent with findings in Transinne.

Of course, such ages obtained close to the weathering front do not preclude older period(s) of weathering for the upper parts of the profile. Interestingly, K–Ar dating of neoformed Mn oxides found in deposits at the base of dissolution pockets within Dinantian limestones at Rechain, just north of NE Ardenne (Fig. 5.1), yielded ages between 88 and 95 Ma (Cenomanian to Santonian; Demoulin et al. 2010). These Late Cretaceous ages are in good agreement with the topographic position of the samples, situated very close beneath the trace of the pre-Cretaceous erosion surface emerging from below the nearby Cretaceous cover of the Herve Plateau. These deposits thus represent the westernmost extension of the continental sediments making the basal Hergenrath Member of the Late Cretaceous cover in the area, corresponding to a wealden-type lowland sedimentation contemporaneous with the deep weathering that took place farther inland (Demoulin et al. 2010).

Morialmé

The Morialmé weathering profile is located in the "Le Faya" quarry, at 260 m asl near Morialmé, in the Entre-Sambre-et-Meuse (ESEM). The quarry cuts the southern flank of an anticline and is mined for the brick industry. The saprolite was developed on micaceous slates of the Famennian Etroeungt and Ciney Formations, weathering processes being mainly kaolinite neoformation in the upper part of the profile and swelling clay formation in its lower part (Barbier et al. 2012). Although the degree of weathering is highly variable inside the quarry (Barbier et al. 2010), the top of the weathering profile still displays the alloterite (Barbier et al. 2012), i.e., the upper part of the profile that has undergone structure collapse and compaction (Wyns et al. 1999), showing that the profile top was quasi unaffected by erosion at the time of its burial under fluviatile sediments in the Late Paleocene (Schuler et al. 1992) to Early Eocene (Barbier et al. 2012) (Fig. 5.8). Two samples of hollandite from the lower middle part of the profile yielded K–Ar ages of 53.3 ± 2.3 Ma and 58.2 ± 2.4 Ma (Barbier et al. 2010) (Table 5.1), confirming that weathering was still active just prior to this fluvial sedimentation episode at the Paleocene-Eocene transition. This sequence of events is also consistent with the Selandian age assigned by Demoulin (1995) to the erosion surface preserved in the western part of the Ardenne and ESEM, which continued to evolve during the Thanetian where it was not drowned by the Early Thanetian Sea.

Beez

Cut into a bench at ~160 m asl along the northern valley side of the Meuse valley close to Namur, the Beez quarry exposes a cryptokarst developed in Visean limestones. In this karstic setting, possibly in relation with the circulation of acid fluids originating from pyrite leaching and oxidation in the overlying Namurian black shales, the weathering paragenesis includes neoformed halloysite, gibbsite, Fe and Mn (hydr)oxides, gypsum, and jarosite. Based on data from neighbouring cryptokarsts and saprolites, Bruyère et al. (2003) pointed to a possible Miocene age of this weathering. New K–Ar analyses performed on jarosite yielded ages of 16.5 ± 0.8 and 16.8 ± 0.5 Ma that confirm the Miocene timing of weathering in Beez (Table 5.1). One notes that these ages are somewhat younger than those obtained at the base of the Transinne and Bihain profiles. However, the weathering history at Beez might have responded mainly to local geochemical (leaching of iron sulphides) and drainage conditions (karstic vertical transfer, possibly linked to the Meuse valley downcutting) and have a more limited climatic and geodynamic meaning than that assigned to the weathering mantles dated elsewhere in the Ardenne. Moreover, the unknown original depth of the cryptokarst makes it very difficult to relate this weathering episode with any contemporaneous topography.

Summarizing, the available age data for weathering mantles and residues in the Ardenne identify periods favourable to deep weathering that are in excellent agreement with those recognized in neighbouring areas throughout the Mesozoic and the Cenozoic, namely during the Early and early Late Cretaceous, the Late Paleocene, and the Early Miocene plus, to a lesser extent, the Late Permian-Early Triassic. As we will show in the next section, these periods fit

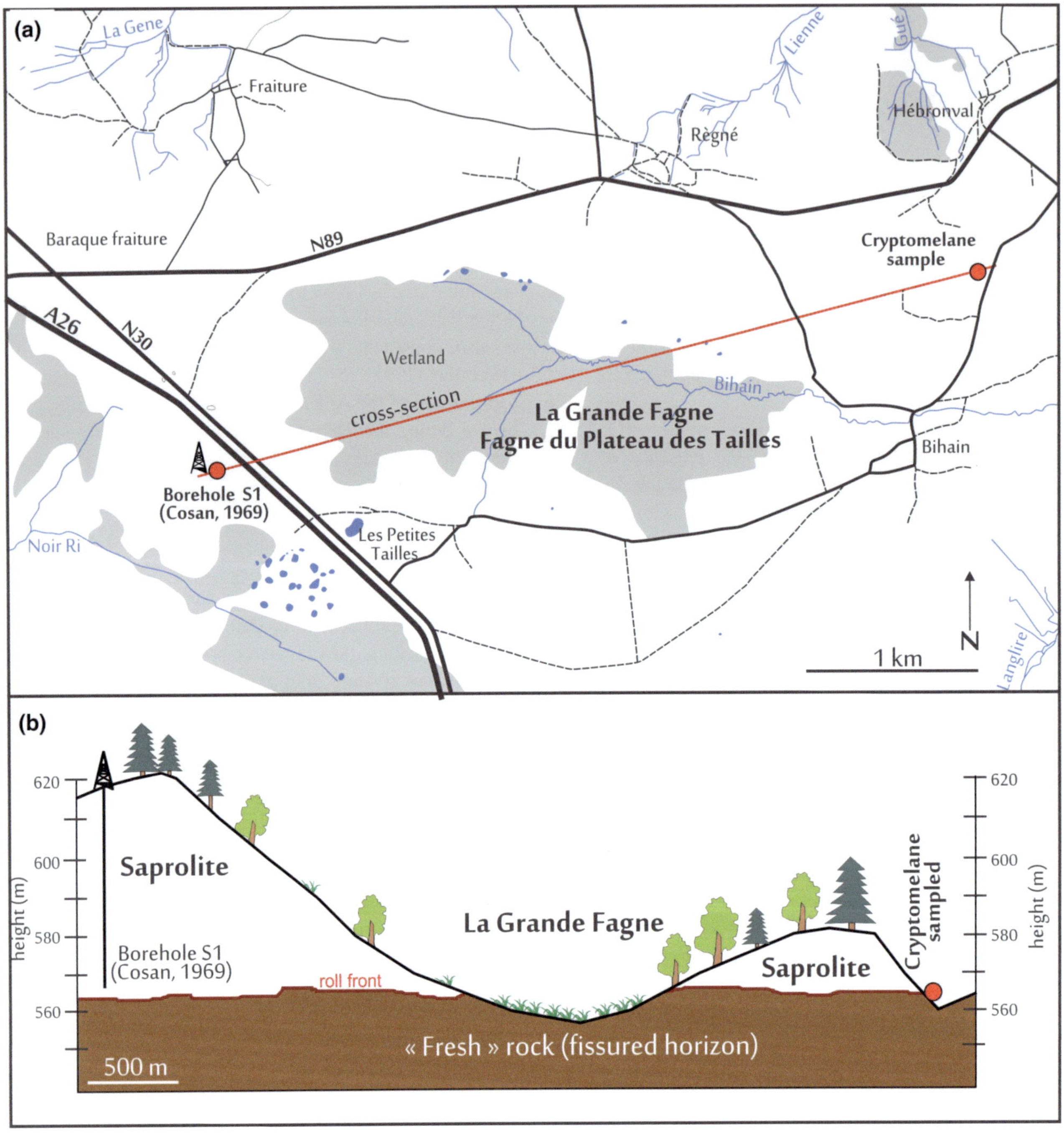

Fig. 5.5 **a** Map of the Bihain site in the Plateau des Tailles (see location in Fig. 5.1), one of the few areas of central Ardenne-Eifel whose summits still bear witness to the pre-Senonian surface in a region where the Late Cretaceous topography was possibly never covered by marine transgressions. **b** Weathering profile across the Plateau des Tailles. Note the location of the dated saprolite close to the weathering front (roll front) at 565 m asl, i.e., ~90 m below the summit surface of Cretaceous age

also very well with the periods of planation and surface development inferred in the Ardenne from independent data, weathering and planation going hand in hand under conditions of warm humid climate and relative tectonic stability in regions where mean elevation is high enough to create the vadose zone required for effective leaching. It remains of course a possibility that further dating of weathering products may point to other phases of weathering, whose climatic/geodynamic meaning will have to be examined carefully in any case, avoiding to put too much emphasis on isolated occurrences linked to specific environments (e.g. karstic setting, water table fluctuation zone).

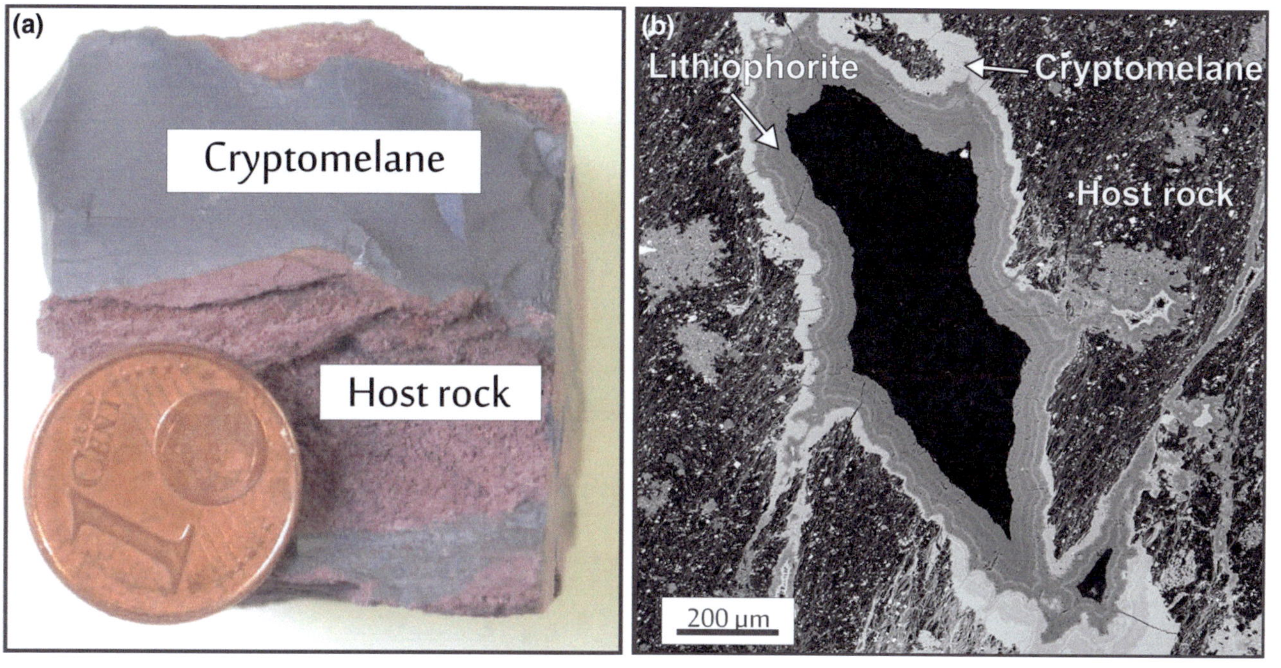

Fig. 5.6 **a** Sample of cryptomelane of the Bihain quarry (collection of the Laboratory of Mineralogy, University of Liège, courtesy A.-M. Fransolet). The host rock is an alternation of Ordovician quartzite and phyllite. **b** SEM-EDS (Scanning Electron Microscopy with Energy Dispersive X-ray Spectroscopy) view of the mixture cryptomelane-lithiophorite in Bihain

5.3 Erosion Surfaces in the Ardenne and Oesling

5.3.1 Geographical and Geological Setting

Beyond being famous for its unique periglacial features (see Chap. 7), the Hautes Fagnes Plateau of NE Ardenne (E Belgium) exposes remarkably well-preserved remnants of erosion surfaces of various Mesozoic and Cenozoic ages. Located at the tip of the recently uplifted Paleozoic Ardenne (e.g. Demoulin et al. 2009), this old erosional landscape of stepped surfaces dominates by ~400 m the Pays de Herve, the northern foreland of the massif, to which it connects by a narrow margin of moderately steep slopes drained by the Vesdre River flowing at its foot. While all slopes are covered by dense forests, the top of the plateau above 550 m bears extended moorland that enhances the flat character of the inherited topography, which is still only marginally affected by the Plio-Quaternary incision of rivers. This presence of large peat bogs is a specificity of the Hautes Fagnes Plateau, which is otherwise similar to all plateau areas of the Ardenne (and the nearby Eifel as well).

The Ardenne–Oesling massif belongs mostly to the Rheno-hercynian zone of the Variscan fold-and-thrust belt of middle Europe, the northern limit of the present-day plateau corresponding more or less to the Variscan front (Fig. 5.9). Structures inherited from the Caledonian orogeny and deformed again in Variscan times crop out along the axis of the uplifted massif, and especially in its northeastern part, resulting in a structurally complex basement wherein longitudinal ENE–WSW folds and thrust faults are cut by numerous NW–SE to NNW–SSE striking normal faults (Demoulin and Ek 1995). The Hautes Fagnes Plateau pertains almost entirely to the Cambrian-Ordovician Stavelot inlier, one of these inherited Caledonian structures aligned along the axis of the Ardenne anticlinorium and essentially made of variably erodible phyllites and very resistant quartzites. To the south of this anticlinorium, the bulk of the Ardenne Plateau corresponds to the Neufchateau synclinorium, which mainly exposes fairly homogeneous Lower Devonian slates and where marked lithological contrasts arise only locally from intercalations of conglomerate or arkose formations. Such arkoses crop out for instance along the eastern limit of the Stavelot inlier, in the SE of the Hautes Fagnes area. To the north of the Ardenne anticlinorium, the northern part of the Ardenne allochthon corresponds to the Dinant synclinorium, where the slaty bedrock of the Ardenne *s.s.* gives way to alternating Middle Devonian to Lower Carboniferous sandstones, shales and limestones. These rocks define the particular landscapes of the Famenne, a wide depression carved in Fammenian shales, and of the Condroz plateau, with alternating elongated ridges on sandstones and valleys cut in the limestones (Fig. 5.9). Scattered over the Condroz ridges and the flat summits of W and NE Ardenne, few remnants of loose sediments, mainly marine sand and clay-with flints, also attest that the massif

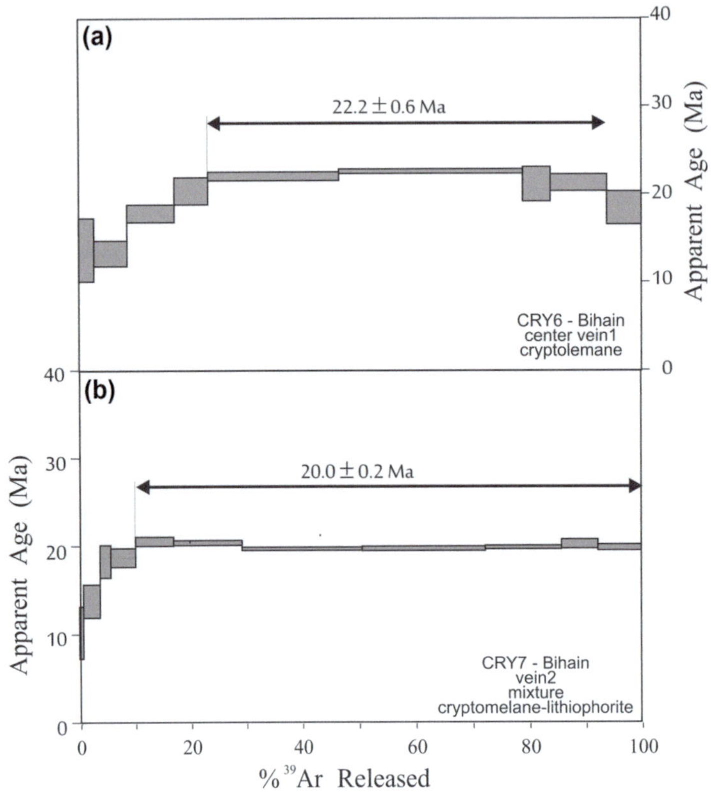

Fig. 5.7 ^{39}Ar–^{40}Ar age spectra of the Bihain cryptomelane. **a** ^{39}Ar–^{40}Ar age spectrum obtained for an homogeneous vein of cryptomelane, yielding a calculated plateau age of 22.2 ± 0.6 Ma (70, 6% of ^{39}Ar released). **b** ^{39}Ar–^{40}Ar age spectrum of a mixture of cryptomelane and lithiophorite, yielding a calculated plateau age of 20.0 ± 0.2 Ma (89, 8% of ^{39}Ar released)

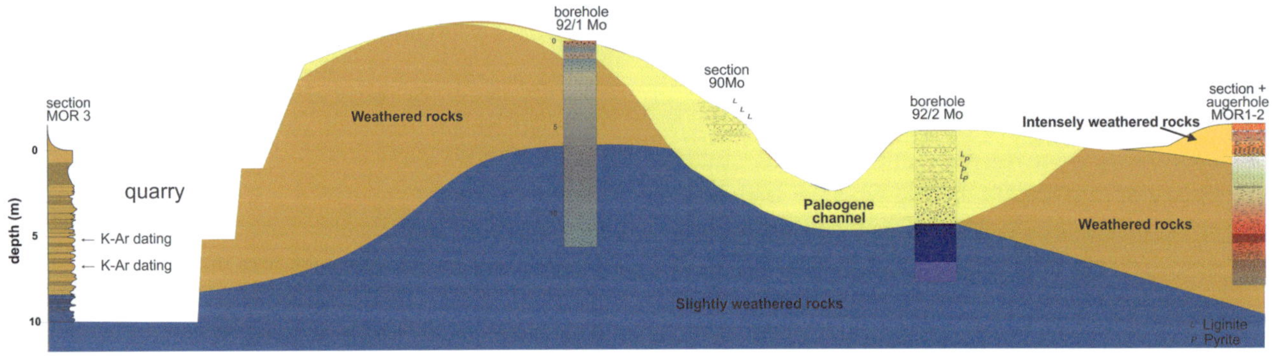

Fig. 5.8 The weathering profile at Morialmé (see location in Fig. 5.1), with location of the samples for K–Ar dating on hollandite

was partly drowned by marine transgressions coming either from the north or the west during the Cretaceous and the Paleogene (e.g. Demoulin 2003; see also Chap. 2).

Morphologically, the Ardenne represents a western annex to the Rhenish shield in the form of an E–W elongated relief whose flat summits gradually descend from elevations above 600 m in the east to 500–600 m in the central domain and below 400 m west of the Meuse valley. As a consequence, the transition from the massif to its N and S forelands, rather sharp in the east, is progressively less conspicuous westward and, would geology not witness its limit, the Ardenne goes imperceptibly into the Mesozoic cover of the Thiérache to the west. Except its SE confines drained by the Sûre River towards the Mosel and the Rhine, the Ardenne Plateau is drained by rivers of the Meuse basin that have incised 100- to 150-m-deep Plio-Pleistocene valleys through it (history of

Table 5.1 K–Ar dating results on hollandite in Morialmé and jarosite in Beez (see location in Fig. 5.1)

Location	Sample	K_2O (%)	Ar (%)	^{40}Ar (10^{-6} cm^3/g)	Age (Ma ± 2σ)
Morialmé	08mori-oxmn1	0.52	17.38	0.90	53.3 ± 2.3
	08mori-oxmn2	0.66	13.00	1.25	58.2 ± 2.4
Beez	Jar-Beez-A	3.39	67.50	1.85	16.5 ± 0.8
	Jar-Beez-B	1.92	18.25	1.05	16.8 ± 0.5

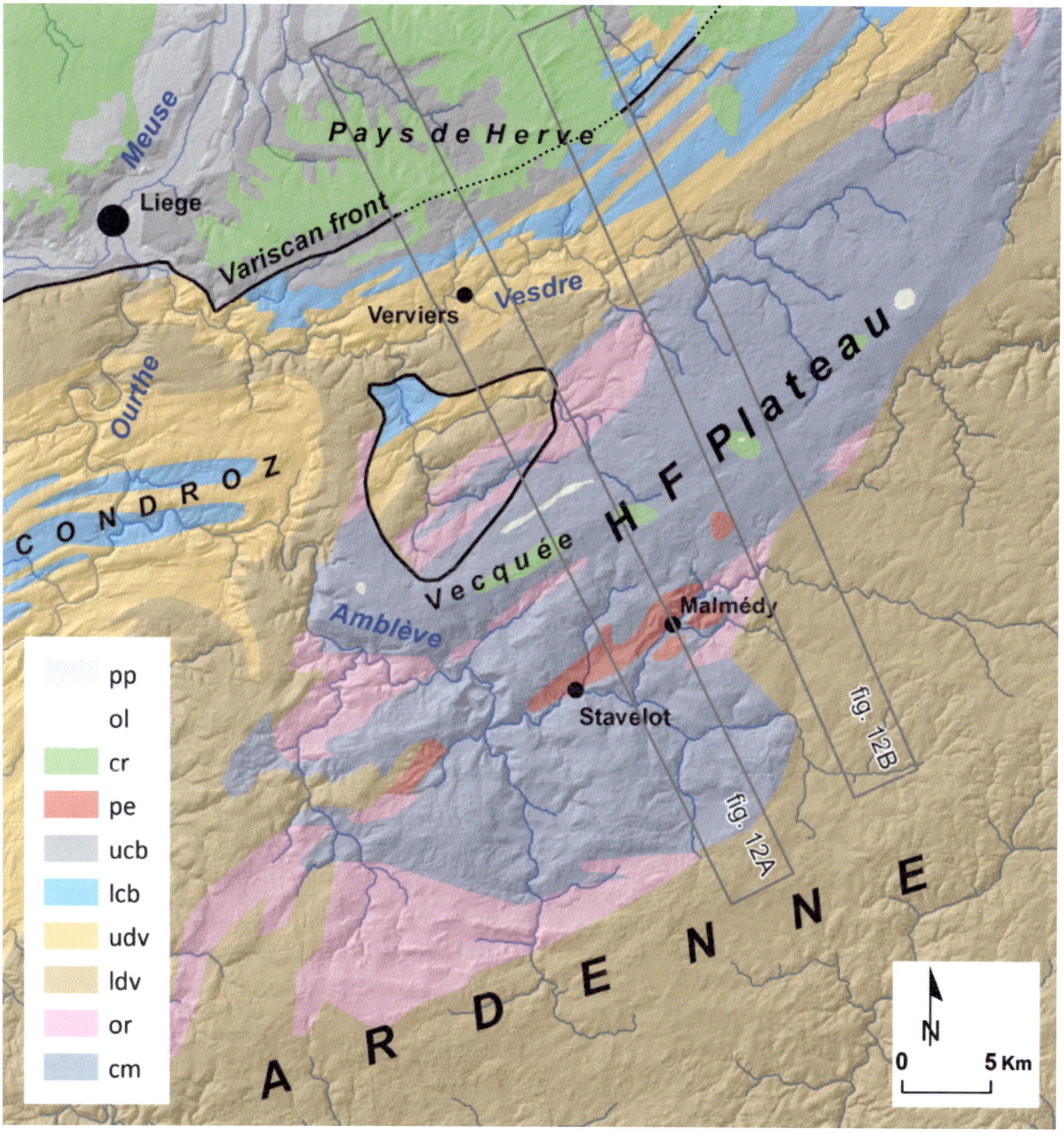

Fig. 5.9 Simplified geological map of the Hautes Fagnes Plateau and surrounding areas in NE Ardenne (see location in Fig. 5.1) draped on a hillshade of the SRTM 3″ digital elevation model. *cm* Cambrian or Ordovician. *ldv* Lower Devonian. *udv* Middle and Upper Devonian. *lcb* Lower Carboniferous. *ucb* Upper Carboniferous. *pe* Permian. *cr* Cretaceous. *ol* Oligocene. *pp* Plio-Pleistocene

the incision is dealt with in Chap. 10). Culminating at ~700 m asl, the Hautes Fagnes Plateau corresponds to the highest part of the massif.

5.3.2 Shape and Characteristics of an Erosion Surface

Being remote descendants of etchplains formed under warm-humid conditions, the Ardennian erosion surfaces display a variable mix of extremely flat areas and more undulating topographies (Fig. 5.10). In principle, active etchplains are essentially flat and remarkably horizontal as long as a deep weathering mantle covers them uniformly, leaving only large-scale inselbergs emerge abruptly from the surrounding surface and dominate it often by 100 or more meters (Thomas 1994). However, when climatic conditions turn drier or cooler, it is common that the etchplain is stripped of part or all of its weathering mantle and that the weathering front becomes exposed. Being more sensitive to lithological and joint density contrasts, this front usually presents irregularities at all scales that appear then at the surface, forming topographic bumps from m- to dm-sized whalebacks to large subdued hilly reliefs up to a few tens of meters in height. This is the large-scale shape most usually shown by the Ardennian surfaces, in agreement with the patchy preservation of their associated weathering mantles, whose stripping probably mainly resulted from the drastic slowdown of weathering imposed by generally cooler conditions since the Oligocene. Recognising the weak relief currently displayed by the erosion surfaces is important in paleolandscape reconstructions because it prevents one from the temptation of distinguishing supposed surfaces vertically too close to each other.

5.3.3 Identifying an Erosion Surface

The most obvious criterion of surface identification is the overall geometry of individual levelled elements and ensembles of elements in the present landscape. However, it relies on the assumption that the envelope of part of the topography may correspond to one particular ancient erosion surface, which is sometimes questionable in regions with a complex history involving many episodes of planation and intervening deformation and degradation. Nevertheless, the reconstructed surfaces display in general, and especially in the Ardenne, a large-scale plane shape, more or less tilted, sometimes locally or regionally flexured, and admitting a relief amplitude in the order of 20–30 m (Demoulin 2006). Beyond post-planation deformation, two factors complicate such geometry-based surface reconstructions. The first one relates to the inherent relief of any erosion surface, especially when stripped of their weathering mantle. It implies that the reconstruction allows for a careful balance between topographic variations within a single-stage paleolandscape and close positions of successive surfaces in some areas, especially where differently tilted surfaces are intersecting each other. The second factor refers to the issue of defining the moment when a given surface ceases to evolve as such and begins to degrade, thus giving way to the formation of a new, distinct landscape. This comes to find a clear separation between pure acyclism and clear-cut morphogenetic cyclicity, whereas all transitions actually exist between these end-members in real landscape evolution (Klein 1990).

Facing such difficulties, geometric reconstructions are advantageously supported by the analysis of the so-called correlative deposits. In particular, when geometry allows the connection of the exhumed part of a paleosurface with its counterpart buried as the base surface of a sedimentary cover in the massif's foreland, the very existence of the latter attests an episode of aggradation, continental or marine, that clearly ends the previous planation stage and defines its specific topography. In such cases, there is most often a more or less extended transition zone where scattered remnants of the sedimentary cover confirm the position of the exhumed surface within the uplifted massif, at least if they may be unequivocally assigned to this cover. In any case, as scarcely as it may be preserved, the sedimentary evidence is a key element in the analysis of erosional paleolandscapes. In this respect, next to sediments, the weathering mantle associated with the surfaces may also be helpful for distinguishing planation episodes, thanks to specific features such as depth, degree of preservation, secondary products (e.g. duricrusts) and, of course, dating of weathering profiles.

5.3.4 Erosion Surfaces of the Hautes Fagnes Plateau

The area whose successive paleolandscapes we describe in some detail corresponds to the Hautes Fagnes region s.l., which is made of the Hautes Fagnes massif s.s. (HFM), its western prolongation in the Vecquée ridge (VR), their northern margin, gently sloping toward the Plateau de Herve in the north, and the high plateau area to the SE of the HFM (Fig. 5.11). Mapping of the intricate erosion surfaces will be aided by the analysis of topographic swath profiles (Fig. 5.12).

The most conspicuous surface element is represented by the remarkably flat, almost horizontal plateau surface covering the HFM at >600 m asl and currently occupied by peat bogs and extensive heath that confer to the area its wild character and wide horizons (Fig. 5.10). In the east, this surface is dominated by a few subdued summits rising a few

Fig. 5.10 The pre-Senonian surface as it appears on top of the Hautes Fagnes Plateau, near Baraque Michel, underlain by a widespread peat cover that still locally overlies residual Cretaceous clay-with-flint and sand, and Oligocene marine sands. The faint ~8‰ slope to the north (toward the *left* of the *photograph*) was produced by post-Cretaceous tilting of N Ardenne (essentially in the very Early Paleogene, as all Cenozoic erosion surfaces are horizontal across central and northern Ardenne)

tens of meters above their surroundings. To the NW, it goes rather abruptly in a tilted and dissected surface that makes the margin of NE Ardenne. The forested interfluves materialising the surface remnants slope regularly (by ~4 %) down to ~320 m asl, where they join the upper edge of the incised Plio-Quaternary valley of the Vesdre. On the opposite side of the HFM, steeper slopes rapidly lead down southeastward to another very regular, hardly dissected surface at 560–580 m asl, covered by pasture and interrupted by small wooden areas only where upper valley reaches of the Roer catchment begin to incise. Though essentially horizontal, this 560–580 m landscape element shows imperceptible slopes going up in a residual relief that culminates in the SE at ~690 m asl at the Weisser Stein (Fig. 5.12). On its southwest flank, the HFM displays a stepped descent toward the perfectly horizontal ridge of the Vecquée. Extending at 560–570 m asl for ~15 km to the SW and with a width of 0.5–1.5 km, the VR represents a distinct topographic element in the present landscape of the area. On the north, it is bordered by a ~120-m-high escarpment that links it to a lower surface well preserved on the interfluves between the incised valleys of the Gileppe, the Hoëgne, and the Wayai. This surface, which skirts the western end of the ridge, is observed at elevations around 450 m at its foot, from where it slopes very gently (~0.6 %) toward NW. It includes extended levelled areas at 360–390 m asl between the villages of Theux, Sart-lez-Spa, and Jalhay (Figs. 5.11 and 5.13). On the south of the VR, a smaller, ~60-m-high escarpment leads down to a local intramontane basin whose levelled bottom appears in the form of wide flat interfluves at uniform elevations of 500–510 m asl. Centred on the area where weak rocks of the Permian graben of Malmédy crop out (Fig. 5.9), this erosional feature opens towards the massif's exterior in the

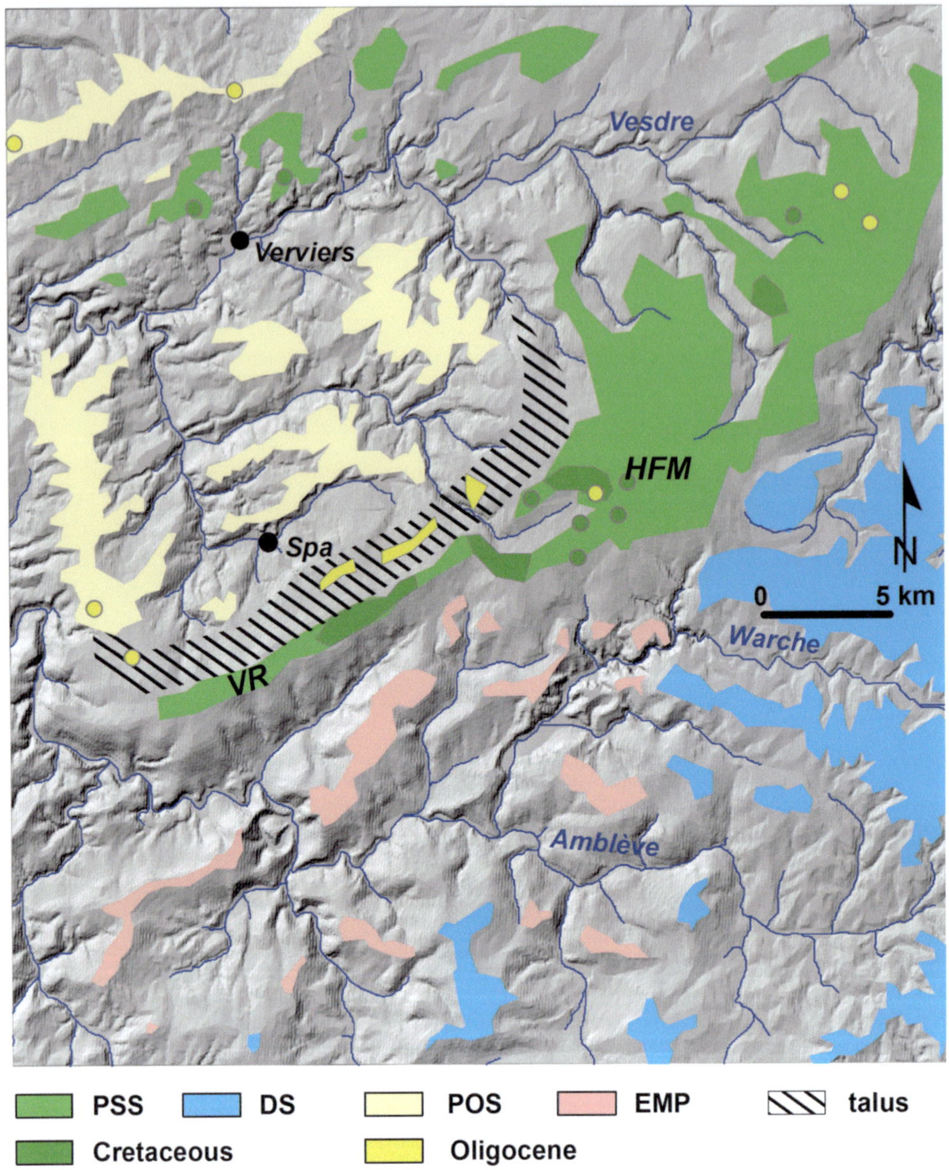

Fig. 5.11 Map of the erosion surface remnants in the Hautes Fagnes area of NE Ardenne, draped on a hillshade of the SRTM 3″ digital elevation model. All flat interfluves and broader plateau areas belong to pre-Quaternary topographies of various ages. *PSS* Pre-Senonian surface. *DS* Danian surface. *POS* Pre-Oligocene surface. *EMP* Early Miocene planation, developed in less resistant Permian conglomerates and Ordovician shales. Talus: ~ 100-m-high erosional riser between the pre-Senonian/Danian and the pre-Oligocene surfaces as it is preserved since it was carved in the regional topography during the Paleogene

west, where it is geometrically connected to the 450-m surface.

In the northern foreland of the massif, the Meso-Cenozoic sedimentary cover of the Pays de Herve offers an opportunity for reconstructing the profile of two paleosurfaces onto which transgressing seas encroached. The first one is the base surface of the Senonian deposits (the clays and sands of the basal Aachen Formation are dated to the Santonian), which shows a slight but consistent tilt of 1.4% towards NNW. Emerging at the topographic surface in the south of the Pays de Herve, it appears there in the form of large exhumed remnants, similarly tilted and preserving sands and clays of early Late Cretacous age in solution pockets of the underlying Dinantian limestones (Demoulin et al. 2010). Prolonged southward toward the Hautes Fagnes summits, the 1.4% tilt angle connects the VR surface and the tilted surface on the northern flank of the HFM with the Senonian base surface in the Pays de Herve (Fig. 5.12). The second marker interface in the Meso-Cenozoic cover is the base surface of the sand formation that accumulated during the last marine transgression that covered the northern part of the Ardenne in the Oligocene (Rupelian?). Again, NNW–SSE sections of the cover indicate a minor (0.6%) but uniform NNW tilt of this interface that can easily be followed across the Vesdre valley to the 360–450-m surface in the massif, north of the VR, which shows exactly the same tilt in the same direction.

Finally, remnants of the Cretaceous sands and clay-with-flints and Oligocene marine sands scattered over the northern margin and summits of the Hautes Fagnes region provide additional evidence of the geometric connections established between the massif's erosion surfaces

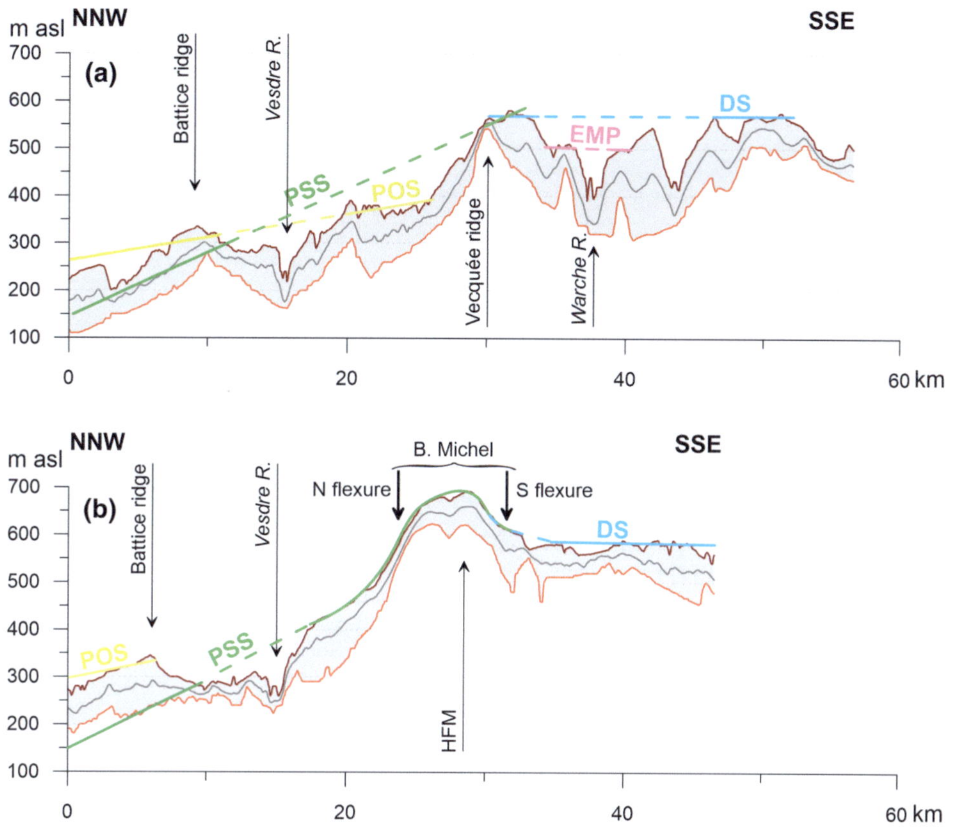

Fig. 5.12 NNW-oriented 4-km-wide swath profiles (see location in Fig. 5.9; *orange, grey,* and *ruby brown lines* are for minimum, mean, and maximum elevations, respectively) showing the geometry of the erosion surfaces across the northern margin of NE Ardenne at the height of the Vecquée ridge (**a**) and the Hautes Fagnes Plateau (**b**), where the general tilt to north is complicated by post-Oligocene uplift of the Baraque Michel block inducing the flexure of the surfaces. *PSS* Pre-Senonian surface. *DS* Danian surface. *POS* Pre-Oligocene surface. *EMP* Early Miocene planation level. The trace of exhumed erosion surfaces obviously follows maximum elevation lines of the swath profiles. By contrast, north of the Vesdre valley, the pre-Senonian surface is still buried under a continuous Upper Cretaceous cover. Height exaggeration ×33

and their buried counterpart in the Pays de Herve. Discontinuous clay-with-flint, often overlying Aachen sand and gravel, are still widespread over the VR and the HFM and are found locally on the sloping surface leading up to the latter, especially southeast of Eupen (Fig. 5.11). Elsewhere, beyond rare occurrences of isolated reworked flints lying around on the 500-m surface south of the VR, the 360–450-m surface in the north and the 560–580-m surface in the south are completely devoid of Cretaceous sediments. By contrast, sparse deposits of Oligocene sands are mainly encountered all over the 360–450-m surface, while rarer remnants of this cover also exist next to, and sometimes over, Cretaceous sediments on the HFM summit and northern slope and on the VR.

Integrating all these observations, one obtains the following consistent reconstruction of the successive paleosurfaces whose fingerprint is still visible in the present landscape (Fig. 5.11). The oldest preserved surface in the Hautes Fagnes area is pre-Senonian in age and includes the levelled summits of the region (VR and HFM) and also the tilted surface that makes the northern margin of the HFM and links it to its foreland east of a line joining Verviers to Malmédy (Fig. 5.12). The unity of this surface made of distinct topographic elements is demonstrated by the Cretaceous deposits that all of them still expose. However, the pre-Senonian surface underwent strong post-planation deformation during the Cenozoic, mainly in the form of a marked differential uplift of the HFM. Its overall geometry shows that the HFM block uplift was limited by flexures on its northern and southern sides (Fig. 5.12b), while a series of NNW-striking Variscan faults, reactivated in normal mode, accommodated the ~100 m height difference with the adjacent Vecquée ridge to the west.

Even if they share the VR summit (Fig. 5.12), the 560–580-m surface to the SE of the HFM represents a planation episode different than that attested by the pre-Senonian surface. Indeed, with the exception of few residual reliefs such as the Weisser Stein, whose summits are probably hardly degraded isolated remnants of the pre-Senonian surface, the remarkable horizontality of the 560–580-m surface

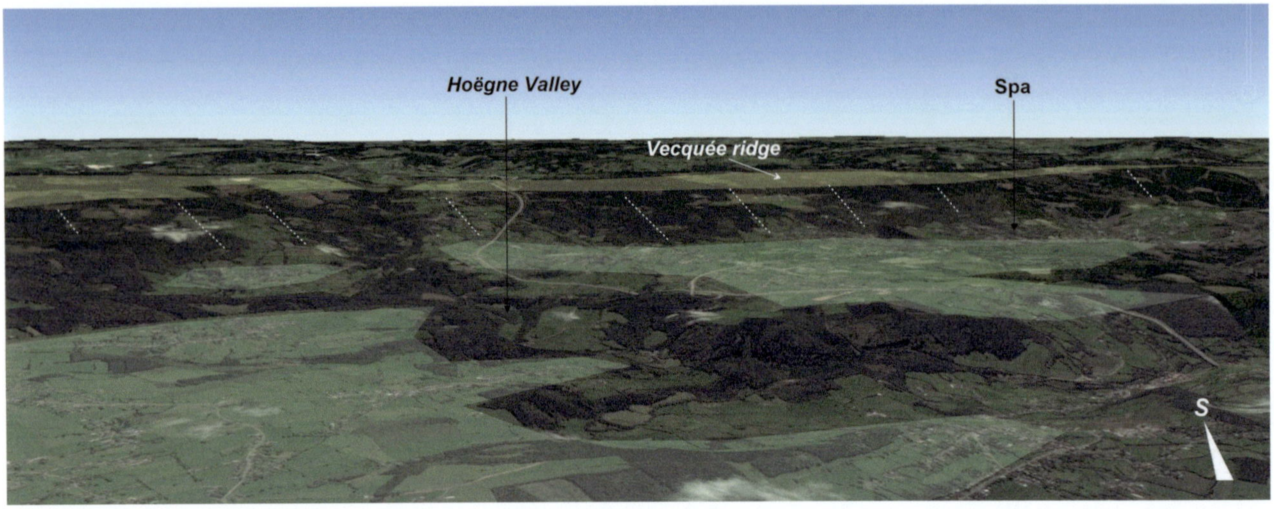

Fig. 5.13 S-looking oblique *aerial* view of the pre-Oligocene surface (in *light green* transparency) developed at 350–400 m asl north of the Vecquée ridge, which represents the pre-Senonian surface, exhumed and reexposed since the Danian (in *light yellow* transparency). The erosional scarp between both surfaces (*dotted white lines*) clearly appears as a regional morphological feature. In the background, beyond the barely visible Early Miocene planation level developed in the Stavelot-Malmédy area, south of the Vecquée, the uniform *horizon line* evidences the well-preserved Danian surface of central Ardenne. *Scale* the visible length of the Vecquée ridge is ~20 km. Elevation factor 2 (© Google Earth 2015)

across the whole central Ardenne and nearby Eifel shows that it resulted from low-angle bevelling of the slightly tilted previous (pre-Senonian) topography, the intersection between both surfaces occurring along the VR (Fig. 5.12a). This interpretation is further supported by the absence of Cretaceous sediments on the 560–580-m surface south of the VR, where their complete removal (their former presence in this area being only proved by isolated clay-with-flints remains on the SE flank of the Weisser Stein) had to precede regradation of their base surface (regradation being the process of tending towards a new grade after a perturbation; Fairbridge 1968).

Finally, the last envelope surface of the topography locates a third nested paleolandscape that is chiefly associated with the 360–450-m surface developed in the north and west of the Hautes Fagnes region and prolonged inside the massif by the 500-m intramontane basin south of the VR. However, whereas remnants of the Oligocene sand cover scattered over the surface north of the VR attest that it was developed before the Oligocene transgression on the massif, the absence of corresponding deposits in the intramontane basin despite drowning of the VR by the Oligocene sea strongly suggests that this inward extension of the 360–450-m surface was carved in later times.

Separation between the different surfaces occurs in two ways. While the 560–580-m and, north of the Vesdre valley, the 360–450-m surfaces both cut at low angle the tilted older pre-Senonian surface, a well-developed 60- to 150-m-high erosion scarp, making the northern flank of the VR, separates them from each other (Fig. 5.12a). The erosional nature of this scarp, as opposed to the tectonic flexures limiting the uplifted HFM, is demonstrated by the fact that N–S sections north of the VR show no equivalent to this flexure in the older, uniformly tilted pre-Senonian surface (Fig. 5.12). The two types of surface separation evidence distinct erosional mechanisms. Low-angle bevelling is typical of acyclic morphogenesis where surfaces temporally distinguished by their associated sedimentary evidence gradually regrade their predecessor under etchplanation regime in response to slow, low amplitude deformation. By contrast, erosional scarps signal morphogenic cyclicity, implying a combination of etchplanation and scarp retreat in response to more rapid, larger tectonic deformation with a significant component of en-bloc uplift.

5.3.5 Dating a Surface

It is clear from the Hautes Fagnes example that geometric relationships between paleosurfaces and their links to buried base surfaces within foreland sedimentary covers already provide powerful tools for unravelling the chronology of long-term geomorphology in erosional settings. Taking into account the various types of discontinuity that give rhythm to such a "surface stratigraphy" further allows inferences about the nature of the often tectonic triggers of morphogenesis.

The paleolandscape chronology is also aided by the analysis of correlative deposits preserved on the erosion surfaces. In the Ardenne–Oesling, the oldest sediments of interest in this respect correspond to the band of Bunter sandstones (Lower Triassic) cropping out at the southern edge of the Oesling, and continuing eastward inside the massif into the N–S Eifel zone. More to the west, the Mesozoic deposits directly resting on the southern border of the Ardenne are younger, mainly Liassic, and ending with Turonian and Coniacian chalks at the limit between W Ardenne and Thierache (Fig. 5.14). The equivalent first post-Variscan sediments present on the northern side of the massif are Upper Cretaceous sands and chalks (weathered to clay-with-flint). Their outcrop zone is restricted to the Hautes Fagnes region in the NE (Bless and Felder 1989), where they nevertheless covered large areas inside the massif as witnessed by the clay-with-flint retrieved at Dalhem, east of the Weisser Stein (Fig. 5.14). Later, further seas which encroached upon the Ardenne are recorded during the Early Thanetian in the west, where Upper Thanetian continental sands also sparsely cover the western Rocroi Plateau, and during the Rupelian, when marine sands were deposited not only in the Hautes Fagnes but also over the whole Condroz Plateau and in the ESEM region (Fig. 5.14).

Finally, as indicated in the first part of this chapter, dating techniques applied in the past two decades to minerals neoformed during weathering brought significant improvement in the appraisal of the age of erosional paleolandscapes in uplifted massifs, and especially in the Ardenne, by incorporating in the discussion quantitative information about the age of the associated weathering mantles. The next section brings a brief overview of the current understanding of the long-term geomorphic evolution in the Ardenne–Oesling at the light of this diverse, recently enlarged deal of evidence.

5.3.6 The Whole Picture: Stepped Surfaces of the Ardenne–Oesling

During Mesozoic and Cenozoic times, the Ardenne–Oesling underwent many phases of subaerial denudation, only interrupted by limited marine transgressions. The seas that drowned the margins of the massif came from E to SE in Triassic and Liassic times, from W during the Eocene and from N in the Early Oligocene. Only the Late Cretaceous sea probably submerged larger parts of central Ardenne.

Seven generations of paleolandscape may be recognized (Demoulin 1995, 2003) (Fig. 5.15). A pre-Triassic topography, mainly observed in W Eifel, where it displays a fairly animated relief (Junge 1987), also emerges from below the Triassic cover to form the slopes of the SE margin of the Oesling. It is continued westwards by the diachronic, so-called post-Hercynian peneplain, which constitutes the southern margin of the Ardenne, with interfluves tilted by 1–3% to the south. In NE Ardenne, following a longer phase of post-Variscan denudation, the oldest paleosurface preserved as part of the modern landscape dates back to the Late Cretaceous. As stated above, remnants of this pre-Senonian surface are also preserved as the highest summits, at 650–700 m elevation, in the heart of the massif (Baraque Michel, Baraque Fraiture, Weisser Stein, Schneifel). Then, after the Late Cretaceous regression, acyclic erosion of the Ardenne–

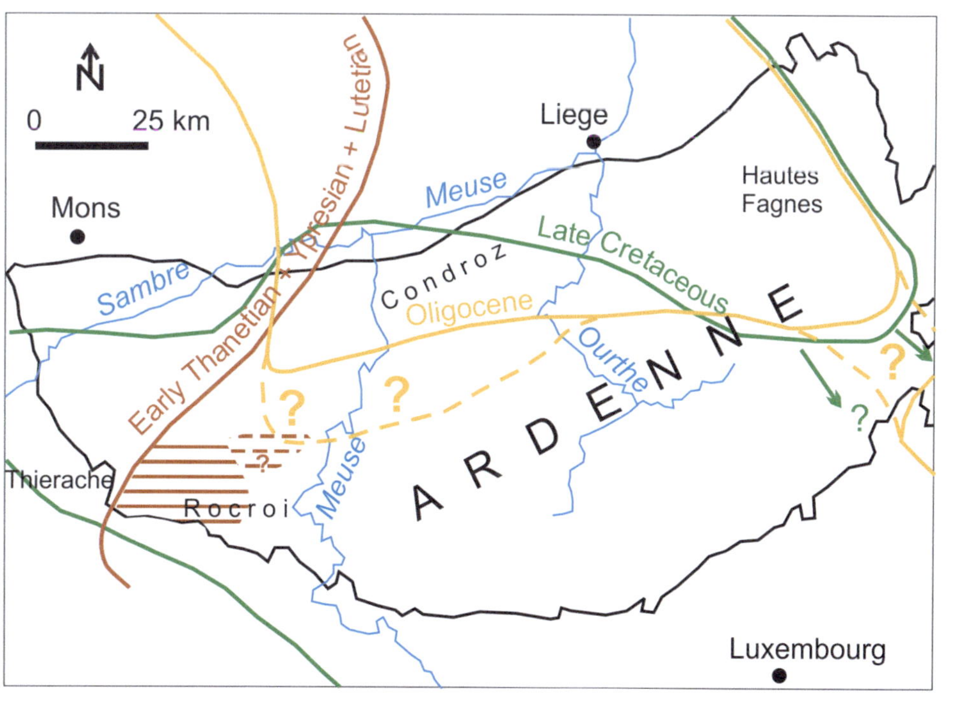

Fig. 5.14 Map of the Mesozoic and Cenozoic marine transgressions over the Ardenne massif (modified after Demoulin 1995). *Horizontal brown hatching* denotes the area of occurrence of Upper Thanetian continental deposits

Oesling removed its Cretaceous cover almost completely and caused an extensive erosion surface to develop over northern and central Ardenne by regrading the pre-Cretaceous landscape during the Danian. Corresponding to the '*surface supérieure*' of Macar (1938), this Danian surface is now observed at altitudes of 560–580 m, going down slowly in the south to reach ~500 m asl along the southern border of the massif. In contrast with previous surfaces, it developed independently of any marine ingression and its age is primarily inferred from geometrical considerations and the derived "surface stratigraphy". Another information is, however, provided by the Early Cretaceous radiometric ages obtained for the upper part of the weathering profile at Transinne, which roughly locates the weathering front at (modern) 450 m asl at 130 Ma, i.e., ~80 m lower than the trace of the Danian surface in the same location, suggesting that the Danian regradation of the pre-Cretaceous topography was minimal in this area.

The Danian surface represents the end product of the long-lasting acyclic evolution of the Ardenne–Oesling under conditions of prevailing tectonic stability during the Mesozoic. From the Middle Paleocene onwards, rather than regrading the older surfaces more or less uniformly and cutting them at low angle, new paleosurfaces developed at their expense by creating 50–170 m-high scarps that retreated inward, progressively nibbling the higher topography and producing a stepped landscape in response to more perceptible tilt of the massif. This happened so first on the western side of the massif, where the Danian landscape was replaced by a lower erosion surface during the

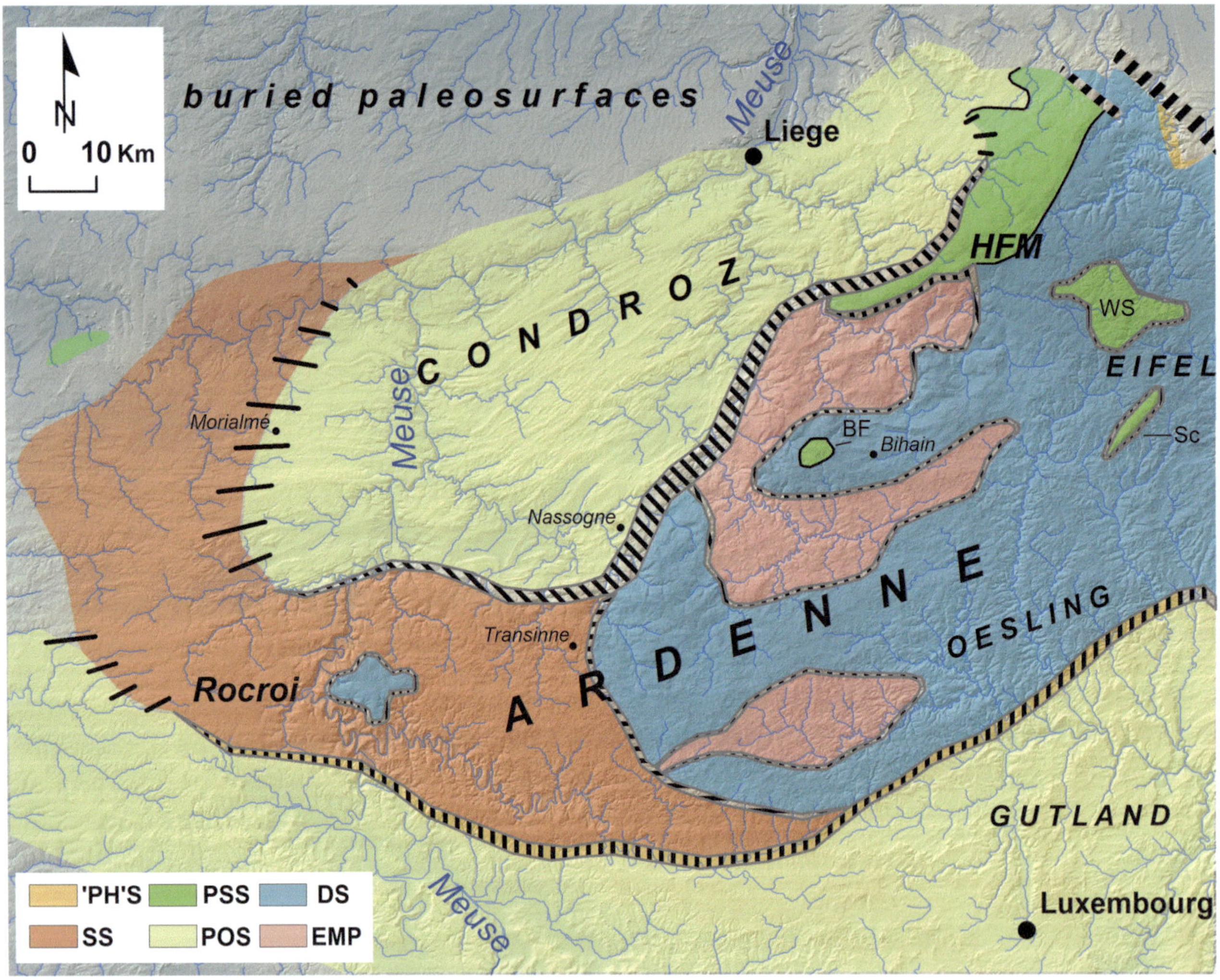

Fig. 5.15 Map of the erosion surfaces of the Ardenne and its margins, draped on a hillshade of the SRTM 3″ digital elevation model. 'PH'S. Post-Hercynian surface (including its more specific pre-Triassic variant in Luxembourg and north Eifel). *PSS*. Pre-Senonian surface. *DS*. Danian surface. *SS*. Selandian surface. *POS*. Pre-Oligocene surface. *EMP*. Early Miocene planation level. *Thin hatching* denotes erosional scarps between surfaces. In the west, the scarps between the Selandian and pre-Oligocene surfaces are progressively less conspicuous, finally giving way to a regradational type transition between the two surfaces (*widely spaced black lines*). *Bold hatching* (*in top right corner* of the map) is for fault scarps. *BF* Baraque Faiture. *WS* Weisser Stein. *Sc* Schneifel

Selandian, until this surface was buried by the Early Thanetian Sea in the west. In the east, it probably remained emerged and continued to evolve and proceed inward for some time, as attested by the 55 Ma radiometric age of the Morialmé weathering profile and the presence of Upper Thanetian continental sands scattered on the Rocroi Plateau (Voisin 1981) (Fig. 5.15). At its eastern limit, east of the Meuse valley, the Selandian surface abuts the Danian one through a ~70-m-high gently sloping erosional scarp.

Then, another phase of massif uplift caused new surfaces to develop along the northern and southern sides of the Ardenne–Oesling, news scarps separating them from the inner Selandian and Danian topographies. Such a Late Eocene surface (usually called 'pre-Oligocene') formed in N Ardenne, erasing the former landscape in part of the ESEM, in the Condroz, and along the NW slopes of the Hautes Fagnes before being sealed by the sand cover abandoned by the Rupelian sea. The scarp leading to the Danian surface is up to 170-m-high near Nassogne (Fig. 5.15). Likewise, south of the massif, active denudation and scarp retreat produced another well-developed surface with elevations around 300–420 m asl in the NE Paris basin and the Gutland area, prolonging eastward in the Moselle trough. While the age of this surface, Upper Eocene to Miocene, is still debated (Baeckeroot 1942; Quesnel 2003; Demoulin 2006), its antiquity is not disputed, identifying the southern margin of the Ardenne –Oesling as a Tertiary erosional feature and disqualifying the assumption that it primarily resulted from the Plio-Quaternary differential uplift of the massif.

Finally, while the Paleogene sand covers were progressively removed from the younger surfaces, the Oligo-Miocene evolution of the massif's interior was restricted to the development of intramontane planation basins of limited extent within the Danian topography, owing to climatic conditions generally less favourable to chemical weathering and etchplanation. These basins formed preferentially on weaker bedrock, in relation with the main drainage axes of the massif, and open all to the west, where their base level corresponded to the top surface of the sand covers of the pre-Oligocene and Selandian surfaces (Fig. 5.15). Their Early Miocene age is consistently supported by the absence of Paleogene marine deposits and the evidence of a last phase of active bedrock kaolinization around 20 Ma from K–Ar and Ar–Ar dating of cryptomelanes at the base of the weathering mantle at Transinne and on the Plateau des Tailles (Bihain).

5.4 Erosion Surfaces, Tectonic Uplift, and Denudation Rates in the Ardenne–Oesling

Although paleosurface reconstruction is now often deemed superseded by developments in the thermochronological and cosmogenic nuclide approaches of long-term landscape evolution, the recent progress in the dating of weathering products allows for sounder such reconstructions, with important implications regarding the underlying tectonic evolution of the Ardennian erosional setting and the associated denudation volumes.

Beyond local tectonics such as the HFM block uplift, broad-scale differential deformation between the originally almost horizontal surfaces of different ages suggests that every particular uplift episode of the massif during the Cenozoic did not exceed a few hundred meters with respect to the forelands. For example, the difference of tilt between the pre-Senonian and Danian surfaces in NE Ardenne indicates a ~400-m-high broad upwarping of central Ardenne in the Early Paleocene, which however never raised the massif's heart to such elevations, as surface regradation more or less kept pace with uplift. The varying tilt angle of the external facets of the acyclic surface developed over Ardenne–Oesling, systematically larger for older facets, attest that this slow doming of the massif was a background trait of its tectonic behaviour during the Mesozoic.

The change for cyclic morphogenesis during the Selandian was probably related to more focused uplift of the massif, highlighted especially in N Ardenne by the contrast between horizontal topography in the massif's interior and abruptly, though weakly, tilted surfaces close to its margins. At least in the north, probably in relation with the dawn of the Lower Rhine rift activity, subdued doming thus gave way to more en-bloc, multiphased uplift. The cumulated heights of the erosional scarps that were consequently produced between the successive surfaces amount to ~200 m from Selandian to Rupelian times, while the surfaces' deformation points to an additional few tens of meters of Neogene marginal tilt and the Plio-Quaternary component of en-bloc uplift may be estimated from river incision in the order of 150 m (see Chap. 11).

Paleosurface analysis and the dating of weathering products are also useful tools to assess the plausibility of long-term landscape evolution and denudation rates derived from thermochronology and cosmogenic nuclide data in regions of low to moderate uplift and elevation. In the Ardenne, thermochronological data are so far limited to fission track (FT), mainly on apatite (Glasmacher et al. 1998; Xu et al. 2009; Bour 2010) and led to a variety of interpretations. From the stand-alone analysis of their FT data, Xu et al. (2009) infer that slow exhumation prevailed in the Ardenne from 230 to 45 Ma, as indicated by cooling rates of 0.1–0.3 °C/Ma. Based on the usually used thermal gradient of 30 °C/km, this corresponds to denudation rates of 4–13 m/Ma. Then, from 45 Ma onwards, they model a phase of faster cooling (0.7–1.1 °C/Ma) that translates in denudation rates of 20–30 m/Ma, and suggest that an exhumation of 0.9–1.3 km would have occurred since the Middle Eocene. However, in a thermal history modelling that does not call for a thick Upper Cretaceous cover over the Ardenne, this km-scale amount of Cenozoic exhumation is in contradiction with field data attesting the preservation of Lower Cretaceous weathering products and Upper Cretaceous sediments on the Ardennian surfaces. In order to overcome such discrepancies between field data and FT inferences, Bour (2010) introduces in the modelling of his own FT data constraints imposed by the measured ages of the weathering mantle and the timing of marine transgressions upon the massif's margins. Consequently, following a long stay from 230 to 90 Ma in the [10–50 °C] domain not resolved by FT but most probably with insignificant denudation rates, his best-fit thermal histories display a heating episode indicating that the Upper Cretaceous sea would have buried the N Ardenne under a more than 1 km-thick sediment cover, succeeded by rapid cooling during the Cenozoic, which Bour (2010) interprets as the result of tectonic inversion and massif uplift. However, though consistent with Lower Cretaceous kaolinic weathering in central Ardenne and the observed burial under Upper Cretaceous chalks along the massif's northern margin, this thermal history faces another difficulty because the Thanetian marine cover directly resting on the Paleozoic basement of the ESEM's western confines and the 55-Ma age of weathering products at Morialmé imply that this thick Cretaceous cover should have been removed very rapidly in the beginning of the Paleocene, at rates of 100–150 m/Ma. Though not impossible per se, such rates would imply comparatively high Paleogene rock uplift rates in the Ardenne, which no other evidence supports so far. Independent information about the volume of the Late Cretaceous Ardennian cover is thus required, which might perhaps be derived from estimates of the fraction of reworked Cretaceous material included in the Selandian and Thanetian sediment covers surrounding the massif. While the apparent inconsistencies between paleosurface and paleoweathering observation on one hand, FT modelling results on the other hand, might refer to the use of inappropriate T°-depth profiles in the latter, they especially stress the importance of the former data type and the need for further efforts towards dating of weathered material, sediments, and exposed surfaces in slow-evolving erosional settings, which the Ardenne massif is highly representative of.

Likewise, cosmogenic radionuclide (CRN) studies have yielded a lot of denudation estimates in uplifting massifs, and especially in the Ardenne (Schaller et al. 2002, 2004; Sougnez et al. 2011) and the Rhenish shield (Meyer et al. 2010), which, though indicative of shorter term denudation, are often compared to the long-term evolution of the massif. These estimates are obtained from measurements of cosmogenic ^{10}Be concentration in the quartz sand fraction of river bed loads, assumed to carry information on the average denudation rate in the river's catchment upstream of the sampling point. Most CRN estimates from Ardennian rivers fall in the range 20–80 mm/ky, being thus meaningful at best for the last 50 ky. However, as secular equilibrium is assumed in the calculation of CRN denudation rates and the climatic conditions of the last 50 ky are indeed representative of those of at least the last 1 My, this suggests that a rock slice 50–200 m-thick would have been removed from the massif just during the Quaternary, a figure hardly compatible with the preservation of >30-Ma-old loose sediments and kaolinized rocks over many levelled interfluves. Moreover, Demoulin et al. (2009) showed that these rate estimates are much closer to the rates of valley incision since 0.7 Ma in the Ardenne, concluding that the CRN rate estimates in the massif are not representative of average denudation because the river bed load does not sample the extended horizontal remnants of the paleosurfaces where virtually no erosion takes place currently. Here again, the geomorphological reconstructions of paleolandscapes appear thus as a welcome counterweight to sometimes hasty conclusions derived from new techniques.

5.5 Conclusion

Praised by writers, celebrated by painters, the Ardenne's scenic landscapes have also been studied for more than a century by geoscientists. While several local names were chosen already in the nineteenth century, and are still often in use, to describe stages, especially of the Devonian System, of the international stratigraphic chart, geomorphologists studied the Ardennian relief as a well-preserved example of the long-term geomorphic features of temperate latitudes' old massifs, highlighting their typical evolution most probably under former warm, often wet, climatic conditions. Within the Ardenne, the Hautes Fagnes area, with its wild beauty, is certainly one of the best places to observe the general character of extended remnants of etchplains and unravel the complex relationships between regional tectonic deformation, continental weathering and relief development, and episodic marine transgressions leading to the present layout of variously intersecting or stepped paleosurfaces. The sections offered by the Transinne quarry and the Bihain outcrop across the deeply kaolinized bedrock of west- and north-central Ardenne are also reference places for the easy observation they allow of the thick profile produced by long-lived chemical weathering, indispensable to etchplanation. In other words, the Ardenne–Oesling is not only a world-class area for its contribution to the history of the understanding of long-term landscape evolution in temperate regions (i.e. the typical setting of the Davisian "normal" geomorphic evolution) but also a remarkable example of how the geomorphic landscape record may be as powerful a tool as the geological rock record for reconstructing the long-term history of regions where continental, tectonically stable or weakly active regimes prevailed over 10^7–10^8 yr timescales. Moreover, the recent results presented in this overview reveal new perspectives for future research and advances, especially in the chronology of the Ardennian evolution and its meaning as a piece in the history of the large-scale foreland domain of the Alpine collision zone in Europe. Meanwhile, every rambler walking through the Ardenne plateau's wild scenery and deep forested valleys cannot keep from having a strong feeling of the grandeur and long time scale of the landscapes in which humans pass so quickly.

Acknowledgements Thanks are due to G. Feraud (University of Nice) for Ar–Ar dating.

References

Alexandre J, Thorez J (1995) Au secondaire et au tertiaire, l'Ardenne tropicale. L'altération des roches et les climats anciens. In: Demoulin A (Ed) L'Ardenne. Essai de géographie physique, Dépt de Géographie physique et Quaternaire, University of Liège, pp 53–67

Baeckeroot G (1942) OeslingetGutland. Morphologie du bassin ardennais et luxembourgeois de la Moselle. A. Colin, Paris, p 396

Baele J-M, Quesnel F, Dupuis C (2016) Silcrete in northern France and Belgium: a terrestrial record of surface environments under the influence of pyroclastic ash-falls. 5th International Geologica Belgica Meeting, Mons, 26–29 Jan 2016, p 40 (Abstract Book)

Barbier F, Prognon C, Quesnel F, Dupuis C, Yans J (2010) Dating and weathering characterization of the Morialmé quarry (Entre-Sambre-et-Meuse, Belgium). 4th French congress on stratigraphy, Paris, p 18 (Abstract Book)

Barbier F, Quesnel F, Dupuis C, Yans J (2012). The late paleocene-early eocene interval as a potential period for weathering in Western Europe: the case of the Morialmé section (Belgium). 4th International Geologica Belgica Meeting, Brussels, 11–14 Sept 2012, p 59 (Abstract book)

Baulig H (1926) Le relief de la Haute Belgique. Ann de Géog 35:206–235

Belanger I, Delaby S, Delcambre B, Ghysel P, Hennebert M, Laloux M, Marion J-M, Mottequin B, Pingot J-L (2012) Redéfinition des unites structurales du front varisque utilisées dans le cadre de la nouvelle Carte géologique de Wallonie (Belgique). Geol Belg 15:169–175

Biagioni C, Capalbo C, Pasero M (2013) Nomenclature tunings in the hollandite supergroup. Eur J Mineral 25:85–90

Bless MJ, Felder PJ (1989) Note on the Cretaceous of Hockai (Hautes Fagnes, NE Belgium). Ann de la Soc géol de Belg 112(1):47–56

Boulvain F, Belanger I, Delsate D, Dosquet D, Ghysel P, Godefroit P, Laloux M, Roche M, Teerlynck H, Thorez J (2000) New lithostratigraphical, sedimentological, mineralogical and palaeontological data on the Mesozoic of Belgian Lorraine: a progress report. Geol Belg 3:3–33

Bour I (2010) Histoire thermique des massifs ardennais et bohémien. Conséquences sur la dynamique de l'Europe de l'Ouest au Méso-Cénozoïque. PhD thesis, University of Paris-Sud. http://geologie-et-alpes.pagesperso-orange.fr/tfa/texte/somthese.htm, 364 p

Bruyère D, De Putter T, Perruchot A, Dupuis C (2003) Neogenesis of halloysite in cryptokarstic environments (Beez, Belgium): chemical modeling. In: 10th Conference of the European clay groups association, Modena, 22–26 Jun 2003, p 48 (Abstract Book)

Cosan Y (1969) Etude géologique de sondages au Plateau des tailles. Mémoire du Service Géologique de Belgique, p 32

Davis WM (1896) The Seine, the Meuse and the Moselle. Natl Geogr Mag 7(89–202):228–238

Davis WM (1899) The peneplain. American. Geologist 23:207–239

Dejonghe L, Boni M (2004) The "Calamine-type" zinc-lead deposits in Belgium and West Germany: a product of Mesozoic paleoweathering processes. Geol Belg 8:3–14

Demoulin A (1995) Les surfaces d'érosion méso-cénozoïques en Ardenne-Eifel. Bull de la Soc géol de Fr 166:573–585

Demoulin A (2003) Paleosurfaces and residual deposits in Ardenne-Eifel: historical overview and perspectives. Géol Fr 2003 (1):17–21

Demoulin A (2006) La néotectonique de l'Ardenne-Eifel et des régions avoisinantes. Mémoires de la Classe des Sciences, Acad R de Belg 25:252 p

Demoulin A, Ek C (1995) La naissance de l'Ardenne. Mise en place du massif paléozoïque de l'Ardenne. In: Demoulin A (ed) L'Ardenne. essai de géographie physique, Dépt de Géographie physique et Quaternaire, University of Liège, pp 15–30

Demoulin A, Hallot E, Rixhon G (2009) Amount and controls of the quaternary denudation in the ardennes massif (Western Europe). Earth Surf Proc Land 34:1487–1496

Demoulin A, Quesnel F, Dupuis C, Gerrienne P, Yans J (2010) Cenomanian sands and clays north of the Vesdre valley: the oldest known Cretaceous deposits in eastern Belgium. Geol Belg 11:241–256

De Putter T, Ruffet G, Yans J, Mees F (2015) The age of supergene manganese deposits in Katanga and its implications for the Neogene evolution of the African Great Lakes Region. Ore Geol Rev 71:350–362

Dill HG, Hansen B, Keck E, Weber B (2010) Cryptomelane: a tool to determine the age and the physical–chemical regime of a Plio-Pleistocene weathering zone in a granitic terrain (Hagendorf, SE Germany). Geomorphology 121:370–377

Dupuis C (1979) Esquisse paléogéographique du Nord et du Nord-Ouest du Bassin de Paris au Paléocéne et à l'Eocène inférieur. Incidences structurales. C.R. Acad Sci Paris 288:1587–1590

Dupuis C, Charlet JM, Dejonghe L, Thorez J (1996) Reconnaissance par carottage des paléoaltérations kaolinisées mésozoïques de la Haute-Ardenne (Belgique). Le sondage de transinne (194E-495): premiers résultats. Ann de la Soc géol de Belg 119:91–109

Dusar M, Lagrou D (2007) Cretaceous flooding of the Brabant Massif and the lithostratigraphic characteristics of its chalk cover in Northern Belgium. Geol Belg 10:27–38

Fairbridge R (1968) Gradation. In: Fairbridge R (ed) Encyclopedia of earth science. Geomorphology, Springer, Berlin, pp 485–486

Fransolet A-M (1979) Occurrences de lithiophorite, nsutite et cryptomélane dans le Massif de Stavelot, Belgique. Ann de la Soc géol de Belg 102:303–312

Glasmacher U, Zentilli M, Grist A (1998) Apatite fission track thermochronology of Paleozoic sandstones and the hill-intrusion, Northern Linksrheinisches Schiefergebirge, Germany. In: van den Haute P, de Corte F (eds) Advances in fission-track geochronology, Solid earth sciences library vol 10, Springer, pp 151–172

Gosselet J (1888) L'Ardenne Mémoire pour servir à l'explication de la carte géologique détaillée de la France. Baudry et cie, Paris, 889 p

Gullentops F (1954) Contributions à la chronologie du Pléistocène et des formes du relief en Belgique. Mém de Inst de Géol de Univ de Louvain 18:123–252

Hautmann S, Lippolt HJ (2000) 40Ar/39Ar dating of central European K–Mn oxides—a chronological framework of supergene alteration processes during the neogene. Chem Geol 170:37–80

Jérôme A (1907) De la découverte d'un gisement notable de kaolin en Ardenne, Bull de la Soc belg de Géol XXI:217–222

Junge H (1987) Der Einfluss von Tektonik und eustatischen Meeresspielgelschwankungen auf die Ausbildung der Reliefgenerationen im Norden der Eifeler Nord-Süd-Zone. Z für Geomorphol Suppl Bd 65:35–84

King LC (1947) Landscape study in southern Africa. Proc Geol Soc S Afr 50:23–102

Klein C (1990) L'évolution géomorphologique de l'Europe hercynienne occidentale et centrale. Aspects régionaux et essai de synthèse. Mémoires et Documents de Géographie CNRS, Paris, 177 p

Macar P (1938) Contribution à l'étude géomorphologique de l'Ardenne (note préliminaire). Ann de la Soc Géol de Belg 61:B187–217

Mees F, Stoops G (1999) Palaeoweathering of lower Palaeozoic rocks of the Brabant Massif, Belgium: a mineralogical and petrographical analysis. Geol Mag 34:349–367

Meyer H, Hetzel R, Strauss H (2010) Erosion rates on different timescales derived from cosmogenic ^{10}Be and river loads: implications for landscape evolution in the Rhenish Massif, Germany. Int J Earth Sci 99:395–412

Nahon D (1991) Introduction to the petrology of soils and chemical weathering. Wiley, New York, 313 p

Quesnel F (2003) Paleoweathering and paleosurfaces from northern and eastern France to Belgium and Luxembourg: geometry, dating and geodynamic implications. Géol Fr 2003(1):95–104

Quesnel F, Dupuis C, Yans J, Wyns R, Farjanel G, Théveniaut H, Voisin L, Lacquement F, Vergari A, Baele JM, De Putter T, Colbach R, Muller A, Vandycke S, Giot D, Brulhet J (2002) Paléosurfaces et paléoaltérations dans le Nord de la Fr et en Belg. Géologues 133–134:40–43

Schaller M, von Blanckenburg F, Veldkamp A, Tebbens L, Hovius N, Kubik P (2002) A 30,000 yr record of erosion rates from cosmogenic ^{10}Be in Middle European river terraces. Earth Planet Sci Lett 204:307–320

Schaller M, von Blanckenburg F, Hovius N, Veldkamp A, Van den Berg M, Kubik P (2004) Paleoerosion rates from cosmogenic ^{10}Be in a 1.3 Ma terrace sequence: response of the River Meuse to changes in climate and rock uplift. J Geol 112:127–144

Schuler M, Cavelier C, Dupuis C, Steurbaut E, Vandenberghe N, Riveline J, Roche E, Soncini M-J (1992) The Paleogene of the Paris and Belgian Basins. Standard stages and regional stratotypes. Cah Micropaléontol 7:29–92

Sougnez N, Vanacker V, Kubik P (2011) Reconstruction of Quaternary landscape evolution based on ^{10}Be denudation rates and tectonic uplift data in a moderate uplifted region (Ardennes Massif, Belgium). XVIII INQUA-congress, Bern, Poster, p 683

Théveniaut H, Quesnel F, Wyns R, Hugues G (2007) Palaeomagnetic dating of the "Borne de Fer" ferricrete (NE France): lower Cretaceous continental weathering. Palaeogeogr Palaeoclimatol Palaeoecol 253:271–279

Thiry M, Quesnel F, Yans J, Wyns R, Vergari A, Théveniaut H, Simon-Coinçon R, Ricordel C, Moreau MG, Giot D, Dupuis C, Bruxelles L, Barbarand J, Baele JM (2006) Continental France and Belgium during the Early Cretaceous: paleoweatherings and paleolandforms. Bull de la Soc Géol de Fr 177:155–175

Thomas M.F (1994) Geomorphology in the tropics. A study of weathering and denudation in low latitudes. Wiley, Chichester, 460 p

Vasconcelos PM, Renne PR, Becker TA, Wenk HR (1995) Mechanisms and kinetics of atmospheric, radiogenic, and nucleogenic argon release from cryptomelane during ^{40}Ar/^{39}Ar analysis. Geochim Cosmochim Acta 59:2057–2070

Voisin L (1981) Analyse géomorphologique d'une région-type: l'Ardenne occidentale. Service de reproduction des thèses, Université Lille 2, 883 p

Wayland EJ (1933) Peneplains and some other erosional platforms. Ann Rep Bull, Protectorate Uganda Geol Surv 1:77–79 Department of Mines

Wyns R, Gourry JC, Baltassat JM, Lebert F (1999) Caractérisation multiparamètre des horizons de subsurface (0–100 m) en contexte de socle altéré. PANGEA 31(32):51–54

Xu C, Mansy JL, van den Haute P, Guillot F, Zhou Z, Chen J, De Grave J (2009) Late- and post-Variscan evolution of the Ardennes in France and Belgium: constraints from apatite fission-track data. Geol Soc Lond Spl Publ 324:167–179

Yans J (2003) An overview of the saprolites of Belgium and their potential kaolinitic supplies to Mesozoic and Cenozoic sediments. Géol Fr 2003(1):33–37

Yans J (2013) Gestion durable des ressources minérales wallonnes: pistes de réflexions en vue d'une meilleure intégration de la problématique. Actes du 1er congrès interdisciplinaire du développement durable, Namur, pp 195–206

Yans J, Dupuis C (2005) Timing of saprolitisation in the Haute-Lesse area (Belgium). Geophys Res Abst 7:7064

A Unique Boulder-Bed Reach of the Amblève River, Ardenne, at Fonds de Quarreux: Modes of Boulder Transport

6

Geoffrey Houbrechts, François Petit, Jean Van Campenhout, Etienne Juvigné and Alain Demoulin

Abstract

The Amblève valley at Fonds de Quarreux displays one of the most spectacular incised landscapes of the Ardenne massif. Cut into highly resistant Cambrian quartzites and surrounded by steep wooded hillslopes, a 3-km-long gorge is cluttered with hundreds of huge quartzite boulders up to several metres in size. These boulders originate from the locally outcropping Cambrian Venne Formation and reached the riverbed through periglacial slope processes. Up to 2-m-large boulders have been transported by the river over distances of several kilometres, some of them being observed 90 km downstream in the Meuse terraces of the Campine Plateau. We show that such transport of very large boulders over long distances cannot occur through purely fluvial processes and conclude that ice rafting was the most likely transportation mode, active during Pleistocene cold periods and probably supplemented by other ice breakup-related processes. We also briefly present nearby sites of geomorphological value within the Amblève catchment.

Keywords

Ice-rafted river deposit • River competence • Paleohydrology • Amblève river • Ardenne

6.1 Introduction

Boulder-bed channels are scarce in the Ardenne, a Paleozoic massif with moderate elevations not exceeding ~700 m asl. Most rivers flow mainly over Weichselian sheets of coarse gravel (median particle diameter D_{50} between 2 and 10 cm) containing only locally sparse boulders of mostly <50 cm diameter (Houbrechts et al. 2006). In general, these boulders do not reveal any systematic in-channel arrangement and are very rarely displaced (Houbrechts et al. 2015). Concentration of blocks and boulders larger than 1 m is observed in only few, mostly short, stream reaches where rivers cut across resistant rock units that deliver massive blocks upon hillslope or channel erosion. Among the most resistant rocks of the Ardennian basement are Emsian conglomerates (e.g. at Roche à Frêne along the Aisne River), Lochkovian conglomerates and arkoses (e.g. crossed by the Chavanne at Bra) and, especially, the quartzites of the Stavelot massif (La Venne Formation, Revin Group, Cambrian), which are crossed by several rivers (e.g. the Amblève, Hoëgne, Helle, Soor) (Fig. 6.1). Several boulder-bed sectors are present in the Amblève between Montenau and Ligneuville, Bellevaux and Coo, and downstream of Targnon (Fig. 6.1). They occur wherever the river flows over resistant Revinian rocks of the Stavelot massif and its Lochkovian rim, block density varying with the degree of resistance of the basement, the

G. Houbrechts (✉) · F. Petit · J. Van Campenhout · A. Demoulin
Department of Physical Geography and Quaternary, University of Liège, Sart Tilman, B11, 4000 Liège, Belgium
e-mail: G.Houbrechts@ulg.ac.be

F. Petit
e-mail: Francois.Petit@ulg.ac.be

J. Van Campenhout
e-mail: Jean.VanCampenhout@ulg.ac.be

A. Demoulin
e-mail: ademoulin@ulg.ac.be

E. Juvigné
University of Liège, Liège, Belgium
e-mail: ejuvigne@skynet.be

A. Demoulin (ed.), *Landscapes and Landforms of Belgium and Luxembourg*,
World Geomorphological Landscapes, DOI 10.1007/978-3-319-58239-9_6

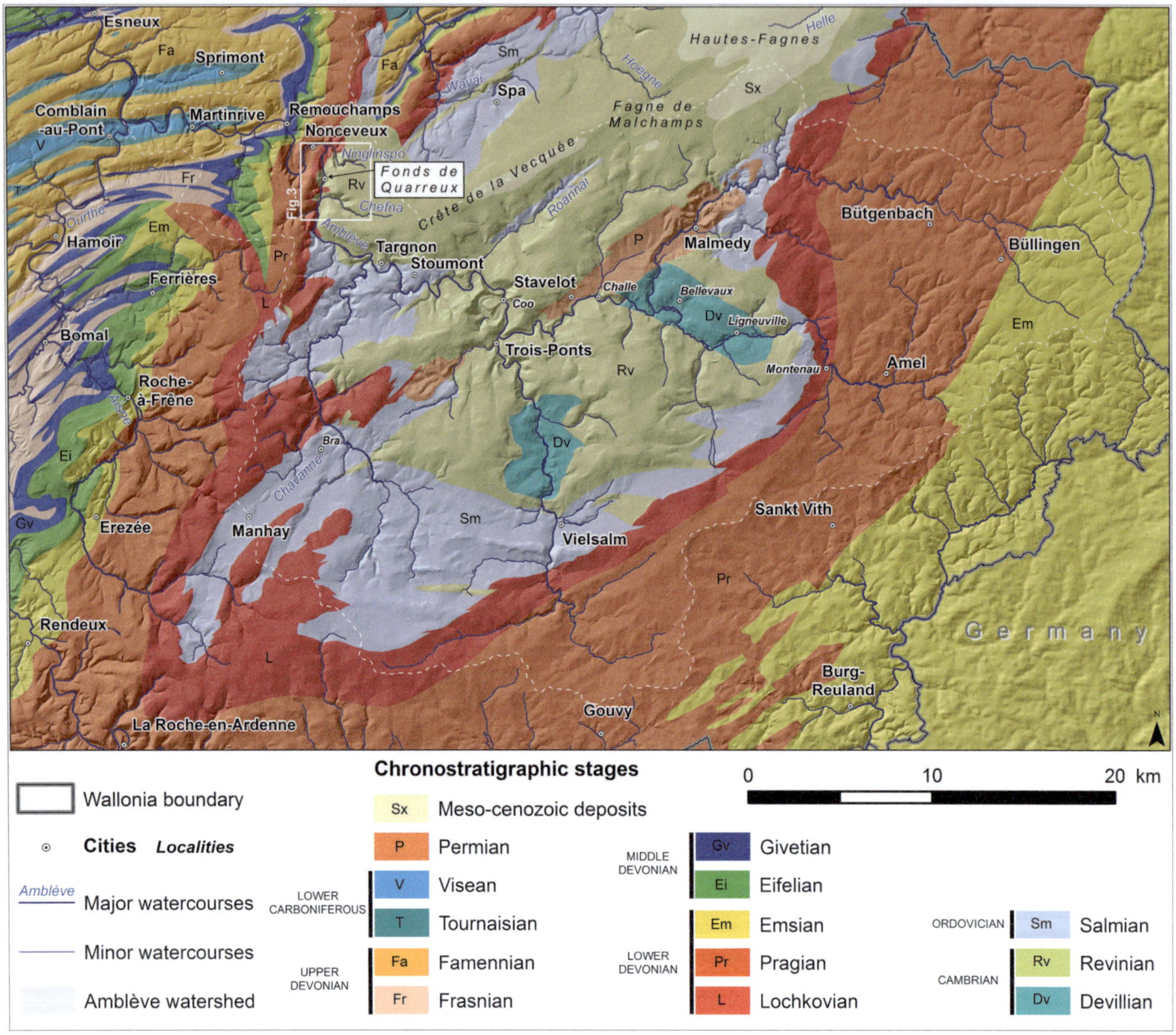

Fig. 6.1 Geology of the Amblève catchment and the Fonds de Quarreux area, draped on the hill-shaded SRTM 1″ DEM. All place names cited in the text are located on this map. *Sources* IRSNB-SGB (1902) Cartes géologiques de la Belgique, 1/40.000. Asselberghs (1946) Carte géologique de l'Eodévonien de l'Ardenne et des régions voisines. Fourmarier (1958) Carte géologique 148 Louveigné-Spa, 1/25.000. Asselberghs and Geukens (1960) Carte géologique 159 Harzé—La Gleize, 1/25.000. Geukens (1963) Carte géologique 160 Stavelot—Malmedy, 1/25.000. Legrand (1965) Carte géologique 233 Saint-Vith—Schönberg, 1/25.000. Vandenven (1990). Geukens (1999) Carte géologique du Massif de Stavelot

delivery by tributaries, and the distance between channel and hillslope foot. Downstream of Targnon, the proximity of the quartzites of the Vecquée ridge favours boulder accumulation, which peaks at Fonds de Quarreux (Fig. 6.2). There, hundreds of quartzite boulders larger than 1 m lie in the Amblève channel over ~3 km. The wild, almost mountainous scenery that results from boulders cluttering the channel and causing multiple small cascades has led to include the Fonds de Quarreux in the *Exceptional Heritage of Wallonia* since 1970, as is also the nearby tributary valley of the Ninglinspo. It also always captured the popular imagination, a local legend telling that the boulder accumulation in the valley bottom results from the brutal destruction by the Devil of a mill he had built on the hill, because he was angry of having been tricked by the wife of the miller with whom he had passed a pact (La Garde 1858).

Geologists have also been intrigued by this boulder-bed stream sector since the nineteenth century. While Dumont (1847) had suggested the boulder accumulation bore witness to rockfall, strong current or glacier activity, Dewalque (1885), based on the observation of striated blocks in the Stavelot area, favoured glacial transport. However, careful examination of such blocks by Stainier (1895) and Van den Broeck (1900) convinced them of the fluvial and/or

Fig. 6.2 Photograph of the Amblève channel at Fonds de Quarreux showing the accumulation of large boulders and the direct connection between hillslopes and channel's banks

differential weathering origin of the striae. In order to explain the presence of big boulders in the Meuse terrace deposits downstream of the Ardenne, Lohest (1891) monitored the mobility of a set of boulders in the Amblève for several years. He noted that boulders of 0.5 m^3 (equivalent diameter ~ 1 m) did not move in spite of high discharges and river ice breakup at the end of the harsh 1890–91 winter (Flammarion 1891). This ice breakup had caused ice jams and the accumulation of numerous ice blocks over the floodplain, where they locally piled up to several metres. Having found only a few small (<0.25 m^3) rock boulders stuck in the ice, Lohest (1891) concluded that the transport of big boulders such as those observed in the Meuse terraces is impossible under the present climate.

After Rahir (1909) had suggested the boulders originated from direct erosion of the valley bottom, the next step in their understanding came with Macar (1946, 1969), who identified their source in the nearby Vecquée ridge, which they would have descended through gelifluction, embedded in a fine clayey matrix in which some of them are still found along the ridge slopes. This clearly points to the periglacial environment in which their hillslope transport occurred. More recently, Collard (2002) mapped block occurrences in the Fonds de Quarreux valley bottom and on the valley slopes of the Amblève and the Chefna, a tributary that joins it just upstream of the boulder-bed sector. Based on this map, Collard et al. (2012) confirm that the block concentration in the Fonds de Quarreux has an origin in the local hillslopes. They also estimate the competence of the Amblève and Chefna and show that both rivers are presently unable to transport such boulders, whose displacement they explain through ice rafting under periglacial conditions. By contrast, although acknowledging the theoretical possibility of boulder transport by ice rafting processes, De Brue et al. (2015) favour a mixed mode of transport in which hydraulic transport and catastrophic flood surges related to ice jam release might be equally important.

6.2 Geographical and Geological Settings

The Fonds de Quarreux valley sector of the Amblève extends from the Chefna confluence to that of the Ninglinspo, between the villages of Stoumont and Remouchamps

(Fig. 6.3). With a length of 88 km and a catchment of 1070 km^2, the Amblève is the largest tributary of the Ourthe, which it joins near Comblain-au-Pont (96 m asl), coming from a source located at 600 m asl south of Büllingen, in E Ardenne. Its roughly E–W course first crosses Lower Devonian rocks before entering the Cambro–Ordovician Stavelot massif down to the Fonds de Quarreux area. There, it follows the N–S oriented boundary between the Stavelot massif in the east and the Dinant synclinorium in the west before leaving the Stavelot massif to take anew an E–W direction across the Devonian and Carboniferous rocks of the Dinant synclinorium.

In the Fonds de Quarreux, the two sides of the valley display contrasted rock types (Fig. 6.3). In the east, the La Venne Formation (Revin Group, Cambrian) characterized by thick beds of very resistant quartzite (Lamberty and Marion 2014) covers the largest part of the valley slopes, where less resistant phyllites and subordinate quartzites of the Cambrian La Gleize Formation also crop out. These intensely folded and faulted rocks of the Stavelot massif are unconformably overlain by Lower Devonian formations that crop out mainly west of the valley. With the exception of the basal conglomerates (Collard et al. 2012), the Devonian silicoclastic rocks are generally less resistant to erosion. North of the Chefna confluence, the La Venne quartzites are present on both sides of the valley, causing its abrupt narrowing and steepening, with slopes up to 30° and extensive quartzitic outcrops especially on the western valley side (Fig. 6.4). Culminating at 357 m asl, this side is however shorter than the opposite one, whose 17°–22° slopes go up into a broad convexity leading to a remnant of the pre-Oligocene erosion surface at ~420 m, which in turn merges into the summit surface of the Vecquée ridge (Demoulin 1995; see also Chap. 5).

6.3 The Fonds de Quarreux Boulder-Bed Stream Reach

Collard (2002) mapped all boulders larger than 1 m in the channel of the Amblève. Starting at the Chefna confluence, his map extends over 2.4 km. It could not be drawn farther downstream because most blocks have been used there to protect the piles of a railway bridge and prevent lateral erosion of the right bank. Many blocks are therefore buried in the bank and a reliable counting of block density is no longer feasible. Based on the data of Collard, we calculate the number of blocks per 100-m-long channel segments and consider its alongstream evolution (Fig. 6.5). In addition, we measured the size of the largest boulders in each segment and show the alongstream variations in maximum boulder size (D_{max}) also in Fig. 6.5.

At the height of the Chefna confluence and down to 0.8 km, the boulder frequency remains low on average (mostly below 25 boulders/100 m) while the maximum length of the b-axis varies randomly but never exceeds 4 m. Then, both frequency and maximum size abruptly increase from 0.9 km to reach peak values of 235 boulders/100 m and a D_{50} of 5.9 m at 1.2 km before decreasing regularly down to 2.4 km, where values close to those observed at the Chefna confluence are found again. This evolution clearly locates the provenance area of the big boulders between 0.9 and 1.2 km. Downstream of the latter point, the block frequency f decreases as a power law of distance x from the source of material

$$f = 55.2x^{-2.3} \qquad (6.1)$$

x being expressed in km, which yields an excellent coefficient of determination of 0.98 (n = 14). Maximum boulder size first decreases more rapidly before reaching a first plateau between 1.7 and 2.1 km ($D_{50} \approx 2.15$ m) and, after two probable random higher values, another slightly lower one between 2.6 and 2.9 km ($D_{50} \approx 1.8$ m). These observations are consistent with the conclusions of other studies of bedload displacement (Houbrechts et al. 2011) and provide insights into the competence of the responsible transport process. Additional size measurements made more downstream at Nonceveux and Remouchamps (4.8 and 9.3 km, respectively) yielded maximum b-axis length of 1.8 and 1.9 m, showing that the 1.8 m competence plateau is stable at least down to Remouchamps.

Hundreds of Revinian quartzite boulders similar to those of Fonds de Quarreux are also present in the Meuse terrace deposits downstream of Liege, and especially in the Main Terrace of the Campine Plateau, located 90 km downstream of Fonds de Quarreux (De Brue et al. 2015). Based on a set of 536 boulders, a longest b-axis of 1.95 m has been measured there, again very close to the plateau value obtained in the downstream sector of the Fonds de Quarreux.

6.4 Modes of Boulder Transport

6.4.1 Provenance of the Blocks

The boulders in the Amblève channel at Fonds de Quarreux are exclusively Revinian quartzites of the La Venne Formation. They clearly have not been plucked off the channel bottom as the bedrock is systematically hidden beneath a gravel sheet on which the boulders rest. Most boulders are fairly well rounded, essentially through abrasion by the suspended load during floods rather than through own displacement. The graphs of Fig. 6.5 demonstrate that they do not originate from the Chefna catchment but come from the hillslopes of the Amblève valley where the river cuts across the La Venne Formation. Indeed, the sector with highest boulder concentrations and largest boulders (0.8–1.7 km) exactly coincides with this zone, where numerous quartzite beds crop out to form impressive walls rising from the hillslopes (Figs. 6.3 and 6.4). Moreover, owing to the narrowness of the valley, the channel is almost continuously connected with the base of the valley sides in this sector. However, despite the number of boulders accumulated in the valley, and even if we take account of that fraction of the quartzite that was delivered from the hillslopes in the form of readily transportable smaller blocks and gravels, a rough estimate of the rock volume removed through valley incision leaves no doubt that the total number of large boulders that reached the valley bottom since incision began has been much larger than what is observed today, evidencing that many boulders were indeed transported over long distances. We evaluate hereafter candidate processes that might have been responsible for this transport.

6.4.2 Hydraulic Transport Under Temperate- and Cold-Period Hydrodynamic Conditions

Several approaches may be used to determine the hydrodynamic conditions required to put bedload particles into motion, namely evaluating (1) critical flow speeds (Hjulström 1935; Sundborg 1956, 1967), (2) critical shear stresses using, e.g. the Shields parameter (Shields 1936), or (3) critical unit stream power, with various empirical relations derived for specific grain size ranges and bed forms (e.g. Petit et al. 2005a; Ferguson 2005; Mao et al. 2008; Gob et al. 2010; Houbrechts et al. 2015). Based on either mean bed

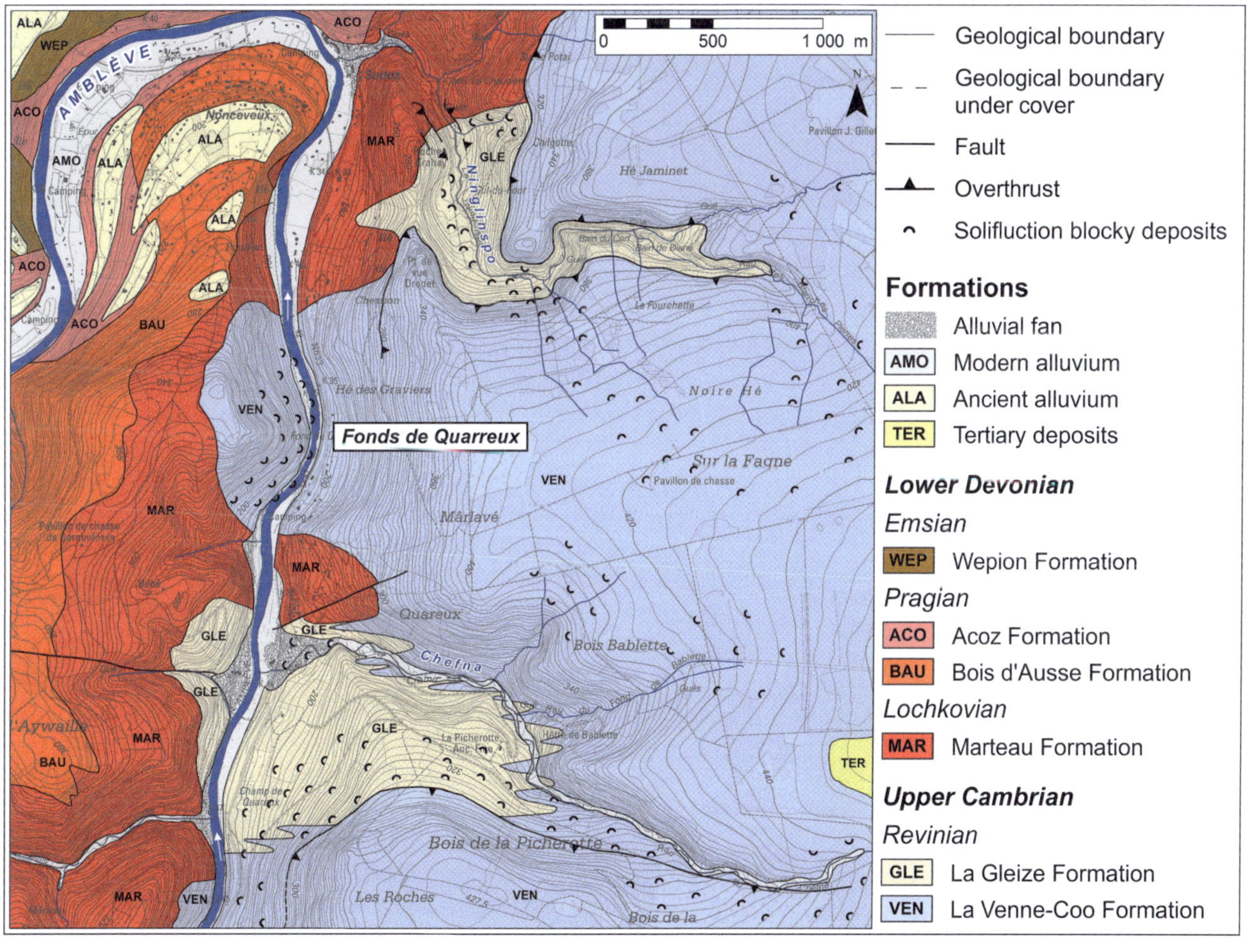

Fig. 6.3 Zoom on the geology of the Fonds de Quarreux area illustrating the relation between outcrop of the resistant quartzites of the La Venne Formation (in *blue*) on both sides of the valley and marked narrowing of the floodplain (see location in Fig. 6.1). *Sources* Marion (2016) Geological map of Wallonia, 49/3–4 & 49/7–8, interim document (personal communication). National Geographic Institute (2001) Topographic map 1/10.000, 49/7 N. National Geographic Institute (2000) Topographic map 1/10.000, 49/3S

Fig. 6.4 Ancient postcard of the Fonds de Quarreux. The photograph was taken at the beginning of the twentieth century. Note the steep cleared hillslopes of the left bank, scattered with outcrops of resistant quartzites and directly joining the channel

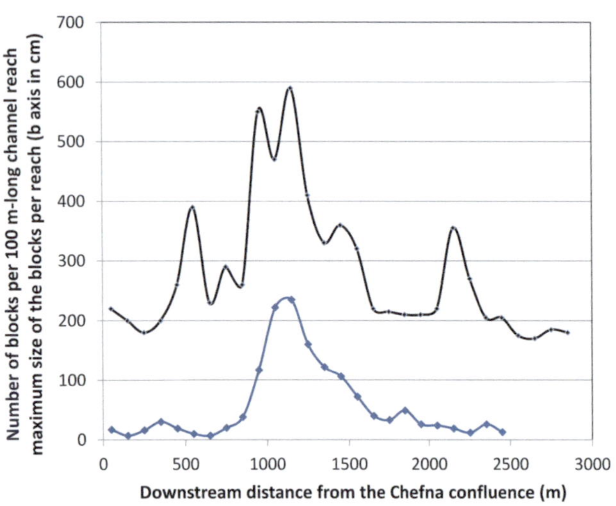

Fig. 6.5 Alongstream variations of the concentration of boulders larger than 1 m in the channel (in *blue*) and the size of the largest boulders (in *black*). Data per 100-m-long channel segments starting upstream at the Chefna-Amblève confluence

gradient or a variety of gradients, water and particle density values, and protrusion ration values (D_i/D_{50}, with D_i = size of the particle to move and D_{50} = median size of the bedload), Collard et al. (2012) and De Brue et al. (2015) estimated critical shear stress and unit stream power values required to move the boulders of the Fonds de Quarreux, showing that the largest boulders (b-axis of several metres) cannot be displaced by fluvial processes. They calculated that, under present climatic conditions, the largest blocks that may be transported by exceptional, 100-year or more, floods are ~0.5 m in size.

However, these estimates did not account for the high variability in channel gradient, channel width, and floodplain width that the presence of a lithologically controlled knickpoint in the Fonds de Quarreux entails. Therefore, we recalculate the unit stream power for each successive 100-m-long channel segment between the Chefna and Ninglinspo confluences, and for various characteristic discharges, namely bankfull discharge Q_b, 100-year flood discharge Q_{100}, and maximum expected discharge Q_{max} under temperate and periglacial climates. We then compare the obtained values with the critical unit stream power required to move the boulders.

The width variable that appears in the equation of unit stream power ω

$$\omega = (\rho g Q S)/W \qquad (6.2)$$

where ρ = water density, g = Earth gravity, Q = discharge, S = channel gradient, W = flow width, has been taken as bankfull channel width in the calculation of $\omega(Q_b)$ and floodplain width for the estimates of $\omega(Q_{100})$ and $\omega(Q_{max})$. Field measurement of Q_b at the nearby site of Martinrive (Fig. 6.1) yielded a value of 140 m^3/s (Petit et al. 2005b). Based on a 41-year-long series of daily data (SPW-DGO2, 1975–2016), a Q_{100} value of 360 m^3/s has been estimated by the Gumbel method (Gumbel 1960). As for Q_{max}, we calculate it as the modified Myers probable maximum flood (Jarvis 1926; Pardé 1960, 1962) although Commons (1961) has disputed the validity of the exponent on A

$$Q_{max} = CA^{0.5} \quad (6.3)$$

where A = drainage area and C = Myers coefficient. In the present conditions of an oceanic *Cfb* climate (see Chap. 3), the highest Myers coefficients recorded for exceptional flood events in the Hautes Fagnes area, ~25 km to the NE of the Fonds de Quarreux, are in the range 16–18 (Pissart 1961; Bravard and Petit 1997). However, they refer to very small catchments (<40 km^2) undergoing intense rainfall of short duration and limited areal extent. Here, we use a value of 18 for the Amblève, keeping in mind that the conditions are different, with floods mostly caused by long-lasting winter rainfall and/or spring snowmelt (Pauquet and Petit 1993). Applying the Myers formula to the Martinrive site thus yields Q_{max} = 580 m^3/s, to which the Gumbel method associates a several kyr-long return period based on the available 41-year-long time series.

As boulder transport may have occurred also during the cold periods, we also consider potential flood intensities under periglacial conditions. Rapid spring melting of the snow accumulated along the winter months, possibly concurrent with heavy spring rainfall, probably caused higher floods than those recorded today. Though concrete data are missing in the Ardenne, one may obtain insight in their intensity from observations in present-day rivers pertaining to the corresponding hydrological regime. The Weichselian Amblève catchment may be approximated by the lowland nival regime of Pardé (1933) that is currently found, e.g. in Sweden, Poland, Russia, and northern Canada. If we use a Myers coefficient value of 50, i.e. the upper bound of the 30–50 range of estimates for rivers in such areas (Pardé 1960; Frécaut 1971), we get Q_{max} = 1600 m^3/s.

Gradient values have been extracted from the LIDAR DEM of the Walloon Region (1-m resolution) for 500-m-long channel reaches in order to avoid large local gradient variations caused by a sensing carried out during a low water period. In the Fonds de Quarreux area, the 0.4% gradient near the Chefna confluence increases first downstream to 1.1% where the lithological knickpoint disturbs the Amblève long profile. Downstream of this knickpoint, the channel gradient decreases back to values around 0.22% near the Ninglinspo confluence (Fig. 6.6). Likewise, the floodplain width, averaged by 100-m-long segments, varies between 43 and 184 m as a function of the resistance to erosion of the rocks cut by the river. As for the channel width, it is affected by the presence of low terraces, alluvial fans at confluences, and colluviation at the base of the valley slopes and varies in the 27–57 m range. Based on this set of data, we calculate unit stream power values that strongly vary along the Fonds de Quarreux sector, with maxima at 1.1 km where the La Venne quartzites determine greatest narrowing of the valley and steepening of the channel gradient (Fig. 6.7). There, our estimates of Q_b, Q_{100}, and Q_{max} lead to unit stream power estimates of 540 W/m^2 for bankfull discharge, 900 W/m^2 for a 100-year flood,

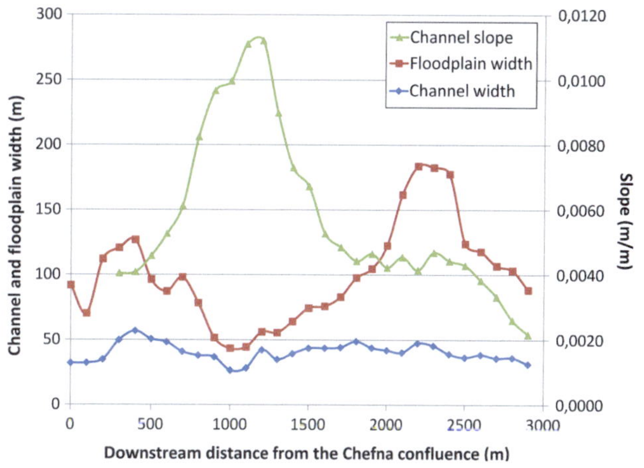

Fig. 6.6 Alongstream variations of channel gradient (*green*), and channel (*blue*) and floodplain (*red*) widths. The *curve* starts upstream at the Chefna-Amblève confluence

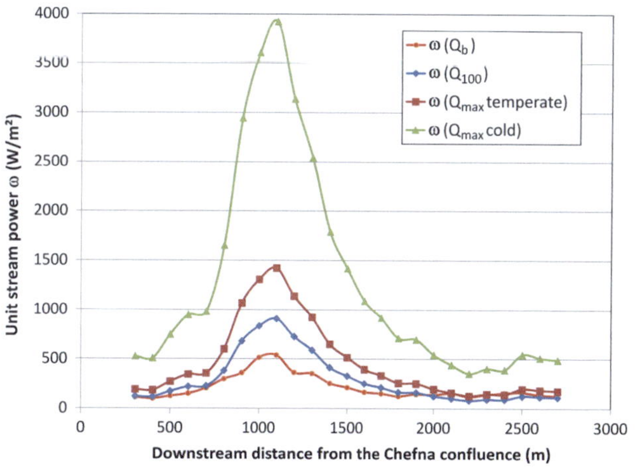

Fig. 6.7 Alongstream variations of estimated ω for Q_b (in *light red*), Q_{100} (in *blue*), and Q_{max} (temperate in *dark red*, cold periods in *green*). Data per 100-m-long channel segments starting 250 m downstream of the Chefna-Amblève confluence

1400 W/m^2 for the probable maximum flood (also called 'project flood') in the present climatic conditions, and 3900 W/m^2 for the same under periglacial conditions.

In order to determine whether these values were high enough to put the boulders into motion, we then estimate the critical unit stream power ω_{crit} needed to allow such a transport. Simple relations have been proposed, which link ω_{crit} only to the size D_i (in mm) of the mobilized particles (Williams 1983; Costa 1983)

$$\omega_{\text{crit}} = a(D_i)^b \quad (6.4)$$

where a and b are constants. Gob et al. (2010) estimated $a = 0.025$ and $b = 1.65$ for blocks ≥ 1 m in boulder-bed streams of the French Central Massif. Applied to the Amblève, these values imply critical unit stream powers $\omega_{\text{crit}} = 700$, 2200, and 7000 W/m^2 to move blocks with 0.5-, 1-, and 2-m-long b-axes, respectively. Comparing these critical powers to the largest unit stream power values calculated at 1.1 km for Q_{100} and $Q_{\max}$ in the Fonds de Quarreux, the present (interglacial) 100-year and probable maximum floods would not be able to displace 1-m and larger boulders. Only the cold period probable maximum flood could have displaced boulders up to 1.4 m in size in the steepest and narrowest sectors of the river, but was still insufficient to move the many boulders ≥ 2 m currently found in the valley bottom. However, this is a matter of fact that many boulders close to 2 m in size are observed downstream in the Amblève, for instance at Nonceveux and Remouchamps, and farther, down to the Campine Plateau in N Belgium. This implies that processes other than purely fluvial must be called on to explain such large distance transport.

6.4.3 Boulders as Ice-Rafted Material?

Several authors have observed and described the transport of ice-rafted blocks in rivers (Lyell 1868; Dionne 1979; Kempema and Ettema 2011). Dionne (1979) mentioned blocks up to 5 m scattered along the banks of the Saint Lawrence estuary and originating from more than 25 km upstream. Likewise, Cailleux and Hamelin (1967) observed ice-rafted blocks of more than 5 m abandoned on the shores of the Bic area of the same region.

Several ice-forming processes concur in forming large sheets of river ice (New Brunswick Subcommittee on River Ice 2011), namely

- border ice, which develops vertically and laterally from the river banks
- frazil ice, which appears as tiny particles in supercooled water. The ice particles aggregate in buoyant flocs that form ice pans at the water surface, which can in turn concentrate in favourable areas (e.g. a constricted channel reach) and coalesce into a continuous ice cover. In turbulent waters, frazil ice may also adhere to the channel bed and form anchor ice that may lift and take bed particles downstream upon thawing
- thermal ice cover, which develops in slowly flowing streams from ice crystals forming on the surface. The initially thin ice cover grows downward. For instance, in central New Brunswick, thermal ice covers typically thicken to between 40 and 80 cm (New Brunswick Subcommittee on River Ice 2011) under winter conditions slightly milder than those experienced by the Ardenne in the Last Glacial Maximum.

Owing to Archimedes' principle and the density of ice (0.92) lower than that of water, ice rafts can bear and carry significant amounts of sediment. It is readily shown that the theoretical maximum rock volume V_{r} that an ice raft of volume V_{i} can transport is

$$V_{\text{r}} = \frac{(\rho_{\text{w}} - \rho_{\text{i}})}{(\rho_{\text{r}} - \rho_{\text{w}})} V_{\text{i}} \quad (6.5)$$

With ρ_{r}, ρ_{i}, and ρ_{w} are density of rock, ice, and water, respectively. For example, a 1-m^3 ice block is able to support $\sim$0.05 m^3 of quartzite ($\rho \approx 2.7$), weighing approximately 135 kg and representing a single spherical block of 0.46 m diameter. If we take as a reference example the largest boulders found downstream of the Fonds de Quarreux, down to the Campine, i.e. blocks of $\sim$2 m in b-axis, 3.5 m^3 and $\sim$9500 kg in approximate volume and weight (De Brue et al. 2015), we find that an ice raft of $\sim$70 m^3 was required to bear one such block, at the exclusion of any other rock material.

Another constraint for ice rafting transport is that the ice sheet should not be broken through by the loading quartzite boulder, which essentially depends on ice thickness. Based on full-scale observations, Gold (1971) showed that the upper bound for failure of ice cover under a concentrated load is given by the relation

$$P = 350\,h^2 \quad (6.6)$$

with P = breakthrough load in kN and h = ice thickness in m, while the lower bound for successful use of ice cover to support a load corresponds to the relation

$$P = 1750\,h^2 \quad (6.7)$$

thus providing a realistic empirical range for the required minimum ice raft thickness to bear a boulder. Theoretical considerations allowed Sodhi (1995) to reach similar conclusions. According to these relations, the minimum ice thickness for successful ice rafting of our reference boulder is therefore between 0.3 and 0.6 m.

In the present climate, Ardennian rivers may freeze during the coldest winters, such as those of 1890–91 (Lohest 1891), 1908, 1917, 1927–28, 1956, 1963, 1985. Such very cold winters were probably much more frequent during the Little Ice Age, especially in the Seventeenth century (e.g. Robinson 2005). Observations show that ice starts to form after about ten days of continuously negative temperature (Fig. 6.8). When a continuous ice cover has been formed, ice breakup, especially following fast thawing, may produce very large ice blocks that ice jam may cause to land in the floodplain (Fig. 6.9). For instance, a 48 m^3 piece of ice sheet (12×5 m^2 surface $\times$ 0.8 m thickness) was driven out of the Ourthe channel and abandoned in the floodplain at Tilff during the ice breakup of 1956 (ACT 2016). According to the figures given above, it would have been able to carry a block of $\sim$2.4 m^3 (1.65 m in diameter, $\sim$6500 kg). However, we note that, in the Ourthe, the top of such a block lying on the channel bottom is about 1.5 m below the water surface and thus would not have been reached by the bottom of a 0.8-m-thick ice cover. Therefore, whereas ice rafts might occasionally entrain boulders up to 1.5–2 m in size in the shallower stream reach of the Fonds de Quarreux, this process cannot currently be held responsible for mobilizing them in the deeper channels of the Lower Amblève, the Ourthe, and the Meuse. Moreover, as shown by Fig. 6.9, the large pieces of ice cover needed to entrain big boulders are very rapidly broken into smaller pieces unable to bear them, which strongly limits the effectiveness of the process under

Fig. 6.8 The Amblève channel frozen in the boulder-bed sector of the Fonds de Quarreux on February 1, 2006

Fig. 6.9 Ice breakup in the Ourthe between Jupille and Marcourt on March 9, 1947 (*Photograph* M.B. Mormont)

the present temperate climate and suggests that the largest ice-rafted blocks of ~0.25 m^3 observed by Lohest (1891) are much closer to the actual ice rafting competence.

The conditions of ice rafting were clearly different during the cold periods. Based on estimates of the mean annual air temperature (MAAT) at sea level in our regions during the Younger Dryas (Isarin 1997), one derives an MAAT of ~ −3.5 °C in the Fonds de Quarreux during this period, which implies a still lower MAAT during the Last Glacial Maximum. This allowed in principle the formation of ice covers much thicker than today, probably freezing the entire channels during each winter. However, the periglacial conditions of the time also entailed very high sediment loads and the replacement of the interglacial meandering channels by braided channels, whose depths certainly did not exceed 1–2 m, and probably still less in the Fonds de Quarreux area. Indeed, at places where the floodplain width is <50 m, considering the channel width to depth ratio control on braiding (Engelund and Skovgaard 1973; Fredsoe 1978; Fukuoka 1989), channel depth is expected to have been <1 m, which in turn limited the potential thickness of ice rafts. However, such shallow depths allowed winter ice covers to enclose more readily and actively displace blocks resting on the channel bottom (at least those not exceeding ~2 m in size) everywhere in the drainage system down to the Campine.

6.4.4 Remaining Issues and Other Potential Transport Modes

The above reasoning has only considered the ability of ice rafts to overcome the gravity control on boulder transportation. Another constraint might in fact have strongly limited the efficiency of ice rafting, namely the necessity for an ice sheet nearly as thick as the channel in which it flows (especially in the Amblève), and thus unable to lift the boulder significantly, to overcome the resistance of the gravel sheet in which this boulder is partly embedded before entraining it. In such a case, either the ice has enclosed not

Fig. 6.10 Valley bottom of the completely frozen Ninglinspo at the place called 'Hermes bath', February 10, 2012

only the block but also sediments of the gravel sheet, thus reducing the portability of the block, or the ice boulder ensemble has to scrape the upper part of the gravels, an effect that has to be quantified prior to any evaluation of the effectiveness of ice rafting in the Fonds de Quarreux.

These uncertainties suggest that the system probably is in the grey zone where both fluvial and ice rafting processes are at the limit of their domain of effectiveness and that boulder transport was further promoted by other processes. Among these, De Brue et al. (2015) mention that, during ice breakup, ice jams may form, which more or less completely block sediment and water until pressure building and/or thermal weakening cause them to fail in succession, inducing catastrophic flood surges. The latter are very powerful and greatly increase the river competence because they combine extremely high water discharges and the punching effect of large ice masses impacting the boulders protruding from the bedload top surface (Shen et al. 2000; Turcotte et al. 2011).

Another noteworthy intriguing observation is that the many boulders found in the Meuse Main Terrace of the Campine Plateau, dated around 725 ka by $^{10}Be/^{26}Al$ (Rixhon et al. 2011), were necessarily transported before the Mid-Pleistocene climatic deterioration, i.e. in a time when milder cold periods did not favour neither effective ice rafting nor exceptionally high Q_{max}. Furthermore, the Ardenne had not yet undergone uplift acceleration and river gradients were on average lower than today, which further limited maximum unit stream power. Although field evidence from such remote times is barely available to trace back other potential boulder transport processes, we envision two possible research avenues unexplored so far in the Ardenne. A first one should deal with the behaviour of less stable boulders on a bare bedrock channel when the rivers, especially the Amblève, were actively incising to form their successive terraces. Valuable results should also arise from exploring how the bridge scouring effect can create a complex pattern of vortices around a boulder, removing the gravels that support the boulder on its downstream face and destabilizing it (e.g. Romero et al. 2012).

6.5 Other Sites of Geomorphic Interest in the Amblève Valley Near the Fonds de Quarreux

Before closing this chapter, we briefly describe two sites of interest in the vicinity of the Fonds de Quarreux. A first one is the already mentioned valley of the Ninglinspo, which joins the Amblève just downstream of the boulder-bed stream reach. Registered in the Exceptional Heritage of Wallonia since 1949, the Ninglinspo has all characteristics of a mountain bedrock stream with picturesque cascades and plunge pools (Rixhon and Juvigné 2010) (Fig. 6.10). Born from the confluence of two brooklets coming from the Vecquée ridge, it flows down the steep slopes of the ridge from 306 m asl to 152 m where it joins the Amblève after a 2.5-km-long course at mean gradient of 6.2%. The headwaters of the Ninglinspo are thus located on the very old topography of the Vecquée ridge (Fig. 6.11), inherited from a Late Cretaceous erosion surface (see Chap. 5) and where some of the lowest-lying lithalsas have been mapped (see Chap. 7).

Another much visited site of geomorphic interest is a cascade of the Amblève at Coo, 20 km upstream of the Fonds de Quarreux. Though only 12 m high, it is the highest cascade of Belgium, caused by a meander cutoff. This cutoff was in large part natural but it might be that the monks of the abbey of Stavelot-Malmedy accelerated the process or perfected its result, however, in an unknown way and at an unknown time. According to Rahir (1909), small cracks would have appeared naturally in the ground at the cutoff

Fig. 6.11 WNW-looking view of the wooded Vecquée ridge and, at its foot, the E–W course of the Amblève valley. The Fonds de Quarreux lie behind the far end of the visible part of the ridge

point, which one of the last Prince-Abbot of the Stavelot-Malmedy principality, Jacques de Hubin, ruling from 1766 to 1787, made people enlarge in the second half of the eighteenth century. Nevertheless, the more reliable description of the cascade given in a regional work by the Baron of Pöllnitz (1734) and accompanied with a lithography showing the two separate chutes (Fig. 6.12) presents this "spectacularly wild" cascade as perfectly natural half a century before the purported intervention of the monks. This work, which was acclaimed throughout Europe, remarkably describes the lifestyle of the high and educated society that came at Spa to take the waters and attests that the Coo site was already highly touristic in the eighteenth centuries. Pöllnitz also mentions that a mill existed in the first half of the century, taking advantage of the chutes. Cosyn and Cosyn (*sd*) confirms this, writing that the mill is cited as soon as in 1492, thus confirming the older age and likely natural origin of the cascade.

Fig. 6.12 Lithography by M.A. Xhrouet showing the full development of the Coo cascade already in the first half of the eighteenth century, taken from Pöllnitz (1734)

6.6 Conclusion

The boulder-bed stream reach of the Fonds de Quarreux is essentially inherited from the Middle and Late Pleistocene cold periods. Indeed, periglacial conditions prevailed when big boulders of Cambrian quartzite came down from the hillslopes of the Amblève valley to clutter its bottom and its channel, always close to the hillslope base in this narrow sector. Transportation of these boulders is attested by their presence in the Meuse terraces as far downstream as in the Campine. We have shown that purely fluvial transport processes could displace only boulders up to 1–1.5 m in size during the largest floods of the cold periods and that the boulders up to 2 m found in the Campine required other transport processes, among which ice rafting during cold periods was most likely predominant, probably supplemented by catastrophic flood surges induced by ice jams during ice breakup. The morphometric characteristics of the Amblève channel probably determined the maximum size of ice rafts, which in turn limited the transport to boulders ≤ 2 m, a value defining the competence of the process. However, these transport processes require very cold conditions that were probably not met during transportation of the blocks now found in the Main Terrace of the Meuse, suggesting that other processes probably also played a role.

References

ACT (Action Culturelle Tilffoise) (2016) Il y a soixante ans, hiver 1956, souvenirs. Contact 2016(1):11–13

Asselberghs E (1946) Carte géologique de l'Eodévonien de l'Ardenne et des régions voisines, Mémoire de l'Institut de Géologie de l'Université de Louvain 16:598p

Asselberghs E, Geukens F (1960) Geological map 159 Harzé-La Gleize, 1/25.000

Bravard JP, Petit F (1997) Les cours d'eau. Dynamique du système fluvial. Armand Colin, Paris, 222 p

Cailleux A, Hamelin L (1967) Périglaciaire actuel sur le littoral du Bic (Bas-Saint-Laurent). Cahiers de Géographie du Québec 11 (23):361–378

Collard S (2002) Contribution à l'étude géomorphologique du bassin de la Chefna et de la vallée de l'Amblève dans les Fonds de Quarreux. Unpublished M.Sc. thesis, University of Liege, 187 p

Collard S, Juvigné E, Marion JM, Mottequin B, Petit F (2012) L'origine des mégalithes du Fond de Quarreux (Ardenne, Belgique). Bull Soc Géog Liège 58:33–51

Commons G (1961) The Myers formula and Myers rating compared with actual floods. J Geophys Res 66:4322–4323

Costa J (1983) Palaeohydraulic reconstruction of flasflood peaks from boulder deposits in the Colorado Front Range. Geol Soc Am Bull 94:986–1004

Cosyn P, Cosyn M (sd) La vallée de l'Amblève. Guides Cosyn, Brussels, 112 p

De Brue H, Poesen J, Notebaert B (2015) What was the transport mode of large boulders in the Campine Plateau and the lower Meuse valley during the mid-Pleistocene? Geomorphology 228:568–578

Demoulin A (1995) Les surfaces d'érosion méso-cénozoïques en Ardenne-Eifel. Bull Soc Géol Fr 166:573–585

Dewalque G (1885) Sur la présence de stries glaciaires dans la vallée de l'Amblève. Ann Soc Géol Belgique 12:157

Dionne JC (1979) Les blocs d'estran à Saint-Fabien-sur-Mer, estuaire du Saint-Laurent, Québec. Marit Sed 15:5–13

Dumont A (1847) Mémoire sur les terrains ardennais et rhénan de l'Ardenne, du Rhin, du Brabant et du Condros. Première partie: Terrain Ardennais. Mém Acad Royale Sciences, Lettres, Beaux-Arts Belg 20, 163 p

Engelund F, Skovgaard O (1973) On the origin of meandering and braiding in alluvial streams. J Fluid Mech 57:289–302

Ferguson R (2005) Estimating critical stream power for bedload calculations in gravel-bed rivers. Geomorphology 70:33–41

Flammarion C (1891) L'hiver de 1890–1891. L'Astronomie 10:43–57

Fourmarier P (1958) Geological map 148 Louveigné-spa, 1/25.000

Frécaut R (1971) La Moselle et son basin. Contribution à l'hydrologie et à la dynamique fluviales en milieu tempéré océanique. Service de Reproduction des Thèses d'Etat, Univ Lille III, 19, 840 p

Fredsoe J (1978) Meandering and braiding of rivers. J Fluid Mech 84:609–624

Fukuoka S (1989) Finite amplitude development of alternate bars. In: Ikeda S, Parker G (eds) River meandering, American Geophysical Union, Water Resources Monographs 12, pp 237–265

Geukens F (1963) Geological map 160 Stavelot-Malmedy, 1/25.000

Geukens F (1999) Geological map of the Stavelot Massif

Gob F, Bravard JP, Petit F (2010) The influence of sediment size, relative grain size and channel slope on initiation of sediment motion in boulder bed rivers. Earth Surf Proc Landf 35:1535–1547

Gold L (1971) Use of ice covers for transportation. Can Geotech J 8:170–181

Gumbel E (1960) Statistics of extremes. Columbia University Press, New York, 375 p

Hjulström F (1935) Studies of the morphological activity of rivers as illustrated by the River Fyris. Bull Geol Inst Univ Uppsala 25:221–527

Houbrechts G, Hallot E, Gob F, Mols J, Defechereux O, Petit F (2006) Fréquence et importance du charriage dans les rivières du Massif ardennais. Géographie physique et Quaternaire 60(3):241–251

Houbrechts G, Levecq Y, Vanderheyden V, Petit F (2011) Long-term bedload mobility in gravel-bed rivers using iron slag as a tracer. Geomorphology 126:233–244

Houbrechts G, Levecq Y, Peeters A, Hallot E, Van Campenhout J, Denis AC, Petit F (2015) Evaluation of long-term bedload virtual velocity in gravel-bed rivers (Ardenne, Belgium). Geomorphology 251:6–19

IRSNB SGB (1902) Geological maps of Belgium, 1/40.000

Isarin R (1997) Permafrost distribution and temperatures in Europe during the Younger Dryas. Permafrost Periglac Process 8:313–333

Jarvis C (1926) Flood flow characteristics. Trans Am Soc Civil Engrs 89:985–1104

Kempema E, Ettema R (2011) Anchor ice rafting: observations from the Laramie River. River Res Applic 27:1–10

La Garde M (1858) Le Val de l'Amblève. Histoires et scènes ardennaises. Tome second, Schnée, Brussels, 225 p

Lamberty P, Marion JM (2014) Les formations cambro-ordoviciennes du Massif de Stavelot et le conglomérat permien (Poudingue de Malmedy) sur les cartes Harzé – La Gleize et Stavelot – Malmedy. Livret-guide de l'excursion du Comité de Suivi du 5 novembre 2014, Carte géologique de Wallonie, 20 p

Legrand R (1965) Geological map 233 Saint-Vith-Schönberg, 1/25.000

Lohest M (1891) Sur le transport et le déplacement des cailloux volumineux de l'Amblève. Ann Soc Géol Belgique 18:107–111

Lyell C (1868) Principles of geology, 9th edn. Appleton and Co, New York, 834 p

Macar P (1946) Principes de géomorphologie normale. Vaillant-Carmanne, Liège, 304 p

Macar P (1969) Actions périglaciaires et évolution des pentes en Belgique. Biuletyn Peryglacjalny 18:137–152

Mao L, Uyttendaelle G, Iroumé A, Lenzi M (2008) Field-based analysis of sediment entrainment in two high-grade streams located in Alpine and Andine environments. Geomorphology 93:368–383

Marion JM (2016) Geological map of Wallonia 49/3-4 & 7-8, interim document (personal communication)

National Geographic Institute (2000) Topographic map 1/10.000 49/3 S

National Geographic Institute (2001) Topographic map 1/10.000 49/7 N

New brunswick subcommittee on River Ice (2011) New brunswick river ice manual. Environment Canada NB, Inland Waters Directorate, Department of Environment, 33 p

Pardé M (1933) Fleuves et rivières. Armand Colin, Paris, 224 p

Pardé M (1960) Sur la puissance des crues au Japon, à Madagascar et à La Réunion. Int Assoc Sci Hydro Bull 5:38–51

Pardé M (1962) Sur les crues exorbitantes qu'éprouvent certaines rivières des Etats-Unis, notamment au Texas. Int Assoc Sci Hydro Bull 7:17–33

Pauquet A, Petit F (1993) Evolution et fréquence des inondations de l'Ourthe inférieure. Bull Soc belge d'Etudes géogr 62(2):361–375

Petit F, Gob F, Houbrechts G, Assani A (2005a) Critical unit stream power in gravel-bed rivers. Geomorphology 69:92–101

Petit F, Hallot E, Houbrechts G, Mols J (2005b) Evaluation des puissances spécifiques de rivières de moyenne et de haute Belgique. Bull Soc Géogr Liège 46:37–50

Pissart A (1961) Les inondations dans la région de Verviers-Eupen. Etude préalable à un aménagement du territoire. Bull CEBEDEAU 123:62–75

Pöllnitz KL. (Baron of) (1734) Les amusements des eaux de Spa. Tome second, P. Mortier, Amsterdam, 515 p

Rahir E (1909) L'Amblève et l'Ourthe. J. Lebègue, Brussels, 306 p

Rixhon G, Juvigné E (2010) Periglacial deposits and correlated processes in the Ninglinspo valley (Ardenne, Belgium). Geologica Belgica 13(1):49–60

Rixhon G, Braucher R, Bourlès D, Siame L, Bovy B, Demoulin A (2011) Quaternary river incision in NE Ardennes (Belgium)—insights from $^{10}Be/^{26}Al$ dating of river terraces. Quat Geochronol 6:273–284

Robinson P (2005) Ice and snow in paintings of little ice age winters. Weather 60(2):37–41

Romero M, Canedo N, Segales R (2012) Scouring processes around a semi-circular nose pier with attack angles. In: Murillo Muñoz R (ed) River flow 2012. CRC Press/Balkema, Leiden, pp 951–958

Shen H, Su J, Liu L (2000) SPH simulation of river ice dynamics. J Comput Phys 165:752–770

Shields A (1936) Anwendung der Aehnlichkeitsmechanik und der Turbulenzforschung auf die Geschiebebewegung. Mitt. der Preuss. Versuchsanstalt für Wasserbau und Schiffbau 26, Berlin

Sodhi D (1995) Breakthrough loads of floating ice sheets. J Cold Reg Eng 9(1):4–22

Stainier X (1895) Stries pseudo-glaciaires en Begique. Bull Soc belge Géol Paléontol Hydrol 10:212

Sundborg A (1956) The river Klarälven, a study of fluvial processes. Geogr Ann 38:127–316

Sundborg A (1967) Some aspects on fluvial sediments and fluvial morphology: 1. General views and graphic methods. Geogr Ann 49A:333–343

Turcotte B, Morse B, Bergeron N, Roy A (2011) Sediment transport in ice-affected rivers. J Hydrol 409:561–577

Van den Broeck E (1900) Contributions à l'étude des phénomènes d'altération dont l'interprétation erronée pourrait faire croire à l'existence de stries glaciaires. Bull Soc belge Géol Paléontol Hydrol 13:328

Vandenven G (1990) Explications de la carte géologique du Synclinorium de l'Eifel (Région de Gouvy - Sankt-Vith - Elsenborn). Annales de la Société géologique de Belgique 113:103–116

Williams G (1983) Paleohydrological methods and some examples from Swedish fluvial environments, I. Cobble and boulder deposits. Geogr Ann 65A:227–243

The Periglacial Ramparted Depressions of the Hautes Fagnes Plateau: Traces of Late Weichselian Lithalsas

7

Alain Demoulin, Etienne Juvigné and Geoffrey Houbrechts

Abstract

Once thought to be of human origin, ramparted circular or elongate depressions that densely cover many, but not all, summit surfaces of the Hautes Fagnes area, NE Ardenne, have been recognized since the 1950s to represent relics of periglacial mounds. We examine first the regional distribution of these periglacial ramparted depressions (PRDs), whose location seems controlled mainly by elevation and lithologic factors. Geomorphological characteristics of the PRDs and trenching data obtained from several ramparts consistently show that these landforms initially developed as lithalsas through accumulation of segregation ice in the mineral soil, whose later thaw produced the presently observed ramparted depressions. ^{14}C dating of, and pollen data from, peat layers pre- and postdating the formation of the periglacial mounds, further supported by the presence of the 12.9-ka-old Laacher See tephra in the underlying peat, point to a period of rapid formation during the second half of the Younger Dryas. Mean annual air temperature and mean temperature of July derived for the Hautes Fagnes area ($\sim$600 m asl) during the Younger Dryas from a compilation of NW European pollen data are in excellent agreement with the temperature values observed in the western Labrador Peninsula where lithalsas are currently active, showing that PRDs are valuable paleoclimate indicators.

Keywords

Periglacial geomorphology • Periglacial ramparted depression • Lithalsa • Younger Dryas • Paleoclimatic indicators • Hautes Fagnes

7.1 Introduction

At the turn of the twentieth century, it was seriously debated whether the Ardenne had been covered by an ice cap during the Quaternary glacial periods (Dewalque 1885; Lohest 1896; Stamm 1912). However, a few decades later, with the rise of periglacial geomorphology, the main features on which this assertion relied, namely 3- to 10-m-thick mud bodies occurring in many valleys flowing down the Hautes Fagnes Plateau and showing an accumulation of large blocks at their surface, were identified as periglacial debris flows (Fourmarier 1933; Pissart 1953). From this time, observations of periglacial forms (cryopediments) and sedimentary structures (cryoturbations, ice wedge casts, stratified slope deposits) multiplied rapidly (see a synthesis in Pissart 1995), yielding strong support to the idea that periglacial conditions had reigned in Belgium during the glacial periods. In this context, studying ramparted circular depressions that occur in dense fields over many flat summit surfaces of the Hautes Fagnes area and that were previously believed to attest

A. Demoulin (✉) · G. Houbrechts
Department of Physical Geography and Quaternary, University of Liège, Sart Tilman, B11, 4000 Liège, Belgium
e-mail: ademoulin@ulg.ac.be

G. Houbrechts
e-mail: G.Houbrechts@ulg.ac.be

E. Juvigné
University of Liège, Liège, Belgium
e-mail: ejuvigne@skynet.be

A. Demoulin (ed.), *Landscapes and Landforms of Belgium and Luxembourg*,
World Geomorphological Landscapes, DOI 10.1007/978-3-319-58239-9_7

prehistoric man action (Bouillenne and Bouillenne 1937), Pissart (1956) was the first to suggest that they were relics of periglacial mounds, which he initially interpreted as pingos.

Here, we describe these very particular and highly rare landforms in their geological and paleoenvironmental setting and expose the current understanding of their forming mechanisms. Referring to their modern analogues, we then underscore their use as paleoclimatic proxies, before briefly showing how new observations (re)opened important questions about their history and the conditions required for their formation. Following a recent use (e.g. Bryant and Carpenter 1987; Ross et al. 2011), we call this relict morphology of thawed ground-ice mounds now identified as lithalsas (Pissart 1998) periglacial ramparted depressions (PRDs) in order to distinguish them from non-ramparted depressions of simple thermokarstic origin (French 2007). The remarkable state of preservation of lithalsa-derived PRDs in the Hautes Fagnes, the fundamental multidisciplinary research conducted there for 60 years, and the rarity of these forms elsewhere in the world have ensured worldwide geomorphological recognition to the region.

7.2 Geographical and Geological Setting

PRDs have been mapped essentially in two areas of NE Ardenne, namely the Hautes Fagnes Plateau (Hohes Venn in German, culminating at 694 m asl at Botrange) and its western continuation by the Vecquée ridge, and the smaller Baraque de Fraiture, or Tailles, massif (with a maximum elevation of 652 m asl), some 25 km south of the Vecquée (Fig. 7.1). They are typically found on extended flat or very gently sloping (<2°) summit surfaces that make the bulk of the upland topography, incised by the drainage network especially along its margins, and correspond to remnants of a pre-Cretaceous erosion surface partly regraded in the Early Paleocene (see Chap. 5). In the Hautes Fagnes, this surface truncates the Revinian quartzites and phyllites of the core of the Caledonian Stavelot massif, while it cuts across Middle to Late Ordovician shales, and Lochkovian coarse quartzites of the Variscan envelope of this massif at Baraque de Fraiture. In both areas, a thick kaolinic weathering mantle of Early Cretaceous to Middle Miocene age (Yans 2003) is still preserved over large parts of the surface, with thicknesses ranging from 1–2 to several tens of metres (see Chap. 5). Its overall clayey texture tends to include a fine sand component in proportion with the abundance of quartzite layers, while it generally gives way to a mixture of clay and more or less fragmented rocks where the mantle thins, thus exposing the weathering front, or where thick quartzite layers predominate. Locally, the Mesozoic weathering mantle is still overlain by a few metres of clay-with-flint, in situ residues of the Upper Cretaceous chalk cover that once buried the erosion surface (Fig. 7.1). Moreover, 10- to 25-m-deep boreholes in various locations have repeatedly shown that the weathering products are mixed in the upper metres with a predominant silt fraction attesting the presence of reworked loess deposits (Pissart 1974; Juvigné and Pissart 1979). Finally, owing to their mostly impervious substrate and high rainfall (P_{ann} > 1400 mm in the Hautes Fagnes), the flat top of the Baraque de Fraiture and, still more, the summit surfaces of the Hautes Fagnes are extensively covered by Holocene ombrotrophic peat bogs in the form of either a thin blanket bog or more localized active raised bogs.

7.3 The Ramparted Depressions: Characteristics and Distribution

The most typical PRDs are almost perfectly circular depressions occurring on quasi-horizontal surfaces (<1°) (Fig. 7.2). Their diameter varies between 30 and 90 m and the peripheral ~10-m-wide rampart generally rises 1–3 m above the current surface of the depression and the surrounding ground. However, the overall high density of forms is often such that they are contiguous or coalesced into bilobate or more irregular depressions enclosed by sinuous ramparts (Fig. 7.3). On surfaces sloping between 1° and 2°, the circular forms give way to slope parallel elongate depressions up to 500–600 m in length (Fig. 7.3). Their rampart, poorly developed at the upslope end, well-defined and sinuous along both sides of the depression, and sharply closed at its downslope end, suggests a downslope development of the forms, including the possible coalescence of a string of smaller individuals.

Beyond their topographic signature, PRDs are also highlighted in the field as well as on aerial photographs and satellite images by the contrasted vegetation that grows on the surrounding ground, the ramparts, and in the depressions respectively. Within a moor landscape dominated by *Molinia caerulea* and dotted with shrubs, ramparts appear densely covered by varied species of *Vaccinium* and by *Calluna vulgaris* attesting the drier character of their mineral substrate. As for the depressions, if not completely filled, they generally host more or less floating *Sphagnum* mats and *Eriophorum* sp., however, rapidly replaced by the invading *Molinia* as soon as a depression is drained. In any case, sounding the ground of a PRD with a rod shows that it is made of a several metres thick peat fill.

The spatial distribution of PRDs in NE Ardenne is mainly known from field investigation and mapping from aerial photographs. However, in a ground-penetrating radar survey covering flat summit surfaces of the Baraque Michel area (Fig. 7.1) whose topography betrays only few PRDs, Wastiaux et al. (2000) revealed that innumerable similar forms existed beneath a thin continuous peat layer levelling the

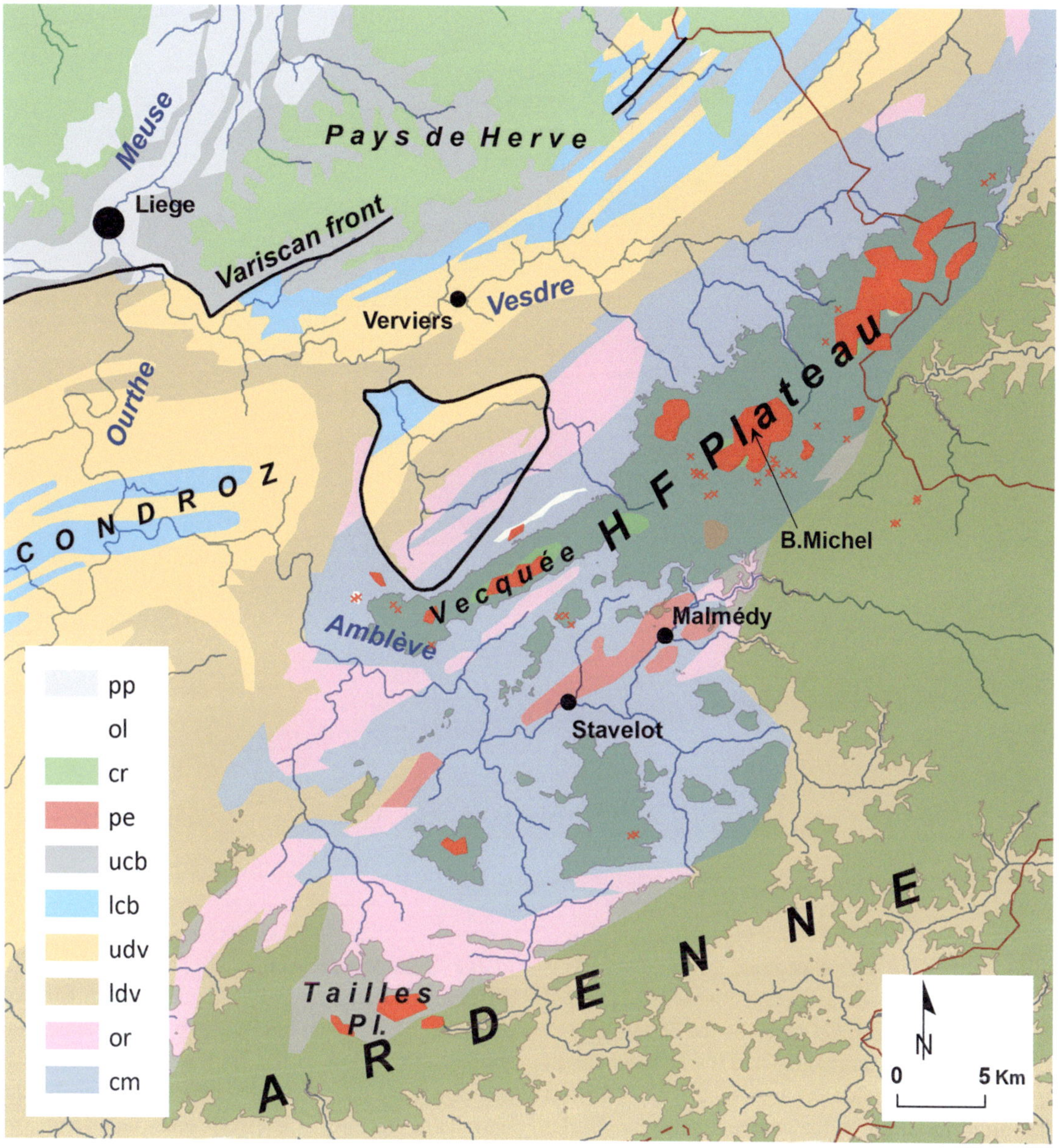

Fig. 7.1 Map of PRD fields (*in red*) and isolated forms (*red crosses*) in NE Ardenne, highlighting their distribution with respect to geology (*cm* Cambrian. *or* Ordovician. *ldv* Lower Devonian. *udv* Middle and Upper Devonian. *lcb* Lower Carboniferous. *ucb* Upper Carboniferous. *pe* Permian. *cr* Cretaceous. *ol* Oligocene. *pp* Plio Pleistocene) and elevation (all areas above 500 m asl are outlined and appear in *greenish* transparency)

surface, making their density as high there as in the NE part of the massif. Moreover, the forests covering an important fraction of the region had so far largely prevented any exhaustive PRD mapping in these areas. The Lidar DTM of Wallonia recently made publicly available (SPW 2015) should soon remedy this lack of information (Fig. 7.3) but, although preliminary analysis already confirmed that many PRDs remain to be mapped, this is not expected to change the fundamental characteristics of their distribution, except maybe with respect to the minimum elevation above which periglacial mounds could form.

The PRDs' spatial distribution seems to be controlled by two main factors. First, in the Hautes Fagnes and at Baraque de Fraiture, but also in the very few other nearby places where they have been observed, they overwhelmingly occur above 500 m asl (Fig. 7.1). Only on benches interrupting the northern slope of the Vecquée ridge have a few small PRDs been mapped at 420–450 m elevations. However, the altitudinal control cannot account for the total lack of PRD on the many flat surfaces extending above 500 m asl in the Ardenne's interior and in the nearby Eifel. Therefore, Pissart (e.g. 2009) strongly insisted on the additional control of the

Fig. 7.2 Oblique aerial view of PRDs in the Brackvenn area (see location on Fig. 7.3) showing circular forms on a quasi-horizontal surface. Note a coalesced form on the *right* of the photo

lithology. Indeed, with the exception of a few PRDs occurring on Pragian shales and sandstones, all recorded forms are located in the outcrop areas of the Revinian 3–5 (interlayered phyllites and quartzites) and the Lochkovian of the Baraque de Fraiture zone (coarse sandstones). Pissart (2009) also drew attention to the indirect meaning of elevation, which, beyond a possible climatic effect, determines the presence of a weathering mantle: all remnants of the extensively weathered pre-Cretaceous and Early Paleocene erosion surfaces are situated above 500 m asl. within the Ardenne.

7.4 Trenches in PRD Ramparts

7.4.1 Trench Description

Several trenches were cut across PRD ramparts in the '70s in order to examine their internal structures and unravel the mechanisms at work in the formation and evolution of the forms that finally evolved in PRDs. Since then, one trench unveiling the downslope end of the rampart of an elongate PRD has been preserved for research and didactic purposes in the Konnerzvenn (Fig. 7.3) and serves as a reference (Pissart and Juvigné 1980). Another trench across the rampart of a circular form has recently been cleaned again with a similar aim 700 m to the southwest in the same area. We describe the Konnerzvenn trench, which nicely displays all significant features observed elsewhere and, moreover, includes a peat layer at the base of the rampart, thus offering an opportunity for age determination (Fig. 7.4).

The local substratum is exposed in the bottom of the trench as a homogeneous bluish-grey clayey silt including rock fragments up to 50 cm in size (*a*, Fig. 7.4). In the internal part of the rampart, slight tongue-shaped colour variations of this material suggest outward displacement. Pertaining also to the original ground is then an overlying peat layer that shows a maximum thickness of ~0.3 m (*b*, Fig. 7.4). This peat layer is sloping outward (with respect to the central depression) and shows characteristic ramp and flat deformations. Though gradually thinning and finally fading out toward the depression, it is clearly overturned at its upper end. This peat is in turn overlain by a 0.2-m-thick package of thin layers of alternating grey silt and dark silty peat that show similar sloping and undulating deformation and inward overturn (*c*, Fig. 7.4). Finally, the bulk of the

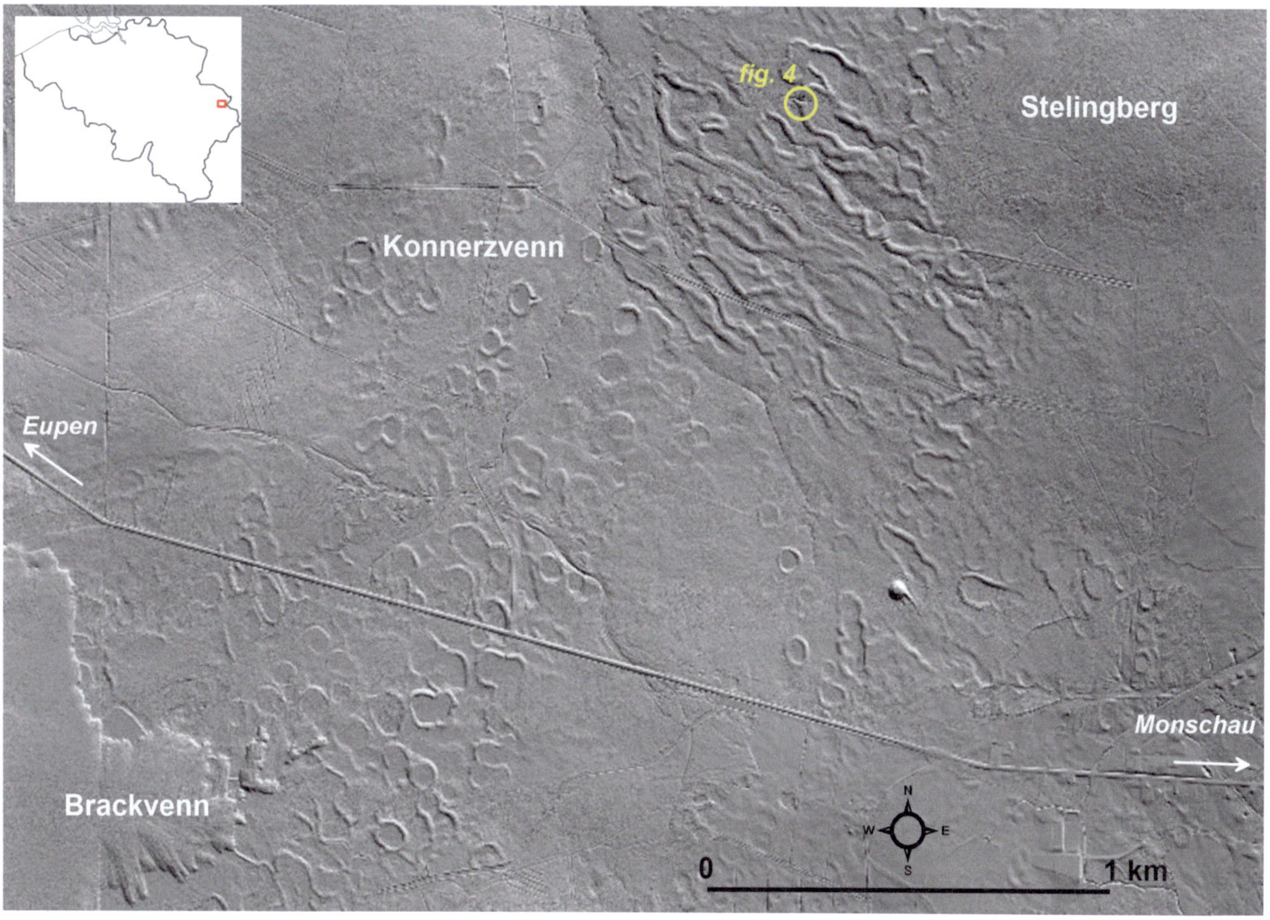

Fig. 7.3 Excerpt of the Lidar DTM of the Walloon Region in the Hautes Fagnes area *east* of Eupen (SPW 2015). Dense fields of round PRDs cover the flat surfaces of the Brackvenn and Konnerzvenn, while elongate forms are characteristic of the gentle footslopes linking the Stelingberg summit (660 m asl) to the subhorizontal Konnerzvenn surface *east* of the nascent creek of the Getzbach

rampart is made of ~2 m of grey to beige clayey silt with a varying part of quartzite gravels and blocks, and showing only faint disordered discontinuities (*d*, Fig. 7.4). On top of it, a thin, poorly developed Holocene soil (*e*, Fig. 7.4) ends up with an O horizon that rapidly thickens toward the depression and goes into a peat fill reaching a maximum depth of 4.2 m (Pissart and Juvigné 1980) (*f*, Fig. 7.4). Grain size (with a mode around 30 μm), heavy mineral assemblages, and the up to 11% carbonate content of the material beneath the bottom peat layer bear witness to the eolian origin and Late Weichselian age of the silt fraction in the shallow substratum and the reworked sediments of the rampart.

7.4.2 Palynostratigraphy

Although no palynological analysis was performed in the peat filling of the Konnerzvenn PRD, equivalent studies were carried out in several other forms of the Hautes Fagnes (e.g. Bouillenne and Bouillenne 1937; Florschütz and Van Oye 1938; Slotboom 1963; Woillard 1975) and Baraque de Fraiture regions (e.g. Mullenders 1963), bringing a consistent picture of their palynostratigraphy. Since the pioneering study of Bouillenne and Bouillenne (1937) in the Baraque Michel area, it was recognized that peat accumulation in the depressions began systematically with the earliest Holocene (Preboreal). However, this timing was long questioned because cold phase pollens were often observed in peaty silts at the very base of the sequence (Florschütz and Van Oye 1938; Woillard 1975) and suggested an end Late Glacial starting time of the filling. The debate was closed when proofs were provided of the presence of Younger Dryas deposits within the PRDs' ramparts (see below), and it is now admitted that the Late Glacial pollens at the base of the filling are reworked, having descended with their host material from the top of the rampart when the depression was created.

As already stated, the value of the Konnerzvenn trench especially lies in the presence of a peat layer at the base of

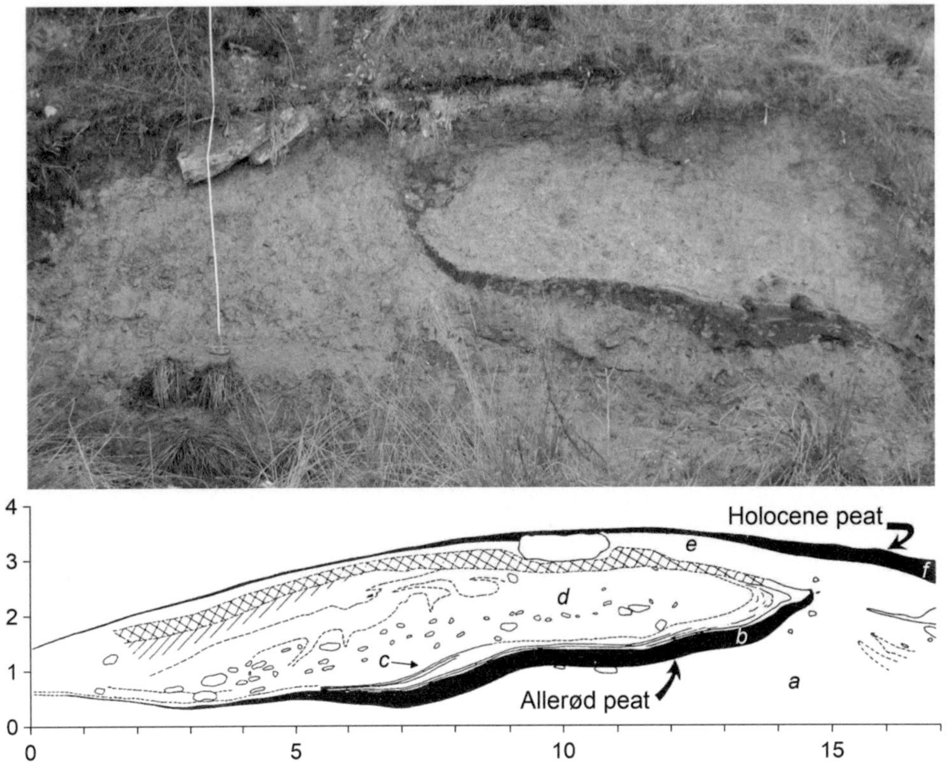

Fig. 7.4 Trench in the downstream part of an elongate periglacial rampart of the Konnerzvenn (see location of the PRD on Fig. 7.3). *Top* View of the southern wall of the trench, refreshed in September 2015 (*Photograph* E. Juvigné). *Bottom* The northern wall of the trench as exposed during initial trenching in 1978 and published by Pissart and Juvigné in 1980. Note in particular the Allerød peat layer predating the periglacial ground deformation and featuring a ramp/flat geometry and overturning of its raised end thinning to the right toward the depression. *Thin lines* above this peat layer, also overturned, correspond to parallel thin peat beds interlayered with clayey silts. *The dashed line* and hatched and cross-hatched areas delineate zones of differently coloured gravelly clayey silt. The big quartzite block on *top* of the trench is typical of the surface material covering the area. The Holocene peat thickens toward the centre of the depression

the rampart. Palynological analyses of this peat identified a vegetational sequence encompassing the period from the Bølling to the beginning of the Younger Dryas (Damblon and Bastin 1986; Damblon 1994), (Fig. 7.5). In this sequence, from bottom to top, a peak of *Selaginella selaginoides* marks the cooler and wetter transition between Bølling and Allerød, followed by the predominance of *Pinus* during the Allerød before the latter species suddenly regresses at the benefit of the *Cyperaceae* at the beginning of the Younger Dryas. Important new findings have recently been brought by Damblon et al. (2014), who examined the pollen and plant remain content of three thin silty peat laminae above the main peat and concluded that they clearly differ from that of the underlying layer and attest to the local vegetation during the early Younger Dryas (Fig. 7.5).

7.4.3 Age Determination

The peat layers in the Konnerzvenn trench offer the additional opportunity of ^{14}C dating of the rampart. First ^{14}C age estimates of the base and top of the main peat were obtained very early (Pissart and Juvigné 1980), additional dating being performed in the following decades. Summarized in Pissart (2000), they yield an age between 15,442 and 13,924 cal BP (from two estimates with their 68% confidence interval, using the IntCal13 curve) for the bottom of the layer, with a most probable value around 14,050 cal BP consistent with the pollen data. Three dates obtained at the height of a thin tephra layer situated $\sim$9 cm below the top of the peat consistently point to a calibrated age in the range 13,036–12,582 cal BP, while two samples taken at the top of the layer give slightly diverging numerical ages of 12,711–12,520 and 12,240–12,016 cal BP. Overall, this age data set is internally coherent and also perfectly matches the palynological results. Heavy mineral analysis of the tephra demonstrated that it originated from the Laacher See eruption in eastern Eifel (Pissart and Juvigné 1980), thus providing an additional chronological tie (Fig. 7.6a). Indeed, Van den Bogaard (1995) gives a $^{40}Ar/^{39}Ar$ numerical age of 12,900 $\pm$ 560 year BP for the Laacher See tephra, which fits well in the ^{14}C chronology of the Konnerzvenn trench.

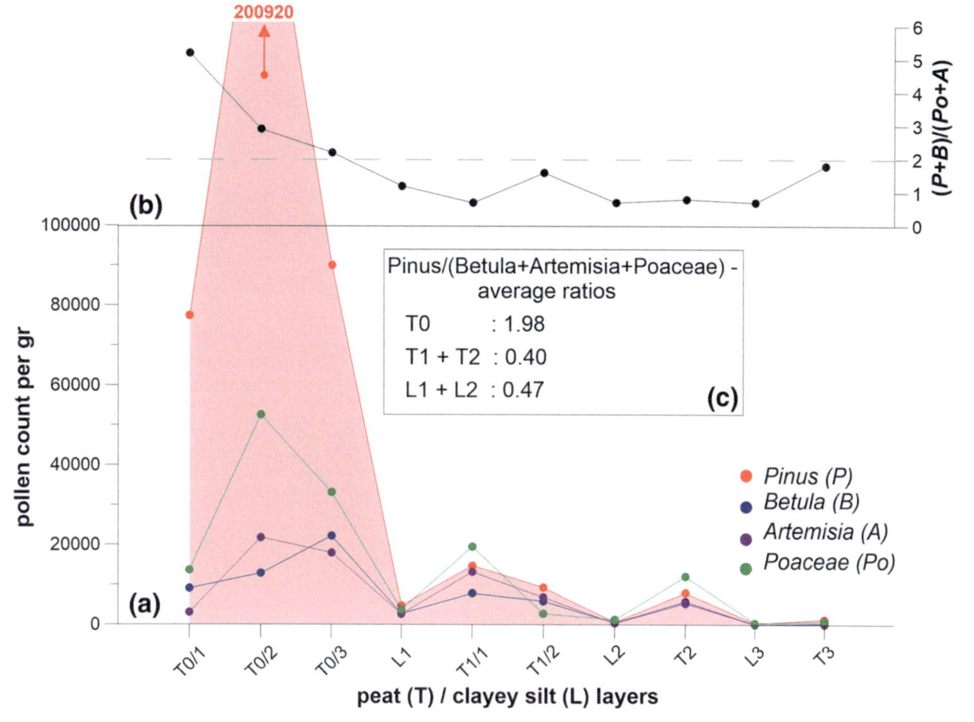

Fig. 7.5 Pollen data from the peat layers deformed in the rampart of the PRD described in the Konnerzvenn (see Fig. 7.4). **a** *Pinus*, *Betula*, *Artemisia*, and *Poaceae* counts in the main Allerød peat T0 and the thin overlying peat (T1, T2, T3) and silt (L1, L2, L3) layers. **b** Ratio between arborean (*Pinus*, *Betula*) and non-arborean (*Artemisia*, *Poaceae*) pollen contrasting the milder climate in which the main peat layer accumulated and the generally colder climate that followed. **c** Ratio between *Pinus* and the other counted pollens. The totally different ratios observed in T0 versus T1 and T2 show that the latter have not been reworked from T0. Their ratio is instead very close to that of L1 and L2

Recently, Juvigné and Streel (2007) produced a new set of ^{14}C ages not only for the thin peat laminae above the main layer but also for the top of the latter. Two clear laminae and a tenuous upper zone of enrichment in organic matter are successively dated to 12,652–12,377, 12,382–12,126 and 12,383–11,986 cal BP (Fig. 7.6b), and two dates agree on an age of 12,779–12,691 cal BP for the top of the thick peat. While the internal consistency of these ages is quite good, they show discrepancies with those previously published by Pissart (2000) for the top of the peat layer, possibly due to laboratory biases.

7.5 Interpreting the Data

7.5.1 Mechanism of Formation

The outward sloping of the sediment layers in the Konnerzvenn trench shows that the PRD's central depression has taken the place of a formerly upheaved mound. Moreover, the overturned upper end of the peat layer and the thin alternate silty/peaty laminae indicates that the slopes thus created were steep enough to allow mass transport to downbend them (Fig. 7.4). Finally, the ramp and flat geometry displayed by the deformed peat layer evidences that episodic lateral growth complemented upheaval of the mound.

The progressive buildup of the mound slopes caused successive sediment transport processes to take place. In the beginning, only runoff, possibly aided by frost creep, was able to occur on the initial gentle ground undulation. It was long held that all thin laminae, both silty and peaty, had been emplaced by runoff (Pissart and Juvigné 1980) but their recently obtained ^{14}C and palynological ages (Juvigné and Streel 2007; Damblon et al. 2014) rather suggest an evolution with alternating silt deposition by runoff under cool episodes and peat accumulation during milder interludes. At a later stage of the mound development, the sharp hinge of the peat overfold and the unstructured appearance of the overlying gravelly/blocky silt mass point to cold conditions and gelifluction as the most probable transport process of a seasonally thawed active layer above the underlying frozen ground of steepened slopes.

The evolution from a mound to the modern depression attests to the former presence of an ice accumulation in the ground, now thawed. While the original hypothesis of pingos formed by injection ice (Pissart 1956) was soon abandoned when it was recognized that the physical conditions

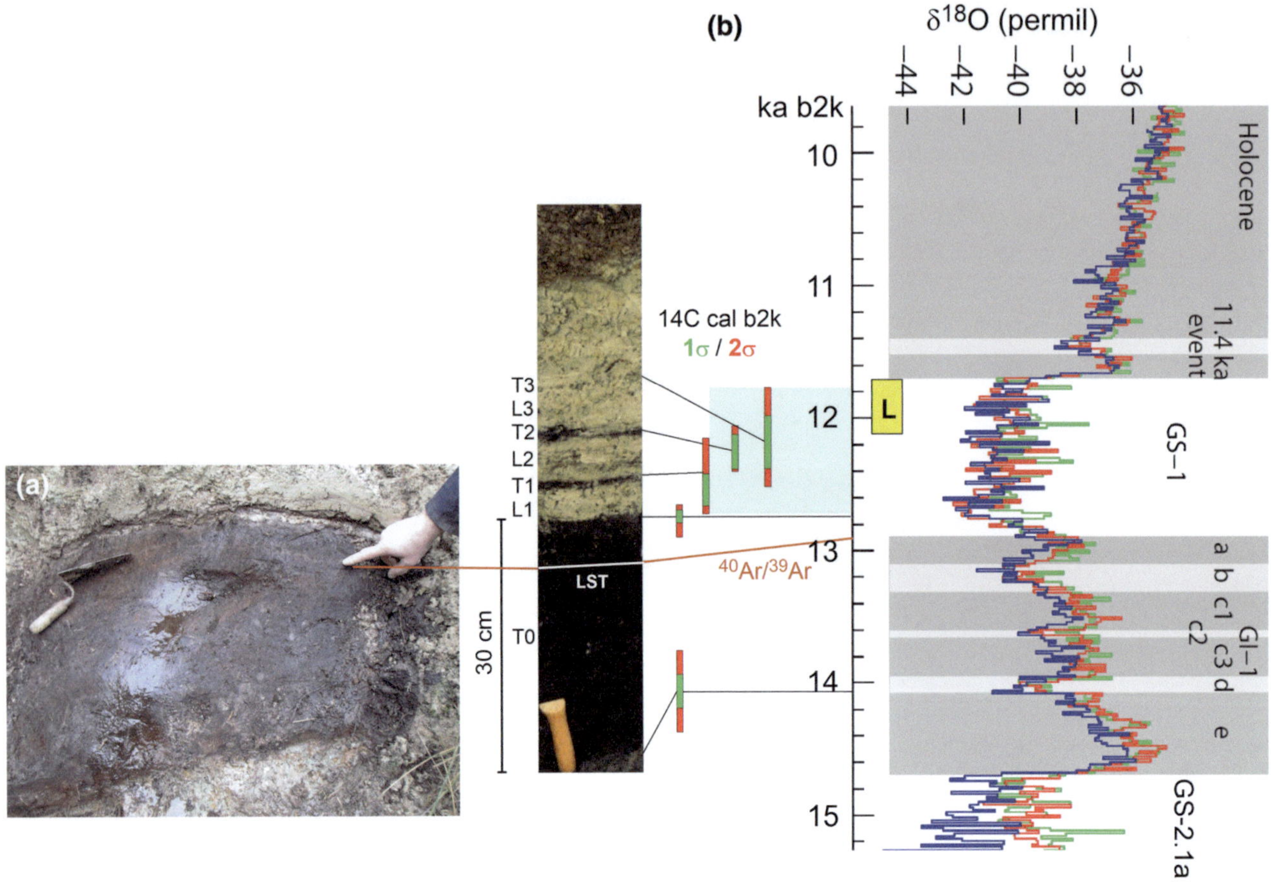

Fig. 7.6 Dating of the rampart of the PRD presented in the Konnerzvenn (Fig. 7.4). **a** Photo of the main peat layer with the Laacher See tephra appearing as a thin beige lamina a few centimetres below the top of the peat (between finger and trowel). **b** Compared with the δ^{18}O curves of three dated ice cores in central Greenland (*blue* NGRIP; *red* GRIP; *green* GISP2) (Rasmussen et al. 2014), the calibrated ^{14}C ages of the peat assign an Allerød (or Greenland interstadial GI-1) age to the main layer, supported by the Ar–Ar age of the Laacher See tephra (LST), and younger ages corresponding to the first half of the Younger Dryas (Greenland Stadial GS-1) to the overlying thin layers (T1, T2, T3). The latter ages suggest that the main phase of lithalsa formation occurred within a few centuries during the second, not necessarily coldest, half of the Younger Dryas (time window noted *L*, *highlighted in yellow*)

for such a process were not met in the Hautes Fagnes area (Pissart 1965, 1974), a large body of evidence (present-day analogues in northern Canada, laboratory experiments, microstructures in the PRD ramparts) now consistently indicates that accumulation of segregation ice was responsible for the development of the frost mounds (Taber 1929; Pissart 2009) (Fig. 7.7). The process of segregating ice in the ground corresponds to the growth of ice lenses in a water-saturated soil by freezing of water drawn from the surrounding substrate through cryosuction. Ice accumulates near the ground surface because water films depleted by freezing at the surface of the forming ice are replenished by capillary water migrating down a thermal gradient through the fine pore and cracks of the nearby unfrozen (though generally below 0 °C) ground toward the growing ice lens. When it comes in contact with the ice at the base of the lens, this water freezes in turn and the process keeps going as long as free water is available, leading to the development of large masses of segregation ice that cause ground heave. Obviously, the first requirement for large-scale ice segregation is thus the juxtaposition of frozen and unfrozen ground, i.e. the presence of discontinuous permafrost where mound growth is related to the development of patches of frozen ground. Other conditions required for the process to work include a slow progression of the freezing front, which is favoured by the exothermic character of the water-to-ice phase change (liberating latent heat of fusion), and a grain size in the clay to silt range, which offers the optimal capillary diameter for water to migrate. In the Hautes Fagnes, the predominant silt fraction of eolian origin in the upper metres of the otherwise clayey cover material makes them especially apt at cryosuction.

Accumulation of segregation ice was identified as the cause of the formation of frost mounds in Finnish Lapland

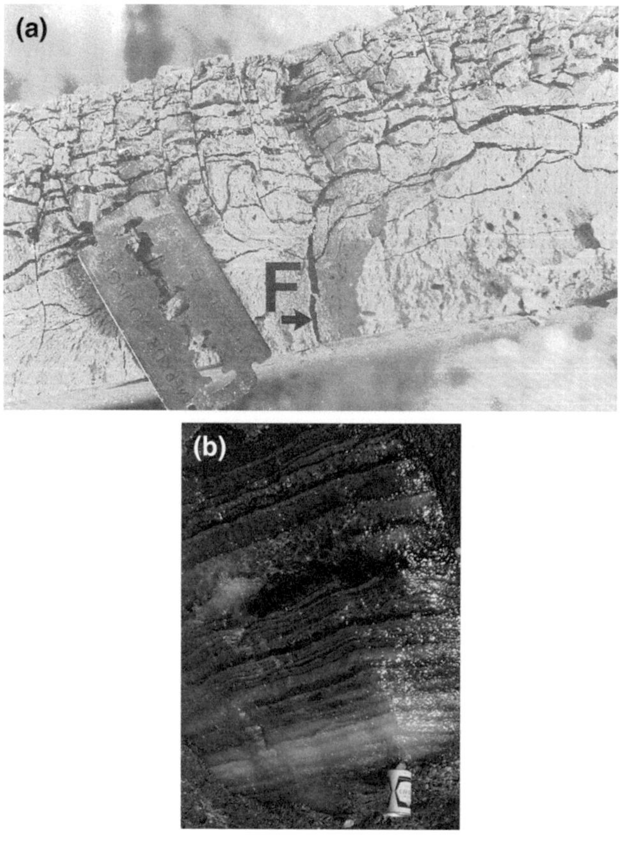

Fig. 7.7 **a** Segregation ice lenses developed upon slow freezing of a silt sample, experiment by Pissart (1987). F. Desiccation crack due to the migration of pore water toward the freezing front. **b** Segregation ice in a frozen soil near Inuvik, northwestern Canada (*Photo* A. Pissart)

by Seppälä (1986). These frost mounds, first described and called palsas by Fries and Bergstrom (1910), display individual and group characteristics very similar to those reconstructed for the cryogenic mounds to which the Hautes Fagnes PRDs bear witness, except that they are developed in peat-covered areas, with a core essentially made of frozen peat. Yet, in the Hautes Fagnes, no such extended peat blanket existed at the time of mound formation (the local peat layer observed in the Konnerzvenn trench is rather an exception and, in any case, disappears toward the centre of the form, where ice accumulated). Consequently, because they result from the same process as palsas but were formed within a mineral substrate, Pissart et al. (1998) called the Hautes Fagnes frost mounds lithalsas, after the name forged by Harris (1993). This difference in the nature of the substrate basically results in a difference in temperature conditions needed for the mounds to form. When peat is present, the contrast between dry peat insulating the ground against summer warmth and heat-conducting wet peat facilitating ground cooling in winter tends to maintain colder soils on average and favours ice segregation for slightly milder air temperatures than in the case of mineral substrates. In other words, lithalsas require harsher conditions than palsas to form.

Finally, would lithalsas have been created solely by vertical upward push of the ground by growing ice lenses, they would have then simply resorbed without leaving traces when their core ice subsequently thawed under warming conditions. The remarkable relict form of the PRDs is therefore by itself clear evidence of the lateral, outward growth component of ice lens and mound, to some extent supplemented by the descent of material on the slopes of the mound.

7.5.2 Timing of Lithalsa Formation and Destruction

The time of lithalsa formation in the Hautes Fagnes was once hotly disputed. Since publication of the first pollen data (Bouillenne and Bouillenne 1937), extensive palynological evidence had been gathered and consistently indicated that the peat fill of the PRDs had started to accumulate in the early Preboreal, after thawing of their frozen core had destroyed the mounds as soon as temperature rose at the end of the Younger Dryas. However, this only defined a minimum age of the lithalsas and authors disagreed on whether they dated back to the Younger Dryas or the Tardiglacial. The debate came to an end when ^{14}C dating of the pre-lithalsa peat layer in the Konnerzvenn trench showed that it had been formed mainly during the Allerød (Pissart and Juvigné 1980), (Fig. 7.6). Furthermore, a heavy mineral assemblage signing the Laacher See tephra ($\sim$12.9 ka) was found not only in this peat layer but also in pre-lithalsa sediments in other trenches across PRD ramparts, strongly suggesting that all observed PRDs derive from a single generation of lithalsas produced during the Younger Dryas in the Hautes Fagnes and Baraque de Fraiture areas. This is also consistent with the similar state of preservation of all forms, whereas older PRDs would be hardly recognizable in the field, if any, due to overprinting by the next generation of cryogenic mounds.

However, Juvigné and Streel (2007) recently claimed that lithalsas were probably also produced during the cold stage that preceded the Bølling-Allerød warming, i.e. around 17–15 ka (Oldest Dryas, or Greenland Stadial 2.1a sensu Rasmussen et al. 2014). Indeed, noting that the Allerød peat layer in the Konnerzvenn trench obviously accumulated in a shallow depression, they argue that this depression attests to the thawing of an older lithalsa. Now, this claim for another generation of frost mounds is not supported by field

Fig. 7.8 Active periglacial mounds in west-central Labrador, close to the Boniface River ~25 km inland from the eastern shore of the Hudson Bay at ~57° 45′N. The photo shows active (*A*) and partly thawed (*T*) lithalsas. Note also the vegetational landscape in which these forms occur (*Photograph* A. Pissart)

evidence: (1) the depth and the diameter of the suspected PRD are much smaller than those of known forms, (2) no trace of the associated rampart is visible in the undisturbed area west of the trench (Fig. 7.3). Moreover, after more than 12 ky of uninterrupted very cold conditions, the Oldest Dryas permafrost was most probably continuous over the Hautes Fagnes area, thus ruling out significant ice segregation at this time.

More enlightening are the ^{14}C ages obtained for the organic laminae overlying the main peat layer (Juvigné and Streel 2007), which shed light on the time needed for a lithalsa to develop. They show that, although limited ice segregation was possibly initiated during cooler, drier time spans within the first half of the Younger Dryas, mound formation was each time aborted by the return of milder, wetter conditions manifested by thin peat accumulations, and it was not until the second, colder half of this stadial (from ~12.2 ka onwards) that the lithalsas really began to grow (Fig. 7.6). This reduces the duration of the period favourable to lithalsa development and aggregation in dense fields to ~0.5 ky. Though apparently short, such a time span is consistent with observations by, e.g. Seppälä (1982) that vertical growth rates of palsas may reach several decimetres per year, with ^{14}C ages of palsas and lithalsas in the modern subarctic regions showing that many of them are only a few centuries old (Allard et al. 1987; Seppälä 2006; Wolfe et al. 2014), and with modelling results indicating that full development of palsas and lithalsas is achieved in a maximum of 200 years (An and Allard 1995).

7.6 PRDs as Paleoclimatic Proxies

Beyond the typical character of these PRD-covered landscapes of the Hautes Fagnes plateau and their geomorphic value, PRDs are also useful indicators of the paleoclimate that reigned over NE Ardenne during the Younger Dryas. Indeed, modern lithalsas are known to develop in very few areas worldwide, mainly northern Canada (Harris 1993; Calmels and Allard 2008; Wolfe et al. 2014) (Fig. 7.8) plus rare occurrences in mountains of central Asia (Wünnemann et al. 2008; Iwahana et al. 2012), which correspond to a narrow range of climatic conditions. Lagarec (1982) noted in Northern Québec that, while palsas chiefly develop in the forest tundra zone, lithalsas are preferentially found in the shrub tundra zone, north of the treeline. Climatic parameters of the latter area meaningful for lithalsa development include mean annual air temperature (MAAT) between −4 and −7 °C, mean temperature of the warmest month between 9 and 10.5 °C, and rather low mean maximum snow thickness of ~0.5 m. While Allard et al. (1987) summarize the required conditions to MAAT <−5 °C and July mean temperature <10 °C, Pissart (2003) slightly tightens the ranges to [−4.3 to −5.8 °C] and [9.5–10.7 °C], respectively. Importantly, he associates the MAAT lower bound with the mapped limit between discontinuous and continuous permafrost and the July 10 °C isotherm with the treeline, thus clearly defining the climatic and ecological conditions needed for lithalsa formation as those of the coldest part of the zone of discontinuous permafrost (Fig. 7.9). This is also in agreement with theoretical optimal ground temperatures between −3 and −8 °C for the growth of ice segregation lenses (Hales and Roering 2007).

Based on these modern observations, Pissart (2003) thus concluded that the Hautes Fagnes and Baraque de Fraiture areas underwent similar MAAT and mean July temperatures when lithalsas developed, i.e. during the colder half of the Younger Dryas (Juvigné and Streel 2007). Taking into account a temperature gradient of 0.6 °C per 100 m of elevation change, these estimates are fully confirmed by partly independent results of Isarin (1997) and Isarin and Bohncke (1999), who give a Younger Dryas MAAT of −2 °C at sea level, yielding a corrected [−5 to −6.2 °C] range for the Hautes Fagnes area, and a mean July temperature of ~13.5 °C corrected for elevation to 9.3–10.5 °C, inferred from a spectrum of periglacial features and pollen and plant macrofossils, respectively (Fig. 7.10).

Finally, an intriguing issue remains: why are summit surfaces of central Ardenne above 500 m asl devoid of PRDs while they are so numerous in the Hautes Fagnes and Baraque de Fraiture massifs? Field evidence contradicts the suggestion by Pissart (1974, 2009) that small perched aquifers in quartzite-derived weathering sands might have provided only in these areas the amount of ground water needed for the buildup of large masses of segregation ice. In this respect, exploring why all PRD areas are aligned on the first high ridgeline that the Ardenne plateau presents to the prevailing winds from WNW and the implications thereof for snow cover and soil saturation during the Younger Dryas might be more promising.

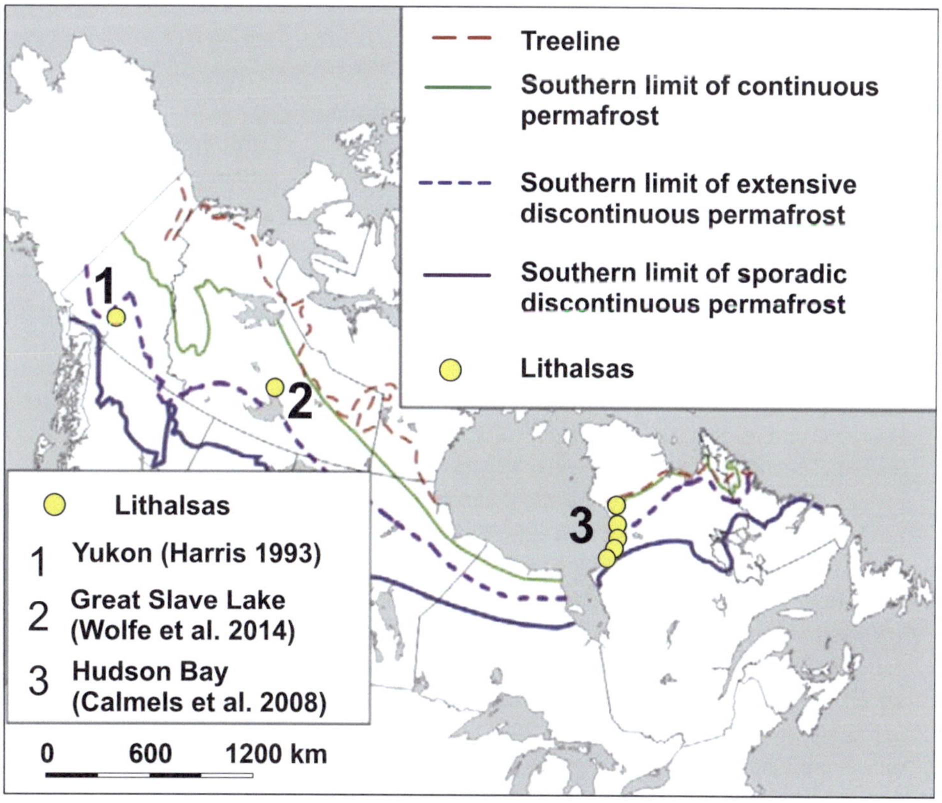

Fig. 7.9 Location of the areas of mapped active lithalsas in Canada with respect to permafrost continuity and summer temperature conditions (*indicated by position of the tree line*) (from Wolfe et al. 2014)

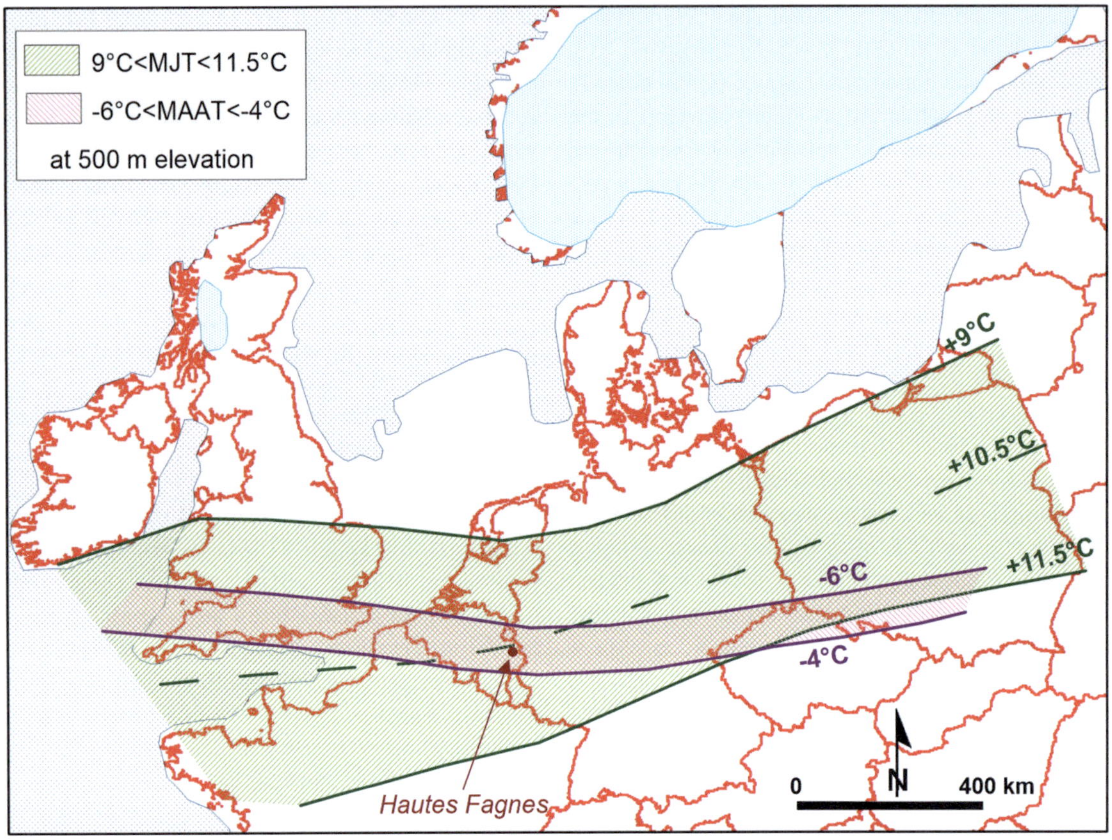

Fig. 7.10 Mean annual air temperature (MAAT) and mean July temperature (MJT) isotherms at 500 m elevation during the Younger Dryas, redrawn after Pissart (2000) from data of Isarin (1997). The Hautes Fagnes lithalsas were formed in the area where the combination of Younger Dryas MAAT (between −4 and −6 °C) and MJT (between +9 and +11.5 °C) was similar to that now observed in northern Canadian regions showing present lithalsa formation. *Light blue dots* ice caps; *middle blue dots* sea ice

7.7 Conclusion: The Geoheritage Perspective

The Hautes Fagnes plateau is the best among the rare places worldwide where PRDs attesting to periglacial environments of the last Glacial are preserved. Moreover, the area hosted a lot of research about not only PRDs but also many other periglacial landforms and highly contributed to their understanding for more than a century. As such, the Hautes Fagnes and their PRDs represent a unique geomorphosite of very high value. This quality led to the extension of the perimeter of the Hautes Fagnes domanial nature reserve first created in 1957 in order to include the zones displaying the most spectacular PRD fields in NE Hautes Fagnes, thus ensuring an adequate frame for their preservation. Since then, large-scale work supported by national and European funds has been achieved with the aim of maintaining and restoring the general character of wet areas of the Hautes Fagnes, which had been largely degraded in the last centuries by agropastoral practices, peat exploitation and, since the nineteenth century, extensive spruce planting. Owing to their real outstanding universal value as natural feature and landscape, an application is currently planned for the inscription of the PRDs of the Hautes Fagnes and the Baraque de Fraiture areas as World Natural Heritages properties of the UNESCO. Meanwhile, it is already not that rare that, walking across the beautiful misty landscapes of the Hautes Fagnes, you meet a naturalist or a geoscientist pursuing research about peat bogs and PRDs.

References

Allard M, Seguin M, Lévesque R (1987) Palsas and mineral permafrost mounds in Northern Québec. In: Gardiner V (ed) International geomorphology 1986. Proceedings of the first international conference on geomorphology, Wiley, Chichester, vol II, pp 285–309

An W, Allard M (1995) A mathematical approach to modeling palsa formation: insights and growth conditions. Cold Reg Sci Tech 23:231–244

Bouillenne R, Bouillenne M (1937) Les viviers du Plateau de la Baraque Michel. Bull de la Soc r des Sci de Liège 12:404–427

Bryant R, Carpenter C (1987) Ramparted ground ice depressions in Britain and Ireland. In: Boardman J (ed) Periglacial processes and landforms in Britain and Ireland. Cambridge University Press, Cambridge, pp 183–190

Calmels F, Allard M (2008) Segregated ice structures in various heaved permafrost landforms through CT scan. Earth Surf Proc Land 33:209–225

Calmels F, Allard M, Delisle G (2008) Development and decay of a lithalsa in Northern Quebec: a geomorphological history. Geomorphology 97:287–299

Damblon F (1994) Les dépôts tourbeux et l'histoire de la végétation sur le plateau des Hautes Fagnes (Belgique). Ann de la Soc géol de Belg 117:259–276

Damblon F, Bastin B (1986) Etude paléoécologique d'une lentille tourbeuse ensevelie dans le rempart d'une palse de la Konnerzvenn (Hautes Fagnes, Belgique). Bull de la Soc R Botanique Belg 120:20–29

Damblon F, Juvigné E, Streel M (2014) Nouvelles données polliniques concernant la formation du lithalse de la Konnerzvenn au cours du Dryas récent (Tardiglaciaire des Hautes Fagnes). Ht Fagnes 3:26–27

Dewalque G (1885) Sur la présence de stries glaciaires dans la vallée de l'Amblève. Ann Soc géol Belg 12:157

Florschütz F, Van Oye E (1938) Over de ouderdomsbepaling van de "vijvers" op het plateau van het Belgisch Hoogveen. Tijdschr K Ned Aardrijksk Genoot 55:454–461

Fourmarier P (1933) Observations au sujet des coulées pierreuses du plateau de la Baraque Michel. Ann Soc géol de Belg 57:B164–B171

French H (2007) The periglacial environment, 3rd edn. Wiley, Chichester 478 p

Fries T, Bergstrom E (1910) Några iakttagelser öfver palsar och deras förekomst i nordligaste Sverige. Geol Fören i Stockh Förh 32:195–205

Hales T, Roering J (2007) Climatic controls on frost cracking and implications for the evolution of bedrock landscapes. J Geophys Res 112:F02033. doi:10.1029/2006JF000616

Harris S (1993) Palsa-like mounds developed in a mineral substrate, Fox Lake, Yukon territory. In: Proceedings of the 6th international conference on permafrost, 5–9 July 1993, Beijing, China, vol 1, pp 238–243

Isarin R (1997) Permafrost distribution and temperatures in Europe during the Younger Dryas. Permafrost Periglac Process 8:313–333

Isarin R, Bohncke S (1999) Mean July temperatures during the Younger Dryas in Northwestern and Central Europe as inferred from climate indicator plant species. Quatern Res 51:158–173

Iwahana G, Fukui K, Mikhailov N, Ostanin O, Fujii Y (2012) Internal structure of a lithalsa in the Akkol Valley, Russian Altai Mountains. Permafrost Periglac Process 23:107–118

Juvigné E, Pissart A (1979) Un sondage sur le plateau des Hautes Fagnes, au lieu-dit "La Brackvenn". Ann Soc géol Belg 102:277–284

Juvigné E, Streel M (2007) Un patrimoine exceptionnel en Europe moyenne: les lithalses des Hautes Fagnes. Hautes Fagnes 4:18–20

Lagarec D (1982) Cryogenetic mounds as indicators of permafrost conditions, northern Québec. In: French H (ed) Proceedings of the 4th Canadian permafrost conference in Calgary, The Roger JE brown memorial volume, Ottawa, Canada, pp 43–48

Lohest M (1896) Des dépôts tertiaires de l'Ardenne et du Condroz. Ann de la Soc géol de Belg 23:37–53

Mullenders W (1963) Note préliminaire sur la palynologie des pingos du Plateau des Tailles (Belgique). Zeitschr f Geom 7(2):165–168

Pissart A (1953) Les coulées pierreuses du plateau des Hautes Fagnes. Ann de la Soc géol de Belg 76:B203–B219

Pissart A (1956) L'origine périglaciaire des viviers des Hautes Fagnes. Ann de la Soc Géol de Belg 79:119–131

Pissart A (1965) Les pingos des Hautes Fagnes: les problèmes de leur genèse. Ann de la Soc géol de Belg 88:277–289

Pissart A (1974) Les viviers des Hautes Fagnes sont des traces de buttes périglaciaires. Mais s'agissait-il réellement de pingos? Ann de la Soc géol de Belg 97:359–381

Pissart A (1987) Géomorphologie périglaciaire. Texte des leçons de la chaire Franqui belge. Lab Géogr Phys Univ Liége 135 p

Pissart A (1995) L'Ardenne sous le joug du froid. Le modelé périglaciaire du massif ardennais. In: Demoulin A (ed) L'Ardenne, Essai de Géographie Physique. Département de Géographie Physique et Quaternaire, Université de Liège, pp 136–154

Pissart A (1998) Les traces de buttes périglaciaires des Hautes Fagnes et le climat du Dernier Dryas (13000 à 11450 ans calendrier BP). Bulletin de la Classe des Sciences de l'Académie. Bull Cl Sci l'Académie. Bruxelles 7–12:395–429

Pissart A (2000) Remnants of lithalsas of the Hautes Fagnes, Belgium: A summary of present-day knowledge. Permafrost Periglac Process 11:327–355

Pissart A (2003) The remnants of Younger Dryas lithalsas on the Hautes Fagnes Plateau in Belgium and elsewhere in the world. Geomorphology 52:5–38

Pissart A (2009) Les viviers des Hautes Fagnes et du Pays de Galles: nouvelles connaissances et hypothèses. BSGLg 53:47–73

Pissart A, Juvigné E (1980) Genèse et âge d'une trace de butte périglaciaire (pingo ou palse) de la Konnerzvenn (Hautes Fagnes, Belgique). Ann Soc Géol Belg 103:73–86

Pissart A, Harris S, Prick A, Van Vliet-Lanoë B (1998) La signification paléoclimatique des lithalses (palses minerales). Biul Peryglac 37:141–154

Rasmussen S et al (2014) A stratigraphic framework for abrupt climatic changes during the last glacial period based on three synchronized Greenland ice-core records: refining and extending the INTIMATE event stratigraphy. Quatern Sci Rev 106:14–28

Ross N, Harris C, Brabham P, Sheppard T (2011) Internal structure and geological context of ramparted depressions, Llanpumsaint, Wales. Permafrost Periglac Process 22:291–305

Seppälä M (1982) An experimental study of the formation of palsas. In: French H (Ed.), Proceedings of the 4th Canadian permafrost conference in Calgary, The Roger JE brown memorial volume, Ottawa, Canada, pp 36–42

Seppälä M (1986) The origin of palsas. Geogr Ann Ser A 68:141–147

Seppälä M, (2006) Palsa mires in Finland. The finnish environment 23| 2006, 155–162

Slotboom R (1963) Comparative geomorphological and palynological investigation of the pingos (viviers) in the Hautes Fagnes (Belgium) and the mardellen in the Gutland (Luxemburg). Z für Geomorphol 7:194–231

SPW (2015) Relief de la Wallonie—Modèle numérique de terrain (MNT) 2013–2014. (http://geoportail.wallonie.be/WalOnMap/)

Stamm K (1912) Glacialspuren im Rheinischen Schiefergebirge. Verh Naturhist Ver Preussischen Rheinlande Westfalens 69:151–214

Taber S (1929) Frost heaving. J Geol 37:428–461

Van den Bogaard P (1995) $^{40}Ar/^{39}Ar$ ages of sanidine phenocrysts from Laacher See tephra (12,900 yr BP): chronostratigraphic and petrological significance. Earth Planet Sci Lett 133:163–174

Wastiaux C, Halleux L, Schumacker R, Streel M, Jacqmotte JM (2000) Development of the Hautes-Fagnes peat bogs (Belgium): new perspectives using ground-penetrating radar. Suo 51:115–120

Woillard G (1975) Recherches palynologiques sur le Pléistocène dans l'est de la Belgique et dans les Vosges lorraines. Acta Geographica Lovaniensia 14:118 p

Wolfe S, Stevens C, Gaanderse A, Oldenborger G (2014) Lithalsa distribution, morphology and landscape associations in the Great Slave Lowland, northwest territories, Canada. Geomorphology 204:302–313

Wünnemann B, Reinhardt C, Kotlia B, Riedel F (2008) Observations on the relationship between lake formation, permafrost activity and lithalsa development during the last 20,000 years in the Tso Kar basin, Ladakh, India. Permafrost Periglac Process 19:341–358

Yans J (2003) An overview of the saprolites of Belgium and their potential kaolinitic supplies to Mesozoic and Cainozoic sediments. Géol Fr 2003–1:33–37

8 Karstic Systems in Eastern Belgium: Remouchamps and Noû Bleû

Alexandre Peeters and Camille Ek

Abstract

Belgium is characterised by a wide variety of rocks, including carbonate formations, which are significantly represented in eastern Belgium. Among those formations, Devonian and Carboniferous carbonates display the most developed karstic features, as illustrated by the two cave systems presented in this chapter. Although both systems are water caves, they differ in many ways. The Remouchamps Cave, developed in the Upper Devonian limestones, is a touristic cave, while the Noû Bleû Cave, developed in the Carboniferous carbonates, is a brand new discovery still under exploration. The two systems are of scientific interest and contain many deposits (speleothems and detrital sediments) providing records of past environmental changes and paleoclimates. Beside the Quaternary dynamics, the present-day dynamics can be highlighted within the karstic basins that encompass the two cave systems. The generation or reactivation of sinkholes and swallow holes is often favoured or triggered by human activity. Furthermore, the two basins offer several good examples of human interaction with karst environments, not only in terms of karstic constraints for humans but also as a resource that must be preserved.

Keywords

Remouchamps Cave • Noû Bleû Cave • Karst dynamics • Speleothems • Sinkholes

8.1 Introduction

Wallonia, the southern half of Belgium, is rich in carbonate rocks, amounting to 12% of the outcrops, but occurring much more widely underground. These carbonate formations are mostly Palaeozoic limestones and dolostones but there are also Palaeozoic calcareous conglomerates and Cretaceous chalks (Boulvain and Pingot 2015; see also Chap. 2).

These carbonate rocks are significantly represented in eastern Belgium (Fig. 8.1). Devonian limestones and dolostones include many pure carbonate rocks (more than 90% calcium and magnesium carbonate). Carboniferous carbonate rocks are also frequently pure. Permian rocks are conglomerates with a calcareous matrix. North of Liège, Cretaceous formations include chalk, with notable karstic features (dolines) but only a few and small caves (see Chap. 11). In the very south of the country (outside the map) there are a few outcrops of Jurassic carbonate rocks, but without notable known caves.

In this chapter two cave systems from the region will be presented in detail. One of them—Remouchamps—is selected because it is large, very popular and includes a 600 m boat trip; the other one—so-called Noû Bleû—because it is a brand new discovery, still currently in progress (2012–2016). Both caves are the resurgences of important karstic basins, fed by numerous swallow holes. The first of these basins is the Sinkhole Dale, resurging at Remouchamps. This 7 km long watershed drains an area of 22 km^2.

A. Peeters (✉)
Department of Physical Geography and Quaternary, University of Liège, Sart Tilman, B11, 4000 Liège, Belgium
e-mail: A.Peeters@ulg.ac.be

C. Ek
Université de Liège, Liège, Belgium
e-mail: camille.ek@ulg.ac.be

A. Demoulin (ed.), *Landscapes and Landforms of Belgium and Luxembourg*, World Geomorphological Landscapes, DOI 10.1007/978-3-319-58239-9_8

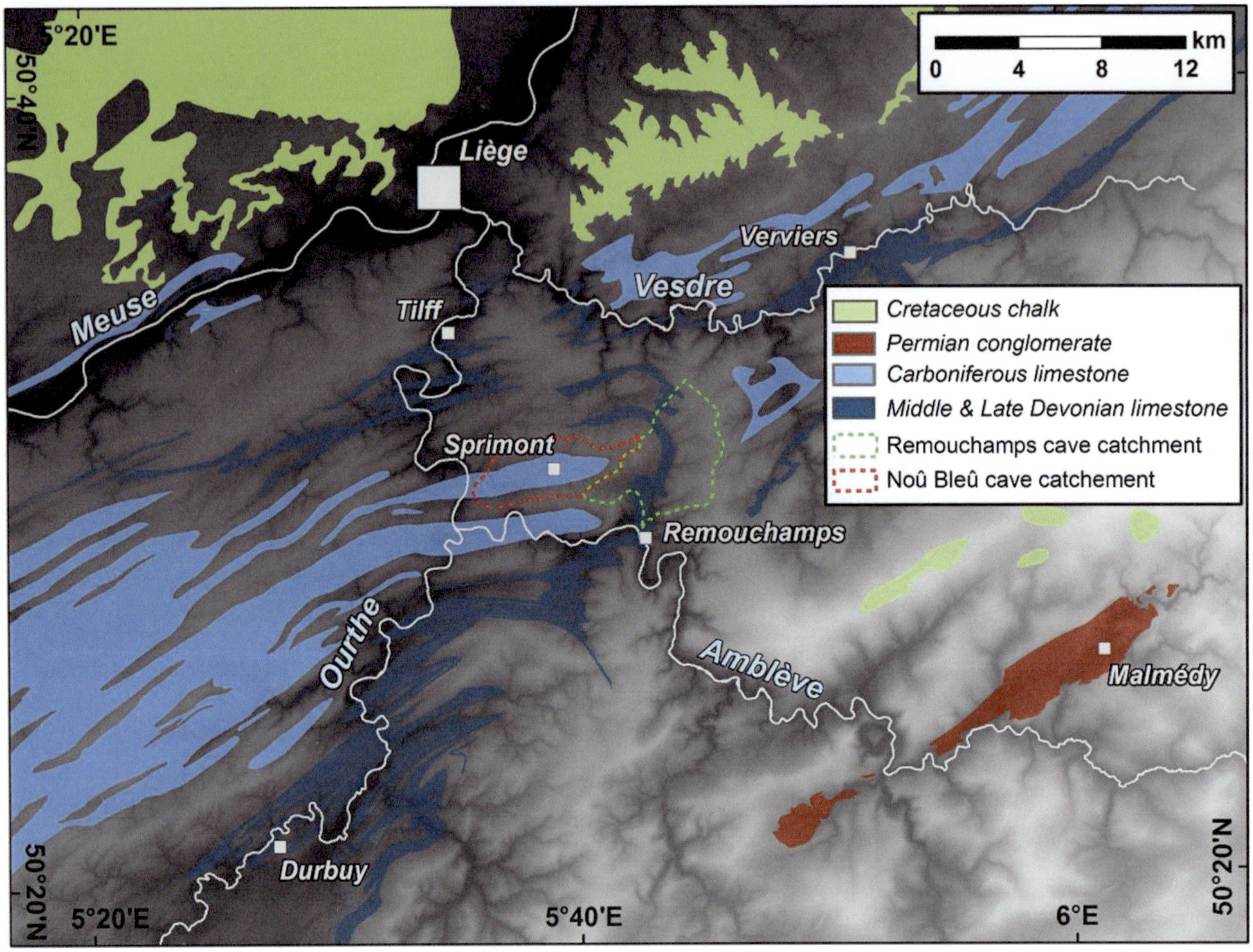

Fig. 8.1 Carbonate formations of eastern Belgium and the two studied catchments: Sinkhole Dale (*green dotted line*) and Sprimont Syncline (*red dotted line*)

It is riddled by hundreds of small dolines and swallow holes, including a few which are very spectacular (Fig. 8.2). The second basin is one of a dry valley extending along a syncline which stretches across the Sprimont municipality. It is 8 km long and collects the water from an area of 18 km^2. A water cave was discovered there by a local caving club, the "*Club de Recherches Spéléologiques Ourthe-Amblève*" (C.R.S.O.A.). This important cave opens in an active quarry, hence some security problems.

8.2 The Remouchamps Cave and Its Drainage Area

8.2.1 The Sinkhole Dale

25 km south of Liège, as the crow flies, the Sinkhole Dale extends N–S in Frasnian (Upper Devonian) limestones. Frasnian limestones have been described in detail by Boulvain et al. (1999). On the eastern side of the valley, they rest on older rocks, mainly shales, sandstones and quartzites. The western flank consists of a thin shaly formation and calcareous sandstones. The limestone strip, here in lowered position and forming a dale, separates the Condroz (west of the depression) from the Ardenne (east of the depression) (Fig. 8.3). The slopes, on both sides, are generally steep and the limestone is deeply entrenched. The bottom displays no talweg.

The small rivulets coming from both sides disappear in small or large swallow holes when reaching the calcareous formation. The surface of the depression displays numerous dolines (sinkholes) and no stream drains the bottom of this dry valley. This leads to disastrous floods when stormy rains occur (Fig. 8.4).

The general dip of the bedding planes is to the west, but secondary folds and numerous faults complicate the geological structure. Nevertheless, most of the rivulets flowing from Ardenne and Condroz run westward—according to the dip—when sinking underground. Consequently, whereas rivulets from Ardenne continue their course straight to west,

Fig. 8.2 The Grandchamps doline and swallow hole. The doline is about 250 m long and 26 m deep. The swallet generally absorbs a few l/s, sometimes less in summer, but can swallow 400 l/s and perhaps more in flood conditions

Condroz tributaries make a U-turn and their underground course passes below their surficial way (Fig. 8.5).

Quaternary and present-day dynamics

During Pleistocene times, periglacial conditions occurred repeatedly, influencing the type and rates of geomorphological processes. Proofs of permafrost were discovered in the Remouchamps Cave 60 m below the surface (Pissart et al. 1988). Solifluction deposits filled the valley and blocked the caves. Some holes were filled up to the roof, as demonstrated by sandstone pebbles still staying in fissures nearby the roof of some caves, like in the Rouge-Thier swallow hole. Some blind valleys heading to swallow holes were totally filled by deposits coming from the slopes. The periglacial periods were thus characterized by the interruption of karstic processes.

At the end of the last periglacial period, frozen ground was thawing so that the Holocene is characterized by the reactivation of perennial runoff. Nowadays, many caves have been unblocked. Underground flow is reborn. New swallow holes frequently open upstream from the old ones, as will be shown for the Noû Bleû area in the next section.

8.2.2 The Remouchamps Cave

8.2.2.1 A Large Cave, a Long Boat Trip

The Remouchamps Cave, a water cave, is drained by a stream which is a tributary of the Amblève River. It is a two level cave: an underground river flows in the lower storey, whereas, about ten meters higher, the upper level is dry (Fig. 8.6a). The total length of all passages is about 3900 m.

The Remouchamps Cave is the resurgence of all the swallow holes riddling the Sinkhole Dale. The underground stream flowing in the cave is deep enough to allow touristic navigation throughout the year, except during heavy floods. The biggest hall of the cave, the Cathedral, is about 60 m long and 35 m high (Ek 1995). But the floor of the hall is resting on a 25-m-thick mass of fallen rocks, so that if the scree was removed the true height of the Cathedral would be about 60 m (Fig. 8.6b). Tourists visiting the cave first walk along the upper level, go to the Cathedral, and come back by the river, sailing in a small boat. The navigation is about 600 m long.

8.2.2.2 A Complex Geological Structure

The cave is developed in a single stratigraphical unit of the Frasnian limestones: the Lustin Formation (Coen 1970; Ek 1970b). This $\sim$100-m-thick formation consists of biostromal limestone with stromatoporoids and corals, and bedded micritic limestones. The shaly Aisemont Formation surmounts the Lustin Formation. These shales clearly appear at the very end of the cave, in the so-called Shale Chamber (shown in red on the geological map of Fig. 8.7).

Below this shaly formation, the Lustin Formation is divided in three parts. The Lower part (formerly F2d) is a thick biostrome, as is also the Upper one (formerly F2h). Between them, the Middle part (formerly F2efg) does not display massive Stromatopora, but many tabular ones and corals. All the beds are pure (>90%) to generally very pure limestone, except for a few thin argillaceous layers. Thanks to the continuous exposure, the cave map reveals a structural complexity unpredicted by the surface survey. Located at the very eastern end of the Dinant basin, the bedrock dips west. But, as indicated by the structure contours of the map, there are several secondary folds and transverse faults, all Variscan in age (Fig. 8.7). The main structural features are two anticlines, with faulted hinges, separated by a transverse syncline.

The floor of the cave is generally covered by alluvium or fallen blocks, or in some places by artificial embanking, so that geological mapping of the floor is impossible. We have

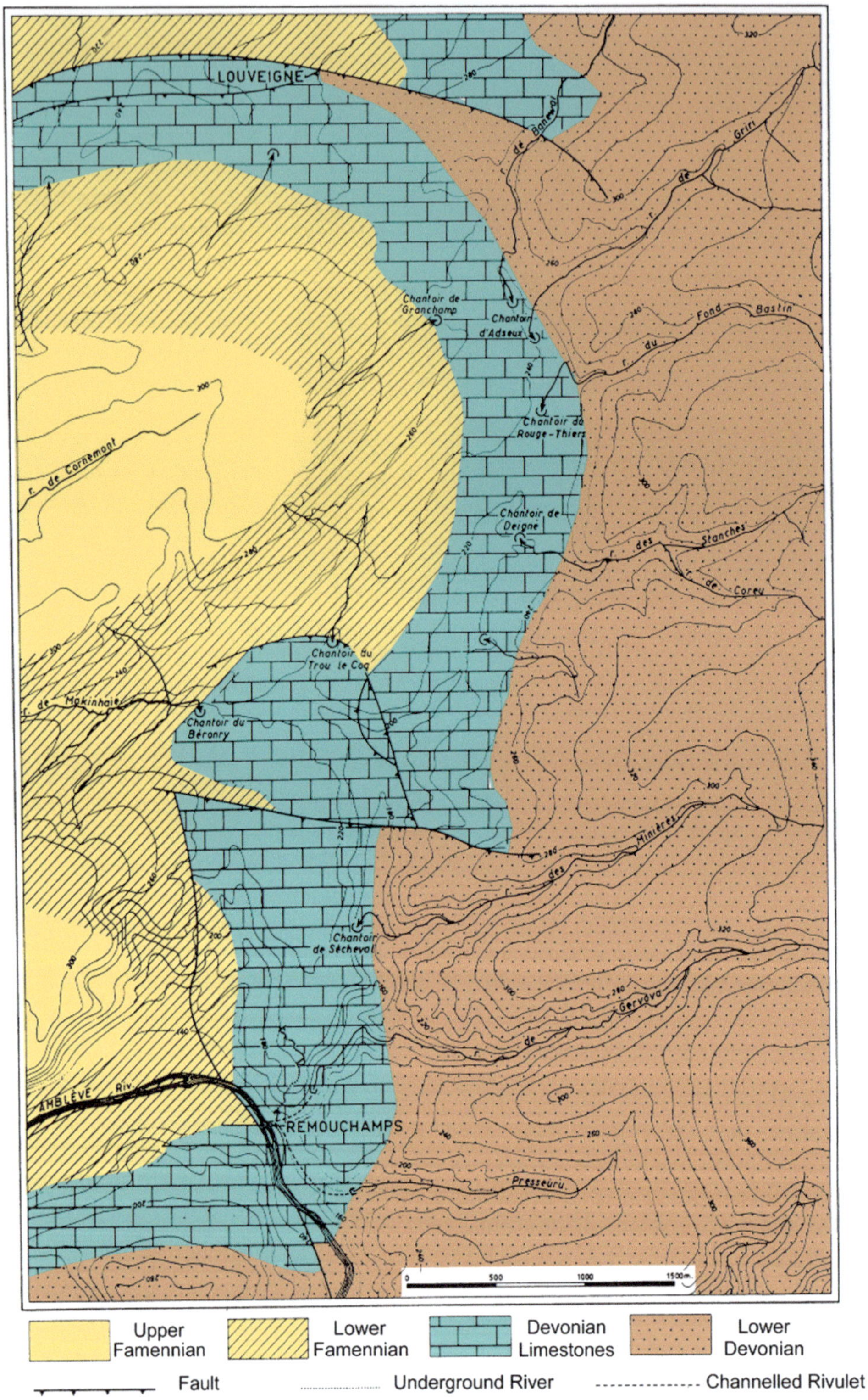

Fig. 8.3 Geological map of the Sinkhole Dale. The Frasnian limestone is surrounded by Famennian shales and sandstones to the west and by Middle and Lower Devonian shales and sandstones to the east. Famennian sandstones generally have a calcareous or dolomitic cement (up to 15%) and their waters are hard while Lower Devonian sandstones are devoid of carbonate and deliver acidic waters

thus used the walls and the roof. As the height of the roof is very variable, the stratigraphic limits look more complicated than they really are. The observed faults are reverse, except for the more northerly—the Faille Cathédrale Nord—which is normal.

8.2.2.3 A Very Varied Morphogenesis

The geological structure has had much influence on the cave morphology. Some beds are more easily dissolved than others, not necessarily because they are purer but, e.g., because of a petrographic structure (porosity, clay

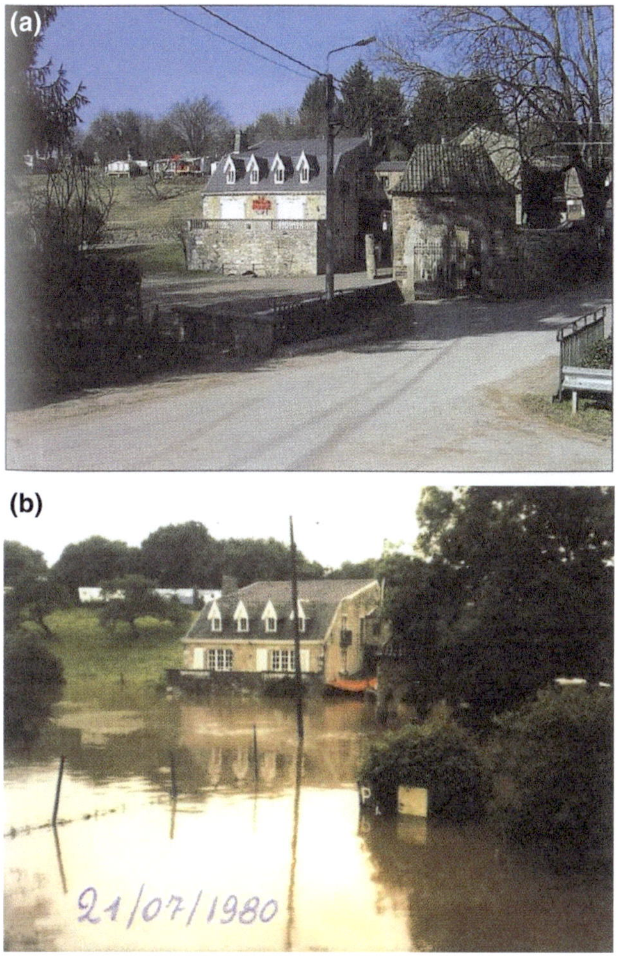

Fig. 8.4 **a** Trou-du-Moulin doline in low water conditions in 2005 (*Photo* JC Schyns). **b** The same in extreme flood conditions in 1980. The ground floor of the house in the background is totally flooded (*Photo* M Etienne)

distribution) that makes the attack by water easier. Generally, dissolution does not begin in a purer bed but along a more open sedimentary joint.

The dip of the beds had in some places a determining effect on the underground stream, some important sumps having formed where the flow direction was parallel to the direction of the dip (Fig. 8.8). In such a situation, the stream saps the rock and displays a trend to siphon. Joints also may affect water circulation: they provide more favourable routes than the bedding planes because they are frequently more open and cleaner.

The Labyrinth displays an example of a combined influence of joints and beddings planes (Fig. 8.9). All passages of the upper level, in this figure, are in the same bed. The direction of all passages is determined by joints. The current situation (last stage of the figure) shows the river flowing in the lower storey, whereas the upper level, dipping 25 W, displays two large galleries and a lot of narrow passages. Originally (stage 1), the river was flowing through a gallery (A on Fig. 8.9), where rounded pebbles can still be found. After the intermediate stage 2, the stream moved westward (stage 3). In this situation, it passed through a sump, due to the westward dip of the bed. During dry periods, when there was no flow anymore, some water could remain in the bottom of the sump. And when the main river, the Amblève, eroded its bed and began to flow 10 m lower, the underground tributary followed this descent by a waterfall through a pit, prepared long ago by dissolution and leading to the developing lower level (B on stage 4). Later on, this lower level developed upstream in the cave, leaving the upper level totally dry.

Though Variscan in age, the faults also affect karstification, favouring important Quaternary breakdowns along their traces. As an example, the biggest hall of the cave corresponds to a normal fault (Cathedral-North Fault), expressing stretching (Ek 1969) (Fig. 8.10).

8.2.2.4 Water Action and Breakdowns: A Long Speleogenesis

The main erosion process responsible for the origin of the present morphology of the cave is certainly concentrated underground flow. The lower level of the cave consists only of the riverbed, making clear that the stream is the agent which carved this whole storey. Most of the passages of the upper level, showing the same morphology, have most probably the same origin.

A fluvial terrace of the Amblève located ~8 m higher than the present floodplain corresponds to the altitude of the upper level of entrance and of the cave (Ek 1957, 1961; Rixhon and Demoulin 2010). As it does not contain minerals of the 74–90 ka-old Rocourt tephra (Juvigné and Gewelt 1988; Rixhon and Demoulin 2010), this terrace is presumably older than 90 ka and so is probably the age of this level of the cave.

Underground fluvial morphology in the Remouchamps Cave is characterized by large and high erosion notches, large meanders and pebble deposits. These features are lacking at some places of the upper level and, whereas it is sure that a river has run along the whole storey, it is not obvious that each place of the upper level owes its morphology to fluvial action. Other water actions, like diffuse flow giving rise to what Bretz (1942) called roof anastomoses and rock pendants, could as well have carved some passages, which were identified as such in the morphogenetic map of the cave (Fig. 8.10). This map provides sufficient evidence that water was the main agent of the cave morphogenesis, although specific processes varied (pressure pipes, phreatic processes, diffuse flow, underground stream, sumps). Even pits and fissures, when visible, attest enlargement by water action. In narrow passages or where water flows very slowly, dominant action is dissolution. Where flow rate or velocity is high, mechanical action

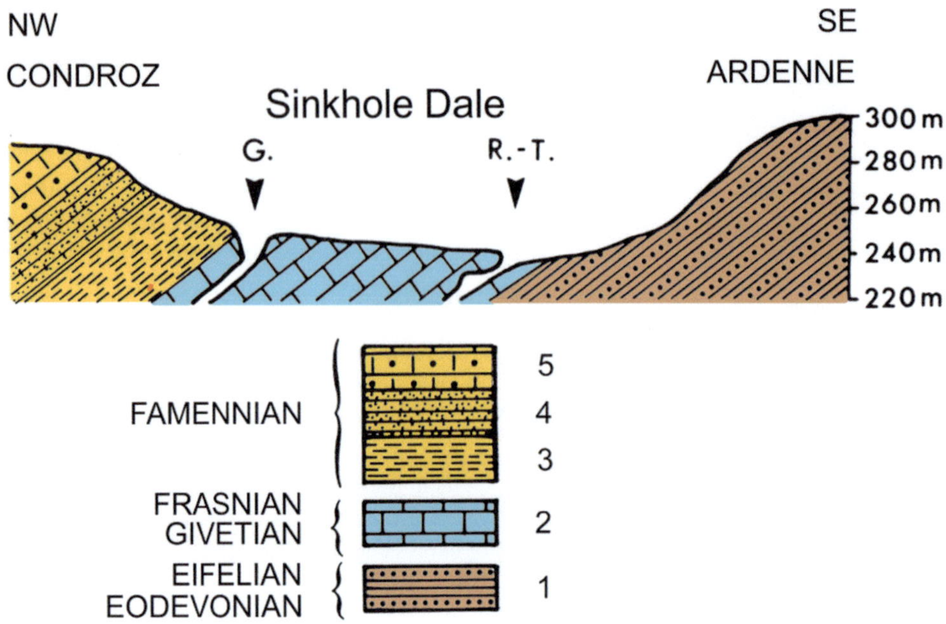

Fig. 8.5 NW–SE cross-section in the Sinkhole Dale (G: Grandchamps swallow hole, see Fig. 8.2; R.T.: Rouge-Thiers swallow hole in the Trou du Moulin doline, see Fig. 8.4). Underground streams run westward. Thus waters of Grandchamps swallow hole run eastward outside and then westward underground

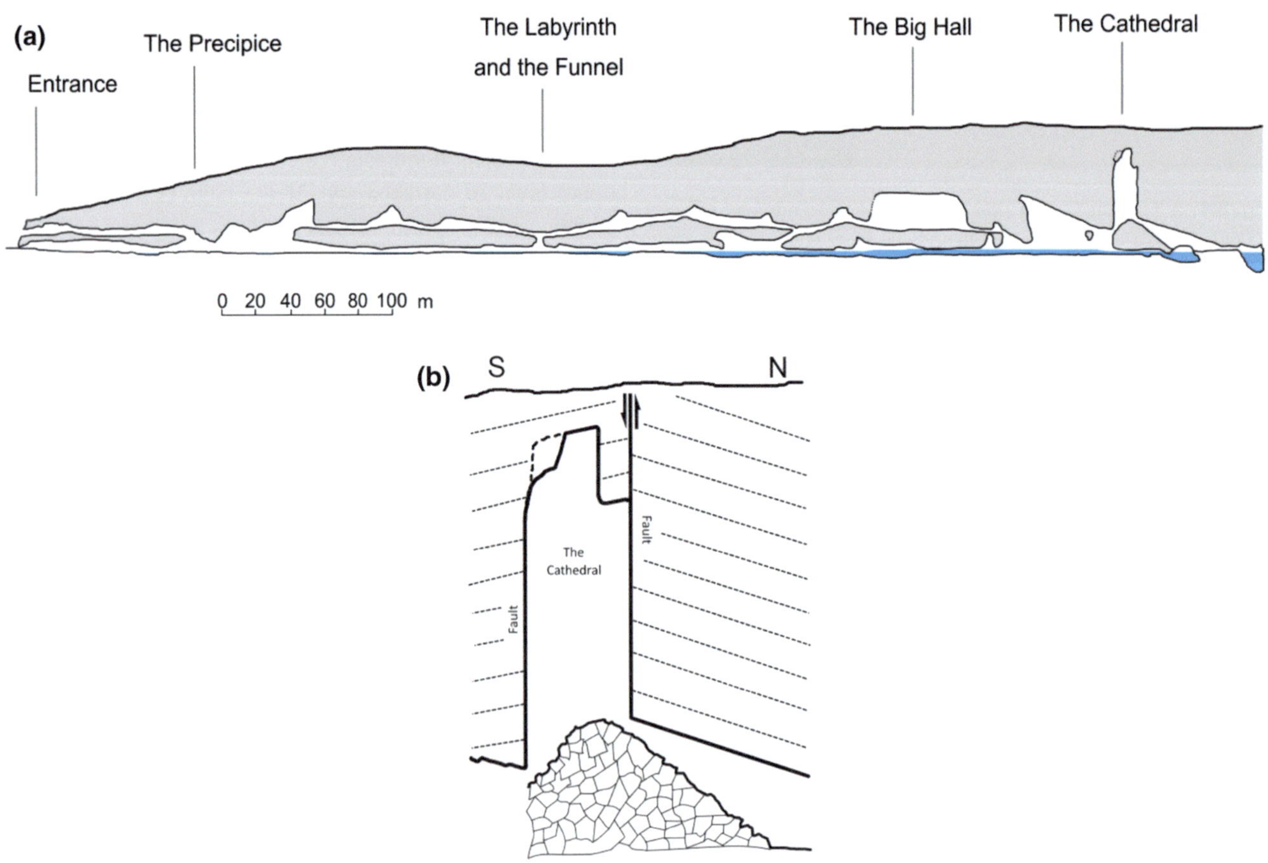

Fig. 8.6 **a** Longitudinal profile (developed) of the two-storey cave of Remouchamps (heights not exaggerated). **b** Cross-section of the Cathedral Hall. The roof is less than 8 m below the surface (modified after Ek 1970b)

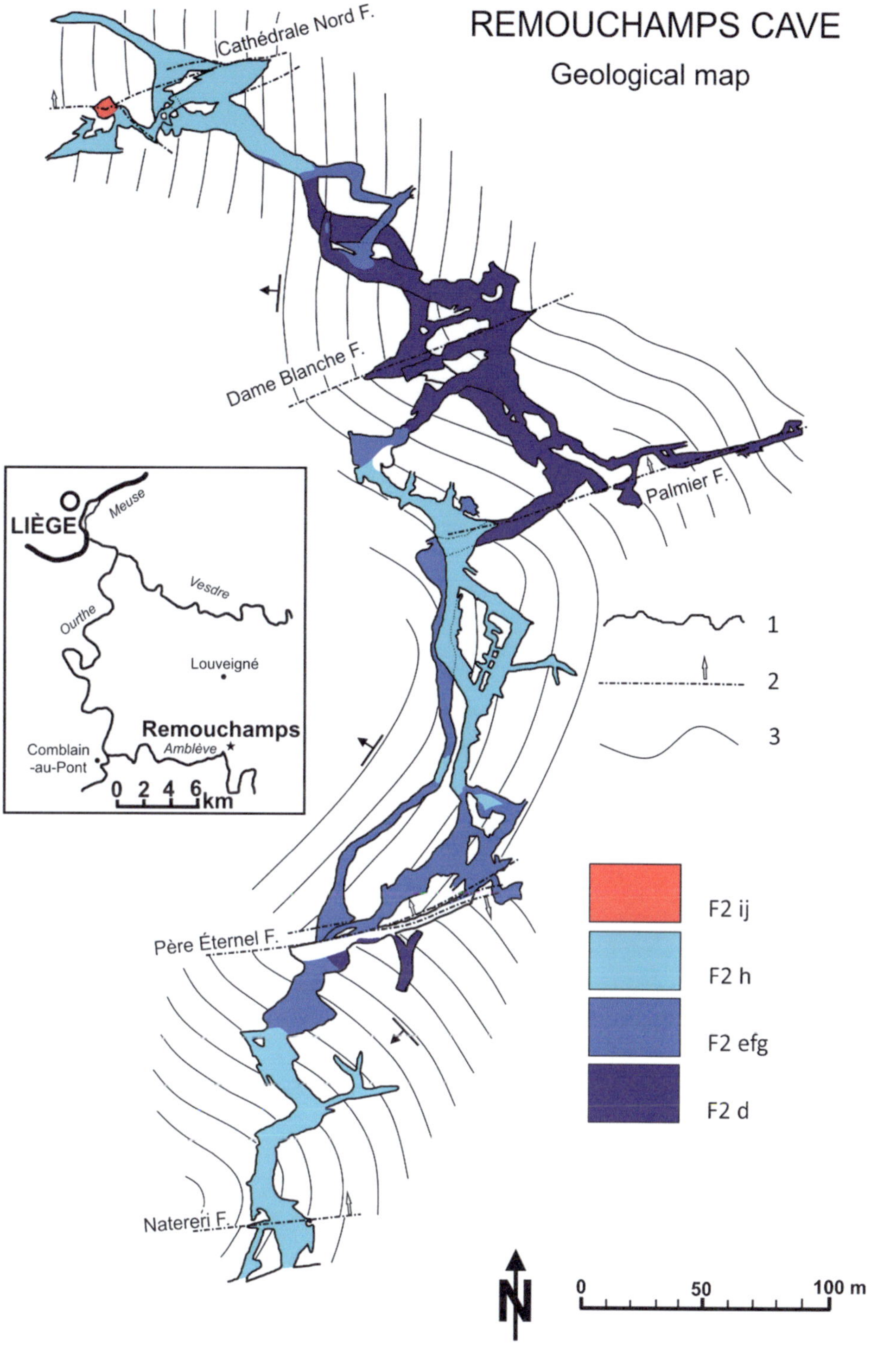

Fig. 8.7 Geological map of Remouchamps Cave. *1* Cave limit. *2* Fault. *3* Structure contour (generalized). *F2ij*. Aisemont Formation (shale). *F2h*. Lustin Formation (*upper part*). *F2efg* Lustin Formation (*middle part*). *F2d* Lustin Formation (*lower part*) (modified after Ek 1970a)

generally becomes predominant. Sumps occur in the riverbed when the stream is not graded (Fig. 8.8). There are three major sumps in the upper level of the cave (Fig. 8.10). Likewise, while the present bed is better graded in the touristic downstream part of the cave, its upstream section is not yet so well graded, with several sumps having been explored by L. Funcken and other divers.

We have noted a few traces of phreatic processes. In a narrow passage of the upper level, there are some boxworks, i.e., networks of thin calcite veins (1–3 mm) protruding from

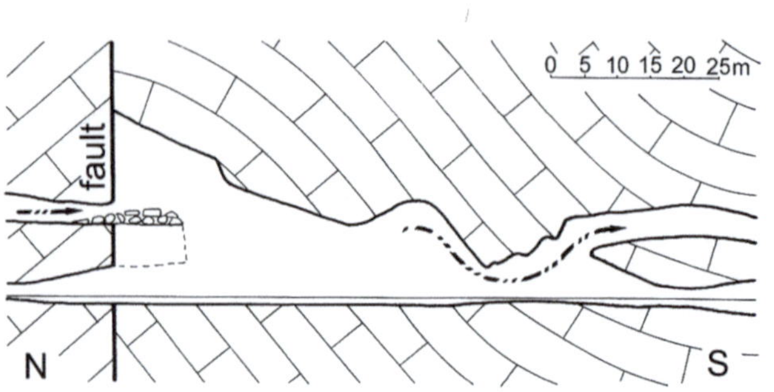

Fig. 8.8 Longitudinal section of a sump in the Remouchamps Cave. Dip and stream flow directions are parallel (modified after Ek 1970b)

the wall by several cm because they are less easily dissolved than the surrounding limestone (Bretz 1942). This kind of very fragile feature cannot withstand a fast flow or a flow including a solid load and it was thus interpreted by Bretz as a sign of standing or very slowly moving water.

Roof and wall breakdowns are a water-independent process involved in the origin of the present morphology of some parts of the cavern (Fig. 8.10). Talus cones are present, and they are generally huge and occur in every large hall. A rockfall needs a preliminary void to occur, so the breakdowns are younger than the water action that created the chambers.

Putting all pieces of evidence together, one can summarise the history of the cave development as follows:

- Fissures and pits (Fig. 8.10) first allowed water to circulate in the bedrock in the very early stages of cavern development.
- Water, descending towards the local base-level, filled voids too scarce for the aquifer recharge; this led to the development of cylindrical pressure pipes, which are marks of a generally saturated aquifer (more incoming water than available voids).
- On the overlying ground surface, swallow holes opened above the Big Hall; although they later filled up and disappeared, they are still evidenced by seismic refraction (Vandenvinne, Girolimetto and Ek, unpublished results) (Fig. 8.11).
- The swallow holes fed a permanent underground rivulet, which progressively became a stream and tended toward a graded profile. This stream was contemporaneous with the 8 m terrace level; it could not grade perfectly, however, and three sumps remain from that time.
- The underground river of the 8 m storey occasionally carried huge amounts of exogenous pebbles and eventually blocked some siphons; this is conspicuous in the so-called Hall of the Virgin.
- When, outside, the Amblève River lowered its bed to the present level, the underground stream progressively

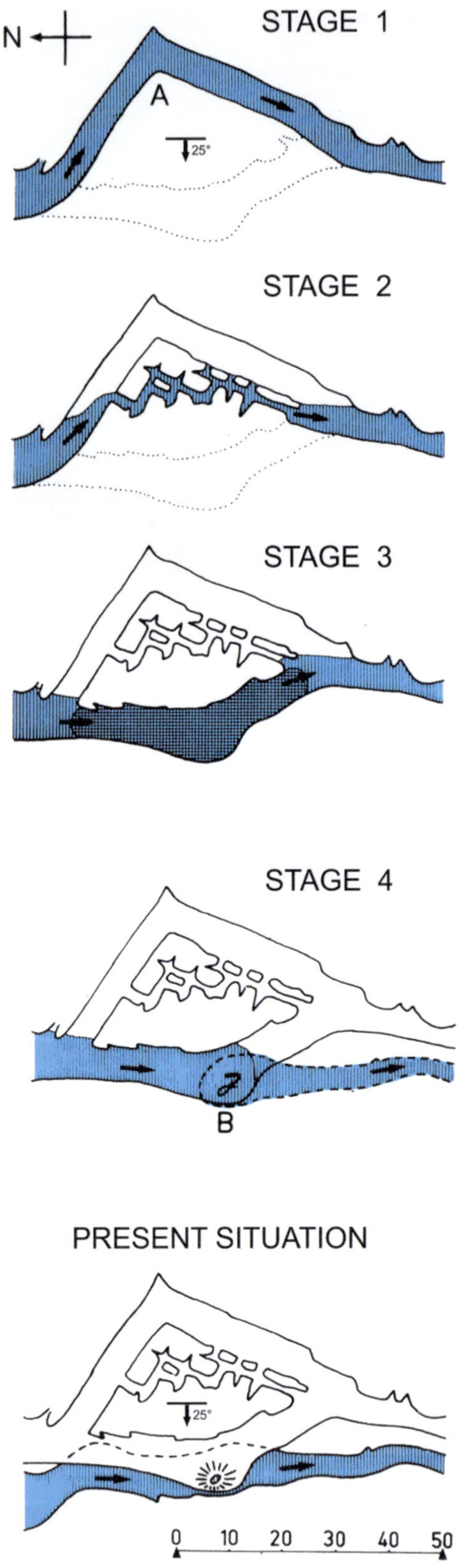

Fig. 8.9 Plan view of the Labyrinth. The present topography of the Labyrinth region of the cave is displayed by the *lowest sketch*: the river (in *blue*) is flowing southward straight on along the lowest level whereas all passages of the upper floor, now dry, are determined by two perpendicular sets of joints

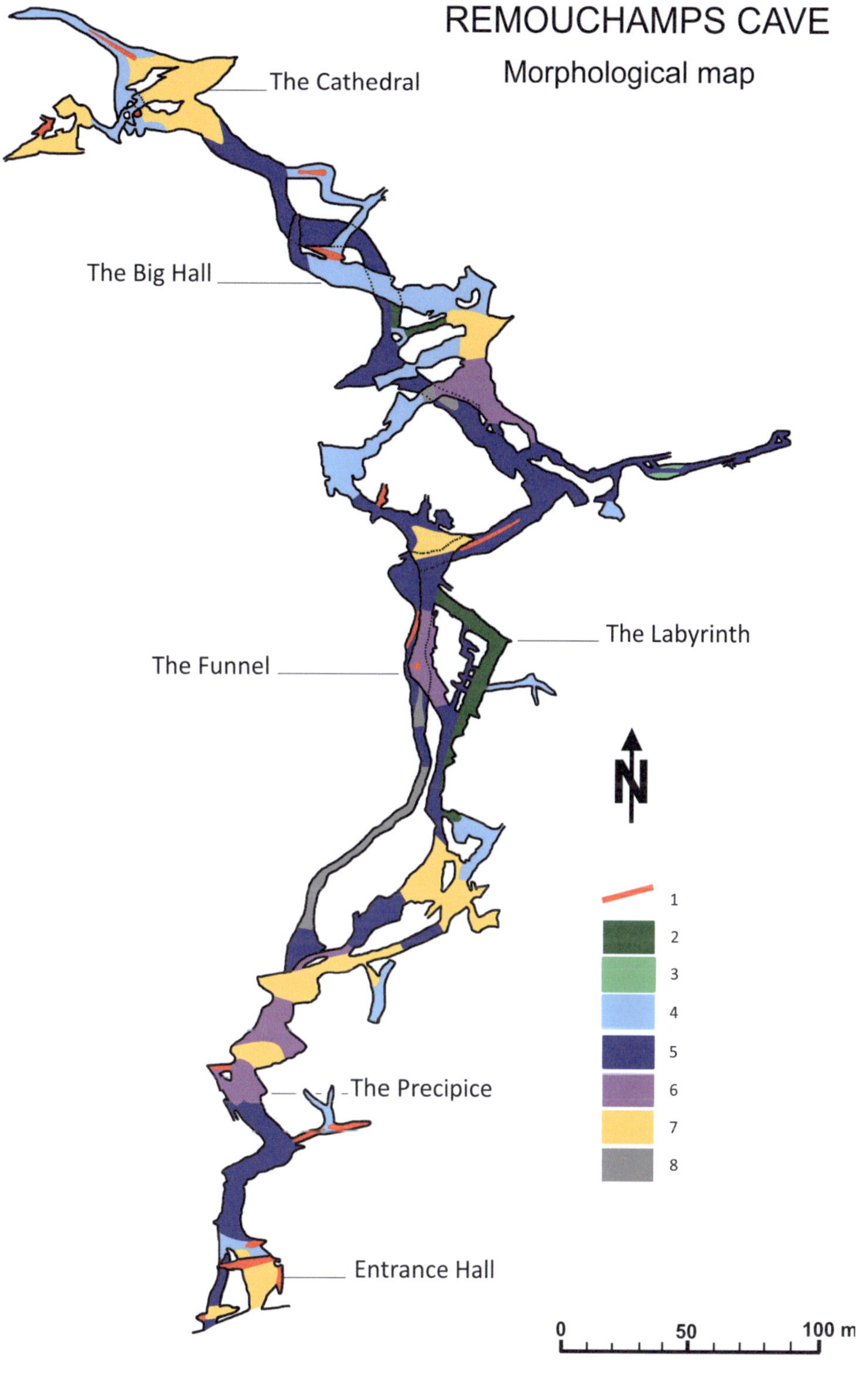

Fig. 8.10 Morphogenetic map of the Remouchamps Cave. *1* Fissure, crack, pit; *2* Pressure pipe; *3* Phreatic processes (very slow move or still water); *4* Diffuse flow, roof anastomoses, unspecified water action; *5* Underground stream; *6* Sump; *7* Breakdown; *8* Anthropogenic digging (simplified after Ek 1969)

joined its level by successive self-captures, or domestic piracy.

- Several breakdowns occurred and created talus cones in the upper level after it has been partly graded.

U/Th dating of a big broken stalagmite within a huge collapse yielded dates of 126 and 95 ka for the bottom and the top of the speleothem, respectively (Gewelt 1985), (Fig. 8.12). The fall of the stalagmite thus occurred at or

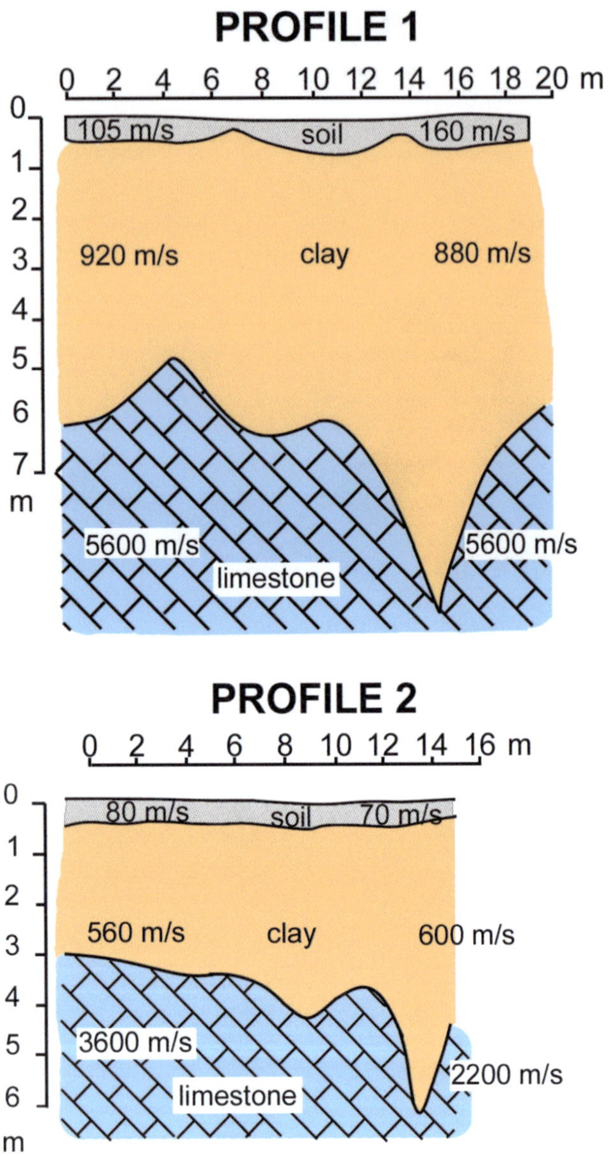

Fig. 8.11 Seismic soundings carried out above the Big Hall (Vandenvinne, Girolimetto and Ek, unpublished results). The analysis of the soundings clearly shows the existence of karstic depressions above the Big Hall, which is the upward terminal of the upper level of the cave. We interpret these depressions as the marks of the swallow holes which once flowed into the cave

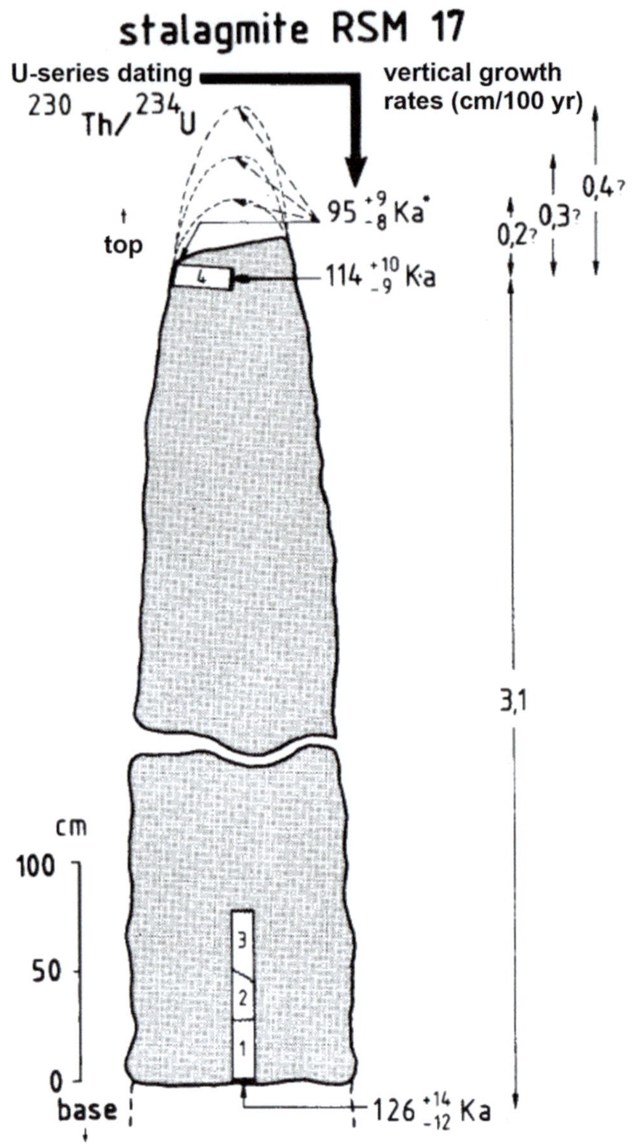

Fig. 8.12 Thorium–Uranium dating of a large broken stalagmite in the Remouchamps Cave (Gewelt 1985). This 4-m-high speleothem began growing 126,000 years ago, and stopped about 95,000 years ago. This corresponds to the Eemian interglacial period. The ending of growth probably was caused by the fall of the stalagmite, swept along by a rockfall

after 95 ka. Even though the speleothem's growth period coincides with an interglacial, consistent with the idea that karstic processes are favoured by temperate climates and hampered by periglacial conditions, it is likely that the collapse, possibly earthquake-triggered, was the cause of the stopping of the stalagmite's growth. However, no other speleothem broken in that time has been retrieved so far in the cave.

8.2.3 Man in the Cave

8.2.3.1 The First Inhabitants: Cave Dwellers

At the very end of the Upper Paleolithic, about 10,000 years ago, a group of hunters of Ahrensburg culture lived temporarily in the cave (Dewez 1974; Ek 1974a, b). They cut and left there a rich flint assemblage, and particularly the so-called Ahrensburg arrow heads (Fig. 8.13a).

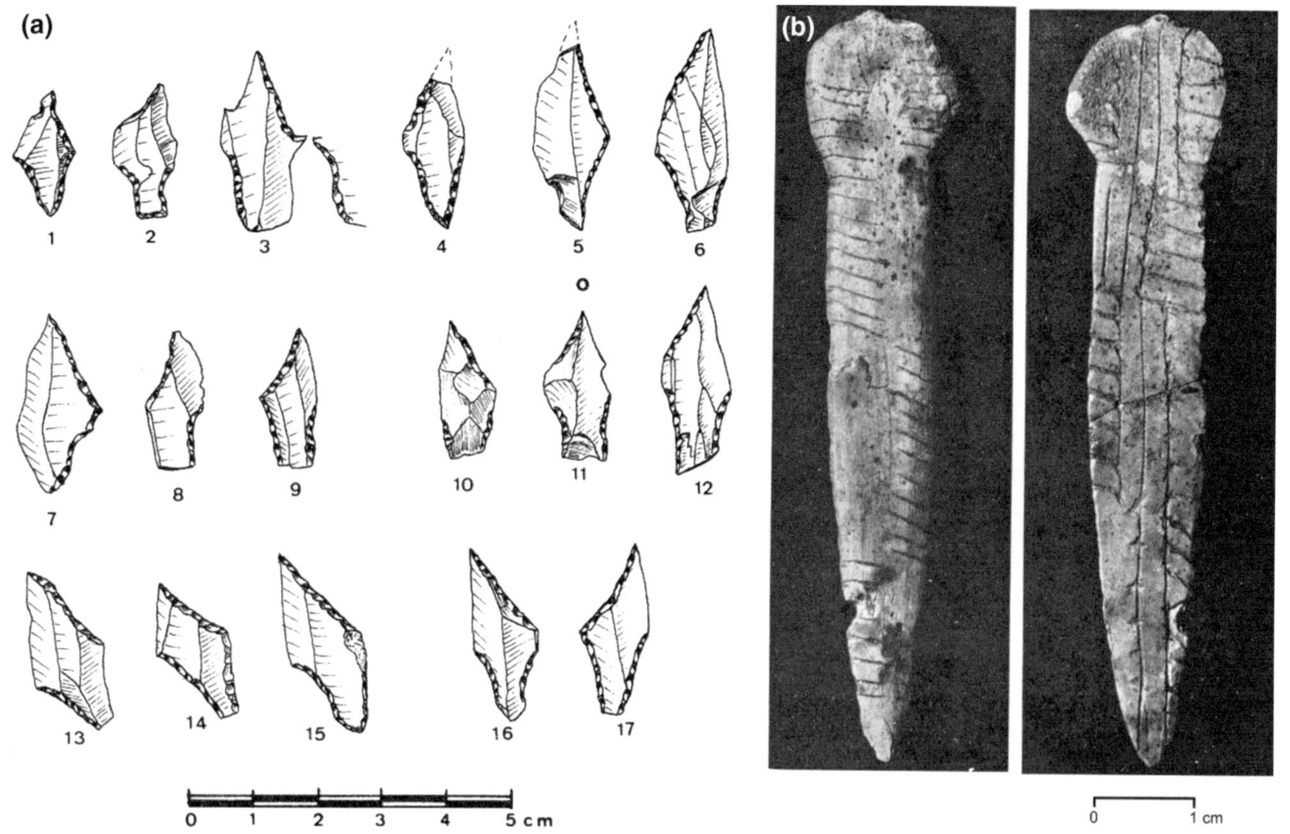

Fig. 8.13 **a** Ahrensburg arrow heads from the Remouchamps Cave (from Dewez 1974). **b** 72 mm long polished and engraved bone blade found in the Remouchamps Cave, assigned to the Ahrensburg culture (from Dewez 1974)

They also left perforated Lutetian shells, probably imported from the Paris basin as necklaces, and a polished and curiously striated bone blade, whose purpose could not be defined (Fig. 8.13b). They ate reindeer, Elaphus stag, Megaceros stag, polar fox, horse, wolf and snow grouse. According to an uncalibrated ^{14}C dating of bones of eaten animals, this group of hunters lived and cut flints in the cave by 8430 ± 170 years B.C. (Dewez 1974). This yields a calibrated age of 12,196 ± 324 year cal BP (CalPal_2007_HULU; Weninger and Jöris 2008), which corresponds to the Younger Dryas, the very last cold impulse of the last glaciation. The very small Ahrensburg arrowheads retrieved in the cave suggest that they sometimes hunted very small animals.

No later prehistoric dwellers are known from the cave but Middle Age occupancy is attested by a lot of coins, particularly from the sixteenth century, a period frequently troubled by the wars of religion and during which the coins were probably lost by refugees.

8.2.3.2 The Explorers

In the nineteenth century, the first explorers were students of the Liège University, among them an Englishman, L. Wilmar (Courtois et al. 1969). Taking no heed of the villagers' fears, they went down a 10 m pit on the 1st of August 1828 (Fig. 8.14), forded the underground stream, and in one night explored more than 200 m of the cave. They were stopped by a cul-de-sac. In September, they were relieved by M. Schols, staff captain in the Belgian army, who mapped the known part of the cave (Schols 1832).

Several other researchers succeeded in dangerous explorations, namely Knight Hoy, general Niellon and Count de Cornelissen in 1834, and later on A. Delhasse in 1851 (Delhasse 1851), E.A. Martel, E. Van den Broeck and E. Rahir from 1898 to 1912 (Van den Broeck et al. 1910). More explorations opened new galleries from 1956 to 1959. A first dive in the upward sump opened the way to upward galleries, and more dives lengthened the cave, doubling the length of the underground stream, from 600 m to more than 1000 m.

8.2.3.3 The Researchers

The Remouchamps cave has always fascinated palaeontologists and archaeologists. Van den Broeck (1902) studied Eocene shells, originating from the Paris basin, and certainly imported by cave dwellers as a necklace. Rahir (1920) discovered hundreds of flint tools, very small, later considered as epipaleolithic by Dewez (1974).

Fig. 8.14 The Remouchamps Cave: "Le Précipice", engraving in Schols (1832)

The cave was carefully studied and described by Van den Broeck et al. (1910) in a very large book, "Les cavernes et les rivières souterraines de la Belgique", the bible of Belgian cavers and speleologists.

But it is mainly during the second half of the twentieth century that geomorphologists and geologists of the University of Liège devoted a lot of work to the cave, linking the levels of the cave with fluvial terraces of the Amblève (Ek 1961), mapping the geology of the cave (Ek 1970a) and analysing the influences of the geological structure on the cave morphology (Ek 1970b). In the 1980s, numerical ages and growth rate estimates were obtained for Late Pleistocene (Eemian) and Holocene speleothems (Gewelt 1985). Ek (1973) analysed waters of the underground river and seeping waters. Stream water analyses showed that stormy rains remove considerable amounts of dissolved calcium carbonate (>600 kg in 24 h in one analysed case), whereas moderate rains dissolve much less (<0.5 kg in 24 h in another reported case) (Ek 1969).

The study of CO_2 in the air above the underground stream showed that the river, coming from the remote upward passages and appearing in the show cave, degasses carbon dioxide very fast. Based on measurements made 30 cm above the water surface, the CO_2 content of the air is of ~6400 ppm, about 16 times the CO_2 partial pressure in open air (Delecour et al. 1968).

Last but not least, the Walloon Commission for Study and Protection of Underground Cavities (C.W.E.P.S.S.) has mapped all karstic features of the basin, and the survey is under permanent updating, because of the activity of some swallow holes (De Broyer et al. 1996). From 1995 to 2005, the Public Service of Wallonia has ordered a geohazard assessment of karstic areas and, as such, Remouchamps is mapped at the scale of 1/10,000.

8.3 The Noû Bleû Cave and Its Drainage Area

8.3.1 The Syncline of Sprimont

The geomorphological landscape of the eastern Condroz displays a succession of ridges and valleys aligned on the WSW–ENE structural grain of the Variscan basement. The ridges correspond to Upper Devonian sandstone anticlines and the valleys to Lower Carboniferous limestone synclines (see Chap. 2). Located 20 km south of Liège, the small town

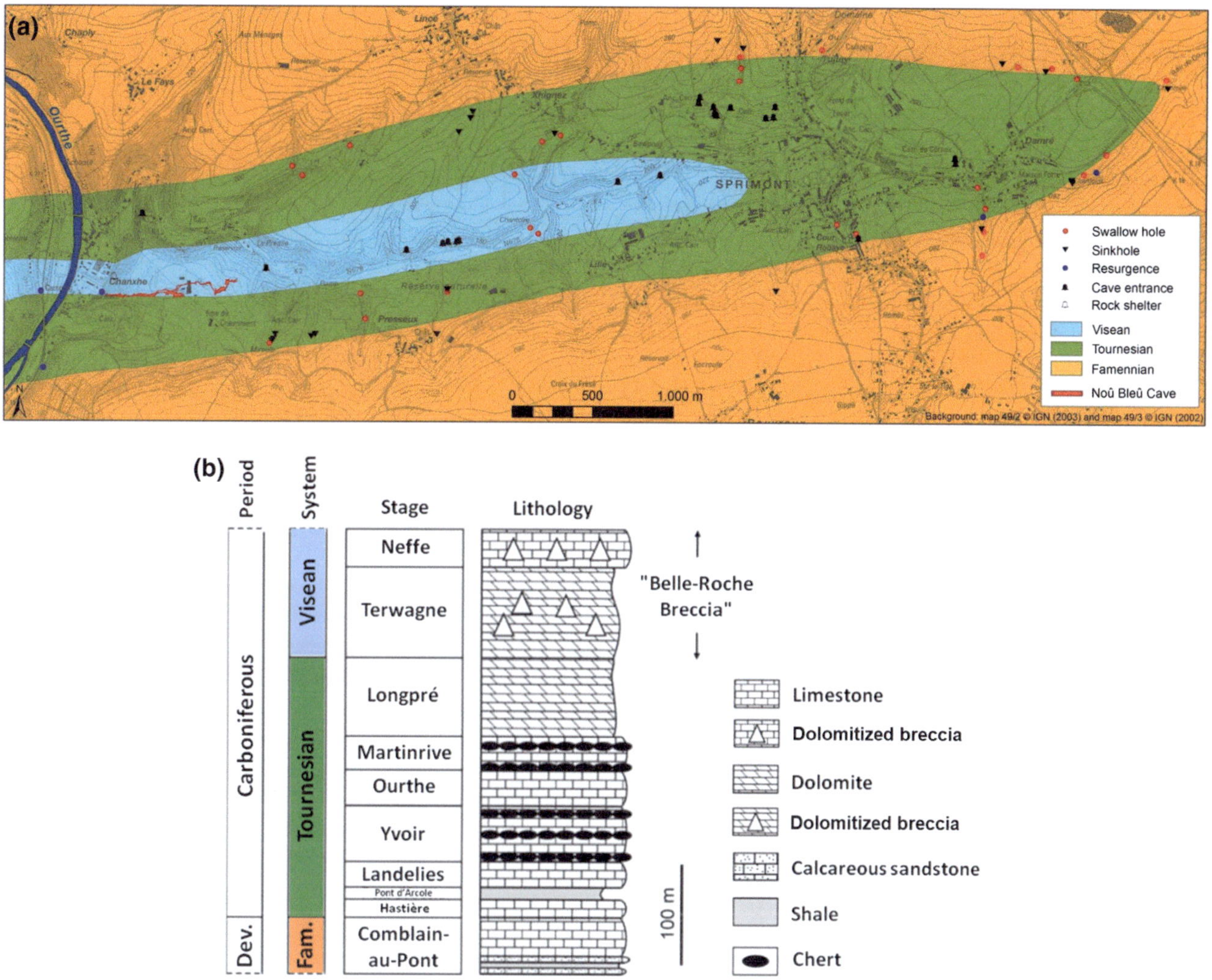

Fig. 8.15 **a** Karst phenomena and simplified lithostratigraphic map of the Sprimont Syncline (Geological data from the Geological map of Wallonia, Tavier-Esneux 49/1-2, by Bellière and Marion 2009, draft version, SPW-DGO3/ULg; Louveigné-Spa 49/3-4, by Mottequin and Marion 2012, modified). **b** Simplified lithostratigraphic column of the Tournaisian and Visean series within the Sprimont Syncline. *Dev* Devonian; *Fam* Famennian (modified after Mottequin and Marion 2012)

Fig. 8.16 **a** Dry swallow hole in the Sprimont Syncline (*Photo* A Peeters). **b**. 3 m diameter cover-collapse sinkhole in the Sprimont Syncline (*Photo* A Peeters). **c** Backfilled swallow hole in the Sprimont Syncline (*Photo* P Labarbe)

of Sprimont lies in such a typical limestone valley. The Syncline of Sprimont is cut transversally by the incised valley of the N-flowing Ourthe (Fig. 8.1). Its 1.2-km-wide, 8.4-km-long eastern part, in which the Noû Bleû karstic system is developed, stretches from Cornemont (270 m asl) in the east to Chanxhe (100 m asl) in the west, where the valley joins the Ourthe. The entire catchment extends over 18 km^2, including the sandstone crests, and is mostly covered by grassland and cropland, with small patches of forest on the steepest slopes and several, partly still active quarries.

8.3.1.1 Geological Setting

The Sprimont Syncline displays several lithostratigraphic units of the Tournaisian and Visean series. The Tournaisian is mainly characterized by carbonate sedimentation (crinoidal limestones, used as building stone under the name of 'Petit Granit') with subordinate shales (Poty et al. 2001; Mottequin and Marion 2012), (Fig. 8.15a, b). The Visean also features carbonate rocks, which are partly dolimitized and brecciated. The "Belle-Roche Breccia" originated in solution collapse of underlying evaporitic layers (Paproth et al. 1983). All these rocks are prone to karstic phenomena.

Moreover, Mottequin and Marion (2012) noted that some Upper Famennian calcareous sandstones of the syncline periphery may also be locally affected by karst phenomena, for instance in the eastern part of the valley (Fig. 8.15a, easternmost swallow hole).

The Sprimont Syncline is characterized by vertical to overturned beds on its south limb, while the north limb dip is ~50°S (Mottequin and Marion 2012). This geometry plays an important role in the formation of underground galleries.

8.3.1.2 Geomorphological Setting

The Sprimont valley has an asymmetrical cross-profile, with steeper south-facing slopes, especially in its central stretch. As this asymmetry does not seem to be structurally controlled, it is hypothesized that it appeared during Pleistocene cold periods due to a different sensitivity of the two valley sides to periglacial slope processes. N–S tributary valleys also display some asymmetry, with steeper west-facing slopes. In this case, it is classically attributed to the periglacial evolution of the valleys during cold periods, when east-facing lee slopes were especially affected by solifluction, which pushed the creek against the opposite valley side and progressively steepened it (Michel 1971).

Many small tributary streams running from the sandstone interfluves disappear underground into swallow holes halfway downslope, as soon as they flow over limestone. Frequently, two or three swallow holes are aligned in the same tributary valley, the most upstream one collecting the stream permanently and the others being active only under high flow conditions. As a consequence, the main valley has no thalweg and the downstream swallow holes are most often dry (Fig. 8.16a). The main valley long profile displays slope breaks, many of them related to the presence of swallow holes. Some of the swallow holes are cave entrances but more such entrances are also scattered over the higher slopes of the valley (Ek et al. 1997). In addition, the Sprimont valley is riddled with up to 10-m-wide, 3-m-deep sinkholes, especially on the valley slopes, more rarely in the valley bottom and on the benches (Michel 1971). Most of them are cover-collapse sinkholes resulting in bowl-shaped depressions (Fig. 8.16b). Similar depressions observed in the tributary valleys are also produced by piping, more or less promoted by dissolution, occurring in the unconsolidated soil cover, whereby a cavity or a small conduit is developed upwards, due to progressive sediment removal by seepage water (Field 1999). Many of these sinkholes and swallow holes have been backfilled by the farmers in grassland and cropland (Fig. 8.16c).

In his description of the 16 caves so far identified in the Sprimont area, Briffoz (1983, 2012) highlighted four types of relationship between gallery morphology and geology: (I) galleries following the bedding planes (thus showing moderate and subvertical gradients in the northern and southern valley side, respectively), (II) galleries developed

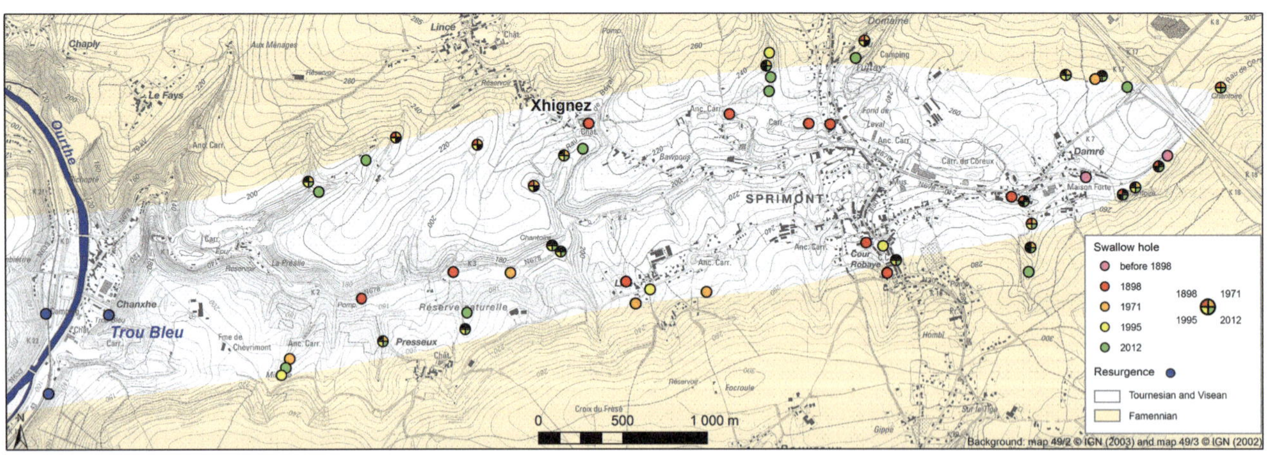

Fig. 8.17 Swallow hole dynamics in the Sprimont Syncline (from Peeters et al. 2012)

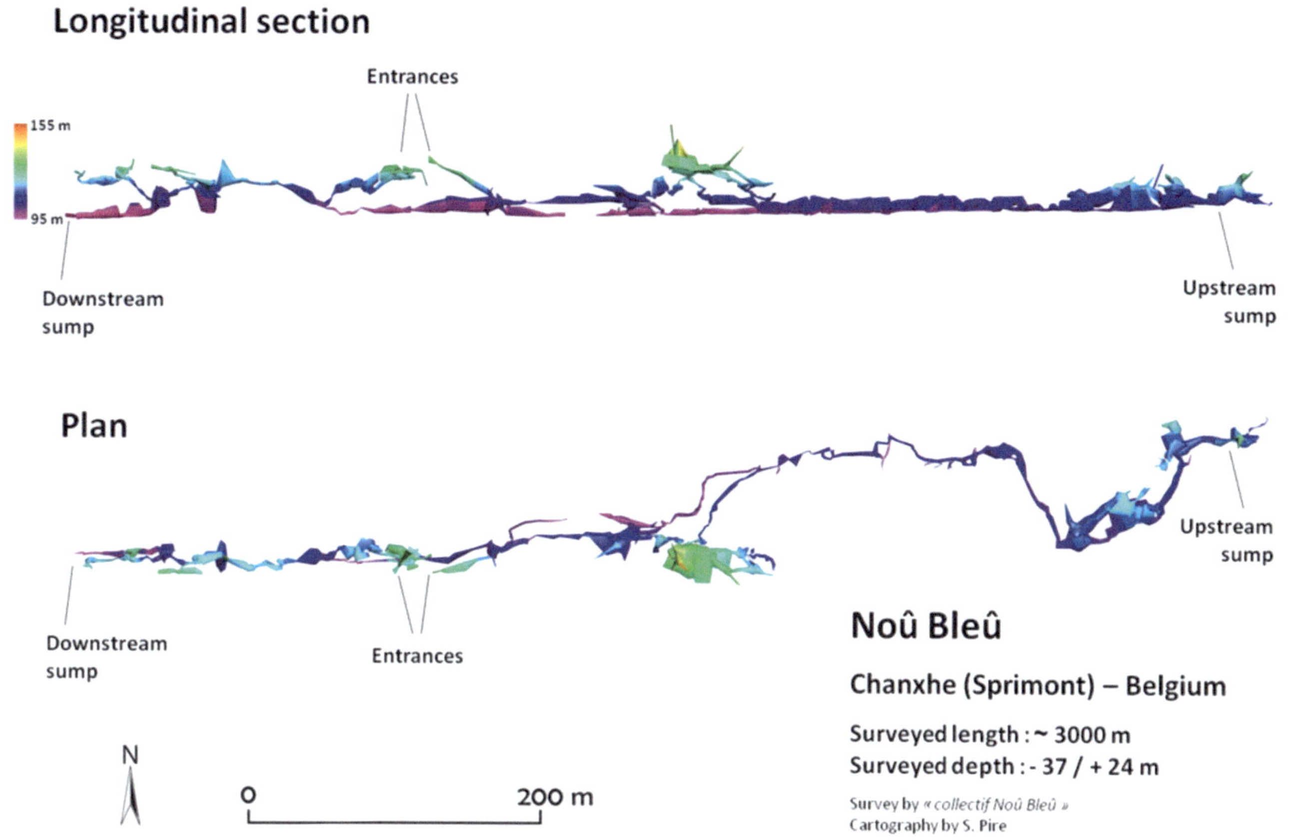

Fig. 8.18 The Noû Bleû Cave survey (April 2015)

along a joint (most pits of the northern valleyside) or (III) at joint intersection, and (IV) subhorizontal galleries with fluvial sediment. Groundwater flow patterns are typical in subhorizontal galleries whereas collapses are common in vertical galleries.

At the downstream end of the valley, outflow in the Ourthe of most waters of the Sprimont syncline occurs through the *Trou Bleu* resurgence located at the foot of the eastern valleyside of the Ourthe, although other resurgences might exist in the Ourthe riverbed (Ek et al. 1997), (Fig. 8.17). The *Trou Bleu* mean discharge is ~0.23 m^3/s (Meus et al. 2014).

8.3.1.3 Karst Dynamics, from Quaternary to Present Day

After solifluction deposits had filled the valleys and clogged the swallow holes during the Weichselian cold period, thus contributing to the interruption of every karstic process (Michel 1971), the Holocene reactivation of perennial runoff reopened some of the filled swallow holes. This dynamics is still going on, with new swallow holes opening upstream of the former ones. Inactive open swallow holes and their evolution were first described by Van Den Broek et al. (1910). Surveys of the valley bottoms (Michel 1971; Ek et al. 1997; Peeters et al. 2012) confirmed the upstream migration of swallow holes, eight of them opening in 72 years at an average distance of 270 m from their downstream predecessor (Fig. 8.17). As an example, the active swallow hole in the middle reach of the main valley moved ~700 m upstream from 1898 to 1995, at a mean rate of 6.7 m/year (Figs. 8.16a and 8.17). However, this natural process is often disrupted by human actions. For example, the Xhignez tributary stream was artificially diverted from its swallow hole in the early twentieth century, causing (re) activation of a downstream swallow hole in the 1950s, which in turn is now migrating upstream (Fig. 8.17).

8.3.2 The Noû Bleû Cave

After the main drain of the Sprimont Syncline had been the object of caving research since the 1960s (Briffoz 1983), an entrance leading to it was discovered in 2012 by the "*Club de Recherches Spéléologiques Ourthe-Amblève*" (CRSOA) in a large limestone quarry. The new cave has been called *Noû Bleû* ('New Blue') in relation to the name *Trou Bleu* of its nearby resurgence. The cave has now been surveyed on more than 3000 m by the joint action of the CRSOA and geoscientists (Figs. 8.15a and 8.18). Two entrances open into the northern wall of the Chanxhe quarry. The depth,

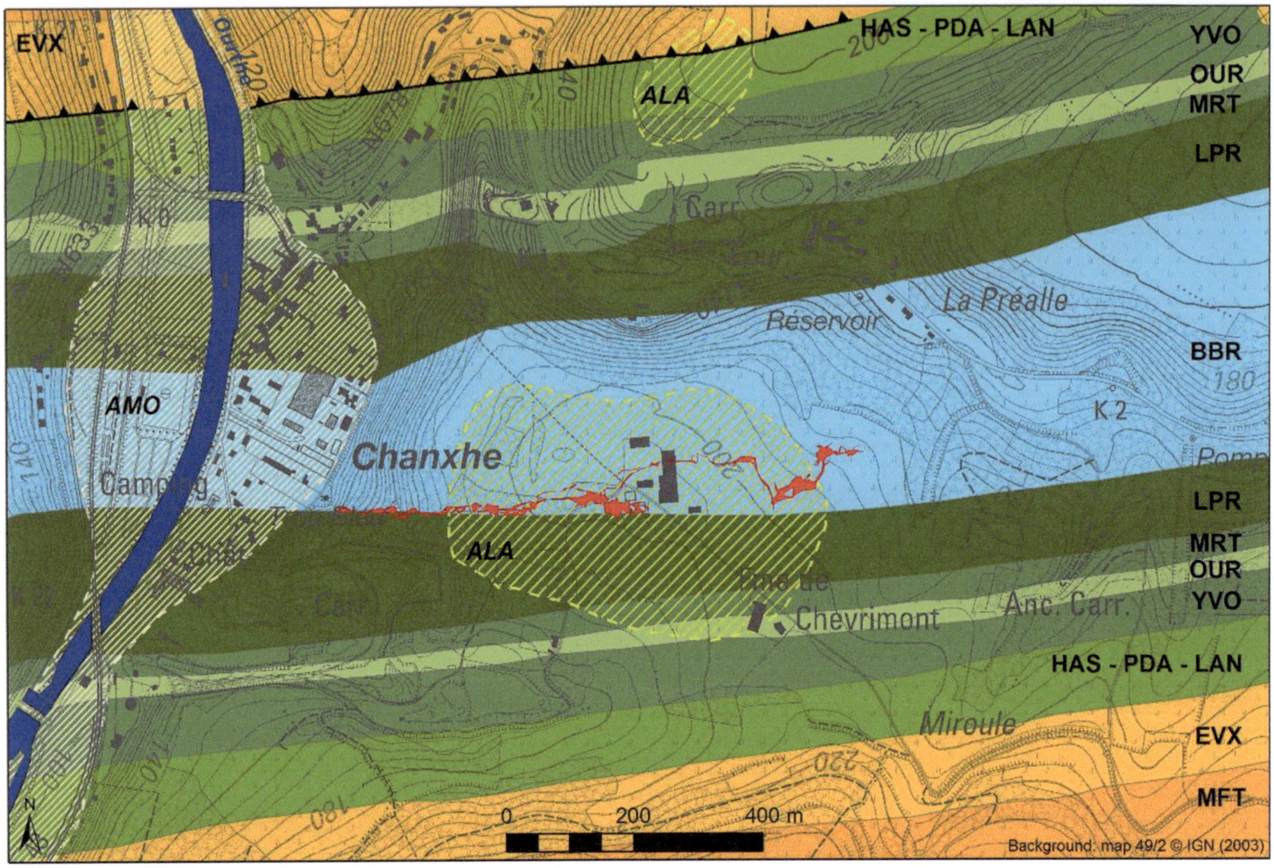

Fig. 8.19 The Noû Bleû Cave developed within the lithostratigraphic units of the Tournaisian and Visean series (Geological data from the Geological map of Wallonia, Tavier—Esneux 49/1-2, by Bellière and Marion 2009, draft version, SPW-DGO3/ULg convention). *AMO* recent alluvium deposits; *ALA* ancient alluvium deposits; *BBR* "Belle-Roche Breccia" Fm; *LPR* Longpré Fm; *MRT* Martinrive Fm; *OUR* Ourthe Fm; *YVO* Yvoir Fm; *HAS-PDA-LAN* Hastière—Pont d'Arcole—Landelies Fms; *EVX* Evieux Fm; *MFT* Monfort Fm)

calculated from the eastern entrance, amounts to 37 m and a maximum height of 24 m has been measured at the level of a chimney. The 900-m-long active part of the cave, travelled by a stream, is limited by two sumps, the downstream one located at only $\sim$20 m from the *Trou Bleu* resurgence. The cave has also several dry upper levels taking the form of either subhorizontal galleries with fluvial sediment or large halls partly filled with collapsed boulders.

8.3.2.1 Geological Setting

Generally, the underground system follows the ENE–WSW axis of the syncline. The largest part of the cave extends in the "Belle-Roche Breccia" Formation (Fig. 8.19), whose limestone and dolomite boulders are particularly well exposed in the active galleries. The downstream half of the system closely follows the faulted contact with the more or less dolomitized Longpré limestones (Fig. 8.20).

The underground drainage is influenced by the geological structure. In the downstream half of the system, the main drain itself more or less follows the faulted contact between the Longpré and "Belle-Roche Breccia" formations. The fault is observed in several places in the active galleries and the dry upper levels. By contrast, the upstream half of the system is entirely developed in the Belle-Roche breccia. Details of the karstic system are strongly controlled by the dolomitized and brecciated nature of the limestone and the density of secondary folds, faults and fractures. In places, the subvertical dip of the beds has led to the development of high galleries (Fig. 8.21a), while in other places subhorizontal bedding favoured the formation of large halls and roof collapses (Fig. 8.21b). The largest discovered hall is 40×25 m, with a maximum height of 15 m.

8.3.2.2 Active River-Run and Dry Upper Galleries

Like Remouchamps, the Noû Bleû system is a water cave characterized by an active level and dry upper levels. Dissolution by the river has resulted in a varied morphology of the active level with alternating high and narrow

Fig. 8.20 Entrances of the Noû Bleû Cave and fault separating the Longpré Formation from the "Belle-Roche Breccia" Formation (*Photo* A Peeters)

high-gradient reaches (Fig. 8.22a) and wider low-gradient reaches (Fig. 8.22b). The stream also runs through sumps and underneath boulders scattered across large halls.

The riverbed mostly exposes a sandy bedload. Likewise, some dry upper levels contain large quantities of sand derived from residues of the Oligocene marine sand cover still present in the watershed (Fig. 8.23a). Some river reaches are also characterized by coarser sediments, namely rounded pebbles including quartz and quartzite and coming from river terrace deposits located above the cave at ~200 m asl (Fig. 8.19, ALA). These pebbles have been injected in the cave through soil piping failures in sinkholes and are now found in the stream at, and just downstream of, the plug injection places (Figs. 8.22b and 8.23b).

Among the dry upper levels, notches (Fig. 8.24) and fluvial sediment accumulation attest the action of an ancient underground river, particularly in galleries more than 8 m above the active level. Other upper levels display large halls partly filled with boulders attesting the predominance of gravitational processes.

8.3.2.3 Speleothem Development

The Noû Bleû is also characterized by its richness in speleothems of extremely diverse types, shapes and colours. Their type and shape diversity evidences that all speleothem-forming hydrologic mechanisms (through flowing, dripping, seeping, pooled and condensation water) were active in the cave (Hill and Forti 1997). Halls, walls and floors are frequently covered by flowstone sheets with various stalactites, stalagmites and columns (Fig. 8.25). Dripping effect is attested by delicate forms, such as fried egg stalagmites lying on a thin base of calcified sediment (Fig. 8.26a). The seeping mechanism is represented by helictites, which grow in every direction by capillarity water seeping through tiny internal canals, and by coralloid aragonite covering the walls (Fig. 8.26b). Rimstone dams of

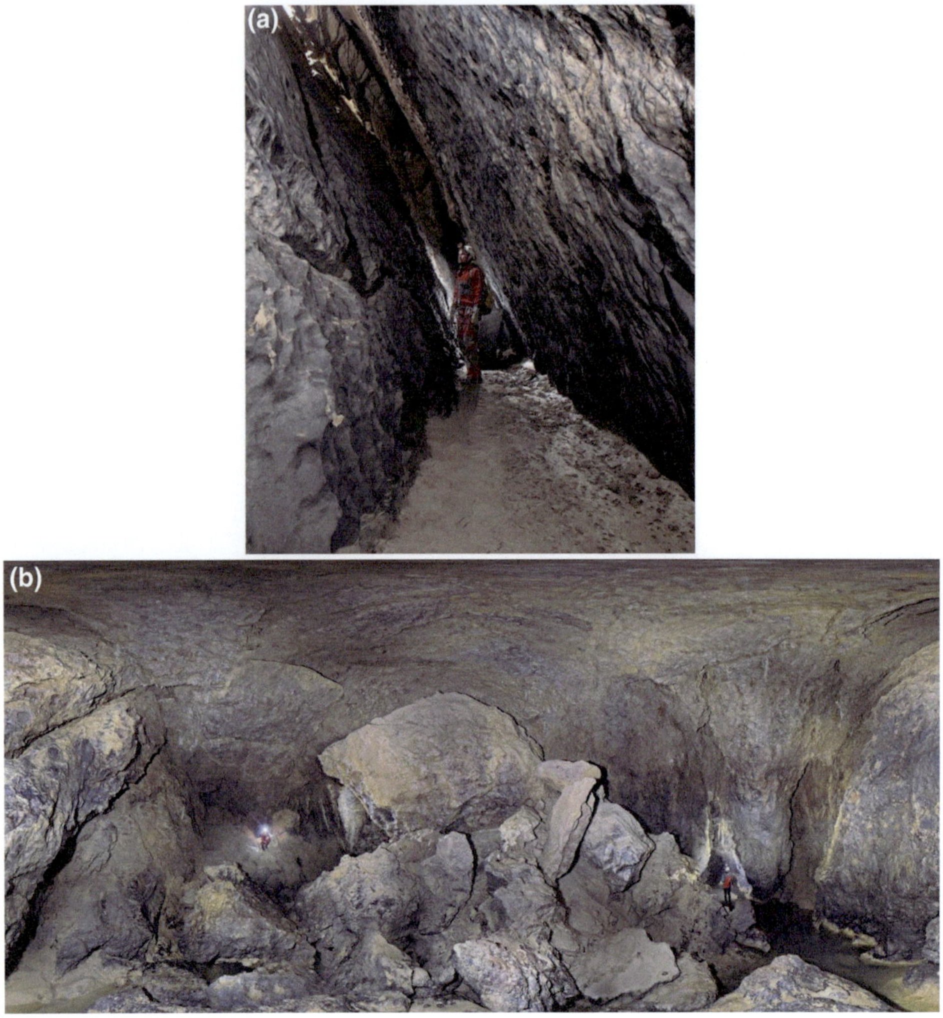

Fig. 8.21 **a** Dry upper level developed along subvertical beds in the Noû Bleû Cave (*Photo* G Rochez). ***b*** Large hall with roof collapse in the Noû Bleû Cave (*Photo* V Gerber)

various size and shape are also common (Fig. 8.26c). Cave pearls were found within a shallow rim pool. Some speleothems may result from the combination of two or more mechanisms, like draperies, formed by water dripping and flowing (Fig. 8.26d). One anthodite, a fragile speleothem composed of aragonite needle clusters radiating outwards from the wall, has also been found (Fig. 8.26e). Finally, the large amount of fine sediment allowed the formation of conulites, i.e., hollow, cone-shaped speleothems that developed in deep conical depressions created in the sand by falling water. Figure 8.26f shows such conulites in their early stage of formation.

In order to preserve the environment and to maintain the speleothems in natural immaculate conditions, measures were taken from the beginning of the cave exploration. The installation of an appropriate mark-up was immediately set up to avoid damage and dirt. Highly fragile reaches are permanently closed while others are regularly cleaned.

8.4 Human Involvements and Constraints

The Remouchamps and Noû Bleû case studies allow one to highlight the human interaction with karst environments, not only in the sense of karst constraints for humans but also in that of karst as a resource that must be preserved (Ek 1985).

8.4.1 Natural and Human-Induced Geohazards

Karst represents multiple geohazards for man-made structures, which are moreover often accelerated or triggered by human actions. Dissolution of karst terrains may induce ground instability problems. For example, the Remouchamps viaduct, which carries a motorway across the Amblève valley, has seven piers which are founded at least partly on limestone. During the excavations for foundations in the late 1970s, the discovery of large open cavities led to

Fig. 8.22 a High gradient reach of the underground stream in the Noû Bleû Cave (*Photo* V Gerber). b Low gradient reach of the same stream (*Photo* G Rochez)

some redesign of the viaduct structure and an extra cost of 15% (Waltham et al. 1986). In the early 1980s, a parcelling project for several properties was established on the land extending over the Remouchamps cave. Based on precise topographic surveys, Ek (1983) demonstrated that the rock thickness was only 8 m between the ceiling of the Cathedral hall and the surface. The presence of faults increased the risk of instability due to roof collapse. The project was therefore abandoned.

Ground subsidence induced by the generation of sinkholes causes engineering and foundation problems. This underlines the need for a proper land use policy plan which would take into account the peculiarities of karst environments. That is why the Service Public de Wallonie (SPW) has compiled a digital karst dataset, which is used to delimit areas of subsidence susceptibility and help the decision makers to manage the land use policy. Mapping karst geohazards implies several issues to be dealt with, including the dynamics of karst. Indeed, the generation or reactivation of sinkholes and swallow holes is often favoured or triggered by human activity. The surveys conducted in the Sprimont syncline for over 100 years have clearly shown that natural process can be accelerated or triggered by the presence of diverted streams or broken water conduits (Peeters et al. 2012).

The presence of swallow holes can also lead to flood geohazards. During intense rainfall events, swallow holes can be blocked (e.g., clogged with logs and other debris), resulting in flooding of the blind valley, which can cause massive destruction of property (as for the Sinkhole Dale, see Fig. 8.4).

Karst aquifers are also extremely vulnerable to pollution due to the rapid transmission of groundwater and contaminants. Consequently, aquifer protection zones are delimited

Fig. 8.23 Clastic deposits in the Noû Bleû Cave. a Sand deposit within a dry upper level (*Photo* V Gerber). b Rounded pebbles deposit in the underground stream (*Photo* JC London)

Fig. 8.24 Upper level (8 m above the underground stream) of the Noû Bleû Cave, with flat ceiling and notch on the walls (*Photo* V Gerber)

by the SPW to avoid spoiling sensitive karst ecosystems (SPW-DGO3 2014).

8.4.2 Karst Environment, a Resource that Must Be Preserved

Relations between karst environment and human activities are quite different for the two studied caves. The Remouchamps Cave, a commercial show cave, is the property of the local municipality and is rented by a private manager through a 25-year lease. Staff is partly paid by tips and the some 100,000 tourists a year make the rent a good business for the municipality and the lessee. The long boat trip at the end of the visit is of course an important source of success. The business has its own souvenir shop and its own cafeteria and restaurant. The positive impact of tourism for the village retail traders and pubs is thus limited, although not negligible. Some stretches of the touristic part of the cave are vulnerable to light pollution, leading to the growth of undesirable lampenflora (Michel 2015). A complete renewal of the lighting is under way to reduce this excessive light pollution.

Officially, the whole cave is classified by the National Commission for Sites and Monuments. No major changes are thus theoretically allowed without permission of this Commission. The "wild" (non touristic) passages of the cavern are listed as "underground cavity of scientific interest" (De Broyer et al. 1996) and, as such, legally protected against pollution and damages. Nevertheless, some pollution of the underground river is caused by upward agricultural activities and camping.

Noû Bleû Cave is quite different. The cave opens in an active quarry and is not touristic at all. It is indeed still under exploration thanks to agreements with the owners and quarry operators. A large part of the cave, including its two

Fig. 8.25 Speleothem formation in "Les Cocognes" gallery (*Photo* G Rochez)

entrances, extends on properties belonging to the quarry owners. They could have decided to close the cave and prohibit its exploration. Instead, they decided to engage in constructive dialogue with the explorers, supported by the Sprimont municipality. Using the precise topographical survey of the cave, the quarry operators could define a buffer zone protection along the boundaries of the cave. On the other hand, the topographical surveys highlighted the danger of extracting rocks on this side of the quarry because it could provide a new outlet for the underground river, causing serious flooding problem within the quarry. Organizing visits of the cave for quarry operators and quarry-workers helped to raise awareness about the importance of preserving and protecting the karst environments. Similarly, conferences for the general public were organized by the explorers.

These two examples emphasize that karst environments are extremely vulnerable to pollution and degradation but on the other hand, they can represent an economical value in terms of touristic activities and extracting rocks.

8.5 Conclusions

The Sinkhole Dale, draining hard waters from Condroz on the west and aggressive waters from the Ardenne on the east, is a unique hydrological case in Belgium. There is no talweg in this valley and swallow holes and sinkholes collect the whole precipitation and divert it underground. A careful mapping of karstic geohazards is thus essential for the administration in charge of land use policy.

The Remouchamps Cave is for numerous tourists from Belgium and neighbouring countries a unique opportunity to discover the underground environment. Hence, a mission of awareness to nature is of great importance as a

Fig. 8.26 Variety of the small-scale speleothem forms in the Noû Bleû Cave. **a** "Fried egg" stalagmite. **b** Coralloid aragonite. **c** Rim pools with candle-shaped stalagmite. **d** 2 m^2 of drapery. **e** Anthodite aragonite. **f** Conulites in the early stage of formation (*Photos* V Gerber)

pedagogic project, and also a duty in order to preserve the natural treasures of the cave, notably by keeping the light to a minimum in view of preserving the speleothems.

The Sprimont Syncline is a very well documented case of karst dynamics. Four surveys in one century give an acute perception of movements of swallow holes and sinkholes. It also emphasizes that karst dynamics is often favoured or triggered by human activity.

The newly discovered Noû Bleû Cave leading to the resurgence of this basin has caused some trouble to the explorers because a large part of the cave, including its two entrances, lies in a private quarry. Fortunately, and this is rare in the history of cave exploration, constructive agreements could be reached between the public administration, the owners, the operators of the quarry and the explorers. The usefulness of the explorers was recognised for the sake of the quarry and the quarry-workers.

References

Bellière and Marion (2009) Geological map of Wallonia (Belgium), scale 1/10.000, sheet Tavier-Esneux 49/1-2 draft version, SPW-DGO3

Boulvain F, Pingot JL (2015) Genèse du sous-sol de la Wallonie. Académie royale de Belgique, Bruxelles, p 208

Boulvain F, Bultinck P, Coen M, Coen-Aubert M, Lacroix D, Laloux M, Casier J-G, Dejonghe L, Dumoulin V, Ghysel P,

Godefroid J, Helsen S, Mouravieff NA, Sartenaer P, Tourneur F, Vangestaine M (1999) Les formations du Frasnien de la Belgique. Mem Geol Sur Belg 44:1–125
Bretz JH (1942) Vadose and phreatic features of limestone caverns. J Geol 50:675–811
Briffoz A (1983) Sprimont: le sous-sol régional à la loupe, Les cahiers de Spéléo Flash N°3. Commission des publications de la Fédération spéléologique francophone de Belgique, Bruxelles
Briffoz A (2012) Descriptif du karst Sprimont-Chanxhe In: Livret-guide de l'excursion de la 16ème édition des Journées de Spéléologie scientifique, Centre belge de l'étude du Karst et Commission scientifique de l'Union belge de spéléologie, 8–9 Dec 2012
Coen M (1970) Stratigraphie du Frasnien de la grotte de Remouchamps. Ann la Soc géol Belg 93:73–79
Courtois H, Courtois J, Denoël R, Ek C (1969) Les explorations de la grotte de Remouchamps, notice historique. Parcs Nationaux 24: 19–29
Delecour F, Weissen F, Ek C (1968) An electrolytic field device for the titration of CO_2 in air. Natl Speleol Soc Bull 30:131–136
De Broyer C, Thys G, Fairon J, Michel G, Vrolix M (1996) Atlas du Karst wallon, province de Liège, 3 vol. Commission wallonne d'Étude et de Protection des Sites souterrains, La Hulpe, p 896
Delhasse A (1851) La grotte de Remouchamps, près de Spa, avec notes historiques et orné d'une vue et d'un plan de la grotte. Bruxelles, p 114
Dewez M (1974) Nouvelles recherches à la grotte de Remouchamps. Bull Soc r belge Anthropol Préhistoire 85:1–161
Ek C (1957) Les terrasses de l'Ourthe et de l'Amblève inférieures. Ann Soc Géol Belg 80:333–353
Ek C (1961) Conduits souterrains en relation avec les terrasses fluviales. Ann Soc Géol de Belg 84:313–340
Ek C (1969) Facteurs, processus et morphologie karstiques dans les calcaires paléozoïques de la Belgique, vol 3. Unpublished PhD thesis, Université de Liège
Ek C (1970a) Carte géologique de la grotte de Remouchamps (Belgique). Notice explicative. Ann Soc Géol Belg 93:287–292
Ek C (1970b) Les influences structurales sur la morphologie de la grotte de Remouchamps (Belgique). Ann Soc Géol Belg 93:293–304
Ek C (1973) Analyse d'eaux des calcaires paléozoïques de la Belgique. Serv Géol Belgique, Prof Pap 18:1–33
Ek C (1974a) Le cadre géologique et géomorphologique In: Nouvelles Recherches à la Grotte de Remouchamps. Bull Soc R Belge Anthropol Préhistoire 85:7–15
Ek C (1974b) Étude sédimentologique dans la Grotte de Remouchamps (avec la collaboration de S. Alexandre-Pyre et E. Juvigné) In: Nouvelles recherches à la grotte de remouchamps. Bull Soc R Belge Anthropol Préhistoire 85:16–41
Ek C (1983) L'eau dans le Vallon des Chantoirs (Remouchamps). Une leçon sur les collaborations nécessaires. La prot Eaux karstiques. Hydrographica 1–2:7–14
Ek C (1985) De la connaissance du milieu karstique à sa gestion. Ann Soc Géol Belg 108:303–304
Ek C (1995) Grottes et rivières des régions calcaires In: Demoulin A (éd) L'Ardenne, Essai de Géographie physique, pp 178–193
Ek C, Michel R, Mousny V, Closson D (1997) The dynamics of the karstic features of Sprimont (Belgium) and its consequences on the land-use planning. In: Proceedings of the 12th International Congress of Speleology, vol 5. Switzerland, pp 13–14
Field MS (ed) (1999) A lexicon of cave and karst terminology with special reference to environmental karst hydrology. Report EPA/600/R-99/006. U.S. Environmental Protection Agency, Washington, DC, p 195
Gewelt M (1985) Cinétique du concrétionnement dans quelques grottes belges: apport des datations 14C et 230Th/234U. Ann Soc Géol Belg 108:267–273
Hill CA, Forti P (eds) (1997) Cave minerals of the world, 2nd edn. Natl Speleol Soc, Huntsville, AL, p 463
Juvigné E, Gewelt M (1988) Téphra et dépôts des grottes: intérêt stratigraphique réciproque. Ann Soc Géol Belg 111:135–140
Meus P, Moureaux P, Gailliez S, Flament J, Delloye F, Nix P (2014) In situ monitoring of karst springs in Wallonia (Southern Belgium). Environ Earth Sci 71(2):533–541. doi:10.1007/s12665-013-2760-x
Michel G (2015) Les enjeux d'une exploitation touristique durable. Remouchamps: perle de l'Amblève ou jungle souterraine? Ecokarst 100:10–11
Michel R (1971) Le Vallon de Sprimont à Chanxhe. Unpublished Ms thesis, université de Liège, p 163
Mottequin B, Marion JM (2012) Lithostratigraphie du Tournaisien et du Viséen (Carbonifère) dans le synclinal de Chanxhe-Sprimont (province de Liège, Belgique) In: Livret-guide de l'excursion de la 16ème édition des Journées de Spéléologie Scientifique, Centre belge de l'étude du Karst et Commission scientifique de l'Union belge de Spéléologie, 8–9 Dec 2012
Paproth E et al (1983) Bio- and Litho-stratigraphic subdivisions of the Dinantian in Belgium, a review. Ann Soc Géol Belg 106:185–239
Peeters A, Ek C, Labarbe P & Michel R (2012) La dynamique du karst dans le synclinal de Sprimont de 1898 à 2012. In: Livret-guide de l'excursion de la 16ème édition des Journées de Spéléologie Scientifique, Centre belge de l'étude du Karst et Commission Scientifique de l'Union belge de spéléologie, 8–9 Dec 2012
Pissart A, Van Vliet-Lanoë B, Ek C, Juvigné E (1988) Des traces de glace de ségrégation dans la grotte de Remouchamps (Belgique). Conséquences en ce qui concerne la sédimentation et la paléoclimatologie. Ann Soc Géol Belg 111:125–133
Poty E, Hance L, Lees A, Hennebert M (2001) Dinantian lithostratigraphic units (Belgium). In: Bultynck P, Dejonghe L (eds) Guide to a rivised lithostratigraphic scale of Belgium. Geol Belg 4:69–94
Rahir E (1920) L'habitat tardenoisien des grottes de Remouchamps, Châleux et Montaigle. Bull Soc Anthropol Brux 35:31–89
Rixhon G, Demoulin A (2010) Fluvial terraces of the Amblève: a marker of the Quaternary river incision in the NE Ardennes massif (Western Europe). Z für Geomorphol 54(2):161–180. doi:10.1127/0372-8854/2010/0054-0008
Schols M (1832) Description de la grotte de Remouchamps, située à deux lieues à l'ouest de Spa. Bruxelles, p 8, one plan, 8 engravings
SPW-DGO3 (2014) État des nappes d'eau souterraine de Wallonie. Service public de Wallonie, DGO 3 (DGARNE), Belgique. Dépôt légal D/2014/11802/11—ISBN 978-2-8056-0142-2
Van den Broeck E (1902) Quelques mots à propos des nouvelles fouilles exécutées dans la grotte de Remouchamps. Bull Soc Anthropol Brux 21:35–44
Van den Broeck E, Martel EA, Rahir E (1910) Les cavernes et les rivières souterraines de Belgique, Bruxelles, 2 vol, p 1592 + 92 + LXVI
Waltham AC, Vandenven G, Ek C (1986) Site investigation on cavernous limestone for the Remouchamps Viaduct, Belgium. Ground Eng 19(8):16–18
Weninger B, Jöris O (2008) A ^{14}C calibration curve for the last 60 ka: the Greenland-Hulu U/Th timescale and its impact on understanding the Middle to Upper Paleolithic transition in Western Eurasia. J Hum Evol 55:772–781

The Karstic System of Han-sur-Lesse

9

Yves Quinif and Vincent Hallet

Abstract

The karstic system of Han-sur-Lesse is the archetype of underground meander cutoff. Located in the "Calestienne", a narrow bench extending along the northern margin of the Lower Devonian Ardenne, it comprises all caves and underground flow paths related to the epigenetic Lesse River across the Givetian limestone anticline of Boine. The Han-sur-Lesse caves result from a two-stage karstogenesis. First, ghost-rock karstification, the residual alterite of which is still observed locally in the system, affected the substratum of a fairly uniform topography under hot and wet climates, probably during the Cretaceous and the Paleocene. Second, as a result of the Ardenne uplift during the Neogene, river incision created a hydraulic potential that led to removal of the alterite by underground water circulation. Currently, the underground circulation short-circuits a meander of the Lesse, from the swallow hole of "Gouffre de Belvaux" to the resurgence at "Trou de Han". Another small underground stream is separated from the main circulation by less permeable nodular shale and argillaceous limestone strata. The overall system is essentially made of horizontal galleries, with large rooms located at structural nodes or produced by the coalescence of superposed galleries. The river accumulated pebbles, sand, and clay in the caves, often capped by flowstones, the whole recording Quaternary climate oscillations.

Keywords

Karstic system • Ghost-rock karstification • Karstic hydrogeology • Han-sur-Lesse • Underground meander cutoff

9.1 Introduction

Developed in Givetian limestones along the northern margin of the Lower Devonian Ardenne, the internationally renowned caves of Han-sur-Lesse form the greatest karstic system of Belgium and represent the archetype of an underground meander cutoff. This group of caves (caves of Han-sur-Lesse, "Père Noël", "Lesse Souterraine", "Trou des Crevés") belongs to the karstic system of the underground Lesse River. Since the beginning of the 1980s, multiple scientific studies, in particular in the fields of hydrogeology, paleoclimatology, underground sedimentology and, of course, karstology, highlighted its geological and geomorphological characteristics, which together allow inferences about the system as part of the regional landscape evolving in the frame of long-term morphostructural and paleoenvironmental changes since the Mesozoic. In many areas, e.g. in western Belgium, it has been shown that the classical speleogenesis may be preceded by an initial process of ghost-rock formation (Quinif 2010; Dubois et al. 2014), a form of partial limestone weathering leaving a residual alterite similar in structure to the isalterite of kaolinic

Y. Quinif (✉)
Géologie fondamentale et appliquée, University of Mons, rue de Houdain, 9, 7000 Mons, Belgium
e-mail: bouqui@skynet.be

V. Hallet
Department of Geology, University of Namur, rue de Bruxelles 61, 5000 Namur, Belgium
e-mail: vincent.hallet@unamur.be

A. Demoulin (ed.), *Landscapes and Landforms of Belgium and Luxembourg*, World Geomorphological Landscapes, DOI 10.1007/978-3-319-58239-9_9

weathering profiles in siliceous rocks (see Chap. 5). This process requires specific conditions combining a warm humid climate and low regional altitudes that point to an activity during the end of the Mesozoic. In western Belgium for instance, the presence of a Cretaceous cover dates the ghost-rock formation in Lower Carboniferous limestones to the period between the Late Jurassic and the Campanian (Quinif 2010). The Han-sur-Lesse karstic system offers a unique opportunity to check this hypothesis and suggests that, if ghost-rock residual alterite ever existed there, its erosion should have occurred as a result of the regional uplift of High Belgium and subsequent river incision, which occurred essentially during the Late Neogene and the Quaternary. River sediments and flowstone capping are present in the two levels of the generally two-storeyed caves. Pollen and isotopic analyses and dating (^{14}C, U/Th, laminae counting) of the speleothems have shed light on the temporal relations between uplift, incision, climate oscillations, and cave development. Moreover, the presence of several great rooms with fallen blocks allows for the study of formation of such voids in possible relation with earthquakes.

In this chapter, we provide an overview of various characteristics of the Han-sur-Lesse karstic system. After having given a panorama of the morphological features and deposits encountered in the caves, we focus on the structure and dynamics of the subterranean water circulation. Then, based on all available data, we show that the long-term evolution of the system fits in the two-stage conceptual frame of karstogenesis first established in western Belgium.

9.2 Geological and Geomorphological Settings

The Han-sur-Lesse area belongs to the Paleozoic massif of the Ardenne *s.l.*, folded and faulted during the Variscan orogenesis (Overlau and Quinif 1979; Delvaux de Fenffe 1985; Dumoulin and Blockmans 2013; see also Chap. 2). Located in between the Lower Devonian core of the high Ardenne to the south and the Upper Devonian/Lower Carboniferous Famenne and Condroz regions to the north, it includes Givetian hills surrounded by depressions carved in Eifelian and Frasnian shales (Fig. 9.1). The geological structure is mainly formed by faulted E–W folds affecting heavily jointed rocks (Fig. 9.2). Over the Mesozoic, the Variscan topography was eroded to a monotonous, low-elevation planated landscape, while the dense vegetation cover developed under Cretaceous warm and humid climates allowed for effective weathering of limestone rocks. Although this is not definitively demonstrated, the Paleogene marine transgressions that deposited sands and clays on parts of the Ardenne plausibly briefly drowned the study area (see Chap. 5). This is to some extent supported by the epigenetic character of rivers such as the Lesse, whose first installation took place, probably in the Oligocene, on either loose cover sediments or a thick weathering mantle, independently of the basement structural grain and the variable resistance to erosion of the rocks. Only recently, during the Late Neogene and, especially, the cold periods of the Quaternary, has differential erosion shaped an Appalachian topography of alternating ridges on hard rocks and depressions in weaker rocks. Depending on the respective contrasting rocks, while the Condroz region displays the classical scheme of Upper Famennian sandstone ridges separated by depressions in Lower Carboniferous limestones, the Han-sur-Lesse area opposes ridges corresponding to Givetian limestones and depressions in Eifelian or Frasnian shales (Poty 1976).

Draining the Lower Devonian siliceous basement of the Ardenne in its upper course, the Lesse River comes across the Givetian limestones of the Calestienne, which form a bench at 280–300 m elevation along the NW margin of the Ardenne massif, before flowing to the NW toward the depression carved in Lower Famennian shales. At Resteigne, the river passes through a first limestone massif where it cuts a deeply incised valley (Fig. 9.1). Then, after having crossed the shaly depression of Belvaux, it flows across the limestone hill of Han (also named the hill of Boine). There, the valley displays a wide but generally dry meander to the east and north of the hill, while the river sinks into the swallow hole of "Gouffre de Belvaux" and comes out at "Trou de Han" after the subterranean crossing of the hill. Likewise, 5 km to the southwest, the Ry d'Ave takes an underground course when traversing the limestone massif of Wellin. From Trou de Han, the Lesse finally takes again a surface course across another shaly depression at Han (Fig. 9.1). The Boine massif, which contains the karstic system of Han, is a W-plunging faulted anticline (Fig. 9.2, Wavreille anticline). Cutting the western part of the karstic system, a 1-km-long NNW-trending, east-dipping fault shows a reverse motion attesting a Late Variscan phase of E–W compression (Delvaux de Fenffe 1985) (Fig. 9.2). Finally, many joints, arranged in two N50°E–N70°E and N140°E–N150°E main systems, play an important role in the karstogenesis (Havron et al. 2004).

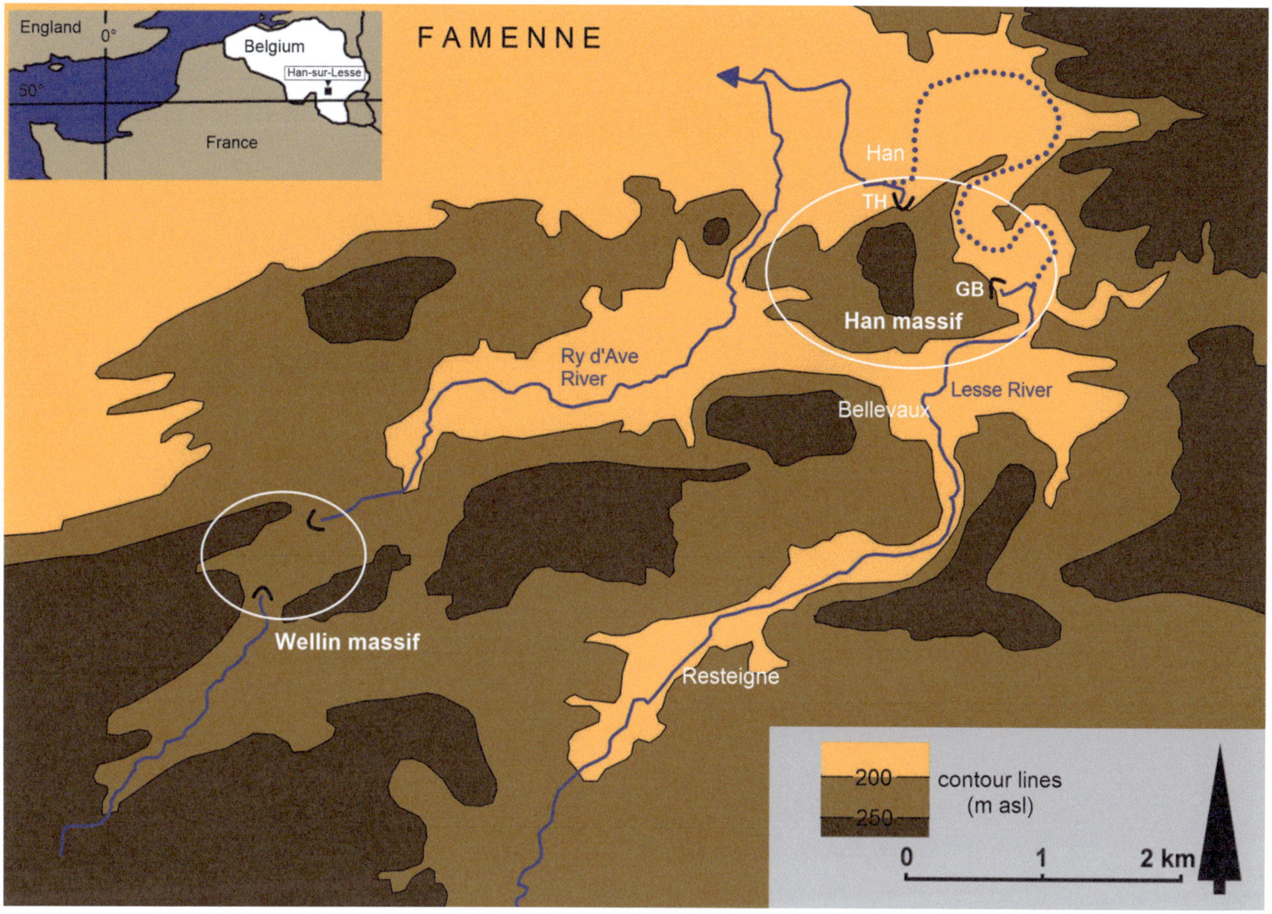

Fig. 9.1 Location of the limestone massifs of the Han-sur-Lesse—Wellin area on a simplified oro-hydrographic map (200 and 250 m contour lines). *Circled in white*, the Han and Wellin massifs include karstic systems related to meander cutoff by the Lesse and Ry d'Ave, respectively. Two other massifs at Resteigne and Grignaux-Turmont (west of Han), though also crossed by these rivers, show no significant karstic development. The *dotted blue line* denotes the abandoned surface course of the Lesse upstream of Han. *GB* Gouffre de Bellevaux. *TH* Trou de Han

9.3 The Karstic Network: Morphology and Deposits

9.3.1 Structure of the System

After a ~1 km surface course across the limestones of the Boine hill's southern flank, the Lesse disappears totally in the swallow hole of "Gouffre de Belvaux" and takes a subterranean course across the rest of the Boine massif (Figs. 9.3 and 9.4). At the surface, the valley continues around the hill as a flat, dry valley floor bordered by steep wooded slopes. The subterranean system of caves that cuts this meander off starts with the "Lesse Souterraine" cave immediately downstream of the "Gouffre de Belvaux", essentially a major gallery travelled by the underground Lesse, and goes downstream in the touristic cave of Han, ending in the resurgence of "Trou de Han" (Fig. 9.3). Two other caves, the smaller "Trou des Crevés" and the "Grotte du Père Noël", are disconnected from the main system. The course of the underground river has been explored and followed in its totality by divers (Quinif and Bastin 1986; Quinif 1987a, b, 1988a). Beyond the "Gouffre de Belvaux", the system includes other swallow holes, located more downstream along the margin of the abandoned valley floor, among which the "Trou d'Enfaule" is still active for discharges >25 m^3/s (Quinif 1988a) in high water periods, and the higher lying "Trou du Salpêtre" is the permanently dry entrance of the touristic underground circuit (Fig. 9.3).

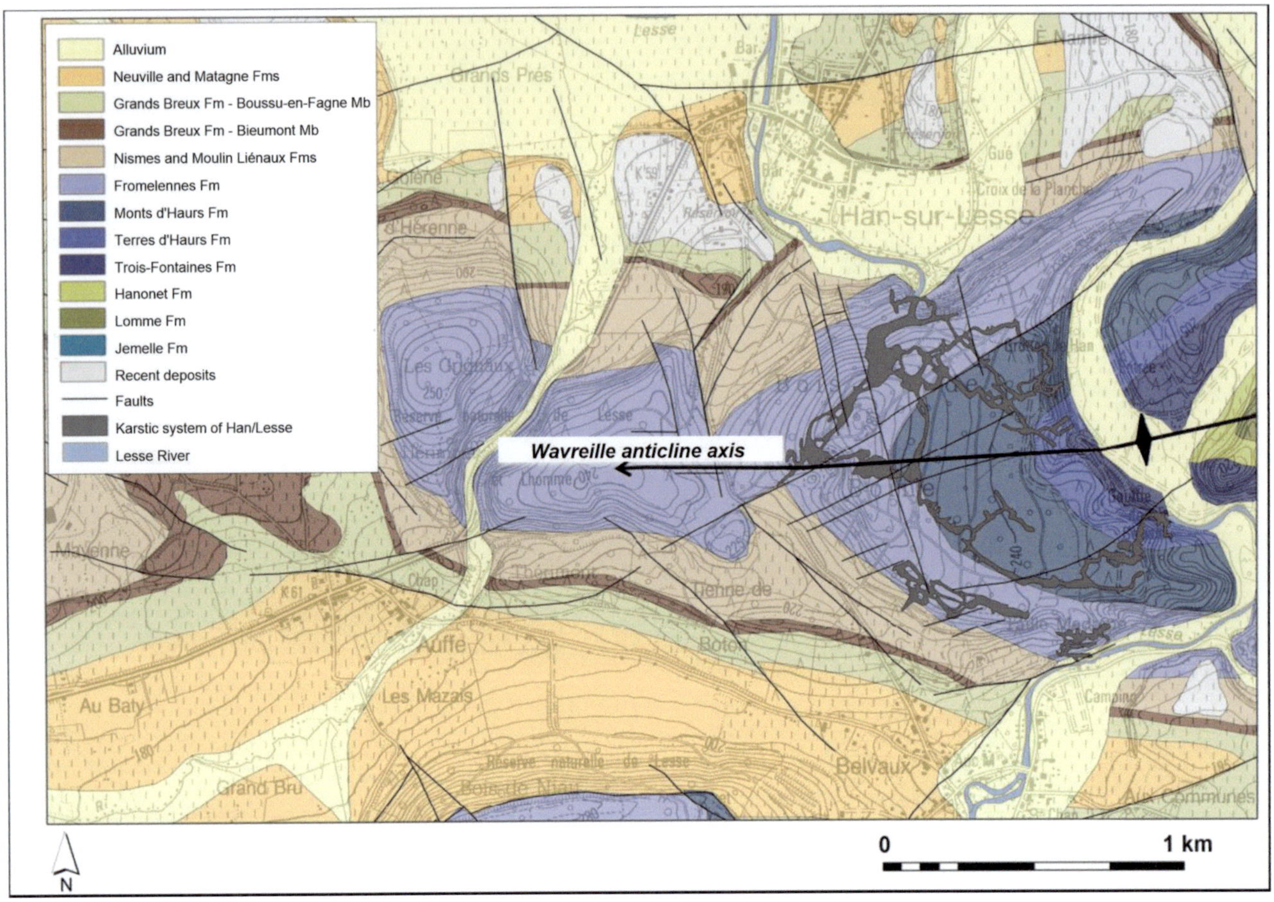

Fig. 9.2 Geological map of the massifs of Boine (in the east) and Grignaux-Turmont (in the west) along the axis of the Wavreille anticline (from Bonniver 2011, modified after Dumoulin and Blockmans 2013). The Givetian limestone formations appear in *blue* tones

These entrances open in distinct gallery systems that finally meet in the "Salle d'Armes" room (Van Den Broeck et al. 1910; Kaisin and de Pierpont 1939). According to Quinif (1988b), the galleries of the Han system extend over a total length of almost 14 km.

From the geological point of view, the karstic system is developed within two main ensembles, the first one encompassing the Lower/Middle Givetian Trois-Fontaines, Terres d'Haurs and Mont d'Haurs Formations, and the second one corresponding to the Upper Givetian Fromelennes Formation. These two ensembles are separated by the watertight Flohimont Member, which marks the base of the Fromelennes Formation and is made of shales and argillaceous limestones. While the "Trou des Crevés" and "Père Noël" caves are developed in the Fromelennes Formation of the anticline's southern flank (Deflandre et al. 1987), a large part of the Cave of Han, including the "Réseau sud" and the exit gallery of the Lesse (Quinif 1987b), lies in the same Formation on the northern limb of the anticline (Fig. 9.2). The rest of the system, i.e. mainly the "Lesse souterraine" cave and the other galleries starting from the swallow holes in the abandoned meander, extends within the Lower/Middle Givetian limestones, on the other side of the Flohimont barrier.

The main galleries, and obviously the "Lesse souterraine", develop at or near the river base level. Upper levels consist only of the "Verviétois" gallery (Fig. 9.5), accessed by the "Trou du Salpêtre" and situated ~15 m above the present Lesse, the upper part of the largest collapse rooms, and some cupola systems. The "Père Noël" cave is the only exception. There, a small water flow is observed only very locally at the base level and the main gallery, including the wide and extended, up to 30-m-high "Bivouac" room, is located higher, while a smaller room has still been found ~20 m above this main level (Verheyden et al. 2008).

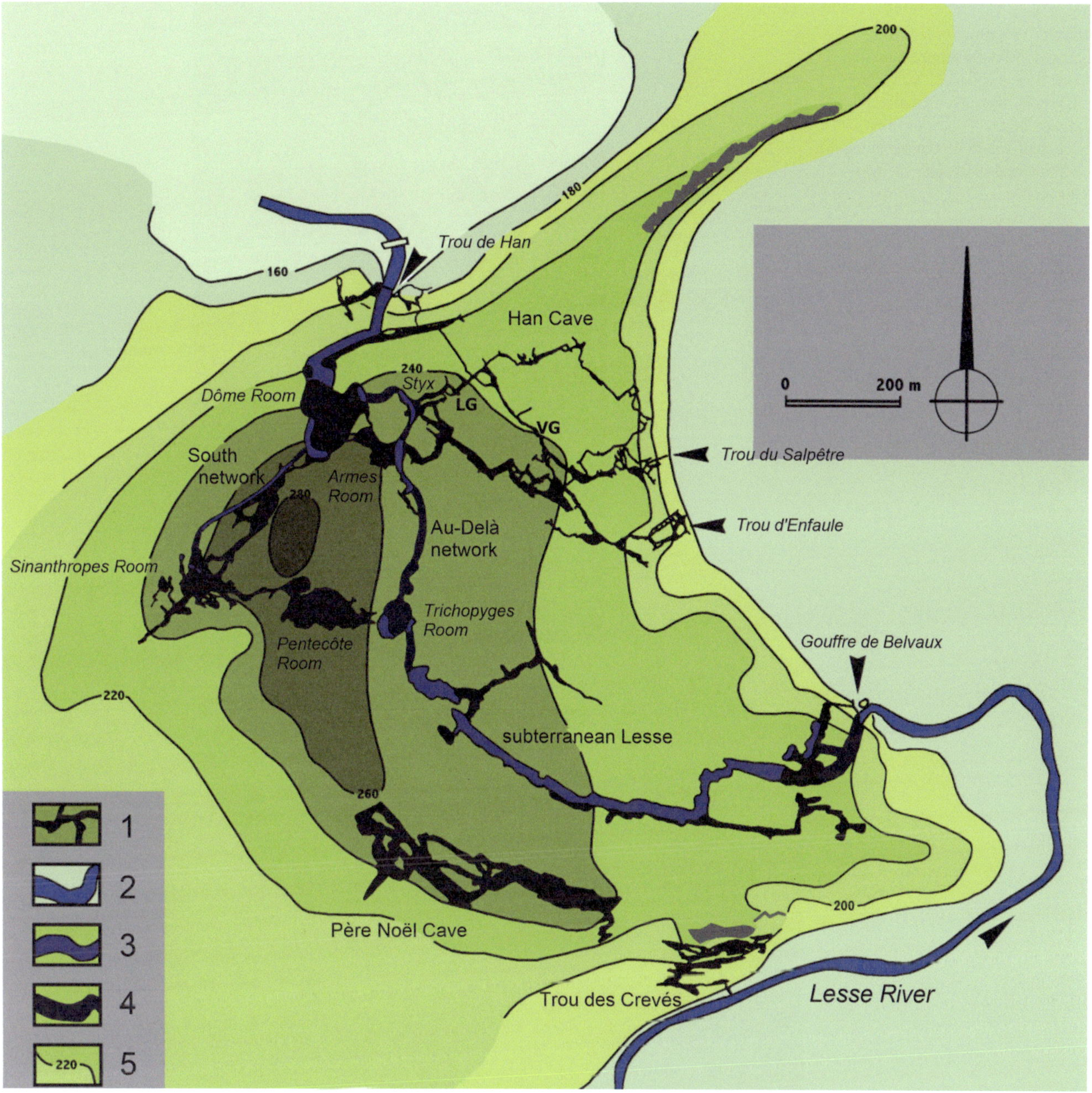

Fig. 9.3 General map of the karstic system of Han-sur-Lesse. *1* Layout of the caves. *2* Subaerial river. *3* Underground river with a free surface. *4* Galleries, including sumps, and rooms. *5* Contour lines of the surface topography. *VG* "Verviétois" gallery. *LG* "Lannoy" gallery. Note the prevailing orientation of galleries following the conjugate N60°E and N150°E joint directions (modified after Quinif 1987)

9.3.2 Morphology and Deposits of the Han Cave

9.3.2.1 From the "Trou du Salpêtre" to the "Styx" Room: A Sedimentary Sequence of the Middle and Upper Pleistocene

The "Trou du Salpêtre" entrance of the cave, an old swallow hole of the Lesse River, is the starting point of the upper level of the karstic system, where the "Verviétois" gallery, connecting successive large ("Scarabées", "Vigneron", "Mystérieuses") and smaller (e.g. "Chaos") rooms, occasionally scattered with fallen roof blocks, is followed by the "Lannoy" gallery leading to the "Salle d'Armes" room (Figs. 9.3 and 9.5). Such galleries and the succession of large rooms they connect are characteristic features of the Han Cave. The higher than large "Verviétois" gallery developed following major N150°E vertical joints.

This upper level contains a sedimentary sequence comprised of a basal detrital unit and a capping stalagmitic complex made of a thick flowstone and stalagmites

(a)

(b)

Fig. 9.4 **a** The "Gouffre de Belvaux" swallow hole, where the Lesse River disappears totally. **b** For flow discharges exceeding ~25 m^3/s, the cave is saturated and overflows. The river reuses its ancient subaerial course along the abandoned meander of the "Chavée"

(Fig. 9.6). The several m-thick basal stratified fine sands become more clayey upwards and end in a thin settling facies, bearing witness to a finishing phase of water flow at this level. Small sandy channels within this unit indicate temporary weak resumption of flow. Except for two dates pointing very locally to MIS 7 at the base of the 1-m-thick flowstone, multiple U/Th dating of the stalagmitic complex yielded a consistent set of ages encompassing the whole marine isotopic stage 5 (~128–70 ka) (Quinif 1991). The colder substages 5.2 and 5.4 are identified by variations in

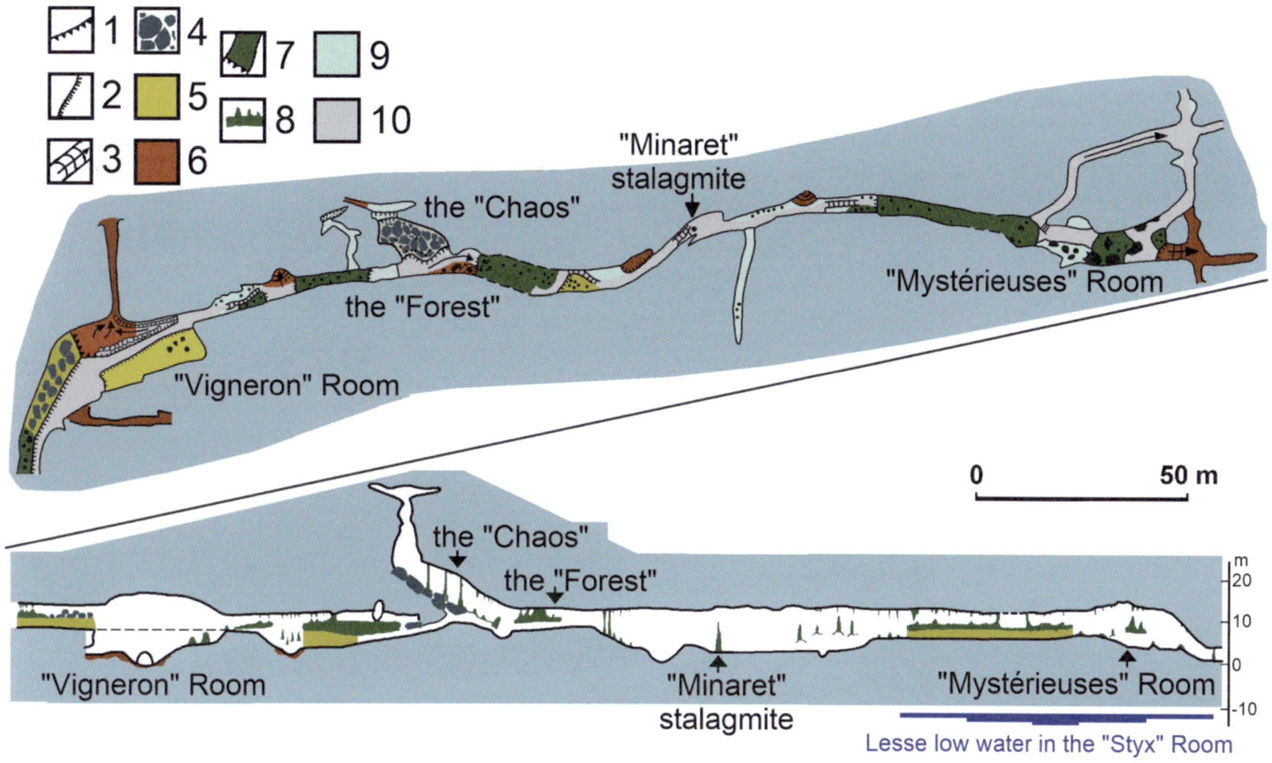

Fig. 9.5 Map (*top*) and longitudinal section (*bottom*) of the "Verviétois" gallery in the Han Cave (the "Trou du Salpêtre" swallow hole is on the *left*, see Fig. 9.3). *1* Escarpment >5 m. *2* Escarpment <5 m. *3* Slope of the ground. *4* Collapsed blocks. *5* Detrital sediment accumulation. *6* Loams deposited during recent high floods. *7* Flowstone. *8* Flowstone and stalagmites in section *9* Limestone bedrock. *10* Touristic path. The successive sections of thick flowstone are at the same elevation and systematically cap a detrital sequence. The deepened parts of the gallery result from either collapses between two superposed karstic levels (e.g. the "Vigneron" room) or decanting into lower voids

the speleothem growth rate without interlayered sedimentation (Quinif and Bastin 1994). The underlying sands and clays are thus attributed to MIS 6.

9.3.2.2 The "Petites Fontaines" Gallery and the Lower Filling

The "Petites Fontaines" is a small dead-end gallery lying a few metres above the current low water level of the underground Lesse, near the exit of the river at "Trou de Han" (Fig. 9.7a). Though at a lower level, the "Petites Fontaines" displays a partial filling fairly similar to that observed in the "Verviétois" gallery, again superposing thick speleothems over several metres of detritic material (Fig. 9.7b) (Blockmans et al. 1999). Here, the basal unit is made of thinly layered clays locally interrupted by centimetric sand layers, evidencing lake deposition occasionally disturbed by temporary water flow. The sedimentological analysis of these deposits assigned them to the Late Glacial (Blockmans et al. 1999), which is confirmed by the 8–9 ka ages of charcoal found at the base of the overlying flowstone (Fig. 9.8). We thus find the same sedimentary sequence in the "Verviétois" and "Petites Fontaines" galleries, except that they refer to two successive glacial interglacial cycles. Detrital sediments are accumulated under quiet water flow to lake conditions during a cold period before being sealed by speleothems that develop in the temperate conditions of the next interglacial. This scheme, also recognized in Remouchamps (see Chap. 8), may be envisioned within the frame of the biorhexistatic theory of Erhart (1967). In this view, speleothem growth is the subterranean counterpart of soil development at the surface under the chemically driven biostatic equilibrium conditions of interglacials (assuming no tectonic interference), whereas accumulation of river sediments in the cave is the subterranean response to their increased erosion at the surface

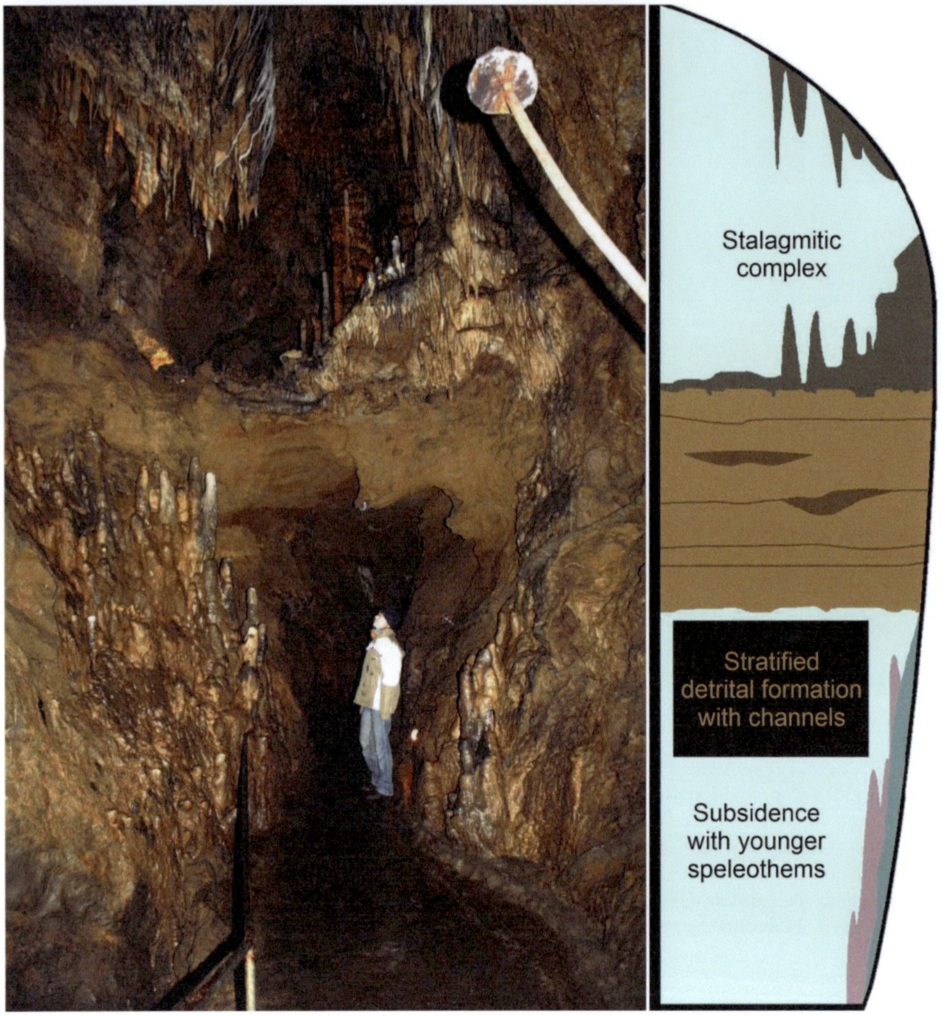

Fig. 9.6 View of the filling in the "Verviétois" gallery observed from a low level caused by subsidence into a lower void. This event may be dated thanks to the speleothems that grow on the flank of the collapse. The flowstone overlying the detrital sequence is dated to marine isotopic stage 5, defining a minimum age of the sediment accumulation

during the mechanically driven rhexistatic episodes induced by the cold periods, which act here as perturbations of the biorhexistatic cycle.

9.3.2.3 The "Dôme" Room

Large collapse rooms are a characteristic feature of the karstic system of Han-sur-Lesse. Most of them are structurally controlled. With the exception of the "Vigneron" room, which results from the collapse of the roof of a lower level, connecting it with the higher lying gallery (Fig. 9.5), they develop in relation with geological discontinuities such as faults, joints, or anticlinal hinges, the initial voids enlarging mainly in the vertical direction by successive breakdowns along bedding planes. In several rooms of the lower level (e.g. the "Dôme", "Antiparos", "Cornet", "Sinanthropes", and "Pentecôte" rooms), the cones of fallen blocks are soaked by the underground water flow.

The "Dôme" room is a representative example of the structure and morphology of such voids (Quinif and Bastin 1987; Fig. 9.9). In its eastern part, it developed within the very brittle argillaceous limestones of the Flohimont Member. In the west, the room is limited by a ~N350°E thrust fault and the associated overturned anticline (Delvaux de Fenffe 1985; Fig. 9.10). Breakdown progression and enlargement of the room occur in the southward direction, up-dip of the stratification, while the Lesse soaks the basis of the large cone of fallen blocks to the N and NE.

9.4 The Subterranean Water Systems

Since 2004, the hydrogeology of the Boine Massif and the Han-sur-Lesse karstic system has been intensively studied (Bonniver 2011; Poulain et al. 2015; Dewaide et al. 2016).

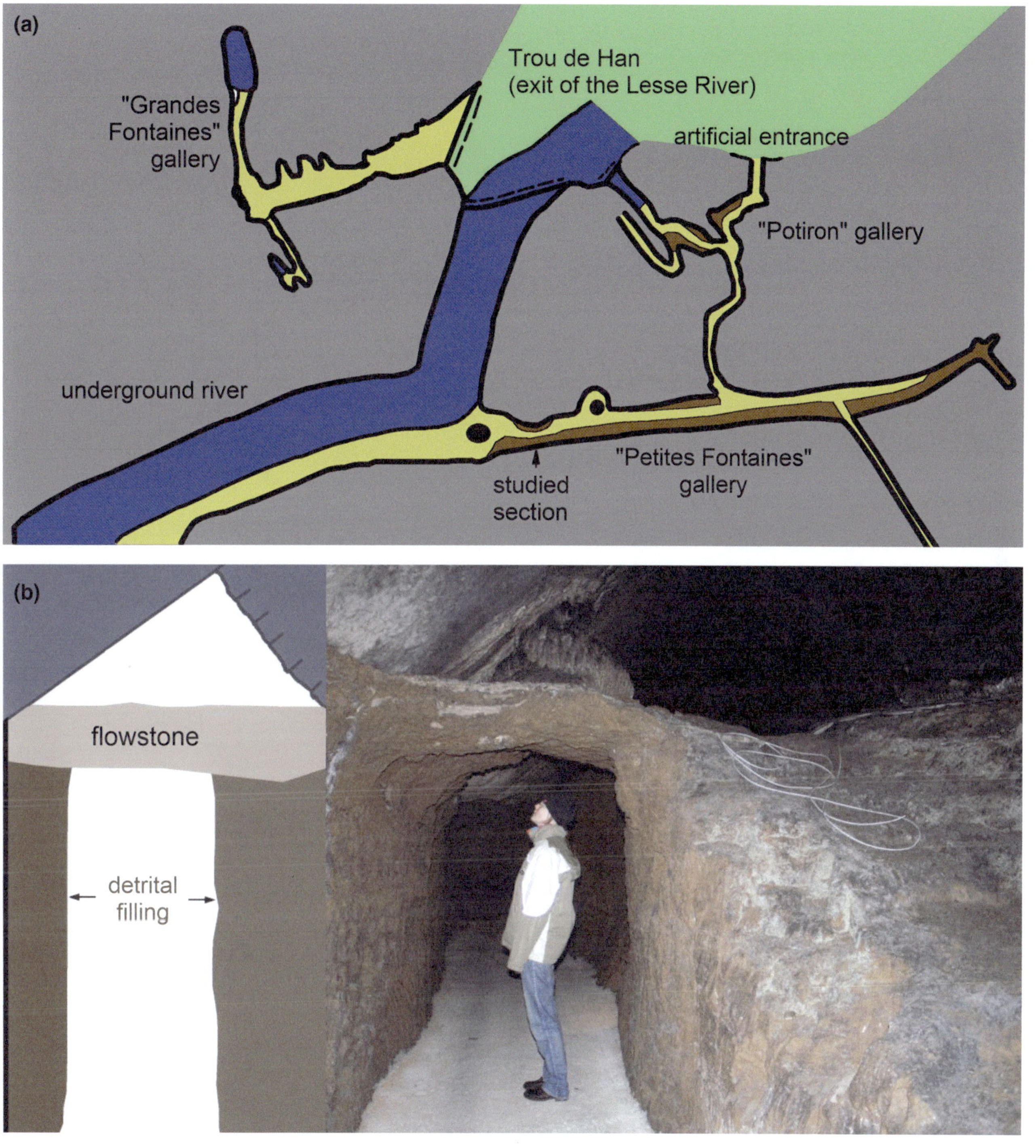

Fig. 9.7 **a** Plan view of the "Petites Fontaines" gallery in its environment close to the exit of the Han Cave. The sedimentary filling appears in *brown*. **b** Sedimentary section observed in the gallery, showing the same superposition of detrital sediments and a thick flowstone also found in the "Verviétois" gallery. Paleoclimatically, this superposition evidences the succession of the cold and warm stages of a glacial–interglacial cycle

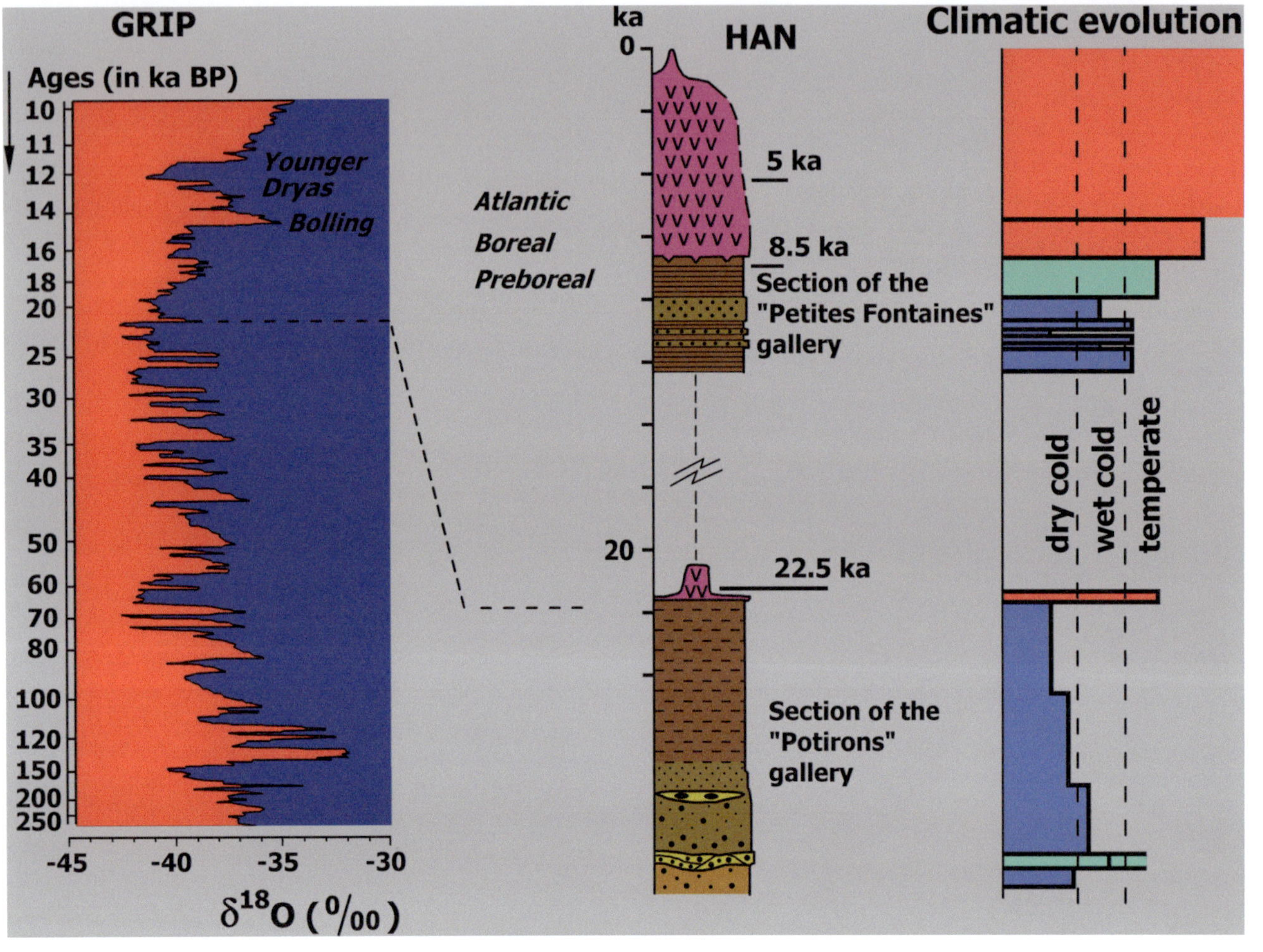

Fig. 9.8 Climatic interpretation of the filling of the "Petites Fontaines" gallery. The stratigraphic column shown in the *centre* is a composite of observations and results in the "Petites Fontaines" and the adjacent "Potirons" galleries. Speleothems are represented in *magenta*, silty (*dashed*) to clayey (*hatched*) sediments in *brown*, sands in *green brown*. The sediments of the "Petites Fontaines" gallery are stratigraphically higher than those of the "Potirons" gallery, both being located in time by dated overlying stalagmites ("Potirons") or flowstone ("Petites Fontaines"). Ages in ka BP are derived from U/Th and pollen analyses on speleothems and radiocarbon on charcoal in the upper fine sediments of the "Petites Fontaines". *Left* excerpt of the $\delta^{18}O$ curve of the GRIP ice core in Greenland sets the climatic background of the sedimentary sequence at Han. The stalagmite developed over the "Potirons" sediments may be linked to a warming peak. *Right* climatic interpretation of the data, evidencing oscillations from dry cold through wet cold to temperate conditions. The *red* colour is for $CaCO_3$ precipitation, *middle blue* for river deposition, *light blue* for quiet water deposition

For example, the groundwater table was monitored across the Boine Massif between 2004 and 2011 and several tracer tests were conducted.

9.4.1 Structure of the Aquifer

Results clearly show that the Boine massif is subdivided in three hydrogeological units that are separated by the ~20 m thick argillaceous limestones and shales of the Flohimont Member, located at the base of the Fromelennes Formation (Figs. 9.2 and 9.11). The three units are the «Trou des crevés (CR)—Père Noël (PN)—Réseau sud (RS)» system (Unit I), the «Lesse souterraine (DA to SA)» system (Unit II), and the «Trou de Han (TH)» system (Unit III), which is the downstream collecting unit of the whole underground system and goes to the "Trou de Han" resurgence (Fig. 9.11). Unit III is limited to the north by the shales of the Nismes Formation.

The piezometric maps (Fig. 9.12) show that, in low water conditions, the water level of Unit I is 2–3 m higher than that of Unit II. The groundwater gradient is ~0.3% to the north in Unit I, against ~0.1% in Unit II. These low gradients confirm the high permeability of the Boine massif in link with its intense karstification. By contrast, in high water conditions, the water level in Unit II is higher than in Unit I,

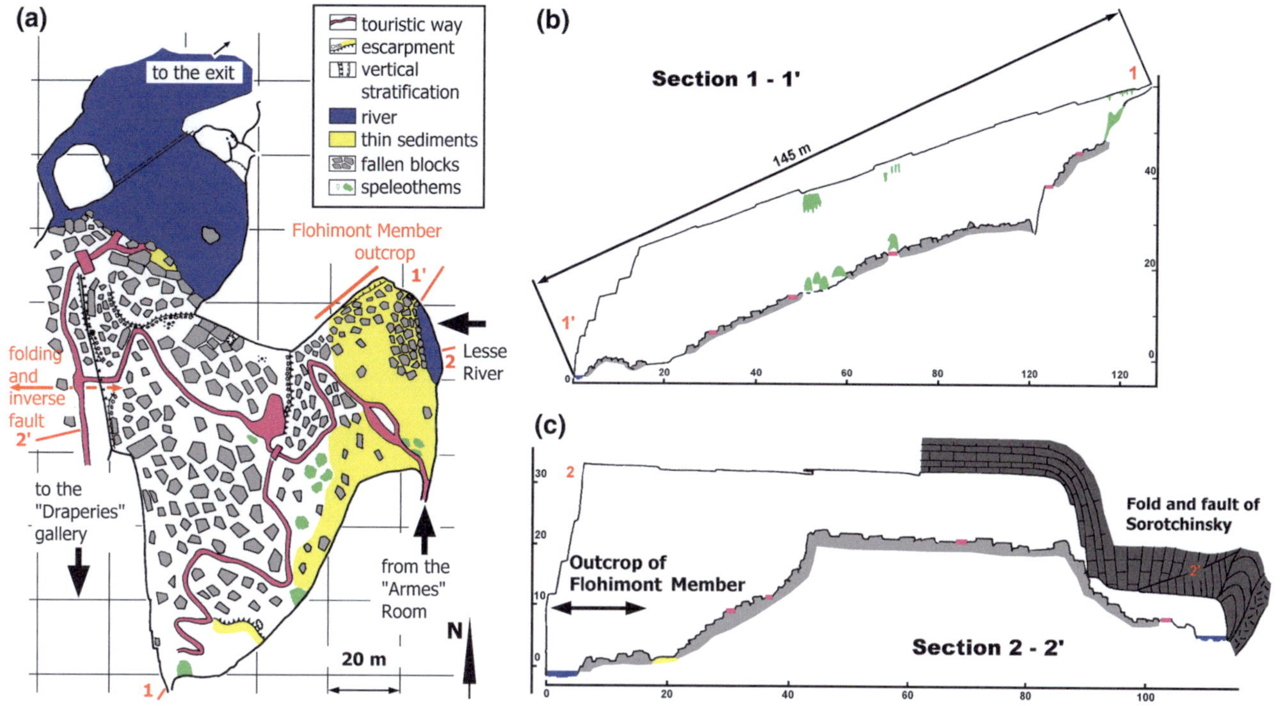

Fig. 9.9 **a** Geomorphic map of the "Dôme" room. This room is located between the argillaceous limestones of the Flohimont Member in the east and the fold and thrust fault of Sorotchinsky in the west (see Fig. 9.2). In the east, blocky accumulation on the sand beach along the river results from a breakdown of Flohimont limestones in 1984. **b** NE–SW section of the "Dôme" Room, following the dip of the strata (see location on the map). **c** ENE–WSW section of the "Dôme" Room, showing the outcrop zone of the Flohimont Member and the geometry of the fold and thrust fault of Sorotchinsky, which control the room position and development (see location on the map)

because of the reactivation of the dry meander in flood events with river discharges exceeding the 25 m^3/s capacity of the "Gouffre de Belvaux" swallow hole. In such events, the subaerial Lesse section acts as an infiltrating river that recharges the aquifer. Low water conditions are by far predominant, as the aquifer is generally drained in less than three weeks after a flooding event. In summer time, the contribution of the Boine massif aquifer to the Lesse base flow is almost nil.

9.4.2 Dynamics of the Underground Lesse

Four tracer tests were performed in various hydrological conditions encompassing discharges between 0.9 and 19.5 m^3/s. The shape of the breakthrough curves shows that transport is mainly by advection (Fig. 9.13). Dispersion effects are low, even if slightly higher in low discharge conditions, and transient storage effects remain poor regardless of flow conditions.

First arrival times of the tracer at the measurement sites (either the "Trou de Han" resurgence or the "Salle d'Armes" room) range from 2.2 h (for Q = 19.5 m^3/s) to 17 h (for Q = 0.9 m^3/s) after injection at the Belvaux swallow hole, corresponding to maximum and minimum flow velocities of 910 and 130 m/h, respectively. Based on the evolution of transfer times with discharge, Fig. 9.14 shows that mean flow velocity increases linearly for discharges up to ~10 m^3/s, then tends to stabilize for discharges over 10 m^3/s. This suggests that the understanding of sedimentation processes within the cave should take account of a maximum underground flow velocity of ~1 km/h.

The underground Lesse consists of a succession of free surface river and siphon segments (see Fig. 9.3). In order to test how flow velocity and transport process behave in these successive sections, a spatial multisampling tracer test was performed during low water conditions (Q_{Lesse} = 0.7 m^3/s). The obtained breakthrough curves confirm that advection is the main transport process, especially in the open surface section, while dispersion and, to a lesser extent, transient

Fig. 9.10 View of the "Dôme" Room, taken from the middle of the room and looking north. People *circled in red* for scale. The stratification of the Fromelennes limestones, which dip gently toward the Sorotchinsky fold hinge in the *right*, is clearly visible

storage occur mainly in the flooded sections (Fig. 9.15). Based on these experimental data, the Lesse karstic network was modelled with the OTIS software (Dewaide et al. 2016; Fig. 9.16). The volume of the "Gouffre de Belvaux" siphon has been estimated to 15,400 m^3 (or 57 m^3/m of section), from which 35% are used as transient storage zone (Fig. 9.16, Box 2). The volume of the second siphon, downstream of the underground Lesse, has been estimated to 31,200 m^3 (or 52 m^3/m of section), from which 11% are used as transient storage zone (Fig. 9.16, Box 4). These volumes are thrice that measured and calculated in the free surface water section (18 m^3/m of section, no transient storage).

9.4.3 Dynamics of the "Trou Des Crevés—Père Noël—Réseau Sud" Network

Two tracer tests were conducted in the system of Unit I. In the first test, fluorescein was injected in the "Trou des Crevés" and a monitoring station was installed in the "Père Noël" cave. In the second test, fluorescein was injected in the "Père Noël" cave and the monitoring station was installed at the "Réseau sud" outlet (Fig. 9.3). The breakthrough curve observed in the "Père Noël" (Fig. 9.17), clearly shows that the transport process within the "Trou des Crevés" and "Père Noël" caves is essentially advective, with low dispersion and almost no transient storage effect. First

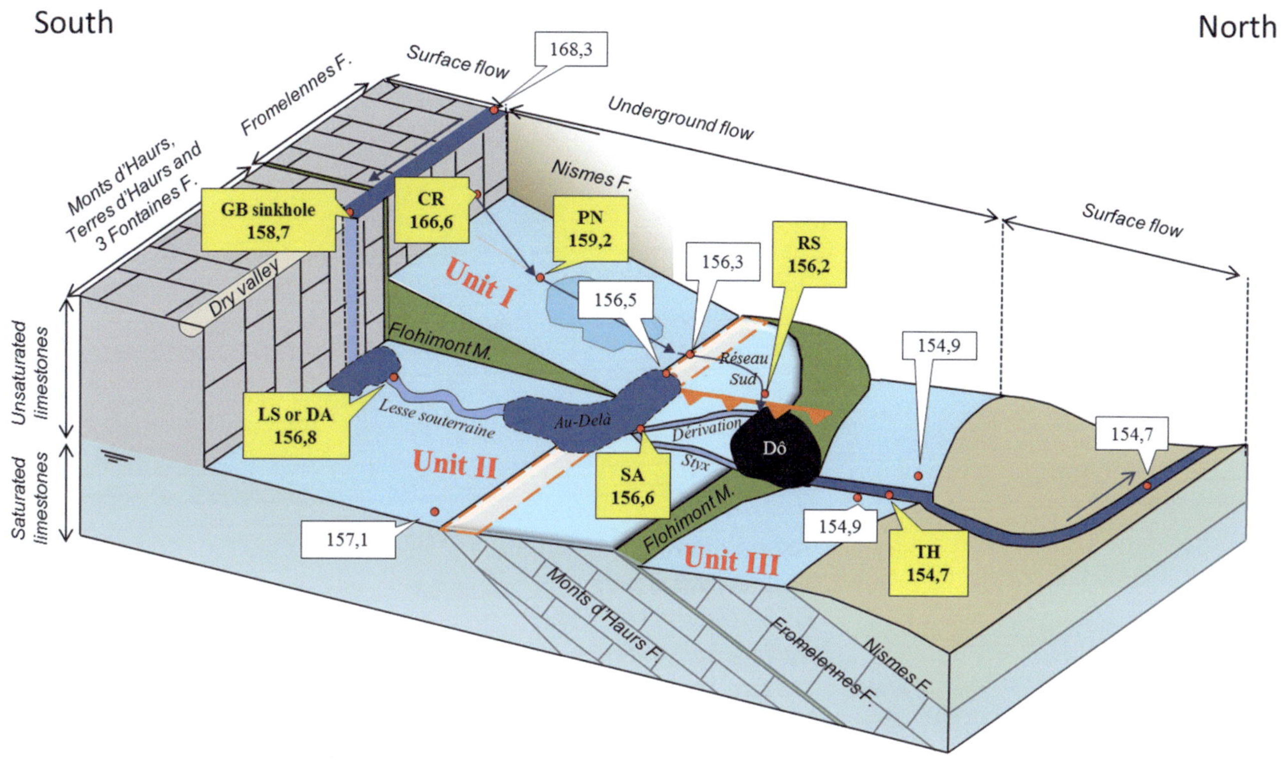

Fig. 9.11 Sketch of the hydrogeological units of the Han-sur-Lesse karstic system, evidencing the role of the argillaceous limestones and shales of the Flohimont Member as a barrier between these units. Water levels are indicated in m asl (modified after Bonniver 2011). *GB* "Gouffre de Belvaux". *LS or DA* "Lesse souterraine" or "Daniel Ameye". *SA* "Salle d'Armes". *TH* "Trou de Han". *CR* "Trou des Crevés". *PN* "Père Noël". *RS* "Réseau sud". *Dô* "Dôme room"

arrival of the tracer in the "Père Noël" was observed ~2.8 h after injection, corresponding to a maximum velocity of 115 m/h. An OTIS model run showed a homogeneous narrow conduit (Fig. 9.18). The breakthrough curve observed in the "Réseau sud" shows a completely different transport process downflow of the "Père Noël" (Fig. 9.19), with strong dispersion and transient storage effect. Observed 75 h after injection in the "Père Noël" cave, the tracer's first arrival indicates a maximum velocity of 13 m/h. Results of an OTIS run point to a ~25 m^2 flow section and a bigger storage zone section of ~40 m^2 (Fig. 9.20).

9.5 Speleogenesis

9.5.1 Ghost-Rock Karstification

This new karstogenetic theory suggests that the origin of many cavities is not directly linked to the present flow dynamics (Dubois et al. 2014; Quinif et al. 2014). The classical view about cave initiation considers that it begins with the enlargement of fissures by chemical dissolution of bedrock, followed by geometrical structuration of the growing system according to the laws of irreversible system dynamics. The fundamental assumption therein is that all products of chemical dissolution, i.e. not only the dissolved species (ions, colloids, molecules) but also the solid residue (e.g. clay, sparitic calcite and quartz grains) leave the system together in a process we call "karstogenesis by total removal".

However, a substantial body of evidence from paleokarsts in the Carboniferous limestones of the Hainaut Province, W Belgium, has shown that, in the first stage of karstogenesis, only the dissolved species actually leave the system. This typically happens during a biostatic stage (Erhart 1967), when flat landscapes provide no significant hydrodynamic potential. Limited water circulation in the phreatic zone does not allow for exportation of the solid particles, which stay in place and constitute a porous residual alterite of low density impeding the formation of voids in the rock. Such alterite

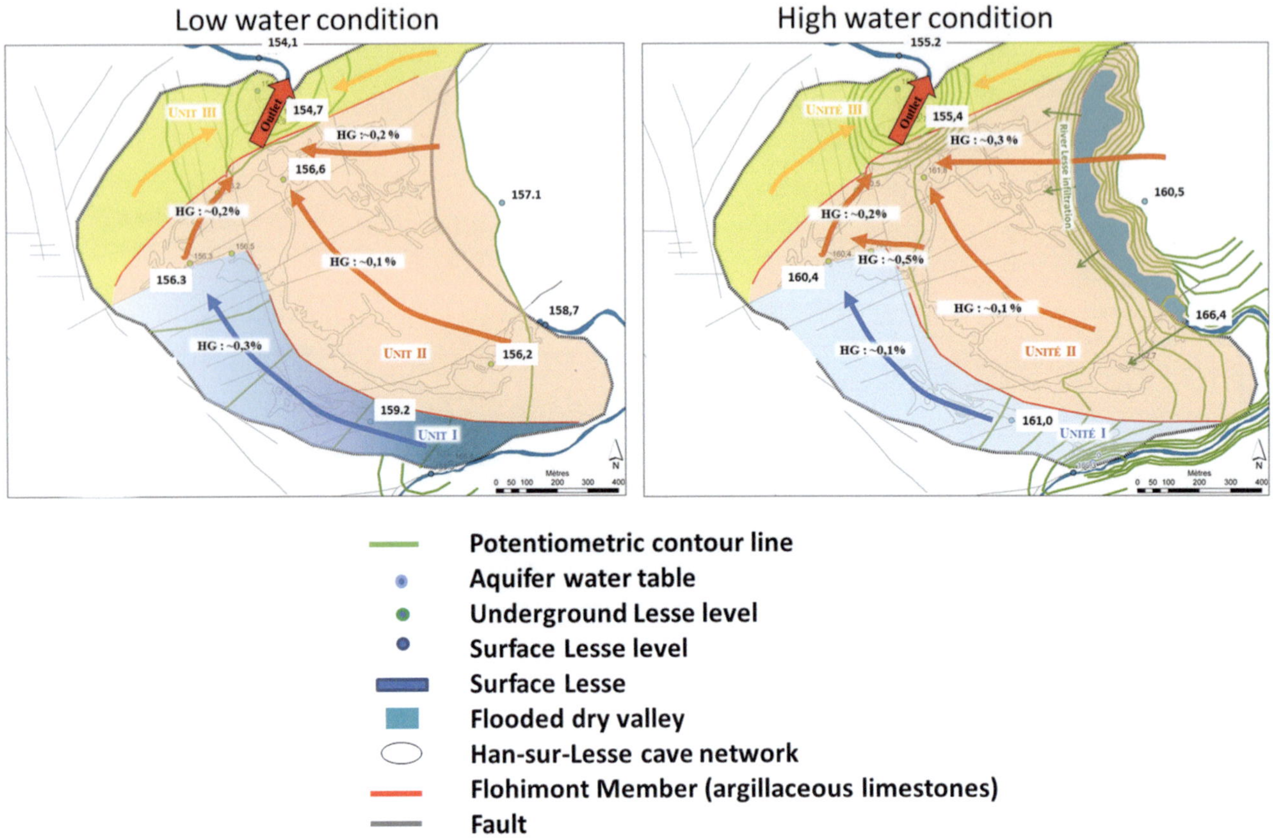

Fig. 9.12 Piezometric maps of the Boine massif at low (*left*) and high (*right*) water conditions

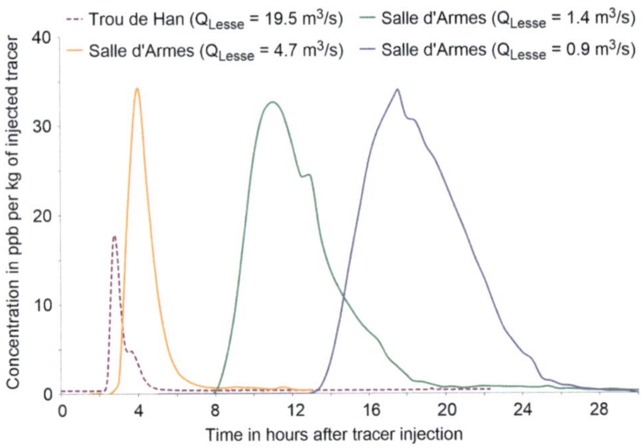

Fig. 9.13 Tracer tests performed in hydrogeological Unit II of the Han karstic system (injection at "Gouffre de Belvaux", measurement at "Salle d'Armes" or "Trou de Han"). Breakthrough curves for different hydrological conditions

bodies within the fresh massif are named ghost rocks and generally form networks whose overall geometry mimics the initial flow pathways through the fresh rock. They thus typically feature alterite chimneys starting from the top and going deep into the limestone massif. Other such bodies are pseudoendokarstic forms totally included in the bedrock, whose shapes may resemble never emptied galleries (Fig. 9.21). The porous, more or less soft material of the residual alterite is made of sparitic calcite grains derived from fossils, calcite veins or clasts, and less soluble to insoluble minerals (clays, quartz). Many ghost rock occurrences exist in western Belgium because this low-elevation region has not undergone any significant change in hydrodynamic potential since the time of ghost-rock karstification. However, if this potential increases for geodynamic (regional uplift, marine regression) or any other reasons (for instance, a climatically induced turn toward rhexistasy), the enhanced underground water circulation, which is still located mainly in the readily infiltrated alterite of the ghost rock network, will remove this soft material from vadose and epiphreatic zones that progressively deepen with uplift, thus opening the cavities formerly filled by the alterite and creating caves in the classical meaning of the term. This erosion process obviously allows then the development of typical

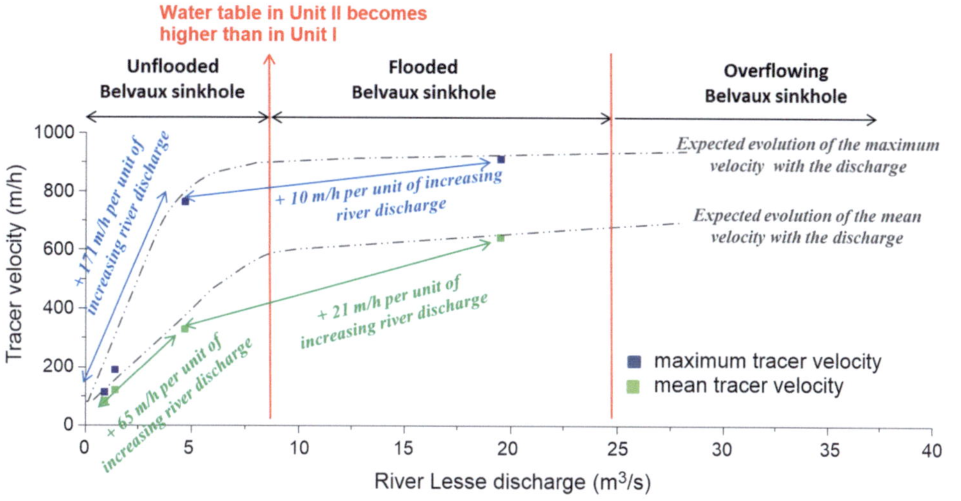

Fig. 9.14 Tracer tests performed in hydrogeological Unit II of the Han karstic system (injection at "Gouffre de Belvaux", measurement at "Salle d'Armes" or "Trou de Han"). Evolution of flow velocities against discharge of the underground Lesse

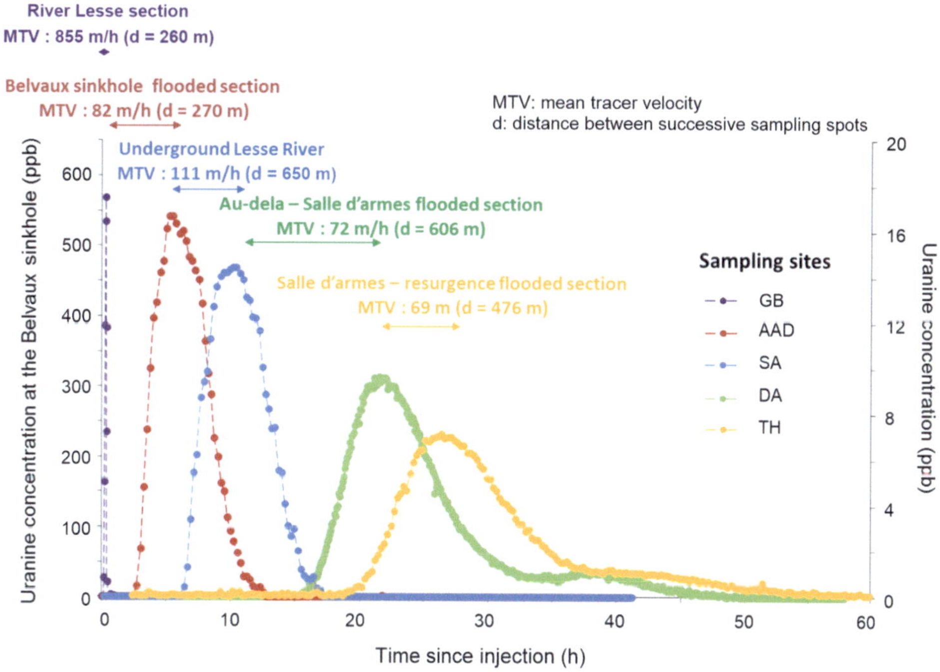

Fig. 9.15 Tracer tests performed in hydrogeological Unit II of the Han karstic system (injection at "Gouffre de Belvaux", measurement at "Trou de Han"). Spatial evolution of the breakthrough curves and flow velocities along the successive sections of the underground Lesse River at low water conditions (Q_{Lesse} = 0.7 m^3/s) (see Figs. 9.3 and 9.11 for location and Fig. 9.11 for site abbreviations)

water flow forms such as potholes, scallops, meander niches, or overdeepenings, whereas other microforms like, e.g. cupolas have also been found in a ghost-rock environment (Quinif 2010; Dubois et al. 2014). In brief, the ghost-rock karstification paradigm highlights that the chemical dissolution and mechanical removal stages of limestone erosion may be disconnected but it makes no prediction about their isolated or common occurrence in each particular case.

9.5.2 Ghost Rocks in the Han System

Finding ghost-rock evidence after large underground flows have travelled through a cave is difficult because the alterite has been more or less completely removed by the hypogeal river. Traces are found only in sheltered areas such as, for example, the highest part of the Antiparos Room of the Han Cave, where the residual alterite of weathered strata has been

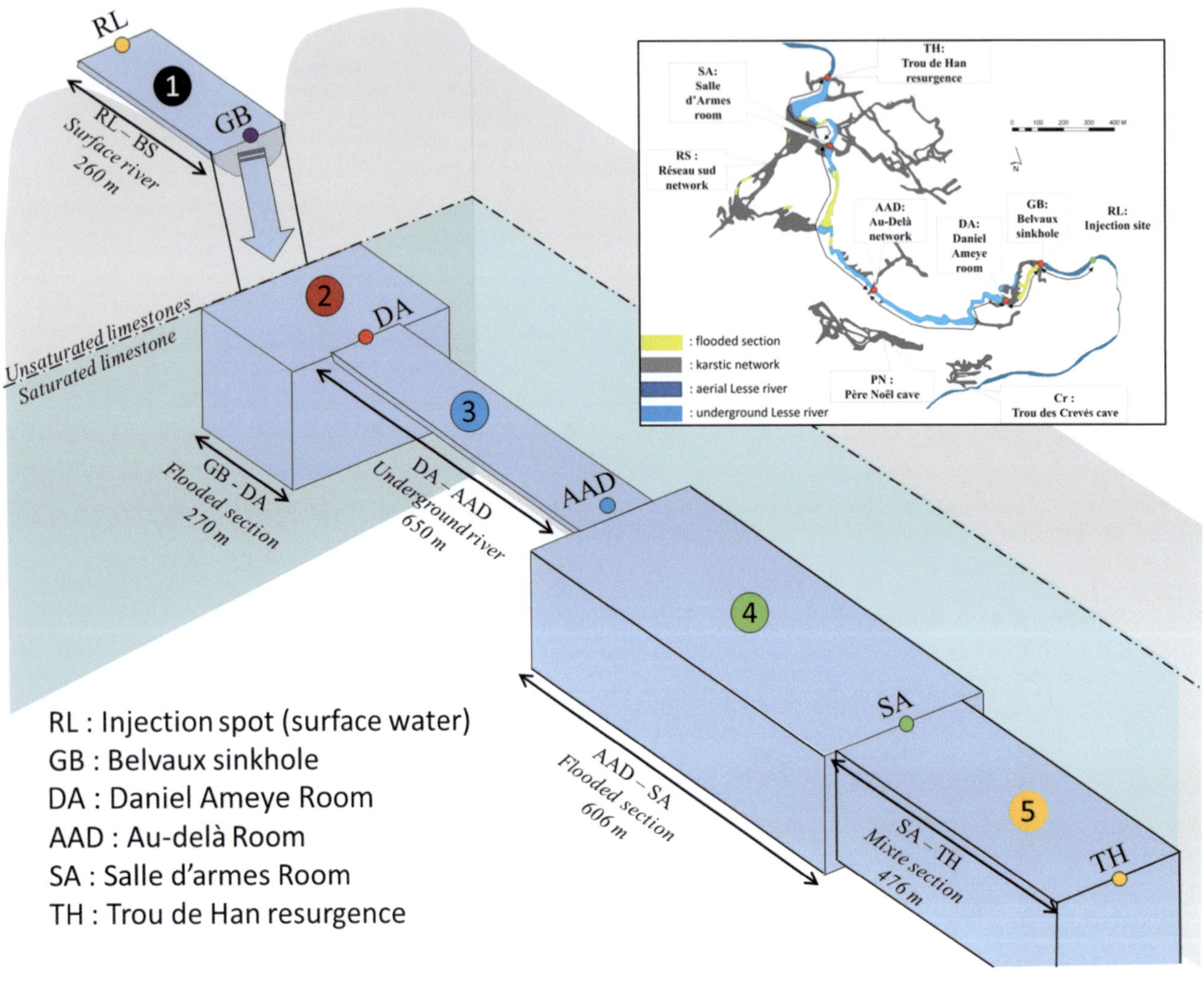

Fig. 9.16 Flow section dimensions along hydrogeological Unit II of the Han karstic system estimated from an OTIS modelling based on the experimental data of the tracer tests (see text for explanation)

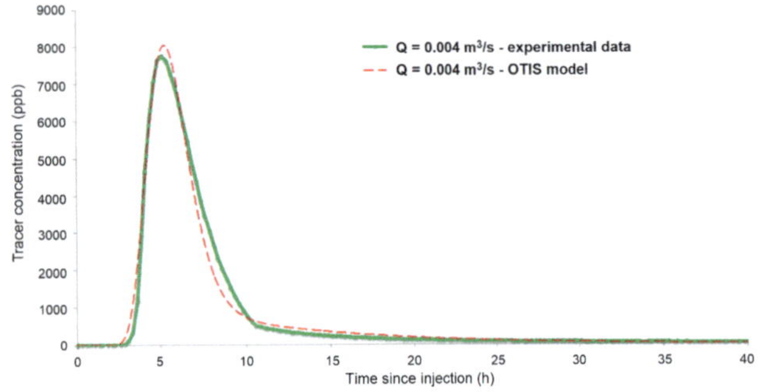

Fig. 9.17 Breakthrough curve obtained from a tracer test performed in the first section of hydrogeological Unit I of the Han karstic system (injection at "Trou des Crevés", measurement at "Père Noël") and best fit OTIS simulation

exposed by the collapse of blocks (Fig. 9.22). This suggests that ghost-rock karstification was indeed the first stage in the evolution of the Han karstic system. In this view, we admit that the structurally controlled geometry of the system, which follows N60°E and N150°E joint directions inherited from Mesozoic and Early Cenozoic extensional tectonic episodes (Fig. 9.3), was foreshadowed by a ghost-rock network that was formed during the Cretaceous and probably also the Paleogene. The warm humid climates of the time in the Han area allowed intense chemical weathering to create a system of "pseudoendokarstic" features preserving a residual alterite in the phreatic zone because the flattened topography and the subsequent absence of hydrodynamic energy robbed close-to-zero water flow in this zone of all mechanical erosive power.

It was only when rivers began to respond by active incision to the Ardenne uplift from Pliocene times onward that a new hydrodynamic potential was created and more efficient groundwater circulation became capable of removing the alterite from the pseudoendokarsts and opening the caves. This was a multistage process progressing downward

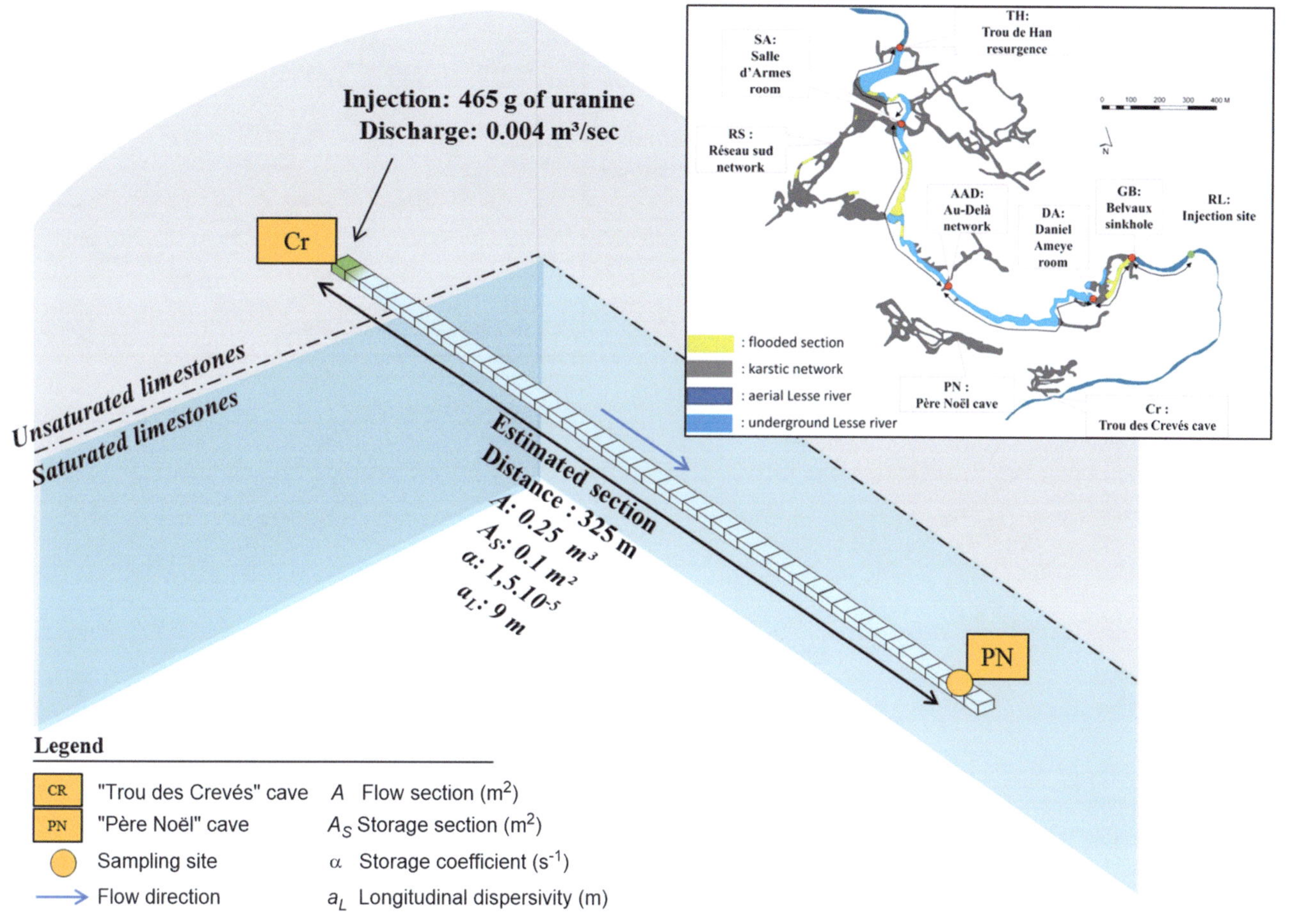

Fig. 9.18 Flow section dimensions along the first section of hydrogeological Unit I of the Han karstic system ("Trou des Crevés"—"Père Noël") estimated from an OTIS modelling based on tracer test data (see text for explanation)

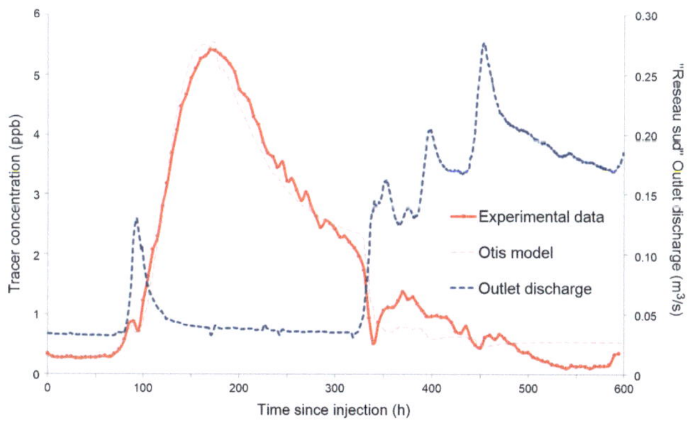

Fig. 9.19 Breakthrough curve obtained from a tracer test performed in the downstream section of hydrogeological Unit I of the Han karstic system (injection at "Père Noël", measurement at "Réseau sud") compared to the best fit OTIS simulation and measured discharge at the outlet of the system

and emptying successive karstic levels at the pace of valley downcutting. Currently, seepage and the underground course of the Lesse continue to shape the cave.

9.5.3 Consequences of Ghost-Rock Karstification

The most important implication of the Han system origin by ghost-rock karstification is that its overall structure is inherited. The Quaternary caves and galleries and the present subterranean water circulation follow a network of channels created in a remote past, when the Cretaceous and Paleogene climate and topography favoured the development of pseudoendokarsts. Therefore, the directions of the galleries are mostly dictated by those of past tectonic episodes, in this case Mesozoic and Early Cenozoic extensional phases. As the location of the present caves depends on that of the zones where ghost rocks developed, the pre-existing system structure has determined the present hydrogeology much more than the opposite. Sedimentological studies have shown that separate connections exist between the "Trou des Crevés"—"Père Noël"—"Réseau sud" sectors (Unit I) on one hand and the "Lesse souterraine"—"Salle d'Armes"—"Dôme Room" axis and the "Trou d'Enfaule" axis on the other hand.

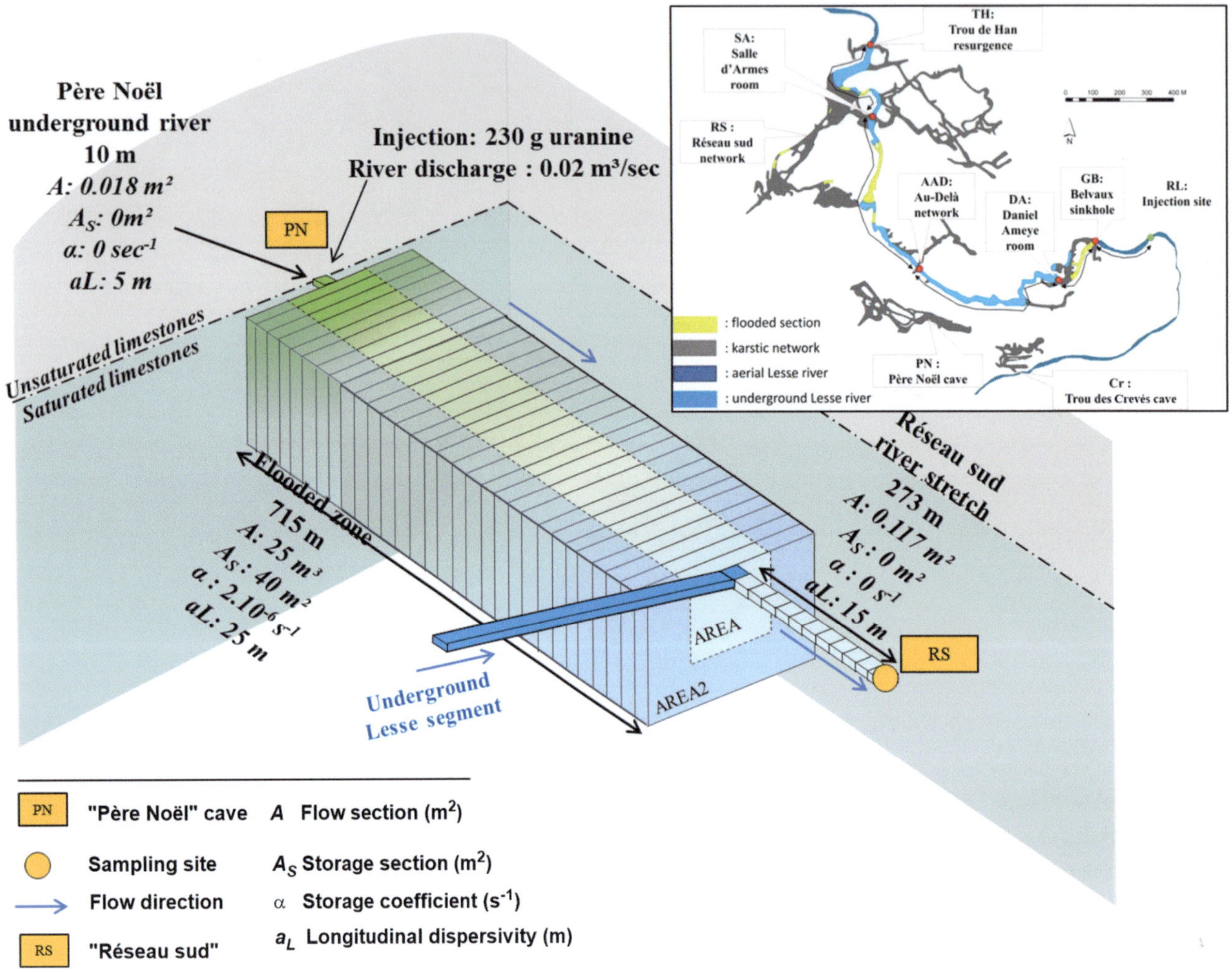

Fig. 9.20 Flow section dimensions along the downstream section of hydrogeological Unit I of the Han karstic system ("Père Noël"—"Réseau sud") estimated from an OTIS modelling based on tracer test data (see text for explanation)

9.6 Conclusion

In this brief overview of the karstic system of Han-sur-Lesse, we showed that this system, developed in Givetian limestones, is the archetype of a subterranean meander cutoff. Secondly, we emphasized how the underground features, their geometry, and the analysis of speleothems and sediments allow for the reconstruction of past environmental and, especially, climatic conditions. Three distinct hydrogeological units have been distinguished, separated by the less permeable shales and argillaceous limestones of the Flohimont Member. While water flow and transport processes are essentially advective in Unit II (with velocities up to ~0.9 km/h in the "Lesse souterraine" system) and the first section of Unit I ("Trou des Crevés"—"Père Noël" connection), dispersion and important transient storage effects in the latter unit between "Père Noël" and "Réseau sud" suggest the existence of a large groundwater volume so far undiscovered.

Finally, scarce but unquestionable traces of ghost rocks in the Han Cave allowed us to reconstruct the history of the karstic system of Han-sur-Lesse within the ghost-rock karstification frame of karstogenesis. This implies recognizing the long geological past of the system, which was initiated by ghost rock formation under warm humid climate and low topography conditions of the Late Mesozoic and Early Cenozoic before being freed from residual alterite and acquiring its present cave status in relation with Plio-Quaternary valley downcutting through the uplifting Ardenne massif. Unknown local conditions during the Mesozoic may explain that ghost rocks (and their evolution into modern caves) intensely developed in the Boine massif but are almost absent in the nearby Resteigne limestone massif. Pre-existence of ghost rock probably also determined the vertical position of the karstic levels, which thus would not have the usually assumed straightforward correlation with river terrace levels. All this attests the high geomorphic and

Fig. 9.21 Typical pseudoendokarst feature in the Lower Carboniferous limestones of the Clypot quarry, Soignies, W Belgium. The crumbly matter within the cavity is not detrital filling but the residual alterite of a ghost rock after partial dissolution of the limestone bedrock. This kind of karst results from chemical action alone, without mechanical erosion component, and occurred under biostatic conditions of warm humid climates and lowland topography

Fig. 9.22 Ghost-rocks in the Han Cave. **a** Limestone layer weathered to *violet-blue* residual alterite still preserving the original structures (hammer for scale). **b** Ghost-rock weathering of the micritic matrix, leaving the sparitic calcite of fossil stromatoporoids and corals intact (*photo* width ~1 m)

karstic value of the Han-sur-Lesse site and thus amply justifies its exploitation as a highly visited (~0.5 million visitors per year) touristic site where a pedagogic approach of the local natural wealth is cleverly combined with adapted leisure activities.

References

Blockmans S, Quinif Y, Bini A, Zuccoli L (1999) Dynamique sédimentaire et paléoenvironnements durant la transition Weichsélien—Holocène à partir des dépôts endokarstiques de la grotte de Han-sur-Lesse (Belgique). Bull Soc Géol Fr 170(6):841–852

Bonniver I (2011) Etude hydrogéologique et dimensionnement par modélisation du système-traçage du réseau karstique de Han-sur-Lesse (massif de Boine-Belgique). Unpublished PhD thesis. University of Namur

Deflandre G, Bastin B, Quinif Y, Gewelt M (1987) La Grotte du Père Noël. Livret-guide du Colloque International de Sédimentologie Karstique (CISK). Centre Belge d'Etudes karstologiques, Han-Sur-Lesse, Belgique, pp 37–44

Delvaux De Fenffe D (1985) Géologie et tectonique du parc de Lesse et Lomme au bord sud du Bassin de Dinant (Rochefort, Belgique). Bull Soc Belge Géol 94(1):81–95

Dewaide L, Bonniver I, Rochez G, Hallet V (2016) Solute transport in heterogeneous karst systems: dimensioning and estimation of the transport parameters via multi-sampling tracer-tests modelling using the OTIS (One-dimensional Transport with Inflow and Storage) program. J Hydrol 540:567–578

Dubois C, Quinif Y, Baele JM, Barriquand L, Bini A, Bruxelles L, Dandurand G, Havron C, Kaufmann O, Lans B, Maire R, Martin J, Rodet J, Rowberry MD, Tognini P, Vergari A (2014) The process of ghost-rock karstification and its role in the formation of cave systems. Earth-Sci Rev 131:116–148

Dumoulin V, Blockmans S (2013) Pondrôme—Wellin 59/5–6. Carte géologique de Wallonie, échelle: 1/25 000. Notice explicative. SPW, DGARNE, p 84

Erhart H (1967) La genèse des sols en tant que phénomène géologique. Masson Ed., Paris, coll. Evolution des sciences, p 90

Havron C, Quinif Y, Vandycke S (2004) Tectonique et karstification. Le cas de la région de Han-sur-Lesse (Belgique). Karstologia 43 (1):19–26

Kaisin F, de Pierpont E (1939) Hydrogéologie des calcaires de la Belgique. Monogr. Sc Nat Soc Scient Brux 54:109

Overlau P, Quinif Y (1979) La région d'Eprave et de Han-Sur-Lesse. Excursions du Colloque franco-belge de karstologie appliquée. Ann Soc Géol Belg 102:168–170

Poty E (1976) L'influence de la structure géologique sur le relief de la Belgique. In: Pissart A (ed) Géomorphologie de la Belgique. Université de Liège, Laboratoire de Géologie et de Géographie physique, pp 107–113

Poulain A, Rochez A, Bonniver I, Hallet V (2015) Stalactite drip-water monitoring and tracer tests approach to assess hydrogeologic behavior of karst vadose zone: case study of Han-sur-Lesse (Belgium). Environ Earth Sci 74:7685–7697

Quinif Y (1987a) Le système karstique de Han-Sur-Lesse. Livret-guide du Colloque International de Sédimentologie Karstique (CISK). Centre Belge d'Etudes Karstologiques, Han-Sur-Lesse, Belgique, pp 19–26

Quinif Y (1987b) Le réseau Sud. Livret-guide du Colloque International de Sédimentologie Karstique (CISK). Centre Belge d'Etudes Karstologiques, Han-Sur-Lesse, Belgique, pp 27–36

Quinif Y (1988a) La structure hydrogéologique du Massif de Boine. Lapiaz hors série "Spécial Han", pp 11–14

Quinif Y (1988b) Une nouvelle topographie de la Grotte de Han. Lapiaz hors série "Spécial Han", pp 15–18

Quinif Y (1991) La série stalagmitique de la Galerie des Verviétois (Han-sur-Lesse, Belgique). Speleochronos 3:29–42

Quinif Y (2010) Fantômes de roche et fantômisation—Essai sur un nouveau paradigme en karstogenèse. Karstologia Mém 18:196p

Quinif Y, Bastin B (1986) Le système karstique de Han-Sur-Lesse (Belgique). Actes 9ème Congr Int Espeleol Barcelona 1:158–161

Quinif Y, Bastin B (1987) La grotte de Han. Livret-guide du Colloque International de Sédimentologie Karstique (CISK). Centre Belge d'Etudes Karstologiques, Han-Sur-Lesse, Belgique, pp 19–26

Quinif Y, Bastin B (1994) Datation uranium/thorium et analyse pollinique d'une séquence stalagmitique du stade isotopique 5 (Galerie des Verviétois, Grotte de Han-sur-Lesse, Belgique). C.R.A. S.Paris, vol 318, série II, pp 211–217

Quinif Y, Baele JM, Dubois C, Havron C, Kaufmann O, Vergari A (2014) Fantômisation: un nouveau paradigme entre la théorie des deux phases de Davis et la théorie de la biorhexistasie d'Erhart. Geol Belg 17(1):66–74

Van Den Broeck E, Martel EA, Rahir E (1910) Les cavernes et rivières souterraines de la Belgique(Ed.) Lamertin, Bruxelles, p 1826

Verheyden S, Genty D, Deflandre G, Quinif Y, Keppens E (2008) Monitoring climatological, hydrological and geochemical parameters in the Père Noël cave (Belgium): implication for the interpretation of speleothem isotopic and geochemical time-series. Int J Speleol 37:221–234

The Picturesque Ardennian Valleys: Plio-Quaternary Incision of the Drainage System in the Uplifting Ardenne

10

Gilles Rixhon and Alain Demoulin

Abstract

As a response to Late Cenozoic tectonic uplift of the Ardenne massif, the Meuse River and its drainage system deeply incised the landscapes of Southern Belgium. Well-preserved terrace staircases flanking the main Ardennian valleys show a break at the height of the Main Terrace Complex, whose younger level (YMT) marks the transition towards the steep-sided lower part of valley transverse profiles. Numerical $^{10}Be/^{26}Al$ dates obtained for the YMT in various places along the lower Meuse—lower Ourthe—Amblève line yield an age around 620 ka for the pulse of tectonic uplift responsible for increased incision rates and fast valley deepening. They also show diachronic abandonment of the YMT, indicating that post-YMT erosion invaded the drainage system through the migration of knickpoints originating in the ~20 m base level lowering initially created at the margin of the en-bloc uplifted region. A study of knickpoints in modern long profiles of Ardennian rivers fully confirms this view. Assimilation of the new data imposes a revised model be proposed for river incision in the Ardenne. Integrating the climatic control with the various modes of tectonic control over river incision, this conceptual model offers a sounder frame of the Quaternary river system evolution in the Ardenne, also explaining observed cases of stream piracy. Basin average denudation rates published for the Ardenne are also briefly compared with valley incision rates.

Keywords

River incision • Cosmogenic nuclide dating of terraces • Knickpoint propagation • Stream piracy • Ardenne • Meuse basin

10.1 Introduction

The spectacular scenery of the valleys of the Meuse (Maas in Dutch and German) and its tributaries, deeply incised in the Ardenne plateau and so poetically described by Hugo (1842), not only inspired world-renowned landscape painters such as J.M. William Turner but also soon attracted fathers of the geomorphological science such as Albert de Lapparent and William. M. Davis. Since the latter's precursory geomorphological study about the formation of the Belgian Meuse valley (Davis 1895), generations of geoscientists have investigated the Quaternary evolution of the drainage network in the Ardenne massif, focusing in particular on the geometrical and morphological correlations between river terrace levels and the sedimentological analysis of terrace deposits (e.g. Macar 1938; Alexandre 1957; Pissart 1961, 1974; Pissart et al. 1997; Rixhon and Demoulin 2010). The transverse profile of all major river valleys cutting into the Paleozoic basement widely exhibits a typical morphological

G. Rixhon (✉)
Institute of Geography, University of Cologne, Zülpicher Strasse 45, 50674 Cologne, Germany
e-mail: grixhon@uni-koeln.de

A. Demoulin
Department of Physical Geography and Quaternary, University of Liège, Sart Tilman, B11, 4000 Liège, Belgium
e-mail: ademoulin@ulg.ac.be

A. Demoulin (ed.), *Landscapes and Landforms of Belgium and Luxembourg*, World Geomorphological Landscapes, DOI 10.1007/978-3-319-58239-9_10

transition between a broad, gently sloping Early Pleistocene valley characterized by extended terrace surfaces and a nested narrow, steep-sided Middle Pleistocene valley with more confined terraces (Fig. 10.1; Van Balen et al. 2000; Rixhon et al. 2011). This transition is materialized by a more or less ubiquitous complex of closely spaced and well-developed terrace levels, the so-called Main Terrace Complex, whose younger level, abbreviated hereafter to YMT, is located directly at the edge of the deeply incised lower part of the valleys (Fig. 10.1). Whilst this morphological hinge is also recognized in the Meuse valley (e.g., Juvigné and Renard 1992; Van Balen et al. 2000), it is particularly conspicuous in its main tributary valleys, such as those of the Ourthe and the Vesdre (Fig. 10.1; e.g., Rixhon et al. 2011). Furthermore, it is also observed in the Rhine drainage system throughout the adjacent Rhenish shield, and especially in the valleys of the Rhine and the Mosel in Germany (e.g. Boenigk and Frechen 2006). It was thus interpreted at the regional scale as the result of a pulse of uplift that affected the whole Ardenne/Rhenish massif for a while around the Early Pleistocene—Middle Pleistocene transition (Van Balen et al. 2000; Demoulin and Hallot 2009).

In the last two decades, many studies based on DEM analysis, numerical modelling, and modern dating methods provided substantial new insights into the spatial and temporal characteristics and the mechanisms of river incision in response to the massif's uplift, which led in particular to the abandonment of the 'YMT floodplain' in the Ardennian Meuse catchment. They encompassed (i) quantification of long-term denudation rates in the Meuse catchment (Van Balen et al. 2000; Schaller et al. 2004; Demoulin et al. 2009), (ii) numerical dating of river terrace sediments by terrestrial cosmogenic nuclides (Rixhon et al. 2011, 2014), and (iii) the modelling of knickpoint migration across the river network (Demoulin et al. 2012a, b; Beckers et al. 2015). The outcomes of these studies, supported by independent morphometric analyses of the drainage system (Sougnez and Vanacker 2011), were far-reaching enough to impose revisiting the accepted understanding of the controls on river incision and developing a new conceptual frame for the Quaternary evolution of the Ardenne (Demoulin et al. 2012b). After a brief presentation of the geological background and an overview of the Ardennian drainage system at the eve of the Quaternary, this chapter thus aims at providing a comprehensive but concise picture of the interplay between tectonics and climate in controlling Quaternary river incision and shaping the Ardenne massif. It focuses first on the morphological and sedimentological characteristics of the main terraces, in particular the YMT, and the recently developed chronological framework. Then, it details the mechanisms at work and how they intermingled to govern incision and denudation in the Ardenne, especially since the beginning of the Middle Pleistocene. Finally, it underlines how, beyond the vertical component of drainage system evolution, planform changes associated to stream piracy were another product of the specific conditions that determined the geomorphological evolution of the massif in this period of time.

10.2 Geological and Hydrographical Settings

10.2.1 Structural and Geodynamical Framework

The Ardenne massif constitutes the western continuation of the Variscan Rhenish massif in south-eastern Belgium, northern France and Luxembourg (Fig. 10.2a). It extends between the Paris basin to the south and the Cenozoic Anglo-Belgian basin to the north. Overall, the structurally complex basement of the Ardenne exhibits longitudinal ENE-WSW folds and thrust faults cut by numerous NW-SE to NNW-SSE striking normal faults (Demoulin and Ek 1995).

A series of Cambro-Ordovician massifs inherited from the Caledonian orogeny and deformed again in Variscan times emerge along the axis of the recently uplifted area, which corresponds to the axis of the Variscan Ardenne anticlinorium. Chiefly made of variably erodible phyllites and resistant quartzites, they are mainly the Rocroi and Stavelot massifs, respectively located in western and north-eastern Ardenne (Fig. 10.2b). Surrounding these Caledonian massifs, the slaty Lower Devonian makes the bulk of the Ardenne anticlinorium and the Neufchâteau synclinorium (Fig. 10.2b). North of the Ardenne anticlinorium, the Dinant synclinorium represents the northernmost unit of the Ardenne allochthon, limited to the north by the Midi-Eifel thrust (Fig. 10.2b). In contrast with the exclusively siliceous rocks of the Lower Devonian Ardenne core, it exposes more varied Middle Devonian to Lower Carboniferous rocks that characterize the lower lying Famenne and Condroz areas. From south to north, a narrow belt of strongly karstified Givetian limestones extends along the northern margin of central Ardenne (see Chaps. 8 and 9), followed by the up to 10-km-wide Fagne/Famenne region, where Frasnian and Famennian shales prone to frost shattering were largely eroded during the cold periods of the Quaternary, creating a 150-m-deep depression between Ardenne and Condroz (Fig. 10.2b). To the north, the latter is then chiefly formed by an alternation of Upper Devonian mica-rich sandstones and Lower Carboniferous limestones cropping out in ENE-WSW trending anticlines and synclines, respectively, and defining a corresponding series of smooth ridges and depressions (Fig. 10.2b). Scattered over the Condroz ridges and the flat summits of W and NE Ardenne, sparse remnants

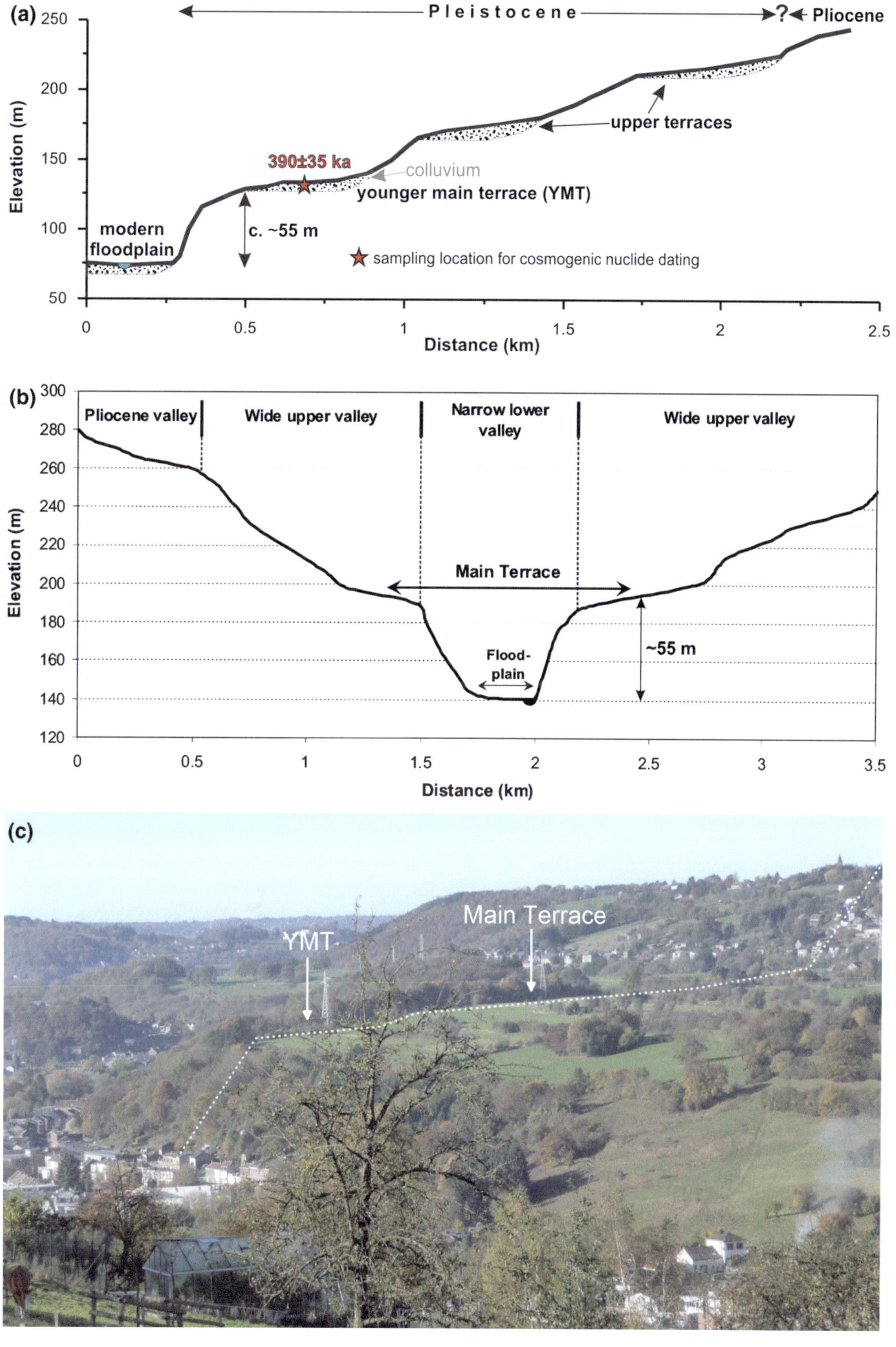

Fig. 10.1 Typical valley cross section of Ardennian (sub-)tributaries of the Meuse as exemplified by the lower Ourthe (**a** adapted from Rixhon et al. 2011) and middle Vesdre (**b** from Rixhon and Demoulin, 2010) valleys and a view of the Vesdre valley downstream of Verviers showing the position of the younger main terrace (YMT) just above the steep-sided lower part of the valley section (**c**). The YMT systematically marks the transition from a broad upper part of the valleys to their deeply incised narrow lower part. The CRN age of 390 ± 35 ka points to the abandonment time of the YMT in the lower Ourthe valley (Rixhon et al. 2011)

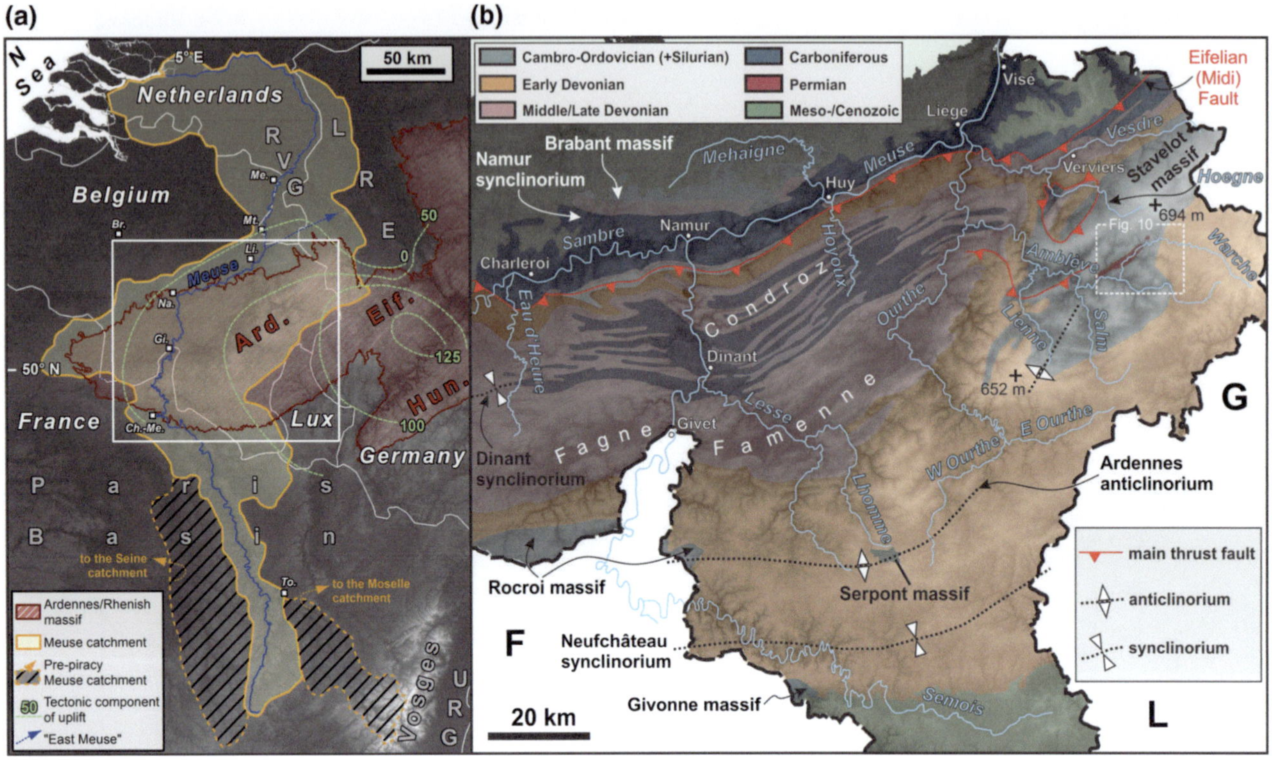

Fig. 10.2 **a** Location of the Meuse catchment within NW Europe. The *white frame* delimits the area detailed in figure. **b** Isolines of the tectonic component of rock uplift in the Ardenne/Rhenish massif since the early Middle Pleistocene are taken from Demoulin and Hallot (2009). *Ard.* Ardenne; *Eif.* Eifel; *Hun.* Hunsrück; *LRE.* Lower Rhine Embayment; *RVG.* Roer Valley Graben; *URG.* Upper Rhine Graben. *Ch-Me.* Charleville-Mézières; *Gi.* Givet; *Li.* Liège; *Me.* Maaseik; *Mt.* Maastricht; *Na.* Namur; *To.* Toul. **b** Simplified geological map of Wallonia, encompassing the Ardennian Meuse catchment in Belgium (Adapted from Bultynck and Dejonghe 2001)

of Upper Cretaceous and Paleogene marine sediments attest that the massif was partly drowned by several Meso-Cenozoic seas and that the subsequent denudation was unable to remove them completely (see Chap. 5).

As a part of the Rhenish shield, the Ardenne experienced late Cenozoic tectonic uplift. Large-scale deformation and present-day elevations of well reconstructed Tertiary erosion surfaces demonstrate that this uplift was primarily of en-bloc type, notably limited to the NE by a narrow marginal flexure and the bounding faults of the Roer Valley Graben, and amounted to ~450 m, locally even 500 m, since the Oligocene. Even though first signs of uplift are detected as early as during the Selandian (~60 Ma), the recent deep incision of the valleys into the massif attest a strong Plio-Quaternary increase in uplift rates. From terrace studies in many rivers of the Ardennian Meuse catchment, it is commonly acknowledged that uplift acceleration occurred stepwise, partly also in a pulsed way. With an estimated rate of 0.3–0.5 mm year^{-1} probably lasting a few 10^4 years, the largest and best documented uplift pulse took place around the Early to Middle Pleistocene transition and was responsible for the abandonment of the YMT floodplain level, before the massif underwent anew relative tectonic quiescence since ~400 ka (e.g., Van den Berg 1996; Quinif 1999; Van Balen et al. 2000; Demoulin and Hallot 2009).

Different possible causes of uplift have been proposed and are still debated: (i) lithospheric thinning underneath the Rhenish massif (e.g. Prodehl et al. 1995), (ii) lithospheric folding as a result of compressive stresses exerted by the Alpine belt on its foreland (e.g. Ziegler and Dezes 2007), (iii) mantle upwelling beneath the Eifel area (Ritter et al. 2001). Recently, Demoulin and Hallot (2009) synthesized geomorphological observations and Demoulin (2011) brought additional morphometric evidence showing that the latest recorded uplift is oldest in the south and youngest in the north. According to them, this suggests that an uplift wave migrated across the massif over the Quaternary, a conclusion that seems to point to lithospheric folding as the most probable cause of uplift of the Ardenne/Rhenish shield.

10.2.2 The Ardennian Meuse Drainage System at the Dawn of the Quaternary

Looking at the main traits of the Ardennian drainage system and, especially, at the course of its trunk stream, the Meuse, one anticipates that it is the product of a long and complicated history. We briefly summarize this story here as a preamble to the discussion of Quaternary river incision.

Although the central Ardenne was and remained constantly emerged since a much longer time, the last sea that drowned a substantial part of the massif during the Rupelian regressed northwards sometime around 30 Ma. It is commonly accepted that the many N- to NNW-striking drainage axes featured by the then-emerged areas bear witness to the consequent river system that was developed at that time (de Heinzelin 1963). Beyond the Hantes and Eau d'Heure-Piéton lines (*a*, Fig. 10.3) in the west, they are the Amblève-lower Ourthe and the Hoyoux-lower Méhaigne axes (*b*, Fig. 10.3), plus the Meuse reach from Dinant to Namur (Fig. 10.3). The last two axes were perhaps simply the prolongation of preexisting rivers (corresponding to the present middle Ourthe and the Meuse reach between Revin and Dinant) that already drained the continental central Ardenne in the Eocene. This brings us to one of the major unsolved issues of the long-term evolution of the Meuse river course: why does the upper Meuse, flowing north in the Paris basin, leave this low-elevation, readily erodible bedrock to cut across the much more resistant Paleozoic basement of the higher uplifted Ardenne? Although superimposition (e.g., Cornet 1903) and antecedence (e.g. Davis 1895; Bustamante and Voisin 1975) have had their proponents to explain this paradox, the capture hypothesis, though being still not definitely demonstrated, prevailed among Belgian geomorphologists in the past half-century, especially because it was vigorously defended by Pissart (1961, 1974). However, the ~50-m-high erosion scarp that Demoulin (1995) has shown to define the southern margin of the Ardenne already in the Oligo-Miocene (see also Chap. 5) makes highly improbable that a capture could have taken place there sometime during the Middle to Late Miocene at the expense of a river flowing in the Paris basin. Moreover, it would be surprising that, empowered by such a capture, the Meuse catchment would then, even if several million years later, be largely beheaded at the benefit of the adjacent Seine and Rhine catchments (Pissart et al. 1997) for allegedly the same reason, namely uplift of the Ardenne (Fig. 10.3). Superimposition of an Ardennian Meuse initially established on Eocene and Oligocene covers (plus the underlying deep weathering mantle, see Chap. 5), followed by antecedence in response to the Quaternary uplift of the massif seems thus equally probable, to say the least.

Another peculiar feature of the Ardennian drainage system is the long straight major drain running along the northern margin of the massif (Figs. 10.3 and 10.4). There, the Sambre and the middle Meuse (between Namur and Liege) collect all the waters flowing down the Ardenne and direct them toward ENE, down to Liege where the Meuse changes its course to flow north. However, in older times, the Meuse kept the same ENE flow direction beyond Liege, passing north of Aachen and accumulating sediments in the Lower Rhine Embayment (the Holzweiler Formation) before joining the Rhine near Mönchengladbach (Westerhoff et al. 2008). As revealed by terrace mapping in the Liege–Maastricht–Aachen triangle (Macar 1938; Felder et al. 1989), the change to its modern course, probably caused by uplift and tilt motion of the shoulder of the Roer Valley Graben, occurred by progressive shifting, initially aided by a capture that took place sometime between 2 and 1 Ma (Juvigné and Renard 1992; Van den Berg and Van Hoof 2001; Westerhoff et al. 2008).

The ENE draining of all Ardennian waters by the so-called Sambre-Meuse axis most probably responded to the regional tilt that developed during the Miocene toward the Lower Rhine Embayment, actively subsiding all along the Neogene. This tilt is likewise evidenced in Middle Belgium by the NE retreat of Late Miocene shorelines and the NE preferred orientation of the consequent drainage network that developed on the emerged land tract when the Diestian sea regressed in this direction (Houthuys 2014; see also Chap. 18). It is commonly believed that the Sambre-Meuse drain and, especially, the Meuse valley segment between Namur and Liege were formed by successive regressive erosion of small streams in this WSW-ENE furrow and captures of Ardennian rivers formerly flowing north or NNW. The origin of this remarkably rectilinear major drainage axis is still unsure. While it has often been suggested that the Sambre-Meuse furrow was the result of a slight reactivation of the Variscan Namur syncline during the Miocene (e.g. Cornet 1899; Fourmarier 1926; Macar 1957), Demoulin (1993) proposed that the main factor controlling the location of the rivers within this furrow was the existence of a shear zone active during the Middle Miocene, which is better supported by the rectilinearity of this 170-km-long line and its spatial coincidence with the North-Artois shear zone (Camelbeeck et al. 2007). In any case, the oldest river terraces preserved along this segment of the Meuse valley contain kieseloolite (silicified oolite) gravels (Fig. 10.5a), a petrographic marker that indicates provenance from the Paris basin and unequivocally relates them to the Kieseloolite Formation, dated from the Late Miocene and Pliocene in the Lower Rhine area (Boenigk and Frechen 2006). This demonstrates that the modern course of the Meuse from the Paris basin down to Liege was already installed at least in the Late Miocene.

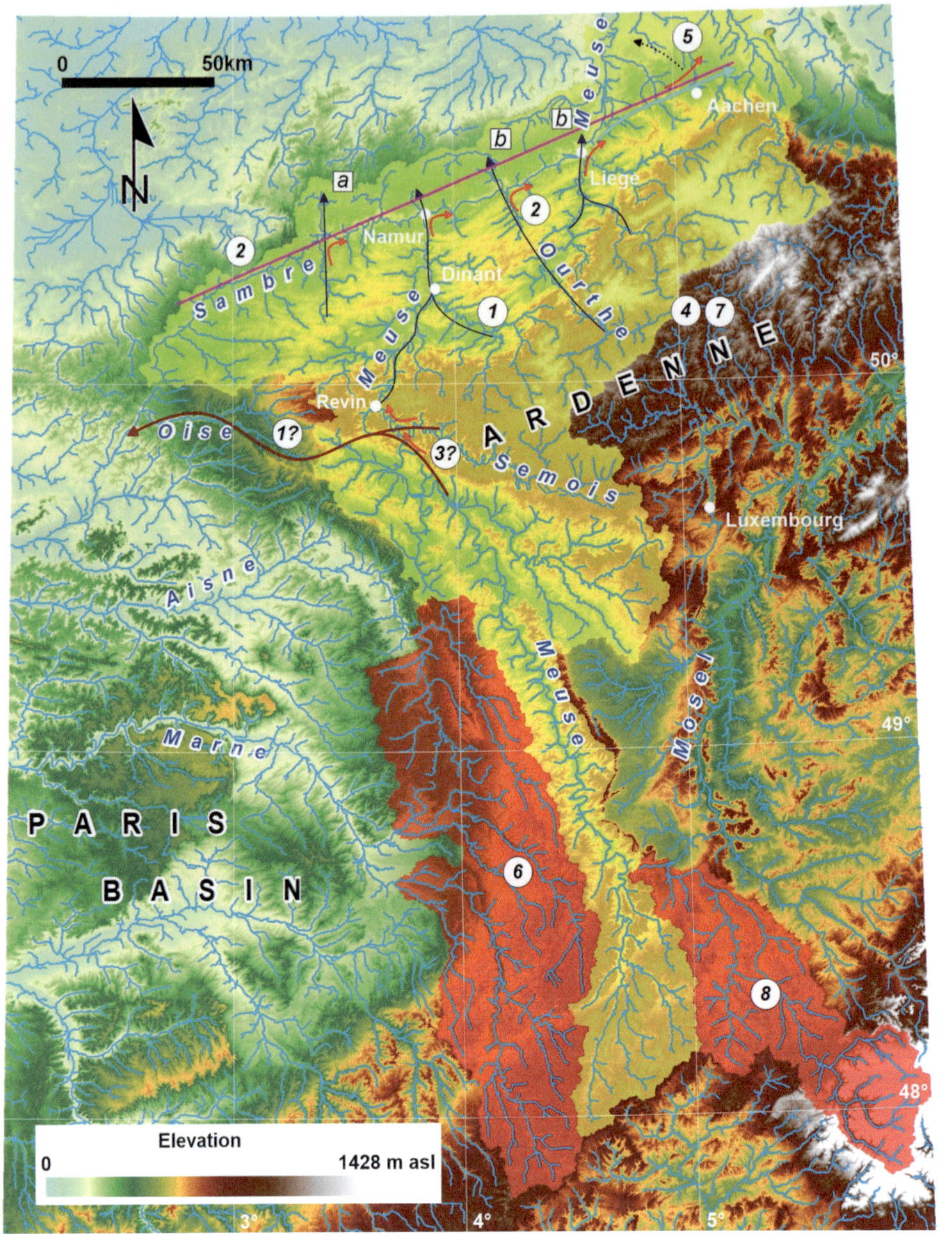

Fig. 10.3 Summary of the Neogene and Quaternary main evolutionary stages of the Meuse catchment (present extension highlighted in *yellow*). *1* N- to NNW-oriented consequent rivers (*thin deep-blue arrows*) drained the Ardenne massif after regression of the Oligocene sea. *a* Eau d'Heure—Piéton axis. *b* Hoyoux—lower Méhaigne and Amblève—lower Ourthe axes. *1?* According to Pissart (1961), the ancient course of the Mesozoic Meuse (i.e. the Upper Meuse, flowing in the Paris Basin) followed the southern margin of the massif and continued in the present Oise (*thick brown arrow*). However, it should be noted that, contrary to Pissart's assumption that a capture was needed to direct the Mesozoic Meuse into the more resistant Paleozoic massif, this ancient course already cut locally into the massif. *2* The Middle to Late Miocene activation of the Sambre-Meuse shear zone (*magenta line*) caused the capture (*red arrows*) of the consequent rivers, now redirected toward the subsiding Roer Valley Graben in the ENE (e.g. Demoulin 1993). *3?* The capture of the Upper Meuse, improbable though claimed by Pissart (1961, 1974) (see text), would have occurred through regressive erosion of an Ardennian river in the Middle to Late Miocene. *4* The Ardenne undergoes a first episode of moderate uplift during the Pliocene-Early Pleistocene. *5* As a consequence of the associated NNW tilt of NE Ardenne, the so-called 'East Meuse' (*light blue arrow*) is diverted toward a more northern course (*red arrow*) that progressively shifts to the NW (*dotted black arrow*) during the Early Pleistocene (e.g. Juvigné and Renard 1992). *6* The Upper Meuse loses a large part of its drainage area (*highlighted in red*) through multiple captures at the benefit of the Seine catchment around 0.9 Ma (Pissart et al. 1997). *7* The Ardenne undergoes a pulse of accelerated uplift at the Early to Middle Pleistocene transition inducing a wave of erosion into the Ardennian drainage system. *8* The last significant capture in the Upper Meuse catchment occurs during the late Middle Pleistocene, when the Upper Moselle is redirected towards the Rhine catchment (Davis 1895; Harmand et al. 1995; Harmand and Cordier 2012)

10.3 The Main Terrace Complex: A Major Geomorphic Marker in the Ardennian Valleys

10.3.1 Morphological and Sedimentological Characteristics

Located approximately at mid-height of the terrace staircases, the Main Terrace Complex (MTC), displaying 2–4 closely spaced terrace levels, constitutes a conspicuous morphological marker not only in many Ardennian valleys but also in the Meuse valley north of the massif. Both its anomalously great lateral development and its location at the sharp hinge between the upper broad and lower narrow parts of valley profiles make it remarkable (Fig. 10.1). However, despite a kilometer-wide three-level MTC (Macar 1938; Juvigné and Renard 1992), the reach of the Meuse valley between Liège and Visé (see Fig. 10.2) shows a less contrasted transverse profile because recently accelerated incision could not prevent this major river to develop wide floodplains during cold stages. The lower Ourthe and the middle Vesdre are certainly among the places where the two-storey valley profile is best expressed in the topography (Fig. 10.1c). There, the MTC, with its YMT lowest level which defines the upper edge of the deeply incised valley, is typically ~300-m-wide, attesting the long period of stability that preceded the onset of the recent episode of incision. The YMT is, therefore, a key feature that has often been used as a major marker in studies of the Quaternary evolution of the Ardennian drainage network (Van Balen et al. 2000; Demoulin et al. 2009; Rixhon et al. 2011).

Being easily identified throughout the Ardennian valleys, the MTC has served as a reference for terrace profile reconstructions for a long time (see the historical review of terrace studies in the Ardennian Meuse catchment in Alexandre and Kupper 1976). A main characteristic of terrace data sets in Ardennian valleys, including the Meuse between Charleville at the southern limit of the massif and Liège at its northern border, is the systematic upstream decrease of the number of terrace levels. Moreover, a careful examination of the representative diagram of Fig. 10.4 shows that this decrease is accounted for essentially by reduced numbers of terrace levels *below* the YMT. In the Meuse valley, the profile reconstruction of Pissart et al. (1997) indicates for example that five post-YMT terraces identified downstream of Liège reduce to three at Namur and only one at Givet, before the YMT tends to merge with the modern floodplain near Charleville (Fig. 10.4). This observation goes hand in hand with that of upstream decreasing relative elevations of the MTC: from 65 m above the modern floodplain of the Meuse near Visé to 60 m at Liège, ~30 m at Namur and ~20 m at Givet. Relative elevations of the MTC in the Ardennian tributaries of the Meuse show a similar trend, decreasing for instance in the Ourthe valley from 55 m at 5 km upstream of the Ourthe-Meuse confluence to 44–48 m in the area of the Amblève-Ourthe confluence and ~34 m at the sharp bend toward NE the river makes in the Famenne (Fig. 10.2b).

If one relies on the widely accepted profile reconstruction of Pissart (1974) for the Meuse terraces, a meaningful feature is the horizontality, or locally even upstream sloping of the MTC levels and all higher terraces between Namur and Liège, i.e., in the river's ENE-striking reach running along the northern border of the Ardenne massif. By contrast, the lower terrace levels remain roughly parallel to the modern long profile in this reach (Fig. 10.4). This profile pattern has both temporal and spatial significance. Spatially, it provides evidence of the geometry of the uplift that deformed the

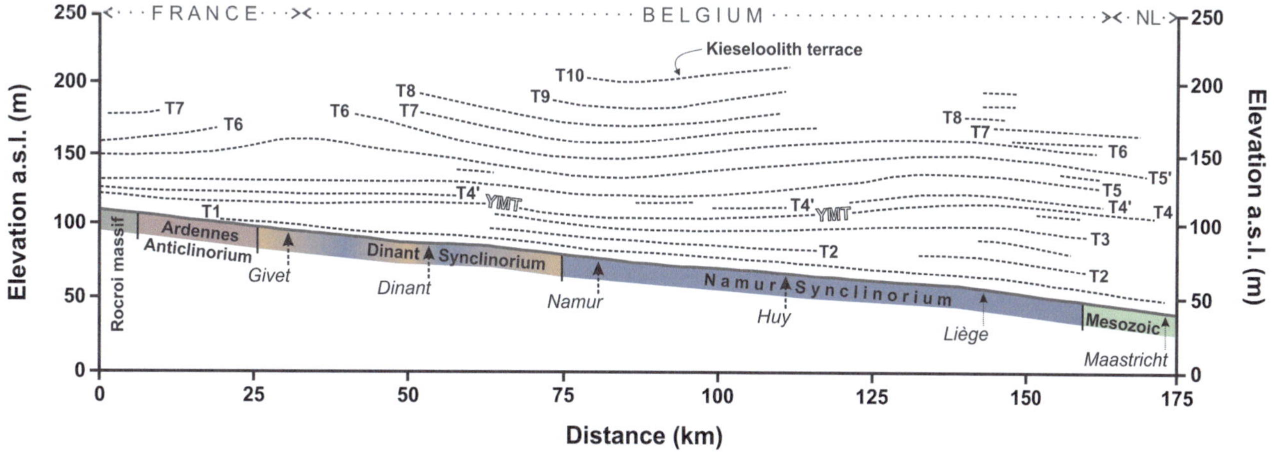

Fig. 10.4 Terrace long profiles of the Meuse in N France and Belgium (adapted from Pissart et al. 1997). Level T4 corresponds to the YMT. Note the profile deformations between Namur and Liege and, for the high terraces, in the Givet area, and the upstream gradual decrease in relative elevation of the profiles

YMT and all higher levels, highlighting the WNW tilting of the margin of the uplifting Ardenne. Temporally, it constrains the time of the deformation, which occurred as a short-lived uplift pulse in the period between abandonment of the YMT and development of the next lower floodplain level.

In the major and middle-sized valleys of the massif, preserved remnants of the MTC usually appear as flat surfaces with minimal transverse slope, extending over a few (in subtributaries) to tens of hectares (in the Meuse valley). Sediment thickness is generally greater in MTC than in other terraces of the same vertical sequence but may vary by a factor 2 or 3 with location in the drainage system, ranging from a mean 7–8 m in the Meuse downstream of Namur to 3–4 m in intra-massif tributaries (Fig. 10.5b). However, anomalously thick alluvial bodies have also been observed in the latter, e.g. >8 m and ~5 m for the YMT deposits in the lower and upper middle Amblève, respectively (Rixhon et al. 2011). Sedimentation patterns typical of braided-river systems along with embedded boulders bigger than 1 m in intermediate axis interpreted as ice-rafted material (see Chap. 6), syngenetic cryoturbations and ice wedge casts are widely observed in the MTC terraces of the Meuse downstream of Liège (Fig. 10.5c, d), including on the Campine Plateau (Macar 1957; Pissart et al. 1969; Juvigné 1992; De Brue et al. 2015). All these elements clearly point to deposition under the periglacial conditions that prevailed in Belgium during the Pleistocene cold periods. In the intra-massif tributary valleys, YMT sediments generally consist of massive bodies of coarse pebbles embedded in an abundant silty-sandy matrix and present only scarce sedimentary structures (Juvigné et al. 2005; Rixhon et al. 2011). They occasionally contain quartzite boulders up to 0.5 m in size and may locally be made of very coarse material (a YMT exposure in the upper middle Amblève revealed a D_{50} larger than 20 cm).

Fig. 10.5 **a** General and closer views of the oldest terrace sediments—the so-called Kieseloolite terrace (T10 in Fig. 4)—overlying Oligocene marine sands atop the left valley side of the Meuse a few kilometres upstream of Liège (*Photographs* G. Houbrechts). **b** View of the Main Teratoprace sediments covered by (mainly) Weichsel loess at Romont, Eben-Emael, close to the Belgian/Dutch border (*Photograph* G. Rixhon). ^{10}Be and ^{26}Al concentration measurements dated the terrace abandonment time to 725 ± 120 ka (Rixhon et al. 2011). **c** Boulder (*a*-axis ~1.25 m) embedded in Meuse terrace deposits of the Campine Plateau north of Liege, generally assigned to the MTC. Such blocks are interpreted as ice-rafted material (*Photograph* from De Brue et al. 2015). **d** Syngenetic cryoturbation features in MTC deposits a few kilometres downstream of Liège (*Photograph* from Cornet 1995)

10.3.2 Age Constraints on MTC and Lower Terraces

Yielding age estimates of the YMT was far from easy until a decade ago because no reliable numerical dating method was available to date fluvial sediments older than half a million years. Supported by sparse measurements elsewhere in the Meuse basin (e.g., Juvigné et al. 2005) and by similar results for the corresponding terraces in the middle Rhine (Koci et al. 1973), paleomagnetic measurements performed in a terrace sequence of the Meuse near Maastricht could at best suggest that the MTC straddled the Matuyama-Brunhes transition and, consequently, that the YMT (called the Sint-Pietersberg 3 terrace in the area of Maastricht) accumulated just after 780 ka (Van den Berg 1996; Westaway 2002). However, the data at hand were differently interpreted by other authors, Van Balen et al. (2000) arguing for instance for a YMT age in the range 0.65–0.72 Ma.

New age data were recently gained from a cosmogenic radionuclide (CRN) study performed by Rixhon et al. (2011) on depth profiles of MTC terraces in the Meuse and its Ardennian tributary system. Jointly modelling ^{10}Be and ^{26}Al depth profiles, they overcame the issue of burial of the fluvial sediments by a loess cover and obtained an age of 725 $\pm$ 120 ka (time of terrace abandonment) for a 5-m-thick terrace remnant exposed in the Romont quarry at Eben-Emael, $\sim$15 km north of Liège (Fig. 10.5b). Careful re-examination of the position of the sampled terrace in the Meuse terrace sequence shows that, contrary to the claim of Rixhon et al. (2011), the Romont terrace pertains to the middle, possibly even the highest level of the MTC, implying an age younger than 725 ka for the YMT, in agreement with the interpretation of Van Balen et al. (2000).

The second Meuse terrace for which numerical ages are available is the Caberg 3 level, at 15–18 m relative elevation in the Liège-Maastricht reach of the river. In the Belvédère site, near Maastricht, the interglacial deposits at the top of the alluvial body have been dated to 250 $\pm$ 20 and 220 $\pm$ 40 ka by thermoluminescence on heated flint artifacts and ESR on mollusc shells, respectively (Huxtable 1993; Van Kolfschoten et al. 1993), while biostratigraphic consideration and amino acid dating rather favoured an MIS9 age ($\sim$340–300 ka) (Meijer and Cleveringa 2009). This Caberg 3 terrace is especially interesting because the disappearance of petrographic and mineralogical markers of the upper Moselle basin (Vosgian granite and basaltic hornblende, respectively) in its deposits helps draw a reliable terrace profile all along the Meuse and constrain the time of the Upper Moselle capture at Toul, in French Lorraine (e.g., Pissart et al. 1997) (Fig. 10.3). Moreover, U/Th dating of a flowstone sealing the last Moselle deposits before capture in a karstic system at the capture point provided a minimal age of 250 ka (Losson and Quinif 2001). All these data seems to be best fitted by a Caberg 3 terrace having developed during MIS8 and 7 (and 9?) and been abandoned at the MIS7-6 transition.

Contrasting with this relative wealth of data in the Meuse valley, the absolute chronology of terrace systems in the Ardennian tributaries of the Meuse was long neglected. Beyond limited tephrochronological hints of the age of a few lowest terraces (Juvigné 1979) and varied time indications gathered in the Belle-Roche cave fill (Figs. 10.6 and 10.7), located just above the YMT in the lower Amblève (Cordy et al. 1993), crucial data have only recently been provided by Rixhon et al. (2011). At Belle-Roche, Amblève gravels occur at the base of the cave fill and are overlain by slope material brought in the cave by mud flows. Supported by the >300 ka U/Th age of a calcitic flowstone sealing the deposits and paleomagnetic measurements indicating normal polarity, the rich and diversified macro- and microfauna included in this material has been dated to 500 $\pm$ 70 ka (Cordy et al. 1993), apparently suggesting a $\leq$500-ka age for the lower lying YMT. Having sampled an $\sim$8-m-thick depth profile in the YMT deposits at Belle-Roche, Rixhon et al. (2011) showed that the ^{10}Be concentration profile displayed two superposed signals, whose modelling allowed them not only to provide an estimate of 222 $\pm$ 31 ka for the time of terrace abandonment but also to date around 550 ka the beginning of the aggradation phase that constructed the YMT floodplain. Moreover, Rixhon et al. (2014) suggested that the gravels in the cave might represent overflow deposits accumulated in relation with the formation of the lower lying YMT, making both gravel bodies pertain to MIS14, in perfect agreement with the MIS13 paleontological age of the slope deposits in the cave.

Still farther-reaching are, however, the overall results that Rixhon et al. (2011) obtained for the YMT remnants they dated all along the Ourthe-Amblève line (Fig. 10.8a): going upstream, (1) the CRN age of the Romont terrace suggests a corresponding age around 620 ka for the time when the YMT was abandoned in the lower Meuse; (2) combined ^{10}Be and ^{26}Al depth profiles at Colonster yield an age of 390 ka for this same YMT in the lower Ourthe; (3) as stated above, the YMT was abandoned around 220 ka at Belle-Roche in the lower Amblève; (4) two sites sampled in the upper middle Amblève, at Stavelot and Lodomez, provided CRN ages in the range 135–140 ka for the abandonment time of the YMT in this part of the valley.

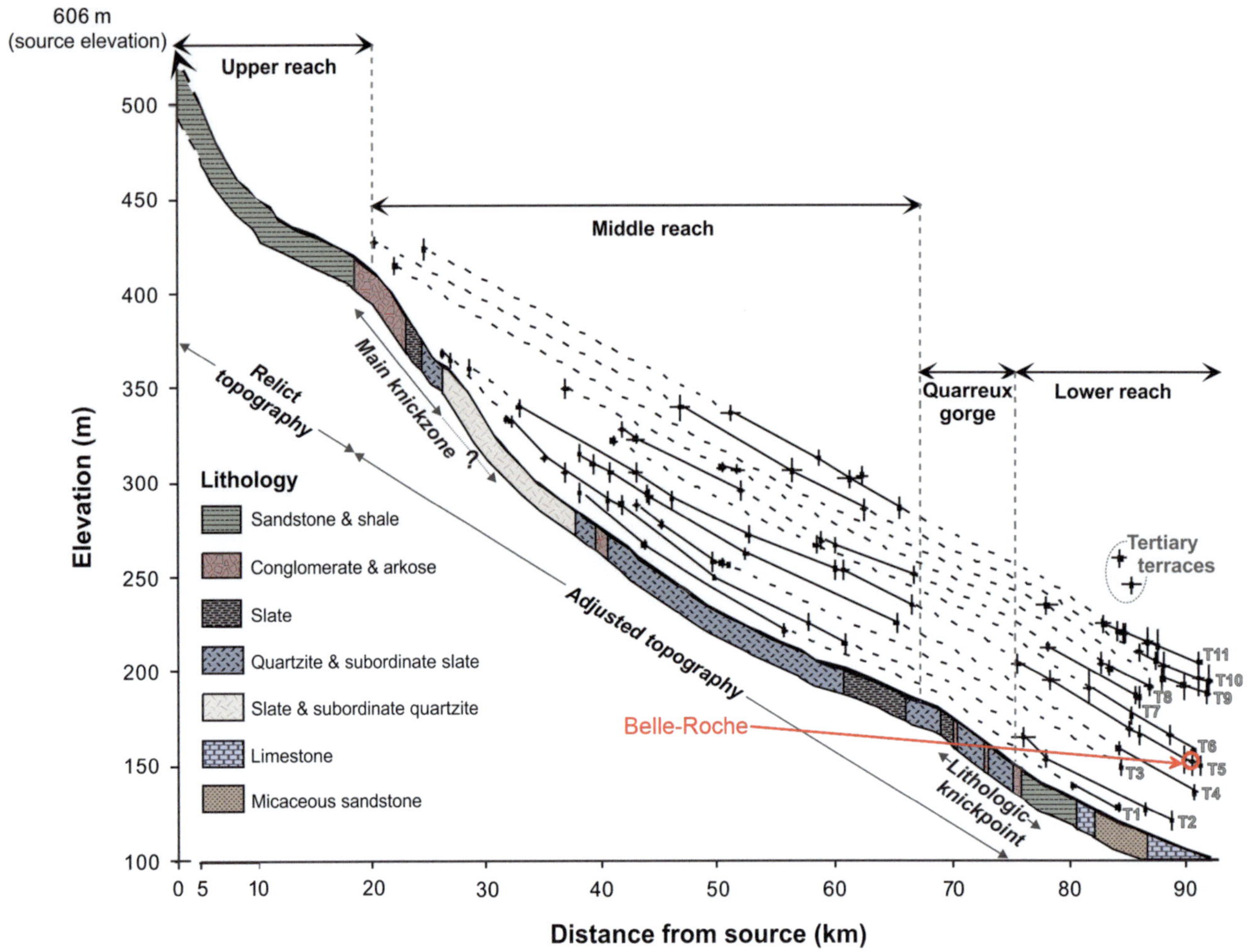

Fig. 10.6 Terrace long profiles of the Amblève River, a main Ardennian sub-tributary of the Meuse, plotted against the present-day river profile with lithology. *Solid and dashed lines* refer to probable and uncertain geometric terrace correlations, respectively. A prominent tectonically controlled knickpoint at km 18 marks the transient boundary between the downstream post-YMT adjusted and upstream relict parts of the profile. The knickpoint in the Quarreux gorge at 70 km is controlled by lithology where the river incises across resistant quartzites (adapted from Rixhon and Demoulin 2010)

10.4 Quaternary River Incision in the Ardenne: A Combination of Tectonic and Climatic Forcing

The main implication of CRN dates decreasing systematically upstream along the lower Meuse—lower Ourthe—Amblève system is that the YMT was abandoned diachronically, due to the propagation of an erosion wave in the Ardenne drainage system. This obvious conclusion cannot be obscured by remaining issues regarding the propagation's varying celerity between the dated sites, which probably relates to the varying duration of the successive periods favourable to incision rather than to other factors invoked by Demoulin et al. (2012b). This important Middle Pleistocene event, which, according to the Romont CRN date, started around 620 ka, has not yet come to an end. Not only does the morphometry of catchment, valley and river profile show that the Ardennian rivers have not yet reached steady-state (Sougnez and Vanacker 2011) but, more importantly, knickpoints are still observed in the upper reaches of many of them at points where the YMT, which converges upstream with the modern floodplain, finally merges into it (Beckers et al. 2015) (Fig. 10.8b, c). As was confirmed by the successful modelling of their migration through the drainage system (Beckers et al. 2015), these lithology-independent knickpoints are tectonically triggered features locating the current position of the erosion wave.

These findings have fundamentally questioned several well-established truths about the Quaternary evolution of the Ardenne and its river system. As most terraces bear witness

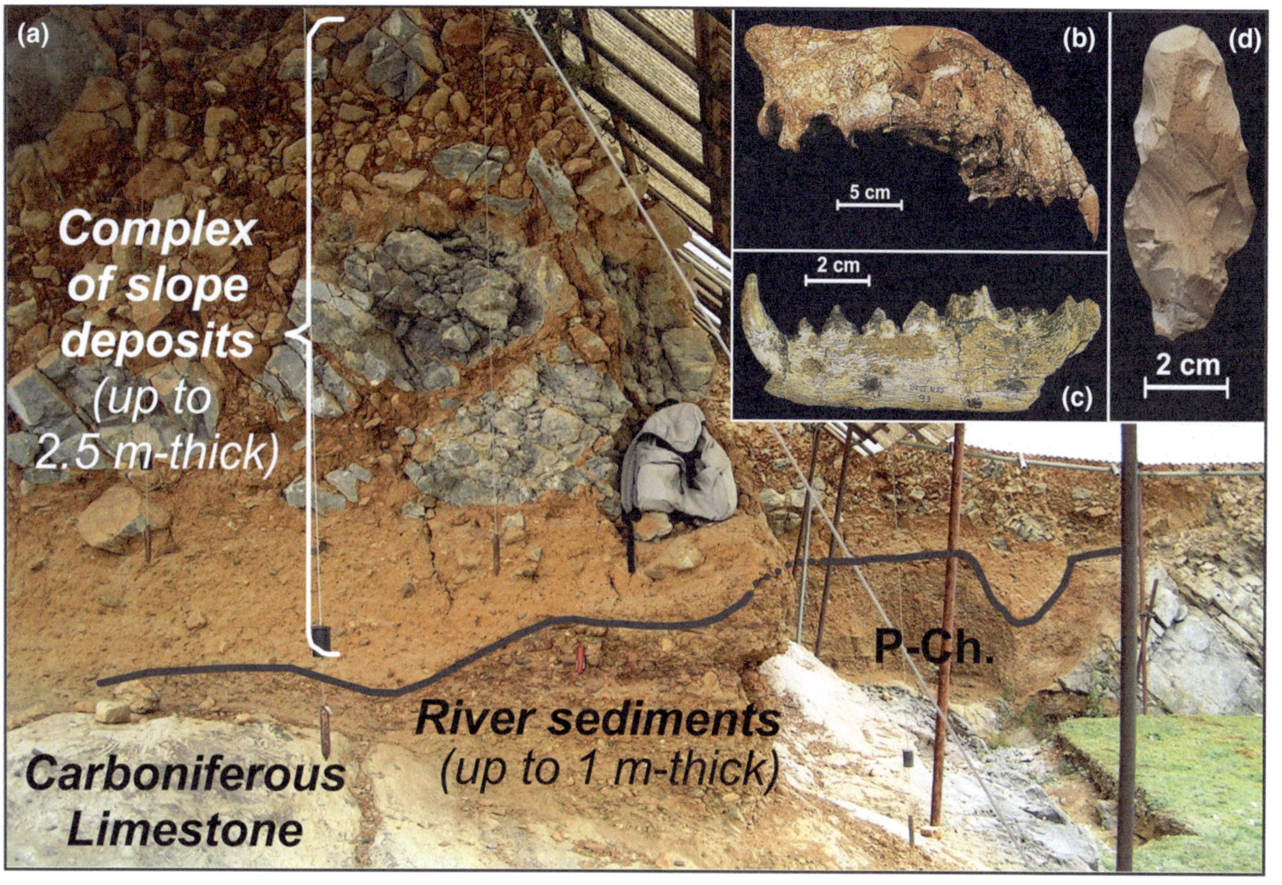

Fig. 10.7 **a** General view of the infill sediments in one gallery of the Belle-Roche palaeokarst in the *Lower* Amblève valley (see position in Fig. 10.6). The sedimentary sequence consists of up to 1 m-thick river deposits of the Amblève at the base, overlain by up to 2.5 m-thick slope deposits (mostly runoff and solifluction products). The latter contain many palaeontological and archaeological remains (**b** skull of *Panthera onca gombaszoegensis*; **c** jaw of *Canis mosbachensis*; **d** flint biface). The very rich palaeofauna assemblage (>50 species, Cordy et al. 1993) has been assigned a biochronological age of 500 ± 70 ka (All *photographs* from Rixhon et al. 2014)

to aggradation under cold-stage conditions, it is widely accepted that each terrace level was representative of one climatic (sub)cycle, the succession of Quaternary glacial-interglacial cycles being responsible for the formation of the terrace staircases (e.g., Van den Berg 1996). This classical scheme of climatically driven alternation of terrace formation and river incision, adopted far beyond the Ardenne area (e.g. Bridgland and Westaway 2008), conveys the implicit assumption that every particular level formed synchronously and thus has the same age everywhere in the drainage system, from the Meuse between Liège and Mastricht down to the tiniest of its Ardennian tributaries. However, the new data show this is incorrect at least for the YMT. Beyond the conceptual change it imposes, the observed diachronism in local river incision also entails that calculations derived from a single age of the YMT throughout the Ardenne to map uplift rates or estimate time-dependent eroded volumes (Van Balen et al. 2000; Meyer and Stets 2007) are basically flawed.

In order to account for the new knickpoint and terrace age information, Demoulin et al. (2012b) proposed a new evolutionary model of river incision in the Ardenne. Its main focus is on integrating the effect of the tectonically triggered longitudinal propagation of a post-YMT erosion wave with that of climatically controlled episodic vertical incision in response to background uplift. Starting from the steady-state reached by the wide YMT valleys at the dawn of the Middle Pleistocene and still attested all over the massif by the barely incised uppermost reaches of the Ardennian valleys, this model acknowledges the uplift pulse recorded by the deformed terrace levels of the Meuse between Namur and Liège and the larger than usual vertical spacing between the YMT and the next lower terrace observed in all valleys. Dated to ~620 ka at Romont, this essentially en-bloc uplift of the Ardenne created a ~20-m-high knickpoint in the Meuse long profile in the vicinity of Maastricht, which in turn initiated the propagation of a wave of regressive erosion in the drainage system. This propagation was, however, also

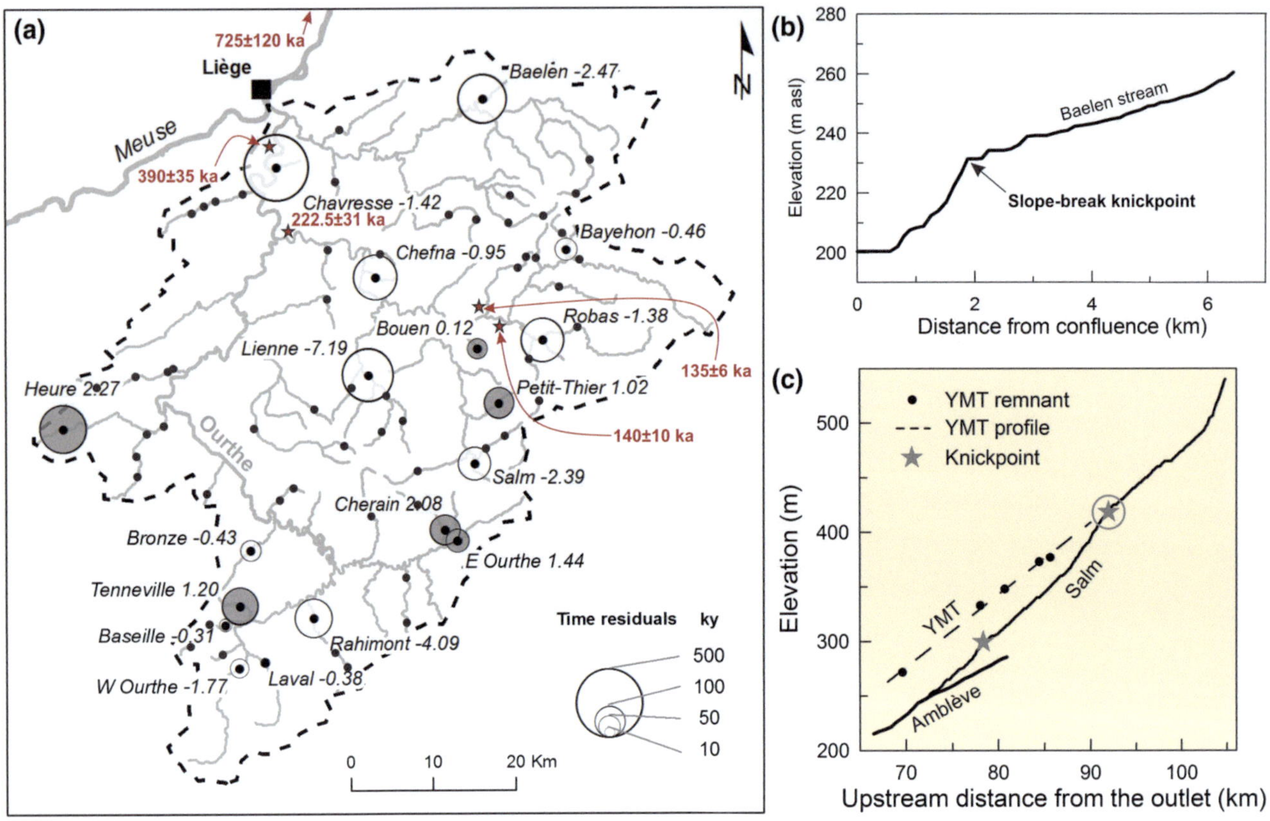

Fig. 10.8 **a** Spatial distribution of knickpoints (*black dots*) in the Ourthe catchment (modified from Beckers et al. 2015). Only circled dots are safely associated to the post-YMT wave of erosion. *Open and solid grey circles* refer to knickpoints delayed with respect to, and ahead of, their modeled position, respectively. Circle size quantifies time residuals whereas distance residuals are given in km by numbers behind every river name. *Red stars* and *red numbers* refer to MTC (north of Liege) and YMT sediments dated by ^{10}Be concentration profiles (Rixhon et al. 2011). **b** Long profile of the Baelen creek displaying a slope-break knickpoint typical of the post-YMT erosion wave (modified from Beckers et al. 2015). **c** YMT profile of the Salm River plotted against the current river profile. The knickpoint marked as a *circled star* indicates the present-day position reached by the post-YMT erosion wave, which is also the point where the YMT profile merges upstream with the modern profile

climatically controlled in the sense that it went on essentially during climatic transitions, when stream power largely outbalanced hillslope sediment delivery, allowing river incision (e.g. Vandenberghe 2003). Owing to the direct dependence of knickpoint migration rate on drainage basin size, the erosion wave was possibly able to travel the whole distance of the Ardennian Meuse valley within one or two such transitions. Moreover, transport-limited conditions rapidly caused diffusive dampening of the knickpoint form in this major valley. However, knickpoint propagation rate decreased significantly when proceeding toward the smaller branches of the Ardennian tributary system, thus extending the drainage network's transient response over multiple glacial–interglacial cycles, up to the present time. In parallel, once a knickpoint had migrated past a valley reach, erosion kept on deepening it, but to a much slower rate in equilibrium with the constant background uplift rate <0.05 mm/year of the last half-million years. Now, this took place after the climatic fashion of ubiquitous vertical erosion, but only in that part of the drainage network in equilibrium with the new background uplift rate downstream of the front of the erosion wave. While this process nicely accounts for the upstream diminishing number of post-YMT terraces in the Meuse catchment, it has a profound impact on the way terrace levels should theoretically be reconstructed and, consequently, on the age of each particular terrace remnant. As no other significant tectonic perturbation occurred after the 620-ka uplift pulse, the post-YMT levels basically display profiles parallel between each other and with the modern floodplain and intersect the YMT profile all the more upstream as they are younger (Fig. 10.9a).

The interaction between the climatic and tectonic controls on river incision is thus twofold. While the classical model only recognized that uplift was required to cause incision under episodically favourable climatic conditions, the new one adds the effect of change in uplift rate, with fast rates

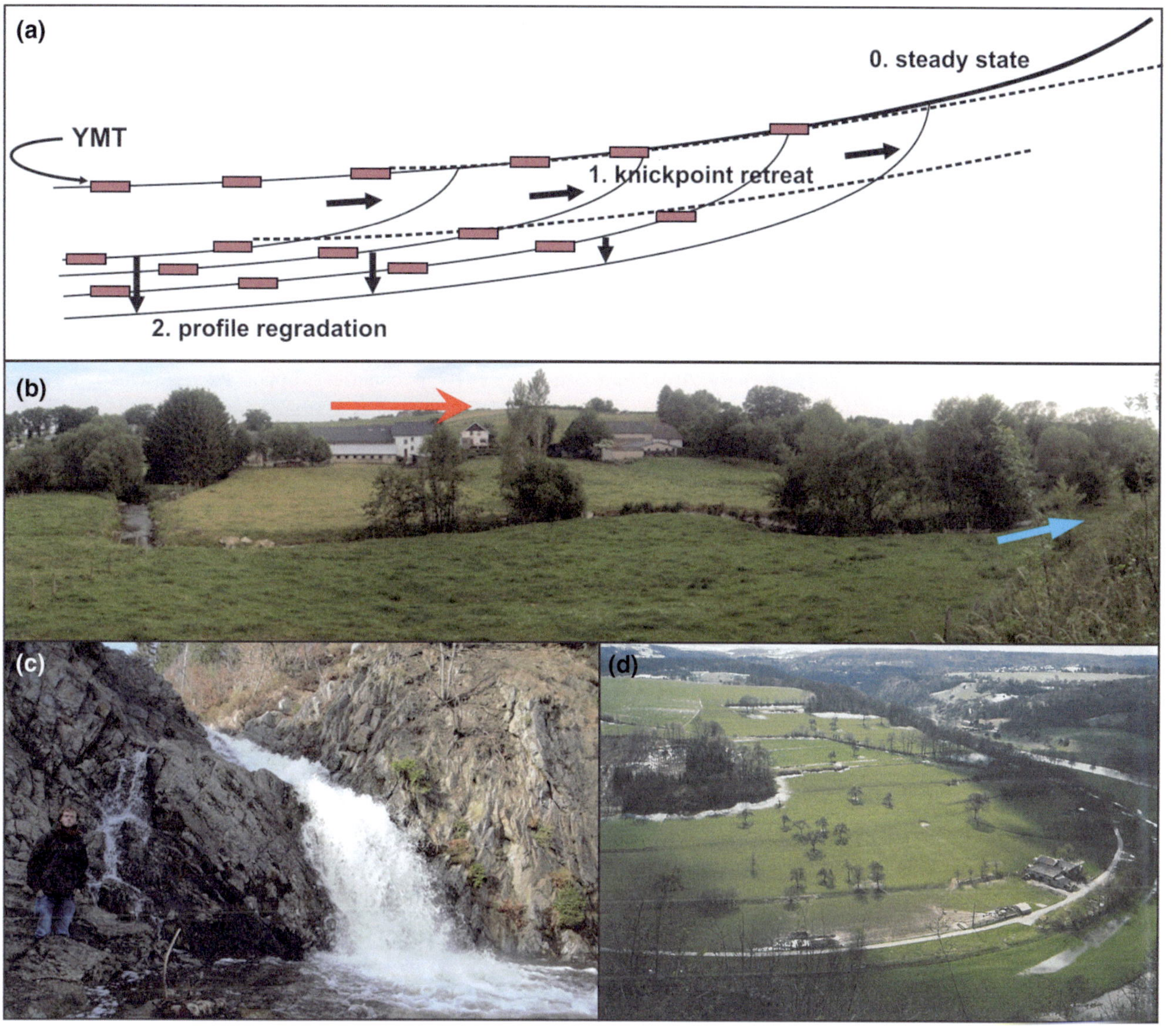

Fig. 10.9 **a** Conceptual sketch of the formation of a time-transgressive terrace (e.g. the Ardennian YMT), showing also the three regions of specific incision behaviour associated with the transient response to an uplift signal (*solid purple rectangles* denote terrace remnants). From source to outlet, in the Ardennian example: *0* upstream of the knickzone, almost no incision occurs in the profile part still in equilibrium with the extremely small pre-event uplift rate; *1* fast vertical erosion is associated with the migrating knickzone; *2* measurable vertical incision ubiquitously lowers the river profile where a new equilibrium has been established with the post-event uplift rate downstream of the knickzone. The height of the propagating knickpoint is given by the difference in elevation between the YMT and the next terrace profile at the point downstream where a base level fall initiated the wave of erosion (from Demoulin et al. 2012b). **b** View of the upper reach of the Amblève valley upstream of the knickzone (flow direction as *blue arrow*). The relict landscape evolves very slowly, in continued steady state with the quasi-zero uplift rate of the Early Pleistocene, with a river developing free meanders in a broad floodplain barely incised into Neogene planation surfaces (*red arrow*) (*Photograph* from Rixhon 2011). **c** Waterfall in the Bayehon stream (person for scale), materializing the present-day position reached by the post-YMT erosion wave in the Ardennian network. Incision occurs here in resistant quartzites of the Cambro-Ordovician Stavelot massif (*Photograph* A. Beckers). **d** View of the Ham meander in the lower Ourthe valley showing continued erosion after the passage of the post-YMT erosion wave. The climate-dependent terraces that form the gentle slope from the YMT flat summit down to the present floodplain are not directly distinguished in the topography, where extended scarplets are not terrace risers but lynchets (*Photograph* from Cornet 1995)

allowing the creation of knickpoints at the edge of the uplifting area in the time span between two erosion phases, thus causing the propagation of erosion waves. During the Quaternary, such a situation occurred only once in the Ardenne, around 620 ka. Since then, at every time until the transient response of the system will be completed, three distinct parts of the Ardennian drainage network display specific behaviours: (1) the front of the erosion wave, where rapid incision results from the passage of migrating knickpoints (Fig. 10.9c), (2) downstream of the front, where river profiles keep on regrading at a slower background pace by ubiquitous vertical erosion (Fig. 10.9d), (3) upstream of the front, where barely changing Early Pleistocene steady-state river profiles still characterize the regions not yet reached by the uplift signal, thus preserving the ancient equilibrium with quasi-zero Early Pleistocene uplift rate (Fig. 10.9b).

Beyond their respective role in river incision, tectonics and climate have also been considered as concurrent factors of massif-wide denudation during the Quaternary. Analysing their CRN-derived denudation rate estimates in the Meuse catchment, which increase from values around 30 mm/ky in the period 1.3–0.7 Ma to 60–80 mm/ky in the Late Pleistocene, Schaller et al. (2004) suggested that the change might result from both the Early Middle Pleistocene uplift acceleration and the climatic deterioration that took place roughly at the same time (Mudelsee and Schulz 1997). Based on geomorphic data, Demoulin et al. (2009) calculated denudation volumes due to various processes in the Meuse catchment upstream of Liege since 700 ka. They showed that river incision (84 km^3 issued from valleys representing 22% of the catchment area) and interfluve denudation (292 km^3 delivered by the remaining 78% of interfluves) were equally effective on average, leading to a regional denudation rate of 27 mm/ky. However, this figure calls for two major comments. First, while river incision is homogeneously governed by tectonically driven knickpoint migration and regressive erosion throughout the catchment, with lateral erosion acting only as a second-order factor of spatial variability, interfluve denudation is spatially very heterogeneous because lithology-dependent sensitivity to climate leads to highly contrasted denudation rates between regions within the massif. Demoulin et al. (2009) provide for instance denudation estimates in the order of 5–10 mm/ky for the flat interfluves of central Ardenne that still preserve remnants of loose sediments of Late Cretaceous and Oligocene age as well as extended patches of Mesozoic and Neogene deep weathering mantles (see Chap. 5), and 150 mm/ky for the lower lying Fagne-Famenne region (Fig. 10.2b), where pure shales are extremely sensitive to frost shattering and underwent very rapid erosion under the periglacial conditions of the cold periods. Second, there is a discrepancy between the geomorphic estimate of 27 mm/ky and an averaged CRN estimate that, based on all values provided by Schaller et al. (2004), is in excess of 40 mm/ky since 700 ka. If significant, this discrepancy might originate in overestimated ^{10}Be rates obtained from samples that did not represent equally all parts of the catchment because of, e.g. inadequacy of the sampled grain size, quasi-zero denudation in some areas, or low to near-zero quartz content of carbonate bedrock. According to Demoulin et al. (2009), the CRN estimates would, therefore, be closer to valley incision rates than catchment-scale denudation.

10.5 Capture Phenomena

The response of the Ardennian drainage network to the 620-ka uplift pulse entailed not only vertical erosion but also planform changes of river courses. We shall not discuss the numerous river captures (upper Aisne, which formerly included also the upper Marne and its tributaries; Aire; upper Moselle) that almost beheaded the Meuse at the benefit of the Seine and Rhine basins (Fig. 10.3), because they occurred outside the Belgium–Luxembourg area dealt with here and, although they are dated to the period 900–300 ka (Pissart et al. 1997; Harmand and Cordier 2012), their relation to the uplift pulse in the Ardenne and NE Paris basin is still more theorized than demonstrated.

However, smaller scale stream piracy has also been observed within the Ardenne as a consequence of the migration of the post-YMT erosion wave. In NE Ardenne, two successive captures affected the Warche and its tributary, the Trô Maret, the first one having taken place between 78 and 50 ka (Juvigné and Schumacker 1985). Flowing westward, the former Warche once made an S-bend to the north and, as proved by petrographically signed sediments, followed the valley of the modern Eau Rouge before joining the Amblève at Stavelot (Fig. 10.10a, b). When travelling knickpoints reached this area, the one migrating in the palaeo-Warche (now Eau Rouge) proceeded much more slowly in the resistant quartzite rocks of the Stavelot massif than another one that propagated in a small tributary of the Amblève draining the weaker Permian bedrock of the Malmédy graben. Consequently, faster regressive erosion in, and lower elevation of, the latter stream allowed it to divert the Warche at Bévercé (Pissart and Juvigné 1982). Importantly, the reconstruction of the long profile of the Warche before capture clearly evidences the knickpoint buried below thick slope deposits in the abandoned valley and the 78–50 ka age of this knickpoint fits perfectly in the series of data points (CRN dates, modern knickpoints in the profiles of the Amblève and the Warche) that attest the propagation of the erosion wave (Demoulin et al. 2012b). After this first capture, the upper Trô Maret continued to flow through the abandoned corridor and the Eau Rouge valley (Fig. 10.10c) but was soon captured in turn by a small creek eroding the part of the abandoned valley that now opened into the incised new Warche.

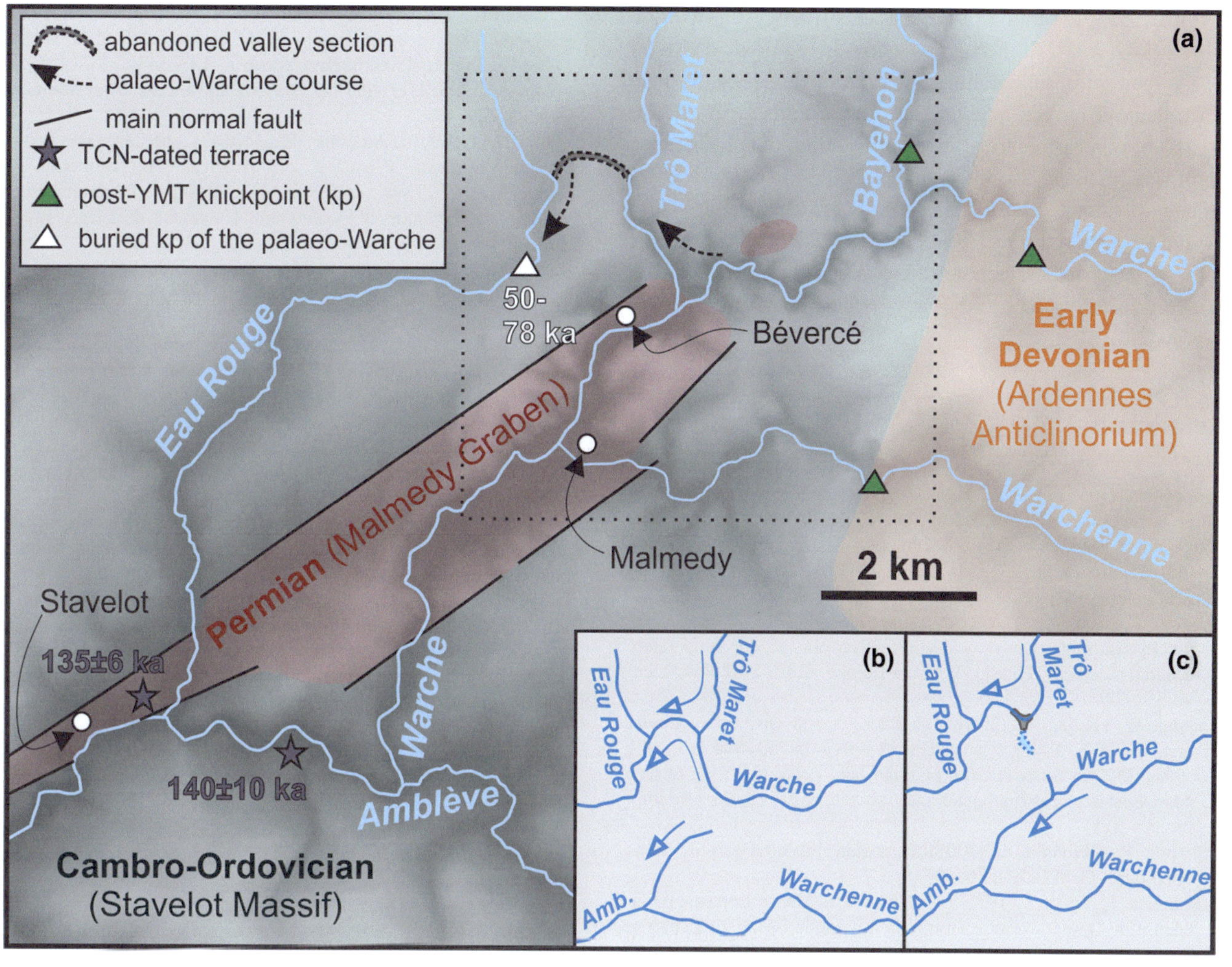

Fig. 10.10 **a** Simplified geological map of the upper Amblève catchment showing the capture of the Warche river and its palaeo-course (modified from Pissart and Juvigné 1982 and Demoulin et al. 2012b). The *white triangle* locates the knickpoint buried in the former course of the Warche through the current Eau Rouge valley (estimated age range according to Demoulin et al. 2012b). The *lilac stars* locate YMT terrace sediments dated by CRN (Rixhon et al. 2011). **b–c** describe the hydrographic network before and after the Warche capture, respectively. Note that a second capture was later responsible for redirecting the Upper Trô Maret toward the new Warche. This second capture occurred through spillover of the small lake that had formed upstream of the alluvial fan accumulated by the Trô Maret in the abandoned valley after the first capture (modified from Pissart and Juvigné 1982)

10.6 Conclusion

The scenic landscapes of the Ardennian valleys with their well-preserved terrace staircases are marked by the dual character of broad Early Pleistocene upper transverse valley sections going into steep-sided narrow Middle Pleistocene lower sections. CRN dating of the younger main terrace (YMT), a prominent marker among the stepped terrace levels, along the lower Meuse—lower Ourthe—Amblève line not only confirmed that the pulse of tectonic uplift responsible for the latest episode of fast river incision in the Ardenne occurred in the Early Middle Pleistocene, sometime around 620 ka, but showed also that the drainage system response took the form of an erosion wave causing the diachronic abandonment of the YMT. This implies a major revision of the classical model of climatically driven river incision and terrace formation in uplifting massifs in order to fully integrate the diverse ways tectonic uplift controls river incision, including in particular the formation of knickpoints in the river long profiles under increased uplift rates and the associated longitudinal mode of migrating incision. Subordinately, planform changes of the drainage network also resulted from the Quaternary uplift of the Ardenne and the NE Paris basin. While stream piracy was caused by changes in the balance between competing major catchments (Seine, Meuse, Rhine) at the regional scale, it was related to differences in knickpoint migration rates between adjacent streams at the local scale.

First numerical ages of Ardennian river terraces have fundamentally changed our understanding of the

Quaternary evolution of the massif's river system but, be that as it may, further dating effort is required to extend the data set in time (dating other terrace levels) and space (obtaining terrace ages in the middle and upper Meuse and in other tributaries) in order to refine the chronology of river incision. In any case, new, still unpublished, CRN ages obtained for the YMT in the middle Ourthe valley and CRN burial ages of cave sediments in karstic systems associated with several terrace levels in the lower Ourthe seem to fully confirm the new model.

References

Alexandre J (1957) Les terrasses des bassins supérieurs de l'Ourthe et de la Lesse. Ann Soc Géol Belg 80:317–332

Alexandre S, Kupper M (1976) L'évolution des rivières. In: Pissart A (ed) Géomorphologie de la Belgique, Lab. de géologie et de géographie physique, University of Liège, pp 51–74

Beckers A, Bovy B, Hallot E, Demoulin A (2015) Controls on knickpoint migration in a drainage network of the moderately uplifted Ardennes Plateau, Western Europe. Earth Surf Proc Land 40:357–374

Boenigk W, Frechen M (2006) The Pliocene and Quaternary fluvial archives of the Rhine system. Quat Sci Rev 25:550–574

Bridgland D, Westaway R (2008) Climatically controlled river terrace staircases: a worldwide Quaternary phenomenon. Geomorphology 98:285–315

Bultynck P, Dejonghe L (2001) Devonian lithostratigraphic units (Belgium). Geol Belg 4:39–69

Bustamante L, Voisin L (1975) La capture de la Meuse Lorraine par la Meuse de Dinant. Remise en question partielle basée sur l'analyse des minéraux lourds. Service géologique de Belgique, 17 p

Camelbeeck T, Vanneste K, Alexandre P, Verbeeck K, Petermans T, Rosset P, Everaerts M, Warnant R, Van Camp M (2007) Relevance of active faulting and seismicity studies to assessments of long-term earthquake activity and maximum magnitude in intraplate northwest Europe, between the Lower Rhine Embayment and the North Sea. In:Stein S, Marzotti S (eds) Continental intraplate earthquakes, Hazard and policy issues, Geological society of America Special paper 425, pp 193–224

Cordy JM, Bastin B, Demaret-Fairon M, Ek C, Geeraerts R, Groessens-Van Dyck MC, Ozer A, Peuchot R, Quinif Y, Thorez J, Ulrix-Closset M (1993) La grotte de la Belle-Roche (Sprimont, Province de Liège): un gisement paléontologique et archéologique d'exception au Benelux. Bulletin de l'Académie Royale de Belgique, Classe des Sciences 6e S 4, pp 165–186

Cornet J (1899) Compte-rendu de l'excursion du 30 avril 1899 à Stambruges, Grandglise, Hautrages. Bull Soc belge Géol 13:125

Cornet J (1903) Etudes sur l'évolution des rivières belges. Ann Soc géol Belg 31:M261–M500

Cornet Y (1995) L'encaissement des rivières ardennaises au cours du Quaternaire. In: Demoulin A (ed) L'Ardenne, Essai de Géographie Physique. Département de Géographie Physique et Quaternaire, Université de Liège, Liège, pp 155–177

Davis WM (1895) La Seine, la Meuse et la Moselle. Ann Géog 4:25–49

De Brue H, Poesen J, Notebaert B (2015) What was the transport mode of large boulders in the Campine Plateau and the lower Meuse valley during the mid-Pleistocene? Geomorphology 228:568–578

De Heinzelin J (1963) Le réseau hydrographique de la région gallo-belge au Néogène. Essais de reconstitution. Bull la Société Belge Géologie Paléontologie Hydrol. 72:137–148

Demoulin A (1993) L'origine de l'axe Sambre-Meuse. Ann Soc Géol Belg 116:29–41

Demoulin A (1995) L'Ardenne des plateaux, héritage des temps anciens. Surfaces d'érosion en Ardenne. In: Demoulin A (ed) L'Ardenne, Essai de Géographie Physique. Département de Géographie Physique et Quaternaire, Université de Liège, Liège, pp 68–93

Demoulin A (2011) Basin and river profile morphometry: a new index with a high potential for relative dating of tectonic uplift. Geomorphology 126:97–107

Demoulin A, Ek C (1995) La naissance de l'Ardenne. Mise en place du massif paléozoïque de l'Ardenne. In: Demoulin A (ed) L'Ardenne. Essai de géographie physique, Dépt de Géographie physique et Quaternaire, University of Liège, pp 15–30

Demoulin A, Hallot E (2009) Shape and amount of the Quaternary uplift of the western Rhenish shield and the Ardenne (western Europe). Tectonophysics 474:696–708

Demoulin A, Beckers A, Bovy B (2012a) On different types of adjustment usable to calculate the parameters of the stream power law. Geomorphology 138:203–208

Demoulin A, Beckers A, Rixhon G, Braucher R, Bourlès D, Siame L (2012b) Valley downcutting in the Ardenne (W Europe): Interplay between tectonically triggered regressive erosion and climatic cyclicity. Neth J Geosci/Geol en Mijnb 91:79–90

Demoulin A, Hallot E, Rixhon G (2009) Amount and controls of the Quaternary denudation in the Ardenne massif (Western Europe). Earth Surf Process Landforms 34:1487–1496

Felder W, Bosch P, Bisschops J (1989) Geologische kaart van Zuid-Limburg en omgeving. Afzettingen van de Maas, Rijks Geologisch Dienst, Harlem, Nederland

Fourmarier P (1926) Les dernières ondulations du sol en Belgique et les terrasses de la Meuse. Société géologique de Belgique, Livre jubilaire, pp 110–114

Harmand D, Cordier S (2012) The Pleistocene terrace staircases of the present and past rivers downstream from the Vosges Massif (Meuse and Moselle catchments). Neth J Geosci 91:91–109

Harmand D, Weisrock A, Gamez P, Le Roux J, Occhietti S, Deshaies M, Bonnefont J, Sary M (1995) Nouvelles données relatives à la capture de la Moselle. Rev Géogr l'Est 35:321–343

Houthuys R (2014) A reinterpretation of the Neogene emersion of central Belgium based on the sedimentary environment of the Diest Formation and the origin of the drainage pattern. Geol Belg 17: 211–235

Hugo V (1842) Le Rhin. Lettres à un ami. *H.L. Delloye*, Paris, 368+653 p

Huxtable J (1993) Further thermoluminiscence dates for burnt flints from Maastricht-Belvédère and a finalised thermoluminiscence age for unit IV Middle Palaeolithic sites. Meded Rijks Geol Dienst 47:41–44

Juvigné E (1979) L'encaissement des rivières ardennaises depuis le début de la dernière glaciation. Z für Geomorphol 23:291–300

Juvigné E (1992) Les formations cénozoiques de la carrière CBR du Romont (Eben/Bassenge, Belgique). Ann Soc Géol Belg 115: 159–165

Juvigné E, Renard F (1992) Les terrasses de la Meuse de Liège à Maastricht. Ann Soc Géol Belg 115:167–186

Juvigné E, Schumacker R (1985) Données nouvelles sur l'âge de la capture de la Warche à Bévercé. Bull Soc géogr Liège 21:3–11

Juvigné E, Cordy J, Demoulin A, Geeraerts R, Hus J, Renson V (2005) Le site archéo-paléontologique de la Belle-Roche (Belgique) dans le cadre de l'évolution géomorphologique de la vallée de l'Amblève inférieure. Geol Belg 8:121–133

Koci A, Schirmer W, Brunnacker K (1973) Paläomagnetische Daten aus dem mittleren Pleistozän des Rhein-Main-Gebietes. Neues jahrbuch für geologie und paläontologie, Monatshefte, pp 545–554

Losson B, Quinif Y (2001) La capture de la Moselle: nouvelles données chronologiques par datations U/Th sur spéléothèmes. Karstologia 37:29–40

Macar P (1938) Compte-rendu de l'excursion du 24 avril 1938, consacrée à l'étude des terrasses de la Meuse entre Liège et l'Ubagsberg (Limbourg hollandais). Ann Soc Géol Belg 61: 187–217

Macar P (1957) Les riviéres synclinales: tectonique active ou tectonique passive? Tijdschr Kon Ned Aardrijksk Gen 74:324–341

Meijer T, Cleveringa P (2009) Aminostratigraphy of Middle and Late Pleistocene deposits in The Netherlands, with notes on published data from the southern part of the North Sea Basin. Glob Planet Change 68:326–345

Meyer W, Stets J (2007) Quaternary uplift in the Eifel area. In Ritter J, Christensen U (eds) Mantle plumes. A multidisciplinary approach. Springer, Berlin, pp 369–378

Mudelsee M, Schulz M (1997) The Mid-Pleistocene climate transition: onset of 100 ka cycle lags ice volume build-up by 280 ka. Earth and Planet Sci Lett 151:117–123

Pissart A (1961) Les terrasses de la Meuse et de la Semois. La capture de la Meuse lorraine par la Meuse de Dinant. Ann Soc Géol Belg 84:1–108

Pissart A (1974) La Meuse en France et en Belgique-Formation du bassin hydrographique. Les terrasses et leurs enseignements. In Centenaire de La Société Géologique de Belgique—L'Evolution Quaternaire Des Bassins Fluviaux de La Mer Du Nord Méridionale, Liège, pp 105–131

Pissart A, Juvigné E (1982) Un phénomène de capture près de Malmedy; la Warche s'écoulait autrefois par la vallée de l'Eau Rouge. Ann Soc Géol Belg 105:73–86

Pissart A, Harmand D, Krook L (1997) L'évolution de la Meuse de Toul à Maastricht depuis le Miocène: corrélations chronologiques et traces des captures de la Meuse lorraine d'après les minéraux denses. Géogr Phys Quat 51:267

Pissart A, Paepe R, Bourguignon P (1969) Dépôts fluviatiles, éoliens et paléosols sur la terrasse de Hermée. Ann Soc géol Belg 92:429–445

Prodehl C, Mueller S, Haak V (1995) The European Cenozoic rift system. In: Olsen K (ed) Continental rifts: evolution, structure, tectonics. Developments in geotectonics, vol 25. Elsevier, Amsterdam, pp 133–212

Quinif Y (1999) Karst et évolution des rivières: le cas de l'Ardenne. Geodin Acta 12:267–277

Ritter J, Jordan M, Christensen U, Achauer U (2001) A mantle plume below the Eifel volcanic fields, Germany. Earth Planet Sci Lett 186:7–14

Rixhon G (2011) L'incision quaternaire des rivières du massif ardennais: Datation de terrasses fluviatiles dans le bassin de la Meuse par les nucléides cosmogéniques terrestres 10Be and 26Al. Editions Universitaires Européennes, Sarrebrück

Rixhon G, Demoulin A (2010) Fluvial terraces of the Amblève: a marker of the Quaternary river incision in the NE Ardenne massif (Western Europe). Z für Geomorphol 54:161–180

Rixhon G, Bourlès DL, Braucher R, Siame L, Cordy JM, Demoulin A (2014) 10Be dating of the Main Terrace level in the Amblève valley (Ardenne, Belgium): new age constraint on the archaeological and palaeontological filling of the Belle-Roche palaeokarst. Boreas 43:528–542

Rixhon G, Braucher R, Bourlès D, Siame L, Bovy B, Demoulin A (2011) Quaternary river incision in NE Ardenne (Belgium)—Insights from 10Be/26Al dating of river terraces. Quat Geochronol 6:273–284

Schaller M, Blanckenburg FV, Hovius N, Veldkamp A, van den Berg MW, Kubik PW (2004). Paleoerosion rates from cosmogenic 10Be in a 1.3 Ma terrace sequence : response of the river Meuse to changes in climate and rock uplift. J Geol 112:127–144

Sougnez N, Vanacker V (2011) The topographic signature of Quaternary tectonic uplift in the Ardenne massif (Western Europe). Hydrol Earth Syst Sci 15:1095–1107

Van Balen RT, Houtgast RF, Van der Wateren FM, Vandenberghe J, Bogaart PW (2000) Sediment budget and tectonic evolution of the Meuse catchment in the Ardenne and the Roer Valley Rift System. Global Planet Change 27:113–129

Van den Berg M (1996) Fluvial sequences of the Maas. A 10 Ma record of neotectonics and climate change at various time-scales. Landbouwuniversiteit Wageningen, p 181

Van den Berg M, Van Hoof T (2001) The Maas terrace sequence at Maastricht, SE Netherlands: evidence for 200 m of late Neogene and Quaternary surface uplift. In: Maddy D, Macklin M, Woodward J (eds) River basin sediment systems: archives of environmental change, Balkema, pp 45–86

Van Kolfschoten T, Roebroeks W, Vandenberghe J (1993) The Middle and Late Pleistocene sedimentary and climatic sequence at Maastricht-Belvédère: the type locality of the Belvédère Interglacial. Meded Rijks Geol D 47:81–91

Vandenberghe J (2003) Climate forcing of fluvial system development: an evolution of ideas. Quatern Sci Rev 22:2053–2060

Westaway R (2002) Long-term river terrace sequences: evidence for global increases in surface uplift rates in the Late Pliocene and early Middle Pleistocene caused by flow in the lower continental crust induced by surface processes. Neth J Geosci 81:305–328

Westerhoff WE, Kemna HA, Boenigk W (2008) The confluence area of Rhine, Meuse, and Belgian rivers: late pliocene and early pleistocene fluvial history of the northern lower rhine embayment. Geol en Mijnbouw/Neth J Geosci 87:107–125

Ziegler P, Dèzes P (2007) Cenozoic uplift of Variscan Massifs in the Alpine foreland: timing and controlling mechanisms. Glob Planet Change 58:237–269

Karst and Underground Landscapes in the Cretaceous Chalk and Calcarenite of the Belgian-Dutch Border—The Montagne Saint-Pierre

11

Luc Willems and Joël Rodet

Abstract

The eastern part of the Hesbaye plateau, which extends into the Dutch Limburg, was excavated by many underground quarries that mainly exploited the Cretaceous chalk of the Maastrichtian. In this area, the Meuse and its tributaries cut into the shallow dipping formations. Thus, entries of quarries usually are on hillsides. Most galleries sink horizontally along calcarenite layers, a coarse chalk also called «Tuffeau of Maastricht». The dimensions and extensions of these networks create real underground landscapes. These galleries intersect thousands of karst features and allow a privileged observation in three dimensions. These karst phenomena have dual characteristics: they mainly grow in a very porous lithology and outside fractures, so a context unfavourable to, a priori, a necessary concentrated weathering. Once dewatered, they cease to function, thus fossilizing a karst system in one of the rarely seen early genetic phases.

Keywords

Chalk karst • Montagne Saint-Pierre • Solution pipe • Blind cave • Underground quarrying

11.1 Introduction

To the North of the city of Liège, the eastern part of the Hesbaye plateau extends into the Dutch Limburg, beyond Maastricht. This is a low rolling plateau with altitudes between 100 and 153 m. Its southeast border is cut by the Meuse and its tributaries that have generated different levels of gravel terraces (see Chap. 12).

This chapter concerns the "Montagne Saint-Pierre" and its surroundings (Fig. 11.1). The Montagne Saint-Pierre (or Sint-Pietersberg in Dutch) is the isolated part of the plateau between the valleys of the Geer and the Meuse rivers that meet in the city of Maastricht. It is crossed from side to side by the Albert canal (accessible to ships up to 9000 T) which connects the industrial area of Liège with the port of Antwerp by the gap of Caster. Where the canal leaves the valley of the Meuse, the Lanaye locks ensure the junction of the Liège industrial basin with the Dutch Meuse, the port of Rotterdam, and the set of channel and river connections with the Rhine, Danube and the Black Sea.

Cretaceous chalk of the Upper Maastrichtian (from 72.1 to 66 Ma) is exposed over a thickness interval of a hundred meters, mainly represented by the Gulpen and Maastricht Formations (Fig. 11.2). The Maastricht Formation consists of macroporous calcarenites rich in flint nodules, subdivided by hardgrounds. The underlying Gulpen Formation consists of very fine calcarenites grading downwards into chalk with numerous flint beds. Pure white chalk and impervious marls, of Campanian age, separate the exposed Maastrichtian carbonates from the underlying deeply weathered and kaolinized Carboniferous limestones (Felder and Bosch 1998). Oligocene sands partly cover the whole (formerly

L. Willems (✉)
Pétrologie Sédimentaire, University of Liège, B20, boulevard du Rectorat 15, 4000 Liège, Belgium
e-mail: l.willems@ulg.ac.be

J. Rodet
UMR 6143 CNRS, Laboratory of Geology, University of Rouen, 76821 Mont Saint Aignan, France
e-mail: joel.rodet@univ-rouen.fr

A. Demoulin (ed.), *Landscapes and Landforms of Belgium and Luxembourg*, World Geomorphological Landscapes, DOI 10.1007/978-3-319-58239-9_11

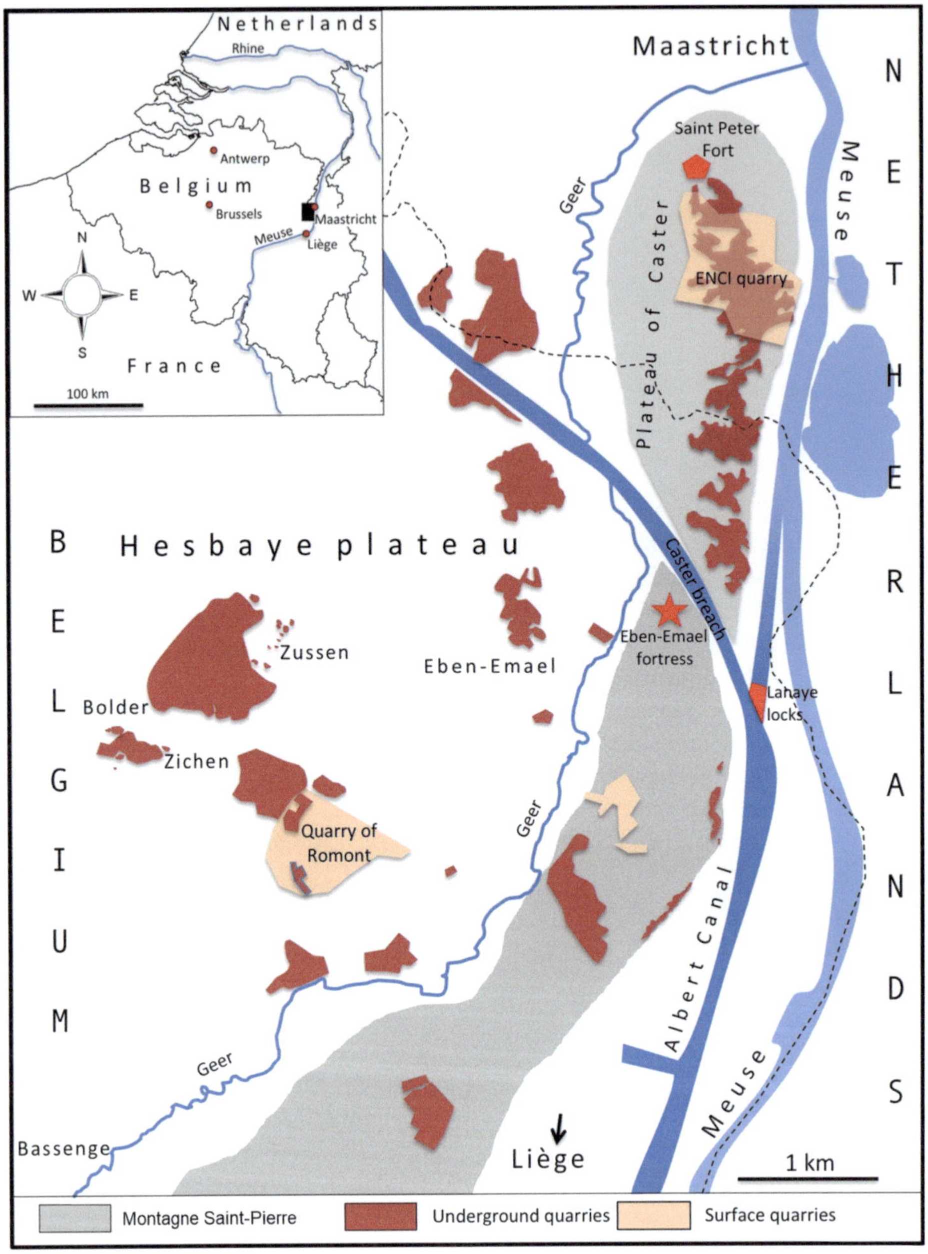

Fig. 11.1 General map of the study area

known as Tongrian sand, in reality, they date from the transition period between the Eocene and Oligocene (Vandenberghe et al. 2004)). During Mio-Pliocene time, the Meuse river drainage system came into existence and progressively incised down to its present level, as testified by peneplain surfaces and gravel terraces. The high terrace of the Meuse river overlying the Oligocene sands is a Middle Pleistocene gravel formation (Sint-Pietersberg terrace, see Chap. 10). These coarse and loamy river deposits armour the underlying Cretaceous chalk or Oligocene layers and lead to a local relief inversion (Juvigné 1976). During the Weichselian glaciation, about ten meters of loess was deposited, burying the pre-existing landscape (Haesaerts 1984).

All these formations were exploited since at least the Middle Ages (Bochman 1985; Breuls 1990; Silvertant 2003) by underground quarries, and since the twentieth century by surface excavations, allowing a privileged observation of the Cretaceous of the region. These quarries intersected

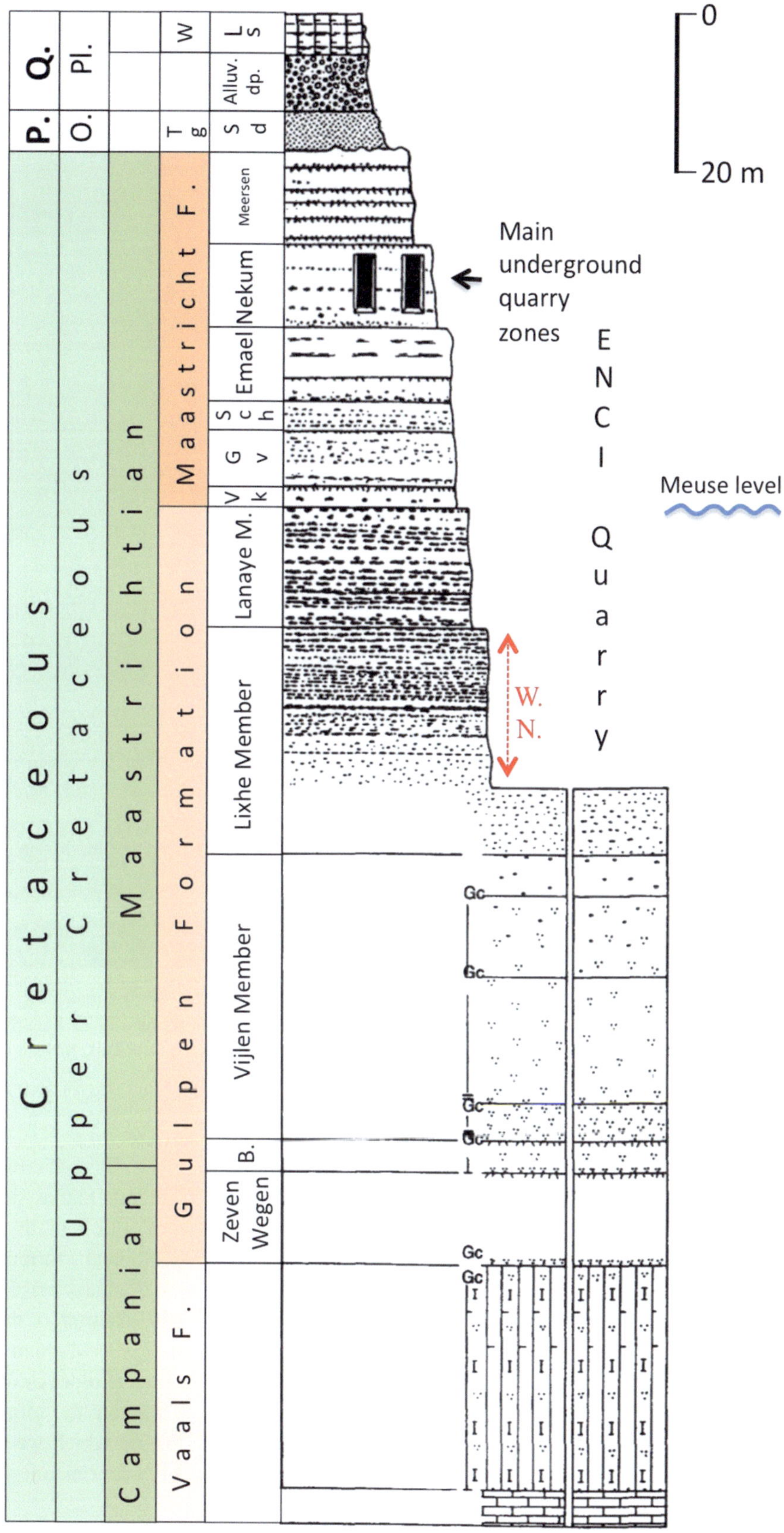

Fig. 11.2 Geology of the Montagne Saint-Pierre (Sint-Pietersberg) near Maastricht, drawn from the ENCI quarry face and from a drilling in the bottom of the same quarry (see Fig. 11.1) (modified after Felder and Bosch 1998). *P* Paleogene; *Q* Quaternary; *O* Oligocene; *Pl* Pleistocene; *Tg* Tongrian; *W* Weichselian; *B* Beutenaken Member; *Vk* Valkenburg Member; *Gv* Gronsveld Member; *Sch* Schiepersberg Member; *Sd* Tongrian sand; *Alluv dp* alluvial deposits; *Ls* loess deposits; *W.N* zone of weathering nodes (see text and Fig. 11.13)

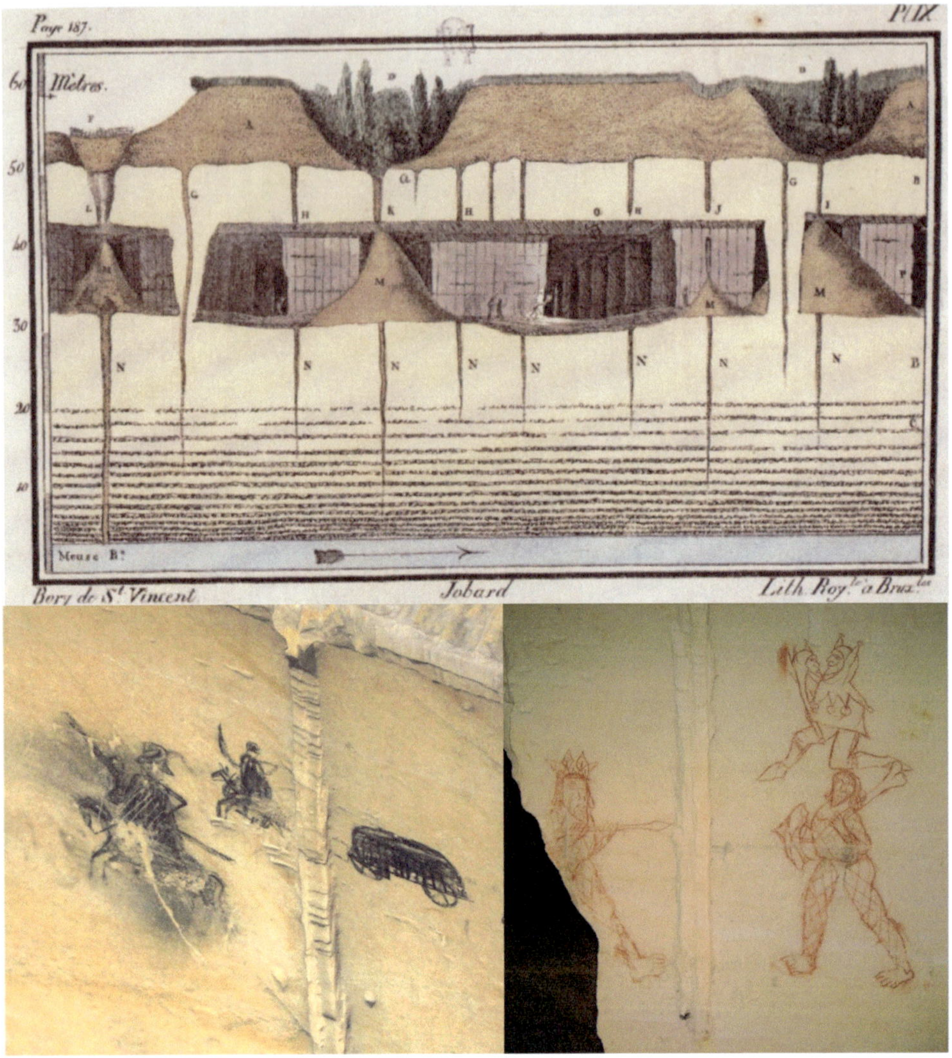

Fig. 11.3 *Top* Schematic cross section of the underground quarries near the city of Maastricht (Bory de Saint Vincent 1819). *Bottom* Old drawings on the walls of underground quarries (Petit Lanaye inf. and Caster quarries, plateau of Caster)

numerous karst systems whose originality lies in their localisation and development fossilized during their early stages.

The first scientific research in the area seems to date from the second part of the eighteenth century, which was a turbulent period for the region, with fighting between the French and Austrian armies for control of the city of Maastricht. Temperature readings in the galleries were first taken in 1782 by Van Swiden (Bory de Saint Vincent 1819). In 1798 the "Histoire naturelle de la montagne de Saint-Pierre de Maestricht" (Natural History of the Montagne Saint-Pierre of Maestricht) was published by Barthélémy Faujas de Saint-Fond. This is the first published description of the underground quarries and their various geological formations. This book also describes the main characteristic fossils, including the Mosasaur, literally «Saurian of the Meuse», at the time seen as a crocodile. In 1813, Mathieu described the "orgues géologiques" (geological pipes or organ pipes) or "puits de terre" (earth pits). These are the solution pipes from the superficial weathering mantle which we will talk about later (Fig. 11.3). In 1814, J.F. Clere mentioned for the first time closed cavities whose existence was doubted by Bory de Saint Vincent a few years later (1819). Their presence will be recognized more than a century later (Calembert 1958; Felder 1974; Smithuysen 1987; de Swart 1993). All these karst features were studied in detail at the beginning of this century (Rodet et al. 2009; Willems et al. 2005, 2007a, b, 2010; Willems and Rodet 2013). We will consider these studies in Sect. 11.3.

11.2 Underground Quarries

In this region, since the Middle Ages, underground quarries primarily operated in the Maastricht Formation and particularly the Member of Nekum. This member is a porous bioclastic grainy limestone, whose content of $CaC0_3$ usually varies between 96 and 98% (Poty and Chevalier 2004). Called calcarenite or «Tuffeau of Maastricht», this coarse chalk is 7–15 m thick in the study area (Fig. 11.2). It is located from a few meters to tens of metres below the plateau surface (Felder and Bosch 1998; Robaszynski 2006). In places, flints are present in the form of nodules or continuous beds, especially in the lower part of the formation where they are recognizable over long distances. Their abundance decreases in the upper part of the Maastricht Formation.

Underground mines consist of more or less subhorizontal straight corridors called streets and avenues, which give a network of galleries of once ~200 km, from which ~80 km are still open today (Fig. 11.4). Originally, these quarries provided building blocks or « limestone sand », derived from the crumbly rock, to improve the farm lands. During the late nineteenth century, some galleries, with vertical access, were dug for phosphate mining (Calembert 1943; Caubergs 1991). Other quarries were opened to supply certain industrial mills with flints to coat the inside of these factories. This was the main objective of these quarries during the twentieth century in the Geer valley. The decline in operating underground quarries for extraction of building blocks continued until the Second World War due to the increasing use of cement. The extraction of sand chalk continued sporadically before disappearing completely in the second half of last century. Underground quarries are no longer used for the extraction of rock but some were converted into mushroom farms, some of which remain active today.

It is surprising to see the use of such a fragile stone like calcarenite for the construction of buildings. In fact, a certain mechanical strength is provided by peripheric induration of the cutting blocks due to a superficial weathering/precipitation process of limestone. The water porosity inside the rock is exuded by block drying. This process produces an impermeable, harder deposit of $CaCO_3$ in contact with air, a few millimetres thick, called «calcin» (cullet) (Froidevaux 2001). However, this building stone remains problematic, because once dried, subjected to friction or cutting, the calcin is removed, exposing the inside of the block to rapid deterioration due to weathering and more specifically the influence of frost (Fig. 11.4).

The size and dimension of the corridors may vary according to the age and extraction techniques and also the kind of product extracted. Corridors, domed halls or rectangular sections can be several meters wide and over 8 m high (Fig. 11.4).

Depending on the location of the underground quarry, only a few metres can separate the gallery roof from the surface, causing stability problems. Some parts of the galleries have already collapsed completely and were sealed by surface deposits. On the surface, this produces closed depressions, sometimes confused with sinkholes, or sometimes more dramatically, sudden collapses (Figs. 11.5 and 11.6). In Hesbaye, it is not uncommon to have buildings or roads heavily impacted by these processes, especially when these underground quarries are located under villages such as Zichen, Zussen and Bolder (e.g. Medaerts 2008; Vital 2008; Vranckx 2010).

11.3 Karst Phenomena

The hundreds of kilometres of underground galleries and some surface quarries allow a privileged observation of karst features in a highly porous lithology, a priori not conducive to a concentrated weathering (measured porosities of 37–44% in calcarenite) (Willems et al. 2010). The shapes generated from the surface (crypto/exokarst) or in the rock, at depth (deep karst), may intersect. The beginning of an ephemeral karst system is quickly inhibited and fossilized due to major dewatering of the geological environment in relation with the Meuse valley downcutting (Fig. 11.7).

11.3.1 Exokarst—Solution Pipes

The exokarst includes forms generated from the surface or close to it, directly from rock dissolution by seepage. Lapiés and sinkholes exist at the border of the Cretaceous and more recent deposits (Rademakers 1998). However, the most

Fig. 11.4 *Top left* Ogival and rectangular gallery sections, Caster quarry. The *horizontal* and *vertical* marks are due to the block extraction technique by sawing. *Top right* Weathered building blocks due to the removal of the "calcin", farm of Caster. *Below* Part of an underground quarry network—Lower Petit Lanaye, Plateau of Caster (base map from De Grood 1967, 1983 in Van Schaïk 1983—modified after Willems et al. 2007)

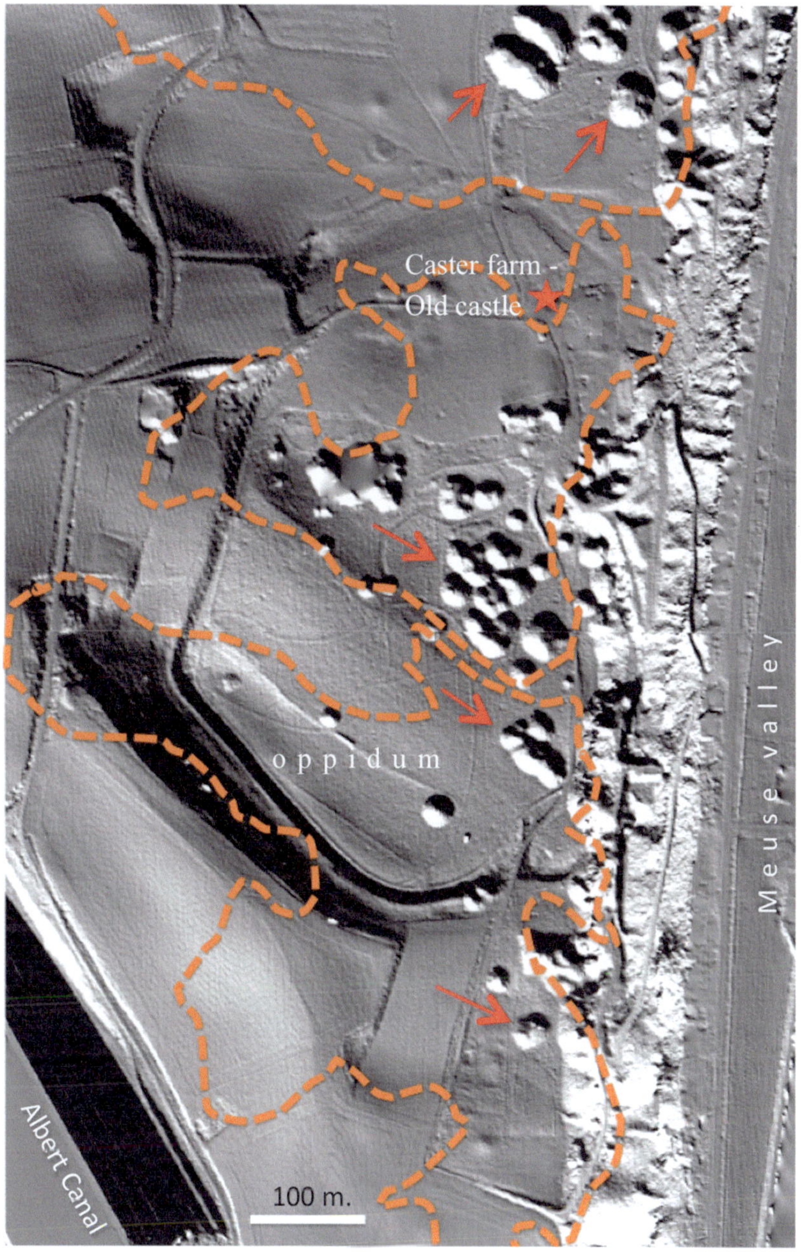

Fig. 11.5 LIDAR DEM image showing the numerous collapses (*red arrows*) on the Caster Plateau (northern part of the Montagne Saint-Pierre). *Orange dashed lines* delimit main underground quarry perimeters (modified after NAVTEQ S.P.W. 2014)

impressive karst features are the thousands of solution pipes which cross the Maastricht Formation (Fig. 11.8).

Known as "geological pipes" or "earth sinks" by the quarrymen, these forms were mentioned for the first time by Mathieu (1813) who described them as "very fat and very long pipes; they are up to two and a half metres in diameter; and their length can range from 60 to 100 m or more (…) These cavities are filled with heaps of stones mixed with the earth… that covers the plateau of the hill." Bory de Saint Vincent (1819), echoing the comments of his predecessor, notices a shorter length than announced for these geological pipes but finds that diameters can reach more than 4 m.

In 1813, Gillet de Laumont proposed an origin through seepage water. Bory de Saint Vincent dismissed the

Fig. 11.6 *Top* Pseudo-doline on the Caster plateau due to the collapsing of an underground gallery, just on the side of the old Middle Age road between Liège and Maastricht. *Bottom left* Collapse in a corn field in the same area (Bassenge, July 2008). *Bottom right* Clogging of a corridor by the drain of an solution pipe (Caster quarry—Plateau of Caster)

hypothesis of a widening of the pipes by the action of animals such as "the mole digs the earth" when the rock was still soft. He complemented the observations made by Mathieu and described the crusting of the walls of these forms, also described much later by Smitshuyzen (1983) as due to a dissolution/precipitation limestone process in contact with the non altered rock.

If seepage water rich in carbonic acid is allowed to explain their formation, Juvigné (1992) noted that these cryptokarst features occur chiefly under fluvial gravels. He hypothesized a location related to a heterogeneous paleo-alluvium, terrace gravels locally creating preferred spots for seepage water.

The thousands of solution pipes intersected by underground galleries show that they develop in the surrounding rock mass, unrelated to fracturing. Moreover, some of these "earth pipes" merge, resulting in their upper part decametric diameters. The development and penetration of these pipes

Fig. 11.7 **a** Large-scale outcrop of the Maastrichtian calcarenites in the open air quarry of Romont showing tens of solution pipes (*vertical brown* deposits in the *upper part* of the wall). **b** Closer view of solution pipes and blind caves cut by the rock extraction

are accompanied by a sinking of the subsurface formations, such as loess, river pebbles and, in places, the underlying Oligocene sand. The lower part of the pipe is sometimes made of highly weathered flint. The passage of infill to the non altered rock is short, a mere few centimetres. Note that in some places, river pebbles (flat and rounded) are racked in a vertical position along the pipes' periphery. Together, these observations suggest rapid genesis of these solution pipes by a very aggressive dissolution. This hypothesis is reinforced by the presence of flint remained overhanging in the vertical conduit, without the overall shape thereof downwardly being affected. To date, many questions remain open about the conditions necessary for pipe formation.

11.3.2 Endokarst: The Blind Caves

The underground and surface quarries have trepanned subhorizontal cavities whose characteristic is to be blind; they have neither upstream nor downstream macroscopic passages. Formed without the influence of fractures, one of the surprising aspects of these karst features is their morphology, characterized by numerous centimetric to plurimetric alveoli in all directions (Figs. 11.7 and 11.9). In some cases, the cave may be formed by the coalescence of metric to plurimetric alveoli. Often, most of them grow parallel to the stratification. Some are located in the floor of the conduit, forming a succession of basins separated from each other by thresholds. The ends of the caves or parts of their walls are often a veritable sponge network. The largest known caves exceed 70 m long (Felder 1974; Smithuysen 1987; Willems et al. 2007a, b, 2010). While the widest known cave to date reaches 10 m (Felder 1974), most do not exceed 2–3 m wide. Without collapse of the roof, the height of the passages rarely exceeds 2 m (Willems et al. 2007a, b, 2010).

These karst features, already mentioned in 1814 by Clere as polymorphic cavities, are isolated and filled with moulds of fossil plants. In the twentieth century, other caves of the same type were recorded in the calcarenite (e.g. Calembert 1958; Felder 1974), being natural cavities brought to light and destroyed by the progression of the working face of the surface quarries. Felder holds that they often develop in favour of fractures and at the intersection of less permeable levels (hardgrounds, flint benches, clay infill levels). He also notes that these are often associated with solution pipes and suggests a genetic link in which water infiltrated via fractures or organ pipes is partially stored up to less permeable levels, generating horizontal conduits, before continuing its descent into the depths. He adopts the "per descensum"

Fig. 11.8 Examples of solution pipes. **a** Large solution pipe, quarry of Romont. **b** Solution pipes cut by a gallery roof of the underground quarry of Caster. **c** Base of a solution pipe showing the short transition between the deposit and the fresh rock, quarry of Romont **d** Fluvial gravels of a Meuse terrace descended in a solution pipe, quarry of Romont

model often used to explain the origin of classical karst phenomena in the more compact, massive and indurated limestone. However, the study of the cavities in the underground quarries of Montagne Saint-Pierre and in the front of the quarry of Romont shows several facts which invalidate this scenario. Thus, the solution pipes may or may not intersect horizontal caves. A genetic connection could really not be established (Willems et al. 2010) (Fig. 11.10). The content of the solution pipe may partially fill the cave and cause a temporary flow on the roof, generating a roof channel (Fig. 11.11). Moreover, the importance of fractures and impervious hardground levels in the inception of the

Fig. 11.9 Alveoli and blind caves in the Maastricht Formation. **a** Nested ceiling pockets in a gallery roof, Zussen quarry. **b** Metric alveoli developed parallel to the stratification (CRSOA cave, Petit Lanaye Inf. quarry—Plateau of Caster). **c** Cave made of connected plurimetric alveoli, Lower Petit Lanaye quarry. **d** Sponge network, Upper Petit Lanaye Quarry, Plateau of Caster. **e** Small blind cave, Zussen quarry

caves is far from proven. Field evidence shows that the cavities often develop outside fractures and without a significant change in the porosity of the surrounding rock. Sometimes, fractures occur at the roof of the passage, without extension into the floor of the cavity, or extend laterally to the conduit. It is likely that such fractures result from rock decompression or are due to the quarrying, well after the establishment of the initial conduit.

If these endokarst features took place in an extremely porous lithology without any of the less permeable levels playing a key role, the presence of small cells testifies to an initial genesis in flooded areas, regardless of surface

Fig. 11.10 Examples of solution pipes intersecting blind caves. **a–b** Blind cave crossed by solution pipes (*red arrows* show the continuation of the pipes under the blind cave floor (cave of Lapin d'avril, quarry of Romont). **c–d** Detail of a partial connection between solution pipes and an horizontal cave (cave of the Two Albert, quarry of Romont). *SP* solution pipe. *c* blind cave. *R* roof channel. *a* lateral alveoli

conditions. The drying out of caves is marked by the birth of ceiling pockets in places of water table variation which temporarily puts under pressure air pockets in the irregularities of the original roof. In some cases, a temporary flow can be set up locally causing vertical erosion. However, the host rock behaves like blotting paper, energy and skill of the flow show a rapid parallel decrease causing a sudden sedimentation (Fig. 11.12).

The hypothesis of spot of alteration along deep water circulation loops (a few hundred meters below the palaeosurfaces) could explain the origin of the alveoli. This hypothesis is also supported by the discovery of weathering

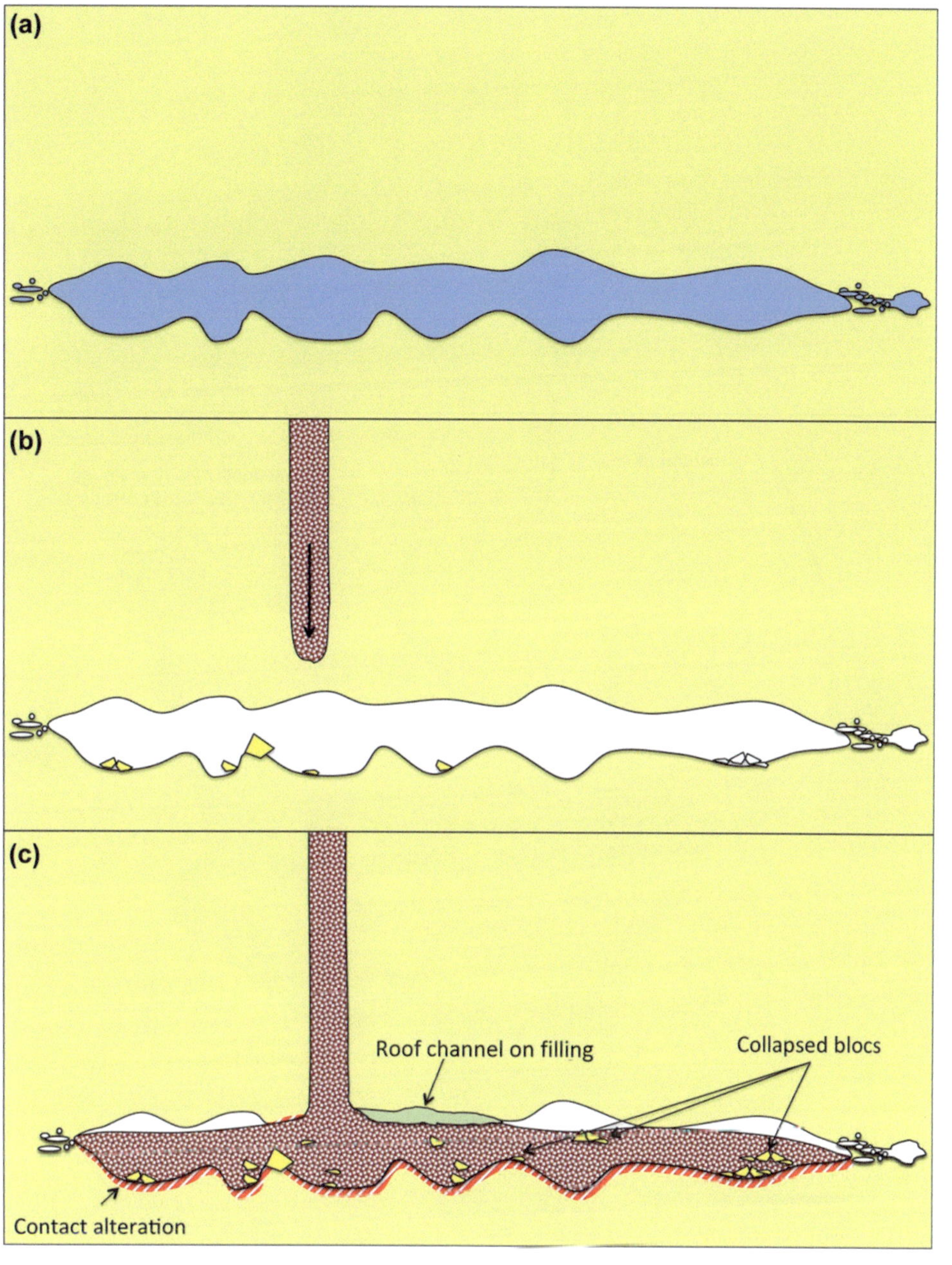

Fig. 11.11 Genesis of a roof channel following intersection of a blind cave by a solution pipe. **a** Formation of a blind cave (evolved from a ghost zone of advanced weathering in the phreatic zone). **b** Water table lowering and dewatering of the blind cave. In parallel, downward development of a solution pipe. **c** Partial filling of the cave by surface material descended in the pipe, leaving limited space for through going water to develop a roof channel

nodes 20 m below the level of the Meuse, in the Gulpen Formation which underlies the Maastricht Formation (Fig. 11.13). By their size and rounded morphology, the nodes of weathered chalk resemble the natural blind caves occurring above in the calcarenite. A further comparison can be made with the is alterites (Campy and Macaire 1989) or "ghost rocks" found inside the Carboniferous limestones of Tournaisian in western Belgium (e.g. Vergari 1998; Quinif et al. 1994; see also Chap. 9). These is alterites result from isovolumetric weathering of bedrock.

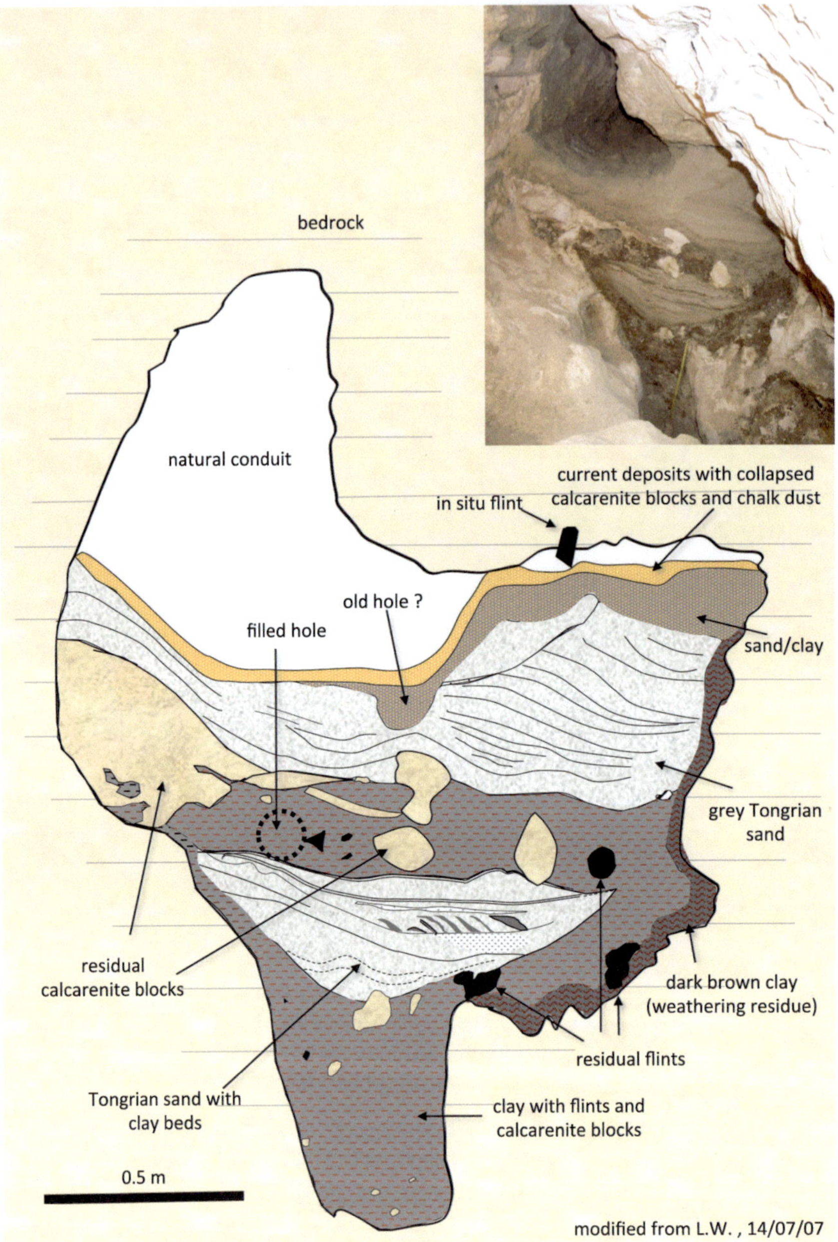

Fig. 11.12 An example of deposits inside a blind cave, CRSOA cave, quarry of Lower Petit Lanaye

11.3.3 Genesis and Fossilization of an Early Karst Network

The various observations presented above clearly show that the solution pipes and the older blind caves, developed deep within calcarenite, are not genetically linked. The dewatering of caves will be accompanied by partial collapse of existing roof. Subsequently, the occurrence of fractures and solution pipe development will be able to intersect these ancient drains. A more or less partial clogging of paleodrains by surface formations (loess, Oligocene sand, paleoterrace deposits) could occur. In some cases, the solution pipes, despite partial release in a cavity, will be able to continue their development to the deep (Willems et al. 2010), demonstrating again the strength of the genetic process. The solution pipes, partially emptied, provide a concentrated vertical transfer of water, which, on coming into the cave, will flow horizontally. While mobilization or remobilization

Fig. 11.13 Weathering nodes 20 m under the level of the Meuse river, inside the ENCI quarry. **a** General view of the quarry in 2005 with the front cut in the Gulpen Formation. **b** Detail of the weathering nodes in the south face of the quarry. **c**, **d** Close views of weathering nodes developed along strata and/or fractures in the east face. Note the temporary spring due to dewatering of the quarry

of the finest elements is possible, it remains limited, water being quickly absorbed by the enclosing calcarenite.

11.4 Conclusion

The network of underground quarries of the Montagne Saint-Pierre and its surroundings represent a remarkable heritage. Historically, they show the link at all times, which unites man with the rock, giving rise to astonishing underground landscapes. Geologically speaking, these networks provide unique access to a set of original karst features, mainly developed in a highly porous lithology. Producing locally aborted karst systems, they give us a glimpse of karstification mechanisms far more complex than those usually discussed.

Acknowledgements This research benefited from the Belgian-French Tournesol program. The authors are very grateful to Martin Andrew Skinner, who friendly translated the text in English, and to the CRSOA (Club de Recherche Spéléologique Ourthe-Amblève) for assisting with the survey work. We also thank E. Juvigné, Camille Ek (University of Liège), John Jagt (Natuurhistorisch Museum Maastricht), Ton Breuls (S.O.K.), S.A. Cimenteries CBR-ENCI Group and the Région Wallonne for their cooperation.

References

Bochman HMA (1985) 16e eeuwsevalsmunters… in een nog oudere groeve. S.O.K. Mededelingen 8:28–33

Bory de Saint-Vincent JB (1819) Sur le plateau de St-Pierre, près de Maestricht. Annales générales des Sciences physiques 1–10: 185–189

Breuls T (1990) Romaans, Romeins of nogouder ? S.O.K. Mededelingen 15:24–29

Calembert L (1943) Les gisements de phosphate de chaux de la Hesbaye. Mémoires de l'Association des Ingénieurs sortis de l'école de Liège 1:1–10

Calembert L (1958) Phénomènes karstiques holofossiles dans le Crétacé supérieur du bassin du Geer. Ann Soc Géol Belgique LXXXI:495–506

Campy M, Macaire JJ (1989) Géologie des formations superficielles: géodynamique, faciès, utilisation. Masson, Paris

Caubergs M (1991) Inventaire de quelques anciennes mines et carrières souterraines de Wallonie. Essai d'archéologie minière, Michel Caubergs

Clere JF (1814) Notice géologique sur l'espèce et la nature du terrain des environs de Maestricht. J Mines 214(36):242–252

de Swart HW (1993) Karst in the Netherlands. Bull Soc Géogr de Liège 29:109–113

Faujas de Saint-Fond B (1798) Histoire naturelle de la montagne de Saint-Pierre de Maestricht. J Jansen, Paris

Felder PJ (1974) Horizontale karstverschijnselen in de groeve Caster van de E.N.C.I. Natuurhistorisch Maandblad 6:112–115

Felder WM, Bosch PW (1998) Geologie van de St Pietersberg bij Maastricht. Grondboor & Hamer 3:53–64

Froidevaux YM (2001) Techniques de l'architecture ancienne: construction et restauration. Mardaga, Bruxelles

Gillet de Laumont F (1813) J Mines 201:202

Haesaerts P (1984) Aspects de l'évolution du paysage et de l'environnement en Belgique au Quaternaire. In: Cahen D, Haesaerts P (eds) Peuples chasseurs de la Belgique préhistorique dans leur cadre naturel, Bruxelles, pp 27–39

Juvigné E (1976) La stratigraphie du quaternaire. In: Pissart A (ed) Géomorphologie de la Belgique. Hommage au Prof. P. Macar, Lab. Géol. And Phys. Geogr, University of Liège, pp 169–179

Juvigné E (1992) Les formations cénozoïques de la carrière C.B.R. du Romont (Eben/Bassenge, Belgique). An Soc Géol de Belgique 115:159–165

Mathieu L (1813) Notice sur les Orgues géologiques de la colline de Saint-Pierre, près de Maëstricht. J Mines 34:197–201

Medaerts V (2008) Vrees voor instorting mergelgrotten Riemst. Het Belang van Limburg, 19/05/2008

Poty E, Chevalier E (2004) L'activité extractive en Wallonie—Situation actuelle et perspectives. Min. de la Région wallonne, Direction générale de l'Aménagement du territoire, du Logement et du Patrimoine, Jambes, 85p

Quinif Y, Vergari A, Doremus P, Hennebert M, Charlet JM (1994) Phénomènes karstiques affectant le calcaire du Hainaut. Bull Soc Bel Géol 102:379–384

Rademakers PCM (1998) Geol Orgenpijpen. Grondboor Hamer 3:71–78

Robaszynski F (2006) Maastrichtian. Geol Belg 9(1-2):63–72

Rodet J, Willems L, Brown J, Ogier-Halim S, Bourdin M, Viard JP (2009) Morphodynamic incidences of the trepanning of the endokarst by solution pipes. Examples of chalk caves in Western Europe (France and Belgium). In: Proceedings of the 15th international congress of speleology, Kerrville, Texas, vol 3, pp 1657–1662, 19–26 July

Smitshuyzen E (1983) Karst in Limburg. S.O.K Mededelingen, mai 2:13–16

Smithuysen E (1987) Een karstgrot in de Sint-Pietersberg. Speel Nederland, 2e jr. N°3:136–139

Silvertant J (2003) De ontstaansgeschiedenis van de gangenstelsels te klein-Ternaiien (België). Natuurhistorisch Maandblad, Dec. Jaargang 92:334–340

Van Schaïk DC (1983) De Saint Pietersberg met aanvullend gedeelte van 1938–1983. Thorn EF and EF

Vandenberghe N, Van Simaeys S, Steurbaut E, Jagt JWM, Felder PJ (2004) Stratigraphic architecture of the Upper Cretaceous and Cenozoic along the southern border of the North Sea Basin in Belgium. Neth J Geosci 83(3):155–171

Vergari A (1998) Nouveau regard sur la spéléogenèse : le « pseudo-endokarst » du Tournaisis (Hainaut, Belgique). Karstologia 31(1):12–18

Vital M (2008) Vrees voor instorting mergelgrotten Riemst. Het Belang van Limburg. 19/05/2008. http://www.hbvl.be/cnt/aid719358/vrees-voor-instorting-mergelgrotten-riemst. Accessed 12 July 2015

Vranckx F (2010) Verzakking doet voetpad instorten. Het Belang van Limburg, 17/08/2010. http://www.hbvl.be/cnt/aid964051/verzakking-doet-voetpad-instorten. Accessed 12 July 2015

Willems L, Rodet J (2013) Endokarsts and cryptokarsts in Cretaceous coarse and highly porous chalk at the Belgian-Dutch border. In: International congress of speleology, Brno, proceedings, vol 3, pp 154–157

Willems L, Rodet J, Fournier M, Masseï N, Laignel B, Dussart-Baptista L, Schyns J-Ch, Dusar M, Ek C (2005) Karst system genesis in the chalk of the lower Meuse (Belgian-Dutch border). In: Proceedings of the 14th UIS Congress, Athens-Kalamos, vol 1, pp 153–157, 23–28 Aug

Willems L, Rodet J, Ek C, Dusar M, Lagrou D, Fournier M, Laignel B, Pouclet A (2007a) Karsts des craies et calcarénites de la Montagne Saint-Pierre (Basse Meuse liégeoise). Bull. Chercheurs de la Wallonie, XLVI:171–186

Willems L, Rodet J, Fournier M, Laignel B, Dusar M, Lagrou D, Pouclet A, Masseï N, Dussart-Baptista L, Compère P, Ek C (2007b) Polyphase karst system in Cretaceous chalk and calcarenite of the Belgian-Dutch border". Z. Geomorph. N.F., Berlin—Stuttgart 51(3):361–376

Willems L, Rodet J, Ek C, St Pirson, Juvigné E (2010) Karsts des calcarénites de la carrière du Romont (Eben—Belgique). Bull des Chercheurs de la Wallonie 3:119–134

12 The Campine Plateau

Koen Beerten, Roland Dreesen, Jos Janssen and Dany Van Uytven

Abstract

Once occupied by shallow and wide braided channels of the Meuse and Rhine rivers around the Early to Middle Pleistocene transition, transporting and depositing debris from southern origin, the Campine Plateau became a positive relief as the combined result of uplift, the protective role of the sedimentary cover, and presumably also base level fluctuations. The escarpments bordering the Campine Plateau are tectonic or erosional in origin, showing characteristics of both a fault footwall in a graben system, a fluvial terrace, and a pediment. The intensive post-depositional evolution is attested by numerous traces of chemical and physical weathering during (warm) interglacials and glacials respectively. The unique interplay between tectonics, climate, and geomorphological processes led to the preservation of economically valuable natural resources, such as gravel, construction sand, and glass sand. Conversely, their extraction opened new windows onto the geological and geomorphological evolution of the Campine Plateau adding to the geoheritage potential of the first Belgian national park, the National Park Hoge Kempen. In this chapter, the origin and evolution of this particular landscape is explained and illustrated by several remarkable geomorphological highlights.

Keywords

Campine plateau • Meuse terraces • Late Glacial and Holocene dunes • Tectonic control on fluvial evolution • Polygonal soils • Natural resources • Coal mining

12.1 Introduction

The Campine Plateau is an extraordinary morphological feature in northeastern Belgium, extending into the southern part of the Netherlands (Fig. 12.1a). It runs from the southeast to the northwest, from an altitude of ca. 100 m a.s.l. (TAW: Tweede Algemene Waterpassing) in the south, to ca. 30 m near the Belgian-Dutch border in the north (Fig. 12.1 b). The polygonal shape of this lowland plateau has attracted a lot of attention from geoscientists during the last 100 years, all the more since it is covered by a thick sheet of coarse fluvial deposits from the Rhine and Meuse (Paulissen 1973, 1983). As these rivers' current channels are located several to tens of km further to the east, at a much lower altitude, the Campine Plateau witnesses a unique episode in the Quaternary evolution of the region. The interplay of fault activity, uplift, weathering, fluvial incision, regressive erosion, substrate characteristics, and aeolian processes created the current shape of the plateau. The plateau can be considered a classic case of relief inversion. The resulting steep bordering slopes are in strong contrast with the otherwise flat landscape of the European sand belt.

K. Beerten (✉)
Engineered and Geosystems Analysis, Belgian Nuclear Research Centre SCK•CEN, Boeretang 200, 2400 Mol, Belgium
e-mail: kbeerten@sckcen.be

R. Dreesen
Geological Survey of Belgium, Royal Belgian Institute of Natural Sciences, Jennerstraat 13, 1000 Brussels, Belgium
e-mail: roland.dreesen@telenet.be

J. Janssen · D. Van Uytven
Werkgroep Geologie, LIKONA (Limburgse Koepel Voor Natuurstudie), Provinciaal Natuurcentrum, Craenevenne 86, 3600 Genk, Belgium

A. Demoulin (ed.), *Landscapes and Landforms of Belgium and Luxembourg*, World Geomorphological Landscapes, DOI 10.1007/978-3-319-58239-9_12

The availability of erodible sand in the surrounding areas promoted the development of aeolian sand sheets and large inland dune areas (Fig. 12.1c) during the (Late) Weichselian when vegetation cover was sparse and the water table low. Furthermore, the poor sandy and stony soils and the deep groundwater table have made the plateau area unattractive for human occupation. Nevertheless, early agricultural activities and Mesolithic settlements seem to have existed close to wetlands (Vermeersch et al. 1974). The collection of heather sods for fertilizing soils (leading to the formation of plaggen soils), a widespread practice in the region, caused severe landscape instability from the fifteenth to the nineteenth century, as a result of which drift sand landscapes developed. Massive pine plantation during the late nineteenth century and first half of the twentieth century led to stabilization of the landscape and destruction of heathland, serving notably the coal mining industry in the Liège and Campine coal fields.

The establishment of pine plantations was a direct result of the need for wood in the galleries of underground coal mines in the nearby Liège area. Later, large amounts of pine wood were used in the Campine Plateau itself, where the geological Campine Basin hosts many coal seams at several hundreds of meters depth. The sandy and dry nature of the soils that developed on Quaternary fluvial and aeolian deposits provided an excellent habitat for pine trees. Similarly, gravel and, especially, sand extraction became widespread economic activities in the plateau region (Gullentops and Wouters 1996). Miocene and Pliocene quartz sand, coarse Rhine sand and loamy Meuse gravels are dug in huge extraction pits, the extent and depth of which can easily be detected on uncorrected digital terrain models, just as the coal spoil heaps. The deep groundwater table, as a result of uplift and river incision, enables dry extraction of some of these mineral resources, in contrast to the much more difficult wet extraction of, e.g., more valuable Meuse gravels in the present-day floodplain.

This chapter provides a state-of-the-art overview on the origin and development of the Campine Plateau, against a background of climate change, tectonic movements and human activities. The link between the current landscape and the overall evolution of the region is highlighted, as well as the geoheritage value of the landscape (see Chap. 24).

12.2 Geographical and Geological Setting

12.2.1 Location of the Campine Plateau

The Campine region is a 4000 km^2 large coversand area in NE Belgium and the southern part of the Noord Brabant province in the Netherlands. The name Campine (Kempen) is derived from Latin "campina", which literally means open space. The location of the Campine Plateau with respect to

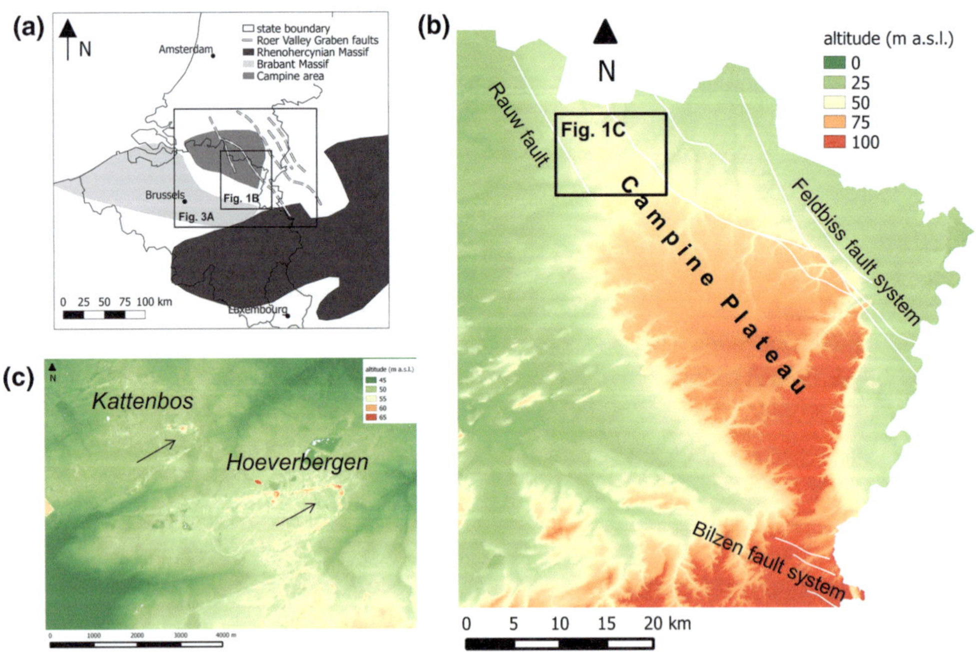

Fig. 12.1 **a** Location of the Campine area in Belgium and the southern Netherlands with indication of Roer Valley Graben faults and Caledonian/Hercynian Massifs. **b** DTM of the Campine Plateau, including Quaternary faults (*white lines*, from G3Dv2; Matthijs et al. 2013). M-G. Meeuwen-Gruitrode. **c** DTM of the parabolic dune areas near Lommel (Kattenbos-Hoeverbergen)

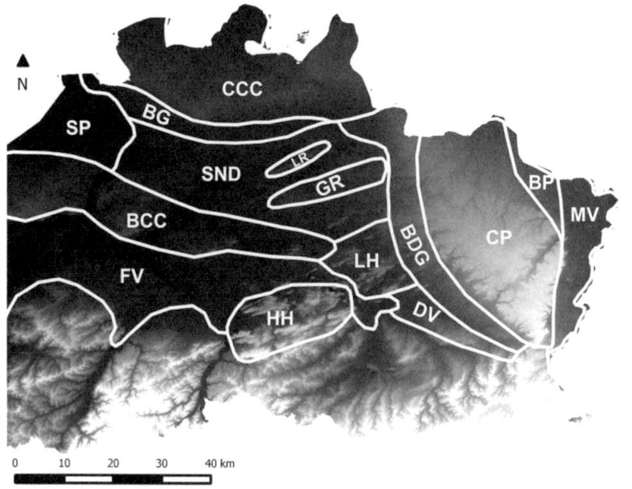

Fig. 12.2 Geomorphological units surrounding the Campine Plateau (from De Moor and Pissart 1992). *CP* Campine Plateau; *BP* Bocholt Plain; *MV* Meuse Valley; *BDG* Beringen-Diepenbeek Glacis; *SND* Schyns-Nete Depression; *LR* Lichtaart Ridge; *GR* Geel Ridge; *FV* Flemish Valley; *DV* Demer Valley; *LH* Lummen Hills; *HH* Hageland Hills; *BCC* Boom Clay Cuesta; *CCC* Campine Clay Cuesta; *SP* Scheldt Polders

other geomorphological entities in NE Belgium is shown in Fig. 12.2 (De Moor and Pissart 1992).

The Campine area is subdivided into several subregions. The Schyns-Nete Depression (SND; Fig. 12.2) in the west, also called the Campine Plain, is an erosive landform characterized by WSW–ENE trending ridges and alluvial plains that follow the strike of the Neogene subcrop (Goossens et al. 1983). The most prominent ridges are the Lichtaart Ridge (LR) and the Geel Ridge (GR). The Campine Plateau (CP) in the east is an accumulative landform, consisting of coarse fluvial deposits from the rivers Meuse and Rhine. These deposits extend towards the north into the Bocholt Plain (BP), which borders the Campine Plateau, north of the boundary faults of the Roer Valley Graben (see Chap. 13). The Beringen-Diepenbeek Glacis (BDG), a cryopediment, connects the SND with the Campine Plateau. Towards the south, the glacis merges into the Demer Valley (DV) and the Lummen Hills (LH), a continuation of the Hageland Hills (HH, see Chap. 14). In the east, the Campine Plateau is bordered by the incised Meuse Valley (MV).

12.2.2 Tectonic Framework

The Campine Plateau is situated in the Campine Basin, which constitutes the southernmost extension of the geological North Sea Basin. It coincides with a deep structural unit, referred to as the Eastern Campine Block (Fig. 12.3a). This structural unit is bounded by deep faults forming the Rauw-Beringen fault system in the (south)west (Fig. 12.3a, RaF fault), the Brabant Massif in the south, and the Feldbiss fault zone in the (north) east (Fig. 12.3a, FF fault) (Dusar and Langenaeker 1992). The thickness of the post-Oligocene sediments range from ca. 200 m on the Western Campine Block over 300 m on the Eastern Campine Block to more than 1000 m in the Roer Valley Graben. Quaternary deposits on the Eastern Campine Block are usually less than 20 m thick, while they may reach up to 100–200 m thickness in the Roer Valley Graben. The area west of the Feldbiss Fault very likely has experienced shoulder uplift during the Middle Pleistocene that is probably associated with accelerated uplift in the Ardenne (Demoulin and Hallot 2009) and/or enhanced subsidence in the Roer Valley Rift System (van Balen et al. 2000).

The morphotectonic map of the Campine Plateau shows the presence of various fault-related escarpments (Paulissen 1997). The most important one is the Bree Fault Scarp, which is the morphological expression of the Feldbiss fault (Fig. 12.3b, see also Chap. 13). To the southeast, this scarp shows a bifurcation known as the Berg Fault Scarp. Towards the northwest, the Bree Scarp splits into the Grote Brogel Fault Scarp, the Reppel Fault Scarp and the Bocholt Fault Scarp, separating the planated fault blocks north of the Campine Plateau, namely the Lommel, Reppel and Kaulille plains, and the Bocholt Plain. The western escarpment partly coincides with the Rauw-Poppel fault (Vandenberghe 1982, 1990; Gullentops and Vandenberghe 1995; Verbeeck et al. 2002), while Neogene and Quaternary displacements are reported along a fault zone inherited from the Caledonian basement directly south of the southernmost plateau tip (Gullentops and Claes 1997) (Fig. 12.1b).

12.2.3 Neogene Evolution

The Campine Block, as part of the Campine Basin, became covered by marine, estuarine, and continental sediments during the late Oligocene and Neogene (Fig. 12.4). This led to overall shoaling of the area (sea level drop) ultimately leading to the establishment of full continental conditions during the Early Pleistocene (Westerhoff et al. 2008). The Neogene sediments consist of the partly continental Lower Miocene sandy Bolderberg Formation, well-known for its high quartz content and lignite layers, the Upper Miocene glauconite-bearing marine Diest Sands, the Upper Miocene glauconitic and slightly clayey marine Kasterlee Sands, and the Pliocene estuarine Mol sands, also known for its high quartz content, and fluvial Kieseloolite Formation. Notably, the white and pure quartz sands of the Bolderberg and Mol formations are intensively quarried, and their preservation is directly related to the Pleistocene evolution of the area, as will be shown below.

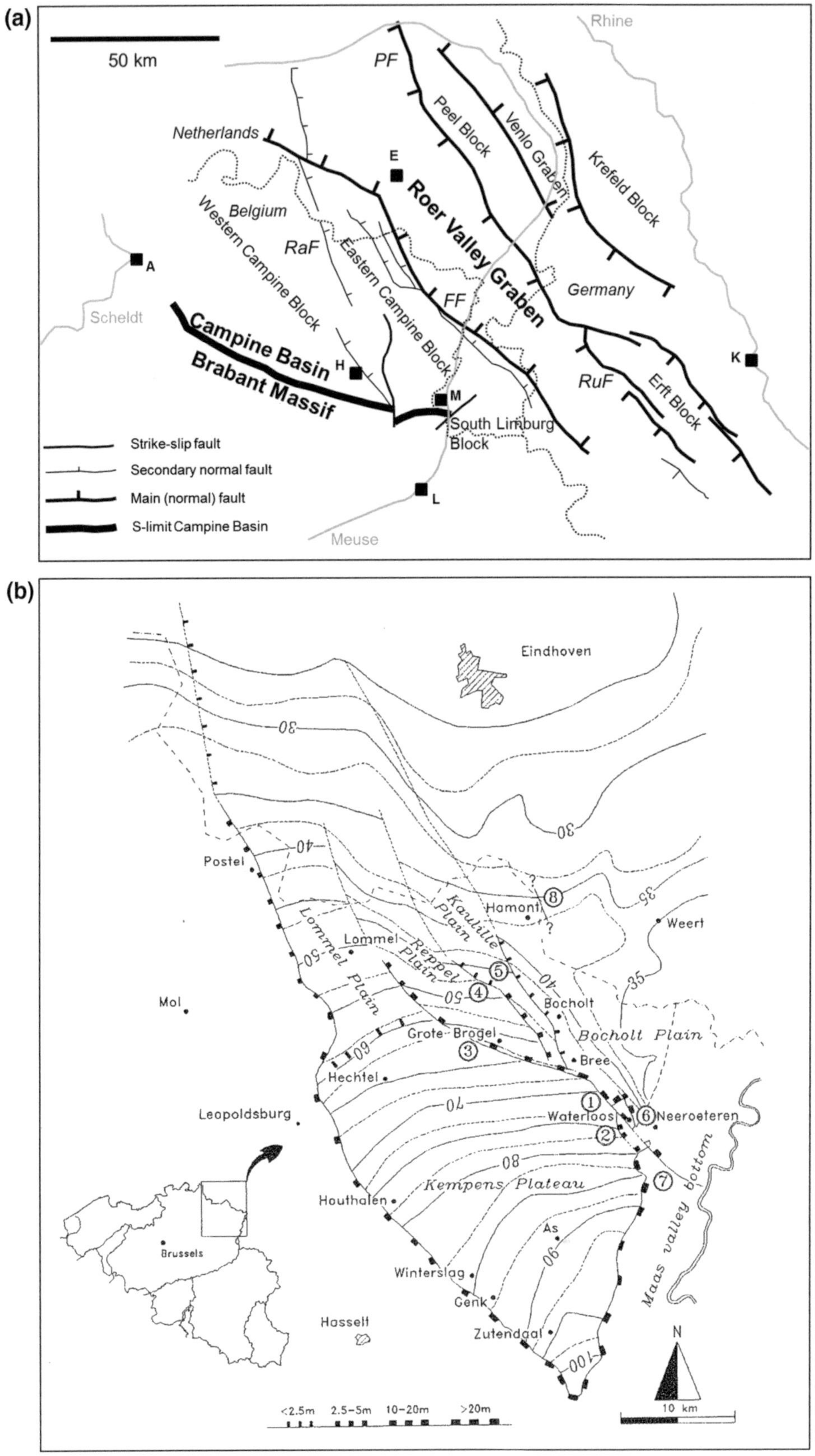

Fig. 12.3 **a** Tectonic framework of the Roer Valley Graben (modified after Geluk et al. 1994; Vanneste et al. 2001). *FF* Feldbiss Fault; *PF* Peel Boundary Fault; *RuF* Rurrand Fault; *RaF* Rauw Fault; *H* Hasselt; *M* Maastricht; *L* Liège; *K* Köln; *E* Eindhoven; *A* Antwerpen. **b** Morphotectonic map of the Campine Plateau highlighting the location of scarps, and notably the following fault scarps: *1* Bree fault scarp; *2* Berg fault scarp; *3* Grote Brogel fault scarp; *4* Reppel fault scarp; *5* Bocholt fault scarp; *6* waterloos scarp; *7* Bichterweerd Scarp; *8* Hamont Scarp. Contour lines in m TAW (Tweede Algemene Waterpassing) (from Paulissen 1997)

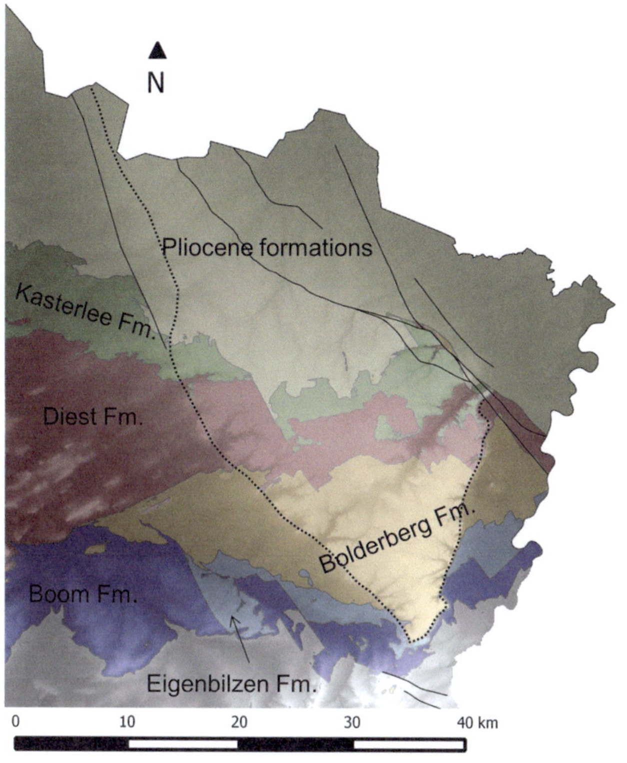

Fig. 12.4 Neogene subcrop map of the Campine Plateau and surrounding areas (from G3Dv2; Matthijs et al. 2013), superimposed on a grayscale DTM. *Dotted line* topographic outline of the plateau; *thin black lines* Quaternary faults. The Pliocene formations comprise the Mol Formation and the top of the Kieseloolite Formation, both consisting of fine to medium sands with intercalated lignite and/or clay layers. The Kasterlee Formation consists of fine glauconitic and slightly clayey sand, the Diest Formation of glauconite-rich coarse indurated (clayey) sand, the Bolderberg Formation of fine to medium pure quartz sand with lignite, the Eigenbilzen Formation of clayey fine sand, and the Boom Formation of (silty) clay

12.3 Origin and Evolution of the Campine Plateau Landscape

12.3.1 Tectonic and Climatic Controls of Early and Middle Pleistocene Fluvial Processes

Following a period of prolonged erosion during much of the Pliocene and Early Pleistocene, sedimentation on the Campine Block resumed when the hydrographic network changed completely around the Early-Middle Pleistocene transition (Juvigné and Renard 1992; Westerhoff et al. 2008). Initially, the Meuse followed an eastern course from Liège to the region north of Aachen where it merged with the Rhine (Fig. 12.5). Shifting tectonic movements along the Feldbiss and Rauw Faults and uplift of the northern margins of the Ardenne-Eifel massif, together with choking by sediments of the area where the Meuse debouched into the graben, caused the river to breach through its northern interfluve and to follow a completely different course over the Eastern Campine Block. At the same time, the Rhine shifted its course as well, flowing over the northern part of the Eastern Campine Block where it merged with the Meuse. Age control is limited, but this drastic event took place in the Early Pleistocene (Juvigné and Renard 1992; Van den Berg 1996), possibly as late as around 1 Ma ago (Westerhoff et al. 2008). Both rivers shifted their course to the east again by 0.5 Ma at the latest (Schokker et al. 2005). Three different lithostratigraphical units cover the plateau: the Lommel Sands, the Winterslag Sands, and the Zutendaal Gravels (Figs. 12.5 and 12.6). The former belongs to the Sterksel Formation while the latter two belong to the Zutendaal Formation (Gullentops et al. 2001). Further to the north, across the Feldbiss fault system, sediments in the Roer Valley Graben are essentially laid down by the Rhine (Bocholt Sands of Sterksel Formation), while the Meuse was still a tributary draining and depositing material (Zutendaal Gravels) on the eastern part of what is now the Campine Plateau (Beerten 2003). However, material with a Meuse signature is also often found within the Bocholt Sands (Beerten 2003).

An important feature of the Zutendaal Gravels is the gradual downstream fining. Near the southern tip of the plateau, the gravel is very coarse, while towards the north, fine gravel and coarse sand become the dominant size fraction. This, together with the conical shape of the plateau and the northward thinning of the gravel deposits led to the view that the plateau constitutes an alluvial fan (Pannekoek 1924). The Meuse deposits of the Campine Plateau are in terrace position relative to other gravel sheets that occur at lower altitude. They were usually linked with the term "Main Terrace" (Pannekoek 1924), according to the large areal extent, or "High Terrace" (Macar 1938), according to the elevated position with respect to the current river course. Paulissen (1973) concluded that the gravels of the Campine Plateau were deposited during the Mindel glaciation, which was correlated with the Elster glaciation in the past. Current correlation schemes correlate the Elster glaciation with Marine Isotope Stage 12, dated to ca. 0.45 Ma (Cohen and Gibbard 2011), while the originally defined Mindel glaciation is probably much older. Age control is poor, but the extensive terrace staircase that developed after deposition of the Zutendaal Gravels already hints towards a relatively old age for the Campine Plateau sediments. Each terrace has been linked with an interglacial–glacial cycle (Vandenberghe 2008), and available palaeomagnetic data suggest that the Main Terrace deposits are Middle Pleistocene in age (van den Berg 1996). This results in age estimates for the Zutendaal Gravels that range between 0.6 and 0.8 Ma, while the Winterslag Sands and Lommel Sands are thought to be slightly younger than 1 Ma (Felder and Bosch 1989; Juvigné and Renard 1992; van den Berg 1996; van Balen et al. 2000; Gullentops et al. 2001) (Fig. 12.6).

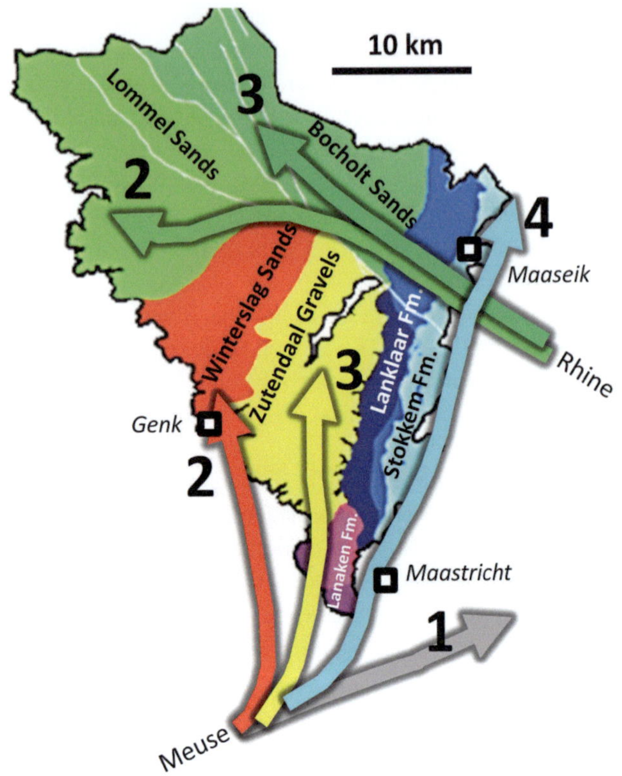

Fig. 12.5 Lithostratigraphic map of the Meuse deposits downstream of Maastricht (modified after Verstraelen 2000; Beerten 2005a, b, 2006). *Arrows* indicate flow direction of the Rhine and Meuse at the time of deposition of the different stratigraphical units. *Numbers* indicate the relative chronology of the various units (from oldest 1 to youngest 4). The entire depositional history is thought to have taken place in the timeframe between ca. 1 and 0.5 Ma (Schokker et al. 2005; Westerhoff et al. 2008). The Winterslag Sands and Zutendaal Gravels are grouped into the Zutendaal Formation (informally referred to as Main Terrace deposits from the Meuse), while the Lommel and Bocholt Sands belong to the Sterksel Formation (Gullentops et al. 2001). The Sterksel Formation consists of Rhine deposits with uptake of material from the river Meuse

12.3.2 The Aftermath: Posterior Development

12.3.2.1 Relief Inversion

The development of fluvial sequences on the Campine Block and in the Roer Valley Graben is clearly associated with the creation of accommodation space, as a result of subsidence. Wide and shallow valleys occupied by braided river channels developed during the Early to Middle Pleistocene transition. The elevated position of these fluvial sequences on top of the plateau, well above the current floodplain of the Meuse, thus needs to be interpreted as a classic case of relief inversion. The erodibility of the coarse grained deposits, ranging from coarse gravel to gravelly sand, is lower than that of the fine sandy Neogene units on which they were laid down. Regions that remained outside the sedimentation realm of the Meuse-Rhine, would thus become preferentially eroded. The western edge of the Campine Plateau can be explained in this way. Sedimentation by the Rhine-Meuse system stopped near the western edge of the Eastern Campine Block, delimited by the Rauw Fault and its southern extension (Fig. 12.3a). The topography of the area west of this fault system was subsequently lowered by regressive erosion from rivers draining the Scheldt basin. This process was accompanied by slope retreat and the development of a pediment, the BDG (Figs. 12.1, 12.2 and 12.6). It resulted in a relief difference of several tens of meters between the plateau and the area west of it.

The effect of regressive erosion in the SND can clearly be observed from the DTM (Fig. 12.1), where the headwaters of the Kleine and Grote Nete can be followed onto the plateau. Enhanced regressive erosion in the Nete basin is probably triggered by erosional events around 450 ka in the southern North Sea basin when the English Channel was opened as a result of a proglacial lake burst (Gupta et al. 2007; Toucanne et al. 2009).

Several uplift records indicate that the southern part of the Campine Plateau experienced around 50 m of uplift since the last 0.7 Ma (van Balen et al. 2000). Around this time, the Meuse was gradually shifting its course to the east again, and started developing the lower part of its extensive terrace staircase (Figs. 12.5 and 12.6). The southern tip of the plateau shows traces of palaeo-meanders that are preserved in the landscape when they laterally carved and incised into older Meuse deposits. An example is the Lanaken terrace (Lanaken Formation), consisting of a relatively thin gravel sheet ca. 10–20 m below the Campine Plateau surface (Paulissen 1973). Subsequent downcutting-infilling cycles were responsible for the development of the Caberg-Pietersem terrace (Lanaken Formation) and Eisden-Lanklaar, Maasmechelen and Geistingen terraces (Lanklaar Formation) (Figs. 12.5, 12.6 and 12.7). The vertical offset between the Campine Plateau Main Terrace and the Saalian Eisden-Lanklaar terrace ranges up to 50 m as a result of fluvial erosion.

The northeastern edge of the Campine Plateau is tectonic in origin. It is known as the Bree Fault Scarp, which is the morphological expression of the Feldbiss fault system (Vanneste et al. 2001; see also Chap. 13). The diffuse continuation of the Meuse deposits north of the Bree escarpment severely complicates their correlation with Rhine deposits in the Roer Valley Graben (Beerten 2003). Nevertheless, it is likely that the Feldbiss Fault caused a cumulative vertical displacement of ca. 25–40 m since deposition of the Zutendaal Gravels (Beerten et al. 1999; Vanneste et al. 2001).

12.3.2.2 Weathering, Soil Formation, and Periglacial Deformation

Several glacial–interglacial cycles occurred after deposition of the Rhine-Meuse sediments on top of the Campine Plateau, leaving sufficient time for post-depositional processes

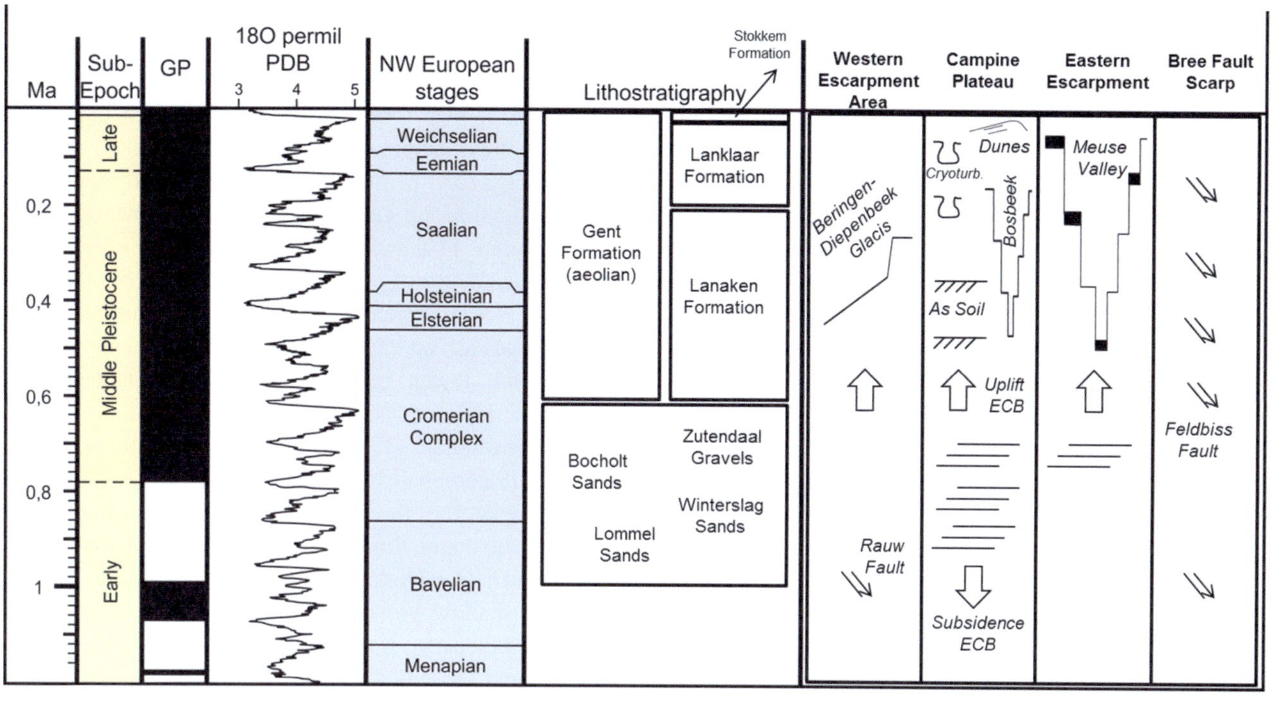

Fig. 12.6 Lithostratigraphic table of the units depicted in Figs. 12.5 and 12.7 and the associated geomorphological, climatic and tectonic development of the area before, during, and after deposition of the Main Terrace deposits (created with TimeScaleCreator-PUBLIC-6.4-21Feb2015). GP refers to the direction of the palaeomagnetic field (*black* normal; *white* reversed). The Early to Middle Pleistocene transition coincides with the Brunhes-Matuyama geomagnetic boundary (~0.78 Ma). *ECB* Eastern Campine Block (see Fig. 12.3a). *Arrows up* refer to the onset of uplift in the region, *arrows down* to subsidence; phases of normal faulting are indicated by *doubled oblique arrows*. *As soil* interglacial weathering soil in Zutendaal Gravels. According to Gullentops et al. (2001), the Meuse deposits on top of the Campine Plateau are younger than the Rhine deposits. This opinion has recently been questioned in the Quaternary subcommission of the National Commission for Stratigraphy. The lateral relationship between the different units is not fully understood yet

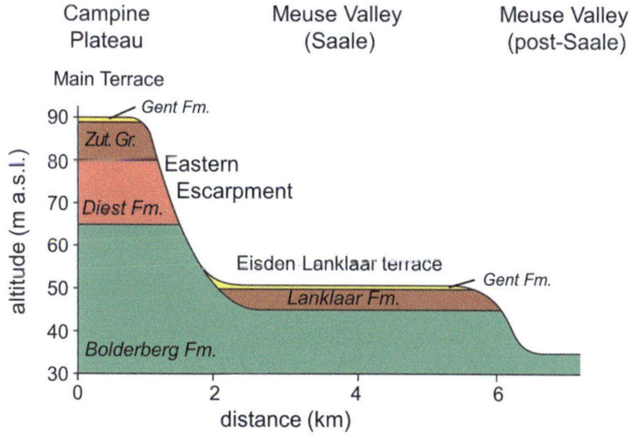

Fig. 12.7 Development of the plateau's eastern escarpment through fluvial erosion and terrace formation. This snapshot shows the situation at the end of the Saalian glaciation or the beginning of the Eemian interglacial (modified after Janssen and Dreesen, 2010). Note the thin aeolian sand layer (Gent Formation) on top of the fluvial terraces

such as weathering, soil development, and periglacial deformation. At some places, the typical red-brown interglacial weathering soil (As Soil; e.g., Gullentops et al. 1993) developed deeply into the gravel deposits (Fig. 12.8). Sufficient precipitation, a deep groundwater table, and the availability of mobile iron are necessary boundary conditions for the development of this soil. Limonite and goethite concretions often give the gravel a cemented appearance, further contributing to its resistance to erosion. This soil is very well developed in the eastern part of the Campine Plateau that more or less coincides with the Zutendaal Gravel outcrop zone (Figs. 12.4 and 12.5). Interestingly, the As Soil is not preserved in the western and northwestern part of the Campine Plateau, or it did not develop there. Several factors might account for this discrepancy. First, the availability of iron could be larger in the Zutendaal Gravels than in other fluvial units of the Campine Plateau, because of the initial presence of now dissolved iron-rich limestones (hematite oolitic ironstones) from Upper Devonian rocks in the south of the Namur syncline, and/or Jurassic minette-type oolitic ironstones from the Lorraine area. Second, the Zutendaal Gravels are more erosion-resistant than the more sandy Winterslag and Lommel Sands, such that the soil may have been eroded from these sandy units. Third, southeast-northwest tilting of the plateau resulted in deep fluvial incision by the Meuse, causing the groundwater table to be significantly lower along the southeastern edge where the Zutendaal Gravels are exposed, in comparison with the sandy facies to the northwest. Finally, the most plausible explanation involves intercalation or superposition of fine-grained sediments, which readily yielded iron in, or on top of, the gravels (M. Dusar, written communication). Where the availability of iron was low, Holocene soil formation processes transformed the top layers of the plateau into a podzol soil, with the typical albic and spodic horizons. In places where the in situ fluvial deposits are overblown by aeolian sand, soil formation took place in the latter. The plateau top layers are heavily deformed by periglacial processes during successive glacials, leading to the development of polygonal soil patterns and cryoturbations (see Sect. 12.4.4).

12.3.2.3 The Development of Dune Landscapes

The Campine Plateau is covered by a discontinuous mantle of homogeneous and fine-grained aeolian sands, which occasionally occur in dry valleys or build up a distinct dune relief. In the revised Belgian Quaternary lithostratigraphy, these aeolian deposits belong to the Opgrimbie Member of the Gent Formation, as defined in the revisited type-locality in the dune complex of Opgrimbie (Derese et al. 2009). Here, and at other places, age control suggests that the aeolian sediments were deposited during the Pleniglacial and Late Glacial under conditions of decreasing landscape stability and increasing soil erodibility associated with declining vegetation during cold interstadials (Kasse et al. 2007). Phases of landscape stability are recorded in these aeolian sands as witnessed by the presence of buried (palaeo)soils which developed during the Allerød and Holocene in response to climatic amelioration and vegetation expansion (Hoek 2001). In many occasions, the sands are blown into parabolic or longitudinal dunes (Gullentops 1957) (Fig. 12.1). A map showing the distribution of dune complexes on the Campine Plateau and surrounding areas is given in Fig. 12.9. Most of them are situated on interfluves between small brook valleys along the (north)western edge of the Campine Plateau, and the Meuse-Scheldt water divide. Famous examples of dune complexes can be found in Lommel (Hoeverbergen and Kattenbos) and near Hechtel. Towards the east, an impressive dune complex can be found between the towns of Gruitrode and Opglabbeek (the "Duinengordel", see also Chap. 24).

The distribution of dune complexes on the Campine Plateau is determined by the availability of erodible sand, and the presence of fluvial processes. The latter hamper the formation of dunes through complex fluvial–aeolian interactions, such as reworking of aeolian material, fluvial erosion, and shallow groundwater tables. In the southeastern plateau area, coarse gravel is outcropping and the lack of suitable sand prevented the development of massive dunes south of the Bosbeek valley. The availability of erodible sand increases towards the northwest, where the sand content of the Early-Middle Pleistocene substrate is high, post-depositional cementation as a result of goethite formation (As Soil, see above) is usually absent, and the deposition of aeolian sand from the north, prior to dune formation

Fig. 12.8 Development of subhorizontal goethite crusts (*arrows*) as part of the As Soil formed in the Zutendaal Gravels (from Dreesen et al. 2005)

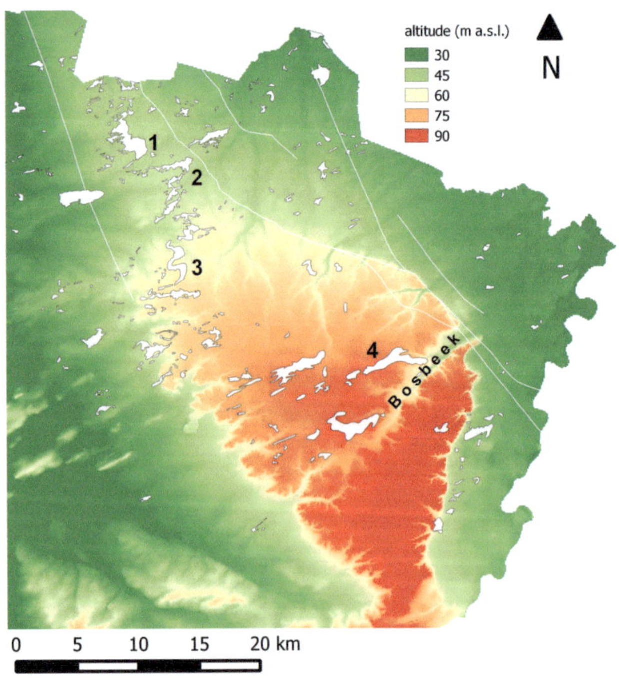

Fig. 12.9 Map of the dune areas (*white spots*) on the Campine Plateau and its surroundings. *1* Kattenbos; *2* Hoeverbergen; *3* Hechtel; *4* Duinengordel. Data from the Quaternary geological map sheets (Beerten 2005a, b, 2006) and the compilation map of Flanders (Bogemans 2005)

during the Late Glacial and Holocene, was more widespread. Finally, dunes are virtually absent in the central-northeastern plateau area; this region is characterized by a relatively dense river network incising deep into the plateau through regressive erosion that is presumably initiated by fault movement along the Feldbiss Fault. Floodplain processes in the valleys, and runoff processes on the interfluves prevented the development of dune systems here.

Forest clearing and intensified land use during the Neolithic gradually destabilized the landscape again, which led to reactivation of aeolian processes and the formation of drift sand dunes (Paulissen 1984). Drift sand dunes can often be interpreted as secondary features associated with the original dune form. Further to the west, off the plateau, the majority of these drift sands is thought to have been deposited during the last millennium (Derese et al. 2010; Beerten et al. 2014). Reforestation and leveling during the late nineteenth and early twentieth century stabilized the landscape again. Today, only few dune areas are still active.

12.3.2.4 The Relation Between Fens and Subsoil Characteristics

The typical fens that are distributed all over the Campine Plateau landscape are rainwater-fed bogs that can be typified according to their position in the landscape (Paulissen 1984) (Fig. 12.10). One type is typical for the southern part of the Campine Plateau, southeast of the Bosbeek valley, where groundwater levels are very low, down to 20 m and sometimes even almost 40 m below the surface. This type of bog developed where aeolian deflation removed any fine sand material that could have been accumulated on top of a hydraulically impermeable clay or silt bed in the top layers of the Zutendaal Gravel. Consequently, a shallow lake could develop in this depression in which peat could accumulate. Another fen type is typical for the northwestern part of the plateau, northwest of the Bosbeek, where groundwater levels are much shallower, less than 5 m deep. Here, the hydraulic barrier developed as an iron-rich B-horizon in response to soil formation processes in blowout hollows. During episodes with elevated groundwater level, bog iron could further develop from precipitation of dissolved iron in seepage water, enhancing the role of the hydraulic barrier.

12.4 Geomorphological Highlights

12.4.1 Escarpments

The shape of the Campine Plateau is determined by three prominent escarpments, with different origins. The southwestern escarpment developed in response to the formation of a pediment. This pediment forms the transition from the Campine Plateau to the river valleys that drain the Scheldt basin (Fig. 12.11). The top of the plateau is studded with residual gravels hampering erosion at this location. The total vertical difference between the plateau and the Demer floodplain that drains the pediment in the west is ca. 50 m. The escarpment itself is ca. 15 m high, while the pediment surface shows a regular dip from ca. 60 to 30 m. The steepest slope is less than 10% over a horizontal distance of 50 m.

The total height difference between the plateau and the Meuse Valley in the east is also ca. 50 m. However, the morphology of the eastern escarpment is completely different from the western one (Fig. 12.11). Much of the height difference takes place over a horizontal distance of less than 1 km, causing the eastern slope to be much steeper than the western one. The steepest slope sections can easily reach 20% over a distance of 50 m. The remaining 5–10 m height difference is accommodated by river terraces that are separated by small escarpments, usually only several m high, until the actual floodplain is reached.

The northeastern escarpment (Bree Fault Scarp) is tectonic in origin, and can clearly be observed near the town of Bree (Fig. 12.11). Its morphology takes an intermediate position in between the western and eastern escarpments. About two-third of the 30 m vertical height difference between the Campine Plateau and Bocholt Plain occurs within a horizontal distance of 1 km. The remaining 10 m is due to the accumulation of slope deposits at the foot of the escarpment. The steepest part of the slope may reach 10–15% over a horizontal distance of 50–100 m.

12.4.2 Development of the Hydrographical Network on the Campine Plateau

The water divide between the Meuse (east) and Scheldt (west) river basins runs through the town of Waterschei (which literally means 'Waterdivide'). It is situated closer to the western boundary of the plateau than the eastern one (Fig. 12.12). Drainage towards the north and northeast developed through small brooks and rivers that penetrated deep into the plateau. Brooks that drain in northward direction have a very low gradient and run in very shallow floodplains. Other brooks and rivers make a distinct bend towards the northeast when passing the Bree Fault Scarp along the northeastern edge of the plateau. Although their upper courses trend N–S, they seem to have been responding to tectonic activity along the Feldbiss fault, finding a new course straight into the sinking Bocholt Plain. The gradient and valley depth are larger than those of the northwards flowing rivers, because of the larger and more abrupt height difference between the Bocholt Plain and Campine Plateau along this fault segment. The most striking feature is the

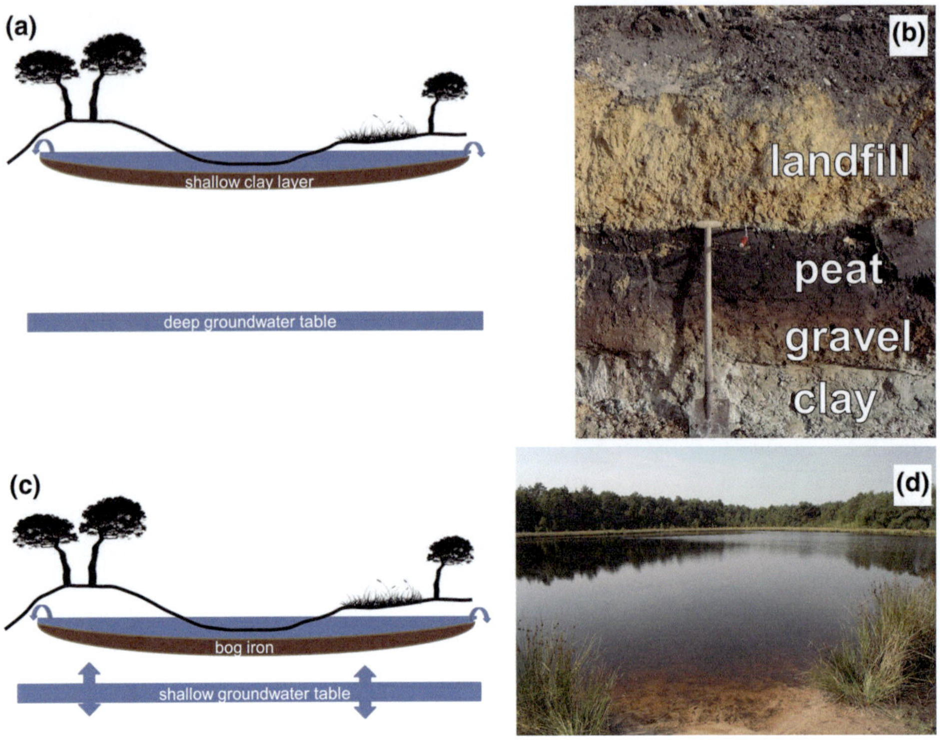

Fig. 12.10 Fen development on the Campine Plateau. **a** Deep groundwater table and a shallow impermeable clay layer. **b** Cross-section through the clay layer and overlying fen deposits of a drained fen (Nieuw Homo, Lanklaar). **c** Shallow groundwater table and impermeable bog iron. **d** Fen developed as a result of the formation of impermeable bog iron (Ruiterskuilen, Opglabbeek). (**a** and **c** from Van Uytven and Dreesen 2015; *photographs* D. Van Uytven)

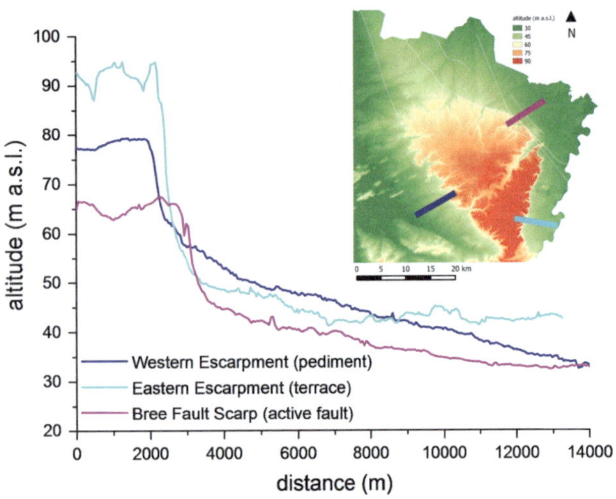

Fig. 12.11 Topographic cross-sections of the escarpments bordering the Campine Plateau on its SW, E and NE sides, respectively. Note the contrasted styles of the escarpments of different origins (SW: cryopedimentation of the Beringen-Diepenbeek Glacis; E: fluvial erosion by the Meuse; NE: Bree fault scarp)

Bosbeek valley, also running southwest-northeast. It has cut through the Zutendaal Gravels and now drains the Neogene substrate (Fig. 12.13).

The Bosbeek valley shows a marked asymmetry in the slopes that border the plateau edges. The evolution of the northwestern slope of the valley (i.e., facing southeast) can be understood in the frame of pediment development under periglacial conditions during the Saalian glaciation (Gullentops et al. 1993). This slope would have functioned as a snow trap for northern winds that would continuously blow in snow from the flat and bare plateau area. Long-lasting snowmelt during summer would have increased active layer processes accelerating slope retreat, the effect of frost cycles breaking up cobbles to transportable dimensions, and the volume of melt water evacuating the debris. These processes finally resulted in the formation of a cryopediment on the west bank of the Bosbeek. The continuous supply of coarse debris from the development of this cryopediment forced the river to flow along the eastern valley side leading to continuous undercutting and erosion of the east bank. This explains why the Bosbeek valley escarpment is about 30 m high along the southeastern valley wall, and only about 10 m along the northwestern valley wall (Fig. 12.13). The northwestern valley edge is drained by perennial tributaries that penetrated into the plateau quite deep, while the southeastern edge is drained by narrow and rather steep dry valleys. This phenomenon is directly related to plateau uplift and fluvial incision by the Meuse, which led to drastic groundwater table lowering along the eastern side of the plateau (Fig. 12.13). As a result, groundwater table is higher along the northwestern valley wall of the Bosbeek compared to the southeastern one. Differences in groundwater table depth also explain why the As Soil developed much more intensely east of the Bosbeek.

The eastern plateau escarpment is drained by deeply incised and very small dry valleys, with a very large gradient. They developed as a response to uplift-triggered incision

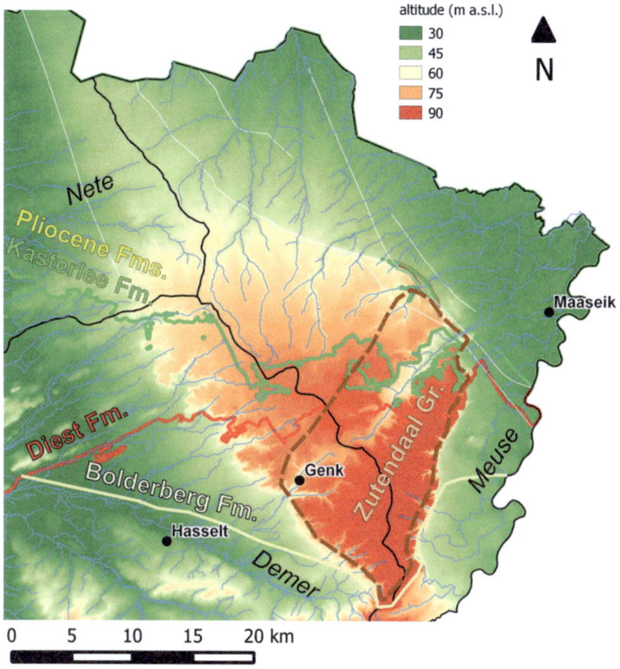

Fig. 12.12 Drainage network of the Campine Plateau, opposing long wide and shallow valleys in the north and west and relatively short narrow deep (and dry) valleys in the east and south. This reflects contrasting conditions of groundwater table (deep in the southeast and shallow in the northwest), and resistance to erosion of the substrate. *Thin white lines* are Quaternary faults and *thin black lines* the limits of the Meuse basin and Nete and Demer catchments (both belonging to the Scheldt Basin) in Belgium. The distribution of Zutendaal Gravels is indicated by a *brown dashed line*, while the southern limits of the Bolderberg, Diest, and Kasterlee and various Pliocene formations are shown as *solid lines of different colors*

of the Meuse river. This led to a significant lowering of the groundwater table, resulting in dry valleys. Their main phase of development is situated in the Saale and earlier (Gullentops et al. 1981). Only in the southern part, where the Boom Clay subcrops below the Zutendaal Gravels, some valleys are drained by perennial groundwater-fed streams (Fig. 12.12). The erosion-resistant role of the Zutendaal Gravels probably hampered the development of these eastern escarpment valleys.

The western escarpment is drained by less deeply incised rivers belonging to the Scheldt basin (Fig. 12.12). They developed as a result of regressive erosion, which finally led to the relief inversion of the Campine Plateau, and penetrated into the plateau edge. This process is also associated with the development of the BDG. In contrast to the valleys draining the eastern escarpment, these valleys are permanently drained by small rivers and brooks. The main reason for this is the relatively high groundwater table compared to the more intensely uplifted eastern part of the plateau. Along this western edge, two trends can be observed in the valley shapes, going from south to north. First, the incision depth and overall gradient decrease in this direction as a result of the differential uplift of the plateau, which is larger in the south than in the north. Second, the width of valleys draining the escarpment increases from south to north, because of the successively less erosion-resistant outcropping sediments, from Zutendaal Gravels in the south to Diest Sands in the middle, and Kasterlee, Mol, and Rhine sands in the north (see also Fig. 12.4). A prominent bend in the western escarpment is visible in the north. The exact cause of this

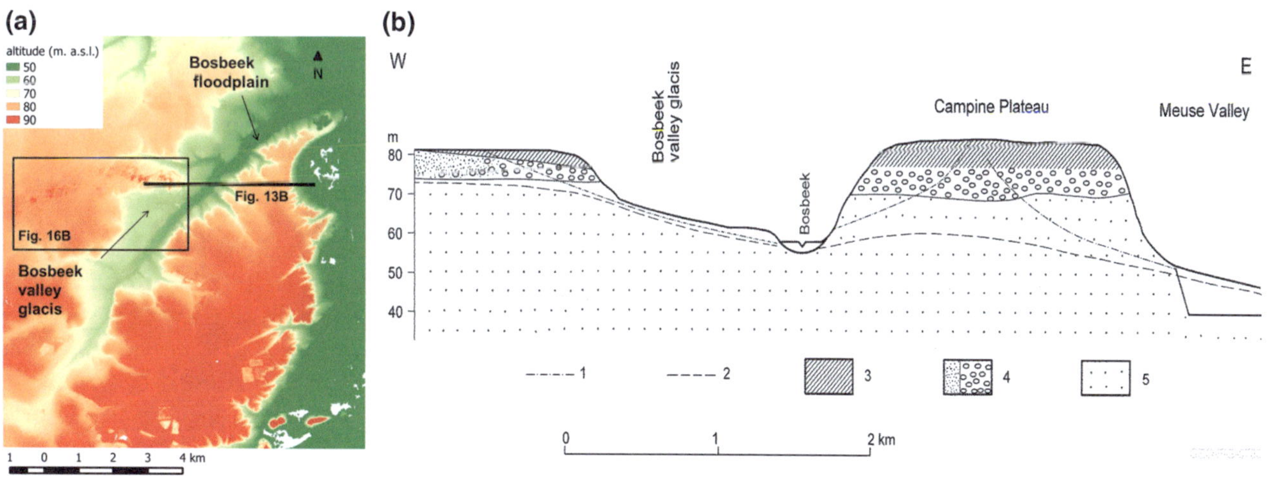

Fig. 12.13 **a** DTM of the Bosbeek valley, showing a steep south-eastern valley side and a smooth northwestern side developed as a cryopediment during the Saalian glaciation. **b** Schematic cross-section through the Bosbeek valley and its divide with the Meuse (redrawn after Gullentops et al. 1993). *1* Projected longitudinal profile of tributaries; *2* present-day groundwater table; *3* as soil; *4* sands and gravels of the Meuse; *5* neogene sands

bend is not clear, but it is probably related to the presence of the Rauw fault, the difference in resistance to erosion between the indurated Diest Sands in the south and the Mol Sands in the north, and the role of regressive erosion in the Nete catchment (see above).

12.4.3 Preserved Natural Resources: Meuse Gravels, Rhine Sands, Mol Sands and Bolderberg Sands

12.4.3.1 Meuse Gravels

The Zutendaal Gravels run in a southwest-northeast direction as a 5 km wide strip on the eastern side of the plateau, eroding the underlying Winterslag Sands and Neogene units. The gravel sheet is up to 15 m thick but is thinning towards the north and south. Extent, thickness, and elevated position of the gravel unit make it a unique morphological feature in Western Europe. The Zutendaal Gravels are characterized by a basal channel-lag deposit consisting of cobbles and larger (ice-rafted) boulders (De Brue et al. 2015) (Fig. 12.14a). The gravel deposits form an aggrading unit, deposited during different aggrading cycles, each cycle consisting of a coarser channel-lag deposit at the base, fining up to fine gravel or even a sand layer or a clay lens at the top (Gullentops et al. 1981; Paulissen 1983). Each cycle signifies deposition in a broad shallow channel. Usually, the gravels are matrix-supported by sand or finer material. They are generally interpreted as a cold-climate braided river deposit. Their characteristics are completely different than those of the Holocene Meuse, which displays graded point bar gravels, several meters thick, capped by silty and clayey floodplain deposits, 2–3 m thick. During deposition of the Zutendaal Gravels, sediment supply was guaranteed by frost weathering during cold periods, under periglacial circumstances. Subsequent snowmelt during early summer caused peak discharge in the tributaries and main channel, promoting the transport of very coarse material.

De Brue et al. (2015) investigated the possibility of a classical fluvial transport mode for the largest boulders—most of which exceed 50 cm and may reach up to 200 cm in intermediate axis and some of them weighting several tons—on the base of empirical hydraulic transport thresholds for gravel bed rivers. Results indicate that hydraulic transport of boulders with intermediate diameters <1 m could have occurred within limited reaches of the palaeoriver, more specifically in the palaeo-Amblève tributary. However, the small slope gradient of the palaeo-Meuse most probably inhibited boulder movement by hydraulic forces only.

Therefore, ice-rafted transport of large boulders is favored here, referring to the existence of recent analogues ("blocs glaciers") in the estuary of the St-Lawrence river of Québec (Dionne 2003) and according to observations and calculations of the transport capacity of ice in the catchment area of the Amblève river by Collard et al. (2012, see also Chap. 6). Furthermore, huge blocks of rock types sensitive to physical weathering, such as phyllites and microgranites (often affected by diaclases), would never have reached the Campine area by normal fluvial transport without breaking into smaller pieces.

The composition of the gravel undoubtedly points to a southern origin, in particular the Ardennes and the Vosges area (northern France). Therefore, the Meuse and its tributaries (including also the Moselle) can be considered as the only transport agent for these gravels. The Zutendaal Gravels display a broad lithological spectrum of weathering-resistant rock types, derived from various Palaeozoic through Meso-Cenozoic geological sources. These include various hard Cambrian quartzites and vein quartz, Lower-Middle Devonian conglomerates and sandstones, Lower-Carboniferous phtanites and silicified oolitic limestones, Cretaceous flints, Tertiary quartz arenites (silcretes) and flint gravels. The striking absence of Palaeozoic and Mesozoic limestone pebbles or boulders is related to chemical weathering before or after transport and deposition of the gravels. Exceptionally, Cambrian-Ordovician phyllites, Late Devonian microgranites (e.g., the "porphyre de Mairupt") and diabases occur within the Zutendaal Gravels as huge ice-rafted boulders. The latter igneous rocks have been derived from dyke swarms in the Rocroi area, SW Ardenne. Finally, pebbles of Palaeozoic granites and Triassic Buntsandstein represent some minor but most characteristic constituents that point to a particular provenance area—the Vosges—and hence transport via the Moselle River (Bosch 1974, 1992; Dreesen et al. 2014). Flint, derived from upstream Cretaceous chalk formations, is also a very typical component.

12.4.3.2 Rhine Sands

The Rhine facies of the Campine Plateau consists of medium to coarse sand, with low clay and silt content, and occasionally containing fine and coarse gravel (Paulissen 1983). Sedimentation took place in wide and shallow channels, probably in a braided river system during glacial stadials or interstadials because a typical fine-grained floodplain facies is absent in these deposits (Fig. 12.14b). Usually, the deposits are arranged in fining upward several-meters-thick sequences. Large boulders and blocks are lacking. Rhine deposits can be found as a deep southwestern bend, covering the western and northwestern part of the Campine Plateau. Along the western edge of the plateau, the sands are eroded and preserved as a residual gravel layer, often containing ventifacts (van Mechelen 1982). Rhine deposits occasionally contain gravel that originates from the Alps, the German middle mountains and the Ardennes-Eifel massif. Gravel, and most notably granule gravel (2–4 mm), is rich in quartz, usually above 60%, while flint fragments are very rare.

Fig. 12.14 Natural resources in the Campine area. **a** Coarse matrix-supported gravel deposited by the Meuse and weathered into the As Soil; the length of the longest axis of the biggest boulder is ca. 0.6 m ('Kikbeek' quarry, Maasmechelen; reproduced from Dreesen et al. 2005). **b** Coarse sand deposited by the Rhine, no obvious weathering ('Blauwe Kei' quarry, Lommel; *photograph* M. Gedeon). **c** White quartz sand and lignite of the Mol Formation (Blauwe Kei quarry, Lommel; *Photograph* K. Beerten). **d** Outcrop in the Bolderberg Formation showing alternations of bleached quartz sand and lignite (Kikbeek quarry, Opgrimbie; reproduced from Dreesen et al. 2005)

These deposits are probably laterally equivalent to the Winterslag Sands, which occupy more or less the central part of the plateau, running southwest-northeast along a 10 km wide zone (Fig. 12.5).

12.4.3.3 Mol Sands

In the northwest, the Rhine sands overlay a thick unit of white to brown relatively pure quartz sands with two distinct lignite layers. These sands and lignite layers are known as the Mol Formation (Fig. 12.14c). They were deposited during the Pliocene in a large estuary fed by the rivers Rhine and Meuse, as attested by the presence of "kieselooliths" (pebbles of silicified oolitic limestone) near the base of the formation. During periods of graben subsidence, thick peat layers developed in a swampy environment, and were transformed into lignites during subsequent burial under quartz sands. The sands in the Mol Formation are fine to medium-grained, while clay particles and glauconite are virtually absent. Post-depositional bleaching of the sand as a result of dissolved organic acids in the infiltrating rain water gave the sands their pureness and white appearance, such that is very suitable in the glass, ceramics, chemical and metallurgical industries (Gullentops and Wouters 1996).

12.4.3.4 Bolderberg Sands

On the southeastern part of Campine Plateau, the Zutendaal Gravel blanket overlays about 40 m of fine-grained white to slightly yellowish quartz-rich sands (Fig. 12.14d). These sands belong to the Lower Miocene (Burdigalian) Bolderberg Formation and correspond to shallow-marine deposits in a North Sea bay bordering the Roer Valley. The source material probably originated from the weathered and therefore quartz-rich material at the margins of the southern North Sea Basin in Belgium and adjacent areas. Their uniform fine grain-size distribution (average 200 μm), good sorting, as well as mineralogical composition (slightly glauconite-bearing) and the presence of herringbone cross-bedding, point to longshore transport, coastal sorting by tidal currents in an estuarine depositional environment (Gullentops 1973). The Bolderberg Sands display a threefold subdivision due to a continental phase interrupting the shallow-marine sand sedimentation. During a short withdrawal of the sea, extensive coastal marshes and back-swamps developed (analogous to the actual Everglades in Florida or the Okefenokee swamps in Southern Georgia, USA) leading to the formation of the so-called Kikbeek Lignite (averaging 3 m in thickness). This lignite is mainly composed of driftwood although swamp trees grew in place, as evidenced by numerous roots that penetrate the underlying sands. This seam is of the same age as the lignite seams of the Roer Valley in the Netherlands (Heksenberg Member of the marine Breda Formation) that correlate well with the well-known thick lignite deposits exploited for fuel in the Garzweiler, Frimmersdorf and Hambach quarries of the Lower Rhine Basin in Germany (Van der Meulen et al. 2009). Early-diagenetic leaching by humic acids originating from the peat affected the underlying sands: so, non-quartz components were dissolved resulting in the creation of the famous Miocene glass-sands or "silversands" of Maasmechelen. These worldwide reputed glass-sands are still extensively quarried on the Campine Plateau because of their pureness (iron content is less than 0.01% Fe_2O_3, Al_2O_3 content less than 0.025%). Moreover, this extraction activity was also responsible for the creation of deep extraction pits, resulting in the creation of important geological outcrops that have been recently reclaimed into highly valuable nature reserves and important geoheritage sites (Dreesen et al. 2005).

12.4.4 Polygonal Soils and Cryoturbations

In the uppermost part of the Zutendaal Gravel, right below the Late Pleistocene-Holocene aeolian sands, spectacular examples of a particular type of cryoturbation occur (Fig. 12.15). In section, these features look like vertical tongues or flames ("involutions") up to 2–3 m in length, whereby the pebble axes are vertically oriented. In plane section (at the surface), these tongues form a characteristic network of large polygons with diameters of less than 1 m up to several meters. The core of these gravel polygons is essentially made of sandy material, and their center is slightly raised (Fig. 12.15b). According to Paulissen (1970), this patterned ground type deformation (called "sorted polygons") is characteristic of periglacial areas undergoing permafrost conditions. These peculiar forms most probably originated within the active layer of the permafrost. Frost heave is generally seen as the main mechanism behind the formation of sorted patterned ground (Pissart 1987; Van Vliet-Lanoë 2014; Yamagishi and Matsuoka 2015). Dependent on grain size (fine sands and silts are most frost-susceptible), freezing rate, and available moisture (ensured where a perched water table is present above the frozen ground during summer), lenses of segregation ice develop gradually within the active layer with seasonal freeze-thaw cycles, causing soil heave and erecting elongated gravel (Pissart 1969). While differential frost heave, mainly related to lateral variations in grain size or soil moisture availability, may be responsible for the polygonal pattern, sometimes also guided by a net of desiccation cracks, the contrast between gravel involutions and cells of finer material is attributed to frost sorting, vertical and lateral, by freezing and thawing. These spectacular polygons bear witness to the harsh (cold and dry) climatic conditions that prevailed in Campine area during the last ice ages, particularly during the Late Weichselian (ca. 20 ka). They are currently characteristic of arctic regions or tundra areas like those located around the Arctic Circle (Canada, Spitsbergen, Greenland).

12.4.5 Late Glacial and Holocene Dunes

Among the most striking landforms topping the Campine Plateau are parabolic and longitudinal dunes. They can easily be identified on DTM images (Figs. 12.1 and 12.9) and historical maps. The dune complex of Hoeverbergen (Fig. 12.1c) was originally studied by Gullentops (1957). It is a parabolic dune, ca. 1 km wide (between tips), ca. 2 km long (along the dune axis) and up to 10 m high. Hand augerings allowed for reconstructing the original surface by mapping the palaeosoil that stabilized the dune, in response to climatic amelioration and vegetation development. The dune itself probably developed during the Late Glacial (Younger Dryas), since it is underlain by a bleached horizon, which is interpreted as the Usselo soil of Allerød/Early Younger Dryas age (Derese et al. 2012). The dominant wind direction during dune development was from the southwest as is evidenced by the steep dip angle (26°) of individual sand beds at the northeastern leeward side of the dune (Fig. 12.16a). The soil that is preserved inside the blown-out surface is described as a hydromorphic soil, such that deflation is supposed to have continued until the water table was reached. The northeastern tip of the dune is blown out at a later stage during landscape development, the blowout sand being deposited towards the east as small isolated dunes 1–2 m high. A similar dune complex is present near Kattenbos and Maatheide (Lommel). The core of the dune at Maatheide is shown to contain the Late Glacial Usselo soil (Derese et al. 2012). Holocene reactivation of the Kattenbos dune as a result of agricultural activity (deforestation, herding, repeated use of tracks for herding, cutting of pieces of heath with roots and humus, called plaggen, as bedding for cattle) caused newly eroded sand to drift onto the older

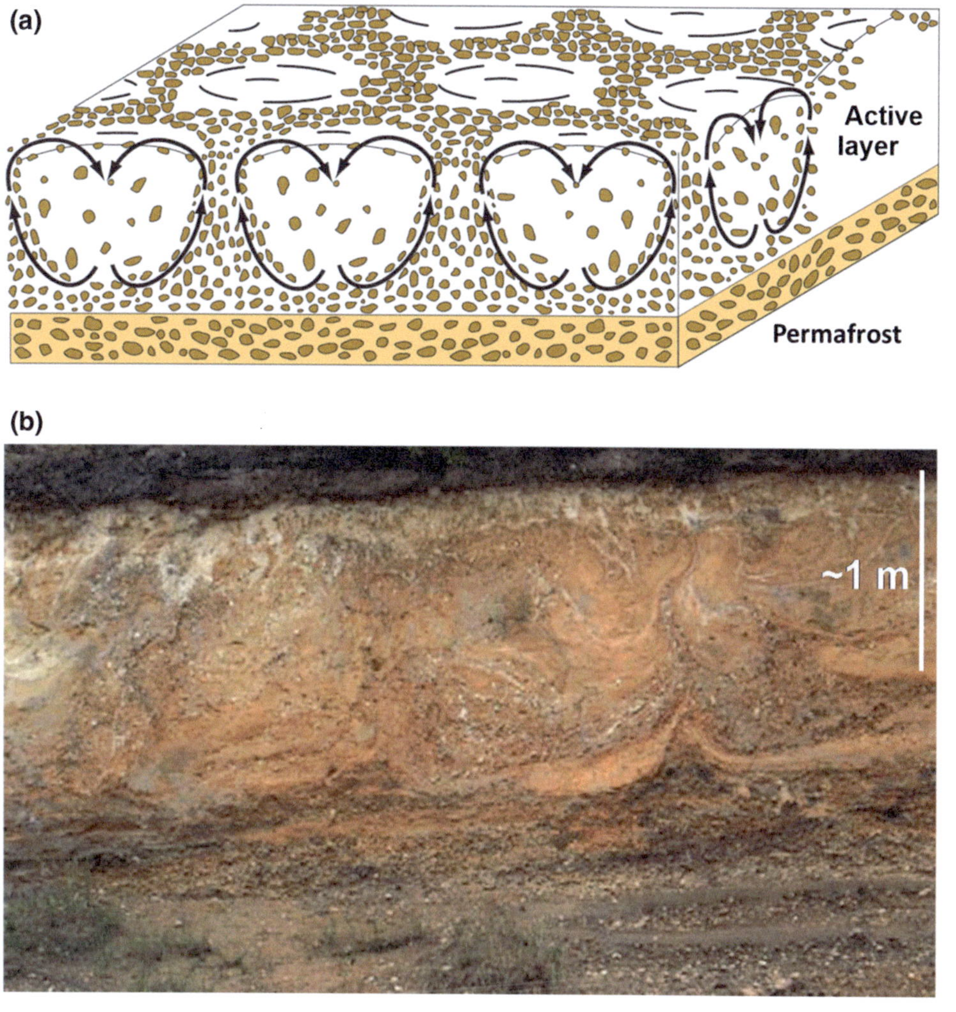

Fig. 12.15 **a** Development of large cryoturbation structures in the active layer above a permafrost table in gravel deposits, according to Paulissen (1970) (modified from Dreesen et al. 2005). **b** Polygonal soil pattern in the top layers of the Zutendaal Gravels, with large-scale involutions and flames containing vertically aligned gravel ('Salamander' quarry, Maasmechelen; reproduced from Dreesen et al. 2005)

dune surface posterior to the development of a podzol soil (Fig. 12.16b).

A different type of dune complex is the one on the west bank of the Bosbeek, called the "Duinengordel" (Fig. 12.16 c). It consists of an array of small parabolic dunes, ca. 200 m wide, 200–400 m long (along the dune axis) and up to 15 m high. The complex itself is about 8 km long, and more or less oriented east-west. Typically, the northeast-facing leeward slope of the individual dunes is much steeper than the southwest-facing slope, indicating that they developed under influence of southwesterlies. The dunes are considered to be relatively young. On the one hand, the absence of Late Glacial palaeosoils in their core suggests that they did not develop from a Late Glacial dune, as is the case in Lommel. On the other hand, traces of soil formation on the dune top are absent, suggesting that they are still active, especially in those areas where vegetation is lacking. A particular feature in this respect is the dune top of the Oudsberg, the highest top of the dune complex (Fig. 12.16d). Aeolian deflation facilitated by over-recreation of the dune led to a system of excavated tree roots. The development of the dune complex is probably related to deflation from ploughed fields south of it.

About 200 circular depressions with an average diameter of 140 m and rims averaging 2.5 m in height have been discovered by using panchromatic and LIDAR-imaging (Laser Imaging Detection And Ranging) in the Bocholt Plain, NE of the Campine Plateau. Field surveys demonstrated that these landforms (with low conservation potential) actually represent blowouts or lunettes (Dusar et al. 2008). Their exact age is still not clear but most probably they were formed as a result of deforestation during the early Neolithicum.

12.4.6 Groundwater Silcretes

About 1.5 m below the base of the Kikbeek lignite seams, ash-white quartz arenitic sandstones occur (the so-called

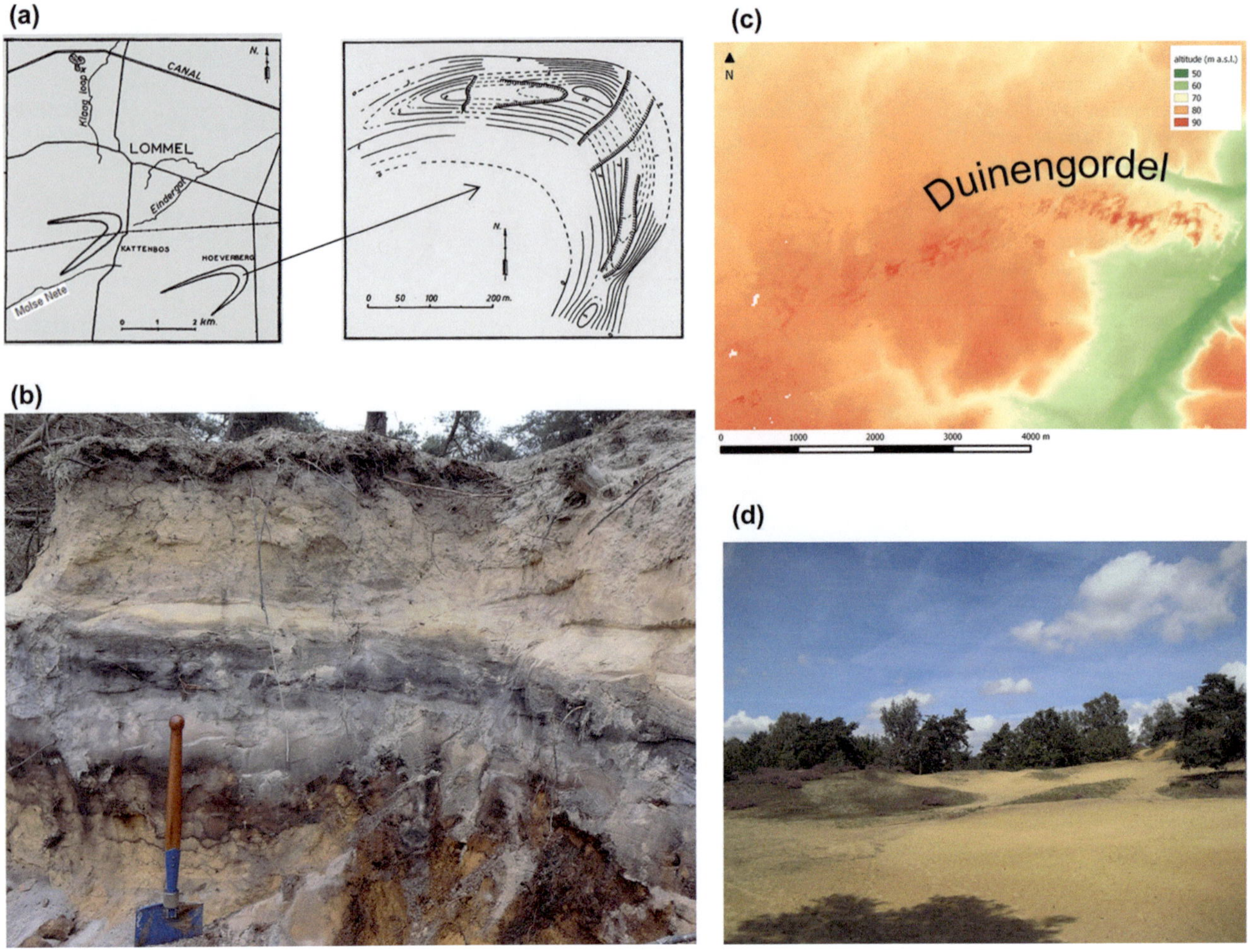

Fig. 12.16 **a**. Detail topographic maps of the Hoeverbergen dune (from Gullentops 1957). **b**. Typical profile showing drift sand migrating over a podzol soil developed on top of the Kattenbos dune near Lommel (*photograph* D. Van Uytven). **c** Detail DTM of the 'Duinengordel' dunes. **d** The Oudsberg, the highest summit of the dune system (*photograph* D. Van Uytven). See also Fig. 12.1c for the location of the Hoeverbergen and Kattenbos dunes

Bolderian Sandstones) in the bleached Bolderberg sand, forming thick plates or thick lenticular beds (Gulinck 1961) that reach volumes of several m^3 (Fig. 12.17). These silcretes often display particular organic or anthropomorphic forms, leading to their naming "*grès mamelonné*" in French. Silicification of the sand, often incomplete, commonly occurs through quartz overgrowth on the quartz grains. The silcrete blocks contain small circular pits at their upper surface and vertical root traces, pointing to the presence of a peat (lignite) layer above. Although different stratigraphic levels display silicification in the subsurface of the province of Limburg, silcretes most frequently formed during continental episodes of the Thanetian (Upper Paleocene) in Belgium and Northern France (Dreesen and Dusar 2012). Silcrete formation is related to post-depositional groundwater circulation and the release of silica accompanying acid leaching of the host material or adjacent strata (Ullyott and Nash 2006). Demoulin (1990) distinguished several types of silicified Tertiary sandstones in Southern Limburg (NL) and at the northern border of the Ardennes (Hautes Fagnes and Pays de Herve). The Miocene quartz arenites of the High Campine area belong to Demoulin's Nievelstein type, typically associated with lignite deposits. The Miocene silcretes show strong analogies with the sarsen stones of southern England and the "Braunkohlequartzite" of the German Lower Rhine area. Moreover, they have been locally used as a vernacular building stone in the province of Limburg (Dreesen et al. 2001). Finally, when some of these huge, massive stones (megaliths) became unearthed through erosion of the enveloping sand, they have been locally used as a grinding stone for flint tools during prehistoric times, such as the sandstone blocks of the Holsteen, in the Zonhoven area (Huyghe 1990).

Fig. 12.17 Blocks of groundwater silcrete indurating Bolderberg sands at the Holsteen, near Zonhoven (Dreesen and Dusar 2012)

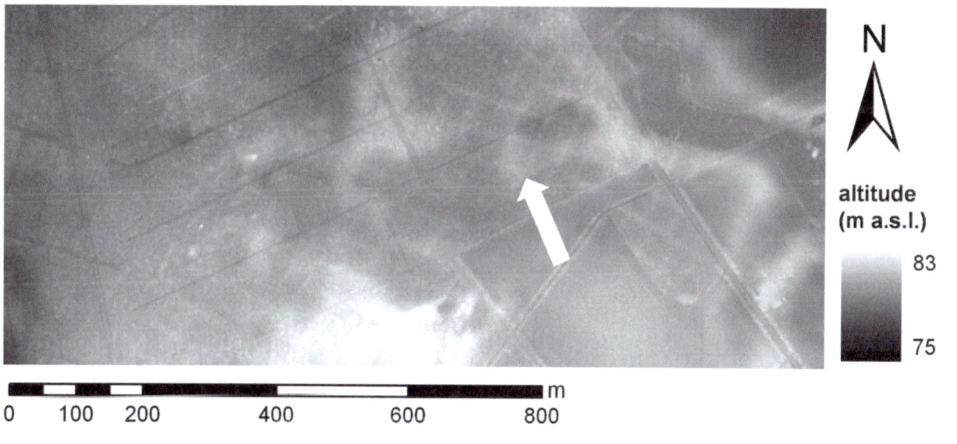

Fig. 12.18 DTM of the Muisven-Ophoven Celtic Field complex (Meeuwen-Gruitrode, see location in Fig. 12.1; modified after Creemers et al. 2011). The squared pattern of the Celtic Field is indicated by an *arrow*

12.4.7 Celtic Fields

Detailed LIDAR images often show a striking pattern of slightly elevated banks, arranged in a multiple crossroad pattern. They are called Celtic Fields and characterized as 'a prehistoric parceling system of which the parcels are completely surrounded by low banks consisting of stones, sand or a mixture of the two, and laid out for agricultural purposes' (Creemers et al. 2011). The banks may only be 20–30 cm high, and the parcels which they surround typically measure 40 × 40 m. A representative example is the Celtic Field of the Muisvenner-Ophovenerheide in Meeuwen-Gruitrode (Fig. 12.18). The fields were used to cultivate, among others, emmer wheat (*Triticum dicoccum*) and gold-of-pleasure (*Camelina sativa*). The term Celtic Fields is somewhat misleading: the Celts were initially thought to have developed this agricultural system, but later investigations showed that the fields date back from as early as the late Bronze Age. The fields were abandoned during the Roman period and developed into heathland until the region became reforested by pine during the late nineteenth and early twentieth century. Their location, as well as that of Mesolithic settlements, seems to be related to the presence of nearby fens and wetlands (see above).

Fig. 12.19 Overgrown coal mining tip ("terril") of Eisden. Groundwater crops out in abandoned gravel pits (*Photograph* M. Bex)

12.4.8 Impacts of Coal Mining on the Geomorphology of the Campine Plateau

Coal mining tips (spoil heaps or waste dumps, named "*terrils*" or "*steenbergen*") represent enormous beacons in the Campine landscape and they are regarded as true sociocultural heritage (Fig. 12.19). They represent silent witnesses of the luxurious tropical rain forests that once ruled the area during the Late Carboniferous, generating numerous coal layers in the deep subsurface of the Campine basin. Their discovery in 1901 meant the beginning of a coal rush. From the first coal production in 1917 until the closure of the last colliery in 1992, over 441 million tons of coal have been produced from seven different collieries, with an equivalent amount of coal waste. The latter mainly consists of mudstone (over 70%) with minor amounts of siltstone, sandstone, coal, and carbonaceous shale. After final closure of the collieries, the coal mining tips and coal sludge basins (liquid waste generated by washing of the crushed coal) were partly reclaimed from 1993 to 2000, because of environmental regulations. Reclaiming measures concentrated on remodeling the dumps, aiming at stabilizing the slopes and avoiding acid coal mine waste drainage. In order to further reduce erosion (e.g., through gully formation and debris sliding) greening of the dumps was carried out by hydroseeding. Due to the latter remodeling, the original outline, shape and location of the dumps have been considerably changed (Fig. 12.20). The coal mining tips originally reached between 70 and 110 m in height. Although they represent enormous reserves of secondary raw materials for the ceramics industry (Dreesen et al. 2008), the ones that have not been remodeled or reclaimed have become nature reserves, hosting thermophilic and halophilic fauna and flora.

Moreover, this deep coal mining activity caused important and extensive ground subsidence. Maps with estimated iso-subsidence lines have been calculated in collaboration with the Campine Collieries between 1981 and 1985 (Vansteelandt 1996) showing subsidence values exceeding 6–8 m (e.g., in the central Zwartberg–Waterschei area). This subsidence is related to the extraction of several superimposed coal layers and to strong groundwater (mine water) pumping (Fig. 12.21). Moreover, with the help of radar interferometry (PS-inSAR) using satellite data from 1992 through 2005, contrasting ground movements have been reported over the abandoned Campine coal mining area (Devleeschouwer et al., 2007). The collieries in the eastern part of the coal basin closed much earlier (1966–1987) than those in the western part (1992). As a result and in contrast with residual mining subsidence in the western part (−4 to −14 mm/year between 1992 and 2000), a distinct uplift has been recorded in the eastern part (+3 to +23 mm/year). This uplift apparently results from poro-elastic rebound of the eastern area when flooding of the underground mine working subsequent to the end of mine water pumping induced a pore-pressure increase. Meanwhile, also the western part of the Campine coal field has reversed to uplift mode (Vervoort 2016; Bejarano-Urrego et al. 2016).

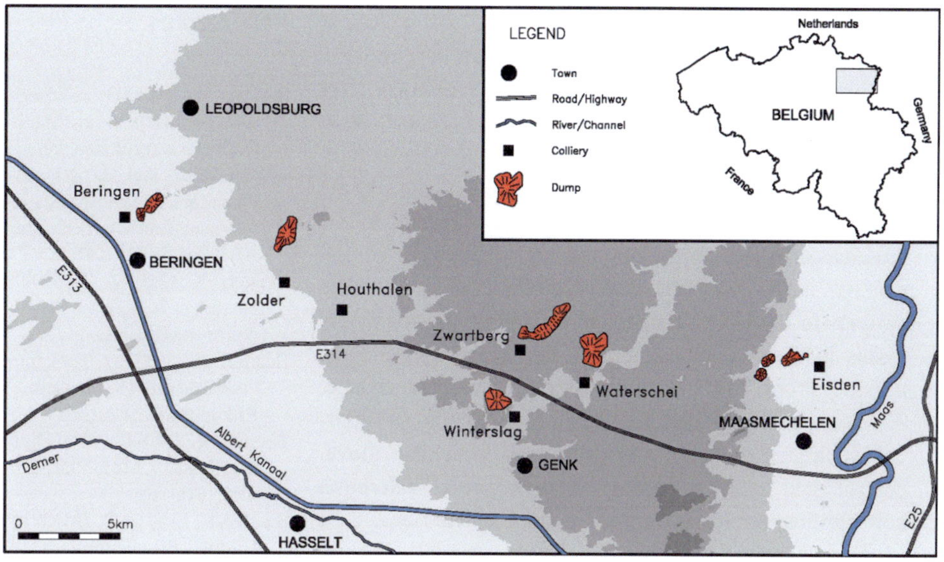

Fig. 12.20 Location map of the remaining spoil heaps (terrils) in the Campine area (modified after Dreesen et al. 2008)

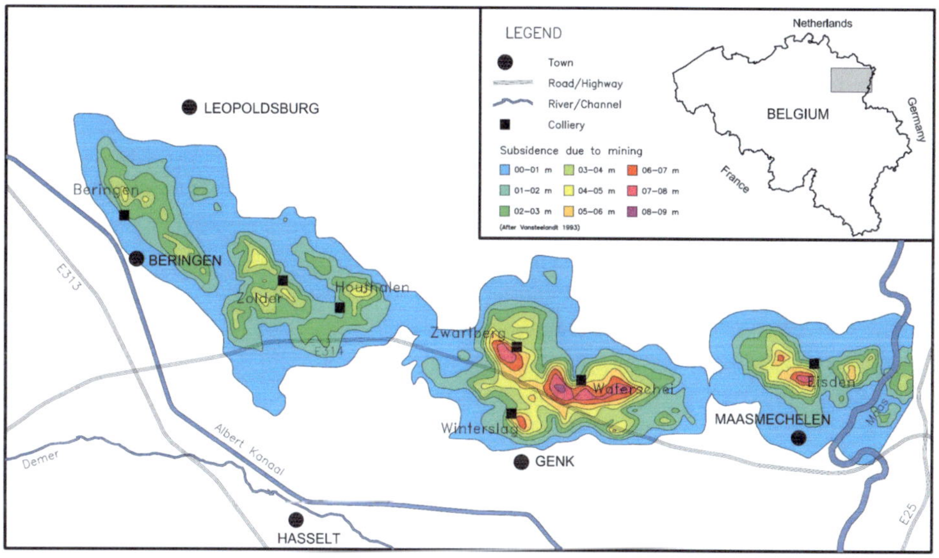

Fig. 12.21 Lines of iso-subsidence above the former coal mining areas in the Campine Plateau (*drawing* J. Matthijs, modified after Vansteelandt 1996)

12.5 Conclusions

In 2011, the Hoge Kempen (High Campine) Rural-Industrial Transitional Landscape (HKRIL) has been nominated for the Tentative list of UNESCO's World Heritage Sites. It bears unique geological and geomorphological witnesses of rather extreme global changes that once affected this area, grading from warm and humid tropical swamps to cold and dry permafrost conditions. Moreover, the effects of dramatic climatological changes that affect our living planet can clearly be demonstrated in natural or artificial outcrops that occur within its geographical boundaries. Abandoned sand and gravel pits in particular represent great geological windows for educational purposes: they offer unique opportunities to display the layer-cake structure and to unravel contrasting climatological conditions and depositional settings in which the exposed sediments and rocks have been formed. Several geotouristic initiatives have been launched by volunteer organizations (e.g., the Geological working group of LIKONA—the Limburgse Koepel voor Natuurstudie) stimulating outreach and triggering public awareness. The latter working group has been involved in the reclamation of abandoned glass sand and gravel pits in the Hoge Kempen National Park, where rather unique geological features are particularly well exposed, including spectacular Pleistocene polygonal soils and gravelly braided river deposits (with ice-rafted boulders up to 2 m in diameter) witnessing the harsh periglacial conditions during the youngest ice age. A particular dune complex represents a key reference section for the Late Glacial sandy aeolian deposits in the southernmost part of the

NW European coversand belt, because of excellent preservation of fossilized soils, rarely seen elsewhere.

Huge ice-rafted boulders embedded in braided river deposits display a broad lithologic spectrum representative of the geology of the upstream area of the Meuse river basin. Below this Pleistocene gravel blanket, pure white quartz sands occur, the origin of which is related to the presence of lignite formed during the Miocene in coastal marshes, analogous to the recent Everglades or Okefenokee swamps in the USA. The lignite as well as the rocks and fossils in the coal mining tips in the direct surroundings provide strong evidence for past global climate change. A selection of ice-rafted boulders, representative of different geological stages, were removed from the gravel deposits and set into a geological "rock garden" offering a new destination for geography school classes and their teachers. A colorful brochure allows the general public to identify pebbles that occur everywhere throughout the National Park and that represent its hallmark (Dreesen 2007). Recently a new geological biking route has been set up, explaining the complex relationship between subsurface, landscape, flora, and cultural history of the High Campine area (Van Uytven and Dreesen 2015).

The Campine Plateau is a unique geomorphological object and one of the best-kept examples in Europe of an enormous alluvial fan-like gravel and sand deposit formed during the Middle Pleistocene. Its present shape is the result of interfering erosional and tectonic processes. The southern tip, the western and northeastern boundaries are all (in)directly fault-related, this particular tectonic configuration controlling the fluvial development of the area around the Early-Middle Pleistocene transition. Once tectonic uplift had raised the Eastern Campine Block, that largely coincides with the Campine Plateau, the fluvial deposits were directly responsible for the preservation of the plateau as a morphological entity in topographic inversion. Aeolian processes actively reshaped the plateau surface during the Late Pleistocene and Holocene. Soil erosion and depletion of nutrients due to intensive grazing had to be countered by massive pine planation on top of the plateau. These could easily thrive on the relatively dry sandy and stony soils. Locally, clay or loam lenses on top of the gravel deposits entail completely different hydrological conditions such that rainwater-fed fens could develop. Coal extraction has led to specific anthropogenic geomorphic features, such as big spoil heaps and subsidence phenomena above the former Campine collieries.

Acknowledgements Michiel Dusar is gratefully thanked for critically reading a late draft of the chapter and helping improve the text.

References

Beerten K (2003) Vroeg- en Midden-Pleistocene sedimentatiegeschiedenis van Maas en Rijn in de Belgische Roergraben – Onderzoeksresultaten van gekernde boringen (Project VLA01-3.1.3.). Ministerie van de Vlaamse Gemeenschap, Brussel. http://www.vlaanderen.be/nl/publicaties/detail/vroeg-en-midden-pleistocene-sedimentatiegeschiedenis-van-maas-en-rijn-in-de-belgische-roergraben. Accessed 5 Feb 2015

Beerten K (2005a) Toelichting bij de Quartairgeologische kaart, kaartblad 26 Rekem. Vlaamse Overheid, Dienst Natuurlijke Rijkdommen, Brussels

Beerten K (2005b) Toelichting bij de Quartairgeologische kaart, kaartblad 10-18 Maaseik. Vlaamse Overheid, Dienst Natuurlijke Rijkdommen, Brussels

Beerten K (2006) Toelichting bij de Quartairgeologische kaart, kaartblad 17 Mol. Vlaamse Overheid, Dienst Natuurlijke Rijkdommen, Brussels

Beerten K, Brabers B, Bosch P, Gullentops F (1999) The passage of the Feldbiss Bundle through the Maas Valley. Aardkundige Mededelingen 9:153–158

Beerten K, Vandersmissen N, Deforce K, Vandenberghe N (2014) Late quaternary (15 ka to present) development of a sandy landscape in the Mol area, Campine region, NE Belgium. J Quat Sci 29:433–444

Bejarano-Urrego L, Verstrynge E, Van Balen K, Wuyts V, Declercq PY (2016) Settlement-induced damage monitoring of a historical building located in a coal mining area using PS-InSAR. In: Proceedings of the CSHM-6: structural health monitoring of new and ageing infrastructure, Belfast, N Ireland, May 26–27

Bogemans F (2005) Technisch verslag bij de opmaak van de quartairgeologische overzichtskaart van Vlaanderen, schaal 1:200.000. Ministerie van de Vlaamse Gemeenschap, Afdeling Natuurlijke Rijkdommen en Energie, Brussels

Bosch P (1974) Diabaas en porfiroïde als zwerfsteen in Zuid-Limburgse Maasterrassen. Grondboor en Hamer 6:110–118

Bosch P (1992) De herkomstgebieden van de Maasgesteenten. Grondboor en Hamer 3:57–64

Cohen K, Gibbard P (2011) Global chronostratigraphical correlation table for the last 2.7 million years. Subcommission on Quaternary Stratigraphy (International Commission on Stratigraphy), Cambridge, England

Collard S, Juvigné E, Marion J-M, Mottequin B, Petit F (2012) L'origine des megalithes du Fond de Quarreux (Ardenne, Belgique). Bulletin de la Société Géographique de Liège 58:33–51

Creemers G, Meylemans E, Paesen J, De Bie M (2011) Laseraltemetrie en de kartering van Celtic Fields in de Belgische Kempen: mogelijkheden en toekomstperspectieven. Relicta 7:11–36

De Brue H, Poesen J, Notebaert B (2015) What was the transport mode of large boulders in the Campine Plateau and the lower Meuse valley during the mid-Pleistocene? Geomorphology 228:568–578

De Moor G, Pissart A (1992) Het Reliëf. In: Denis J (ed) Geografie van België. Gemeentekrediet, Brussel, pp 130–215

Demoulin A (1990) Les silicifications tertiaires de la bordure de l'Ardenne et du Limbourg méridional (Europe NO). Zeitschrift für Geomorpholopgie 34:179–197

Demoulin A, Hallot E (2009) Shape and amount of the quaternary uplift of the western Rhenish shield and the Ardennes (western Europe). Tectonophysics 474:696–708

Derese C, Vandenberghe D, Paulissen E, Van den haute P (2009) Revisiting a type locality for Late Glacial aeolian sand deposition in NW Europe: optical dating of the dune complex at Opgrimbie (NE Belgium). Geomorphology 109:27–35

Derese C, Vandenberghe D, Eggermont N, Bastiaens J, Annaert R, Van den Haute P (2010) A medieval settlement caught in the sand: optical dating of sand-drifting at Pulle (N Belgium). Quat Geochronol 5:336–341

Derese C, Vandenberghe D, Van Gils M, Mees F, Paulissen E, Van den Haute P (2012) Final Palaeolithic settlements of the Campine region (NE Belgium) in their environmental context: optical age constraints. Quatern Int 251:7–21

Devleeschouwer X, Declercq P-Y, Dusar M, Debien A (2007) Contrasting ground movements revealed by radar interferometry over abandoned coal mines (Campine, Belgium). In: Proceedings of FRINGE 2007, workshop, European space agency, Frascati, Italy, vol 12

Dionne J-C (2003) Observations géomorphologiques sur les méga-blocs du secteur sud-est de la batture argileuse de la baie à l'Orignal, au parc du Bic, dans le Bas-Saint-Laurent (Québec). Géog Phys Quatern 57:95–101

Dreesen R (2007) Zoekkaart keien voor het Nationaal Park Hoge Kempen. Regionaal Landschap Kempen en Maasland & LIKONA

Dreesen R, Dusar M (2012) Duivelsstenen in Limburg: zwerfstenen, megalieten of getuigenstenen? LIKONA jaarboek 2011:14–29

Dreesen R, Dusar M, Doperé F (2001) Atlas natuursteen in Limburgse monumenten. Geologie, beschrijving, herkomst en gebruik. Provincie Limburg, Het Groene Huis, Genk, pp 294

Dreesen R, Mareels J, Fries S (2005) De zandgroeve van Opgrimbie: een uitzonderlijk kijkvenster op de geologische geschiedenis van de Hoge Kempen. LIKONA Jaarboek 2005:14–25

Dreesen R, Laenen B, Nielsen P, Matthijs J (2008) Assessment of coal waste dump material as an alternative for primary raw materials in the former coal mining district of the Campine area (N-Belgium). In: 33rd international geological congress, Oslo, Proceedings

Dreesen R, Janssen J, Van Uytven D (2014) Het grind ontleed: samenstelling en herkomst van de stenen in het Maasgrind. LIKONA-jaarboek 24:18–35

Dusar M, Langenaeker V (1992) De oostrand van het massief van Brabant, met beschrijving van de geologische verkenningsboring te Martenslinde. Geol Surv Belg Prof Paper 255, 22 p

Dusar M, Matthijs J, Dreesen R, Poesen J (2008) Cirkelvormige depressies in de vlakte van Bocholt. LIKONA jaarboek n° 17, 20–35

Felder W, Bosch P (1989) Geologische kaart van Zuid-Limburg en omgeving 1:50.000, Afzettingen van de Maas. Rijks Geologische Dienst, Heerlen

Geluk M, Duin E, Dusar M, Rijkers R, Van den Berg M, van Rooijen P (1994) Stratigraphy and tectonics of the Roer Valley Graben. Geol Mijnbouw 73:129–141

Goossens J, Molemans J, Paulissen E, Theuwissen J (1983) De begrenzing van de Kempen. Mededelingen van de Vereniging voor Limburgse Dialect- en Naamkunde 25.153–156

Gulinck M (1961) Note sur le Boldérien d'Opgrimbie (Campine) et remarques sur les grès "erratiques" du Limbourg. Bulletin de la Société belge de Géologie 70:297–302

Gullentops F (1957) Quelques phénomènes géomorphologiques depuis le Pléni-Würm – Dunes paraboliques en Campine. Bulletin de la Société belge de Géologie 66:86–95

Gullentops F (1973) Grainsize and mineralogy of Miocene glass-sands of Maasmechelen, Belgian Limburg. Mededelingen Rijks Geologische Dienst, Nieuwe Serie 23:25–34

Gullentops F, Claes S (1997) The Bilzen fault bundle (NE Belgium). Aardkundige Mededelingen 8:99–102

Gullentops F, Vandenberghe N (1995) Toelichtingen bij de geologische kaart van België, Vlaams Gewest, Kaartblad 17, Mol. Belgische Geologische Dienst, Brussels

Gullentops F, Wouters L (1996) Delfstoffen in Vlaanderen. Ministerie van de Vlaamse Gemeenschap, Brussels

Gullentops F, Janssen J, Paulissen E (1993) Saalian nivation activity in the Bosbeek valley, NE Belgium. Geol Mijnbouw 72:125–130

Gullentops F, Bogemans F, De Moor G, Paulissen E, Pissart A (2001) Quaternary lithostratigraphic units (Belgium). Geologica Belgica 4:153–164

Gullentops F, Paulissen E, Vandenberghe J (1981) Periglacial Phenomena in NE-Belgium. Biuletyn Peryglacjalny, 28:345–365

Gupta S, Collier J, Palmer-Felgate A, Potter G (2007) Catastrophic flooding origin of shelf valley systems in the english channel. Nature 448:342–345

Hoek W (2001) Vegetation response to the 14.7 and 11.5 ka cal. BP climate transitions: is vegetation lagging climate? Global Planet Change 30:103–115

Huyghe D (1990) De "Holsteen" te Zonhoven: geo-archeologie van een prehistorisch landschap. Archeologie in Vlaanderen 1:31–54

Janssen J, Dreesen R (2010) Geologische fietsroute Tussen Kempen en Maas. LIKONA, Limburgse Koepel voor Natuurstudie, Genk 78 p

Juvigné E, Renard F (1992) Les terrasses de la Meuse de Liège à Maastricht. Annales de la Société Géologique de Belgique 115:167–186

Kasse C, Vandenberghe D, De Corte F, Van den Haute P (2007) Late Weichselian fluvio-aeolian sands and coversands of the type locality Grubbenvorst (southern Netherlands): sedimentary environments, climate record and age. J Quat Sci 22:695–708

Matthijs J, Lanckacker T, De Koninck R, Deckers J, Lagrou D, Broothaers M (2013) Geologisch 3D lagenmodel van Vlaanderen en het Brussels Hoofdstedelijk Gewest – versie 2, G3Dv2. Vlaamse overheid, Departement Leefmilieu, Natuur en Energie, Afdeling Land en Bodembescherming, Ondergrond, Natuurlijke Rijkdommen, Brussel, 21 p

MACAR P (1938) Compte rendu de l'excursion du 24 avril 1938, consacrée à l'étude des terrasses de la Meuse entre Liège et 'Ubagsberg (Limburg hollandais). Annales de la Société géologique de Belgique, 61:187–217

Pannekoek A (1924) Einigen Notizien über die Terrassen in Mittel- und Nord-Limburg: Natuurhistorisch Maandblad 13:305–336

Paulissen E (1970) Excursiegids: De geomorfologie van de Limburgse Kempen. De Aardrijkskunde 86:1–48

Paulissen E (1973) De Morfologie en de Kwartairstratigrafie van de Maasvallei in Belgisch Limburg. Verh. Kon. Acad. Wetensch., Letteren en Schone Kunsten v. België. Kl der Wetensch 127:1–266

Paulissen E (1983) Les nappes alluviales et les failles Quaternaires du Plateau de Campine. In: Robaszynski F, Dupuis C (eds) Guides Géologiques Régionaux – Belgique. Masson, Paris, pp 167–170

Paulissen E., 1984. Het fysisch kader van Opglabbeek. In: Molemans J, Mertens J (eds) Opglabbeek, een rijk verleden, Gemeentebestuur Opglabbeek, pp 29–56

Paulissen E (1997) Quaternary morphotectonics in the Belgian part of the Roer Graben. Aardkundige Mededelingen 8:131–134

Pissart A (1969) Le mécanisme périglaciaire dressant les pierres dans le sol. Résultats d'expériences. Comptes-rendus de l'Académie des Sciences de Paris 268:3015–3017

Pissart A (1987) Géomorphologie périglaciaire. Rijksuniversiteit Gent, Faculteit van de Wetenschappen 133 p

Schokker J, Cleveringa P, Murray A, Wallinga J, Westerhoff W (2005) An OSL dated Middle and Late Quaternary sedimentary record in the Roer Valley Graben (southeastern Netherlands). Quartern Sci Rev 24:2243–2264

Toucanne S, Zaragosi S, Bourillet J-F, Gibbard P, Eynaud F, Giraudeau J, Turon J-L, Cremer M, Cortijo E, Martinez P, Rossignol L (2009) A 1.2 Ma record of glaciation and fluvial discharge from the Western European Atlantic margin. Quatern Sci Rev 28:2974–2981

Ullyott J, Nash D (2006) Micromorphology and geochemistry of groundwater silcretes in the eastern South Downs, UK. Sedimentology 53:387–412

van Balen R, Houtgast R, Van der Wateren F, Vandenberghe J, Bogaart P (2000) Sediment budget and tectonic evolution of the Meuse catchment in the Ardennes and the Roer Valley Rift System. Global Planet Change 27:113–129

van den Berg M (1996) Fluvial sequences of the Meuse-a 10 Ma record of neotectonics and climate change at various time-scales. Doctoral dissertation, Wageningen University, 181 p

Van der Meulen M, Westerhoff W, Menkovic A, Gruijters S, Dubelaar C, Maljers J (2009) Silica sand resources in The Netherlands. Neth J Geosci 88:147–190

Van Mechelen J (1982) Geomorfologie van de westelijke rand van het Kempens Plateau en micro-eolisatieverschijnselen op grind. Unpublished master thesis, K.U.Leuven

Van Uytven D, Dreesen R (2015) Geologische fietsroute Hoge Kempen. LIKONA, Genk 144 p

Van Vliet-Lanoë B (2014) Patterned ground and climate change. In Podrovsky O (ed) Permafrost: Distribution, composition and impacts on infrastructure and ecosystems. Nova Science Publ. Inc., pp 67–106

Vandenberghe J (1982) Geoelectric investigations of a fault system in quaternary deposits. Geophys Prospect 30:879–897

Vandenberghe J (1990) Morphological effects of Pleistocene faulting in unconsolidated sediments (Central Graben, Netherlands). Zeitschrift für Geomorphologie 34:113–124

Vandenberghe J (2008) The fluvial cycle at cold-warm-cold transitions in lowland regions: a refinement of theory. Geomorphology 98:275–284

Vanneste K, Verbeeck K, Camelbeeck T, Paulissen E, Meghraoui M, Renardy F, Jongmans D, Frechen M (2001) Surface-rupturing history of the Bree fault scarp, Roer Valley graben: evidence for six events since the late Pleistocene. J Seismolog 5:329–359

Vansteelandt P (1996) Mijnverzakkingen in Limburg. In: Gullentops F, Wouters L (eds) Delfstoffen in Vlaanderen. Ministerie van de Vlaamse Gemeenschap, Departement EWBL, pp 115–116

Verbeeck K, Béatse H, Vanneste K, Camelbeeck T (2002) The use of cone penetration tests in paleoseismological investigations: 2 case studies. Aardkundige Mededelingen 12:9–12

Vermeersch P, Munaut A, Paulissen E (1974) Fouilles d'un site du Tardenoisien final à Opglabbeek-Ruiterskuil (Limbourg belge). Quartär 25:85–104

Verstraelen A (2000) Toelichting bij de Quartairgeologische kaart, kaartblad 34 Tongeren. Vlaamse Overheid, Dienst Natuurlijke Rijkdommen, Brussels

Vervoort A (2016) Surface movement above an underground coal longwall mine after closure. Nat Hazards Earth Syst Sci 16:2107–2121

Westerhoff W, Kemna H, Boenigk W (2008) The confluence area of Rhine, Meuse, and Belgian rivers: Late Pliocene and Early Pleistocene fluvial history of the northern Lower Rhine Embayment. Neth J Geosci 87:107–125

Yamagishi C, Matsuoka N (2015) Laboratory frost sorting by needle ice: a pilot experiment on the effects of stone size and extent of surface stone cover. Earth Surf Proc Land 40:502–511

Morphotectonics and Past Large Earthquakes in Eastern Belgium

13

Kris Vanneste, Thierry Camelbeeck, Koen Verbeeck and Alain Demoulin

Abstract

Tectonic landforms are generally modest in stable plate interiors characterized by low strain rates and rare earthquakes. Nevertheless, specific investigations identified such landforms in Belgium, which is located in the most seismically active region of stable Europe northwest of the Alps. Here, we present two active fault zones among the best documented in continental Europe, and whose geomorphology is related to earthquake activity in eastern Belgium. The 12-km-long Bree fault scarp is readily identified in the flat alluvial landscape of the Campine. It results from the activity since the Middle Pleistocene of the Geleen fault bounding the Roer Valley Graben (RVG) to the west. At its base, a ~1-m-high scarplet corresponds to the surface rupture associated with a Holocene large normal-faulting earthquake. Combined geomorphological and paleoseismic investigations allowed evaluating the average slip rate and the return period of large earthquakes on the Geleen fault during the Late Pleistocene. Extending across a more animated landscape in NE Ardenne, owing also to lower slip rates, the Hockai Fault Zone (HFZ) is morphologically less conspicuous. Microseismicity, geology and geomorphology provide consistent evidence of recent activity of this fault zone. The location of the $M = 6\ ¼$ 18 September 1692 earthquake, the largest historical earthquake recorded in this part of Europe, on the northern HFZ suggests that the fault morphology could have resulted from sporadic bursts of large earthquakes.

Keywords

Morphotectonics • Roer Valley Graben • Geleen fault • Hockai Fault Zone • Paleoseismology • Historical seismicity • Fault activity

13.1 Introduction

The local and regional shape of the Earth's surface results from the interaction between internal forces related to crustal stresses and climate-dependent surface processes (weathering, erosion/deposition). In particular, large earthquakes along active fault zones can leave a lasting imprint in the landscape, depending on the balance between fault slip rate and degradation processes.

The territories of Belgium and Grand-Duchy of Luxemburg are part of a stable continental region far away from tectonic plate limits (Fig. 13.1a). At first glance, despite the

K. Vanneste (✉) · T. Camelbeeck · K. Verbeeck
Section of Seismology—Gravimetry, Royal Observatory of Belgium, Avenue Circulaire 3, Brussels, 1180Belgium
e-mail: kris.vanneste@oma.be

T. Camelbeeck
e-mail: thierry.camelbeeck@oma.be

K. Verbeeck
e-mail: koen.verbeeck@oma.be

A. Demoulin
Department of Physical Geography and Quaternary, University of Liège, Sart Tilman, B11, Liège, 4000, Belgium
e-mail: ademoulin@ulg.ac.be

A. Demoulin (ed.), *Landscapes and Landforms of Belgium and Luxembourg*,
World Geomorphological Landscapes, DOI 10.1007/978-3-319-58239-9_13

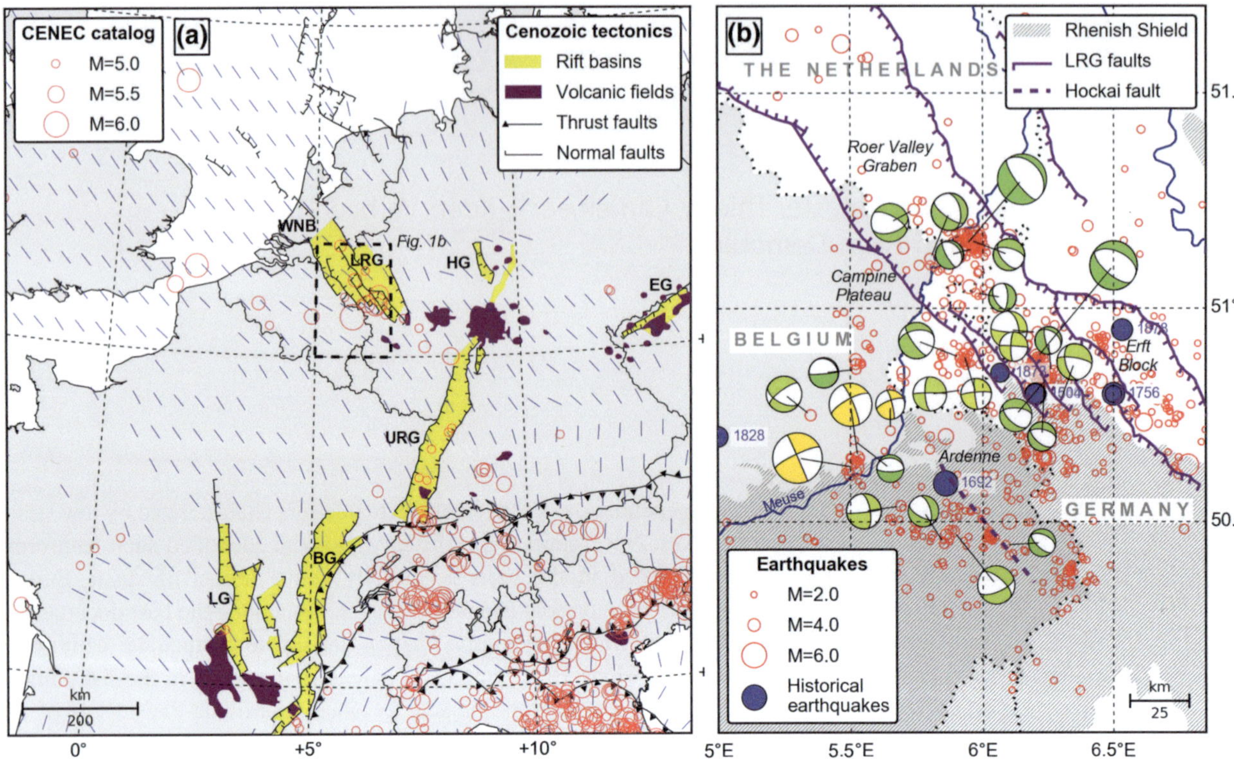

Fig. 13.1 **a** Map of the European Cenozoic Rift System. *BG* Bresse Graben; *EG* Eger Graben; *HG* Hessian Graben; *LG* Limagne Graben; *LRG* Lower Rhine Graben; *URG* Upper Rhine Graben; *WNB* West Netherlands Basin. Modified after Dèzes et al. (2004). *Circles* represent epicentres of earthquakes with $M_w \geq 5.0$ from the CENEC catalogue (Grünthal et al. 2009). *Blue lines* correspond to smoothed S_H orientations from the world stress map (Heidbach et al. 2010). **b** Map showing location of the Roer Valley Graben and Hockai fault in eastern Belgium, along with instrumental and historical seismic activity (ROB earthquake catalogue). Colour of focal mechanisms follows a gradient between pure normal faulting (*green*) and pure strike slip (*yellow*)

proximity of the European Cenozoic Rift System (ECRIS), they do not seem to be an ideal place to look for significant landforms related to active tectonics, except perhaps for the moderate broad-scale deformation of the Plio-Quaternary uplift of the Ardenne (see Chap. 10). Nevertheless, the region is characterized by moderate seismic activity, part of it concentrated in the eastern part of the country and the adjacent regions of the Netherlands and Germany (Fig. 13.1b). The most active area is the Roer Valley Graben (RVG). It is characterized by moderate but ongoing seismic activity, with at least six earthquakes with $M \geq 5.0$ since the fourteenth century, the strongest one being the 1756 Düren (Germany) earthquake with an estimated magnitude around 5¾. During the past century, two $M = 5.3$ earthquakes struck the region, at Euskirchen (Germany) in 1951 and Roermond (the Netherlands) in 1992 (Camelbeeck et al. 2007).

On the occasion of the 13 April 1992 Roermond earthquake, deformations caused by seismic shaking were observed at the ground surface, motivating the search for active, potentially seismogenic faults, their possible footprint in the land morphology and evaluation of the associated seismic hazard. In the frame of these studies, several paleoseismological surveys have been conducted in NE Belgium since the 1990s, which disclosed a record of surprisingly (for the time) large paleoearthquakes caused by seismic ruptures on the Geleen fault, the western border fault of the RVG in the Belgian Campine (e.g. Camelbeeck and Meghraoui 1996, 1998) (Fig. 13.2). A succession of large Middle Pleistocene earthquakes formed a 12-km-long NNW-trending scarp along the fault, the Bree fault scarp, which represents a gentle but noticeable 15–20-m-high geomorphic marker. The surface rupture associated with the most recent Holocene large earthquake is clearly visible as a small 0.5–1-m-high scarplet at the base of the higher main scarp.

However, the $M_w \approx 6¼$, 18 September 1692, Verviers earthquake, which is the strongest known earthquake in this part of Europe (Alexandre et al. 2008), is located outside the RVG. This seismic event is now commonly recognized as one of the two most violent historical earthquakes that struck the NW foreland of the Alpine chain, second only to the Basel earthquake in 1356. The Verviers earthquake likely occurred along a reactivated Variscan structure called the Hockai Fault Zone (HFZ), which shows present-day microseismic activity and whose Quaternary activity is revealed by scattered

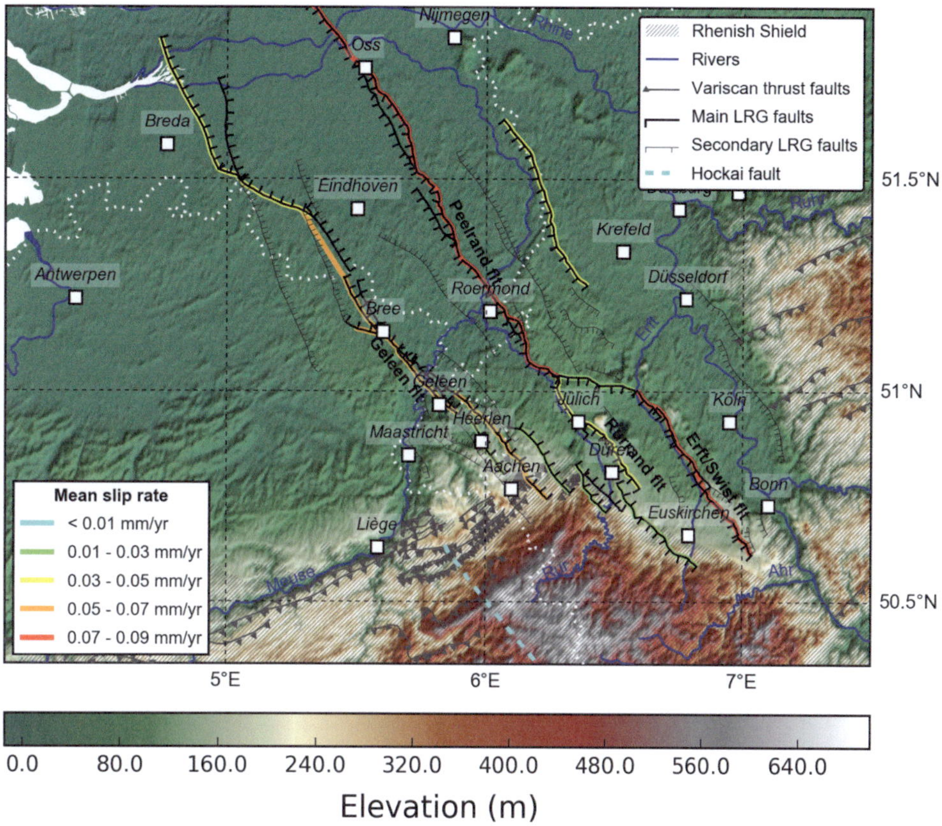

Fig. 13.2 Morphotectonic map showing the main faults of the Lower Rhine Graben and the Hockai fault zone, *coloured* according to their long-term slip rate (Vanneste et al. 2013). Topography from ASTER Global Digital Elevation Model (GDEM) v2. ASTER GDEM is a product of NASA and METI

geomorphological indices. Like the RVG, this area thus appears to have preserved the morphological fingerprint of active faulting with a normal component, with the difference that, here, the fault morphology points to lower slip rates but longer preservation in hard bedrock.

In this chapter, we present an overview of the knowledge on the Geleen fault and the HFZ, which are two of the best documented case studies of morphological evidence related to the occurrence of large earthquakes in tectonically stable continental Europe.

13.2 The Western Border Fault of the Roer Valley Graben in the Belgian Campine

13.2.1 Geographical and Geological Setting

The RVG is the central graben of the Roer Valley Rift System (RVRS), also commonly denoted as Lower Rhine Graben or Lower Rhine Embayment The RVRS is the northwestern branch of the Rhine Graben system, which is in turn part of the ECRIS that extends from west of the Alps to the North Sea (Fig. 13.1a). The ECRIS developed by northward-propagating graben formation during the Cenozoic in response to an extensional stress regime in the foreland of the Alpine collision (Ziegler 1990). In the RVRS, rifting began in the late Oligocene (Demyttenaere and Laga 1988; Geluk et al. 1994). The RVRS consists of a series of grabens and horsts that are separated by NW-SE trending normal faults. This system straddles the border area of the Netherlands, Germany, and Belgium, and extends over a distance of ~200 km along strike between the city of Bonn in the SE and the central Rhine–Meuse delta in the NW, and over a distance of up to 75 km across strike (Fig. 13.2). The Campine region, in NE Belgium, is the country's only area that is directly affected by the RVRS, more specifically by its western border fault system.

13.2.1.1 Geometry of the RVRS

The central and northern parts of the RVRS (mainly NW of the Meuse River) have a relatively simple geometry, with the central RVG flanked by parallel, synthetic faults on either side. This graben is ~145 km long, trends NW-SE, and has an asymmetric geometry (Fig. 13.2). To the NE, it is bordered by a relatively straight and narrow fault zone, the Peelrand fault, which has a throw of 400–800 m at the base of the Miocene (Geluk et al. 1994). To the SW, it is bordered by a broader fault zone, the Feldbiss Fault Zone (FFZ), which consists of a number of shorter faults showing a left-stepping pattern. The throw of the base-Miocene horizon on the FFZ is ~400 m (Demyttenaere and Laga 1988). The central sector of the RVG is between 21 and 30 km wide. To the NW, the graben widens and gives way to the West Netherlands Basin (Fig. 13.1a). This widening is accompanied by a westward step of ~15 km of the FFZ, branching of the Peelrand fault into a number of secondary faults, and decreasing throw on both border faults. To the SE of the intersection with the Meuse River, the geometry of the RVRS becomes more complex as it intersects the Variscan thrust front. The FFZ splits into a number of smaller faults stepping progressively towards the graben (Fig. 13.2). At about the same latitude, the eastern border fault bifurcates into the Rurrand fault and the Erft fault, the latter stepping ~15 km to the east. As a result, the RVG narrows, and finally disappears ~20 km SE of the intersection with the Variscan thrust front (Fig. 13.2). The main subsidence is transferred to the adjacent Erft Block, the tectonic block in between the Rurrand and Erft faults, which develops into a half graben and continues for another 30 km SE before dying out in the Rhenish Shield.

13.2.1.2 Neogene Sediments and Depositional Environment

From the Miocene onwards, sedimentation in the Roer Valley Rift System became increasingly restricted to the RVG (Geluk et al. 1994). Outside the central graben, Neogene deposits are maximum 200 m thick, whereas inside the graben, the thickness increases from 500 m in the west to 900 m in the east, locally up to 1000 m in a depocentre in front of the Peelrand fault (Geluk et al. 1994; Worum et al. 2005; Cloetingh et al. 2006). Quaternary deposits in the graben are generally more than 100 m thick, and locally up to 200 m (De Mulder et al. 2003), compared to generally less than 15 m on the western graben flank (Beerten et al. 2005b). The depositional environment of the central RVG evolved from marine (continental shelf) in the Early Miocene, over marine-deltaic (alternating fluvial-deltaic deposition and widespread peat formation in coastal marshes and shallow lakes) in the Middle Miocene through Pliocene, to fluvial in the Pleistocene (Verbeek et al. 2002). Two major rivers, the Rhine and the Meuse, currently flow into the area. The Meuse River drains from the Ardenne and flows more or less straight across the central RVG, while the Rhine River drains from the more distant Alps, and flows further east, almost entirely outside the RVRS. Nearly all surface deposits in the rift system are fluvial terraces of the two rivers, deposited mainly in a periglacial environment during glacial periods (Paulissen 1973; Houtgast et al. 2002).

13.2.2 Morphotectonics

The Roer Valley Rift System is one of the only regions to date in stable continental Europe, where earthquakes can be linked to active faults that can be recognized in the landscape. In the past two decades, paleoseismological investigations have demonstrated that traces of Late Pleistocene and Holocene earthquakes have been recorded in the morphology and shallow subsurface (see Sect. 13.2.3). Bounding the RVG to the SW, the Feldbiss Fault Zone runs NW-SE through Belgian Limburg (Fig. 13.3). Between the Meuse River, defining the Dutch/Belgian border in the SE, and the city of Bree in the NW, it consists of two subparallel

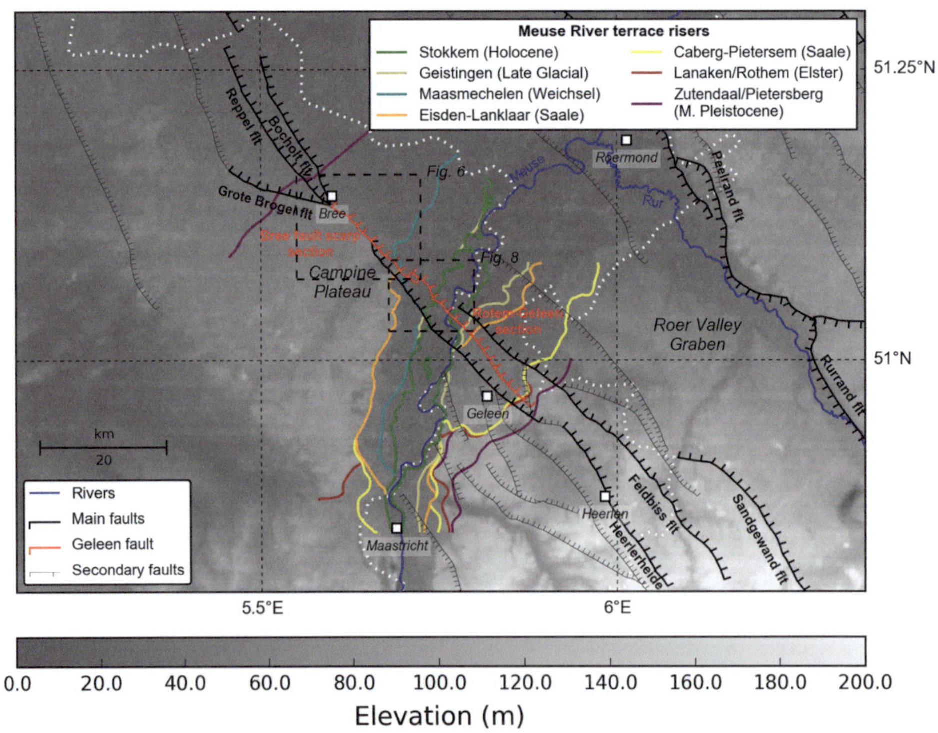

Fig. 13.3 Map of the south-western border fault system of the Roer Valley Graben, showing the two sections of the Geleen fault intersecting the Middle Pleistocene to Holocene sequence of Meuse River terraces (Beerten et al. 2005a; Felder et al. 1989; Houtgast et al. 2002). Topography from ASTER Global Digital Elevation Model (GDEM) v2. ASTER GDEM is a product of NASA and METI

faults, the Rotem fault and the Neeroeteren fault, merging in a single strand near the village of Opitter. The Rotem fault corresponds to the Heerlerheide fault in the Netherlands (Felder et al. 1989), while the Neeroeteren fault, previously thought to connect with the Feldbiss fault in the Netherlands (Paulissen et al. 1985), is in fact continued by the Geleen fault (Beerten et al. 1999). We adopt hereafter the names used in the Netherlands. Seismic reflection data indicate that the Geleen fault has the largest throw of these two faults (Demyttenaere and Laga 1988). NW of Bree, it splits into at least three faults, the Grote Brogel, Reppel and Bocholt faults, among which tectonic displacement is distributed more or less equally (Fig. 13.3).

13.2.2.1 Interplay Between Faulting and Fluvial Dynamics

The current morphology of the Belgian Campine reflects the interplay between active normal faulting and fluvial dynamics. Extensive fluvial terraces have been deposited by the Meuse and Rhine Rivers, mainly during glacial periods. It is suspected (e.g. Gullentops et al. 2001) that activity of the western border faults has attracted both rivers to the area during the Early to Middle Pleistocene, resulting in a thick sequence of fluvial deposits, collectively referred to as the 'Main Terrace' (Paulissen 1973; see also Chap. 10). In the late Early Pleistocene (1 Ma–770 ka), the Rhine deposited coarse, gravelly sands (Sterksel Formation, consisting of the Bocholt Sands in the RVG and Lommel Sands west of the RVG) (Fig. 13.4). In the Cromerian (770–330 ka), the Meuse River deposited a large fan of coarse gravel (Zutendaal Formation) in an area west of the RVG, corresponding to the present-day Campine Plateau, and flowed into the Rhine River, which was still draining the RVG at that time (Beerten 2015). Fault activity thus determined in a general way where river terraces were deposited, and may even have driven shifts of river courses (Kemna 2008). However, long-term fault slip rates (and hence, vertical displacement rates) remained below 0.1 mm/year for all but the most active fault in the RVRS (Vanneste et al. 2013), which is lower than aggradation rates during active phases of terrace deposition. Consequently, the areal extent of individual terraces is scarcely determined by fault activity, as illustrated by the present course of the Meuse, which is perpendicular to both borders of the RVG.

On the other hand, the present elevation of these terraces is the result of fault activity since their deposition. This is for instance evidenced by the age-altitude record of a flight of Meuse River terraces in the footwall of the FFZ in Dutch Limburg (Van den Berg 1994). Note however that the changes in altitude between successive terraces represent footwall uplift and cannot be simply equated with the long-term vertical displacement rate of the FFZ, which can be obtained by comparing terrace altitudes on either side of

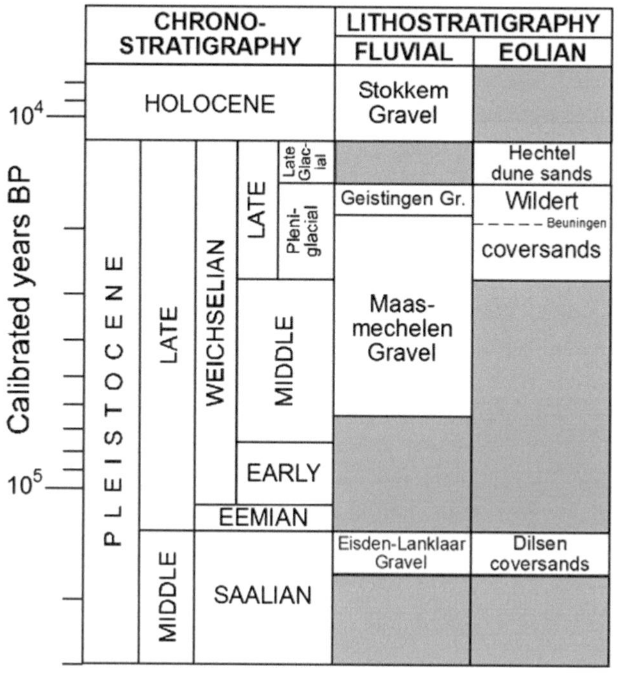

Fig. 13.4 Simplified chronostratigraphic and lithostratigraphic table for fluvial and eolian deposits in the Belgian Campine and Meuse River valley. Modified after Beerten et al. (2005a) and Gullentops et al. (2001). See Chap. 12 for the pre-Saalian stratigraphy

the fault. Based on this terrace record, it could be inferred that around 3 Ma ago, the average uplift rate accelerated from 0.003 to 0.06 mm/year, and has remained relatively constant throughout the Pleistocene (Van den Berg 1994).

13.2.2.2 The Geleen Fault in Belgian and Dutch Limburg

Due to concurrent fault activity and fluvial incision and deposition, the geomorphic expression of faults in the Roer Valley Rift System is determined mostly by the age of the fluvial terraces exposed at the surface. A unique example is the Geleen fault, the main fault of the FFZ and the only fault in the RVRS that intersects a sequence of fluvial terraces with ages spanning the last ~1 Ma (Fig. 13.3). As a consequence, its topographic offset changes along strike, reflecting the different amounts of cumulative vertical offset experienced by the fluvial terraces it runs through. The fault runs NW–SE over a distance of ~27 km between the cities of Bree (Belgium) and Geleen (the Netherlands). Based on geomorphic expression, we can discern three sections: the northern and southern sections are well visible in the landscape, whereas the central section is faint. Along with the Grote Brogel fault, the northern section of the Geleen fault defines the NE edge of the Campine Plateau, an elevated area covered by the Main Terrace of the Meuse (Paulissen 1973). Between Bree and Neeroeteren, the fault appears in the landscape as a very linear, 11.5-km-long and

15–20-m-high scarp, separating the Campine Plateau from the Bocholt Plain (part of the RVG) to the NE. This tectonic escarpment, the most prominent geomorphic feature in the Belgian part of the RVG, has been coined 'Bree fault scarp' (Camelbeeck and Meghraoui 1996), and is described in more detail in Sect. 13.2.2.3.

South of Neeroeteren, the less straight and steep edge of the Campine Plateau turns to a N-S direction, where it becomes an erosional feature determined by the Meuse incision. The Geleen fault continues SE, and its central section between Neeroeteren and Born (the Netherlands) traverses the Meuse River valley, successively intersecting the Weichselian Pleniglacial Maasmechelen terrace, the Holocene floodplain and the Late Glacial Geistingen terrace. In this area, the Geleen fault is associated with a subtle, less than 1-m-high scarp barely visible in the landscape and preserved in only a few places. East of the Meuse valley, the southern section of the Geleen fault runs through successively older Meuse terraces spanning the approximate 130–850 ka interval (Houtgast et al. 2002) (Fig. 13.3). Scarp height increases accordingly, but no longer in a systematic way because the fault gradually dies out and strain is transferred on the overlapping Feldbiss fault (Houtgast et al. 2002). The relation between scarp height of the Geleen fault and terrace age is shown in Fig. 13.5, allowing inferences about the long-term vertical displacement rate of the fault over different periods of time. Based on displaced terraces in Dutch Limburg, a vertical displacement rate of 0.04–0.05 mm/year can be inferred for the overlapping Geleen and Feldbiss faults, which has been essentially constant over the past 1 Ma (Houtgast et al. 2002).

13.2.2.3 The Bree Fault Scarp

The Bree fault scarp (Figs. 13.6, 13.7a) is the best example of an active tectonic lineament in Belgium. Its topographic offset of 15–20 m records continuing activity of the Geleen fault during and after deposition of the Middle Pleistocene Zutendaal Gravel, the base of which shows a total vertical displacement of ~40 m (Beerten et al. 2013). The Campine Plateau to the SW is the uplifted part of this terrace. In the Bocholt Plain, the depression on the downthrown side NE of the scarp, the Zutendaal gravels are replaced with Bocholt sands, due to either erosion (Paulissen 1973) or a sharp lateral transition between Meuse and Rhine sediments (Beerten 2015). In front of the Bree fault scarp, these fluvial gravels are covered by a thick package of eolian sands (referred to as 'coversands') deposited at the end of successive glacial periods. The most recent coversands are the Wildert (~27–15 ka) and Hechtel (~15–12 ka) dune sands (Fig. 13.4). Locally, the Bree fault scarp is intersected by small valleys, such as that of the Itterbeek, where it displays a reduced topographic offset in Holocene alluvial sediments.

Northwest of Bree, the Bree fault scarp splits into three smaller escarpments corresponding to the Grote Brogel, Reppel and Bocholt faults (Fig. 13.3). While the two NNW-trending easterly scarps are only 2–5 m high, the Grote Brogel fault scarp appears to be continuous with the Bree fault scarp, though showing a large bend to the WNW-ESE. Its height decreases from 10 m near Bree, to

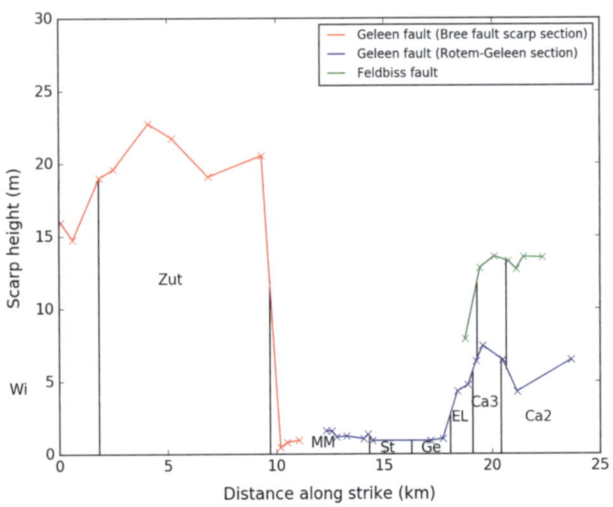

Fig. 13.5 Plot showing along-strike variation of vertical topographic offset (equivalent to fault scarp height) across the Geleen fault and the overlapping part of the Feldbiss fault in relation to the youngest fluvial terrace affected (see Fig. 13.3). *Zut* Zutendaal terrace. *Ca* Caberg-Pietersem terrace. *EL* Eisden-Lanklaar terrace. *MM* Maasmechelen terrace. *Ge* Geistingen terrace. *St* Stokkem terrace

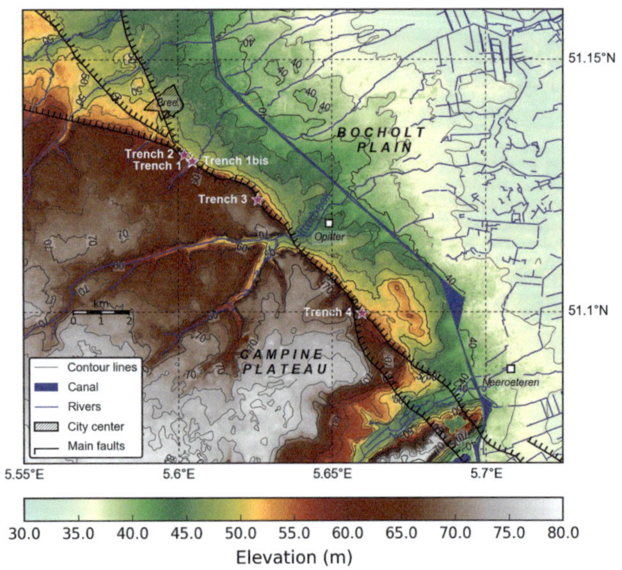

Fig. 13.6 Morphotectonic map of the Bree fault scarp section of the Geleen fault between Bree and Neeroeteren, showing location of the paleoseismic trenches. Topography from DTM-1:10.000 of the Belgian National Geographic Institute

Fig. 13.7 Photos of the Bree fault scarp. *Top* in background, compound fault scarp intersecting the Middle Pleistocene Zutendaal terrace between Bree and Opitter. *Bottom* in the middle ground, single-event scarp intersecting the Wildert coversands, which fill a small valley perpendicular to the fault near Bree

less than 5 m near Grote Brogel. The three scarps disappear gradually towards the northwest.

Careful field observation supported by levelling and aerial photograph analysis evidenced a frontal scarp running along the base of the Bree fault scarp (Camelbeeck and Meghraoui 1998). This scarplet shows a vertical offset of 0.5–3 m across Late Weichselian coversands and a left-stepping en-échelon pattern that suggests a small component of left-lateral motion. As it neither corresponds to any lithological contrast, nor can be of erosional, gravitational or differential settling origin, this frontal scarp has been interpreted as the result of recent coseismic displacements, thus representing a primary geomorphic expression of active faulting. It is best preserved in a small perpendicular valley just south of Bree (Fig. 13.7b). Based on degradation modelling of this frontal scarplet, Camelbeeck et al. (2001) concluded that morphometrically derived scarp ages can only be calculated for the last 14–19 ka (i.e. post Last Glacial Maximum) because the surface expression of older paleoearthquakes in unconsolidated sediments was erased by the strong erosive phase that occurred at the end of the glacial period. Using information from paleoseismic trenches for a 1-m-high scarp, they estimated the mass diffusivity constant between 0.002 and 0.010 m^2/year. However, human perturbations of the surface evidence of recent faulting hampered morphological dating of the scarps in many sites.

13.2.2.4 The Geleen Fault in the Belgian Meuse Valley

Southeast of Neeroeteren, the Bree fault scarp disappears as the Geleen fault enters the Meuse valley (Fig. 13.8). The oldest surface deposits in the valley bottom are late Pleniglacial Wildert coversands overlying middle to late Weichselian fluvial gravel (Maasmechelen gravel). The topographic offset associated with the Geleen fault is therefore not larger than many other landforms such as sand dunes, terrace risers or local drainage features. In combination with centuries-long human modification of the landscape, this explains why the surface trace of the fault has long gone unnoticed in this area. The Geleen fault was first recognized in the valley thanks to electrical soundings that showed a step-like displacement of 8–10 m at the base of the Saalian Eisden-Lanklaar terrace (Paulissen et al. 1985). This buried step was called the Bichterweert scarp. Later, GPR and electric resistivity tomography imaged a fault displacing shallow deposits (including the top of the Maasmechelen Gravel) at places where a subtle <1-m-high scarplet can be observed (Vanneste et al. 2002, 2008a) (Fig. 13.9). These observations confirmed the link between the Bree fault scarp

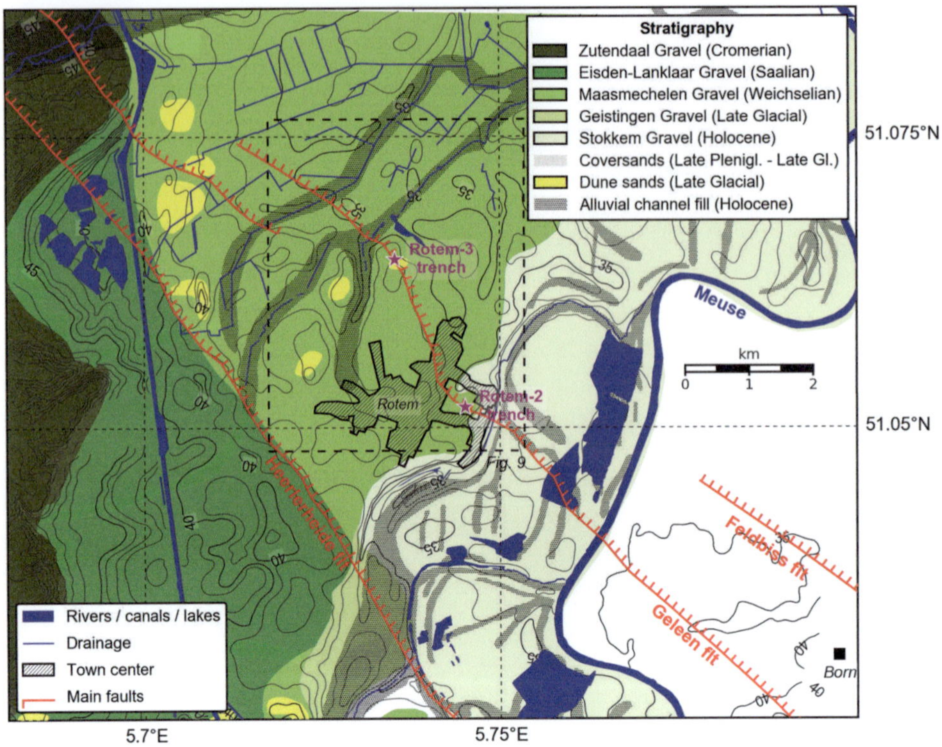

Fig. 13.8 Geomorphic map of the Rotem-Geleen section of the Geleen fault in the Belgian Meuse River valley, showing location of the paleoseismic trenches near the village of Rotem. Stratigraphy redrawn from Beerten et al. (2005a) and Paulissen (1973)

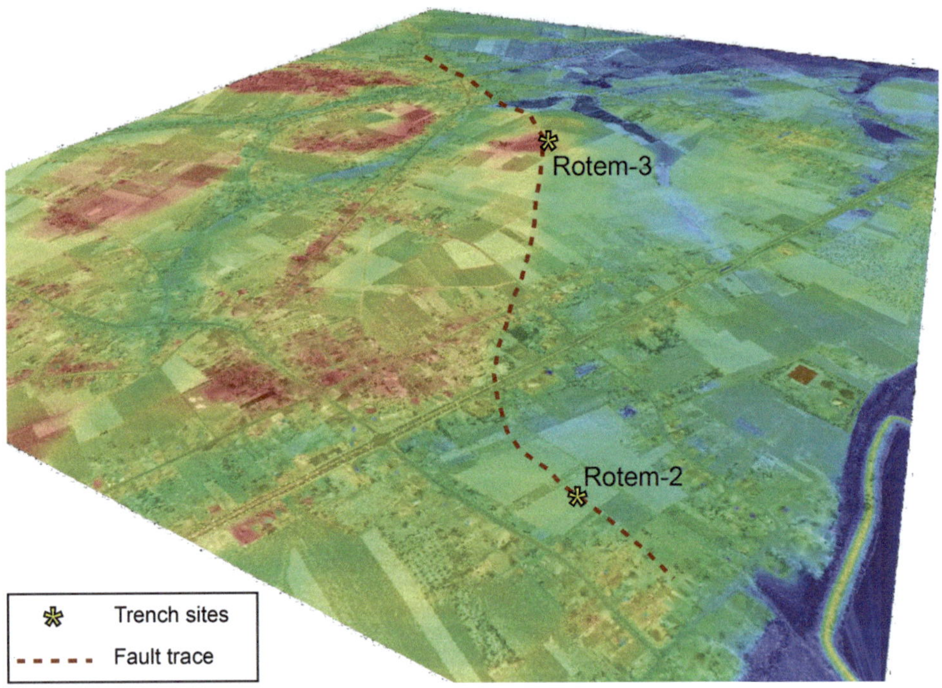

Fig. 13.9 LiDAR-based digital terrain model looking NW and showing the trace, marked by a subtle scarp, of the Rotem-Geleen section of the Geleen fault in the Belgian Meuse River valley near Rotem (see Fig. 13.8 for location). Topographic range of the colour map is approximately 4.5 m (higher in *red*, lower in *blue*). LiDAR data were kindly made available by Agentschap Geografische Informatie Vlaanderen (AGIV)

and the Geleen fault east of the Meuse but also showed previously unrecognized complexity, notably with a ~500-m-wide left stepover north of the village of Rotem, associated with two sharp ~30° bends in between which the fault trends NNW-SSE for ~2 km (Figs. 13.8 and 13.9). This stepover separates two fault sections of 11.5 km (Bree fault scarp) and 16.5 km (Bichterweert scarp and its SE continuation in the Netherlands). It is not yet clear whether the stepover represents a segment boundary (Camelbeeck et al. 2007), but its limited width makes this unlikely.

13.2.3 Active Faulting and Paleoseismology

In 1996, the Royal Observatory of Belgium (ROB) started a pioneering programme to identify and study active faults in the Roer Valley Rift System, focusing on the western border faults of the RVG in Belgium. Before this time, the general consensus was that faulting in the RVRS is largely aseismic (Ahorner 1966, 1975, 1983). The successful identification of past surface-rupturing earthquakes requires pinpointing the 'event horizon' (Pantosti et al. 1993), typically associated with stratigraphic evidence such as upward termination of fault splays and fissures, colluvial wedges resulting from degradation of a fault scarp, angular fault zone unconformities, stepwise downward increasing fault offsets, and buried soils preserved on the downthrown side of a fault (McCalpin 1996).

Since 1996, seven paleoseismic trenches have been investigated across the Geleen fault in Belgium, providing the first evidence for Holocene/Late Pleistocene surface-rupturing earthquakes with magnitude larger than ~6.3 in stable continental Europe. Parallel research was conducted on the Dutch part of the Geleen and Feldbiss faults in Dutch Limburg, which led Houtgast et al. (2003, 2005) to a different interpretation.

13.2.3.1 The Bree Fault Scarp

Between 1996 and 2000, five trenches were excavated at four different sites across the Bree fault scarp (Camelbeeck and Meghraoui 1996, 1998, 2001; Meghraoui et al. 2000; Frechen et al. 2001) (Fig. 13.6). The main results are summarized from two sites in contrasted settings, respectively located in a small transverse valley close to Bree (trenches 1 and 1bis) and an interfluve near Opitter (trench 4). The former site allows characterizing the most recent faulting event (MRE), while the latter covers a longer time span with multiple faulting events.

At site 1, a 1- to 1.5-m-high scarp runs through silty sand filling the valley bottom. This scarplet is cut by a small creek incised in the sands on the upthrown fault side. Trenches 1 and 1bis (Fig. 13.10a) exposed poorly stratified silty sands interlayered with two well-defined pebble beds. These pebble beds are sharply displaced by a single fault splay. The upper pebble bed shows a vertical offset of 0.5–0.65 m across the fault, reaching ~ 1 m if warping is included. A gley soil that developed in the coversands above this pebble bed is preserved only in the hanging wall. In trench 1, the fault is truncated by a silty-sandy soil discernible below the plough layer. This soil thus represents the event horizon of the MRE. While an initial radiocarbon age of the MRE between AD 610 and AD 890 (Camelbeeck and Meghraoui 1996; Camelbeeck et al. 1998) was probably biased by contamination, IRSL luminescence dating in trench 1bis yielded an average age of 16.8 ± 1.8 ka for three samples below the upper pebble bed, and an average age of 15.4 ± 2.3 ka for four samples above this bed (Frechen et al. 2001) (Fig. 13.10a), indicating that the dated sands belong to the Late Weichselian Pleniglacial Wildert coversands (Fig. 13.4) and have not been reworked by the local creek. These ages further allow correlating the pebble bed with the Beuningen horizon, a desert pavement dating from the Last Glacial Maximum (~ 14–19 ka BP, Kolstrup 1980). Unfortunately, the event horizon identified in trench 1 could not be observed in trench 1bis. Instead, a decayed plant root piercing the fault plane without being displaced yielded a ^{14}C age of 1020–1400 cal BC, indicating that the fault has not moved since that time. The dating results from trenches 1 and 1bis thus do not allow constraining the age of the MRE better than sometime between 15.4 ± 2.3 ka and 1020–1400 cal BC.

Trench 4 (Fig. 13.10b) exposed a well-defined single fault splay near the base of the main escarpment, juxtaposing deeply weathered cross-bedded gravelly coarse sand with erosional channels in the footwall against various units of lighter coloured silty sand with varying amounts of gravel in the hanging wall. The gravelly sands in the footwall correspond to the Middle Pleistocene Zutendaal Gravel, and are thrown down several metres below the trench bottom in the hanging wall. The sands in the hanging wall are mostly coversands of various ages, partly reworked. The top shows a characteristic succession of structureless silty coversand overlying an irregular pebble bed and laminated coversand. The two upper units also extend on the footwall, where they cover the Zutendaal Gravel. They show an abrupt vertical displacement of 0.12 m across the fault. However, taking into account their overall flexuring, the vertical offset amounts to 0.80 m, in better agreement with the height of the frontal scarp. Although it was not possible to pinpoint the event horizon, secondary soil horizons ('Bt bands') within the uppermost coversand unit appeared to be displaced, suggesting that the faulting event probably occurred in the late Holocene. The lower laminated coversand extends only a few metres onto the footwall before being eroded by the pebble bed, and both its base and internal stratification show a displacement of 0.45 m. There is thus a stepwise increase in displacement below the pebble bed, indicating that the latter represents the event horizon of a second faulting event with a vertical offset of 0.33 m. No colluvial wedge is observed. IRSL dating yielded ages between 9.6 ± 1.9 and 11.4 ± 1.8 ka for the silty sand above the event horizon and of 12.0 ± 1.6 and 22.2 ± 2.5 ka for the lower laminated sand (Frechen et al. 2001). The age of the silty sand was further confirmed by ^{14}C dating at 8450–8160 cal BC and 11,200–10,650 cal BC. The pebble bed defining the penultimate event horizon is therefore interpreted to correlate with the Beuningen horizon.

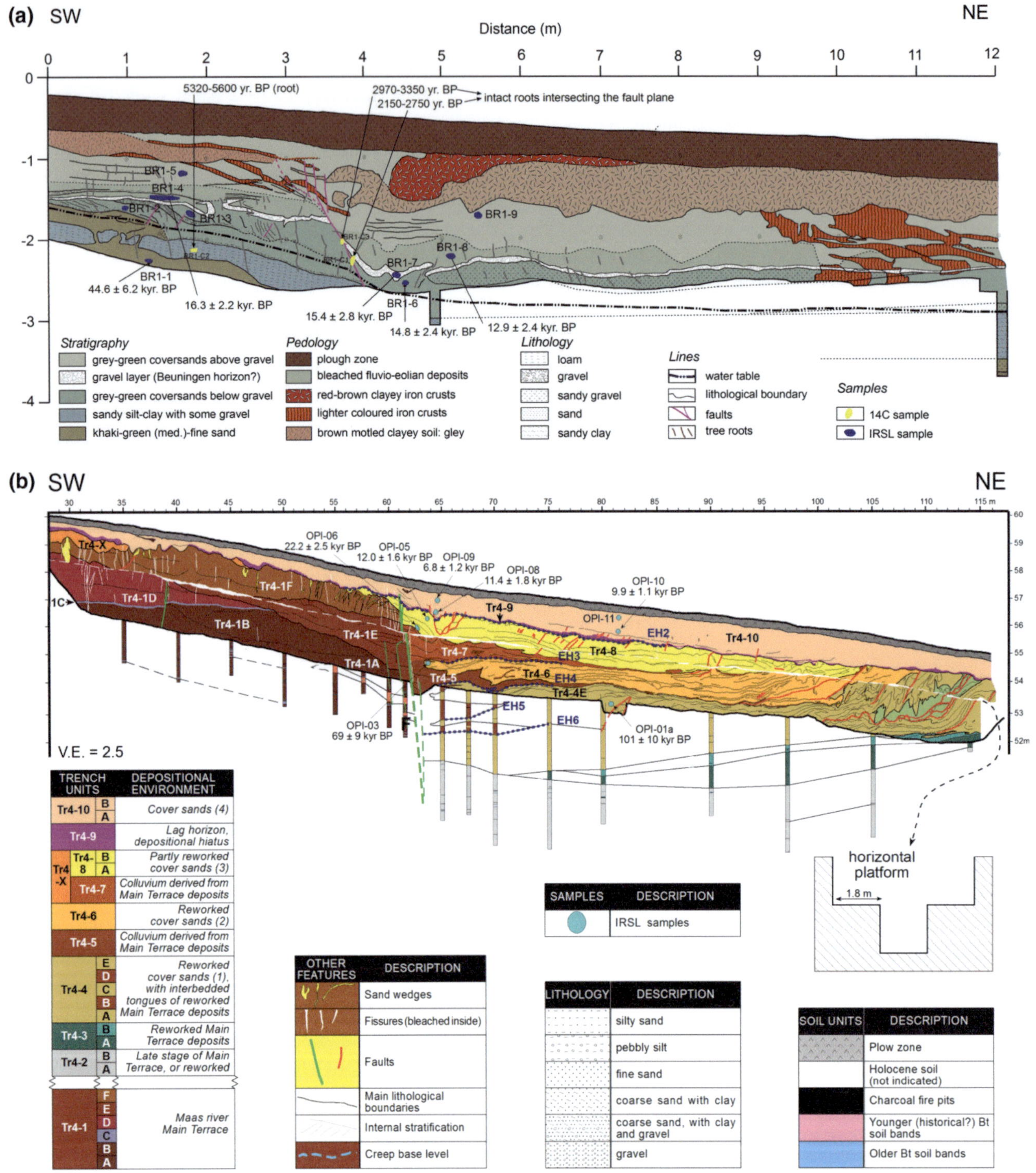

Fig. 13.10 **a** Paleoseismic log of trench 1bis across the Bree fault scarp section of the Geleen fault near Bree, showing faulted Late Glacial coversands and the pebble bed interpreted to correlate with the Beuningen horizon. The event horizon could not be identified due to soil development. Redrawn from Camelbeeck et al. (2007). IRSL ages from Frechen et al. (2001). **b** Paleoseismic log of trench 4 across the Bree fault scarp section of the Geleen fault near Opitter, showing Zutendaal Gravel deposits in the footwall juxtaposed against Weichselian coversands and footwall-derived colluvium in the hanging wall. Unit Tr4-9 is a pebble bed interpreted to correlate with the Beuningen horizon. *Blue dashed lines* indicate event horizons EH2 to EH6 (EH1 could not be recognized). Redrawn from Vanneste et al. (2001). IRSL ages from Frechen et al. (2001)

Deeper in the hanging wall, two older units of reworked coversand containing a significant amount of gravel are separated from each other and from the upper coversand by two wedge-shaped units of dark brown clayey coarse sand. These wedges are up to 1 m thick close to the fault, but rapidly taper downslope. Although their material is hardly distinct from the Zutendaal Gravel deposits in the footwall, the wedges are distinguished by their subhorizontal internal stratification. Thin fine sand layers in the upper wedge can be correlated with subvertical coversand-filled frost wedges at the top of the Zutendaal Gravel in the footwall. Progressive downslope bending of these frost wedges towards the fault is unequivocal evidence that the wedge-shaped units in the hanging wall originated by downslope transport of footwall material, strongly suggesting that they are colluvial wedges accumulated under periglacial conditions in response to scarp formation by surface-rupturing earthquakes (Vanneste et al. 2001). Two more wedges were encountered in boreholes below the trench bottom. Altogether, these colluvial wedges thus record four faulting events with individual displacements between 0.3 and 1 m. IRSL dating of the lowermost coversand in the trench indicates that three wedges are younger than 101 ± 10 ka (Frechen et al. 2001). Trench 4 thus revealed the longest record of surface faulting to date along the Bree fault scarp, with six successive paleoearthquakes, five of which occurred after the Eemian interglacial.

13.2.3.2 Trenches in the Meuse Valley

Two trenches were excavated in the Meuse valley, both situated on the Weichselian Maasmechelen terrace, respectively east and north of the village of Rotem (Fig. 13.8). The first one (Rotem-2), located close to the limit with the Holocene floodplain exposed coarse sandy gravel, showing a complex internal structure with several erosional surfaces and sand lenses. This gravel is overlain by a 1.40–1.80-m-thick sequence of silty sands and sandy silts whose lower part consists of laminated yellowish silty sands similar to the sand in the lenses in the gravel below, while the upper part is siltier and structureless. These sands (~17–12 ka) most likely correspond to the upper part of the Wildert coversands, grading upwards into latest Pleniglacial to Late Glacial loess. The second trench (Rotem-3, Fig. 13.11) was situated on the NNW-trending section of the Geleen fault close to the step-over with the Bree fault scarp. Here, coarse gravel is only present at the bottom of the footwall, covered by a sequence of horizontally laminated sands alternating with cross-stratified coarser sands and silt bands. These sands, interpreted as Wildert coversands partly reworked by local drainage, are 1.5–2-m-thick in the footwall. They are truncated by a thin, continuous pebble bed, similar to the Beuningen horizon in the trenches along the Bree fault scarp. Above this pebble bed, sands with interlayered silts, 1.60-m-thick in the footwall, are interpreted as equivalent to the upper sands in trench Rotem-2. In both cases, a 1-m-thick banded Bt soil developed in these

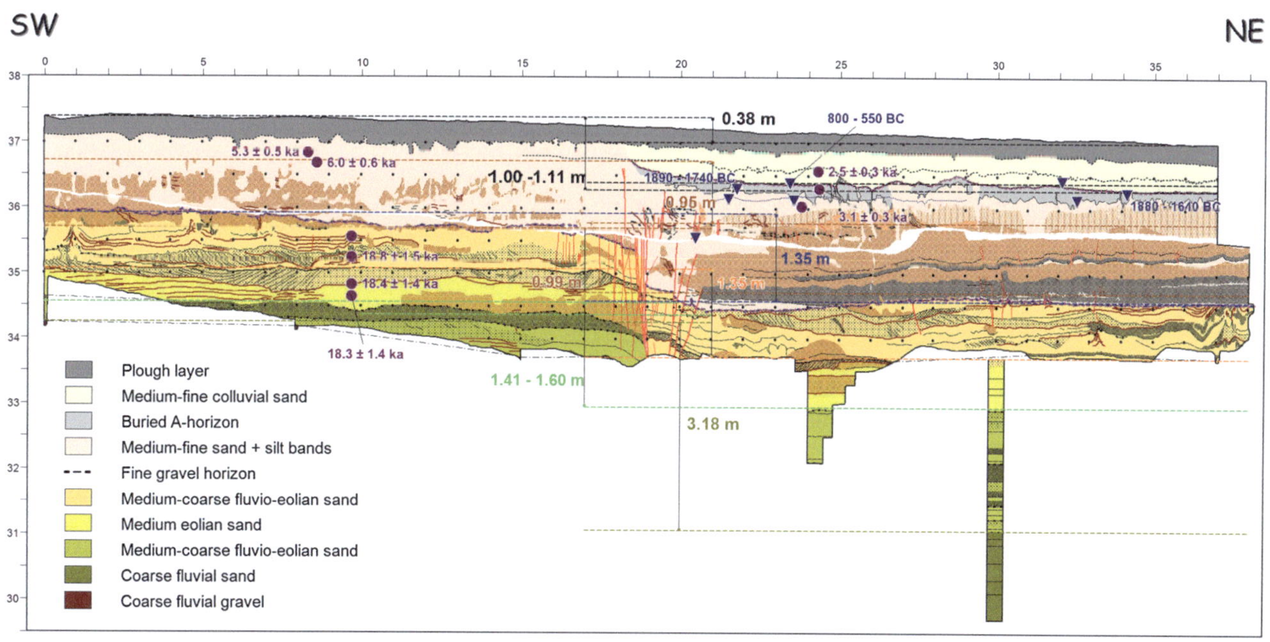

Fig. 13.11 Paleoseismic log of trench Rotem-3 across the Rotem-Geleen section of the Geleen fault in the Belgian Meuse River valley. *Pink dashed lines* indicate event horizons of the two most recent surface-rupturing earthquakes. The most recent event is evidenced by a buried soil horizon overlain by colluvium in the hanging wall, the penultimate event by a discrete increase in offset (from ~1.0 to ~1.4 m) below a pebble bed interpreted to correlate with the Beuningen horizon. *Purple circles* correspond to OSL samples, *blue triangles* to radiocarbon samples. Redrawn from Vandenberghe et al. (2009)

coversands. The gravel in trench Rotem-2 is likely contemporaneous with the Beuningen horizon, whereas the basal gravel in trench Rotem-3 should correspond to an older phase of the Maasmechelen terrace.

The trenches in the Meuse valley allowed better identification and dating of the most recent faulting event. In both trenches, the entire sequence, including the post-depositional soil, is thrown down by a well-defined subvertical fault. In trench Rotem-2, gravel and coversand show a 0.75–1 m vertical offset. In trench Rotem-3 (Fig. 13.11), the Beuningen pebble bed and the younger coversands are offset by ~1 m. Although field determination of the upward termination of the fault splays was difficult, thin sections revealed that a faint horizon in the hanging wall corresponds to the limit between in situ soil below and colluvium derived from this soil above, and hence represents the event horizon (Vanneste et al. 2008b). The coseismic nature of this event is furthermore attested by associated liquefaction features (sand and gravel dikes). In trench Rotem-3, the Beuningen pebble bed truncates older soft-sediment deformation features, while several lower stratigraphic levels show an incremental offset of 0.4 m with respect to the layers above (Fig. 13.11). This indicates that the Beuningen pebble bed defines the event horizon of a second paleoearthquake, similar to the observation made in trench 4 along the Bree fault scarp. The age of the MRE was here consistently constrained by OSL (2.5 ± 0.3 to 3.1 ± 0.3 ka BP) and ^{14}C (785–1765 cal BC). The penultimate event was dated between 15.9 ± 1.1 and 18.2 ± 1.3 kyear BP by OSL (Vandenberghe et al. 2009).

13.2.3.3 Concluding Overview of 120 kyear of Surface-Rupturing Earthquakes Along the Geleen Fault

Paleoseismic investigations in the Belgian Campine provide information on the timing and recurrence of large earthquakes on one of the border faults of the RVG, the Geleen fault. The length of the paleoseismic record depends on the age of the geomorphic surface of the investigated site. The youngest surfaces are formed by Late Weichselian Wildert coversands covering either the Maasmechelen terrace or the Beuningen pebble horizon. At these sites, the fault scarp is subtle, and only the most recent one or two faulting events are recovered in paleoseismic trenches. Elsewhere, where the fault affects the Middle Pleistocene Zutendaal Gravel, the scarp is more pronounced, and the paleoseismic record is much longer.

Along the Bree fault scarp, up to six paleoearthquakes were identified in one trench, five of which occurred since the Eemian interglacial. The three most recent paleoearthquakes could be correlated along the entire scarp, each causing vertical displacements of 0.5–1.0 m. Though not well constrained, the MRE most likely occurred during the (late?) Holocene. Strong indications for the coseismic nature of faulting were found in the form of colluvial wedges and various types of soft-sediment deformation (Vanneste et al. 1999). However, the paleoseismic studies also faced some problems that are directly or indirectly related to the slow rate of deformation: evidence for the MRE is situated at shallow depth and obscured by soil development, as a result of which the event horizon could not be identified; the tectonic signal is overprinted by a strong climatic signal (transition from periglacial to temperate conditions); dating resolution rapidly decreases for older events, etc.

In the Meuse valley, the age of the two latest paleoearthquakes was better constrained. Thin sections evidencing the contact between colluvium and in situ soil allowed identifying the event horizon of the MRE more reliably. OSL dating provided the best constrained ages of the two latest events on the Geleen fault, between 2.5 ± 0.3 and 3.1 ± 0.3 ka BP and between 15.9 ± 1.1 and 18.2 ± 1.3 ka BP, respectively. The interval between the two events has a two-sigma range of 11,800–16,800 year, which is currently our best estimate of the recurrence interval on any fault in the RVRS.

The section of the fault we investigated in the Meuse valley is separated from the Bree fault scarp by a 500-m wide stepover. The ages and amount of displacement determined for the two paleoearthquakes in the Meuse valley are in relatively good agreement with those obtained in the trenches along the Bree fault scarp. This raises the possibility that the Geleen fault defines a single, 27-km-long rupture segment, which would be capable of producing M 6.7 earthquakes. This is also suggested by the rather large offset (~1 m) observed for the MRE close to the stepover, whereas decrease in offset is generally expected towards segment boundaries. However, the stratigraphic and dating resolutions are not sufficient to exclude the possibility of separate ruptures of the two fault sections in somewhat smaller ($M \sim 6.3$) earthquakes within a short span of time. As the Geleen fault represents only ~5% of the more than 550 km of active faults in the RVRS, paleoseismic investigations on the other faults in the system are a necessity if we want to extend our knowledge of the long-term seismic activity beyond the limits of historical and instrumental earthquake catalogues, improve our understanding of fault behaviour and interaction, and ultimately the assessment of seismic hazard.

13.3 Active Tectonics Outside the Roer Valley Graben: The Hockai Fault Zone

13.3.1 Geographical and Geological Setting

The Hockai Fault Zone (HFZ) (Ahorner 1983; Demoulin 1988) is a major reactivated fault system of eastern Belgium that runs in a SSE-NNW direction across the NE part of the Ardenne massif and its northern foreland in the Pays de

Herve, parallel to, but ~25 km away from, the border faults to the west of the southern part of the RVG (Fig. 13.2). Geologically, its existence is attested as far back in time as the basal Frasnian, when it belonged to a limit of tectonic blocks called the Verviers-Trier lineament (Dvorak 1973; Thorez and Dreesen 1986). In any case, it formed then one of the main systems among the numerous steeply dipping faults that were active in the area during the Stephanian—early Permian phase of wrench-induced collapse of the Variscan orogen, chiefly accommodating dextral shear at that time (Ziegler and Dèzes 2006). Later, the HFZ was episodically reactivated during the Mesozoic and the Cenozoic. In particular, the faults limiting a small graben developed along the HFZ in the Pays de Herve displaced the basement by up to 300 m horizontally and 48 m vertically, while the Mesozoic cover underwent 23 m of vertical motion (Ancion and Evrard 1957). Since the Oligocene and the opening of the Lower Rhine Embayment (RVRS), the HFZ is believed to be acting essentially as a normal fault (with a possible subordinate strike-slip component), behaving like the faults inside this rift segment in response to the stress field imposed upon NW Europe mainly by the Alpine collision.

At shallow depths, the HFZ cuts across varied bedrock. In the south, its surface trace is first manifested by a sharp S-bend in the southern edge of the Caledonian Stavelot massif, opposing Lower Devonian arkoses and shales on its eastern wall and Upper Cambrian quartzites and phyllites on the western wall. Then, it runs across the Stavelot massif itself, where it also crosses the eastern part of the small Permian graben of Malmédy, putting a clear-cut limit to the bulk of the graben fill and raising the base of the few Permian remains to the east. To the north, the HFZ is continued through the Devonian and Carboniferous sandstones, schists and limestones of the Vesdre massif and the Pennsylvanian schists of the Herve massif. In this area, as stated above, it also affects the Cretaceous cover and a parallel fault seems even to have displaced Late Pleistocene loess (Graulich 1959).

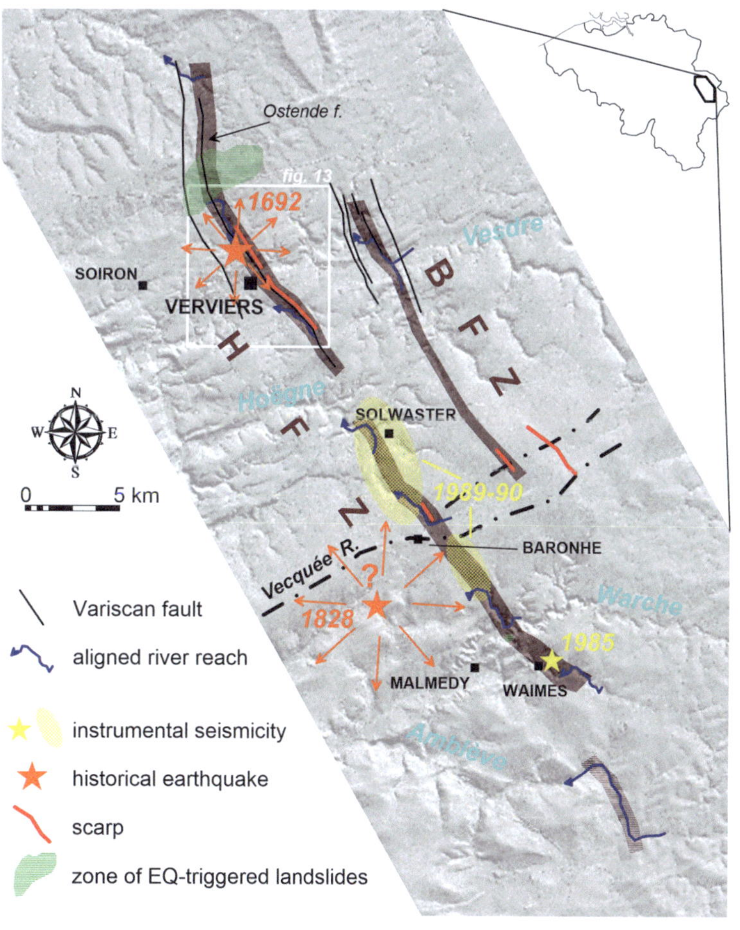

Fig. 13.12 Extent and general layout of the Hockai Fault Zone as mapped from geomorphological, seismological and geological evidence. *Dark brown areas* denote the morpho- and seismotectonic trace of faults. Historical and instrumental seismicity as well as a dated landsliding event support recent activity of the fault zone. The degree of scarp freshness points to Plio-Quaternary fault slip. *HFZ* Hockai Fault Zone. *BFZ* Baelen Fault Zone

13.3.2 Morphotectonics

Geomorphological evidence allows the HFZ to be mapped on a total length of 42 km, from the upper Amblève near Amel in the south to a few kilometres farther than the Battice ridge in the north (Demoulin 2006) (Fig. 13.12). It also suggests that the fault zone is divided in several segments separated by changes of strike, en-échelon stepovers, or gaps, where no sign of faulting has been observed. In particular, such a ~5-km-long gap, probably superposed to a right stepover, makes a main separation between the well-documented northern and southern parts of the HFZ, east and north of Verviers, and between Solwaster and Waimes, respectively. Noteworthily, the Baelen fault zone, a parallel feature ~5 km east of the HFZ, though smaller, also displays a geomorphic signature attesting its Cenozoic, if not even Plio-Quaternary, activity. We present hereafter the various landforms that support the HFZ mapping and allow first inferences about its recent activity.

13.3.2.1 Scarps

Fault scarps provide the best morphotectonic evidence of slip on a normal fault. However, owing to the low slip rates of the faults involved in the evolution of the Lower Rhine Embayment, we cannot expect these scarps to be more than a few tens of metres high, even though they may be preserved for a very long time in the Paleozoic basement. Indeed, the climatic conditions prevailing during the Quaternary also favoured rapid degradation of newly created steep scarps even in many Paleozoic rocks, whose pre-existing fracturation (especially in the fault zone) led to efficient gelifraction. Therefore, fault scarps, in fact rather subdued fault line scarps, are in general not so conspicuous in the landscape as to be easily identified in the hodgepodge of steep slopes related to fluvial erosion or to lithological contrasts that characterize the animated topography of the northern margin of the Hautes Fagnes plateau.

Linear features can therefore be interpreted as fault scarps only if they are definitely not controlled by river incision or lithology, and especially when they cut interfluves and valleys at right angle. Such probable fault scarps are observed at different scales along the HFZ. On one hand, tens of metres high, up to km-long scarps mark the surface trace of the fault zone and bear witness to its longer term (10^5–10^7 years) activity; on the other hand, smaller, often discontinuous and steeper scarps a few metres in height, which might reveal coseismic surface ruptures, are indicative of much more recent (10^2–10^5 years) fault movement.

The most continuous tectonic scarps may be followed across all interfluves running to the WSW between the Vesdre and small creeks' valleys, just to the north and east of the city of Verviers (Fig. 13.13 and 13.14). Together, they define a ~7-km-long linear fault segment. To the SE of Verviers, the Sècheval scarp is the highest, with a height of 40 m and a slope of 16° to the west. The height decreases rapidly from the centre of the scarp towards its ends. Interestingly, the geomorphological trace of the HFZ disappears where the new geological map of Wallonia also stops mapping any fault south of the Sècheval scarp (Laloux et al. 1996). To the north, a 150-m-wide left relay ramp leads to another, less prominent, scarp reaching 20–25 m in height at Stembert and displaying slopes of 7–10°. North of the Vesdre valley, the mean height of the successive scarps keeps diminishing progressively: 25 m at Bôtister, where the scarp slopes by ~13° to the west, 12 m at Mont-Dison, with a 4° slope, before it dies out (except perhaps where its topographic expression is locally exacerbated along the valley side of the Dison creek). In any case, 2 km farther north, the flat summit of the major Battice ridge displays no measurable vertical offset near the zone where the HFZ is assumed to cross it.

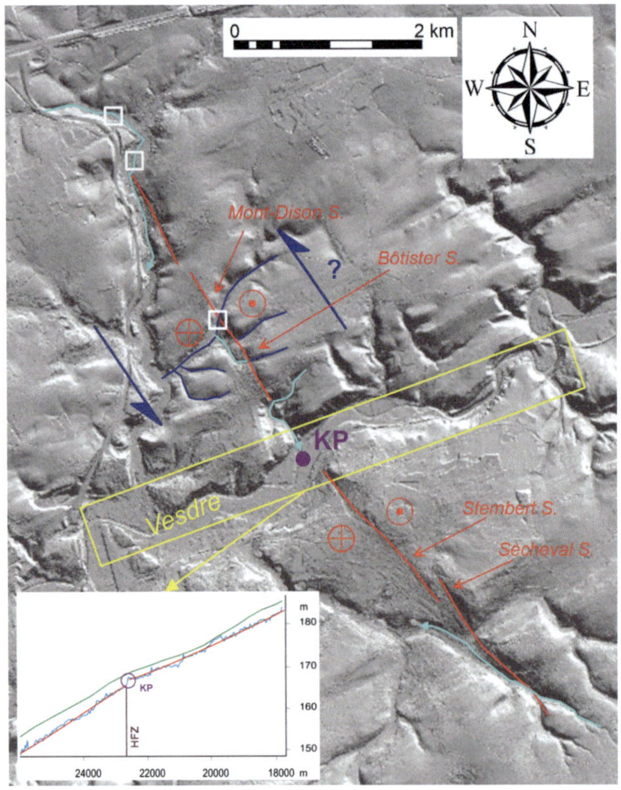

Fig. 13.13 Excerpt of the shaded LIDAR Digital Terrain Model of the Walloon region zooming on the Verviers segment of the HFZ (see Fig. 13.12). Aligned scarps (in *red*) and a knickpoint-type discontinuity (KP) in the Vesdre long profile (*inset*: Vesdre channel bottom profile in *blue*, smoothed in *red*; floodplain top profile in *green*) evidence the vertical component of fault slip in the morphology. While several stream reaches (*light blue*) exploit the weakened substrate of the fault zone, other creeks (*dark blue*) are laterally offset at the crossing of the fault, suggesting a component of left-lateral slip. *White squares* locate electrical resistivity tomographies with steps in the bedrock-sediment contact that might suggest recent (Holocene?) fault displacement

Fig. 13.14 Photos of the fault scarps along the Verviers segment of the HFZ. **a** Looking south across the Vesdre valley, Stembert scarp southeast of Verviers. **b** Looking north across the Vesdre valley, Bôtister scarp north of Verviers. **c** Looking farther north, Mont Dison scarp across the next interfluve. See scarp location in Fig. 13.13. *White dashed lines* enhance topography across the scarps

Beyond these large features, the Verviers fault segment, whose seismogenic character is highly probable from historical data (see below), shows also two small scarps that might have been associated to Middle or Late Quaternary fault activity. The first one has a length of 100 m, a maximum height of 5 m, and a maximum slope of 30°. It locally defines the steep lower part of the Bôtister scarp. This scarplet is developed where the fault cuts across homogeneous, rather resistant silt- and sandstones and it disappears laterally where these rocks give way to less resistant shales and nodular limestones. However, the downslope dip direction of the strata suggests that, rather than having directly caused the development of a lithological scarp, bedrock variations might actually have controlled the differential preservation of a recent scarplet of true tectonic origin. Nevertheless, a strictly erosive origin is equally plausible because it seems surprising that surface ruptures would have repeatedly affected only this small length of the fault line to create the scarplet. Another scarplet was also mentioned by Demoulin and Pissart (1999) close to the northern end of the Verviers segment. Extending to ~800 m in length, it is perfectly aligned with the fault trace and shows a ~2-m-high vertical offset.

Except across the main ridge prolonging the Hautes Fagnes Plateau to the southwest, reliable fault scarps are rarer along the southern half of the HFZ (Fig. 13.12). The connection between the Hautes Fagnes summit of Baraque Michel and this so-called Vecquée ridge occurs through a first NW-striking scarp that pertains to the Baelen fault zone. Though only gently sloping west ($\sim$2.5°), this 30-m-high scarp is clearly distinct from the otherwise horizontal summit and ridge tops. Three kilometres farther southwest, a second scarp of similar height and aspect exists at the HFZ—Vecquée intersection. As the Hoëgne River has deeply cut through the crushed rocks of the fault line, this scarp is only deduced from the difference in elevation of the horizontal ridge on each side of the valley. However, the ridge's general morphology suggests that its slope is not significantly different from that of the other scarp. Moreover, sparse observations of scarplets have been tentatively interpreted as possible hints of more recent activity of the fault in this area too. One such $\sim$1-m-high scarplet is more or less continuously developed over 1200 m in the bottom of the Hoëgne valley at Parfondbois-Solwaster (Demoulin 2006). However, despite preliminary geophysical reconnaissance, it remains unclear whether this feature is of tectonic or erosive origin. Likewise, Lecocq (2011) identifies a faint scarp parallel to the HFZ on the ridge at Baronhé, 3 km to the west of the main fault trace. While the tectonic origin of this scarplet is supported by its location in the surface prolongation of a seismologically mapped fault (see below), its gentle slope to the east suggests that it might represent the topographic expression of a minor, antithetic element of the fault zone.

Although all aligned large and small scarps consistently delineate the HFZ over a distance of 20 km and attest its activity as a fault with a significant normal component, the variable degree of freshness of their shape allows for much less convincing inferences about their age. It is obvious that the large scarps cut in the resistant quartzites of the Vecquée ridge required a very long time to be degraded down to slopes of 2°–3° and their evolution most probably encompasses the whole Neogene. By contrast, the scarplets point to much younger events because such small features cannot preserve slopes higher than 20° for more than a few 10^4 years, but their tectonic origin has still to be demonstrated, let alone their seismogenic character. In between, owing to their undoubted causal link with the HFZ, the large scarps around Verviers offer a more concrete, though not yet realized, opportunity for dating the activity of this fault segment. However, deriving ages from the scarp morphology is not that straightforward here because (1) the scarps integrate multiple events and (2) they are cut in coherent rocks. Making the assumption of dynamic equilibrium between scarp slope and constant long-term slip rate, one might get rate evaluations from a diffusive approach, but overcoming issue (2) is more problematic because reliable soil production rates are crucial in modelling weathering-limited systems and they probably varied significantly under the changing climates of the Plio-Quaternary. In this respect, Bovy et al. (2016a, b) recently modelled the evolution of soil thickness on Ardennian hillslopes during a glacial-interglacial cycle and provided first estimates of the diffusivity and soil production rate parameters for this area that are promising for dating purposes in the near future.

13.3.2.2 Hockai Fault Zone and Drainage Network

Drainage systems have two virtues in morphotectonic studies, on one hand passively adapting to, and revealing, the underlying structures, on the other hand recording active tectonics in their deformed components. Both virtues may be exploited here to further document the surface trace of the HFZ and analyse its recent activity. The passive exploitation of the HFZ crushed rocks ('passive' being meant here from the fault's standpoint) is observed all along its mapped trace, with river courses locally deflected by the fault and up to 3.5-km-long river reaches lodged in the damage zone (Fig. 13.12). The northward deflection, for varying distances, of all westward flowing Ardennian streams crossing the fault (with the exception of the Vesdre but including the paleo-Warche before capture, see below) most probably reflects the north tilting of the whole area during the Neogene rather than betraying any dextral shear component of the HFZ. In particular, complex capture phenomena bear witness to the role of the HFZ in facilitating the erosion of small streams competing with larger rivers. The story of the most spectacular case of the Warche—Trô Maret double capture, south of the Vecquée ridge, is told by Rixhon and Demoulin in Chap. 11 of this book. Briefly, while the paleo-Warche followed the trace of the HFZ, it was captured between 74 and 90 ka and deviated to the southwest by a small stream rapidly eroding the Permian conglomerates of the Malmédy graben. Then, owing to the presence of highly fractured rocks in the HFZ, a very small tributary of the new Warche was able to incise a $\sim$110-m-deep, 2-km-long gorge into the resistant quartzites of the Cambrian Stavelot massif and caused a further capture of the upper Trô Maret within a few tens of thousands of years. A comparable, though much older, event occurred also on the northern side of the Vecquée ridge where, for similar reasons, the upper Hoëgne was captured by a small stream eroding up the northern margin of the massif during the Pliocene (Demoulin 1986).

Plan view features of the drainage system that would suggest active deformation by the HFZ are rare, the only significant example being provided by the systematic offset of several creeks at their crossing of the Verviers segment north of the Vesdre valley, pointing to a possible 100–150 m left-lateral displacement of the fault (Fig. 13.13). More meaningful information may be expected from the analysis of the terrace profiles of the Vesdre valley, which flows from east to west, perpendicularly to the HFZ. Two recent studies of the Vesdre terraces based, respectively, on field observation (Petermans et al. 2004) and extraction of terrace data from a DEM (Demoulin et al. 2007) reached fairly similar conclusions about deformation across the HFZ, namely suggesting that the 2–2.5-Ma-old highest terrace level would have been vertically offset by 10–14 m (Demoulin et al. 2007) to 15–20 m (Petermans et al. 2004), in agreement with the only slightly larger offsets recorded by the interfluve scarps. The two studies only diverge with respect to the estimated timing of the HFZ Quaternary activity, which relies on their respective reconstruction of the terrace staircase. While Petermans et al. (2004) believed that the HFZ was active through the Quaternary but showed increased slip rate from the Middle Pleistocene, Demoulin et al. (2007) suggested that, as all Middle Pleistocene terrace levels are similarly displaced, after a Pliocene(?)–early Quaternary phase of activity, the HFZ was quiescent during the Middle Pleistocene and that it was reactivated only recently. Interestingly, the modern longitudinal profile of the river also reveals a slight break of slope between two concave sections up- and downstream of the HFZ. According to Demoulin (2006), in the HFZ zone, the bedrock crops out in many places of the bottom of the channel incised into Weichselian gravels thicker in the hanging wall than in the footwall of the fault (i.e. the top surface of the floodplain deposits is undisturbed across the HFZ), indicating that the corresponding ~2-m displacement occurred during the Weichselian. In any case, though still lacking numerical dating, the current understanding of the Vesdre profile deformation across the HFZ leaves little doubt about the fault's Quaternary activity.

13.3.2.3 Earthquake-Triggered Landslides

Beyond direct evidence on the fault, geological information about ruptures having caused violent earthquakes may be gained from a range of indirect effects of ground shaking, and especially landslides and rockfalls. In the Pays de Herve, a set of large landslides occurred on the moderate slopes exposing the Cretaceous clays of the Vaals Formation, in close proximity to the Ostende fault pertaining to the HFZ (Fig. 13.12). Available ^{14}C dating suggests they were initiated around 150 AD. We refer to Chap. 20, where Dewitte et al. provide detailed description and analysis of these landslides and we give here only a brief overview of their characteristics that strongly support a seismic origin (Demoulin et al. 2003): (1) they all lie within 3 km from the active Ostende fault; (2) liquefaction of the Cretaceous Aachen sands was the primary cause of landsliding in the overlying Vaals clays; (3) stability analysis of the concerned slopes shows that their failure required a dynamic trigger; (4) they are large features (up to 22 ha) affecting the upper part of the slopes; (5) they probably occurred simultaneously; (6) no extreme climatic event or wet period during the last two millennia was able to cause more than very limited reactivation of the existing landslides.

However, these landslides are especially important because they raise an intriguing issue about the seismogenic activity of the HFZ. Indeed, the great Verviers earthquake of 1692, which we detail hereafter, occurred at the end of a very wet summer season (likely implying saturated ground) and struck violently the area where the landslides were already present, without being apparently able to cause even their slightest reactivation. Unless evidence of extended landsliding in 1692 is found, this might call into question the validity of one of the two equally sound conclusions of the seismic origin of the existing landslides or the location of the 1692 rupture on the Verviers segment of the HFZ. The high-resolution Lidar DTM of Wallonia (SPW 2014) recently shed new light on this issue by highlighting a large landslide that had so far got unnoticed below its dense forest cover. This fresh-looking ~8-ha landslide affects the southern valley side of the Warche valley in front of the Trô Maret confluence right on the surface trace of the southern HFZ (Mreyen et al. 2016). It has a probable seismic origin, which requires however confirmation by a more complete analysis.

13.3.3 Seismicity

Independent information supporting the morphotectonically inferred Quaternary activity of the HFZ is provided by seismological data. In the case of the HFZ, which has been the focus of seismological observation and research since 1980s, the detailed instrumental record covering the last three decades is completed by extensive studies of the region's historical seismicity that led to the revised attribution of the great earthquake of 1692, ascribing it to the HFZ (Vogt 1984; Alexandre et al. 2008).

13.3.3.1 Instrumental Seismicity

A few microearthquakes having been recorded near Solwaster in the beginning of the autumn of 1989, the ROB deployed a set of mobile seismographs in the area, which allowed the record of a swarm of ~500 microearthquakes in the following six months, with maximum magnitude M_L = 2.4 (Camelbeeck 1993) (Fig. 13.12). Beyond attesting the current activity of the HFZ, this seismic event was meaningful in that, combined with fault plane solutions of the largest events, the location of about 100 hypocentres clearly described the geometry of the fault and the associated mechanism. The ruptured segment extends over 12 km across the core of the Cambrian Stavelot massif, with an azimuth of N330°E and a dip of 70° to the east, steepening to nearly vertical in the upper 6 km of the crust. Most earthquakes occurred at depths between 6 and 9 km, responding to oblique fault slip with a predominant left-lateral strike-slip displacement and a subordinate normal component. However, upward extrapolation of the imaged fault geometry makes it intersect the ground surface 2–3 km to the west of the main trace of the HFZ, where it coincides with the location of the Baronhé scarplet. This lateral position within the fault zone and the east

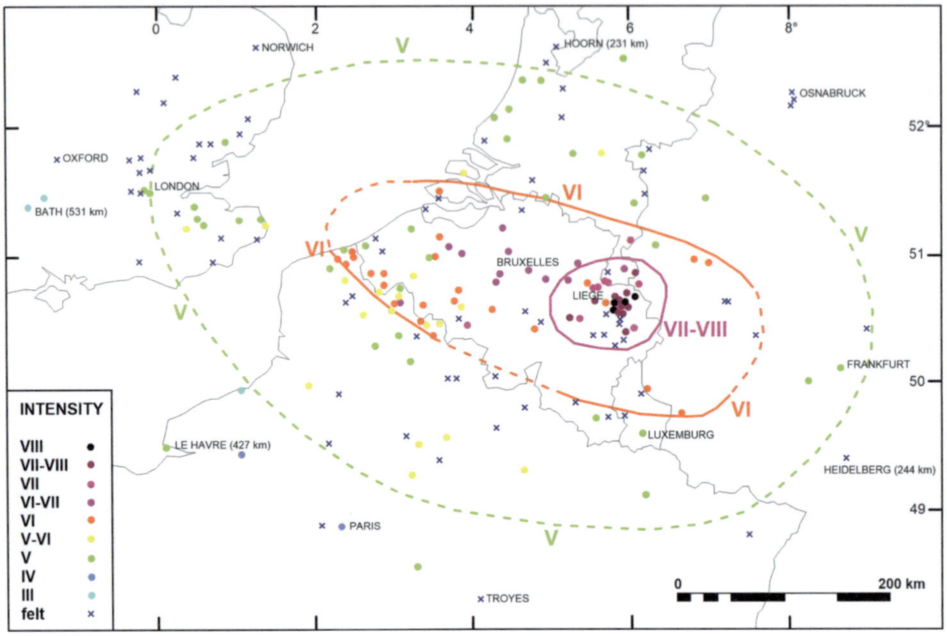

Fig. 13.15 Macroseismic intensity map of the Verviers 1692 earthquake, based on the EMS-98 macroseismic scale adapted to the historical context. The *coloured dots* denote localities where reliable sources allowed an intensity to be estimated (redrawn after Alexandre et al. 2008)

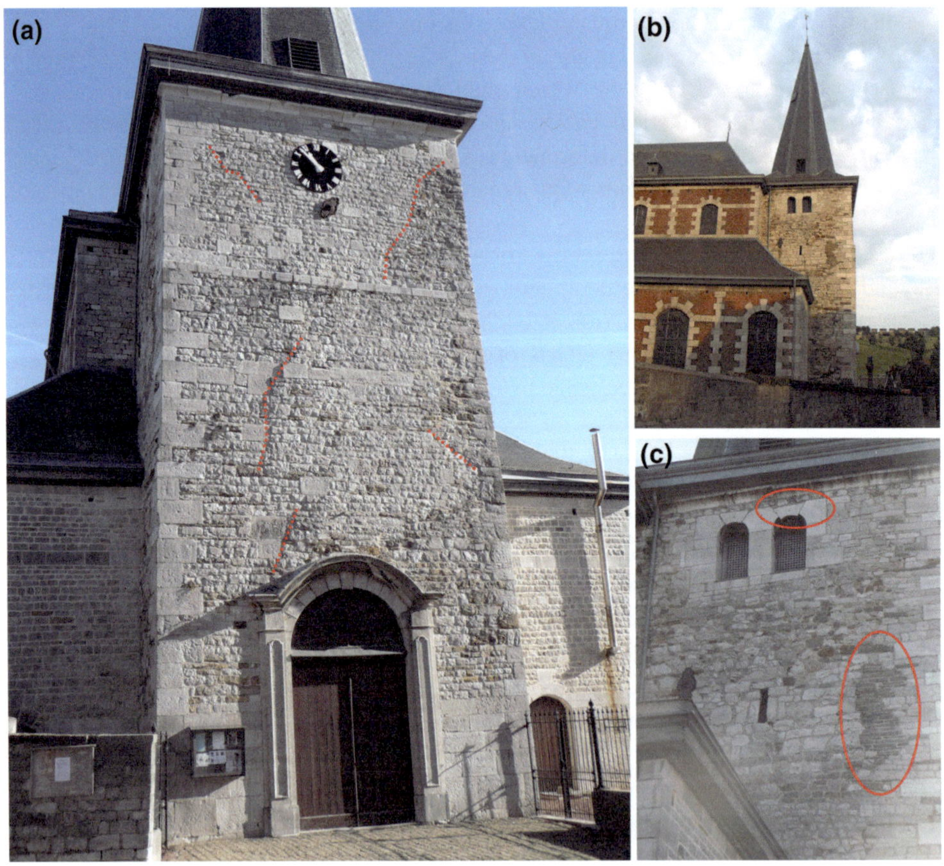

Fig. 13.16 Architectural evidence of the Verviers 1692 earthquake. In the village of Soiron (see location in Fig. 13.12), the church was heavily damaged. While the nave was destroyed by the quake and rebuilt in typical eighteenth century style (**b**), the tower also was severely affected, with diagonal crack formation in the front wall (**a** *red dotted lines*) and damage especially along the *right edge*, where the original squared quoins (still visible along the *left edge*) have been replaced by rawer material during reconstruction (**a**). **c** Close-up of the *upper part* of the *left-lateral wall* showing further damage (different repair material, sagged keystone)

dip of the segment, opposite to the topographic gradient defining the main scarp, strongly suggest that it corresponds to a small antithetic fault.

Incidentally, an isolated $M_L = 2.9$ earthquake was also recorded close to the fault trace between Waimes and Malmédy on 12 May, 1985, confirming that the southern part of the HFZ is currently participating in the regional seismic activity.

13.3.3.2 The Verviers 1692 Earthquake

Though now identified as one of the strongest historical earthquakes north of the Alps, the Verviers earthquake of 18 September, 1692, was poorly studied and mistakenly located by earthquake catalogues in central Belgium or even SE England until recently. After Vogt (1984) had suggested its relocation near the city of Verviers, in eastern Belgium, Alexandre carefully analysed innumerable historical accounts that brought definitive support to this conclusion and strong arguments for a moment magnitude M_w between 6 and 6.3 (e.g. Alexandre et al. 2005, 2008).

Historical documents attesting the effects of the 1692 event in many localities situated within 10 km from the city of Verviers clearly point to a zone of greatest damage centred on this town (Fig. 13.15). A smaller disconnected zone of heavy damage is also identified near Stavelot, ~25 km to the south of Verviers. In these two areas, many monumental buildings—mainly castles and churches—and numerous houses suffered major structural damage often leading to partial or, more rarely, total collapse (Camelbeeck et al. 2014) (Fig. 13.16), and ground shaking was violently felt by people outside the buildings. A frightened crowd gathered in one church of the city was even convinced of having witnessed a miracle in this occasion, a stone statue of the Madonna having purportedly changed her attitude to calm her son's anger, and this made the city a place of vivid pilgrimage (Alexandre and Kupper 1997) until far into the twentieth century. These data locate the macroseismic epicentre of the earthquake a few kilometres north of Verviers (Alexandre et al. 2008), in closest vicinity to the surface trace of the Verviers segment of the HFZ.

Although the limited description of many damages, the lack of information about percentages of damaged houses, and the somewhat ill-defined vulnerability of traditional stone buildings in the area affect intensity assessments, the epicentral macroseismic intensity is reliably estimated to VIII, while the whole Belgian territory and slightly beyond experienced intensity VI or more, and the intensity V area had an average macroseismic radius of 250 km and a ~340 km radius in the ESE-WNW direction of maximum elongation, probably related to ground motion amplification over the London-Brabant massif and its Meso-Cenozoic cover (Nguyen et al. 2004). Although very little information is available for localities where intensity was less than V, we know that the earthquake was felt at least as far as in East Anglia and Oxford in England, Normandy and Champagne in France, N Netherlands, and Heidelberg and Osnabrück in Germany. The most remote mention identified so far of local effects of the earthquake comes from the English town of Bath, 530 km away from the epicentre.

Macroseismic intensity data also offer an indirect means of evaluating an earthquake's magnitude. The magnitude of the Verviers 1692 earthquake has been estimated by comparing its spatial distribution of intensities with the intensity distribution of modern earthquakes, whose magnitude is instrumentally determined. Various methods based either on isoseismal radii or individual intensity data converged in an estimated moment magnitude of 6¼ (Camelbeeck et al. 2007), making the Verviers event an historical equivalent of the paleoearthquakes that ruptured the Bree segment of the Feldbiss fault (Camelbeeck and Meghraoui 1998, Camelbeeck et al. 2000).

Recently, based on historical report analysis, the macroseismic epicentre of another earthquake that struck east Belgium on 3 December, 1828, was located in the Hautes Fagnes area, suggesting that it might have taken place somewhere on the southern half of the HFZ (Knuts et al. 2015). Intensity data distribution suggests a magnitude $M_w = 4.2 + 0.4/-0.2$.

13.4 Synthesis: Seismotectonics in Eastern Belgium

In this chapter, we presented and analysed the Bree fault scarp and the HFZ as two remarkable morphotectonic features. We conclude by focusing on their importance in the understanding of the seismotectonic context of this part of stable continental Europe.

13.4.1 The Bree Fault Scarp

Earthquake fault plane solutions in the RVG show mostly dip-slip normal faulting along NW-SE striking faults (Ahorner 1975; Camelbeeck and van Eck 1994; Camelbeeck et al. 2007). This mechanism is in agreement with a NW-SE maximum horizontal compressive stress inducing tensional forces perpendicular to the graben axis, which explains the extensional crustal deformation manifested in landforms and the geological record. Earthquake locations also attest to this close relationship between seismic activity and the graben structure.

The Bree fault scarp and its prolongation in the Meuse Valley bottom are the morphological expression of the Geleen fault, part of the fault system bounding the RVG to the west. This 15- to 20-m-high escarpment defines the border between the uplifted Campine Plateau, covered by the Main Terrace of the Meuse, and the subsiding RVG, where the Main Terrace equivalent, dated 300–770 ka, is downthrown by ~40 m and is buried below Late Pleistocene alluvial and eolian sediments. Despite its limited amplitude, it has all characteristics of active normal fault scarps.

The ~1–m-high frontal scarplet is clearly identified in many places along the main escarpment. Trench observations show that the amount of Late Holocene slip, marked by a single step in the stratigraphy, is fairly similar to the scarp offsets, suggesting they were produced by the surface rupture of a large earthquake on the Geleen fault.

Based on trench data, Camelbeeck et al. (2007) provided an estimation of the fault slip rate and return period of large earthquakes along the Bree fault scarp. Considering only the two most recent complete earthquake cycles that can be correlated along the entire scarp, average return period is 13.7 ± 7.8 kyear and fault slip rate 0.050 ± 0.036 mm/year. Using the trench 4 record, average return period is estimated to 22.7 ± 4.3 kyear and fault slip rate to 0.031 ± 0.012 mm/year over the last 100 kyear, i.e. figures not significantly different from the shorter term estimates.

The maximum credible earthquake, that is, the largest potential earthquake along a fault or in a specific region within the current tectonic setting, should correspond to that associated with the largest seismogenic source in this region (Camelbeeck et al. 2007). The magnitude of this earthquake is defined as M_{max}, which is an important parameter in seismic hazard studies, and may be constrained for individual faults using paleoseismic data in combination with empirical relationships between magnitude, fault length and amount of fault slip (Wells and Coppersmith 1994). Camelbeeck et al. (2007) discussed various plausible hypotheses for the Bree fault scarp and its SE extension resulting in M_{max} values ranging from 6.3 ± 0.3 to 6.7 ± 0.3.

13.4.2 The Hockai Fault Zone

The seismic activity in the western part of the Rhenish Massif, west of the RVG, is less important than in the graben itself. The HFZ is the only unquestionable alignment of earthquake epicenters suggesting the existence of an important seismogenic structure in this area. Therefore, if we except this fault zone, the regional seismic activity can be understood in terms of tectonic strain release in a 3-D inhomogeneous crustal volume, lending support to the HFZ as seismogenic source of the large 18 September 1692 earthquake.

Earthquake mechanisms in this part of the Ardenne range from pure normal faulting on NW-striking faults to quasi pure strike-slip faulting on NNW-SSE or ENE-WSW faults (Camelbeeck et al. 2007). Based on this variety of mechanisms, Camelbeeck and van Eck (1994) suggested a change of the tectonic stress field with respect to that in the RVG, with a 30° counterclockwise rotation of the horizontal maximum and minimum stresses. This N290°E orientation of S_{Hmax} only slightly greater in amplitude than the vertical stress component is in agreement with the dominant left-lateral strike-slip mechanism observed during the 1989–1990 seismic sequence on the HFZ (Camelbeeck 1993).

Surprisingly, this is also consistent with the long-term displacement on the Ostende fault, a part of the northern HFZ that displays a post-Paleozoic left-lateral movement with a significant normal slip component (Ancion and Evrard 1957). However, not only would not it make much sense to derive average slip rates from post-Paleozoic movements that probably concentrated in specific tectonic phases (for instance, most of the 300 m left-lateral displacement noted on the Ostende fault was possibly accommodated during wrench tectonic phases in the Stephanian–Early Permian and/or Late Jurassic–Early Cretaceous; Ziegler 1990; Ziegler et al. 2006; Ziegler and Dèzes 2007) but there is also no clear evidence that these very long-term horizontal and vertical movements were systematically coupled into oblique slip rather than pertaining to distinct times of fault activity. Restricting such rate considerations to the post-Oligocene deformation of the erosion surface offset by ~30 m between the Vecquée ridge and the Baraque Michel high is hardly more meaningful but, even if we make the reasonable assumption that most of this displacement occurred in Plio-Quaternary times, when the NE Ardenne was uplifted, this yields a very low vertical fault slip rate of ~0.006 mm/year. Finally, the shortest term view we can obtain from the highest terrace of the Vesdre, for which the maximum estimate of vertical offset is 15–20 m since the end of the Tertiary, gives the same figure. As for horizontal slip rates, no reliable data are available so far to provide estimates.

Such low slip rates explain why the HFZ is less conspicuous in the morphology than the Bree fault scarp. Moreover, even within a given tectonic phase, the seismic activity of plate interiors is episodic in nature (Stein 2007). Consequently, considering the present earthquake activity rate on the HFZ, the recurrence of an $M \approx 6.2$ earthquake such as that of Verviers in 1692 should be in the order of 250 kyear. Though highly unlikely, a rupture involving the whole length of the HFZ (42 km) might produce an M 7.0 earthquake.

References

Ahorner L (1966) Erdbeben und jüngste Tektonik im Braunkohlenrevier der Niederrheinischen Bucht. Zeitschrift der deutschen geologischen Gesellschaft 118:150–160

Ahorner L (1975) Present-day stress field and seismotectonic block movements along major fault zones in Central Europe. Tectonophysics 29:233–249

Ahorner L (1983) Historical seismicity and present-day microearthquake activity of the Rhenish Massif, Central Europe. In: Fuchs K, von Gehlen K, Mälzer H, Murawski H, Semmel A (eds) Plateau Uplift. Springer, Heidelberg, pp 198–221

Alexandre P, Kupper JL (1997) Le tremblement de terre de 1692 et le miracle de Notre-Dame des Récollets à Verviers. Feuillets de la Cathédrale de Liège, 28–32, p 43
Alexandre P, Kusman D, Camelbeeck T (2005) Le tremblement de terre du 18 septembre 1692 dans le nord de l'Ardenne (Belgique). Impact sur le patrimoine architectural. In: Levret A (ed) Actes des VIèmes Rencontres du Groupe APS "Archéosismicité et Vulnérabilité. Environnement, bâti ancien et société", Perpignan, 4–5 Oct 2002, pp 1–10
Alexandre P, Kusman D, Petermans T, Camelbeeck T (2008) The 18 September 1692 earthquake in the Belgian Ardenne and its aftershocks. In: Fréchet J, Meghraoui M, Stucchi M (eds) Historical seismology. Interdisciplinary studies of past and recent earthquakes. Springer, Berlin, pp 209–230
Ancion C, Evrard P (1957) Contribution à l'étude des failles Monty, Mouhy et d'Ostende dans la partie occidentale du massif de Herve. Ann Soc géol Belg 80:B477–B488
Beerten K (2015) Quaternary stratigraphy of the Belgian part of the Roer Valley Graben: lithology, petrology and depositional environments. In: Belqua 2015 annual meeting, Brussels, Belgium
Beerten K, Vandenberghe N, Gullentops F, Paulissen E (2005a) Technisch verslag bij de Quartairgeologische kaart van België. Vlaams Gewest, Kaartblad Rekem (26), p 52
Beerten K, Vandenberghe N, Gullentops F, Paulissen E (2005b) Technisch verslag bij de Quartairgeologische kaart van België. Vlaams Gewest, Kaartblad Maaseik (10–18), p 48
Beerten K, Brabers P, Bosch P, Gullentops F (1999) The passage of the Feldbiss Bundle through the Maas Valley. Aardkundige Meded 9:153–158
Beerten K, De Craen M, Wouters L (2013) Patterns and estimates of post-Rupelian burial and erosion in the Campine area, north-eastern Belgium. Phys Chem Earth 64:12–20
Bovy B, Braun J, Demoulin A (2016a) A new numerical framework for simulating the control of weather and climate on the evolution of soil-mantled hillslopes. Geomorphology 263:99–112
Bovy B, Braun J, Demoulin A (2016b) Soil production and hillslope transport in mid-latitudes during the last glacial-interglacial cycle: a combined data and modelling approach in northern Ardennes. Earth Surf Proc Landf (in press)
Camelbeeck T (1993) Mécanisme au foyer des tremblements de terre et contraintes tectoniques: le cas de la zone intraplaque belge. Unpublished PhD thesis, Université catholique de Louvain, p 343
Camelbeeck T, Meghraoui M (1996) Large earthquakes in northern Europe more likely than once thought. EOS Trans, Am Geophys Union 77:405–409
Camelbeeck T, Meghraoui M (1998) Geological and geophysical evidence for large palaeoearthquakes with surface faulting in the Roer graben (northwestern Europe). Geophys J Int 132:347–362
Camelbeeck T, Van Eck T (1994) The Roer Valley Graben of 13 April 1992 and its seismotectonic setting. Terra Nova 6:291–300
Camelbeeck T, Alexandre P, Vanneste K, Meghraoui M (2000) Long-term seismicity in regions of present day low seismic activity: the example of western Europe. Soil Dyn Earthq Eng 20:405–414
Camelbeeck T, Martin H, Vanneste K, Verbeeck K, Meghraoui M (2001) Morphometric analysis of normal faulting in slow-deformation areas: examples in the Lower Rhine Embayment. Neth J Geosci 80(3–4):95–107
Camelbeeck T, Vanneste K, Alexandre P, Verbeeck K, Petermans T, Rosset P, Everaerts M, Warnant R, Van Camp M (2007) Relevance of active faulting and seismicity studies to assessments of long-term earthquake activity and maximum magnitude in intraplate northwest Europe, between the Lower Rhine Embayment and the North Sea. In: Stein S, Mazzotti S (eds) Continental intraplate earthquakes: science, hazard, and policy issues. Geological Society of America Special Paper 425, pp 193–224
Camelbeeck T, Alexandre P, Sabbe A, Knuts E, Garcia Moreno D, Lecocq T (2014) The impact of the earthquake activity in Western Europe from the historical and architectural heritage records. In: Talwani P (ed) Intraplate earthquakes. Cambridge University Press, pp 198–230
Cloetingh S, Cornu T, Ziegler P, Beekman F, Environmental Tectonics (ENTEC) Working Group (2006) Neotectonics and intraplate continental topography of the northern Alpine Foreland. Earth-Science Reviews 74:127–196
De Mulder E, Geluk M, Ritsema I, Westerhoff W, Wong T (2003) De Ondergrond van Nederland. Wolters-Noordhoff, Groningen
Demoulin A (1986) Un phénomène de capture dans les Hautes Fagnes: la Hoëgne à Hockai. Bull Soc belge d'Etudes géographiques 1986–1:45–51
Demoulin A (1988) Cenozoic tectonics of the Hautes Fagnes plateau (Belgium). Tectonophysics 145:31–41
Demoulin A (2006) La néotectonique de l'Ardenne-Eifel et des regions avoisinantes. Mémoires de la Classe des Sciences, Académie Royale de Belgique, 25, p 252
Demoulin A, Pissart A (1999) Les glissements de terrain du Pays de Herve. Convention 99/42075 du Ministère de la Région Wallonne (DGATLP), unpublished report, p 146
Demoulin A, Pissart A, Schroeder C (2003) On the origin of late quaternary palaeolandslides in the Liège (E Belgium) area. Int J Earth Sci 92:795–805
Demoulin A, Bovy B, Rixhon G, Cornet Y (2007) An automated method to extract fluvial terraces from digital elevation models: The Vesdre valley, a case study in eastern Belgium. Geomorphology 91:51–64
Demyttenaere R, Laga P (1988) Breuken-en isohypsenkaarten van het Belgisch gedeelte van de Roerdal Slenk: eerste resultaten van een seismisch onderzoek in het gebied van Poppel-Lommel-Maaseik. Professional papers. Geological Survey of Belgium, 234, p 20
Dèzes P, Schmid S, Ziegler P (2004) Evolution of the European cenozoic rift system: interaction of the Alpine and Pyrenean orogens with their foreland lithosphere. Tectonophysics 389:1–33. doi:10.1016/j.tecto.2004.06.011
Dvorak J (1973) Die Quergliederung des Rheinischen Schiefergebirges und die Tektogenese des Siegener Antiklinorium. Neues Jb Geol Paläontol Abh 143:132–152
Felder W, Bosch P, Bisschops J (1989) Geologische kaart van Zuid-Limburg en omgeving, schaal 1:50000. Afzettingen van de Maas, Rijks Geologische Dienst
Frechen M, Vanneste K, Verbeeck K, Paulissen E, Camelbeeck T (2001) The deposition history of the coversands along the Bree Fault Escarpment, NE Belgium. Geol Mijnbouw 80:171–185
Geluk M, Duin E, Dusar M, Rijkers M, van den Berg M, van Rooijen P (1994) Stratigraphy and tectonics of the Roer Valley Graben. Geol Mijnbouw 73:129–141
Graulich J (1959) Sur le prolongement méridional du graben de La Minerie (Pays de Herve). Ann Soc Géol Belg 83:55–61
Grünthal G, Wahlström R, Stromeyer D (2009) The unified catalogue of earthquakes in central, northern and northwestern Europe (CENEC)—updated and expanded to the last millennium. J Seismolog 13:517–541. doi:10.1007/s10950-008-9144-9
Gullentops F, Bogemans F, De Moor G, Paulissen E, Pissart A (2001) Quaternary lithostratigraphic units (Belgium). Geol Belg 4:153–164
Heidbach O, Tingay M, Barth A, Reinecker J, Kurfeß D, Müller B (2010) Global crustal stress pattern based on the world stress map database release 2008. Tectonophysics 482:3–15. doi:10.1016/j.tecto.2009.1007.1023

Houtgast R, Van Balen R, Brouwer L, Brand G, Brijker J (2002) Late quaternary activity of the Feldbiss Fault Zone, Roer Valley Rift System, the Netherlands, based on displaced fluvial terrace fragments. Tectonophysics 352:295–315

Houtgast R, Van Balen R, Kasse C, Vandenberghe J (2003) Late quaternary tectonic evolution and postseismic near-surface fault displacements along the Geleen Fault (Feldbiss Fault Zone—Roer Valley Rift System, the Netherlands), based on trenching. Geol Mijnbouw 82:177–196

Houtgast R, Van Balen R, Kasse C (2005) Late quaternary evolution of the Feldbiss Fault (Roer Valley Rift System, the Netherlands) based on trenching, and its potential relation to glacial unloading. Quatern Sci Rev 24:489–508

Kemna H (2008) Pliocene and lower pleistocene fluvial history of the Lower Rhine Embayment, Germany: examples of the tectonic forcing of river courses. Quatern Int 189:106–114

Knuts E, Camelbeeck T, Alexandre P (2015) The 3 December 1828 moderate earthquake at the border between Belgium and Germany. J Seismolog. doi:10.1007/s10950-015-9535-7

Kolstrup E (1980) Climate and stratigraphy in northwestern Europe between 30,000 and 13,000 b.p., with special reference to the Netherlands. Mededelingen Rijks Geologische Dienst 32(15):181–253

Laloux M, Dejonghe L, Geukens F, Ghysel P, Hance L (1996) Limbourg-Eupen 43/5-6. Carte géologique de Wallonie, échelle: 1/25.000. Notice explicative. Ministère de la Région Wallonne, DGRNE, p 192

Lecocq T (2011) Seismic Activity in the Ardenne and its relationship with Active Tectonics. *Unpublished PhD thesis, Université libre de Bruxelles*

McCalpin J (1996) Paleoseismology. Int Geophys Ser 62:588 (Academic Press, San Diego)

Meghraoui M, Camelbeeck T, Vanneste K, Brondeel M, Jongmans D (2000) Active faulting and paleoseismology along the Bree Fault Zone, Lower Rhine Graben (Belgium). J Geophys Res 105B:13809–13841

Mreyen A, Havenith H, Fernandez-Steeger T (2016) Geophysical investigation of landslides and fault scarps in the Hockai Fault Zone, Belgium. Geophys Res Abstr 18:2350

Nguyen F, Teerlynck H, Van Rompaey G, Van Camp M, Jongmans D, Camelbeeck T (2004) Use of microtremor measurements for assessing site effects in Northern Belgium—interpretation of the observed intensity during the $M_s = 5.0$ June 11, 1938 earthquake. J Seismolog 8:41–56

Pantosti D, Schwartz D, Valensise G (1993) Paleoseismology along the 1980 surface rupture of the Irpinia fault: implications for earthquake recurrence in the southern Apennines, Italy. J Geophys Res 98B:6561–6577

Paulissen E (1973) De morfologie en de kwartairstratigrafie van de maasvallei in belgisch Limburg. Verhandelingen van de Koninklijke Academie voor Wetenschappen, Letteren en Schone Kunsten van België, Klasse der Wetenschappen

Paulissen E, Vandenberghe J, Gullentops F (1985) The Feldbiss fault in the Maas valley bottom (Limburg, Belgium). Geol Mijnbouw 64:79–87

Petermans T, Camelbeeck T, Alexandre P, Kusman D, Verbeeck K, Vanneste K, Demoulin A, Nguyen F, Jongmans D (2004) The September 18, 1692 earthquake in the Belgian Ardenne and its geologic context. Final Report, SAFE European Project (EVG1-2000-00023), WP2, p 43

SPW (2014) Relief de la Wallonie—Modèle numérique de terrain (MNT) 2013–2014 (http://geoportail.wallonie.be/WalOnMap/)

Stein S (2007) Approaches to continental intraplate issues. In: Stein S, Mazzotti S (eds) Continental intraplate earthquakes: science, hazard, and policy issues. Geol Soc Am Spec Pap 425:1–16

Thorez J, Dreesen R (1986) A model of a regressive depositional system around the Old Red Continent as exemplified by a field trip in the Upper Famennian "Psammites du Condroz" in Belgium. Ann Soc géol Belg 109:285–323

Van den Berg M (1994) Neotectonics of the Roer Valley rift system: style and rate of crustal deformation inferred from syn-tectonic sedimentation. Geol Mijnbouw 73:143–156

Vandenberghe D, Vanneste K, Verbeeck K, Paulissen E, Buylaert JP, De Corte F, Van den haute P (2009) Late Weichselian and Holocene earthquake events along the Geleen fault in NE Belgium: OSL age constraints. Quat Int 199:56–74

Vanneste K, Camelbeeck T, Verbeeck K (2013) A model of composite seismic sources for the Lower Rhine Graben, Northwest Europe. Bull Seismol Soc Am 103A:984–1007

Vanneste K, Mees F, Verbeeck K (2008b) Thin-section analysis as a tool to aid identification of palaeoearthquakes on the "slow", active Geleen Fault, Roer Valley Graben. Tectonophysics 453:94–109

Vanneste K, Meghraoui M, Camelbeeck T (1999) Late quaternary earthquake-related soft-sediment deformation along the Belgian portion of the Feldbiss fault, Lower Rhine Graben system. Tectonophysics 309:57–79

Vanneste K, Verbeeck K, Camelbeeck T, Paulissen E, Meghraoui M, Renardy F, Jongmans D, Frechen M (2001) Surface-rupturing history of the Bree fault scarp, Roer Valley graben: evidence for six events since the late Pleistocene. J Seismolog 5:329–359

Vanneste K, Verbeeck K, Camelbeeck T (2002) Exploring the Belgian Maas valley between Neeroeteren and Bichterweert for evidence of active faulting. Aardkundige Meded 12:5–8

Vanneste K, Verbeeck K, Petermans T (2008a) Pseudo-3D imaging of a low-slip-rate, active normal fault using shallow geophysical methods: the Geleen fault in the Belgian Maas River valley. Geophysics 73:B1–B9

Verbeek J, de Leeuw C, Parker N, Wong T (2002) Characterisation and correlation of tertiary seismostratigraphic units in the Roer Valley Graben. Geol Mijnbouw 81:159–166

Vogt J (1984) Révision de deux séismes majeurs de la région d'Aix-la-Chapelle-Verviers-Liège ressentis en France: 1504, 1692. In Tremblements de terre, histoire et archéologie. Actes du Colloque d'Antibes 2–4 Nov 1983, Valbonne, pp 13–21

Worum G, Michon L, van Balen R, van Wees J-D, Cloetingh S, Pagnier H (2005) Pre-neogene controls on present-day fault activity in the West Netherlands Basin and Roer Valley Rift System (southern Netherlands): role of variations in fault orientation in a uniform low-stress regime. Quatern Sci Rev 24:475–490

Wells DL, Coppersmith KJ (1994) New empirical relationships among magnitude, rupture length, rupture width, rupture area and surface displacement. Bull Seismol Soc Am 84:974–1002

Ziegler P (1990) Geological atlas of western and central Europe. Shell Internationale Petroleum Maatschappij B.V

Ziegler P, Dèzes P (2006) Crustal evolution of western and central Europe. In: Gee D, Stephenson R (eds) European lithosphere dynamics, vol 32, pp 43–56 (Geol Soc Mem)

Ziegler P, Dèzes P (2007) Cenozoic uplift of Variscan massifs in the Alpine foreland: timing and controlling mechanisms. Glob Planet Change 58:237–269

Ziegler P, Schumacher M, Dèzes P, Van Wees J, Cloetingh S (2006) Post-Variscan evolution of the lithosphere in the area of the European Cenozoic Rift System. In: Gee D, Stephenson R (eds) European lithosphere dynamics, vol 32, pp 97–112 (Geol Soc Mem)

The Hageland Hills, Legacies of the Depositional Architecture of the Miocene Diest Sands

14

Rik Houthuys and Johan Matthijs

Abstract

The Hageland area is characterized by long, a few tens of metres high hills that form a remarkable array of parallel topographic features. The existing model stating that the Hageland hills are remnants of offshore tidal sandbanks formed during deposition of the Miocene Diest Formation should be abandoned. Also the role of ironstone beds as the primary cause of the hills is questioned. Instead, the hills are thought to be determined by elongated parallel coarser grained sand bodies that were embedded in a more abundant facies of slightly clayey, bioturbated fine sand inside the Miocene Diest Formation. A sedimentary model is presented that interprets the elongate coarse sand bodies as the fill of a series of parallel, linear depressions, each time scoured by locally constricted flow during the continuous lateral fill of a tidal marine embayment. The coarse sand bodies are thought to have been sculpted out as positive relief features much later, during Pleistocene uplift of the area. Their fabric and higher permeability are put forward as important causes of primary preservation during the landscape morphogenesis. The ironstones can only have formed after initial appearance of the long hills. Their presence may have contributed to preserve some hills during further uplift and erosion.

Keywords

Hageland hills • Diest Formation • Tidal scour-and-fill architecture • Tidal sandbank • Differential erosion • Neogene iron sandstone

14.1 Hageland with Its Typical Elongated Hills

Some remarkable long hills appear east of the wide Dijle river plain at Wezemaal, between Leuven and Aarschot (Fig. 14.1). The hills are arranged as parallel ridges. They are a few km long and less than 1 km wide (Fig. 14.1a) and have steep (Fig. 14.1b), often wooded slopes. The top surfaces are flat (Figs. 14.1 and 14.2) and either horizontal or faintly sloping. A gradual convex or sometimes abrupt transition to the flanks surrounds the flat top. The hills are covered by sandy soils and have long been used as extensive pasture land and heath. Many outcrops of brown iron-cemented sandstone can be found in the flanks and road cuts of the hills. The stone has been quarried and used as a building material for centuries (De Clercq et al. 2014). Abandoned quarries dot the flanks of the hills. Some hills where outstanding natural and historical values have been preserved are now protected as nature and heritage reserves, such as Wijngaardberg (Natuurpunt Oost-Brabant 2014) and Kesselberg (Goolaerts 2015).

Hills like those near Wezemaal typify the region named Hageland. The geographical region Hageland is the area between the cities of Leuven, Diest, Aarschot and Tienen (Fig. 14.3). The special geomorphology characterized by

R. Houthuys (✉)
Geoconsultant, Suikerkaai 8, 1500 Halle, Belgium
e-mail: rik.houthuys@telenet.be

J. Matthijs
VITO, Boeretang 200, 2400 Mol, Belgium
e-mail: johan.matthijs@vito.be

A. Demoulin (ed.), *Landscapes and Landforms of Belgium and Luxembourg*,
World Geomorphological Landscapes, DOI 10.1007/978-3-319-58239-9_14

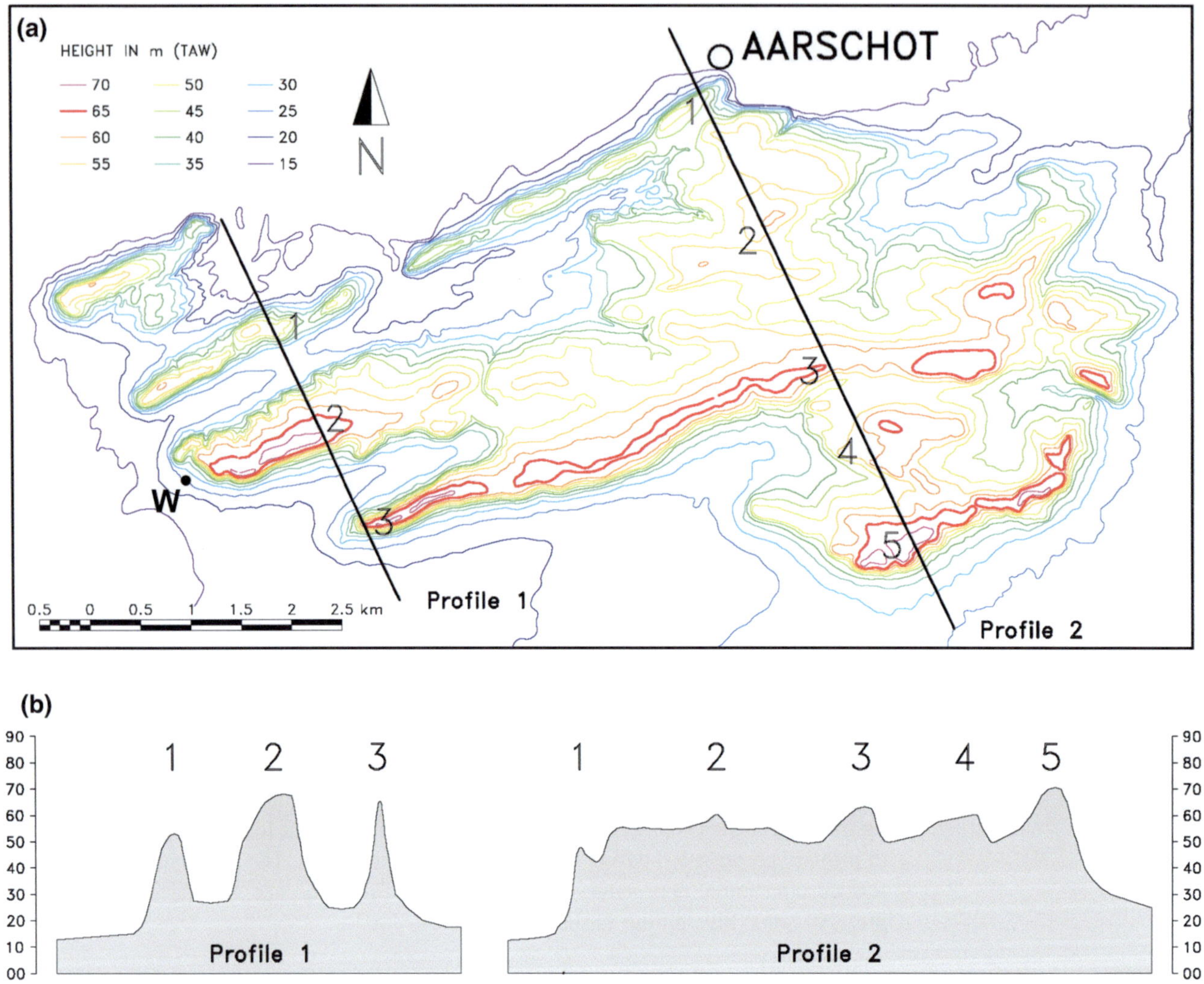

Fig. 14.1 Typical topography of a part of the Hageland area between Wezemaal (*W*) and Aarschot (see Fig. 14.3 for location). **a** Elevation map showing *coloured contour lines* in 5 m steps. Lowest altitude is 15 m near the river Demer in the north and highest is 75 m at the crest of Wijngaardberg (*2*) and Houwaartse Berg (*5*). Source of data: Grid version of the Topographic Map of Belgium at scale 1/10,000, map sheets 24/6, 24/7 and 24/8, NGI 1978–1993). **b** Cross-profiles *1* and *2* in Fig. 14.1. **a** *Note* the strong vertical exaggeration. In *profile 1*, hill *1* is Middelberg, *2* Wijngaardberg and *3* Beninksberg. In *profile 2*, hill *1* is Konijntjesberg, *3* Kratenberg near Nieuwrode, also shown in Fig. 14.4, *4* Benneberg, and *5* Houwaartse Berg

parallel long hills is found in the northern part of the proper Hageland although hills of similar appearance are also found further west (between Leuven and Brussels) and north (southern part of the Kempen (Campine) area). The term "Hageland" is used here in a wider sense to designate the area with elongated hills.

We examine the primary origin of this remarkable ensemble of hills, i.e. only the possible original causes of the topography but not its later geomorphological evolution. It is believed that, once the hills appeared as low positive landforms, further uplift of the country and erosion made the hills stand out more prominently. Steep slopes and large altitude differences are found near the present-day main river valleys and plains that are part of the eastern branches of the "Flemish Valley" (see Chap. 18). This is indicative of the ongoing morphogenic processes, though human activities such as the widespread extraction of iron sandstone also produced many steep slope segments, originating from ancient quarry walls.

14.2 Description of the Hageland Relief

The Hageland hills are characterized by their elongated shape and often stand out as single relief features in a flat surrounding landscape. They all contain outcrops of iron

Fig. 14.2 Aerial photograph of Wijngaardberg taken on 8 March 2015. View from the south. *Two contour lines* have been added to enhance the hill shape

sandstone and consistently show the same geological substrate of Upper Miocene highly glauconiferous Diest Sands (Vandenberghe et al. 2014).

Most of the hills (Fig. 14.3) are between 0.2 and 1 km wide, 0.5 and 5 km long, and 20 and 50 m high, though smaller forms rising only a few metres above the surrounding landscape can also be ranked as Hageland hills (e.g. near Westerlo and Tessenderlo, see Fig. 14.3). While the most prominent hills have steep slopes with values of 20–40% and more (Fig. 14.1), many instances are found of much gentler slopes. Also, the inselberg character is no defining feature. Some hills that sharply emerge from the landscape at one end merge with others at their other extremity in a plateau-like area (Figs. 14.1b and 14.4). Such plateaus are found, e.g. between Brussels and Leuven, around Pellenberg, Scherpenheuvel and Linden.

Although a closer look shows that individual hills may have a range of orientations, the hill crest lines are dominantly oriented WSW-ENE (Figs. 14.1 and 14.5). However, a few ridges, mostly low ones, are almost perpendicular to this direction. Mapping of the Hageland hills relied on (1) their sharp topographic emergence from the surrounding topography, and (2) their independence of the hydrographic network, and thus also included the low hills near Westerlo and Tessenderlo. In total, 275 km of crest lines have been identified.

Figure 14.3 shows the intrinsic tie between the Hageland hills and the outcrop area of the Upper Miocene Diest Sands. These sands are a marine deposit of mostly well rounded medium-size quartz grains and up to 50%, and locally more, transported glauconite pellets (Vandenberghe et al. 2014; Adriaens 2015). Clay is also present in drapes and up to a few centimetres thick layers. Both the glauconite pellets and the clay mineral composition attest the marine provenance of the sedimentary matter (Adriaens 2015).

Without doubt, the Diest Formation plays a key role in the story of the Hageland hills. The dominant WSW to ENE topographic direction is also characteristic of the internal structure of the Diest Sands. Many outcrops show thin and thick cross-bed sets (Fig. 14.6). The great majority of the cross-bed foreset laminae dip to ENE (Vandenberghe and Gullentops 2001), indicating that flow currents were chiefly oriented from WSW to ENE during deposition of the sand. Moreover, the basal surface of the Diest Formation shows long, linear depressions oriented in the same WSW-ENE direction, especially in the Hageland (Fig. 14.7).

14.3 Existing Theories on the Origin of the Topography

14.3.1 Role of the Ferricretes

The Hageland hills have always been considered as a classical example of iron sandstone-capped positive topographic features; this was reported since the nineteenth century (Gullentops 1957; Vandenberghe and Gullentops 2001). Obviously, subhorizontal beds of fairly resistant ferricrete are observed in the hills. They would, on uplift of the area, have protected the present-day hills from erosion.

14.3.2 The Tidal Sandbank Model

Gullentops (1957) correctly stated that the previous theory was insufficient to explain the alternation of narrow hills and

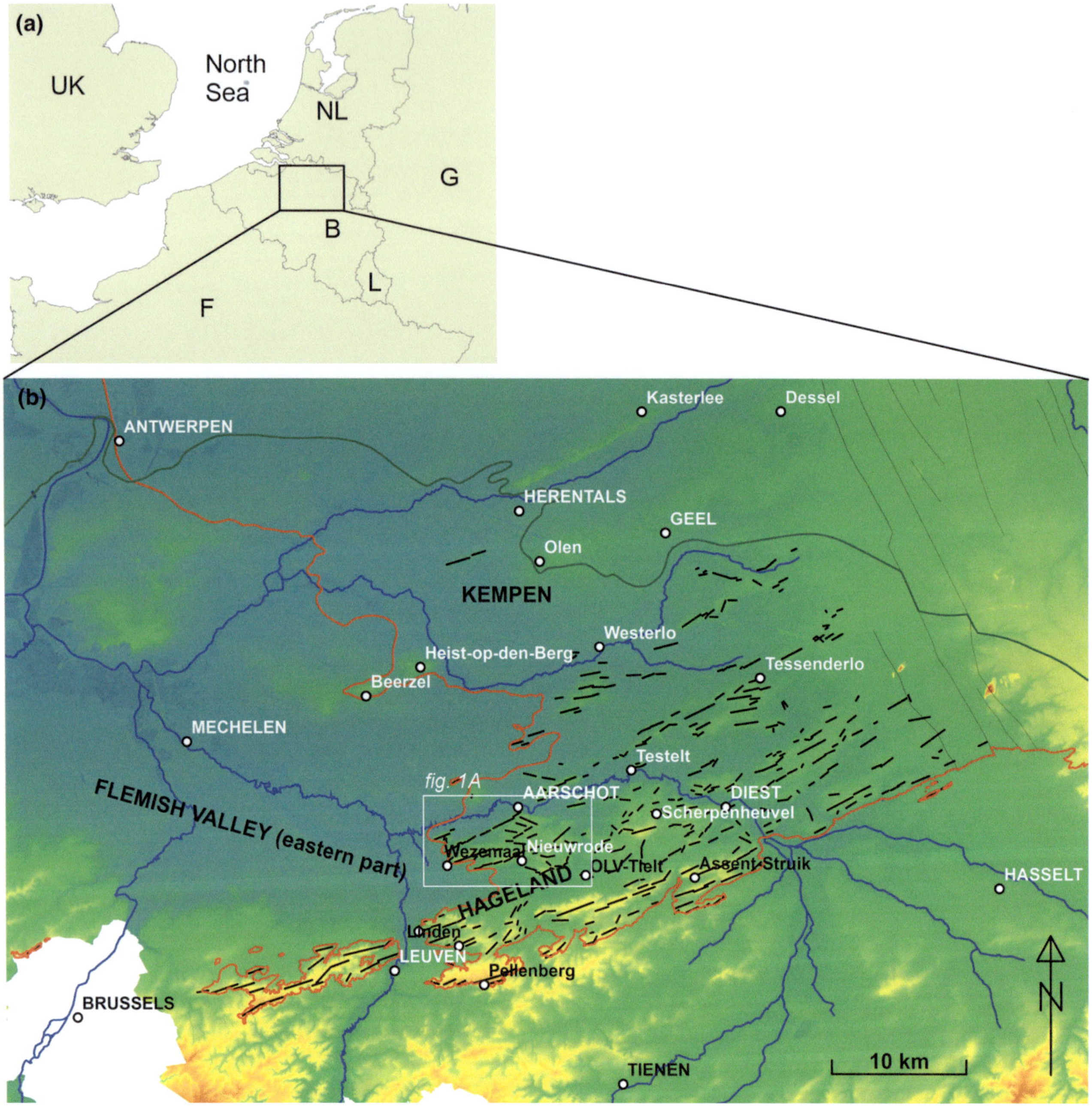

Fig. 14.3 **a** General location of the study area in NW Europe. **b** Elevation map of the Hageland and southern Kempen (Campine) area. Lowest areas in *greenish blue* are at sea level, highest areas in *red* are around 120 m TAW. *Source of data* 5 m digital elevation grid (VMM, Watlab and AGIV (2006) DHM-raster5). *Black lines* indicate crest axes of the Hageland hills. *Red line* is the contour of the Diest Formation outcrop; *dark green line* is the contour of the Kasterlee Formation outcrop. Both formations continue in the subsurface in the north and northeast

wide depressions, oversize with respect to the small creeks that drain them. He proposed instead an explanation inspired by the plan shape of the hill crest pattern and interpreted the hills as remnants of offshore tidal sandbanks, whose submarine relief has been somehow preserved in the present topography (Fig. 14.8). He theorized that at some time, the Diest Sea left the area so suddenly that the sandbank seascape was preserved. In the subsequent subaerial environment, the glauconite present in the sand dissolved and precipitated as iron (hydr)oxides to form resistant sandstone beds, which indeed were able to protect the sandbank relief against erosion during further uplift.

Fig. 14.4 Aerial photo of Nieuwrode, south of Aarschot, taken on 8 March 2015. *Note* the gentle slopes of the long hill, which is connected with steeper hills at its far west end, near Wezemaal. *Two contour lines* have been added to highlight the hill slope

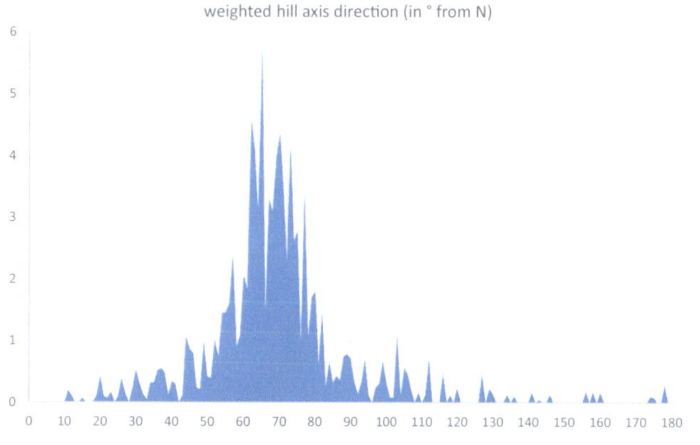

Fig. 14.5 Frequency diagram of structural hill crest axis lines in Hageland, weighted by their length. 90° is east

The tidal sandbank model (Gullentops 1957, 1988; Broothaers 1996; Vandenberghe and Gullentops 2001) invokes the following stages:

- the Upper Miocene Diest Sea transgressed farther SW than the present outcrop area of the Diest Sands, as suggested by similar indurated iron-rich sands on top of the Flemish Hills (as far as Cassel in northern France, more than 100 km west of Hageland), creating a marine connection between the English Channel and the southern North Sea;
- the Diest Formation base surface (Fig. 14.7) was eroded by the tidal currents that imported glauconite-laden sand from the English Channel through this strait and built a topography of shifting tidal sandbanks;
- at the end of the Diest Sands deposition, there was a seascape of linear offshore sandbanks similar to the present Flemish Banks in the southern North Sea (Fig. 14.8);
- the sea level lowered sometime so quickly that this seascape or, at least, the linear banks emerged. The Zanclean flooding event was suggested to have provoked this sudden sea-level drop (see below);
- on initial emersion of the seabed, soil processes under warm conditions during the Pliocene induced iron cementation of the sediments at the top of the emerged sandbanks;
- on further emersion, the ferricretes protected the sandbanks from erosion;
- on further uplift during the Pleistocene, the hills were accentuated by erosion of their surroundings.

Fig. 14.6 Typical Diest Sands cross-bed in a temporary pit west of Leuven. Scraper tool is 40 cm long

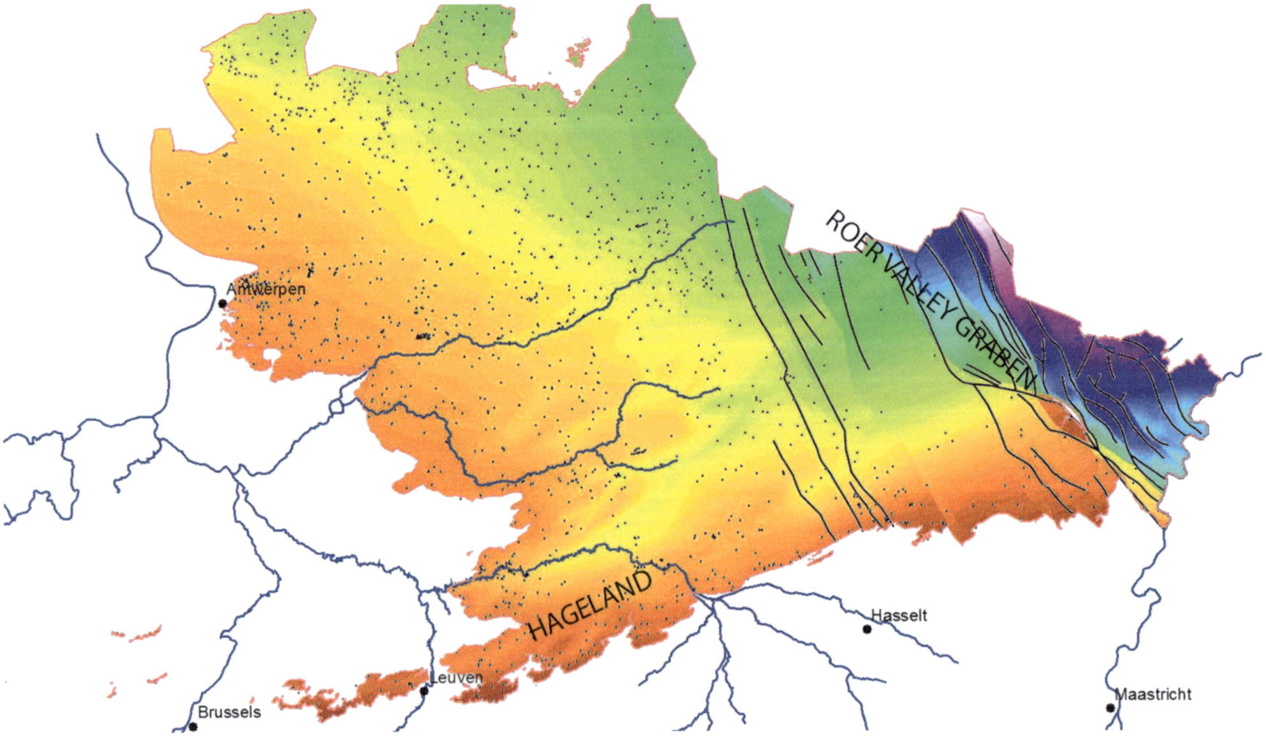

Fig. 14.7 Elevation map of the basal surface of the Diest Formation. *Colours* represent 10 m elevation steps, descending from +100 m in the southwest (*orange brown*) to below −500 m in the northeast (*light purple*). *Dots* represent wells where the base surface has been identified. Source of data: Matthijs et al. (2013), G3Dv2_0203_ Formatie van Diest basis, available on http://www.dov.vlaanderen.be

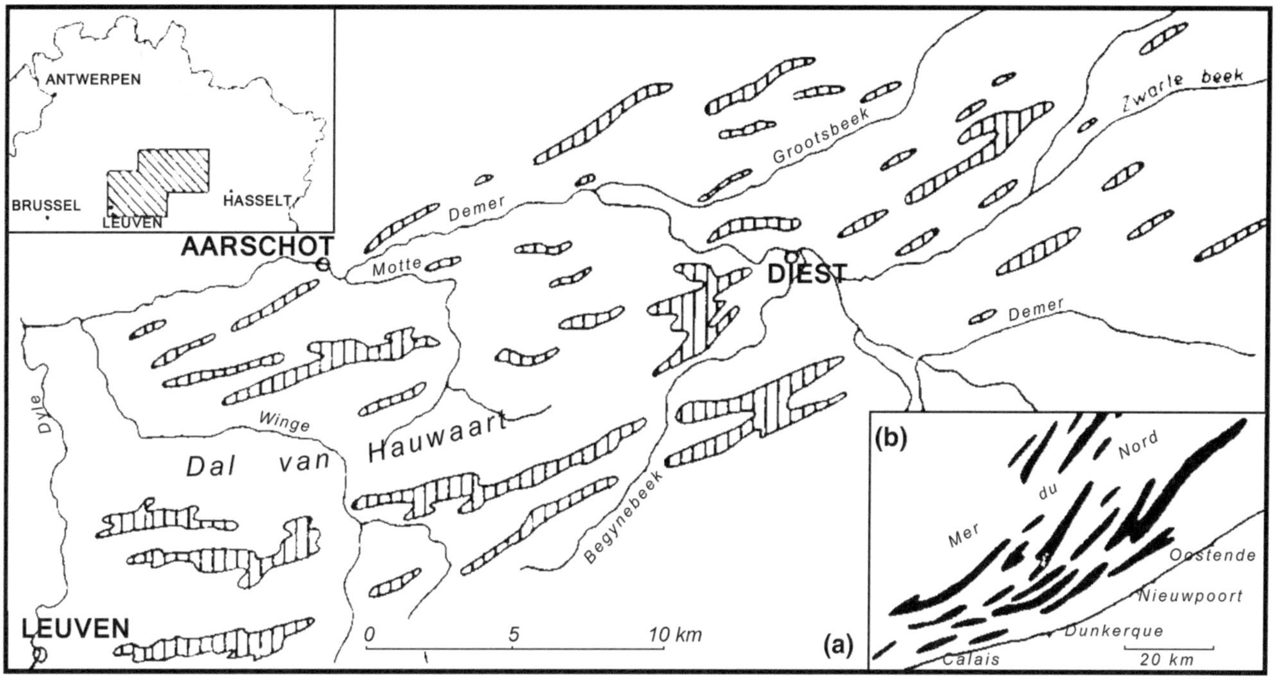

Fig. 14.8 Map of the Hageland hills. *Inset* For analogy, outline of the Flemish Banks on the present-day French–Belgian continental shelf (from Gullentops 1957)

14.4 New Evidence

14.4.1 The Diest Formation Is Absent in the Flemish Hills Area

The sands found on top of the Flemish Hills have been deposited in the shoreface environment of an exposed sea (Houthuys 2014). By contrast, the Diest Sands represent a marine embayment fill (see below). These two environments are incompatible. Instead, Houthuys (2014) suggested an Upper Eocene age for the Flemish Hills Sands, making them ca. 30 My older than the Diest Formation. While the precise age of deposition of the Flemish Hills Sands remains uncertain, the different sedimentary environments and the basin context indeed rule out a coeval Upper Miocene sedimentation. Therefore, a paleogeographic configuration with an E-W coastline parallel to the line of the Flemish Hills as depicted in Broothaers (1996) most likely never existed during the deposition of the Diest Sands.

14.4.2 The Basal Surface of the Diest Formation

The elevation map of the base of the Diest Formation describes an irregular surface dipping ~0.5% to NE and marked by WSW-ENE linear depressions (Fig. 14.7). Vandenberghe et al. (2014) interpreted the depression pattern as the signature of a paleodrainage system of parallel rivers on a relatively flat emerged surface gently sloping to the sea. Indeed, the depressions are perpendicular to the coastline, assumed to coincide with the border faults of the Roer Valley Graben before the Diest marine transgression. While some depressions are well documented (Fig. 14.7), others are more hypothetical, the base surface having been interpolated with the model of a connected channel pattern in mind (Lanckacker, personal communication 2014). However, based on P. Van Calster's data, a detailed map of the Hageland part of the basin revealed a pattern of elongated closed depressions (Houbolt 1982).

14.4.3 The Sedimentary Environment of the Diest Sands

The sedimentary characteristics of the Diest Sands were revised by Houthuys (2014). He identified in the outcrop area a facies suite consisting in "thick wedge-shaped cross-beds" (XT), "tabular descending cross-beds" (XD), "bioturbated cross-beds" (Bx), "bioturbated beds" (B), and "massive breaching beds" (M). Although additional data are needed to support the outcrop evidence, some important deductions about the sedimentary environment can unequivocally be made: (i) the overwhelming majority of foreset directions is to ENE, with a small spread of the data;

(ii) thick cross-beds with long depositional faces can only be the product of very large bedforms formed in relatively deep water (several tens of metres) and very strong flow; (iii) the numerous clay drapes and clay drape packages indicate unsteady currents in relatively confined flow channels where mud can settle during slack waters; (iv) fairly sheltered areas that allowed deposition of slightly clayey fine sand and intense bioturbation were present in the depositional environment; (v) bedding surfaces have a systematic downcurrent dip (i.e. to ENE) and often a transversal dip component pointing to the centre of the hill (Fig. 14.9a); (vi) as the facies are interwoven, they bear witness to a single sedimentary environment; (vii) sedimentation was fast in an environment with slopes steep enough to allow breaching to produce M facies (Van den Berg et al. 2002), and (viii) no sedimentary structures indicative of waves have ever been found.

Reconstruction of the sedimentary architecture is so far partial, restricted to the outcrop-scale. In the longitudinal ENE direction, the beds show downcurrent progradation; in the transverse direction, the bedding indicates a channel fill, while in the vertical direction, a stacked fill is observed (Fig. 14.9a). These elements suggest an environment of eroding channels or linear troughs where filling takes place immediately in the wake of the erosion centre, which migrates downstream (Fig. 14.9b). This mechanism seems to imply that a very high sediment supply drove all other processes involved in the Diest Sands deposition. This is further detailed below.

14.4.4 The Top Surface of the Diest Sands

In the Kempen, the Diest Sand Formation is covered by the Uppermost Miocene marine Kasterlee Formation (Matthijs et al. 2013) (Fig. 14.3), which is also present at Heist-op-den-Berg and Beerzel as an 8–10-m-thick layer at the top of the hills (Fobe 1995; Verhaegen et al. 2014). The contact surface of the Kasterlee and Diest Sands contains rounded and weathered flint pebbles near Heist-op-den-Berg and at Olen (Verhaegen et al. 2014). Near Leuven and in NE Hageland, a residual gravel of well rounded, up to 10-cm-long flint pebbles is present on top of the Diest Sands, marking the base of the Quaternary (Vandenberghe and Gullentops 2001).

The top surfaces of the Hageland hills, generally coincident with that of the Diest Sands, are often fairly flat and the most elevated of them form a relatively flat surface gently sloping to the north and smoothly merging into the Kasterlee Formation base surface. As evidenced by the sedimentary structures in the Diest Sands, this surface truncates the top part of the formation (Houthuys 2014). All this indicates that the top surface of the Diest Sands is a marine abrasion surface formed by wave erosion during the following

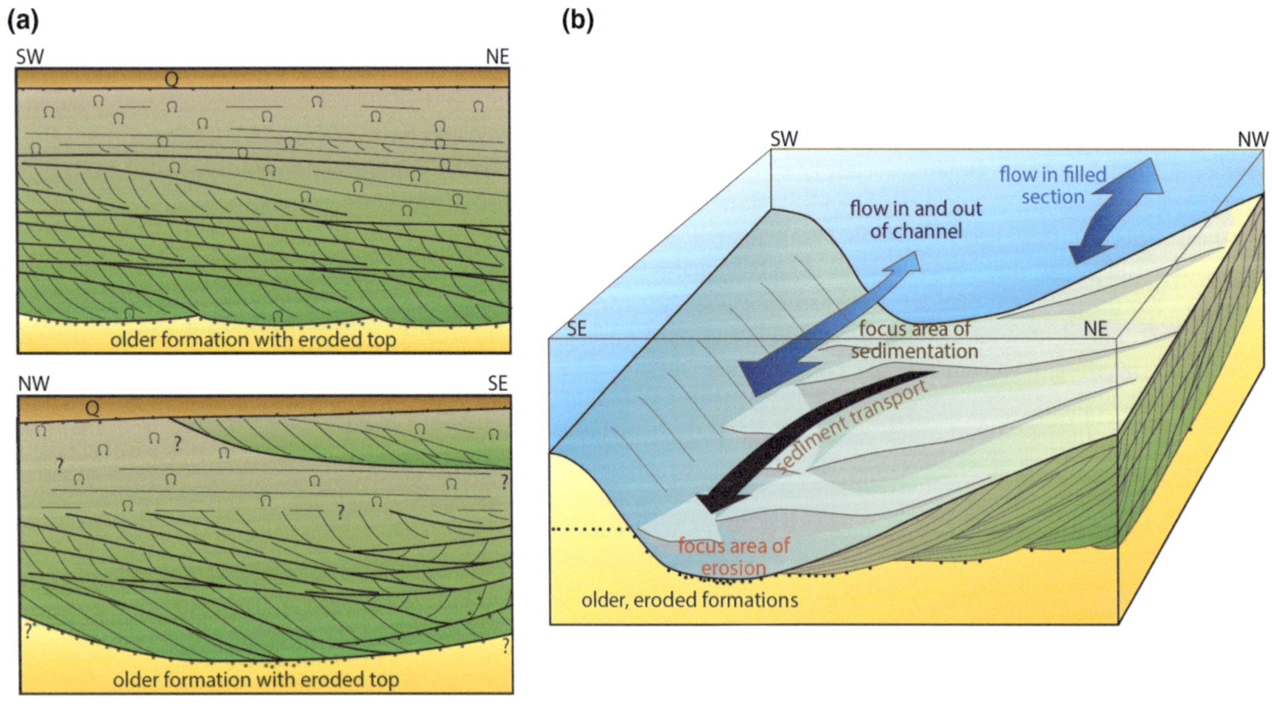

Fig. 14.9 **a** Typical longitudinal (*top*) and transverse (*bottom*) sections through a Hageland hill. Idealized sections of width × height in the order of 50–100 m × 10–20 m. Ω bioturbation. **b** Reconstruction of the sedimentary processes that governed deposition of the Hageland Diest Sands. *Grey-shaded* bedforms are large dunes. They were most probably covered by smaller dunes (*not indicated in sketch*)

transgression, as usual for most Cenozoic marine formations in the southern North Sea basin. It might even be a multi-stage abrasion surface, as the time period between the estimated ages of the Hageland Diest Sands (11–9 Ma, Vandenberghe et al. 2014) and the base of the Kasterlee Formation (7.5 Ma, Louwye and De Schepper 2009) could have accommodated several marine transgression cycles.

14.4.5 On the Iron Sandstones

Iron sandstones are very common in the Diest Sands. The original loose deposit has been cemented by iron minerals of the limonite group through precipitation of iron dissolved in the groundwater. This required oxidizing conditions, and thus emersion. Outcrops of Diest Sands frequently show loosely cemented liesegang rings indicative of slow precipitation in concentric areas according to local chemical gradients (Fig. 14.10a). Though not a prerequisite for iron cementation, clay layers are often found inside and at the bottom of the cemented zones, illustrating the role of perched water tables.

These ferricretes often display capricious shapes, with harder zones jutting out from the outcrop, separated by uncemented or weathered sandy zones (Dusar 2014) (Fig. 14.10a). Though most indurated layers follow the original stratification, some, including concentric, nested spheres ('klapperstenen'), also crosscut the primary layering. However, they generally appear either as thin plates, hard and brittle, often found near the top of the hills, or sheet-like or isolated thick massive or crumbling blocks (Dusar 2014). Only these iron sandstone masses, from ~0.2 m to several metres in thickness, are believed to have had a possible morphogenic role (Fig. 14.10b, c). According to our observations, confirmed by those of others (Van Calster, pers. comm. 2015), such well-developed ferricretes are never found in the deeper subsurface that has always been below the water table. Wells drilled in valleys and throughout the low-lying areas of the Campine never encounter ironstones, apart from local occurrences of siderite (Laga 1972). As observation shows, they also do not occur generally in the centre of plateaus or wide hills but concentrate predominantly along the sides of the hills (Bos and Gullentops 1990; Dreesen et al. 2010; Dusar 2014), often at depths between 10 and 20 m below the hill tops. Outcrops extending over more than a few tens of metres feature multiple (sub)horizontal ferricrete levels (Fig. 14.10b), mostly discontinuous and of varying structure and thickness (Fig. 14.10c).

Iron sandstones are not exclusive to the Diest Sand Formation (e.g. Bos and Gullentops 1990). Famous are the ironcrust-capped Flemish Hills, between Cassel in N France and Flobecq in N Hainaut, and the ferricretes crowning the hills between Geraardsbergen and Kester. While the former indurated sediments are of possible Upper Eocene age, the latter belong to the Middle Eocene Lede Formation (Hout-huys 2014). Ironstone layers locally over 5-m-thick are also found in the Middle Eocene Brussels Sands, often in plateau areas or underneath valley heads, but they had no

Fig. 14.10 **a** Compact ironstone beds (*top*) and Liesegang structures (*bottom*) in the Diest Sands at Pellenberg. **b** *Horizontal*, a few decimetres thick iron sandstone beds at Kesselberg. The slight elevation of the ground surface may be related to the presence of the *upper sandstone bed*. *Note* the presence of a thin inclined iron crust running across the loose sand layer between the *lower* and *middle* horizontal indurated beds. **c** Irregular orange iron sandstone zones in a temporary outcrop of the Diest Sands at Assent Struik. The *top* half metre is late Pleistocene sandy loess. No influence of the indurated zones is observed in the surface topography. Section is oriented WSW (*left*) to ENE

topographic effect. The more or less horizontal development of all extensive ironstones, regardless of the formation in which they were formed, suggests a link with paleowater tables, whose oscillations provided the alternation of oxidizing and reducing conditions needed to produce such thick ferricretes.

While authors often state that conditions warmer than those of today were necessary for the mobilization of iron oxides (Bos and Gullentops 1990; Vandenberghe and Gullentops 2001; Dusar 2014), some observations in the Hageland suggest a fairly recent origin of the iron sandstones. Differential compaction is nowhere observed near thick massive ironstones, indicating that compaction occurred prior to induration. Moreover, while Bos and Gullentops (1990) mention hard crusts formed in a few decades, water seepage at excavation sites in the Diest Sands is often observed to cause the formation of a thin, crumbly crust of iron-cemented sand in only a few weeks. In brief, though no dating is available of the time of duricrusting, there is evidence of a very long period in which it may have occurred, groundwater ferricretes requiring only a continental environment but being not climate-dependent. However, some evidence points to recent induration, which may be going on today.

14.5 Problems with the Offshore Sandbank Model

In view of the above observations, there are several insuperable problems in the offshore sandbank model of the Diest Sands and Hageland hills, which we list hereafter.

14.5.1 The Internal Structure Is not of a Sandbank

Offshore tidal ridges, or sandbanks, are the largest bedforms in seas with strong tidal currents, i.e. near-surface mean spring peak tidal currents well over 0.5 m/s (Belderson et al. 1982). They may reach a few tens of km in length, a few km in width and a few tens of m in height, dimensions that indeed match the Hageland hills. They often occur as groups of parallel linear forms extending over hundreds of square kilometres (see inset in Fig. 14.8), aligned at a small angle to the direction of the peak tidal flow. Actively maintained sandbanks show large dunes (megaripples) on their flanks and top. The crest of these asymmetric bedforms (with lee side slope $\leq 6°$) is barely lower than low-tide level. Observations of recent and well-documented ancient offshore tidal sandbanks consistently point to the following internal sedimentary architecture and structures (Stride et al. 1982; Houthuys 1990; Trentesaux et al. 1999; Reynaud et al. 1999; Mathys 2009; Reynaud and Dalrymple 2012; Michaud and Dalrymple 2016; Leva Lopez et al. 2016):

- sandbanks have either a flat basal surface or developed over an older core of any origin;
- the internal structure predominantly shows cross-bedded sand produced by the dunes that mantle the ridges; the cross-stratification dip tends to change gradually upwards, in accordance with the dunes becoming more parallel to the bank crest when they approach it;
- the master bedding surfaces are parallel to one or both sandbank flanks, and slope at angles of a few degrees; in the case in which both slope directions have been preserved, the left and right flank master beds truncate each other in the middle;
- the base deposits are more bioturbated, finer grained and more poorly sorted than the top beds;
- wave structures, such as flat wave planation surfaces, may be found near the sandbank crest; also internal gravel lags, coarse graded storm beds, or even hummocky cross-stratification may be present;
- ebb- and flood-oriented cross-beds should be more or less equally developed, especially near the top of the form, where cross-beds are short and separated by a multitude of internal bounding surfaces.

None of these features are found in the Diest Sands.

14.5.2 The Flemish Hills Top Sands Should Be Part of the Diest Formation

The paleogeographic configuration compatible with offshore sandbanks similar to the present Flemish Banks (Fig. 14.8) requires that the Flemish Hills top sands are part of the Diest Formation. If they are of a different age than the Diest Sands, as proposed by Houthuys (2014), the coastline configuration needed by the offshore sandbank model is highly unlikely to have occurred.

14.5.3 The Basal Incision of the Diest Sands Is not Compatible with Sandbank Formation

Present-day sandbanks are not found in sea straits, though they do occur in their vicinity. If incision at the base of the Diest Sands is explained by strait erosion, the location of sandbanks inside such a strait setting is problematic as tidal sandbanks are accretionary forms developing on a flat surface or a positive pre-existing form.

14.5.4 A Timing Problem

Tying a supposed post-Diest Sands fast sea-level fall to the event of the Zanclean flood is inconsistent with available ages. The drop in global sea level entailed by the still debated Zanclean flood, which would have ended the Messinian salinity crisis (base level fall of 1.5–2 km in a landlocked Mediterranean Sea) by breaching the Gibraltar land barrier and refilling the Mediterranean basin, has been dated at 5.33 Ma (Krijgsman et al. 1999). By contrast, the Diest Sands are relatively well constrained in time during the first half of the Tortonian (DN8, from 11 to 9 Ma) (Vandenberghe et al. 2014). A Zanclean sea level fall ending deposition of the Diest Sands is therefore excluded. However, this timing problem is not crucial because eustatic sea-level changes in the order of 10 m, corresponding to third- and fourth-order sequences of Vail et al. (1977), occurred throughout the geological times.

14.5.5 The 9-Ma Submarine Topography Is Unlikely to Preserve

Even the quickest $\sim$10-m sea-level drop following deposition of the Diest Sands would have lasted long enough to allow the currents and waves that had built the sandbanks to fill up intervening spaces and flatten the seascape on lowering of the sea level, in accordance with the sequence stratigraphy scheme in which a relative sea level fall causes the lowering wave base to plane off the soft sea floor sediments and create a 'regressive surface of marine erosion' (Catuneanu 2006). In order to get around this problem, Gullentops (1957) invoked fast, quasi syn-regression ironstone formation. However, there is nothing to document this assumption and, as explained above, the iron sandstones are likely to have formed millions of years after deposition of the Diest Sands and, in any case, only when a subaerial topography provided the required oxidizing conditions.

14.6 A New Depositional Model

Houthuys (2014) proposed a sedimentologically consistent depositional model that accommodates the basal incision and the internal architecture of the Diest Sands, based on an analogy with the Brussels Sands depositional history (Hout-huys 2011).

The depositional environmental similarities between the Eocene Brussels Sands and the Miocene Diest Sands are impressive (Houthuys 2014). In the well-exposed Brussels Sands, the fill mechanisms and architecture could be thoroughly documented, showing that the sedimentary system was characterized by an overall eastward lateral accretion and flow constriction occurring on closure of successive pre-existing local depressions (Houthuys 2011). Such a constriction implied strong reinforcement of the flow currents that caused scouring, immediately followed by cross-bedded fill.

The result is an internally differentiated fill style consisting of a background of bioturbated muddy fine sand, in which a series of parallel elongated bodies exist, corresponding to erosive troughs filled with coarser, cross-bedded sand. The troughs have straight, non-meandering axes in relation with the following driving mechanisms:

- a continuous external sand supply, inferred to have included a coastal drift pathway, feeding into the embayment;
- tide action inside the embayment, causing much water to enter it during rising tide, including over the low-tide flats, and leave it at falling tide through increasingly narrower channels at one side of the filling basin.

As local flow constriction becomes more and more dominant when nearing complete fill of the basin, the elongate troughs filled with coarse cross-bedded sand are preferentially found at the downdrift side of the embayment, e.g. to the east in the Brussels Sands case.

Applied to the Diest Sands, this model accommodates key observations (Fig. 14.11), namely (1) the complex erosion history and the shape of the basal surface, (2) the marine nature of the sand fill and (3) the high-stand style of the fill. In contrast to the Brussels Sands, the Diest Sands appear to have undergone a syn-sedimentary tilt towards the Roer Valley Graben. However, this tilt was probably slow enough not to interfere with the concomitant sedimentary processes.

Waiting for a definitive proof that the Brussels Sands depositional model can be applied to the Diest Sands, we note that unknown key elements remain the exact large-scale progradation direction at the basin scale, the valley- or closed-trough-like character of the longitudinal basal depressions, and the precise style of lateral interweaving of the depression fill bodies within the finer, bioturbated sand. If the analogy holds, then:

- no exceptional tectonic event or unrealistic paleogeographic configuration is needed to explain strong incision underneath Hageland, where the parallel troughs would have been created by internal, fill-related erosion cycles;
- this sedimentary process chain created longitudinal internal bodies built of coarser grained, cross-bedded, trough-fill deposits;
- like every sedimentary cycle, sedimentation ended with a complete basin fill, probably followed by the development of a relatively flat surface of marine erosion truncating the upper part of the Diest Sands upon regression.

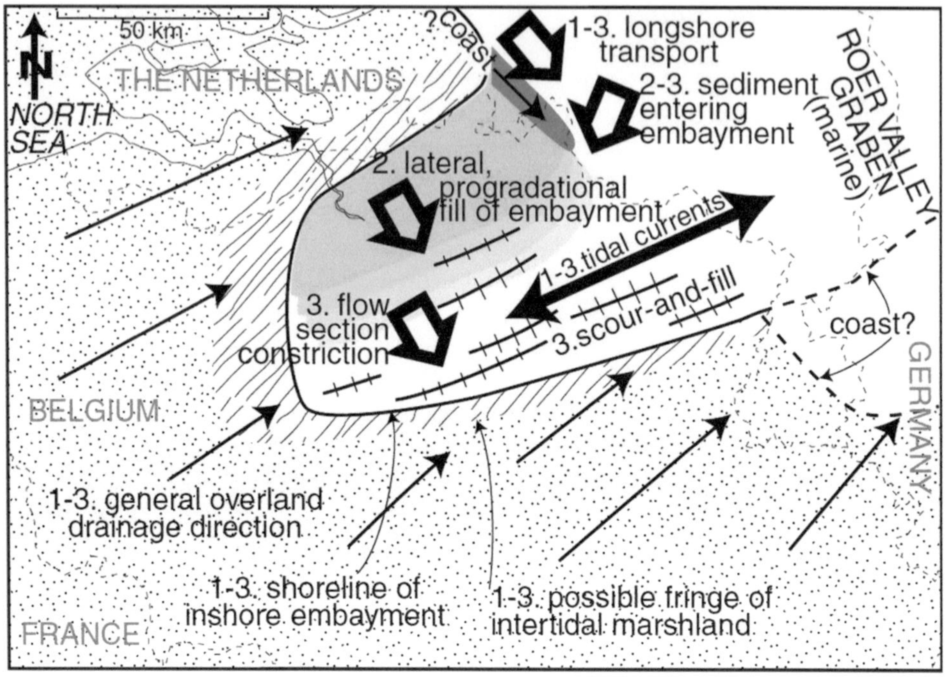

Fig. 14.11 Reconstruction of the paleogeography during deposition of the Diest Sands. Present-day coastline and state boundaries are shown in *thin grey lines* for reference. Stages in the paleogeographical evolution: *1* drowning during relative sea-level rise of a river mouth area and establishment of an inshore marine embayment; *2* first stage of filling, prograding from NW to SE; *3* last filling stages with subtidal flow section constriction and subaqueous formation of scour-and-fill elongate troughs. Stippled area is emerged (from Houthuys 2014)

The differential subsidence during deposition of the Diest Sands also remains to be confirmed. If indeed the NE part of the Diest Embayment experienced stronger subsidence than its SW part during deposition, a decoupling between basal incision and internal incision would be expected in the Diest Formation, with coarse trough fills not necessarily positioned on top of basal incisions.

The model implies a flat, low-lying environment during the Diest Formation deposition, flooded and filled under high sea level conditions. Inferring marine transgressions postdating the deposition of the Diest Sands also implies a significantly younger start of the continental morphogenesis.

14.7 Alternative Explanations of the Development of the Hageland Topography

Beyond purely sedimentary considerations, uncertainties remain about the precise timing and characteristics of the local uplift history, the original extent of the iron sandstones and the conditions and timing of their origin. Regarding the Hageland topography, some new deductions can now be made, which are however necessarily somewhat speculative.

14.7.1 Role of the Iron Sandstones

It seems unlikely that thick ironstones were present in the crucial stage when the topography started to get its grain. Within the flat, low-lying topography that existed prior to significant regional uplift, the Diest Sands lay mostly below the water table, which was unfavourable for iron oxidation and duricrusting anywhere in the area. Ironstone formation required a minimum relief, with an unsaturated zone above an oscillating water table, to take place. Moreover, the repeated iron sandstone levels found at some sites probably reflect a succession of paleo-groundwater tables related either to a stepped uplift history or to perched water tables above clay beds.

Assuming that an iron crust formed very early after regression, one would expect the formation of an extensive sheet-like ferricrete rather than elongated parallel indurated zones that would later turn to hills through differential erosion. In addition, many Hageland plateaus are free of ironstone beds and the isolated hill tops are often underlain by several metres of loose, often clayey sand before compact, thick ironstone layers are encountered in deeper position. Extensive layers of ironstone have nowhere been evidenced underneath hill tops.

Evidence of Kasterlee Sands inside the top part of the hills at Heist-op-den-Berg and Beerzel further discredits the purported relation between iron sandstone and hill formation. Rounded limonite lumps and iron crust fragments have been found at the base of the Latest-Miocene Kasterlee Sands at Beerzel (Verhaegen et al. 2014). They would indicate supply from already emerged and duricrusted Diest Sands around a marine Kasterlee embayment. However, the hills in the Heist-op-den-Berg and Beerzel area have a similar topography to the Hageland Hills, yet contain non-indurated marine sediments (the Kasterlee Sands) significantly younger than the Diest Sands. It seems thus far more probable that ferricretes played no structuring role during the initial stages of definitive emersion of the area.

Once present, the ironstones may have helped to protect locally and temporarily the initial Hageland hills (Fig. 14.10 b). They nevertheless may not be as resistant as they look and their protective efficiency should be put in perspective. For instance, many sunken roads in Hageland show iron sandstone walls suggesting that, if the ferricretes were pre-existing extensive features, they succumbed to a few centuries of passage and runoff. Also, many iron sandstone blocks used in old buildings show significant wear (Fig. 14.12). A critical review of the observations seems to point to a fairly fast ironstone formation (in the order of kyears) at various times since the area is emerged. Bearing in mind the formation of iron crusts is a matter of weeks at excavation sites, one may even wonder whether ironstones in the flanks of sunken roads formed after the latter were incised.

But, regardless of the formation time of the massive iron sandstones, a primitive topography had to exist for such duricrusting to take place, so that the ferricretes cannot have preceded, and have been responsible for, the formation of the hills. At this point, only a few hypotheses independent of the presence of ironstones are thus left to explain these remarkable parallel hills.

14.7.2 Hypothesis 1: Interfluves of a Primary Parallel Overland Drainage System Have Been Preserved

It could be theorized that an overland drainage system with a preferential flow direction to ENE (corresponding to the peak direction in Fig. 14.5) was present after the last marine transgression. After initial uplift and incision, the interfluves would have developed into the typical long hills. In this hypothesis, the internal structure of the Diest Sands is indifferent. Later creation of cross-hill river trunks such as the Dijle, Winge, Motte, Begijnebeek, Demer (Fig. 14.8), river captures and local inversions would have rearranged this original system.

However, the spacing and rhythmicity of the hills seem out of proportion to the intervening small watercourses. Vandenberghe and Desmedt (1979) thought the structures related to the Diest Sands in the eastern part of the Scheldt basin rather imposed river valley directions than vice versa. Also, some typical Hageland hills have a different alignment (Fig. 14.3), barely compatible with the proposed parallel drainage system. Finally, this hypothesis is also weakened by the observation that, on the whole, this parallel arrangement is present only in the rather restricted area where the Diest Sands crop out.

Fig. 14.12 Zoom on the stone front of the Testelt water mill, built in 1608. *Note* wear of the building blocks of local ironstone extracted from the Diest Formation

14.7.3 Hypothesis 2: Denudation Processes Were Guided by Internal Structures

This hypothesis states that the internal structure of the Diest Sands, organized along a preferential direction, has been carved out in the landscape. In other words, the elongated parallel coarser grained, cross-bedded trough fills characterizing the internal architecture of the formation would have been sculpted out as positive relief features during initial uplift and denudation. The regional arrangement of the Hageland hills, their dimensions, and the correspondence between the hill axis direction and the principal internal directions (basal depression long axis, dominant foreset direction) strongly support this hypothesis.

Again, the analogy with the Brussels Sands, where the interpretation is more firmly documented, is striking. There, after removal of a younger cover, differential erosion within the Brussels Sands also allowed internal structures to impose the local topographic grain (Houthuys 2011). Therefore, Hypothesis 2 is a plausible candidate to explain the topographic structure of the Hageland. As grain size is the main differentiating element inside the Diest Sands, we suggest that the more grain-supported fabric of the coarser grained bodies might have resulted in slightly positive relief features after compaction, which in turn would have better resisted to erosion than the surrounding finer clayey sand because their higher permeability favoured infiltration of meteoric water, thus reducing runoff erosion.

14.7.4 Hypothesis 3: Imprint of Directional Structures Inherited from a Younger, Now Eroded Formation

This hypothesis proposes that younger, now eroded Cenozoic deposits were present on top of the Diest Sands in the Hageland. After emersion, they would have imposed topographic directions following mechanisms similar to those exposed in Hypothesis 2 and these directions would have been inherited and preserved within the Diest Sands on further uplift and denudation. Hypothesis 3 has the virtue of accommodating the observation that similarly oriented hills partly cut in younger deposits exist north of the Hageland, notably the hills at Heist-op-den-Berg and Beerzel, covered by Kasterlee Sands, and the long hill between Herentals and Kasterlee, with Poederlee and Kasterlee Sands.

The Kasterlee Formation is made of three stacked members of fairly wide areal extent (Fobe 1995; Verhaegen et al. 2014) but they do not seem to show any spatial internal differentiation, which makes the parallelism of these hills with those of the Hageland very intriguing and depreciates the last hypothesis.

14.7.5 What Happened Once the Direction Appeared in the Topography?

While decisive arguments are currently lacking to choose between the hypotheses presented above, the second hypothesis, i.e. sculpting of internal sedimentary contrasts of the Diest Sands, is suggested to be the most plausible representation of how the Hageland morphology was initiated. Once low embryonic hills had appeared in the emerging landscape, several of the processes discarded above as primary morphogenic causes may have joined forces to perpetuate the differential erosion. The more permeable hills evacuated their share of meteoric water by percolation towards the depressions. Iron sandstone formation at seepage locations may have created an armour that contributed to preserve some hill slopes and make them steeper in a process that may well go on today.

14.7.6 The Story at a Glance

In brief, the Hageland hills were most plausibly created and evolved as follows:

- during deposition, the Diest Sands acquired an internal sedimentary architecture consisting of parallel elongated bodies whose ENE trend was imposed by the lateral fill style that constricted tidal currents; these bodies are cross-bedded, relatively coarse-grained sand fills of longitudinal erosion troughs, embedded in finer grained, clayey and bioturbated sand;
- these coarse sand bodies were sculpted out after first emersion and initial uplift, probably not before the Late Pliocene (depending on the regional uplift history, a currently still unsolved issue); the proposed, but not yet demonstrated, mechanism for differential erosion relies on differential compaction and contrast in permeability between coarse sand and clayey sand;
- ironstone formation started only after an initial topography had formed;
- ironstones may have contributed to the preservation, possibly also the enhancement, of the initial hills during further uplift and associated valley incision.

14.8 Conclusion

The Hageland area has a particular topography: the land surface displays long, parallel hills. The hills are a few tens of metres high and consist internally of an Upper Miocene marine sand deposit, the Diest Sands. Horizontal ironstone beds and variously shaped iron sandstones often crop out in the flanks of the hills.

Building a coherent theory to account for the array of hills is an issue that has boggled the minds of researchers for decades. The present state of knowledge allows discarding the established theory, making two things clear, namely that (i) the morphology of the Hageland hills is not the legacy of an offshore tidal sandbank morphology as the paleogeography and internal structure of the Diest Sands are not compatible with offshore sandbanks, and (ii) massive iron sandstones cannot have been responsible for the creation of the hill array because their formation required an initial topography.

The alternative theory presented here is that the sedimentary architecture of the Diest Sand Formation includes parallel elongated sand bodies and that, a long time after deposition, uplift, probably in the Pleistocene, induced grain-size-controlled differential compaction and erosion that materialized these sand bodies in the topography as positive residual landforms. Some essential elements of this alternative theory are still hypothetical. This paper hopefully will encourage research and new data collection aimed at further unravelling the origin of the intriguing Hageland hills.

Acknowledgements Thanks to Tom De Moor for having performed a photo flight over Hageland on 8 March 2015.

References

Adriaens R (2015) Neogene and quaternary clay minerals in the southern North Sea. Unpublished PhD thesis, KULeuven

Belderson RH, Johnson MA, Kenyon NH (1982) Bedforms. In: Stride AH (ed) Offshore tidal sands, processes and deposits. Chapman and Hall, London, pp 27–57

Bos K, Gullentops F (1990) IJzerzandsteen als bouwsteen in en rond het Hageland. Bull Belgische Vereniging Geol 99:131–151

Broothaers L (1996) Overzicht van de geologie van Vlaanderen. In: Gullentops F, Wouters L (eds) Delfstoffen in Vlaanderen. Flemish authorities, Afdeling Natuurlijke Rijkdommen en Energie, Brussels, Belgium, pp 1–28

Catuneanu O (2006) Principles of sequence stratigraphy. Elsevier, Amsterdam

De Clercq H, Janssens E, Smets S, Verhaert G (2014) Diestiaan ijzerzandsteen. Van Demergotiek tot restauratieproblematiek. Koninklijk Instituut voor het Kunstpatrimonium, Regionaal landschap Noord-Hageland and Vlaamse overheid, ALBON, Contributions of symposium "Omgaan met Diestiaan ijzerzandsteengebruik", 19 Sept 2014

Dreesen R, De Ceukelaire M, De Koninck R (2010) Natuurlijk voorkomen en karakteristieken van inheemse natuursteen in Vlaanderen—Diestiaan ijzerzandsteen in het Hageland. VITO, Mol, internal report commissioned by Flemish Authorities, ALBON

Dusar M (2014) Geowaarden van de Wijngaardberg. Natuur en landschap, special edition with contributions of the symposium held at Wezemaal, 5 oktober 2014. Available via http://www.natuurpuntoostbrabant.be/docs/PDF/Brochure-Wijngaardberg-2015-LR.pdf. Accessed 8 July 2015

Fobe B (1995) Litologie en litostratigrafie van de Formatie van Kasterlee. Natuurw. Tijdschr. 75:35–45

Goolaerts S (ed) (2015) De Kesselberg binnenste buiten. Hona vzw, Leuven

Gullentops F (1957) L'origine des collines du Hageland. Bull Belgische Vereniging Geol 57:81–85

Gullentops F (1988) Neogene (part of Excursion B3: the Cenozoic southern North Sea). In: Herbosch A (ed) IAS 9th European regional meeting, excursion guidebook. Belgian Geological Survey, Leuven, Belgium, pp 255–260

Houbolt JJ (1982) A comparison of recent shallow marine tidal sand ridges with Miocene sand ridges in Belgium. The ocean floor. Wiley, New York, pp 69–80

Houthuys R (1990) Vergelijkende studie van de afzettingsstruktuur van getijdenzanden uit het Eoceen en van de huidige Vlaamse Banken, vol 5. Leuven University Press, Aardkundige mededelingen, p 137

Houthuys R (2011) A sedimentary model of the Brussels Sands, Eocene, Belgium. Geol Belg 14(1/2):55–74

Houthuys R (2014) A reinterpretation of the Neogene emersion of central Belgium based on the sedimentary environment of the diest formation and the origin of the drainage pattern. Geol Belg 17 (3/4):211–235

Krijgsman W, Hilgen F, Raffi I, Sierro F, Wilson D (1999) Chronology, causes and progression of the Mesinian salinity crisis. Nature 400:652–655

Laga PG (1972) Een fossielhoudende zandsteen in de Zanden van Diest te Olmen (Antwerpse Kempen). Bull Belg Ver Geol 81:251–254

Leva Lopez J, Rossi VM, Olariu C, Steel RJ (2016) Architecture and recognition of ancient shelf ridges; an example from Campanian Almond Formation in Hanna Basin, USA. Sedimentology 63:1651–1676

Louwye S, De Schepper S (2010) The miocene-pliocene hiatus in the southern North Sea Basin (northern Belgium) revealed by dinoflagellate cysts. Geol Mag 147(5):760–776

Mathys M (2009) The quaternary geological evolution of the Belgian continental shelf, southern North Sea. Unpublished PhD thesis, Ghent University, Belgium

Matthijs J, Lanckacker T, De Koninck R, Deckers J, Lagrou D, Broothaers M (2013) Geologisch 3D lagenmodel van Vlaanderen en het Brussels Hoofdstedelijk Gewest—versie 2, G3Dv2. Research project by VITO commissioned by Flemish Authorities, Departement Leefmilieu, Natuur en Energie, Afdeling Land en Bodembescherming, Ondergrond, Natuurlijke Rijkdommen, VITO-report 2013/R/ETE/43

Michaud KJ, Dalrymple RW (2016) Facies, architecture and stratigraphic occurrence of headland-attached tidal sand ridges in the Roda formation, northern Spain. In: Tessier B, Reynaud JY (eds) Contributions to modern and ancient tidal sedimentology: proceedings of the Tidalites 2012 conference, IAS Spec Pub 48, Wiley, Chichester. doi:10.1002/9781119218395.ch17

Natuurpunt Oost-Brabant (2014) De Wijngaardberg in Wezemaal, een erfgoed- en natuurmonument van Europees belang. Natuur en landschap, special edition with contributions of the symposium held at Wezemaal, 5 Oktober 2014

Reynaud J-Y, Dalrymple RW (2012) Shallow-marine tidal deposits. In: Davis RA, Dalrymple RW (eds) Principles of tidal sedimentology. Springer, Dordrecht, pp 335–369

Reynaud J-Y, Tessier B, Proust J-N, Dalrymple RW, Marsset T, De Batist M, Bourillet J-F, Lericolais G (1999) Eustatic and hydrodynamic controls on the architecture of a deep shelf sand bank (Celtic Sea). Sedimentology 46:703–721

Stride AH, Belderson RH, Kenyon NH, Johnson MA (1982) Offshore tidal deposits: sand sheet and sand bank facies. In: Stride AH (ed) Offshore Tidal Sands, Processes and deposits. Chapman and Hall, London, pp 95–125

Trentesaux A, Stolk A, Berné S (1999) Sedimentology and stratigraphy of a tidal sand bank in the southern North Sea. Mar Geol 159:253–272

Vail PR, Mitchum RM, Thompson S (1977) Seismic Stratigraphy and global changes of sea level. Part 3: relative changes of sea level from Coastal Onlap. In: Payton CE (ed) Seismic stratigraphy—applications to hydrocarbon exploration, vol 26. American Association of Petroleum Geologists Memoir, pp 63–81

Van den Berg J, van Gelder A, Mastbergen D (2002) The importance of breaching as a mechanism of subaqueous slope failure in fine sand. Sedimentology 49:81–95

Vandenberghe J, De Smedt P (1979) Palaeomorphology in the eastern Scheldt basin (Central Belgium), the Dijle-Demer-Grote Nete confluence area. Catena 6:73–105

Vandenberghe N, Gullentops F (2001) Kaartblad 32 Leuven. Toelichtingen bij de geologische kaart van België, Vlaams Gewest. Belgian geological survey and Flemish Authorities, afdeling Natuurlijke Rijkdommen en Energie

Vandenberghe N, Harris WB, Wampler JM, Houthuys R, Louwye S, Adriaens R, Vos K, Lanckacker T, Matthijs J, Deckers J, Verhaegen J, Laga P, Westerhoff W, Munsterman D (2014) The implications of K–Ar glauconite dating of the Diest Formation on the paleogeography of the Upper Miocene in Belgium. Geol Belg 17(2):161–174

Verhaegen J, Adriaens R, Louwye S, Vandenberghe N, Vos K (2014) Sediment-petrological study supporting the presence of the Kasterlee formation in the Heist-op-den-Berg and Beerzel hills, southern Antwerp Campine, Belgium. Geol Belg 17(3/4):323–332

15 Gullies and Closed Depressions in the Loess Belt: Scars of Human–Environment Interactions

Jean Poesen, Tom Vanwalleghem and Jozef Deckers

Abstract

This chapter illustrates how the study of the characteristics and spatial pattern of closed depressions (CDs) and several types of gullies in old forests may help unraveling human–environment interactions in (pre)historical times. The Meerdaal Forest in central Belgium, being continuously a forest for at least eight centuries and most probably already from the fifth century AD onwards, offers a unique opportunity to study such geomorphic features as it has very well conserved a pre-medieval landscape with 71 CDs and 252 gully channels. A systematic and detailed investigation of these geomorphic features, using process-based knowledge of soil erosion processes, reveals that these features result from a period with intense agricultural activities. The CDs are man-made features, most probably ancient quarries made to extract calcareous loess which was then used to lime nearby cropland on the flatter parts so as to improve its fertility. During high-magnitude rainfall events on these bare cropland areas that were cleared, sufficient runoff could then be produced to scour gully channels on sloping parts of the surrounding forest. Dating of both CDs and gullies in this old forest indicates that these features result from an agricultural land use phase between the eighteenth century BC and the third century AD. The gullies in Meerdaal Forest are rather unique as they are, to our knowledge, the only known gullies dating from the Roman period or before. These gullies teach us that intense soil erosion is not only an important soil degradation process occurring in recent times but that in historical times also periods with significant soil erosion have occurred. As such Meerdaal Forest is an important geomorphosite with a high geovalue.

Keywords

Geomorphosite • Geovalue • Geoconservation • Soil erosion • Soil fertility

15.1 Introduction

The interactions between humans and their environment, an important topic in geography, has been studied thoroughly only since 1945. Few detailed field observations and measurements have been made before that date. As a consequence, one gets the impression that the impact of humans on the landscape, and in particular on soil erosion rates induced by agricultural activities, has never been as high as during the last decades. Recent geomorphologic studies in old forests reveal however that this impression is wrong. During the past 2000 years human activities during

J. Poesen (✉)
Department of Earth and Environmental Sciences, KU Leuven, Celestijnenlaan 200E, 3001 Leuven–Heverlee, Belgium
e-mail: jean.poesen@kuleuven.be

T. Vanwalleghem
Hidrología e Hidráulica Agrícola, Univeristy of Cordoba, Campus Universitario Rabanales, Ctra Madrid km 396, 14071 Córdoba, Spain
e-mail: tom.vanwalleghem@uco.es

J. Deckers
Division of Soil and Water Management Department of Earth and Environmental Sciences, KU Leuven, Celestijnenlaan 200E, 3001 Leuven–Heverlee, Belgium
e-mail: seppe.deckers@kuleuven.be

A. Demoulin (ed.), *Landscapes and Landforms of Belgium and Luxembourg*,
World Geomorphological Landscapes, DOI 10.1007/978-3-319-58239-9_15

particular periods provoked intense soil erosion in the loess belt, resulting in scars in the landscape, i.e., gullies and closed depressions. These features can still be observed clearly in old forests, such as the Meerdaal Forest in central Belgium. This chapter illustrates how the study of these geomorphologic features can contribute to a better understanding of human–environment interactions in historical times.

15.2 Geographical Setting

Figure 15.1 shows the extent of the loess belt in Belgium as well as the location of the Meerdaal Forest. Today, most of this belt is under cropland, given the high soil quality (i.e., large water-holding capacity) of the loess-derived soils. However, these soils rank among the most erodible soils in the world (Poesen 1993). When covered by forests, mean annual soil loss by water erosion processes in the loess belt is quite low, typically <0.1 ton/ha/year (Maetens et al. 2012). In contrast, when deforested, and particularly when topsoil organic matter content drops below 2.5%, mean rates of soil loss by water erosion may easily go up to 10 ton/ha/year and more. On top of this, soil erosion by direct human activities adds to this soil degradation process. Such activities comprise soil tillage (soil translocation by tillage induces a mean soil loss rate of 8 ton/ha/year), extraction of root and tuber crops (inducing a mean soil loss rate of several tons/ha/year) and soil quarrying (Poesen et al. 2001). A combination of these natural and human-induced erosion processes has resulted in a mean soil loss rate of 26 ton/ha/year as well as peculiar geomorphic features in the landscape (Fig. 15.2), such as ephemeral gullies (concentrated flow erosion channels that are annually erased by the farmers) (Fig. 15.3), bank gullies (eroded channels due to runoff crossing earth banks) (Fig. 15.4), lynchets

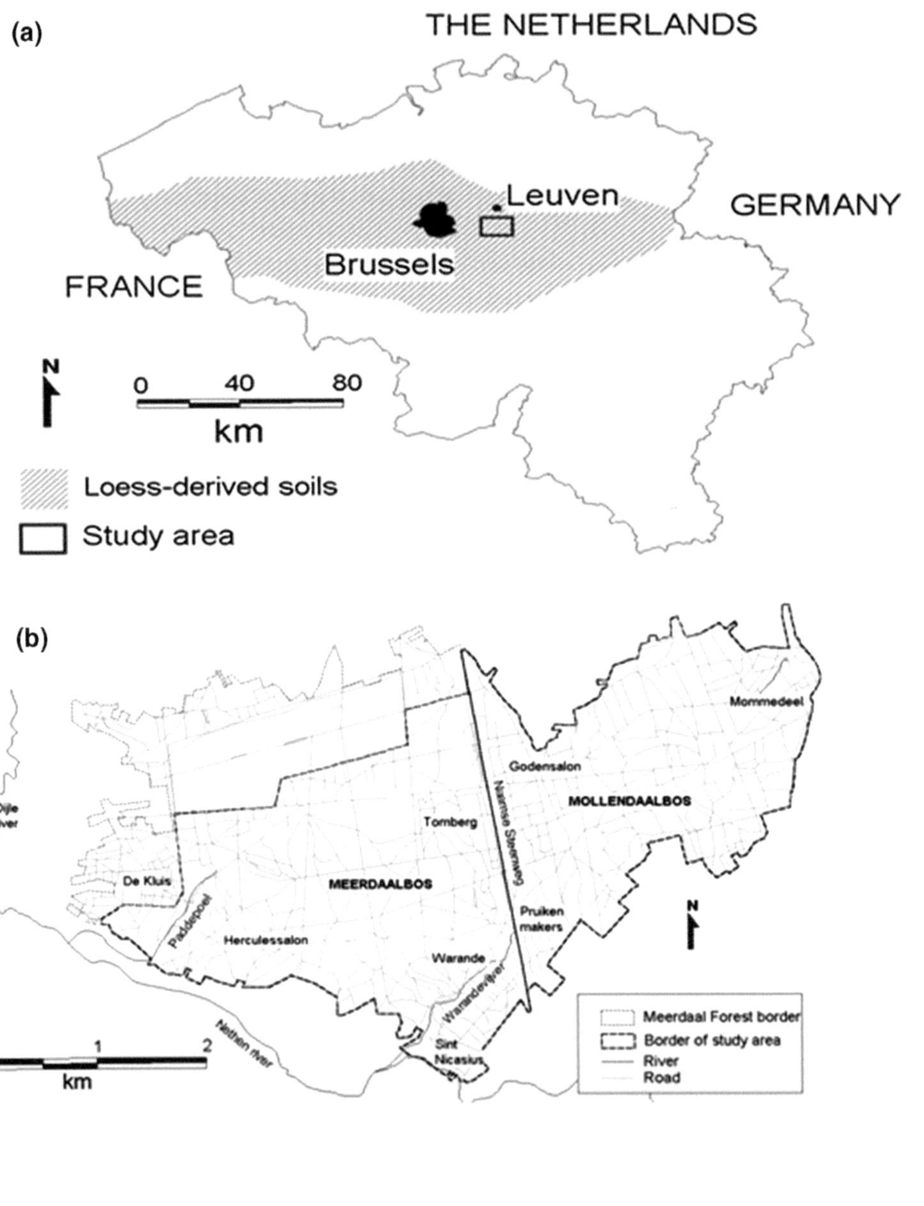

Fig. 15.1 **a** Location of the Meerdaal Forest within the Belgian loess belt. **b** Map of Meerdaal Forest, with indication of important locations (Vanwalleghem 2004)

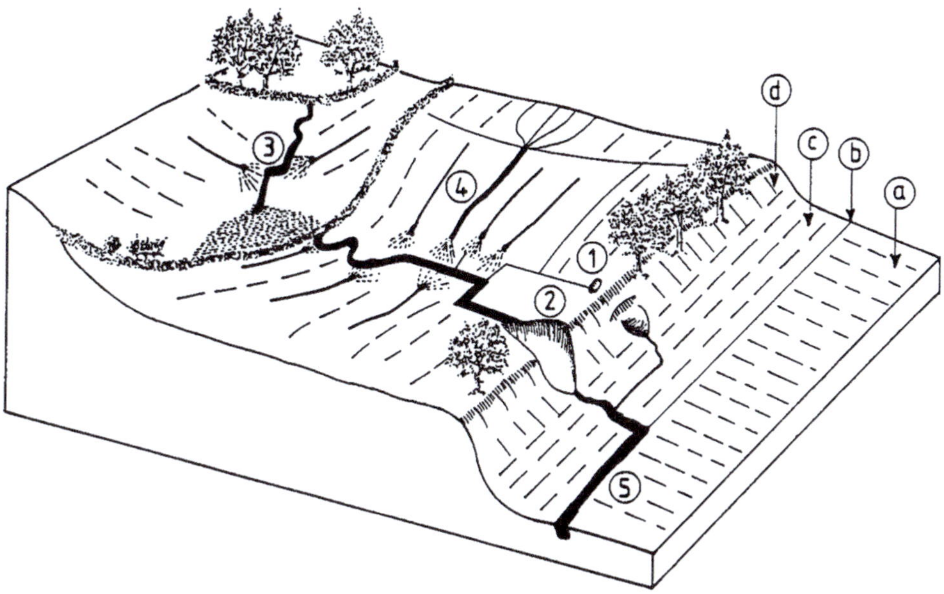

Fig. 15.2 Sketch illustrating various erosion features in agricultural lands of central Belgium. *1* pipe inlet; *2* bank gully; *3* ephemeral gully in valley-bottom; *4* ephemeral gully in valley-side; *5* ephemeral gully in linear landscape element; **a** tillage direction; **b** limit of headland; **c** headland; **d** bank (lynchet) (Poesen et al. 1996)

Fig. 15.3 Ephemeral gully channel formed in cropland of central Belgium (Bertem, May 2009). This erosion process contributes to ca. 50% of total soil loss by water erosion in the Belgian loess belt

(earth banks created by water and tillage erosion as well as by sediment deposition during water erosion and soil tillage) (Fig. 15.5), and road gullies (sunken lanes, or permanent (gully) channels that are used as access roads) (Fig. 15.6). There is a general belief that all these soil erosion processes were either absent or at least much less intense in historical times, when agriculture was much less mechanized and less intense.

Within the loess belt, there are very few ancient forest areas. One of these is the Meerdaal forest (Fig. 15.1). This forest was never entirely cut since the production of the Ferraris map in 1771. Based on older historical documents, one can even state that it was never completely cut since 1367. Miradal, the oldest name that refers to a forest on this spot in Belgium, dates from 1146 (Baeté et al. 2009). Within this forest, several archaeological sites from prehistoric and Roman times are present (Fig. 15.7). Based on these archaeological remains, no conclusion about the extent of the human occupation during these periods can be drawn. Was this forest at those times entirely transformed in

Fig. 15.4 Bank gully erosion channel formed in the bank of a sunken lane (visible on the *right*) (Bertem, May 2009)

Fig. 15.5 Earth bank (lynchet) in cropland formed by a combination of erosion and sediment deposition by tillage and water erosion (Bertem, 25 December 2014)

cropland? Was it only used to harvest wood and other forest products or for cattle grazing? A detailed study of the geomorphic features, namely gullies and closed depressions, may help answering these questions.

The unique history of the Meerdaal Forest makes this part of the loess belt an interesting landscape archive for the study of human–environment interactions dating from more than eight centuries ago. In contrast, in the surrounding cropland areas, various soil erosion processes during the Holocene have resulted in a mean soil profile truncation ranging between 0.3 m (on slopes <3%) and 1.6 m (on slopes >8%) (Verstraeten et al. 2009). As a consequence,

Fig. 15.6 Road gully (sunken lane) in cropland (Bertem, 25 December 2014)

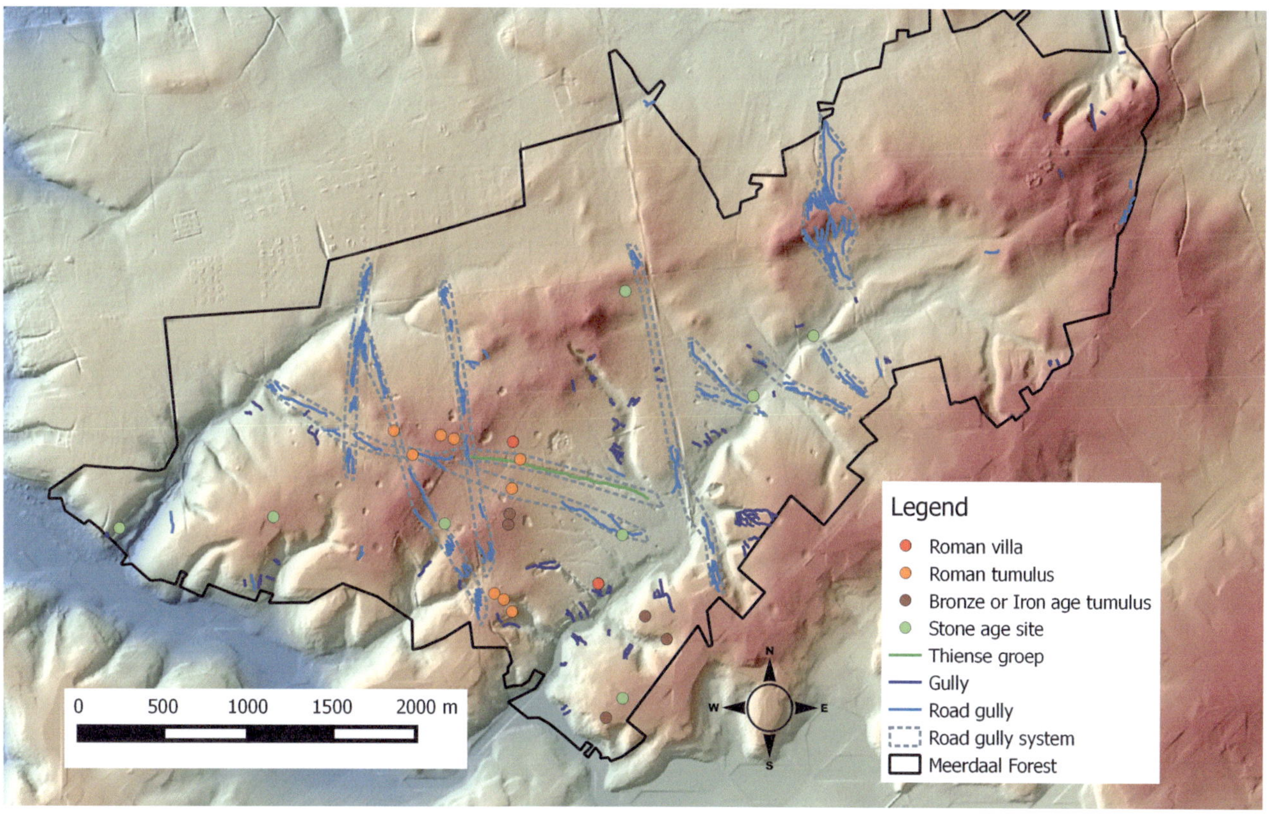

Fig. 15.7 Spatial distribution of the different types of incision (channels) within Meerdaal Forest, with indication of the identified road systems and the location of archaeological sites (Vanwalleghem et al. 2003)

many traces of former human occupation have been wiped out or are buried under colluvial or alluvial deposits. As reported above, the Meerdaal Forest has conserved scars in the landscape (gullies and closed depressions) resulting from human occupation more than eight centuries ago. An analysis of their characteristics and spatial distribution will increase our understanding of the history of this landscape and the role humans have played.

15.3 The Meerdaal Forest

The Meerdaal Forest is a 17 km^2 forest, located ca. 25 km east of Brussels between 50° 47′ 10″ and 50° 49′ 10″N and 4° 39′ 25″E (Fig. 15.1). It is one of the largest ancient forests in the Belgian loess belt. Ancient forests in Belgium are defined as forests that have existed continuously since at least 1775, which correspond to the publication of the first area-covering map, the Ferraris map (Hermy et al. 1999). Because it was owned since the twelfth century by local noblemen who used it for hunting purposes, the Meerdaal Forest is one of the few forest complexes that survived the important medieval clearing phases in Flanders between the second half of the twelfth and the beginning of the thirteenth century (Vanwalleghem 2004).

However, several studies mention the presence of archaeological sites and strange geomorphic phenomena (such as gullies and old sunken lanes) in Meerdaal Forest and other forests which were considered undisturbed (e.g., the Zonian Forest, east of Brussels) (e.g., Gullentops 1992; Vanwalleghem 2004; Baeté et al. 2009). Several hypotheses were put forward to explain their formation. Some researchers saw in the complex pattern of gullies and old sunken lanes clear evidence of Neolithic earth works for a religious purpose. Others believe all observed incisions were natural features or eroded paths. An attempt to investigate and date these phenomena in more detail was however never made (Vanwalleghem 2004).

This turns Meerdaal Forest into an interesting paradox of an apparently pristine, undisturbed area but with many peculiar landforms as relics of past environmental changes. Outside of the forest, a detailed analysis of the relation between past environmental changes and the landscape is difficult because geomorphic phenomena induced during past erosion phases have been erased by significant soil erosion. However, under forest, important past environmental changes have left a trace in the geomorphology. These traces have been accumulating in the "memory of the landscape," which makes Meerdaal Forest a unique study area.

The Meerdaal Forest is located on a large plateau east of the river Dijle and is dissected by three main NE–SW-oriented valleys. Altitudes vary between 35 m a.s.l. and 102.5 m a.s.l. Most of the area (80%) has slopes ranging between 0 and 0.08 m m^{-1}, although steep slopes up to 0.58 m m^{-1} occur. Mean annual precipitation is between 770 and 820 mm (observed at the nearby Beauvechain and Uccle stations) and is well distributed over the year. Annual evapotranspiration rate is ca. 550 mm. Hence, the annual precipitation surplus is ca. 220–270 mm. This implies a lixiviating soil regime, which has an important impact on soil development.

The Meerdaal Forest is part of the Belgian loess belt (Fig. 15.1a). The lithology consists of Pleistocene loess deposits of varying thickness overlying Tertiary marine sands. The loess deposits generally increase in thickness from the northern part of the study area, which is in the transition zone with the sandy or loamy plains to the north, to its southern part dominated by silt loam soils. Loess thickness also depends on the local slope aspect. Slopes oriented to the N, NE, and E nearly always have a markedly thicker loess layer than slopes oriented to the S, SW, and W (Goossens 1988). Especially in the northern part, but also on locally eroded hilltops in the rest of the forest, the Tertiary, and deposits outcrop.

The loess-derived soils occupy the largest part (87%) of the Meerdaal Forest. The major soil-forming processes in loess-derived soils are decalcification and clay eluviation–illuviation. In the literature, the timing of these processes in Belgium is often disputed. Some authors argue that the main clay illuviation phase took place during the Holocene, while others propose that the main soil formation took place during the Tardiglacial (15,000–11,700 cal. year. BP) (Vanwalleghem 2004). The interaction between the different parent materials and these soil-forming processes has resulted in a very varied soilscape. The soils in Meerdaal Forest can be subdivided into eight soil types based on the Belgian soil classification system. Soil classification according to WRB (ISSS working group 1998) is given between brackets.

Soil type 1 are well-drained (loamy) sandy soils with a spodic horizon (Podzols, Dystric Cambisol, or Arenosols) that occur on outcrops of Tertiary sand. Soil type 2 is well-drained (sandy) loam soils with an argic horizon (Haplic Luvisol). This soil type is a typical loess-derived soil found in agricultural land, but occurs in only 2% of the forest and exclusively at the borders of the Meerdaal Forest that have been deforested for some time. Soil type 3 is well-drained silt (loam) soils with a mottled argic horizon (Haplic Luvisol). Soil type 4 is well-drained silt (loam) soils with a strongly mottled argic horizon and albeluvic tonguing (Retisol). Soil types 3 and 4 represent a typical undisturbed loess-derived soil profile. Soil type 5 is soils with a sandy or clayey substrate at shallow depth (any soil type) and is usually associated with sandy soils (soil type 1) on Tertiary outcrops. Soil type 6 is poorly drained soils (Gleysols or Gleyic Luvisol). Finally, soil type 7 is alluvial and colluvial soils (Fluvisols or Regosols), mainly present in the three main valleys. The last group, soil type 8 is artificial soils that occur very locally.

Soil types 3 and 4 are the most important types in Meerdaal Forest. Together, they occupy nearly 50% of the Meerdaal Forest area, mainly in the southern part where there is a sufficiently thick loess cover. These soils have a moderate to high potential fertility but, in their natural, undisturbed condition, they are both chemically and physically infertile (Langohr 1990). Their pH is usually between 3.5 and 4.5, which means that they need fertilization prior to, e.g., cereal cultivation. Recently reforested soil profiles also have such a low pH because the soil pH, even after intense fertilization, returns to its original low value within a few decades (Bossuyt 2001).

Several pits were dug to compare soil profiles known to have undergone human influence at the forest border with undisturbed profiles in the middle of the forest. No morphological, chemical, or physical disturbance indicative of past land use was found in the undisturbed profiles (Vanwalleghem 2004).

In the loess-derived soil types 2, 3, and 4, the original, in situ calcareous loess can still be found if the loess deposit is thick enough. Decalcification depth in undisturbed forest soils ranges between 1.14 and 5.10 m, with an average value of 2.48 ± 0.63 m (n = 184; Vanwalleghem et al. 2010).

15.4 Gullies

Of all water erosion processes, gully erosion produces one of the most spectacular geomorphic features. Within Meerdaal Forest, several linear incisions, or gullies, are observed (Gullentops 1992) (Fig. 15.7). Gullies are defined as channels eroded during concentrated (overland) flow and having a cross section >900 cm^2, or one square foot (Poesen et al. 2003, 2006). In total, 252 such incisions have been mapped in Meerdaal Forest (Vanwalleghem et al. 2003).

Based on their spatial occurrence and their characteristics, they can be classified in two groups. A first group consists of linear incisions that most probably developed from rural roads (Fig. 15.7). Continuous trampling and compaction by humans, animals, and carts reduced the infiltration capacity of the topsoil so that during intense rain events, Hortonian runoff was produced. Once concentrated, this runoff cut erosion channels on slopes exceeding 3–4%. This process still occurs today in cropland areas and explains the formation of many road gullies (sunken lanes, Fig. 15.6) in central Belgium. The road gullies in Meerdaal Forest are typically located along rectilinear north-south and northwest–southeast transects (Fig. 15.7).

A second group of incisions is small and large gullies that seem to be randomly distributed in the forest (Figs. 15.7, 15.8, and 15.9). Mean length and mean depth of these gullies (n = 43) equals 64 m (σ = 50 m) and 1.26 m (σ = 0.98 m), respectively. The largest gullies have depths exceeding 6 m (Fig. 15.9). As to their origin, several hypotheses have been put forward. A first hypothesis states that these gullies developed in former cropland. Similar gullies still form occasionally today on cropland (Fig. 15.10). Several authors described similar features in old forests of Germany and France (e.g., Vogt 1953; Bork et al. 1998). After the cropland was reforested, these gullies were very well conserved as erosion was halted completely under the protective forest cover and the gullies were left "frozen" in the landscape. A second hypothesis states that all these gullies developed under forest during very intense rain storms. These rain storms should have been much higher or more intense than normal as runoff generation or gully incision is not observed in today's forests. A third hypothesis states that these gullies are rather old and were formed during the Pleistocene in cold periods, as shown in the Zoniën Forest by Van Vliet and Langohr (1981) based on the study of soil profiles.

In order to reconstruct the conditions in which the gullies developed, several techniques were applied. A first approach simulated the location of (ephemeral) gullies, assuming that the entire area was deforested and using a topographical threshold index that combines the critical soil surface slope and the corresponding drainage area (Poesen et al. 2003) needed to cut an ephemeral gully channel in cropland (Vanwalleghem et al. 2003). Compared to forest, this index is lowest for gullies forming in cropland (Poesen et al. 2003; Torri and Poesen 2014). It was calculated for every pixel of the digital elevation map of the Meerdaal forest. Next, critical topographic index values were determined for ephemeral gullies in nearby cropland. Pixels in the forest with an index value exceeding this critical value were interpreted as spots in the landscape with a high probability of being scoured by concentrated flow into an (ephemeral) gully channel, obviously for the condition that these pixels and the corresponding catchments were entirely located under cropland. A comparison of the predicted (ephemeral) gully pixels with the mapped gully pixels in the Meerdaal forest reveals that there is some correspondence between both only for the southern part of the forest. For many pixels elsewhere in the forest, predicted gully pixels do not coincide with mapped gullies. Furthermore, the areas showing best correspondence in the south contain a thicker loess cover and are potentially more fertile, while the northern part of the forest is sandier and hence less suitable for agriculture. Moreover, Figs. 15.7 and 15.8 reveal that most of the observed gully channels occur on fairly steep slopes in the southern part of the forest, where also most of the known archaeological sites are located (Vanwalleghem et al. 2003). As the simulation works well for cropland, we conclude that the observed gully channel pattern in the forest can be explained by assuming

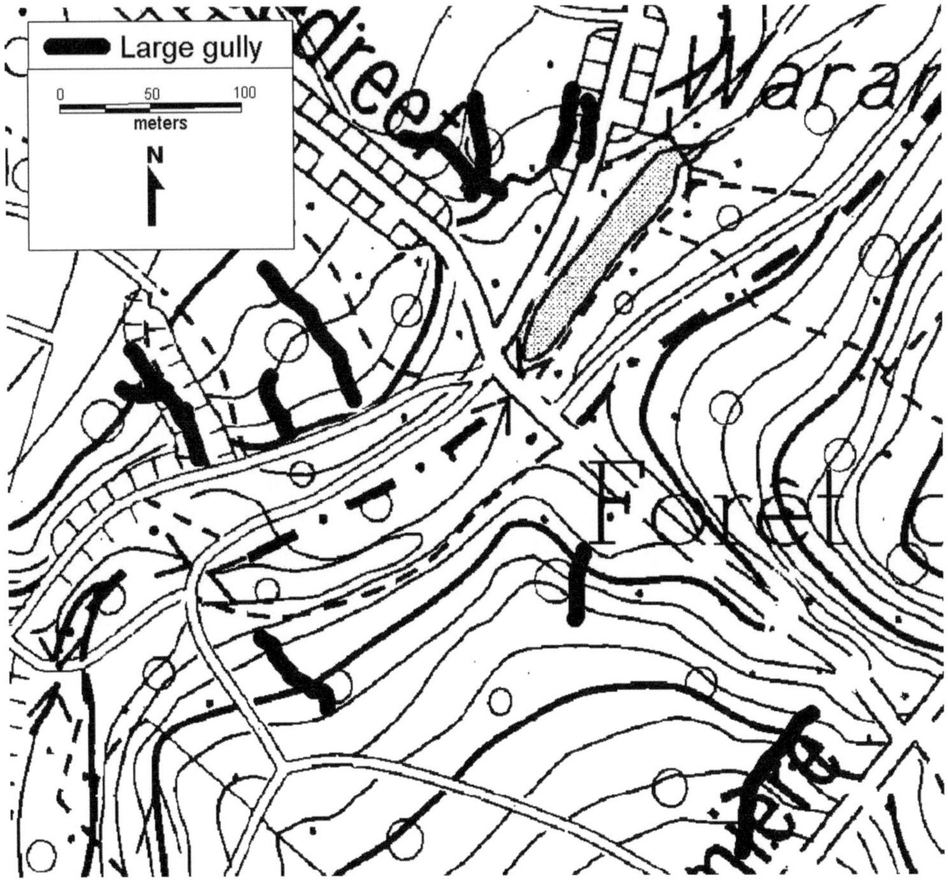

Fig. 15.8 Detail of the spatial distribution of the mapped gullies (only large gullies are shown here) (Vanwalleghem 2004)

Fig. 15.9 Old large gully channel in Meerdaal Forest (December 2009)

Fig. 15.10 Deep gully developed in cropland (Letterhoutem, 29 January 1993). This gully formed at the end of the winter on cropland with a compact topsoil resulting from potato harvesting. Calcareous loess is present at shallow depth in the soil profile explaining the rapid and deep incision by concentrated runoff

that during some period(s), there was locally cropland present in the south of the Meerdaal Forest, whereas the remaining part was left under forest. Similar to the situation in cropland today, large and deep gullies could then form on steep slopes (typically exceeding 8%; Vanwalleghem et al. 2005a) with bare soils during intense rains, where sufficiently large runoff discharge could produce gullying. Under a permanent forest cover with a thick leaf litter, runoff production under the current climatic conditions is far too small to form gully channels.

These results were corroborated by those of a second approach, in which an assessment was made of the peak runoff discharges (Q_p) in loam to sandy loam soils with a channel width similar to those observed in the Meerdaal Forest. Using measured channel bottom widths (W_{bottom}) of the old gullies as input, peak flow discharges can be estimated for these gullies using an empirical equation (Nachtergaele et al. 2002). The old gullies in Meerdaal Forest have an average W_{bottom} ranging between 1.1 and 1.5 m. Corresponding calculated Q_p values range between 0.04 and 0.07 m^3/s (Vanwalleghem et al. 2005b). Next, rainfall intensities (I) were deduced from Q_p using the rational formula. By simulating various land use scenarios via their corresponding runoff coefficient (C) values, I values, and concentration time (T_c) could be calculated for each land use class. Using I, T_c, and intensity-duration-frequency tables for central Belgium, the recurrence interval (RI) of the rain events needed to erode the observed gully channels was assessed (Fig. 15.11). Although analysis of historical documents indicates that Meerdaal Forest has probably been under forest since at least the Middle Age, it is unlikely that the old gullies originated under forest vegetation or even degraded forest vegetation, since rain events with RI in excess of 200 years are needed to cause gullying in such land use scenarios (Fig. 15.11). Cropland is the only land use that provides acceptable values of 11–128 years for RI (Vanwalleghem et al. 2005b) (Fig. 15.11). Moreover, analysis of the relation between soil surface slope and drainage area at the gully heads in Meerdaal Forest reveals that this relation typically corresponds to that for modern gullies that develop in cropland (Torri and Poesen 2014; Schotmans et al. 2015).

Dating of the sediment infill of two gullies using charcoal and artifacts indicates two major soil erosion phases in the Meerdaal forest, one in the Middle Bronze Age and a second one in the Roman period (Vanwalleghem et al. 2006) (Fig. 15.12). Moreover, remnants of a Roman villa have been reported near a gully. Most probably, this villa and the associated cropland were abandoned in the third century AD, as it happened also to other Roman villas in central Belgium. Human occupation and agriculture ceased after this period, as no indication of Merovingian or Medieval settlements were found in the forest. Based on field observation of present-day forest regrowth in this region, forest regrowth is known to have been rather rapid (<5 years), which explains the well-preserved gully microtopography under the protective forest cover during the following centuries. Under cropland however, gullies may rapidly disappear by sediment infilling if the contributing area produces large sediment yields (Vanwalleghem et al. 2005a).

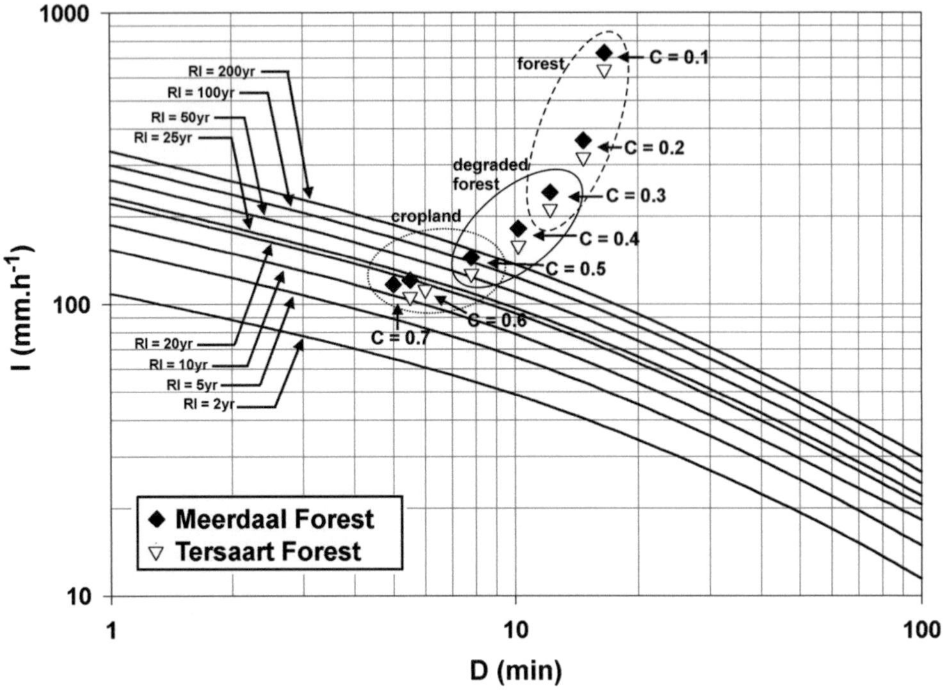

Fig. 15.11 Rain intensity-duration-frequency graph (KMI 2001) with indication of rainfall conditions (intensity *I*, duration *D*, and recurrence interval *RI*) needed to incise the observed gullies in two old forests (Meerdaal Forest and Tersaart Forest) under different land use scenarios, simulated using their corresponding runoff coefficient (*C*; $0.1 \leq C \leq 0.7$). The data points corresponding with the lowest runoff coefficient ($C = 0.05$) were not shown on the graph for practical reasons (Vanwalleghem et al. 2005b)

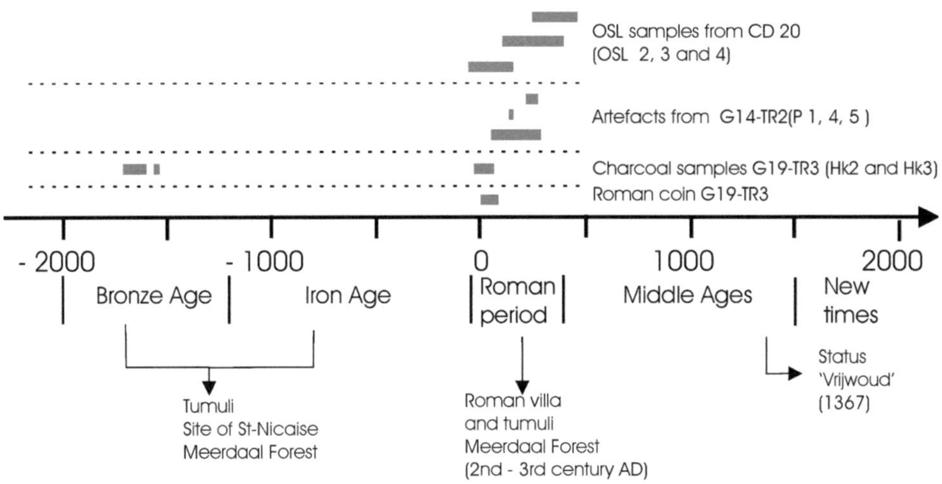

Fig. 15.12 Overview of the dates and datable material obtained from the gullies (*G*) and closed depressions (*CD*) in Meerdaal Forest. Time axis is in calendar years (Vanwalleghem et al. 2008)

15.5 Closed Depressions

A second geomorphic feature typical of Meerdaal Forest consists in the closed depressions (CDs) (Fig. 15.13). CDs are typically circular forms where water and sediment eroded from the surrounding soil surfaces are trapped. They have been reported in several other regions of the European loess belt and are attributed either to natural processes, such as dissolution of subsurface horizons, piping or thermokarst, or to human intervention (e.g., quarrying, bomb craters). Previous CDs studies mainly focused on cropland areas where, however, only few and largely infilled CDs remain.

A systematic survey of the Meerdaal Forest allowed 71 CDs to be mapped (Fig. 15.14). Their mean diameter (*D*), surface area (*A*), maximum depth (d_{max}), and average sidewall slope (*S*) are as follows: $D = 28.8$ m ($\sigma = 13.8$ m), $A = 780$ m^2 ($\sigma = 790$ m^2), $d_{max} = 2.03$ m ($\sigma = 0.94$ m), and $S = 28.9\%$ ($\sigma = 10.9\%$). In contrast, 30 CDs observed in the nearby cropland area have significantly different characteristics, namely $D = 94.7$ m ($\sigma = 29.9$ m), $A = 7100$ m^2 ($\sigma = 4300$ m^2), $d_{max} = 1.10$ m ($\sigma = 0.76$ m), and $S = 2.9\%$ ($\sigma = 2.6\%$) (Vanwalleghem et al. 2007) (Fig. 15.15). The comparison of their morphology indicates that CDs under forest are significantly deeper, with steeper sidewalls and a

Fig. 15.13 Deep (ca. 3.0 m) closed depression in Meerdaal Forest (December 2009). *Note* person as scale

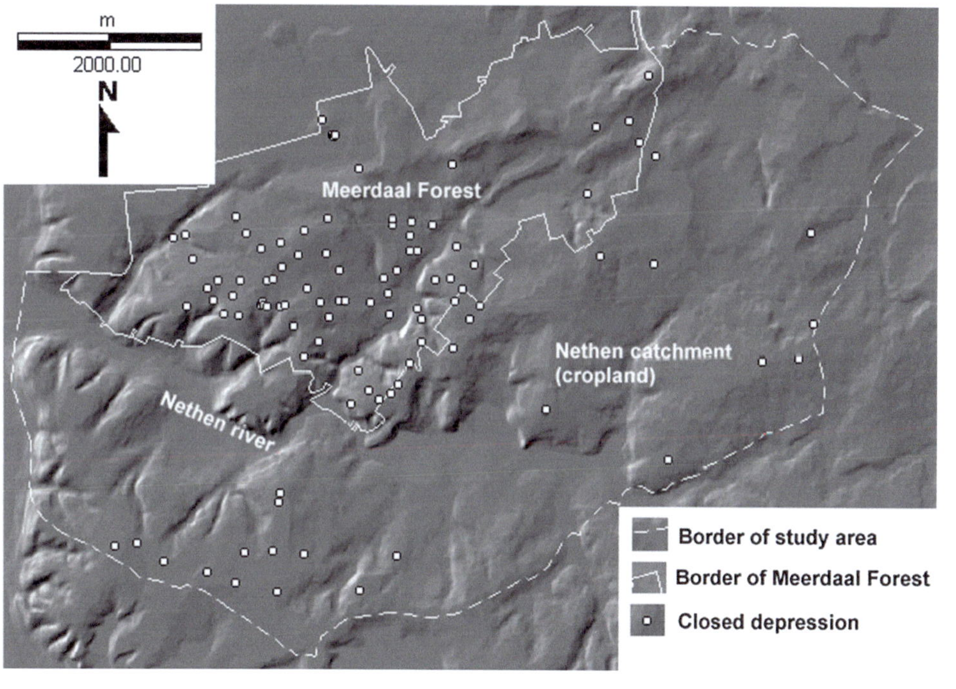

Fig. 15.14 Location of the closed depressions under forest (Meerdaal Forest) and under cropland (southern part of study area) (Vanwalleghem et al. 2007)

smaller surface area than those under cropland subject to more active erosional/depositional processes. By comparing CDs that had been under cropland for different time intervals, the rate of their morphological evolution could be reconstructed (Vanwalleghem et al. 2007).

Taking into account the respective size of the two surveyed study areas (i.e., 13 km^2 for the forest and 29 km^2 for the cropland) reveals a significant difference in CD density, with 5.3 CDs km^{-2} under forest, and only 1 CD km^{-2} for the cropland area (Vanwalleghem et al. 2007). A similar CD

Fig. 15.15 Shallow (<1.5 m deep) closed depression in cropland near Meerdaal Forest (December 2009). *Note* person at the *right* as scale

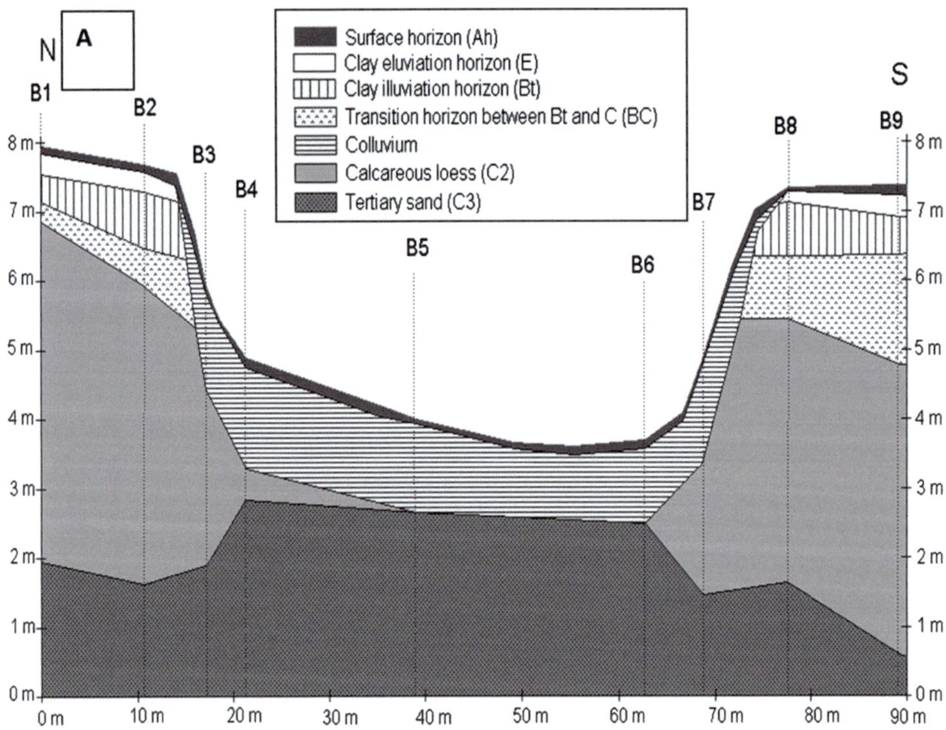

Fig. 15.16 Detailed topography and stratigraphy of a closed depression in Meerdaal Forest based on soil augerings (Vanwalleghem et al. 2007)

density of 0.6 CD km^{-2} was also reported in a dominantly cropland area elsewhere in the loess belt (Gillijns et al. 2005).

In order to unravel their origin, soil augerings and trenches were made within and outside two CDs in Meerdaal Forest (Fig. 15.16). A detailed analysis of the soil stratigraphy confirmed their origin as former quarries. Most probably, calcareous loess, with a $CaCO_3$ content between 10 and 15% (Goossens 1987), was excavated, since this 2–5-m-thick soil horizon was completely absent within the CDs. Dating of the infilling of one CD by optically stimulated luminescence (OSL) shows that the CD filled in between the first century BC and the fourth century AD (Fig. 15.12).

Quarrying of calcareous loess to improve soil fertility by increasing soil pH is a practice that existed already before the Roman invasion (Lindemans 1952) and lasted up to the introduction of mineral fertilizers around 1930. The undisturbed forest soil in the loess belt is rather acidic ($3.5 < pH < 4.5$) and less suitable for crop production (Vanwalleghem et al. 2007). Calcareous loess is an ideal fertilizer in these conditions. Moreover, the clay-rich Bt horizon ("terre-à-brique") could also be used for brick making. In total, 120,000 tons of soil have been quarried from the CDs in Meerdaal Forest, which corresponds to 92.3 ton per ha of forest. Recent studies elsewhere in Belgium, Germany, and France also point to a similar anthropogenic origin of closed depressions (see Kołodyńska-Gawrysiak and Poesen 2017).

15.6 Linkage Between Gullies and Closed Depressions

The analysis of these two distinct geomorphic features, gullies, and CDs, in an old forest allows one to reconstruct land use history in a distant past. Applying rare events logistic regression to the data resulted in two multivariate models that effectively simulate the probability of occurrence of CDs and permanent gullies under forest (Vanwalleghem et al. 2008). The obtained logistic regression models show that both anthropogenic and physical factors are important to describe the spatial distribution of both geomorphic features. For the prediction of the occurrence of closed depressions, anthropogenic factors are most important however.

The probability of finding a CD is larger close to reported prehistoric and Roman archaeological sites, on well-developed, undisturbed Luvisols and on northeast- to southeast-facing slopes. By contrast, the local slope gradient of the soil surface is the most important factor in the multivariate model for predicting the presence of the permanent gullies. The occurrence probability of permanent gullies is also larger on northeast- to southeast-facing slopes. The anthropogenic factor expressed by the distance to the closed depressions is significant as a univariate model. This suggests that a critical slope gradient is a prerequisite for gully incision, but gully formation still needs to be triggered by a human-induced land use change. It can therefore be concluded that both the gullies and the closed depressions in the Meerdaal Forest are related to past human activities, either directly, through quarrying of calcareous loess or indirectly, through runoff production on bare cropland.

In conclusion, both gullies and CDs in this old forest point to the facts that, at their time of formation, agricultural activities took place in their vicinity and that this forest was an important transit area (judging from the road gullies). The dating of CDs coincides with that of sediment deposits in nearby, human-induced gullies that were attributed to an agricultural land use phase between the eighteenth century BC and the third century AD (Fig. 15.12). A most probable scenario is that only parts of Meerdaal Forest have been deforested during this period to establish cropland mainly on the flatter plateau areas. In order to enhance crop productivity, nearby soil pits, which would later become CDs, were dug in order mainly to quarry calcareous loess for liming the land and hence improving soil fertility and crop production. In addition, the clay-rich Bt horizon was possibly used as construction material for houses. During intense rain storms with a return period of 10–25 years or more, these cropland areas, in periods of low vegetation cover, produced large volumes of runoff that caused significant gully erosion on steep slopes (>10%). This former land use has left significant traces also in the vegetation of Meerdaal Forest. Vegetation analysis in 31 carefully selected plots revealed persistent effects of a Roman land use at the individual plant species level, at the species group level, and at the community level. Indications were found that the vegetation on formerly cultivated and fertilized sites still has a more eutrophic character and hence, contains more typical forest species (Vanwalleghem et al. 2004).

15.7 The Geovalue of Meerdaal Forest

Although the results discussed above show that the Meerdaal Forest is far from being a natural, undisturbed area, the geomorphic features found in this old forest are rare and possibly unique in Belgium and even in northwestern Europe. Even if a (pre)Roman land use phase occurred, the Meerdaal Forest is one of the oldest forests known today in central Belgium. Therefore, several arguments can be put forward that stress the large value of this area and the need for geoconservation.

Sometimes called the "forgotten half of nature conservation" because much of the focus in nature conservation is on biodiversity, geoconservation aims to preserve the natural diversity of significant geological, geomorphologic and soil features and processes (i.e., the nonliving components of the natural environment), and to maintain natural rates and magnitudes of change in these features and processes (Sharples 2002). An absence of understanding and information on the geovalue is one of the main reasons for this lack of consideration (Gray et al. 2004). This is perfectly illustrated in an assessment of the economic value of the Meerdaal Forest by Moons et al. (2000), where they only mention that "variability in topography is positively evaluated by visitors" and they include the "prevention of soil erosion" as only contributor to the "geo-section" of the nonuse value of Meerdaal Forest.

The geovalue of the Meerdaal Forest can be defined as the threefold value conferred by its geodiversity (Sharples 2002). The first value is the intrinsic value, meaning that the geomorphic phenomena in Meerdaal Forest have value simply by existing, free of human valuation, or, in other words, being a good example of their type. A second value component comprises the ecological value or the importance in maintaining the natural systems and processes, both biotic and abiotic, of which it is part. In Meerdaal Forest, this is represented by the old forest plants, which thrive on old, undisturbed loess soils only. The last class of value, also called the geoheritage, includes values that have a direct significance to humans. In the case of the Meerdaal Forest, the geomorphic phenomena, but also the quasi-undisturbed soil profiles, provide scientific evidence documenting the early land use history and soil-forming processes. It is therefore an important place for research and education. The incised gully systems and, to some extent, the closed depressions largely contribute to a diversified landscape and can therefore be considered as enhancing the aesthetic aspect of the Meerdaal Forest. In order to stimulate the (geo)-touristic attraction of the Meerdaal Forest, a hiking trail passing by some representative geomorphic and archaeological sites, illustrated with information panels and an excursion guide including a map, has recently been elaborated upon (Baeté 2014).

Within Meerdaal Forest, more than 5 CDs occur per km^2 whereas in the adjoining cropland only one CD per km^2 can still be observed (Fig. 15.14). If one assumes that the CD density was originally the same within and outside the forest area, one can conclude that ca. 80% of the old CDs (and gullies) have disappeared through various erosion processes and sediment deposition. As such, Meerdaal Forest plays a very important role in geoconservation.

15.8 Conclusions

This chapter illustrates that the study of geomorphic features in old forests may be very rewarding in that it allows for a better understanding of human–environment interactions in a distant past. The Meerdaal Forest is rather unique because it has been continuously forested for at least eight centuries and most probably already from the fifth century AD onwards. As such, it has very well conserved a pre-medieval landscape with its microtopographic features, i.e., closed depressions and gullies. For geoscientists, the Meerdaal Forest offers a particular window on the past, allowing them to study human–environment interactions that took place more than 800 years ago. Most gullies in Meerdaal Forests are rather unique in that they are the only known gullies most probably dating from the Roman period or before. Together with the CDs, they point to a different land use that triggered these geomorphic features. The gullies in Meerdaal Forest teach us that intense soil erosion is not only a significant soil degradation process occurring in the last decades but that in historical times also periods with very significant soil erosion occurred, as has also been shown elsewhere, e.g., in France (Vogt 1953) and in Germany (Bork et al. 1998). As such, Meerdaal Forest must be considered as an important geomorphosite with a high geovalue.

References

Baeté H (2014) Miradal. Ontdek je erfgoed in Heverleebos en Meerdaalwoud. Vrienden van Heverleebos en Meerdaalwoud, Heverlee, 72 pp

Baeté H, De Bie M, Hermy M, Van den Bremt P (2009) Miradal. Erfgoed in Heverleebos en Meerdaalwoud. Davidsfonds, Leuven, p 263

Bork HR, Bork H, Dalchow C, Faust B, Piorr H-R, Schatz T (1998) Landschaftsentwicklung in Mitteleuropa. Klett-Pertes, Stuttgart, p 328

Bossuyt B (2001) Plant species and soil dynamics across ancient-recent forest ecotones: consequences for ecological restoration. Unpublished Ph.D. thesis. Fac. for applied Biological Sciences, KU Leuven, pp 182

Gillijns K, Poesen J, Deckers J (2005) On the characteristics and origin of closed depressions in loess-derived soils in Europe—a case study from central Belgium. CATENA 60:43–58

Goossens D (1987) Sedimentatiemechanismen bij natuurlijke stofdeeltjes in de lucht. Unpublished Ph.D. thesis. Department of Geography, KU Leuven

Goossens D (1988) Scale model simulations on the deposition of loess in hilly terrain. Earth Surf Proc Land 13:533–544

Gray M, Jungerius PD, van den Ancker J (2004) Geodiversity and geoheritage of soil protection. In: Advice to the working groups towards a thematic strategy for soil protection. European Union Soil Thematic Strategy, p 10

Gullentops F (1992) Holocene soil erosion in the loess belt of Belgium. Acta Geogr Lovaniensa 33:671–684

Hermy M, Honnay O, Firbank L, Grashof-Bokdam C, Lawesson J (1999) An ecological comparison between ancient and other forest plant species of Europe, and the implications for forest conservation. Biol Cons 91:9–22

ISSS working group RB (1998) In: Deckers JA, Nachtergaele FO, Spaargare, OC (eds) World reference base for soil resources: introduction, 1st edn. ISSS, ISRIC and FAO. Acco, Leuven, p 165

KMI (2001) Extreme precipitation in Flanders, new IDF-curves based on long-term precipitation measurements. Ministerie van de Vlaamse Gemeenschap, Brussels (in Dutch)

Kołodyńska-Gawrysiak R, Poesen J (2017) Closed depressions in the European loess belt—natural or anthropogenic origin? Geomorphology 288:111–128. http://dx.doi.org/10.1016/j.geomorph.2017.02.004

Langohr R (1990) The dominant soil types of the Belgian loess belt in the early neolithic. In: Cahen D, Otte M (eds) Rubané et Cardial, E.R.A.U.L. 39, Liège, pp 117–124

Lindemans P (1952) Geschiedenis van de landbouw in België. Deel 1. De Sikkel, Antwerpen

Maetens W, Vanmaercke M, Poesen J, Jankauskas B, Jankauskien G, Ionita I (2012) Effects of land use on annual runoff and soil loss in

Europe and the mediterranean: a meta-analysis of plot data. Prog Phys Geogr 36:599–653

Moons E, Eggermont K, Hermy M, Proost S (2000) Economische waardering van bossen: een case-study van Heverleebos-Meerdaalwoud. Garant, Leuven, p 356 (in Dutch only)

Nachtergaele J, Poesen J, Sidorchuk A, Torri D (2002) Prediction of concentrated flow width in ephemeral gully channels. Hydrol Process 16(10):1935–1953

Poesen J (1993) Gully typology and gully control measures in the European loess belt. In: Wicherek S (ed) Farm land erosion in temperate plains environment and hills. Elsevier, Amsterdam, pp 221–239

Poesen J, Vandaele K, van Wesemael B (1996) Contribution of gully erosion to sediment production in cultivated lands and rangelands. IAHS Publ. 236:251–266

Poesen JWA, Verstraeten G, Soenens R, Seynaeve L (2001) Soil losses due to harvesting of chicory roots and sugar beet: an underrated geomorphic process? CATENA 43(1):35–47

Poesen J, Nachtergaele J, Verstraeten G, Valentin C (2003) Gully erosion and environmental change: importance and research needs. CATENA 50(2–4):91–133

Poesen J, Vanwalleghem T, de Vente J, Knapen A, Verstraeten G, Martinez-Casasnovas JA (2006) Gully erosion. In: Boardman J, Poesen J (eds) Soil Erosion in Europe. Wiley, Chichester, pp 515–536. ISBN 0-470-85910-5

Schotmans J, Poesen J, Vanmaercke M (2015) Hoe zijn ravijnen in oude bossen ontstaan? Studiegeval Neigembos. Aardrijkskunde 2015:101–110

Sharples C (2002) Concepts and principles of geoconservation. Published electronically on the Tasmanian parks and wildlife service website (version 3)

Torri D, Poesen J (2014) A review of topographic threshold conditions for gully head development in different environments. Earth Sci Rev 130:73–85

Van Vliet B, Langohr R (1981) Correlation between fragipans and permafrost with special reference to silty weichselian deposits in Belgium and Northern France. CATENA 8:137–154

Vanwalleghem T (2004) Environmental change and geomorphic response in old forests of the loess belt. Unpublished PhD. thesis, Department geography, KU Leuven

Vanwalleghem T, Van Den Eeckhaut M, Poesen J, Deckers J, Nachtergaele J, Van Oost K, Slenters C (2003) Characteristics and controlling factors of old gullies under forest in a temperate humid climate: a case study from the Meerdaal Forest (Central Belgium). Geomorphology 56(1–2):15–29

Vanwalleghem T, Verheyen K, Hermy M, Poesen J, Deckers J (2004) Legacies of Roman land-use in the present-day vegetation in Meerdaal Forest (Belgium)? Belg J Bot 137(2):181–187

Vanwalleghem T, Bork HR, Poesen J, Schmidtchen G, Dotterweich M, Nachtergaele J, Bork H, Deckers J, Brüsch B, Bungeneers J, De Bie M (2005a) Rapid development and infilling of a buried gully under cropland, central Belgium. CATENA 63(2–3):221–243

Vanwalleghem T, Poesen J, Van Den Eeckhaut M, Nachtergaele J, Deckers J (2005b) Reconstructing rainfall and land use conditions leading to the development of old gullies. The Holocene 15(3): 378–386

Vanwalleghem T, Bork HR, Poesen J, Dotterweich M, Schmidtchen G, Deckers J, Scheers S, Martens M (2006) Prehistoric and Roman gullying in the European loess belt: a case study from central Belgium. Holocene 16(3):393–401

Vanwalleghem T, Poesen J, Vitse I, Bork HR, Dotterweich M, Schmidtchen G, Deckers J, Lang A, Mauz B (2007) Origin and evolution of closed depressions in central Belgium, European loess belt. Earth Surf Proc Land 32:574–586

Vanwalleghem T, Van Den Eeckhaut M, Poesen J, Govers G, Deckers J (2008) Spatial analysis of factors controlling the presence of closed depressions and gullies under forest: application of rare event logistic regression. Geomorphology 95:504–517

Vanwalleghem T, Poesen J, McBratney A, Deckers J (2010) Spatial variability of soil horizon depth in natural loess-derived soils. Geoderma 157:37–45

Verstraeten G, Rommens T, Peeters I, Poesen J, Govers G, Lang A (2009) A temporarily changing holocene sediment budget for a loess-covered catchment (central Belgium). Geomorphology 108:24–34

Vogt J (1953) Erosion des sols et techniques de culture en climat tempéré maritime de transition. Rev Géomorphol Dynam 4: 157–183

River Landscapes in the Dijle Catchment: From Natural to Anthropogenic Meandering Rivers

16

Gert Verstraeten, Bastiaan Notebaert, Nils Broothaerts, Jef Vandenberghe and Paul De Smedt

Abstract

In this chapter the changing river landscapes in the Dijle catchment since the Late Glacial are discussed in terms of climatic and anthropogenic changes. Climatic changes in the Late Glacial triggered the incision of large meanders in Weichselian braided-river deposits in the downstream part of the study area, which still form one of the most prominent landforms in this region. During the early Holocene period, peat accumulation, which initiated in the now-abandoned meanders, characterized floodplains lacking a clearly defined river channel. The landscape was covered with a deciduous forest, while floodplains were mainly covered with an alder carr forest. Although agriculture started in the region in the Neolithic ($\sim$6200 cal BP), it is only from the Bronze Age onwards that anthropogenic deforestation on the hilly loess slopes caused an increase in soil erosion and sediment delivery. In the upstream part of the Dijle catchment, this resulted in accelerated floodplain sedimentation that halted peat growth and covered the Late Glacial landforms entirely. Rivers developed again a meandering channel pattern with much smaller amplitude compared to the Late Glacial meanders, whilst also levees and colluvial fans were formed. During the past 250 years, many of these channels were artificially or naturally cut off, and they remain visible in the current floodplain as oxbow lakes or shallow depressions. In this chapter we demonstrate how this evolution is the result of a change from a climate-dominated to a human-dominated fluvial landscape.

Keywords

Central Belgium • Dijle valley • Late glacial and Holocene floodplain evolution • Anthropogenic influence • Holocene environments • Free meanders

16.1 Introduction

Contemporary fluvial landscapes of northern and central Belgium are characterized by low energy meandering river channels with a relatively low sinuosity. Floodplains are relatively wide and nearly flat and most rivers can be considered to be under-fit compared to the dimensions of the floodplain. Many of these valley systems exhibit landscape features that testify to both climate- and human-induced changes over the past 15,000 years. As many other streams in the Scheldt River Basin, the River Dijle originates in the hilly central Belgian loess belt and flows in a northerly direction towards the low-lying and flat Flemish Valley (see

G. Verstraeten (✉) · B. Notebaert · N. Broothaerts
Department Earth and Environmental Sciences, Division of Geography and Tourism, KU Leuven, Celestijnenlaan 200E, 3001 Leuven–Heverlee, Belgium
e-mail: gert.verstraeten@kuleuven.be

N. Broothaerts
e-mail: nils.broothaerts@kuleuven.be

B. Notebaert
FWO Fund for Scientific Research, Flanders, Belgium

J. Vandenberghe
Department of Earth Sciences, VU Amsterdam, De Boelelaan 1085-1087, 1081 HV Amsterdam, The Netherlands
e-mail: jef.vandenberghe@vu.nl

P. De Smedt
Heverlee, Belgium
e-mail: dspaul@telenet.be

A. Demoulin (ed.), *Landscapes and Landforms of Belgium and Luxembourg*, World Geomorphological Landscapes, DOI 10.1007/978-3-319-58239-9_16

Chap. 18) (Fig. 16.1). The difference in setting between the hilly loess area and the northern Flemish valley results in contrasting landscapes: in the loess area the river flows through a very flat floodplain in an otherwise hilly region, while in the downstream parts the river valleys themselves exhibit often the most prominent landscape forms. Today, these floodplains have an important environmental and ecological value and are often considered as one of the main natural landscapes in northern Belgium. In this chapter, we will focus on the typical landforms of the floodplains in these two contrasting regions. We will show how first climatic changes have shaped the floodplains, whereas during the past few thousand years unintentional human action, through deforestation and consequent soil erosion, can be considered as the main geomorphic agent. Floodplains are studied using detailed LIDAR data (1 m resolution; AGIV DHM II), extensive coring datasets, floodplain cross sections, palynological analysis and detailed soil maps.

16.2 Geographical Setting

In this chapter we focus on the main floodplain of the River Dijle, downstream of the city of Wavre up to a few kilometres east of the city of Mechelen (Fig. 16.1). From a geomorphic point of view, this study area falls apart in two main regions, namely the central Belgian loess belt for the upstream sections and the Flemish Valley (see Chap. 18) for the downstream sections. The substratum of the central Belgian loess belt is composed of Cenozoic marine deposits (mainly sands in the concerned area), covered with a loess layer of variable thickness (locally absent and up to a few metres). Topographically, it is an undulating plateau with deeply incised, broad river valleys. At the downstream end of this region, i.e. near Leuven, the floodplain is up to 1 km wide and lies approximately 50–70 m below the level of the loess plateau. There is a long history of deforestation and agriculture in this region, starting in the early Neolithic, at least ca. 6200 cal a BP (Crombé and Vanmontfort 2007; Vanmontfort 2007). From the Bronze Age onwards, this region has been under intense human impact (Broothaerts et al. 2014c). At present, large parts of the plateau are used for cropland cultivation, although some historical forests are still present. In contrast, the considered floodplains are mainly used for grassland and plantation forests, while also many fish ponds are present.

A few kilometres north of Leuven, the river enters the "Flemish Valley". Here, repeated erosion and deposition cycles have shaped a large depression, which was, for its major part, filled with sandy fluvial deposits during the Weichselian, giving the landscape its flat character. After the confluence with the River Demer near the village of

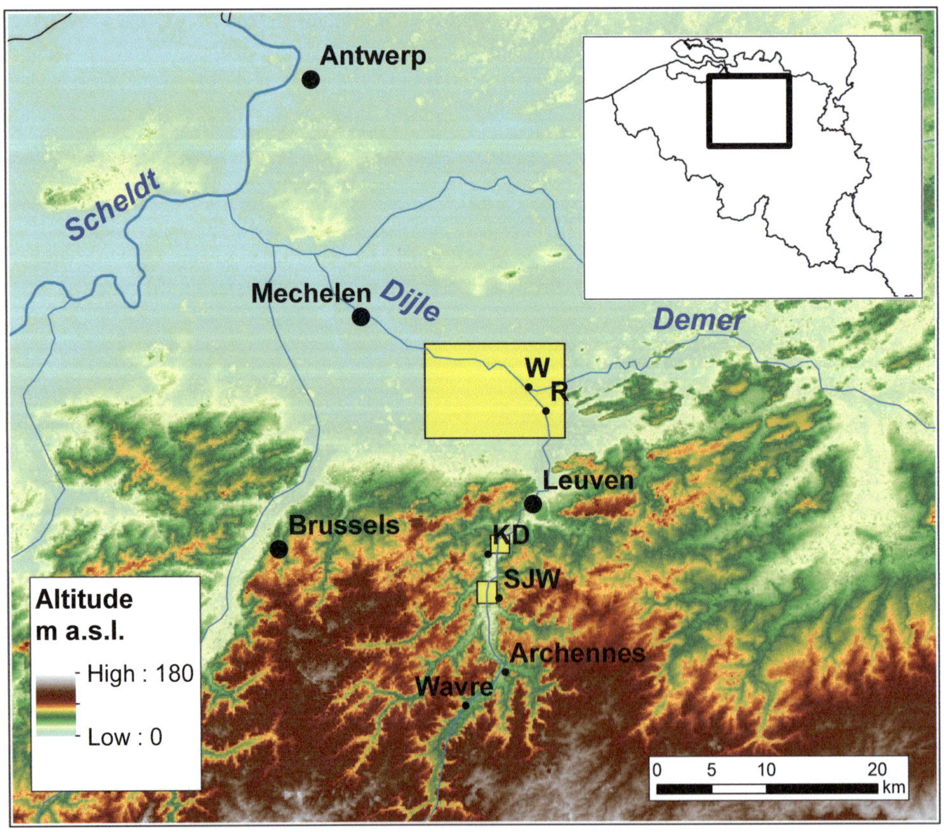

Fig. 16.1 Location map of the study areas in central Belgium. The *yellow squares* indicate the position of the study areas discussed in detail, namely the Dijle–Demer confluence area (see Fig. 16.2) near Werchter (*W*) and Rotselaar (*R*) and the upstream Dijle valley near Korbeek-Dijle (*KD*; Fig. 16.9 c, d) and Sint-Joris-Weert (*SJW*; Fig. 16.9 a, b)

Werchter, the river takes a western direction towards the River Scheldt. Slightly more to the northwest (Tremelo–Keerbergen area), a former Grote Nete course dissecting the resistant Boom Clay substratum joined the Dijle–Demer ('Houtvenne gully') until the Weichselian Pleniglacial (Vandenberghe 1977; Vandenberghe and De Smedt 1979). The W–E trending axis of the Dijle–Demer represents the far end of the eastern branch of the Flemish Valley and reflects the structural influence of the outcropping Cenozoic layers with different erosion resistance, which caused a subsequent valley system to be established in a period of pre-Holsteinian sea-level low-stand (Vandenberghe 1977; Vandenberghe and De Smedt 1979). In this confluence area, a complex system of Late Quaternary floodplains has formed, which is one of the focal points of this chapter. The River Demer has its headwaters also in the loess region but, due to a different morphological setting, a larger part of the catchment drains more gently sloping sandy regions. Here, anthropogenic impact on land cover has long been less intense compared to the loess regions, although cropland was probably also dominant for much of the area since the Middle Age.

16.3 The Dijle–Demer Confluence Area

The flat landscape of the confluence region of the Dijle and Demer is dominated by several late Quaternary fluvial landforms. Figure 16.2 shows a detailed LIDAR-based elevation model of the confluence area, combined with a geomorphic map. Large valleys are recognizable for the Dijle and Demer rivers. These valleys are incised in Weichselian fluvial sandy deposits formed by a wide braided-river system that was present in the entire Flemish Valley in that time (see Chap. 18). The southern edges of this Flemish Valley can locally be recognized easily in the landscape as a series of topographic steps that can reach a height of 5–7 m over a distance of 100–200 m (Fig. 16.2). Pleistocene sandy and gravelly fluvial terraces are locally found on the interfluves south of this Flemish Valley (1, Fig. 16.2) (Vandenberghe and De Smedt 1979).

Developed after a main incision phase at the beginning of the Late Glacial (Vandenberghe 2001), the Holocene floodplains are situated within the Flemish Valley fill. Two parallel E–W oriented incised valleys can be recognized on the LIDAR DEM, with remnants of the Weichselian braided-river deposits in topographic inversion in between (Fig. 16.2). At the transition from the present floodplain to the Weichselian terrace levels, a pattern of large inherited meander forms can be observed (Figs. 16.2 and 16.3). The northern floodplain is characterized by large elongated palaeo-meanders, whilst in the southern valley the palaeo-meanders are of smaller amplitude. Based on the differences in morphological properties of both valleys, it is considered that the northern valley (3 and 5, Fig. 16.2) is a palaeo-Demer floodplain, whilst the southern valley west of the present-day confluence was incised by the palaeo-Dijle river (4, Fig. 16.2). The exact course of these palaeo-channels could not be reconstructed due to subsequent incision and sedimentation in the valley axis. Nevertheless, an almost continuous series of meanders has been preserved for the northern floodplain both up- and downstream the Dijle–Demer confluence. Meander form is variable, from very high amplitude (>1 km) and moderate wavelength (500–800 m) (6, Fig. 16.2) to, in general, more limited amplitude (250–500 m) and higher wavelength (1–2 km) geometries (7, Fig. 16.2). Based on wavelength and amplitude, it is estimated that annual peak flood discharges 10–15 times higher than today were required to form such large-sized meanders. Such high peak discharges were probably of very short duration and can be related to periods of major snowmelt across a permafrost landscape (Vandenberghe 2001; Vandenberghe and Woo 2002).

During the Younger Dryas, a narrow dune belt was formed to the north of the palaeo-Demer valley (9, Fig. 16.2). At several locations, peat of Allerød age could be observed at the base of the dunes, which also locally underwent later erosion by the large palaeo-meanders (De Smedt 1973; Vandenberghe and De Smedt 1979), indicating that the latter were active after the Allerød. Dune sand mineralogy points towards river sands as their source. It has therefore been suggested that these are typical river dunes originating from the deflation of sands from the Weichselian river floodplain to the south. In this region, the dunes are often rather low (≤ 3 m), with a poorly expressed morphology. Also along the Scheldt, Grote Nete, Maas and Leie rivers such river dune complexes formed during the Late Glacial (Gullentops 1957; Paulissen and Munaut 1969; Vandenberghe 1983; Verbruggen et al. 1991; De Moor and Pissart 1992).

Recent coring profiles and radiocarbon dates from the floodplain near Rotselaar (8, Fig. 16.2) provide more data on the formation of the large palaeo-meanders and the floodplain evolution during the Holocene. At the base of the floodplain transects (Fig. 16.4), loamy sands to coarse sands are interpreted as reworked Weichselian fluvial sands deposited as bedload material during meander formation. Pollen from the top 5 cm of this sandy layer were dated by radiocarbon at 11,588 ± 123 cal BP (Fig. 16.5), which indicates that these basal sands were deposited towards the end of the Younger Dryas Period. At Rotselaar but also in other meanders, they are overlain by fine sediments, mainly Early Holocene silts (De Smedt 1973) (Fig. 16.4), an age confirmed by the ^{14}C date of 10 504 cal BP obtained for the top of the silt layer, which evidences a fast meander fill rate. This silt deposition phase was followed by a phase of peat accumulation with local clay deposition but no

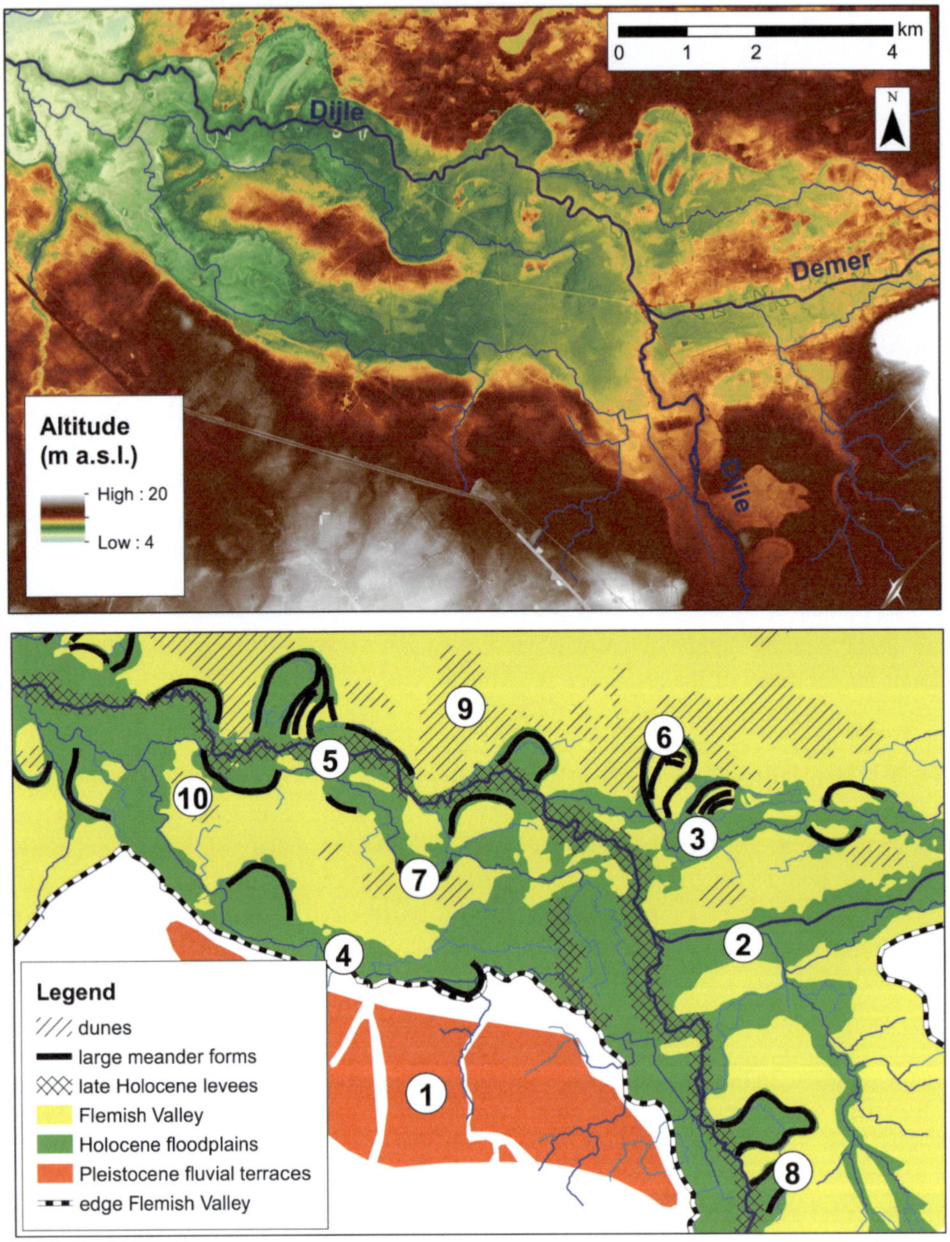

Fig. 16.2 Digital elevation model (LIDAR DEM; AGIV DHMII, 1 m resolution) (*top*) and geomorphological map of the Dijle–Demer confluence area (*bottom*). The geomorphological map is based on field data from De Smedt (1973) and a detailed analysis of LIDAR data and soil maps. Numbers refer to morphological units (see text for details): *1* Pleistocene fluvial terraces; *2* and *3* Southern and northern branches of the Demer floodplain; *4* and *5* Southern and northern branches of the Dijle floodplain; *6–8* Late Glacial meander forms; *9* Younger Dryas inland dunes; *10* Holocene dunes

contemporaneous channel sediments, which suggests either that the large meanders were cut off from the active low-amplitude meander channel, or a lack of a well-defined river channel. At Rotselaar, the top of the peat layer is dated at 2742 cal a BP (Broothaerts 2014). In addition, pollen data confirm the radiocarbon dates (Fig. 16.5). The classical vegetation stages of the Early Holocene are recognized in the pollen of the silty and peat deposits, with an initial dominance of *Betula* (birch) followed by a *Pinus* (pine) dominance and, finally, the appearance of *Corylus* (hazel), *Quercus* (oak) and *Ulmus* (elm). Due to low sedimentation rate, the very low temporal resolution of the pollen data rules out any detailed conclusion for the Middle and Late Holocene.

The peat accumulation phase was followed by a new period of sediment deposition that continued until (sub)recent times. Based on pollen assemblages, De Smedt (1973) dates this transition around 2000–3000 cal a BP, which is consistent with the radiocarbon age of the top of the peat layer at Rotselaar. During this last phase, mainly silts were deposited, with a gradual coarsening of the sediments over time (De Smedt 1973). As exemplified by the Rotselaar transects, the overbank silts near the floodplain edges are intercalated with coarser colluvial deposits originating from erosion of the local Weichselian river sands. During this period, river channels were formed again and the Dijle formed levees. However, such levees are lacking for the River Demer. The transition from peat to silt accumulation is

Fig. 16.3 Satellite image from a part of the Dijle–Demer confluence area. The present River Dijle is clearly visible in the southwest. It is surrounded by cropland situated on the elevated levees from the late Holocene. A nineteenth century anthropogenic cut-off meander can be seen in the southern, central part of the image (*mm*). In the northern part, relicts of two Late Glacial meanders can be seen (*LGm*). They are characterized by woodland and pasture. The meander to the right (corresponding to number 6 on Fig. 16.2) is very elongated and a series of fish ponds nowadays dot the palaeo-channel course. There is a striking adaptation of landscape organization (land use, field boundaries) to the microtopography (see Fig. 16.2). Note also that the palaeo-meanders are nearly not built, which contrasts with the intervening Weichselian terrace and the Late Glacial dune landscape to the north (*Source* Google Earth, image taken 18/4/2015 © Aerodata International Surveys)

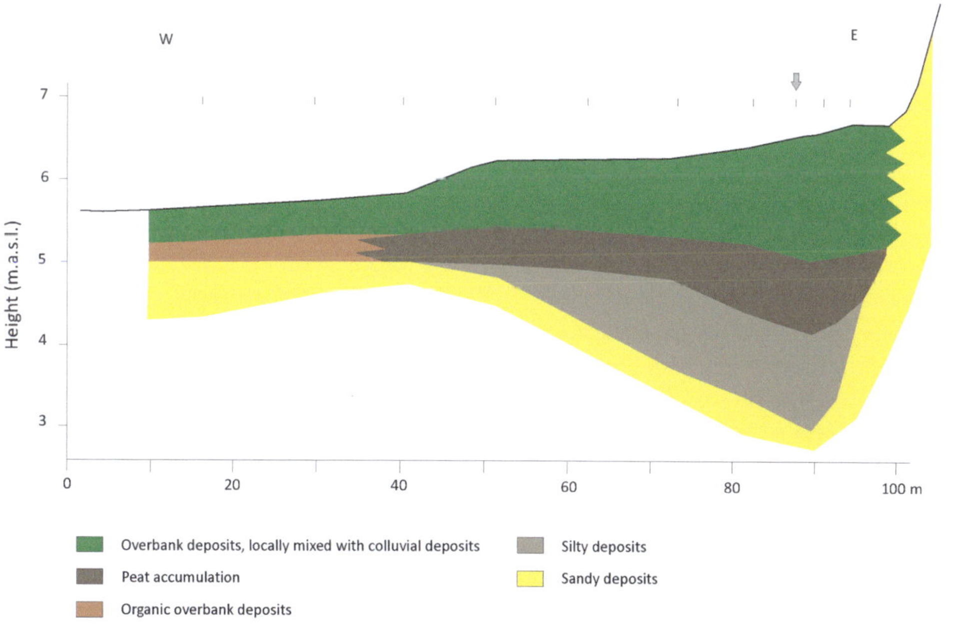

Fig. 16.4 Cross-section through one of the Late Glacial meander forms in the Dijle floodplain near Rotselaar (8, Fig. 16.2) based on coring data, with indication of the different lithostratigraphical units (Broothaerts 2014). *Small vertical bars* at the *top* indicate coring positions, *arrow* indicates the location of the pollen diagram (Fig. 16.5). After formation of the large meander, it was filled in with silty deposits followed by peat accumulation between 10 504 cal a BP and 2742 cal a BP. On *top* of the peat layer, a combination of overbank silty deposits and colluvial sandy deposits is found

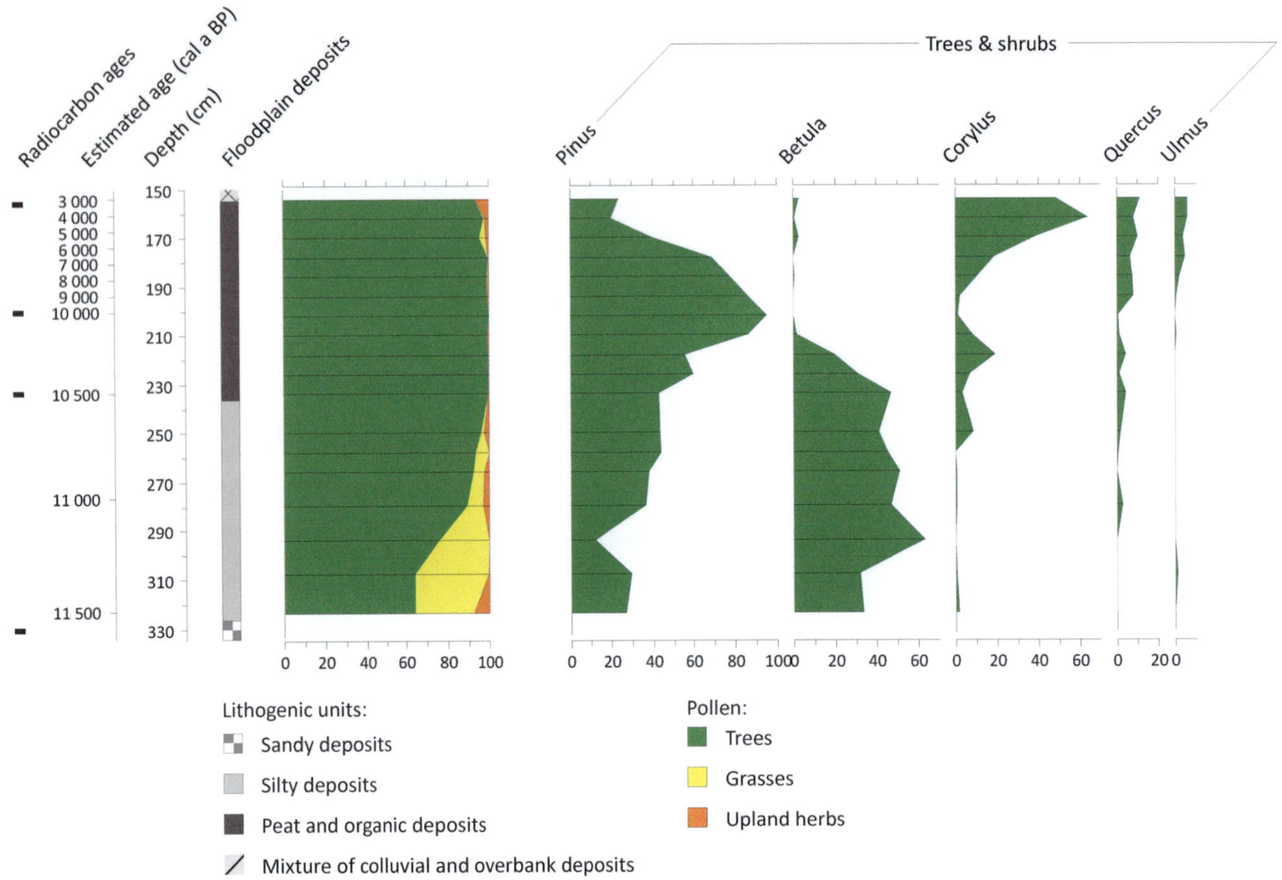

Fig. 16.5 Simplified pollen diagram from the Dijle floodplain near Rotselaar (see Fig. 16.4), based on Broothaerts (2014). The estimated ages are based on a linear interpolation between the four indicated radiocarbon ages. Note that the upper 1.5 m of the core is not displayed. The diagram shows the typical vegetation succession of the Early Holocene, with a *Betula* dominance followed by a *Pinus* dominance and later appearance of *Corylus*, *Quercus* and *Ulmus*

a result of anthropogenic deforestation in the headwaters (see below). As the Demer drains to a large extent relatively gently sloping sandy areas that are less susceptible to soil erosion, it is also characterized by a much lower supply of suspended load. This explains the coarser grained nature of the Demer floodplain deposits and the absence of extended levees along the River Demer.

During the Holocene period, not only the character of the floodplain deposits changed, but also the river system showed an alternating occupation of the northern and southern valleys. The sediment record shows how the Demer occupied first the northern valley branch (3, Fig. 16.2), then shifted sometime during the middle Holocene to its southern branch (2, Fig. 16.2) (De Smedt 1973). Downstream of the confluence, the river system made an opposite movement: first it occupied the southern valley branch (4, Fig. 16.2) before shifting to the northern branch (5, Fig. 16.2). The shift of the Demer is still poorly dated and understood. The shift downstream of the confluence took place sometime during the past 2000–3000 years, caused by the large sediment input and floodplain aggradation related to the anthropogenic deforestation during this period. This triggered the formation of an alluvial fan north of Leuven, where the Dijle leaves the hilly loess area and enters the Flemish valley, causing the river to change its course until its bed became fixed by a natural levee (Vandenberghe and De Smedt 1979). The LIDAR image shows that many floodplain lakes and small depressions occur in the present Demer and Dijle floodplains (e.g. 5, Fig. 16.2), which are the result of natural and anthropogenic meander cutoff. A comparison with historical maps demonstrates that most of them were cut off during the nineteenth century. In the same period, also artificial levees were created next to the channel, hampering further channel migration and preventing flooding. As such, the contemporary channels are disconnected from the floodplains, and floodplain deposition has been very limited or even absent over the last century. In recent times, renewed interest in natural flood management has been prominent in Flanders. There are now many plans to re-activate the nineteenth century meanders along the River

Demer, whereas levees will be lowered locally along the River Dijle and large controlled flood basins will be constructed.

In addition, Late Holocene (probably Medieval) deforestation resulted in the formation of some small dune complexes which rest locally on top of the Late Glacial dunes or on the sandy Weichselian braided-river deposits and coversands (De Smedt 1973). In contrast to the older dunes, they lack a well-developed podzol soil horizon. Such Holocene dunes are found, e.g. on the western part of the interfluve between the two parallel valley branches of the Dijle (10, Fig. 16.2).

16.4 The Dijle River in the Loess Area

In the loess area, south of Leuven, the River Dijle forms a 1-km-wide flat floodplain within a hilly landscape and has been studied in detail using a large dataset of hand corings, palynological data and an analysis of historic maps and LIDAR data (e.g. Mullenders and Gullentops 1957; De Smedt 1973; Notebaert et al. 2009a, b, 2011a, b; Broothaerts et al. 2014a, b).

The coring data allow the reconstruction of an undulating pre-Holocene topography at the bottom of the floodplain fill (Fig. 16.6). Pre-Holocene sediments often consist of coarse to medium sand deposited during the last glacial period by braided rivers. One OSL date indicates a Pleniglacial age (Notebaert et al. 2011b). Locally, Holocene floodplain deposits also cover in situ deposited Weichselian loess. Thanks to the high coring density, cross sections also evidence 2- to 3-m-deep small depressions that bear witness to older channels in this pre-Holocene topography (Fig. 16.6). Precise age information is lacking, but position and morphology of the channels suggest formation by river incision at the transition from cold to warm conditions, thus to some extent equivalent to the large Late Glacial meanders found in the lower part of the catchment.

Corings along 25 floodplain transects and pollen samples from six of these sites were used to reconstruct the Holocene floodplain evolution in relation to increasing human impact (Notebaert et al. 2011a; Broothaerts et al. 2014c; Notebaert et al. 2011b). The Holocene deposits have a twofold structure, with 1- to 4-m-thick organic rich deposits at the base, covered by 2- to 6-m-thick clastic deposits (Fig. 16.6). The organic material is mainly peat and organic silt that started to accumulate in the abandoned channels, i.e. the deepest parts of the floodplain, from 11,588 ± 123 cal BP (Fig. 16.5). This shows that, here also, the channels became inactive at least at the end of the Younger Dryas to Early Preboreal. By around 9500 cal a BP, peat covered over 90% of the contemporary valley (Broothaerts et al. 2014a). By that time, the floodplain lacked clear channel deposits, suggesting diffuse water transport in a marshy environment where organic material accumulated. Pollen data (Fig. 16.7) show that the Dijle floodplain consisted of strongly vegetated marsh areas dominated by an alder carr forest (Fig. 16.8), which is interpreted as the climax floodplain vegetation during interglacials. Human impact was low during these periods, as shown by regional pollen data. During the Middle Holocene, the Dijle catchment was mainly covered by a deciduous forest dominated by *Corylus*, *Tilia* (linden) and *Quercus*. From the beginning of the Bronze Age onwards (ca. 3900 cal a BP), the vegetation gradually changed from a

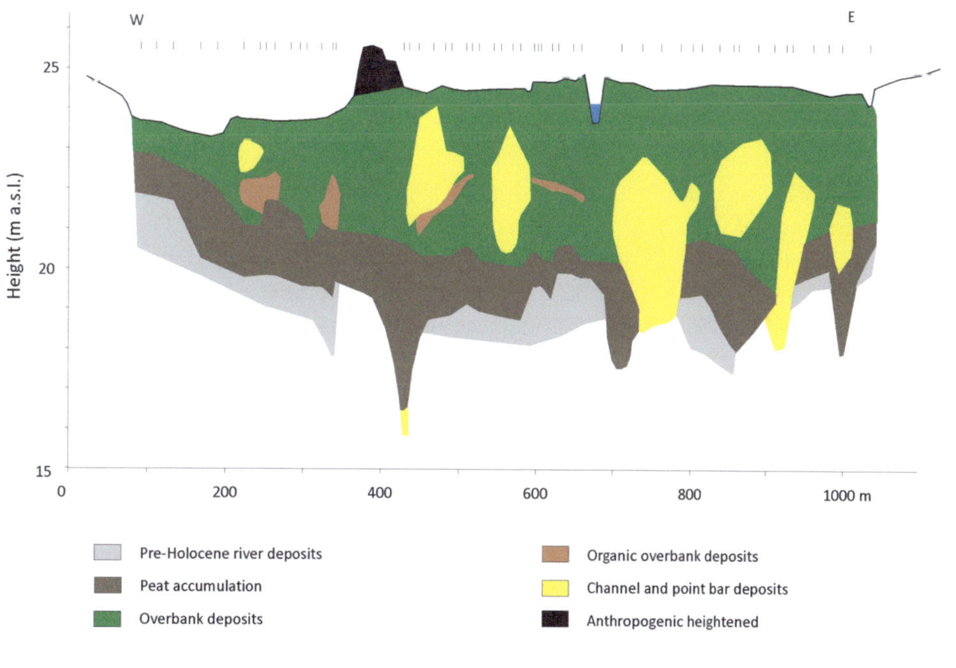

Fig. 16.6 Cross-section through the Dijle floodplain near Korbeek-Dijle (see location on Figs. 16.1 and 16.9d), indicating the main lithostratigraphical units (Notebaert et al. 2011a). It shows that Early to Mid-Holocene peat accumulation was followed by the deposition of overbank silts from ca. 2750 cal a BP. Due to the discontinuity of the transect (see Fig. 16.9d), the section crosses the sandy deposits of the meander belt several times

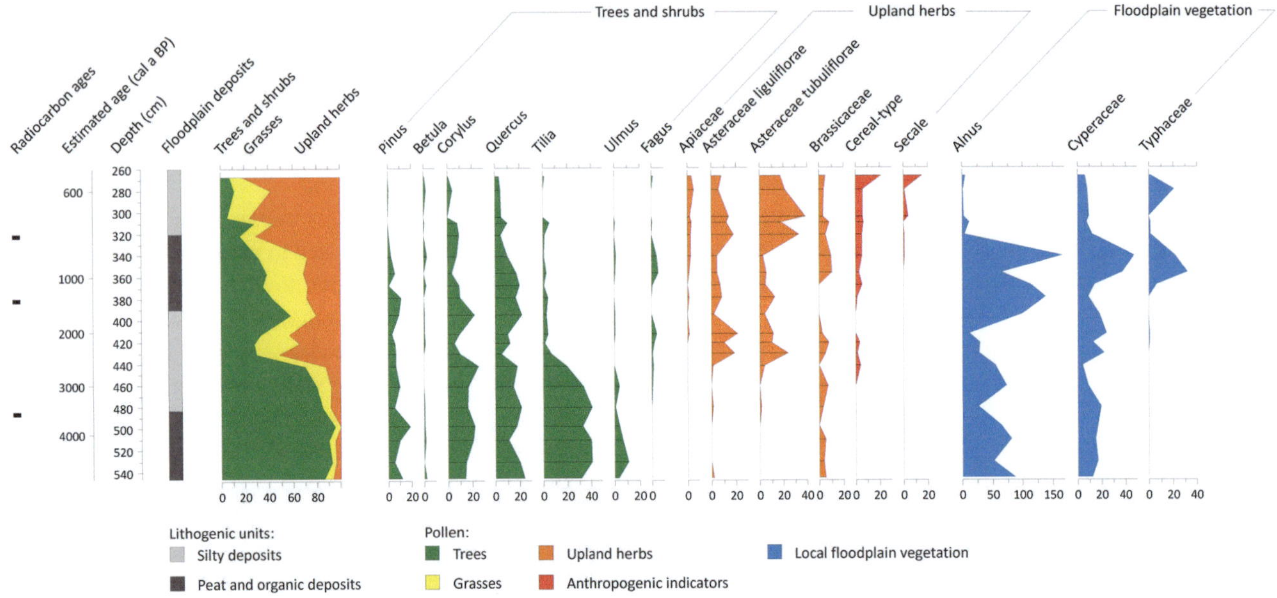

Fig. 16.7 Simplified pollen diagram from Archennes, a typical floodplain site of the loess area (see Fig. 16.1). The estimated ages are based on a linear interpolation between the three available radiocarbon ages. Note that the upper 2.5 m of the core are not displayed. The pollen data show a gradual opening of the landscape starting around 3000 cal a BP. The natural deciduous forest is replaced by a cultural landscape. At this site, vegetation changes in the floodplain occurred rather late, the alder forest being replaced by open floodplain species during the last 1000 years (Broothaerts et al. 2014c)

deciduous forest towards an intensively cultivated landscape. The fraction of forest cover decreased and more upland herbs, *Poaceae* and anthropogenic indicators such as cereals and weeds appeared (Fig. 16.7). This greater human impact at catchment scale caused an increase in overland flow and soil erosion and, consequently, an increase in sediment input that interrupted the peat accumulation in the floodplain system (Broothaerts et al. 2014b). The top of the peat layer is diachronic and ranges between 6500 and 500 cal a BP. Peat accumulation was replaced by sediment deposition first in the floodplain centre, probably near a newly formed channel, while peat growth continued in more distal parts (Fig. 16.8). As high sediment delivery persisted, the entire floodplain was finally dominated by deposition of clastic sediment from 500 cal a BP. Simultaneously with the increase in sediment input, floodplain vegetation changed towards a more open one, dominated by *Poaceae*, *Cyperaceae* and *Typhaceae* (Figs. 16.7 and 16.8). Together with the silty overbank deposits, also sandy river bed and bar deposits were formed, whose position points towards the development of a meandering river system, whereby new well-defined channels eroded the underlying peat layer at some places (Fig. 16.6). A small but readily traceable organic layer dated to ca. 1500–1000 cal a BP is observed in the overbank deposits in many places of the main valley. It most likely coincides with a phase of decreased sediment deposition due to a decrease in human impact (Fig. 16.7).

Due to the thick Holocene peat and sediment cover, the modern fluvial landscape does not show any imprint of pre-Holocene topography but is entirely the result of sub-recent sedimentation processes. The current channel has a meandering pattern of 50–200 m amplitude and 50–100 m wavelength (Figs. 16.9 and 16.10). The sedimentary record shows that Late Holocene meandering affected only a limited part of the floodplain. Channel and (point) bar sediments are only found along a narrow stretch around present and cut off meanders. The floodplain topography is dominated by natural levees of up to 1.5 m in height (1, Fig. 16.9), flanked by backswamps (2, Fig. 16.9) (Notebaert et al. 2009a). Colluvial fans were formed where large tributary dry valleys join the main valley (3, Fig. 16.9). These fans overlie the Early- to Mid-Holocene peat layer, indicating that they are also linked to the anthropogenic deforestation of the past millennia. They often have an elongated form along an artificial drainage channel that leads water and sediment into the centre of the floodplain. A detailed survey using LIDAR data has demonstrated that a pattern of old meandering channels can still be observed in the floodplain (4, Fig. 16.9) (Notebaert et al. 2009a). The most recent cut off meanders have become oxbow lakes, while older ones are filled depressions with a depth of 0.1 to more than 1 m. A comparison with historical topographical maps shows that most of these meanders were cut off in the eighteenth or nineteenth century, but some cut-offs also predate the oldest

Fig. 16.8 Conceptual model of the response of the floodplain morphology to increasing human impact (Broothaerts et al. 2014b). The model shows how the natural landscape (**a**, ~6000 cal BP) is gradually replaced by a cultural landscape inducing changes in the floodplain morphology. Initial local erosion results in a limited sediment supply and sedimentation is first limited to an area near the river channel (**b**). With increasing deforestation, sediment supply increases and larger parts of the floodplain are covered with sediment (**c**), until the entire floodplain is covered with a thick sediment layer in the last 1000 years (**d**). Meanwhile, also floodplain vegetation changed from an alder forest to a more open landscape

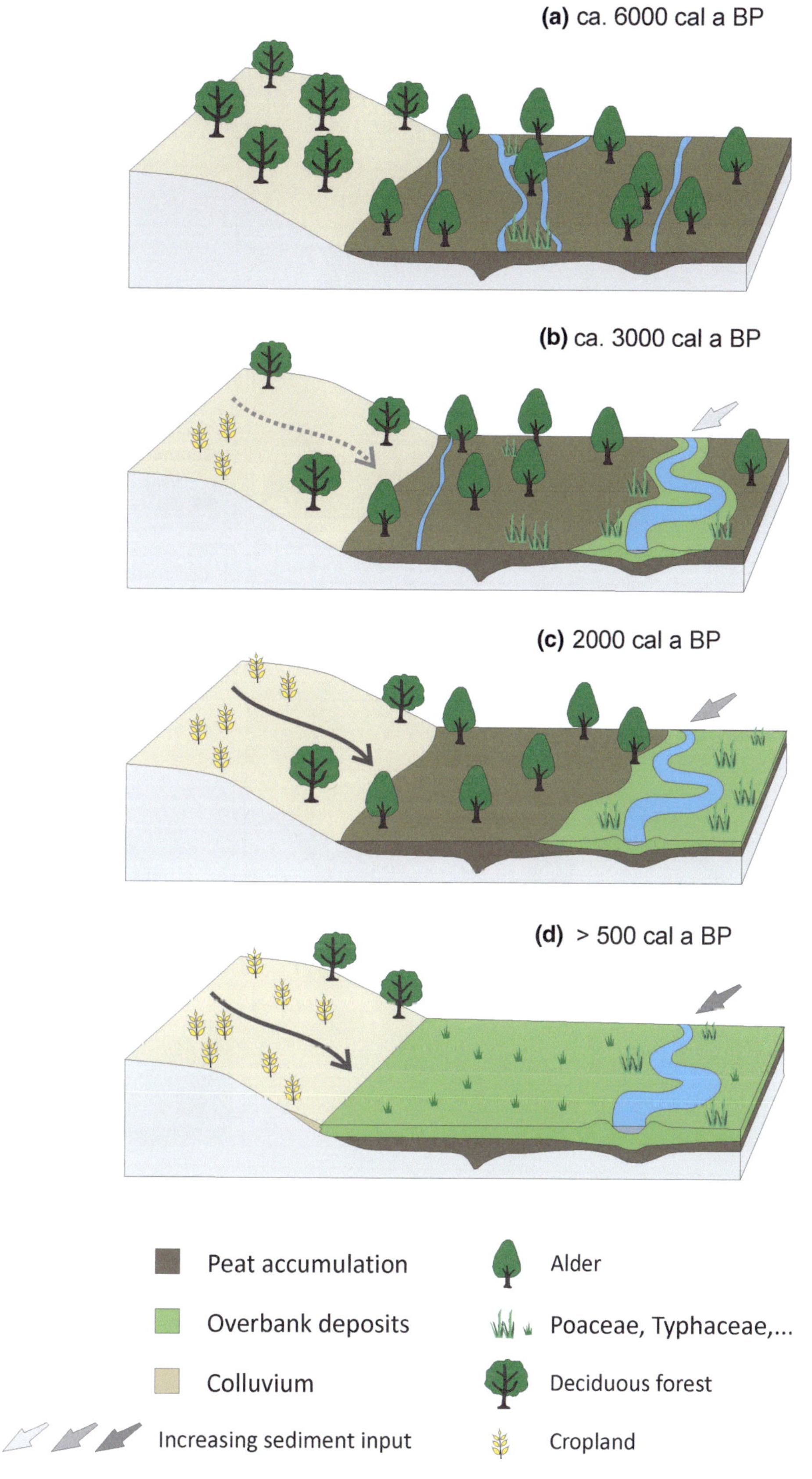

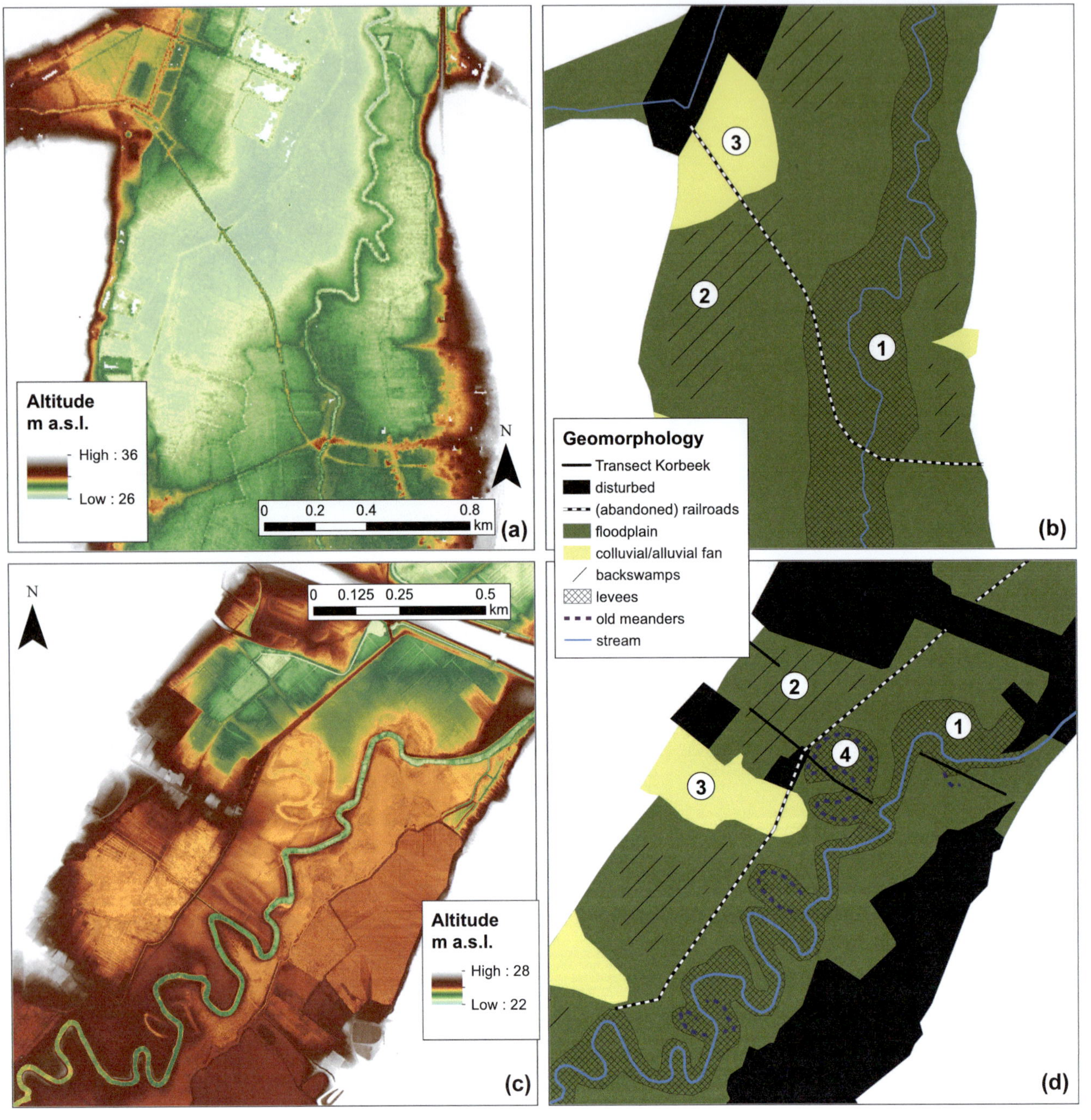

Fig. 16.9 DEM (LIDAR DEM; AGIV DHMII, 1 m raster) and geomorphological maps of the Dijle floodplain near Sint-Joris-Weert (**a**, **b**) and Korbeek-Dijle (**c**, **d**), based on Notebaert et al. (2009a). Numbers refer to morphological units (see inserted legend and text for details). *1* Levees; *2* backswamps; *3* fans; *4* old channels. The position of the former channels is confirmed by coring data and by a comparison with historical maps. Some of the older meanders were cut off before the oldest available map (1775 AD) was published. The position of the coring cross sections at Korbeek-Dijle (Fig. 16.6) is indicated by *black lines* on the geomorphological map d

Fig. 16.10 Aerial photograph of the Dijle floodplain near Korbeek-Dijle, south of Leuven. View is towards the south. The photo covers the southern part of the area shown in Fig. 16.9 c, d. The meandering Dijle is the indirect result of anthropogenic activities through soil erosion and consequent floodplain sedimentation. Land use in the floodplain is now dominated by pasture, while small forest plantations (often poplar trees) and large fish ponds are also present. (© Wim Robberechts & co; http://www.panoramio.com/photo/44794046)

maps from 1775 AD (Notebaert et al. 2009a). Contrary to the northern study site, the River Dijle has no artificial dikes south of Leuven.

16.5 Conclusions

Floodplain landscapes of central Belgium provide a testimony of both climatic changes and anthropogenic land cover changes of the last glacial-interglacial cycle. The downstream parts of the floodplains in the Demer–Dijle confluence area of the Flemish Valley bear many traces of the important climatic changes at the last glacial-interglacial transition. The Weichselian sandy braided-river deposits were incised at the beginning of the Late Glacial and large, often elongated meanders formed in the context of snowmelt-related high peak discharges. These palaeo-meanders are still among the most prominent landforms in this flat area (Fig. 16.3). The effects of these climatic changes are no longer visible in the floodplains of the loess area, mainly because of the thick Holocene sediment cover.

During the Early Holocene, water and sediment transport decreased dramatically. The Late Glacial meanders were filled up, first with reworked Weichselian fluvial sediments, followed by peat accumulation covering the entire floodplain. During this period, no well-defined channel was present. Pollen data indicate that the landscape was covered with a deciduous forest, while floodplains were mainly covered with an alder carr forest. From the Bronze Age onwards, anthropogenic deforestation caused an increase in soil erosion in the hilly loess region and, consequently, an increase in floodplain sedimentation and a stop in peat growth. Rivers developed again a meandering channel pattern, although of much smaller size compared to that of the late Younger Dryas meanders (Fig. 16.10). This same sediment wave was responsible for the formation of levees and colluvial fans that shape the current floodplain topography. During the last 250 years, many channel segments were artificially or naturally cut off, and they remain visible as oxbow lakes or shallow depressions in the current floodplain. Artificial dikes were constructed along the channels in the lower floodplains, fixing the channel position and decoupling the river channel from its floodplain. Over the last 2000–3000 years, the increased sediment deposition also resulted in shifts in river position in the downstream part of the catchment.

This evolution shows how fluvial landscapes evolved from climate- to human-dominated landscapes. Widespread deforestation in the Dijle catchment during the last 4000 years triggered the formation of an anthropogenic fluvial system, which is further artificialized by direct technical interventions mainly since the nineteenth century AD.

References

Broothaerts N (2014) Sensitivity of floodplain geoecology to human impact in the Dijle catchment. KU Leuven, Belgium

Broothaerts N, Notebaert B, Verstraeten G, Kasse C, Bohncke S, Vandenberghe J (2014a) Non-uniform and diachronous Holocene floodplain evolution: a case study from the Dijle catchment, Belgium. J Quatern Sci 29(4):351–360

Broothaerts N, Verstraeten G, Kasse C, Bohncke S, Notebaert B, Vandenberghe J (2014b) From natural to human-dominated floodplain geoecology–a Holocene perspective for the Dijle catchment, Belgium. Anthropocene 8:46–58. doi:10.1016/j.ancene.2014.12.001

Broothaerts N, Verstraeten G, Kasse C, Bohncke S, Notebaert B, Vandenberghe J (2014c) Reconstruction and semi-quantification of human impact in the Dijle catchment, central Belgium: a palynological and statistical approach. Quatern Sci Rev 102:96–110

Crombé P, Vanmontfort B (2007) The neolithisation of the Scheldt basin in western Belgium. In: Whittle A, Cummings V (eds) Going over: the mesolithic–neolithic transition in North–west Europe, vol 144. Proceedings of the British Academy Press, pp 263–285

De Moor G, Pissart A (1992) Het reliëf. In: Denis J (ed) Geografie van België. Gemeentekrediet, Brussel, pp 130–215

De Smedt P (1973) Paleogeografie en Kwartair-Geologie van het confluentiegebiede Dijle–Demer. Acta Geogr Lovan 11:141

Gullentops F (1957) Quelques phénomènes géomorphologiques depuis le Pléni-Wurm. Bull Soc Belge Géol 61(1):86–95

Mullenders W, Gullentops F (1957) Palynologisch en geologisch onderzoek in de alluviale vlakte van de Dijle te Heverlee–Leuven. Agricultura 5:57–64

Notebaert B, Verstraeten G, Govers G, Poesen J (2009a) Qualitative and quantitative applications of LIDAR imagery in fluvial geomorphology. Earth Surf Proc Land 34(2):217–231. doi:10.1002/esp.1705

Notebaert B, Verstraeten G, Rommens T, Vanmontfort B, Govers G, Poesen J (2009b) Establishing a Holocene sediment budget for the river Dijle. Catena 77(2):150–163. doi:10.1016/j.catena.2008.02.001

Notebaert B, Houbrechts G, Verstraeten G, Broothaerts N, Haeckx J, Reynders M, Govers G, Petit F, Poesen J (2011a) Fluvial architecture of Belgian river systems in contrasting environments: implications for reconstructing the sedimentation history. Neth J Geosci-Geol Mijnbouw 90(1):31–50

Notebaert B, Verstraeten G, Vandenberghe D, Marinova E, Poesen J, Govers G (2011b) Changing hillslope and fluvial Holocene sediment dynamics in a Belgian loess catchment. J Quat Sci 26(1):44–58. doi:10.1002/Jqs.1425

Paulissen E, Munaut A (1969) Un horizon blanchâtre d'age Bølling à Opgrimbie. Acta Geogr Lovan 7:65–91

Vandenberghe J (1977) Geomorphology of the Zuiderkempen Belgium, vol. 39. Verhandelingen van de Koninklijke academie voor wetenschappen, Letteren en schone kunsten van Belgie, p 140

Vandenberghe J (1983) Late Weichselian river dune formation Grote Nete valley, central Belgium. Z Geomorphol NF 45:251–263

Vandenberghe J (2001) A typology of Pleistocene cold-based rivers. Quatern Int 79(1):111–121

Vandenberghe J, De Smedt P (1979) Palaeomorphology in the Eastern Scheldt basin (Central Belgium). The Dijle–Demer–Grote Nete confluence area. Catena 6:73–105

Vandenberghe J, Woo MK (2002) Modern and ancient periglacial river types. Prog Phys Geogr 26(4):479–506

Vanmontfort B (2007) Bridging the gap. The Mesolithic–Neolithic transition in a frontier zone. Doc Praehist 34:105–118

Verbruggen C, Denys L, Kiden P (1991) Paleo-ecologische en geomorfologische evolutie van Laag-en Midden-België tijdens het Laat-Kwartair. De Aardrijkskunde 3:357–376

The Scheldt Estuary: An Overview of the Morphodynamics of Intertidal Areas

17

Lennert Schepers, Tom Maris, Patrick Meire and Stijn Temmerman

Abstract

Along the Scheldt estuary, intertidal areas such as intertidal flats and intertidal marshes have important functions including flood water storage, water quality regulation and provision of important habitats. The functioning of intertidal marshes is determined by short-term geomorphic processes: the marshes are regularly flooded and sediments are deposited on the marsh surface, a process through which marshes gain elevation. The spatial patterns and rates of sedimentation on marshes are mainly governed by temporal and spatial variations in tidal characteristics and flow patterns, which are strongly influenced by marsh vegetation, leading to a micro-topography typical for tidal marshes. Strong human modifications of the Scheldt estuary by embankments, channelization and dredging have been increasing the tidal range, especially in the more upstream regions. As part of an integrated management plan for the Scheldt estuary—the so-called Sigmaplan in Flanders, new intertidal areas have been created and more than 2000 ha are planned to be ready by 2030. This intertidal flat and marsh creation is realized in some cases as flood control areas with a reduced tidal exchange between the estuary and the flood control area, where the geomorphologic processes are comparable to natural marshes, although we note some important differences.

Keywords

Scheldt River • Estuarine environments • Intertidal area dynamics • Estuary management • Human impact

17.1 Introduction

The riverbanks of the Scheldt estuary are occupied by intertidal flats and marshes (Fig. 17.1). These are the areas that periodically flood and drain during flood and ebb tides, respectively.

L. Schepers (✉) · T. Maris · P. Meire · S. Temmerman
Department of Biology, University of Antwerp, Campus Drie Eiken, Universiteitsplein 1, B-2610Wilrijk, Belgium
e-mail: lennert.schepers@uantwerpen.be

T. Maris
e-mail: tom.maris@uantwerpen.be

P. Meire
e-mail: patrick.meire@uantwerpen.be

S. Temmerman
e-mail: stijn.temmerman@uantwerpen.be

The lower areas, called intertidal flats, are flooded every semi-diurnal high tide. These areas are too dynamic (strong tidal currents and waves, intense sediment dynamics) or the inundation time is too long to support vegetation establishment. Higher areas have shorter hydroperiods (shorter inundation time and higher frequency) and are characterized by a dense vegetation cover of flood-tolerant plant species (Fig. 17.2). These areas are called intertidal marshes. The lower intertidal marshes are flooded every tide, while the highest marshes are only inundated during spring tides, i.e. the highest water levels of the fortnightly spring-neap tidal cycle (Fig. 17.3). Intertidal marshes are flooded and drained by channels, also called tidal creeks or tidal channels. The intertidal flats and marshes have important functions: they serve as natural flood control areas, they are sinks for nutrients, organic matter and sediments and sources of

A. Demoulin (ed.), *Landscapes and Landforms of Belgium and Luxembourg*,
World Geomorphological Landscapes, DOI 10.1007/978-3-319-58239-9_17

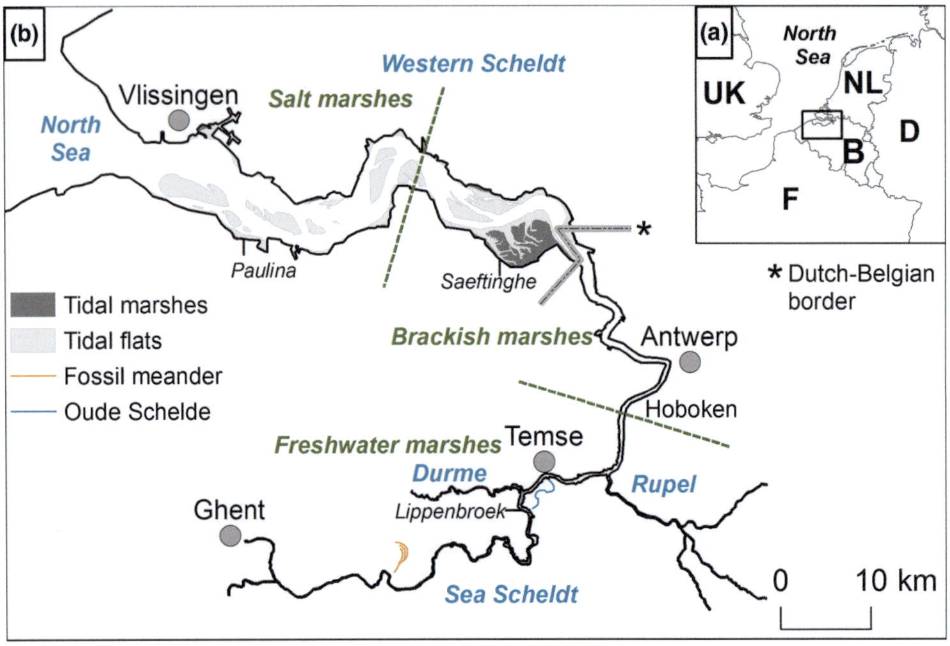

Fig. 17.1 **a** Situation of the Scheldt estuary within NW-Europe, **b** overview of the Scheldt estuary with all the major intertidal flats, marshes and field sites and place names mentioned in the text

oxygen and they provide habitats for many species. These environments are in a geomorphic view very dynamic, with several processes even measurable during one single tide.

The historical evolution and especially anthropogenic influences such as construction of large-scale embankments and dredging have been altering the Scheldt estuary, leading to a rise of the tidal range (i.e. the difference between low and high tide water level). As a safety measure, several large flood control areas with new intertidal habitats have been (or are currently being) installed, which offer unique possibilities to study the geomorphological processes in intertidal areas that are in an early stage of geomorphological and ecological development. The Scheldt estuary, although nowadays strictly confined by historical land reclamation and dike building, still accommodates remains of a scenic natural habitat surrounded by a highly populated and built-up area of Belgium and the Netherlands.

In this chapter, we will discuss the current knowledge of geomorphic processes in the natural and newly created intertidal areas, especially the intertidal marshes, together with their effect on the functioning of these intertidal landscapes along the Scheldt estuary.

17.2 Geographical Setting

The estuary of the river Scheldt extends from the mouth at the North Sea near Vlissingen (NL) (0 km) to Ghent (160 km from the estuary mouth), where sluices impede further inland propagation of the tidal wave (Meire et al. 2005) (Fig. 17.1). Along this stretch, the salinity gradually decreases in landward direction, and therefore the intertidal marshes can be differentiated into salt, brackish and freshwater marshes (Fig. 17.1b, see Meire et al. (2005) for more information). The Saeftinghe brackish marsh, with an area of around 3000 ha is the biggest intertidal marsh area of the estuary and the southern North Sea coastal area (Fig. 17.2b). The other remaining marshes are small and scattered fringing the riverbanks. The Western Scheldt, the Dutch part of the Scheldt estuary near the mouth of the Scheldt River, is a multichannel system with extensive intertidal flats bordering the channels (Fig. 17.1b). On the Belgian territory, the tidal part is called the Sea Scheldt and consists of a single channel river with intertidal flats next to the subtidal channel.

17.3 Functions of Intertidal Flats and Marshes

17.3.1 Attenuation of Floods, Waves and Erosion

Intertidal areas and especially intertidal marshes play an important role in the water management of the estuary. They act as natural flood control areas, creating additional water storage capacity at extreme high water levels, thereby reducing peak water levels within the estuary and reducing the flood risks for nearby urban areas, such as Antwerp. The flood storage capacity of marshes is especially important at the Saeftinghe marsh and the Belgian part of the tidal river. Model analysis shows for example that the flood storage capacity of the Saeftinghe marsh decreased the water level by almost 20 cm in Antwerp during a storm tide in 2013 (Smolders et al. 2015). Furthermore, marshes strongly

Fig. 17.2 **a** View of the Paulina salt marsh at low water, **b** aerial picture of the Saeftinghe brackish marsh (https://beeldbank.rws.nl, Rijkswaterstaat/Joop van Houdt) and **c** view of intertidal flats and freshwater marshes from the Sea Scheldt close to Lippenbroek

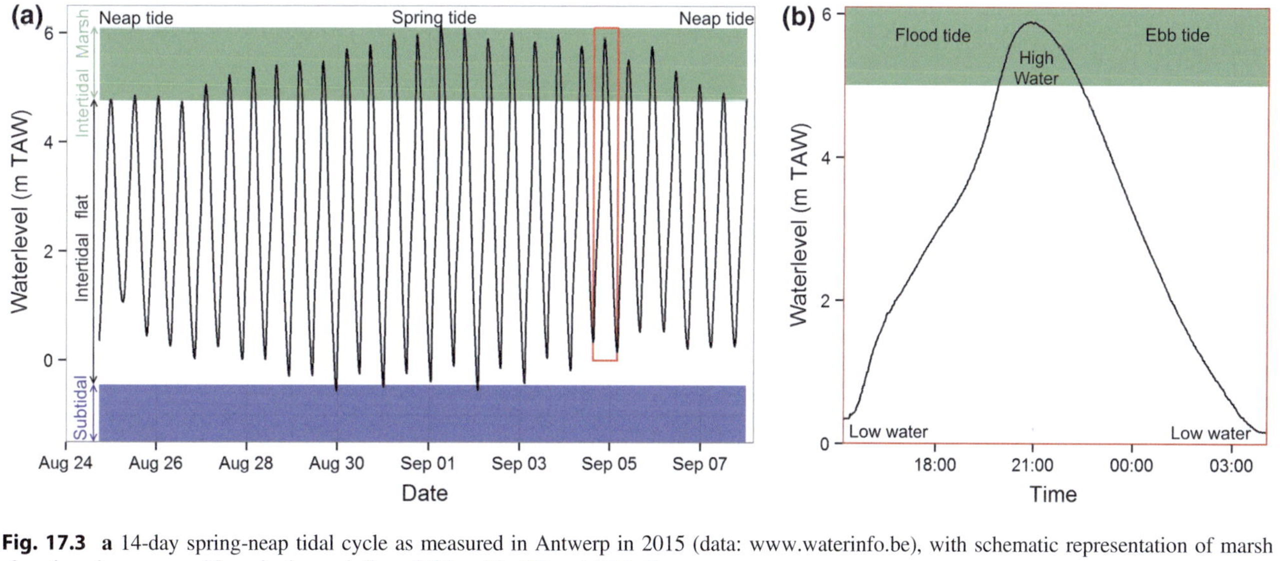

Fig. 17.3 **a** 14-day spring-neap tidal cycle as measured in Antwerp in 2015 (data: www.waterinfo.be), with schematic representation of marsh elevations in *green* and **b** a single semi-diurnal tide with different tidal phases

attenuate waves and tidal currents (Möller and Spencer 2002; Widdows and Brinsley 2002; Bouma et al. 2005a, b; Möller 2006; Van der Wal et al. 2008; Möller et al. 2014). As such they protect against or reduce erosion (Widdows and Brinsley 2002; Neumeier and Ciavola 2004; Neumeier and Amos 2006; Van der Wal et al. 2008; Möller et al. 2014). Dikes, artificially constructed levees for flood protection, need less maintenance if they are protected by marshes at the riverside and are less prone to dike breaches during storm floods (Möller 2006; Möller et al. 2014).

17.3.2 Filter Function

Marshes have an important filter function for the Scheldt water. The river water is heavily loaded with sediment, nutrients (e.g. nitrogen) and pollutants (e.g. heavy metals), originating from soil erosion, agricultural and industrial activities in the catchment of the Scheldt River, and from landward tidal pumping of sediments that are as such trapped within the estuary. High nutrient loads result in eutrophication with episodic anoxic water conditions, strongly limiting biological activity. Luckily, the anoxic episodes have become less common the last 10 years. Marshes, on the contrary, are sources of oxygen and silica and they trap nutrients (Gribsholt et al. 2005; Struyf et al. 2007), pollutants (Du Laing et al. 2007; Teuchies et al. 2013) and sediments (Temmerman et al. 2004b; Van Damme et al. 2009). Silica is an important building block of phytoplankton (diatoms), the primary producers that form the foundation of the estuarine food web. Hence the marshes can be considered as natural filters for the Scheldt River, which may contribute to reducing the costs for dredging and removing nutrients and pollutants. A similar filter functioning is also expected in realignment areas (i.e. de-embankments, with a landward retreat of the dike) along the Scheldt, as well as in the flood control areas with a reduced tide (areas with a sluice through the dike, that allows controlled tidal flooding of the land).

17.3.3 Habitat Function

Intertidal areas comprise important habitats for the ecosystem of the Scheldt estuary. Their nature value is acknowledged by several national and international directives (e.g. EU habitat and bird directives). For example, the Scheldt estuary is an important area for water birds in NW-Europe. Situated along the East Atlantic Flyway, 80 different species of waterfowl and waders are observed (Ysebaert et al. 2000; Van Den Bergh et al. 2005). The tidal wave reaches 160 km upstream making the Scheldt estuary one of the few remaining estuaries in Europe with a full salinity gradient from saltwater over brackish to freshwater marshes and associated plants and fauna (Van den Bergh et al. 2001). Biodiversity decreases with salinity. The salt marshes consist of a limited number of halophytes, plant species adapted to tolerate salinity such as glasswort (*Salicornia* sp.) and common cordgrass (*Spartina anglica*). These species are low herbs and grasses with a maximum vegetation height of 0.8 m (Temmerman et al. 2003b). Brackish marshes have a higher biodiversity, but freshwater marshes host the most diverse species composition, from herbs up to different trees such as willow (*Salix* sp.). Biodiversity and species composition not only change with salinity, but also with flooding frequency. Pioneer species may colonize the higher parts of the intertidal flat whereas the high marsh zones are vegetated by climax species (Van den Bergh et al. 2001). In salt marshes, glasswort (*Salicornia* sp.) and common cordgrass (*S. anglica*) are pioneer species, whereas *Elytrigia atherica* is the climax vegetation that is present on the high salt marshes and on higher channel-bordering levees, but absent on lower elevations. In brackish marshes, seaside bulrush (*Bolboschoenus maritimus*) occurs on the lower elevations, while common reed (*Phragmites australis*) is less salt tolerant and only present on the higher elevated areas. *Schoenoplectus* sp. are the pioneer species of the freshwater marshes. At higher elevations, common reed (*P. australis*) can reach exceptional heights of up to 4 m. Willow (*Salix* sp.) is the climax species in the freshwater zone of the Scheldt estuary.

Also small variations in the micro-topography of the marsh surface lead to distinct plant communities. These species patterns are also visible on Fig. 17.2b, where distinct patches of darker and lighter green vegetation are present, as a result of different vegetation communities (i.e. different species compositions). Hence the geomorphological changes influence habitat shifts in marshes. The geomorphic processes play an important role in the sustainable ecological management of the Scheldt estuary and affect the functioning of the marshes (see Sect. 17.7.4).

17.4 Pleistocene and Holocene Changes in the Scheldt River

During the Saale glacial period, the period of maximum ice sheet extent in North Europe and lowest sea level that lasted until ca. 130 ka BP, the main rivers in Flanders followed a northwest direction past Ghent, eroding a deep valley, called 'The Flemish Valley' (see Chap. 18). During the Weichselian, the last glacial period that lasted until ca. 12 ka BP, the Flemish Valley was gradually filled by mostly sandy deposits originating from a braided river system and covered with eolian sediments (De Moor 1983; Bogemans 1993). This resulted in a relatively flat landscape, only dissected in certain zones by the rivers (Bogemans et al. 2012). Near the end of the Weichselian, a sand ridge was formed that blocked the drainage of the rivers in the Flemish Valley to the northwest, forcing the main rivers to shift their course more eastwards, through the 'Hoboken gap', cutting through the cuesta of resistant Oligocene clay, towards the north (Kiden 1991, 2006). This is visible in Fig. 17.1b between the Rupel confluence and Hoboken where the Scheldt River follows a straight course towards the north. Near the end of the Weichselian the Scheldt River shifted also from a braiding to a meandering river, forming extensive fossil meanders, such as at Overmere Donk (Fig. 17.1b). The size of the fossil meanders indicates that discharges were 3–5 times larger than those of the contemporary Scheldt river (Kiden 1991). During the Late Dryas fluvial processes interacted with eolian processes, resulting in local dune formation in the river floodplain (Bogemans et al. 2012).

The first period of the Holocene was characterized by diminishing fluvial activity. It was a remarkable period of river stability with (almost) neither new river channel formation nor lateral erosion (Kiden 1991; Bogemans et al. 2012). The paleochannels and floodplains were gradually filled with gyttja, peat and organic to peaty clay (Kiden 1991). The tidal influence began ca. after 400 AD at the Belgian border. Brackish water intrusion stopped peat growth and clay was deposited. In Antwerp, peat could continue to grow until 500–600 AD, but the tidal influence in the more inland part of the river was still very limited until 1100 AD (Kiden 2006; Coen 2008). A nice example of this is an old meandering part of the Scheldt that was cut off in the thirteenth century AD. This cut-off part, located on the right bank between the Durme tributary and Temse (Fig. 17.1b), called the 'Oude Schelde' (Old Scheldt), has an average width of 75 m, a dimension that indicates limited or no tidal influence yet in the thirteenth century AD (Kiden 2006).

The tidal range and local high water level in the Scheldt River increased rapidly since about 1000–1100 AD (Fig. 17.4), by natural processes such as the development, widening and deepening of a new, shorter tidal connection to the North Sea (via the Western Scheldt, whereas before that period the connection to the North Sea was via the Eastern Scheldt) and elevation gain by sediment accretion of the intertidal areas, as well as by human impact, namely large-scale embankments of intertidal areas to create agricultural land. Since the middle of the 20th century, land reclamation has been driven by industrial and urban developments (Meire et al. 2005). The land reclamation in the 20th century resulted in a 16% loss of the total surface area of the estuary, mainly reducing the area of mature marshes.

In order to provide safe access to the port of Antwerp and to enable shipping traffic in more upstream areas of the Scheldt, several modifications altered the river system. In parts of the freshwater zone, the river channel was

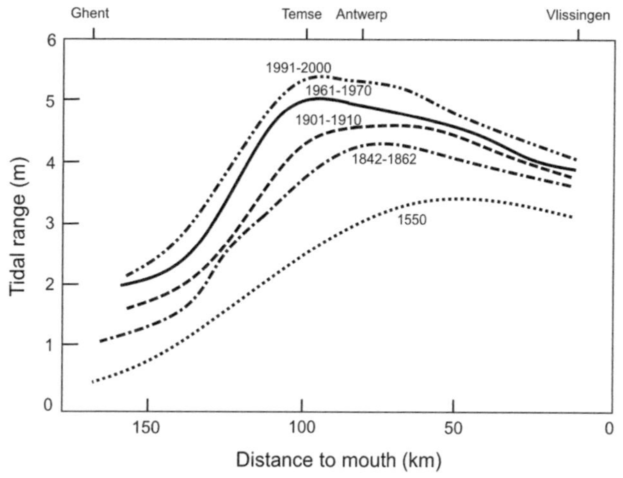

Fig. 17.4 Evolution of the tidal range (difference between high and low tide) in the Scheldt estuary since 1550. Adapted from Kiden (2006) and Van Braeckel et al. (2006)

historically shortened and straightened, thereby reducing the sinuosity of the river (Van Braeckel et al. 2006). In the brackish and salt water zones downstream of the port of Antwerp, maritime access routes are being dredged on a large scale. Also upstream Antwerp the river is dredged for inland shipping. The dredging activities are concentrated on deepening the sills, i.e. shallow areas of the navigation channel. As a consequence, the bottom friction of the channel decreases and the tidal wave can propagate further upstream (Vroon et al. 1997). The average tidal range now increases from Vlissingen (3.8 m) to Antwerp (5.25 m). The maximal tidal range is observed upstream of Antwerp at Temse (5.41 m) and decreases to 2 m around Ghent (data 2001–2010, www.waterinfo.be, and Fig. 17.4). The continued increase of tidal range, tidal discharge and ebb and flood velocities enhanced erosion of the Scheldt river bed, making the river wider and deeper over the last centuries. This led also to an increase of high-dynamic areas. The increase of tidal range raised the high water levels. At the Durme tributary for example, the mean high water level rose 150 cm since 1860 (Van Braeckel et al. 2006) and the inundation risk increased accordingly.

To reduce the flood risk after severe storm floods in 1953 and 1976, the Delta Works and Sigmaplan were initiated along the Dutch (Western Scheldt) and Belgian sides (Sea Scheldt) of the Scheldt River, respectively (Meire et al. 2014). The Dutch Works consisted of conventional engineering solutions such as the cut-off of estuaries by large-scale dams or storm surge barriers (for example the Haringvliet sluices and the Eastern Scheldt storm surge barrier respectively) and by heightening the existing dikes. The more recent Belgian Sigma plan consisted of heightening of the dikes, as well as increasing the water storage capacity of the river by creating new intertidal areas and flood control areas (Table 17.1). This novel approach was adopted to guarantee the safety along with the naturalness and the functioning of the estuarine system. Some of the new intertidal areas are created by managed retreat or realignment of the dikes. Flood control areas (FCA's), on the other hand, are low-lying areas, next to the estuary, which are surrounded by a high dike (Fig. 17.5). The dike at the river side is lowered (the overflow dike, see Fig. 17.5) so that water can flow over the dike during a storm tide, dissipating the flood wave energy and storing a large volume of water. As a result, the water level of the river lowers, reducing the flood risk for inhabited areas. In some FCA's, a controlled reduced tide (CRT) will be introduced: a limited tidal water exchange can enter and exit the FCA-CRT through different sluices (Fig. 17.5). In the FCA-CRT's we expect significant elevation changes, comparable with sedimentation and erosional processes in natural marshes (Cox et al. 2006; Maris et al. 2007).

In total, almost 3000 ha of flood control area (without tides) will be installed and more than 2000 ha of new intertidal areas will be created by 2030 (Table 17.1). The new intertidal areas are either compensation for reclaimed marshes or needed to reach the favourable state of conservation according to the EU Habitat directive.

17.5 Intertidal Flats

Intertidal flats are the intertidal areas that flood every tide, positioned higher in the tidal frame than the subtidal river channel and lower than the vegetated intertidal marshes (Fig. 17.3). Although intertidal flats may look like uniform slopes next to the river channel, surprisingly different morphological features can be distinguished in these dynamic systems.

However, as the Scheldt River is continuously modified by maintenance dredging and sand extraction, the morphology of the tidal river is continuously shifting towards new equilibria. These dynamics, together with natural variations in, e.g. discharge and suspended sediment concentrations, change sedimentation or erosion patterns constantly. This is also evident in monitoring campaigns (e.g. Van Braeckel et al. 2014), where the substrate of the intertidal flat changes from field visit to field visit, and sedimentation or erosion patterns are not consistent between years, places or geomorphologic places (e.g. inner or outer river bend). This uncertainty makes prediction and generalization of the intertidal flat morphological processes very difficult, so we will only give a brief overview of intertidal flat features in the next sections.

Table 17.1 Overview of newly installed flood control areas, flood control areas with reduced tides and realignments (in ha) in 2015 and planned by 2030 in the Scheldt estuary

Year	Non-tidal areas (ha)	Intertidal areas (ha)		Totals (ha)
	Flood control areas	Flood control areas with reduced tide	Realignments	
2015	961	350	243	1554
2030	2760	794	1373	4927

Data Waterwegen & Zeekanaal NV, personal communication, 11/2015

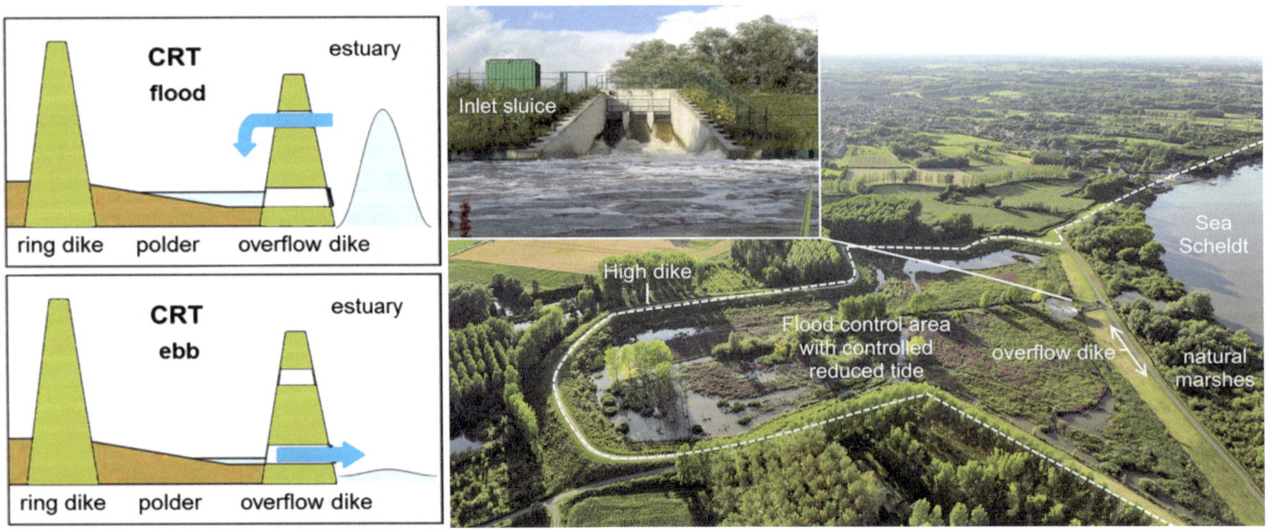

Fig. 17.5 Functioning of a flood control area with controlled reduced tides (FCA-CRT). At flood tide, water enters the area through inlet sluices and at ebb, the water flows back to the estuary through outlet sluices. In the FCA, the tidal range is much reduced compared to the estuary itself. On the right, overview of pilot area Lippenbroek (location see Fig. 17.1b) (*Photo* Vildaphoto/Yves Adams) with inlet sluice (*inset*) after Temmerman et al. (2013)

17.5.1 Mudflat Versus Sandflat

The first difference that is obvious between different intertidal flats is the texture of the sediment, there are sandy and muddy intertidal flats, also called sandflats and mudflats.

The hydrodynamic exposure of a particular place determines the texture of the sediment that is deposited on or transported to the intertidal flats. Areas that experience strong ebb or flood currents or even waves have a sandy texture. Currents and waves keep small particles in suspension, while coarser particles settle, or are transported as bedload. In hydrodynamically sheltered areas, only fine-grained suspended mud (silt and clay sized particles) accumulates.

The differential deposition of sediment texture classes can be observed along an elevation gradient from the lowest intertidal flat areas, close to the subtidal channel to the highest flats at the marsh edge: the lower areas of the intertidal flat experience flooding during almost the whole tidal cycle and hence are generally subjected to stronger flood and ebb tidal currents compared to higher, mudflat areas that only flood at slack high water. This may result in finer sediments deposited at the higher portions of the intertidal flat. Such small changes in sediment texture can be observed at several locations along the Scheldt River (Van Braeckel et al. 2014), although at many locations the difference is negligible.

There is in general also a textural difference evident along the estuarine gradient: the more downstream, the sandier the texture of the intertidal flats. But also here there are many local exceptions in this overall tendency.

The repetition of flood currents, high water slack and ebb currents may lead to the deposition of alternating layers of coarse and fine material on the intertidal flats and marshes, called 'tidal bedding'. During each ebb and flood phase with high flow velocities, coarse sediments are deposited, while during the high water phase, when the flow velocity decreases and changes direction, finer particles are

Fig. 17.6 Different morphological structures of intertidal flats. **a** Tidal flat with tidal bedding visible in the cliff (*white arrow*) and microripples in front of the cliff, **b** Microcliff and **c** meseta morphology caused by *Vaucheria* sp. (*Photos* (c) Research Institute for Nature and Forest (INBO))

deposited. In the same way, also spring and neap cycles or seasonal cycles may be recorded in the sediment. However, the abovementioned tidal bedding will only be observed in areas with continued sedimentation, rather than erosion. Such fine-layered tidal bedding is often visible in intertidal flat or marsh sediments, which can get exposed by vertical cliffs (ranging from a few dm to sometimes more than 1 m in height) originating from wave- and current-induced erosion (Fig. 17.6a).

17.5.2 Erosion on Intertidal Flats

Where erosion is the dominant process on the intertidal flats, different erosional processes can be observed.

When waves or currents erode the intertidal flat deposits or underlying hard substrates, such as clay or peat layers, a variety of geomorphological forms can be observed (De Smedt 1969; Van Braeckel et al. 2014), resulting from differential erosion between harder and softer substrates (Fig. 17.6):

- Microcliffs: small cliffs facing the low water line.
- Micromeseta: small plateaus with eroding edges at both sides.
- Microcuesta: the escarpment facing the marsh edge, the backslope inclining towards the low water line.

During the dry period, water seeping out of the marsh on the mudflat can erode rills and small channels. In general, steeper slopes of the mudflat result in straight, single channels, while meandering, dentritic channels form on gentle mudflat slopes (De Smedt 1969; Van Braeckel et al. 2009). On sandy flats, the steep slopes do not show any geomorphological features such as channels, since water can infiltrate the sandy substrate. On the flat surfaces, however, micro-, meso- or macroripples are often visible, indicators of tidal currents and waves (Fig. 17.6a). The slope of the intertidal flats generally decreases downstream in the estuary, with average slopes >15% towards Ghent to <5% at the Dutch border (Van Braeckel et al. 2014). This corresponds to the decrease of the average width of the river. In the gap of Hoboken, where the river cuts into resistant Oligocene clay (see Sect. 17.4), the slopes are steeper than expected.

A special geomorphological pattern on the intertidal flats may arise by diatoms. Diatoms may accumulate on the intertidal flats, forming a visible, smooth thick brown biofilm on top of the sediment, inhibiting erosion and increasing sedimentation of fine-grained particles (Weerman et al. 2010; van de Koppel et al. 2012). If the diatoms form discontinuous films, erosion of the lower unprotected areas results in a regular pattern with higher diatom-covered hummocks and lower depressions. The same process is also visible when *Vaucheria* sp., a kind of green algae, is present on the intertidal flats (Fig. 17.6c).

It is important to note that only 10% of the riverbanks of the Sea Scheldt are still natural. All the other parts of the Scheldt river banks are reinforced by riprap, on the intertidal flat or at the intertidal marsh edge (Van Braeckel et al. 2009). These structures prevent erosion of the underlying intertidal flats or the marsh cliff (see next section).

17.6 Intertidal Flat—Marsh Edge

The transition from intertidal flat to marshes can be gradual or very abruptly marked by a cliff. While marshes occur on a (gentle) slope towards the mudflat, pioneer species can expand to the lower mudflats, as does for instance the brackish pioneer species *B. maritimus* at the Dutch-Flemish border. In the freshwater marshes, common reed (*Pragmites australis*) is often found in the lower parts of a gentle marsh slope (Van Braeckel et al. 2014). Both species can, under favourable circumstances, laterally expand at rates >2 m/year (Van Braeckel et al. 2014).

Cliffs are characterized by a steep elevation difference between the low intertidal flat and the higher marsh platform and are observed along different salt (van de Koppel et al. 2005; Van der Wal et al. 2008), brackish (Van der Wal et al. 2008) and freshwater marshes (Brys et al. 2005) in the Scheldt estuary. Cliff formation is promoted by differential sedimentation rates on the marsh and the adjacent mudflat, steepening the marsh front and hence making it more vulnerable to wave-induced erosion (Allen 2000; Van de Koppel et al. 2005). At the edge of marsh plateaus, lateral cliff retreat through undercutting can result in marsh rejuvenation (Van de Koppel et al. 2005; Van der Wal et al. 2008) when the hydrodynamic conditions and the elevation of the intertidal flat in front of the cliff are favourable for pioneer species re-colonization. As the marsh further expands over the intertidal flat in front of the cliff, the abandoned cliff may remain visible within the vegetated marsh. Active cliffs and marsh rejuvenation are visible at several marshes along the Western Scheldt such as the Paulina marsh (van de Koppel et al. 2005; Van der Wal et al. 2008). At the Sea Scheldt, however, more than 90% of the river banks are reinforced with riprap (Van Braeckel et al. 2009) and this hinders cliff erosion and marsh rejuvenation.

17.7 Intertidal Marshes

Intertidal marshes are the vegetated areas whose elevation allows flooding only during relatively short periods around high water level (Fig. 17.3). A marsh typically starts with a pioneer species that colonizes the intertidal flat at elevations just below the high water level at neap tides (Fig. 17.3). The vegetation reduces the water flow velocity, thus promoting sedimentation. As the marsh platform accretes, other plant species will also colonize the marshes. The higher the marsh platform rises, the less frequent it will be inundated, reducing accretion, until it only gets flooded by the higher spring tides (Fig. 17.3). However, marshes can also disappear by cliff erosion (see previous section) or by drowning, if they get more and more flooded, for example when relative sea level rise is faster than the sediment accretion rate. In the Scheldt estuary, horizontal expansion or retreat of the marshes is limited to those in the Western Scheldt because most edges at the Sea Scheldt are reinforced with riprap. The geomorphological and sedimentary processes within marshes and newly created flood control areas will be discussed in detail in the next paragraphs. We first focus on vertical sediment accretion on the marsh surface.

17.7.1 Sedimentation in Time and Space

The suspended sediment concentration (SCC), i.e. the mass of dry sediment per volume of water (in g/l), determines the maximum sediment deposition rate that can occur on a marsh surface. The SSC is determined by the incoming river/sea water, although local scouring can entrain sediment from the creek bed and banks. The SSC strongly fluctuates during a spring-neap tidal cycle and throughout the year (Fettweis et al. 1998; Temmerman et al. 2003b). For example, SSC measurements at the Notelaar, a freshwater marsh just downstream Temse (Fig. 17.1b), showed a variation between 10 and 300 mg/l (Temmerman et al. 2003b). In general, the sediment concentration is higher when the marsh is flooded by higher tides (spring tides), due to higher water velocities (Bouma et al. 2005a); additionally, this relationship between SSC and high tide level has a steeper slope in winter, i.e. for a similar high tide level the SSC is higher in winter than in summer (Fettweis et al. 1998; Temmerman et al. 2003b). The total sediment deposited on a marsh surface will follow the tidal and seasonal variations in incoming SCC. As a result, most sedimentation is occurring during the highest spring tides in winter (Temmerman et al. 2003b).

Sedimentation on the marsh surface shows a clear spatial pattern (Fig. 17.7a). A corresponding digital elevation model is shown in Fig. 17.7b. Most of the sediment is deposited in the low-elevated marshes, close to the marsh edge. The lower the marsh elevation, the higher the inundation frequency and hence the higher the sediment deposition rate (Stoddart et al. 1989; Cahoon and Reed 1995; Temmerman et al. 2003b). On the elevated marshes further away from the marsh edge, less sedimentation occurs, the variation being also determined by the distance to the nearest tidal creek. A greater distance from a tidal creek results in lower deposition rates (Reed et al. 1999; Allen 2000; Christiansen et al. 2000; Temmerman et al. 2003b). Sedimentation over marsh platforms is dominated by tidal advective processes (Allen 2000; Temmerman et al. 2003b), hence the spatial patterns of sediment deposition can be explained by the water flow over the marsh. The tidal inundation starts at the lower marsh edge. These are the lowest areas and flooded first. As the water level rises, the tidal channels are filled. Once the tidal channels are full, the water flows from the channels over the surrounding marsh platform. Once the water level exceeds the elevation of the marsh platform, the tidal flow will also directly flow from the Scheldt River over the platform. The sediment-laden water loses progressively more suspended sediment along its flow path from the marsh edge to the vegetated marsh platform (Christiansen et al. 2000; Friedrichs and Perry 2001; Moskalski and Sommerfield 2012). Consequently the largest, heaviest particles (fine sand) are deposited close to the marsh edge and the tidal channels. The remaining fine-grained sediment (silt and clay) is deposited further away on the marsh platform (Allen 2000; Temmerman et al. 2003b, 2005a). When the water level rises higher than the vegetation canopy and the vegetation is deeply submerged, the water will flow as sheet flow from the river over the marsh platform (Temmerman et al. 2005a; Vandenbruwaene et al. 2013).

In FCA-CRT's, the water volume that flows in and out every tide is set by the tidal height in the estuary and sluice configuration, and contrary to natural intertidal marshes, independent of the site elevation. Classical feedback mechanisms, which slow down sedimentation rates when marshes grow higher in elevation, function therefore differently in a CRT (Vandenbruwaene et al. 2011a). A small amount of sedimentation is wanted for the geomorphic development to a marsh in FCA-CRT's, but too much sedimentation will affect the storage capacity (see Sect. 17.7.4). Measurements reveal that the initial sedimentation rate is strongly related to the inundation depth and inundation time, with rates similar to those of natural marshes (Vandenbruwaene et al. 2011a). As sedimentation is related to tidal characteristics, which are regulated by the sluice system, sedimentation rates can be controlled in an FCA-CRT.

17.7.2 Crucial Impact of Vegetation

Marshes have a dense vegetation cover (e.g. Fig. 17.2). In general, if water flows through dense vegetation, the flow velocity is strongly reduced due to the friction exerted by the vegetation canopy on the water flow, thereby promoting sedimentation (Allen 2000; Leonard and Reed 2002; Bouma et al. 2005a; Leonard and Croft 2006; Neumeier and Amos

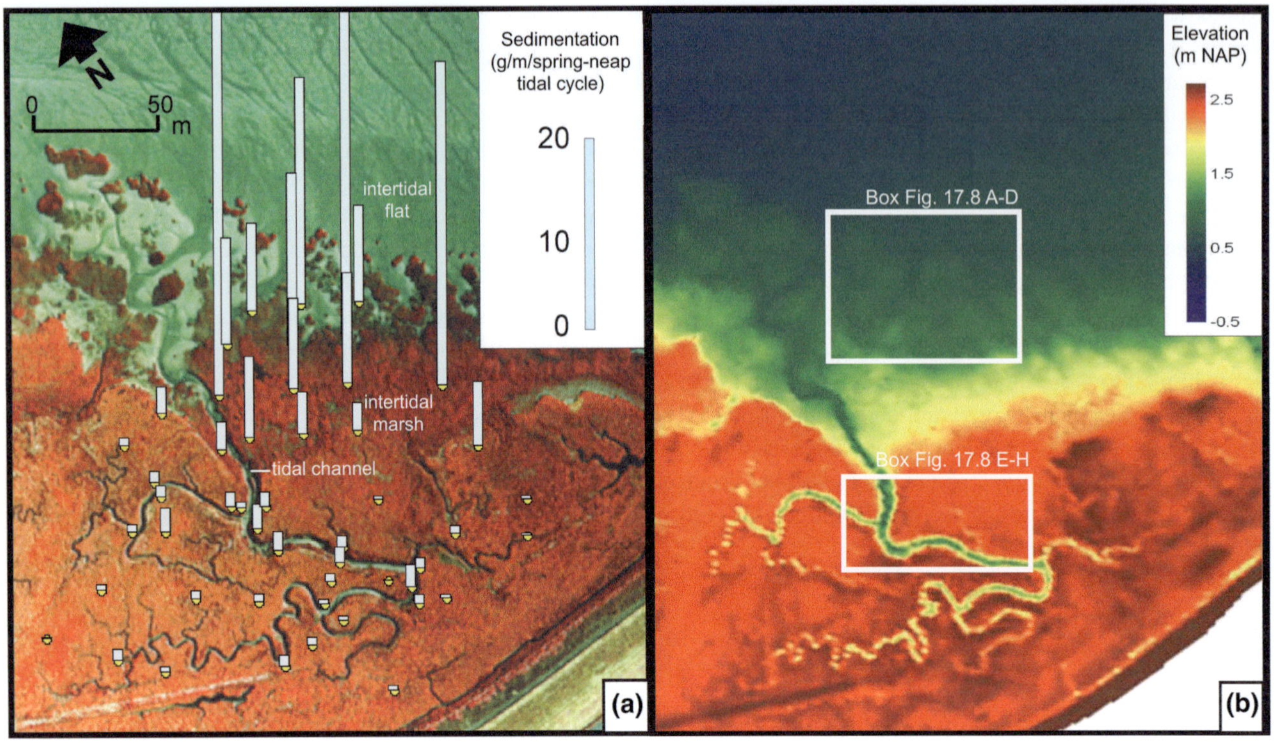

Fig. 17.7 **a** Aerial image of Paulina marsh (see Fig. 17.1 for location). The *bars* represent cumulative sedimentation during a spring-neap tidal cycle (ca. 14 days) on 50 locations. *Reddish colours* represent marsh vegetation, *greenish colours* the intertidal flat and channels. **b** Elevation map of the same area (*green* to *red* represent low to high elevation). *White boxes* correspond to the modelling areas of Fig. 17.8. After Temmerman et al. (2006)

2006). In intertidal marshes, however, the vegetation has a more complex effect that is crucial for erosion and sedimentation. This is illustrated by a computer model (Delft3D) of Paulina marsh (see location in Fig. 17.1b) that incorporates the physical laws of water flow and sedimentation over a marsh surface with vegetation (Temmerman et al. 2005b, 2007). To demonstrate the impact of vegetation, we simulated flow velocities and sedimentation on the same marsh topography, but without vegetation (Temmerman et al. 2012). In other words, we simulated the flow and sedimentation patterns that we should observe if we mowed all the vegetation on an existing marsh platform. The objective of these model simulations is to show the impact of vegetation on the flow and sedimentation patterns on an intertidal marsh by comparing the simulated patterns with presence and absence of vegetation. Of particular interest are the areas on the marsh edge (Fig. 17.8a–d) and along an existing creek (Fig. 17.8e–h):

- At the transition between the unvegetated intertidal flats and vegetated marshes, the flow velocities are strongly reduced within the vegetation by the friction exerted by the vegetation canopy (Fig. 17.8a). In between vegetated areas, however, the flow converges and velocities are enhanced. Apart from the model simulations, this effect is also empirically observed (Vandenbruwaene et al. 2011b Bouma et al. 2013). As a consequence of this flow pattern, the model shows that sediment is deposited within the vegetation, while in between the vegetation, very low sedimentation rates or even erosion can take place (Fig. 17.8c). This effect of vegetation patches on sedimentation within and erosion around the patches has also been empirically observed (Bouma et al. 2007).
- On the mature high marsh with deep tidal creeks, the simulated water velocities are the highest in the creeks (Fig. 17.8e): here the friction is low in absence of vegetation. Once the water level exceeds the level of the creek banks, the creek water starts to flood the high vegetated marsh platform where the flow velocities quickly drop. These simulations are in accordance with field measurements of flow velocity patterns (Temmerman et al. 2012).
- The model simulations show that in absence of vegetation (and hence friction) on the marsh platform, the water will flow easily from the Scheldt river over the unvegetated marsh platform instead of flowing first through the creeks (Fig. 17.8b, F). Therefore the flow velocities in the creeks will be lower. Hence, the flow patterns on vegetated marshes lead to limited sedimentation or erosion in the tidal creeks. In the absence of vegetation,

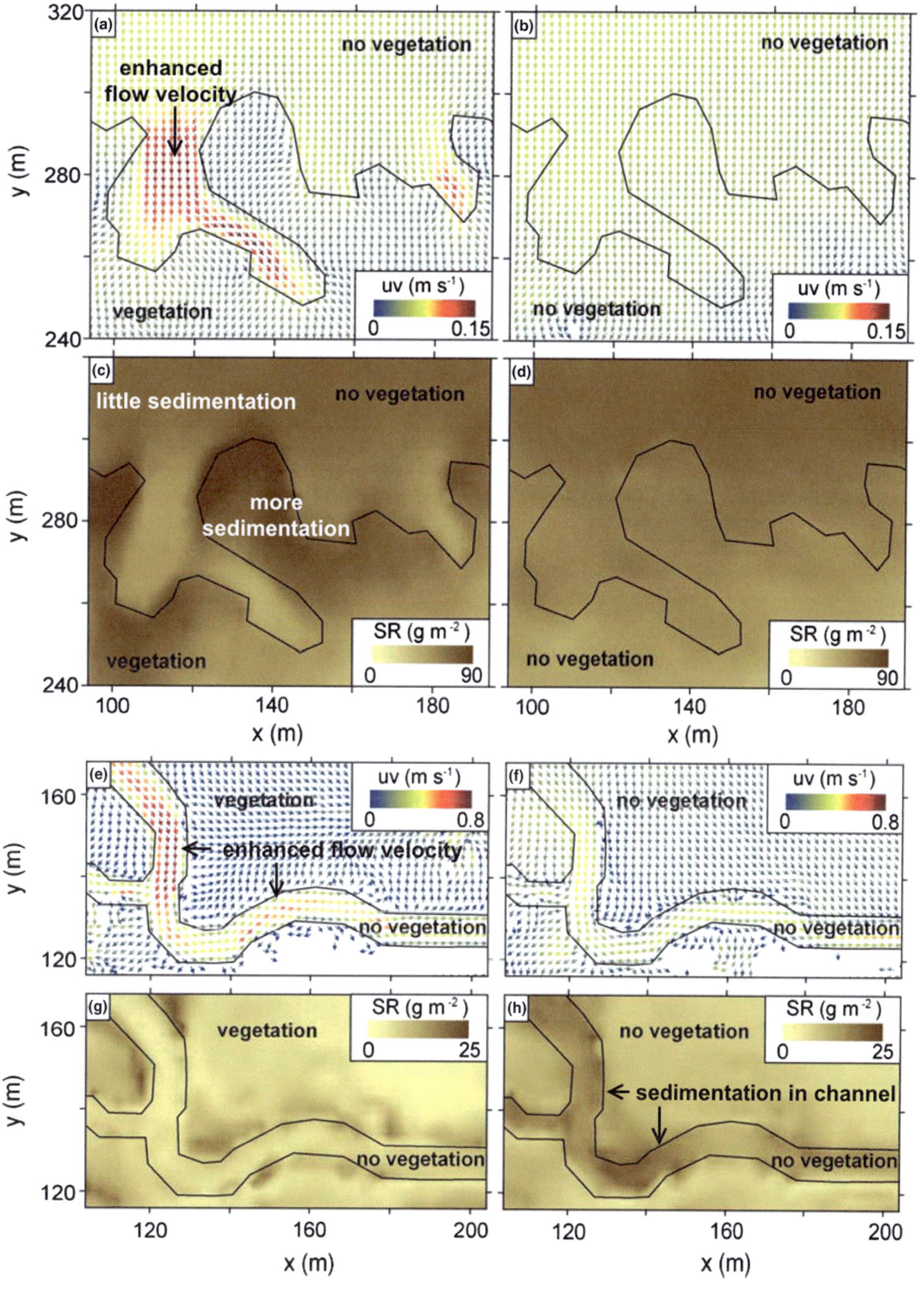

Fig. 17.8 (**a**–**d**) Results of model simulations at the marsh edge (*upper white box* in Fig. 17.7b): flow pattern with (**a**) and without (**b**) marsh vegetation; sedimentation pattern with (**c**) and without (**d**) marsh vegetation. (**e**–**h**) Results of the model simulations at the elevated marsh (*lower white box* in Fig. 17.7b): flow pattern with (**e**) and without (**f**) marsh vegetation; sedimentation pattern with (**g**) and without (**h**) marsh vegetation. Figure after Temmerman et al. (2005b) uv: flow velocity in m s^{-1}; sr: sedimentation rate in g m^{-1}

lower velocities in the creeks could lead to creek infilling and lower sedimentation rates on the marsh platform (Fig. 17.8d, h). This simulated effect of presence or absence of marsh vegetation on flow velocities in the tidal creeks dissecting the marsh are also confirmed by field measurements, where marsh vegetation was experimentally mowed (Temmerman et al. 2012).

These simulations and measurements show that interactions between vegetation, flow hydrodynamics and sedimentation-erosion dynamics play an essential role in the long-term geomorphic development of the intertidal marsh platform, leading to the erosion of tidal creeks in between vegetation patches (D'Alpaos et al. 2006; Temmerman et al. 2007; Vandenbruwaene et al. 2013) and to the development of a levee-basin topography on the vegetated marsh platform (Temmerman et al. 2004a, 2005b; Mudd et al. 2013; Vandenbruwaene et al. 2013, 2015): along the tidal creeks, elevated levees form that are typically 20–30 cm higher than the lower basins further from the tidal creeks. If vegetation is absent such as on intertidal flats, levees are generally not developed along the tidal channels that dissect the surface, while they are typically present along the tidal channels present in vegetated marshes (Vandenbruwaene et al. 2015). This vegetation-induced micro-topography can, in turn, enhance the flow concentration towards the channels (Vandenbruwaene et al. 2015).

In FCA-CRT's, spontaneous channel formation already took place after one year of tidal action, but vegetation did not yet have a significant influence on the channel extension (Vandenbruwaene et al. 2012). Here, channel formation is influenced by the initial topography and the underlying high erosion-resistant polder clay (Vandenbruwaene et al. 2012).

17.7.3 Long Term Effect of Sedimentation

In the previous paragraphs we mentioned that sedimentation on the short time scale of single tides is strongly governed by the inundation frequency, inundation depth and inundation period. On the long term (decades to centuries), the flooding dynamics will change. Marshes build up by sedimentation, but the high water levels also change, e.g. due to relative sea-level rise or construction of embankments (see Sect. 17.4). Relative sea-level rise (RSLR) can cause marshes to become more and more flooded, ultimately drowning them and leading to marsh die-off on a large scale (for example in the Mississippi delta (Penland et al. 2000), Chesapeake Bay marshes (Kearney et al. 1988; Schepers et al. 2017), Venice Lagoon (Carniello et al. 2009)). In other estuaries, however, marshes elevate with a rate similar to RSLR (van der Wal and Pye 2004; Kirwan and Megonigal 2013). It is clear that these different evolutions of marshes with respect to RSLR have important implications for their functioning.

How do the marshes of the Scheldt estuary evolve with changing high water levels? Measurements and coring of an old (>200 year) and a young (ca. 50 year) freshwater marsh (The Notelaar, just downstream of Temse, see Fig. 17.1b) give insight in long-term sedimentation rates representative for the Scheldt River (Temmerman et al. 2003a) (Fig. 17.9):

- Young, low-elevated marshes are frequently flooded by the tides, with a long inundation time and depth, so that the initial sedimentation rate is high (e.g. 6–8 cm year^{-1} between 1950 and 1965). Similar high elevation gains, up to 1.5 m in 15–30 year were also measured in Saeftinghe in the period 1931–1963 (Wang and Temmerman 2013) (see location in Fig. 17.1b). As a

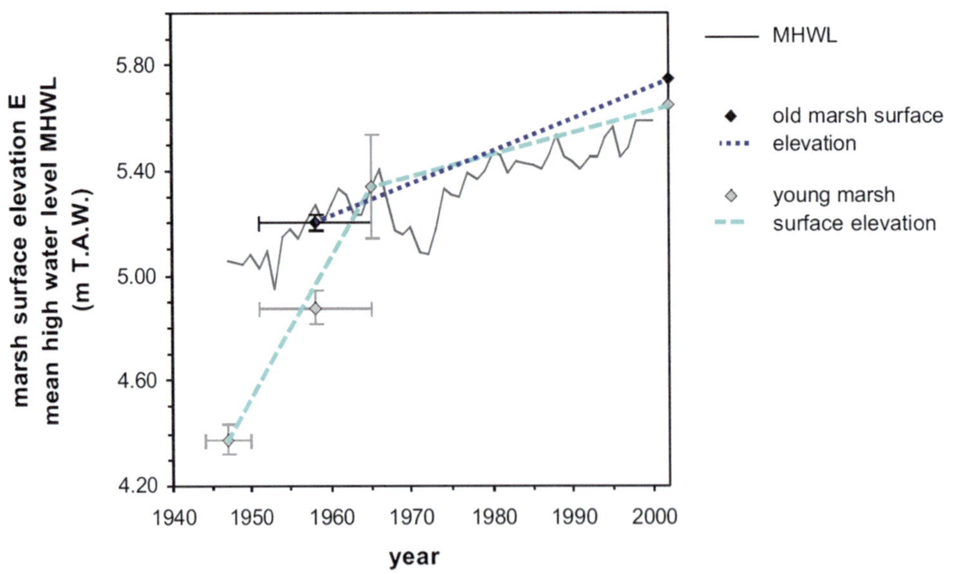

Fig. 17.9 Sedimentation of an old marsh (*dark blue, dotted*) and a young marsh (*light blue, dashed*) of the Notelaar (location just downstream of Temse, Fig. 17.1b) during the last 60 years, dated by sediment cores (after Temmerman et al. 2003a). The *black line* shows the evolution of the mean high water level at the same site

consequence of this elevation gain relative to the mean high water level (MHWL), the flooding frequency, duration and depth decrease gradually. Therefore, the sedimentation rate also decreases. This negative feedback between sedimentation and inundation frequency, inundation time and inundation depth leads to vertical accretion of marsh surfaces up to an equilibrium level, around 10–20 cm above MHWL in the Scheldt estuary (Temmerman et al. 2004a; Wang and Temmerman 2013). As MHWL rises due to RSLR and human modifications to the estuary such as embankment building, the marshes build up and follow the MHWL rise.

- At older, high-elevated marshes (already at the equilibrium level), sedimentation rates are comparable to the rates of MHWL rise (Fig. 17.8). At the Notelaar, this is 1.2–1.8 cm $year^{-1}$.

This explains the generally flat topography of intertidal marshes. However, there is still a micro-topography of levees and basins related to channel systems (see Sect. 17.7.2). Research has shown that once the micro-topography of marshes has established, accumulation on the higher levees and in the lower marsh basins occurs at the same rate, in equilibrium with the rate of MHWL rise (Temmerman et al. 2004a), so the micro-topography of the marshes is preserved with long-term accretion.

These morphodynamics do not only occur in the marshes of the Notelaar but also apply very similarly to other marshes along the Scheldt river (Temmerman et al. 2004b). However, there are also regional differences: young marshes elevate faster in the more inland Sea Scheldt than in the more seaward Western Scheldt, due to higher suspended sediment concentrations in the inland part. Also older marshes rise faster in the Sea Scheldt by a faster MHWL rise rate.

As for the response to increased sea level rise, model results show that the marshes of the Scheldt estuary could keep up with MHWL rise even if MHWL should rise 1.5 times faster in the future than during the last 70 years (Temmerman et al. 2004b). However, if the sediment concentration is halved in this scenario of accelerated MHWL rise, the marshes are no longer able to keep up with it and they will be increasingly flooded, leading to important vegetation changes and maybe even vegetation die-off. Note that in these analyses we do not take into account lateral erosion of the marshes (i.e. cliff erosion).

In realignment areas such as Sieperda marsh (bordering on the east side to Saeftinghe, Fig. 17.1b), sedimentation rates measured close to the Scheldt reach similar high values (up to 3–5 cm $year^{-1}$) as in young marshes, due to the low elevation. This will continue until the high water level is reached (Verbeek and Storm 2001). In the FCA-CRT's, we already mentioned that the sedimentation is related to the inundation depth and inundation time, determined by the initial topography before introduction of the tide. So the lower areas fill up faster than higher elevated marshes that inundate less frequently, leading to a levelling of the topography (not taking into account the formation of tidal creeks and levee-basin micro-topography). In the long term, model analyses show that the marsh elevation–sedimentation feedback (i.e. rising marsh elevation resulting in decreasing inundation depths and hence reduced rate of elevation increase) is absent in FCA-CRT's (Maris et al. 2007; Vandenbruwaene et al. 2011a) because the water volume that is exchanged every tide is not determined by the marsh elevation of the FCA-CRT compared to the water level (as in natural marshes) but depends on (adjustable) sluice dimensions. Hence, in the long term, FCA-CRT's could eventually build up to an equilibrium elevation close to the MHWL of the Scheldt.

17.7.4 Impacts of Geomorphological Processes on the Functioning of Marshes and Flood Control Areas

The water control capacity of intertidal marshes and flood control areas might decline by the natural process of sedimentation and subsequent rise of the marsh elevation. The long-term effect of sedimentation to the water control capacity is dependent on the type of intertidal or flood control area. (i) Natural, low-elevated marshes or realignments elevate rapidly by high, frequent and long inundations. Examples are the young marshes of the Notelaar downstream of Temse, ca. 50 cm in 11 year (1947–1958) (Temmerman et al. 2003a) and Saeftinghe, ca. 150 cm in 32 year (1931–1963) (Wang and Temmerman 2013). This might affect the water storage capacity on natural marshes. However, not only the water storage capacity but also the elevation of intertidal marshes is important for storm tide attenuation, whereby higher marsh elevations provide more storm tide attenuation compared to lower marsh elevations (Smolders et al. 2015). (ii) Flood control areas with a reduced tide store flood water that flows over the overflow dike (Fig. 17.5). Observations from Lippenbroek, an installed FCA-CRT (Fig. 17.5), show that in the lower elevated areas, initial sedimentation rates in the first year after installation reached 13.4 cm $year^{-1}$ (Vandenbruwaene et al. 2011a). As the lowest areas fill up, these high sedimentation rates decrease (currently 6.3 cm $year^{-1}$). Averaged over the whole area, sedimentation rate in the Lippenbroek FCA-CRT is now 2.35 cm $year^{-1}$, comparable with natural marshes. Although sedimentation will continue in the FCA-CRT's, the amount of water (and the amount of sediment) that enters the FCA-CRT can be controlled by the inlet sluice dimensions. In this way, the sluice dimensions can be regulated to keep the flood water storage capacity as high as possible, while letting in enough water for the development of an estuarine ecosystem. (iii) In FCA's without a reduced tide, only during the highest storm tides (one or two times a year) will flood water enter the FCA, depositing very little sediment with no impact on the flood control capacity of the area.

The elevation gain also affects the filter function of intertidal marshes and FCA's. As sedimentation decreases with increasing elevation, also the rate at which sediments are removed from the river water decreases. So we can expect that in a short term, the realignments of marshes and the installation of FCA-CRT's will have a positive effect (due to high sedimentation rates), but eventually this effect will decline.

The filter function of nutrients and pollutants is somewhat more complex and still poorly understood on a longer timescale, since biogeochemical cycling is controlled by sediment structure and sub-surface hydrology (Spencer and Harvey 2012). Nevertheless, marshes and FCA-CRT's remove up to 1 kg of dissolved inorganic nitrogen per tide per ha from the water column, as is observed in the fresh-water marshes of Tielrode (at the mouth of the Durme tributary, see Fig. 17.1b) (Gribsholt et al. 2005) and the FCA-CRT of Lippenbroek.

Finally, also the ecological functioning of marshes will be affected by sedimentation processes. Fast build-up of marsh elevation drives a fast vegetation succession from pioneer species to a climax vegetation (Temmerman et al. 2003a). This is also one of the reasons why low-elevated marshes with pioneer vegetation are scarce in the estuary. Large elevation differences are rapidly levelled out by the inundation–sedimentation feedback, inevitably leading to a loss in habitat diversity and biodiversity. In FCA-CRT's, where the inundation–sedimentation feedback is not present, the inundation frequency and hence the species composition will be determined by the elevation distribution within the area and the sluice dimensions.

17.8 Synthesis and Conclusion

We showed that the geomorphology of intertidal flats and marshes along the Scheldt estuary is very dynamic. The precise processes leading to sedimentation or erosion on intertidal flats are complex, producing a diversity of geomorphic features. We also detailed the morphological developments of intertidal marshes. On a short term and a local scale, the sedimentary processes are determined by the tidal characteristics (flooding frequency, depth and period) and the flow patterns over the marshes, in turn strongly influenced by the marsh vegetation. The resulting micro-topography drives plant species composition, with different species on the slightly higher levees compared to the inner marsh basins.

Extensive embankments decimated the former marsh areas of the Scheldt estuary and, together with dredging and other anthropogenic modifications or natural changes, altered the tidal characteristics. In general, the tides rose significantly in the estuary (Fig. 17.4). In the long term, marshes build up relatively rapidly to an equilibrium elevation around mean high water level. This process led to the few remaining mature marshes that we still see along the river banks of the Scheldt, now usually protected from cliff erosion and marsh rejuvenation by ripraps.

Increased flood risk led to the creation of more than 500 ha of new intertidal areas as an ecosystem-based contribution to flood defense foreseen in the so-called Sigmaplan. In restored marshes with controlled reduced tides, we observed similar processes as in the natural marshes, though some processes are governed by the sluice configuration rather than the elevation of the site. Since the natural process of sedimentation reduces some functions of marshes and flood control areas, continued monitoring is needed in order to follow up and adjust the sluice configuration in order to guarantee an optimal combination of the different marsh functions (flood safety, filtering of the water and valuable ecological habitats).

This research highlights that understanding of present-day geomorphic processes is of utmost importance for a sustainable management of the Scheldt estuary. Hence, a good cooperation between scientists and policy makers is needed to safeguard the estuary as a navigation channel, as a safe work and residential area and as a valuable natural area.

References

Allen JRL (2000) Morphodynamics of Holocene salt marshes: a review sketch from the Atlantic and Southern North Sea coasts of Europe. Quat Sci Rev 19:1155–1231. doi:10.1016/S0277-3791(99)00034-7

Bogemans F (1993) Quaternary geological mapping on basis of sedimentary properties in the eastern branch of the Flemisch Valley. Mem Geol Surv Belgium 35

Bogemans F, Meylemans E, Jacops J, Perdaen Y, Storme A, Verdurmen I, Deforce K (2012) The evolution of the sedimentary environment in the lower River Scheldt valley (Belgium) during the last 13,000 a BP. Geol Belgica 15:105–112

Bouma TJ, De Vries MB, Low E, Kusters L, Herman PMJ, Tánczos IC, Temmerman S, Hesselink A, Meire P, van Regenmortel S (2005a) Flow hydrodynamics on a mudflat and in salt marsh vegetation: identifying general relationships for habitat characterisations. Hydrobiologia 540:259–274. doi:10.1007/s10750-004-7149-0

Bouma TJ, De Vries MB, Low E, Peralta G, Tánczos IC, Van De Koppel J, Herman PMJ (2005b) Trade-offs related to ecosystem engineering: a case study on stiffness of emerging macrophytes. Ecology 86:2187–2199. doi:10.1890/04-1588

Bouma TJ, van Duren LA, Temmerman S, Claverie T, Blanco-Garcia A, Ysebaert T, Herman PMJ (2007) Spatial flow and sedimentation patterns within patches of epibenthic structures: Combining field,

flume and modelling experiments. Cont Shelf Res 27:1020–1045. doi:10.1016/j.csr.2005.12.019
Bouma TJ, Temmerman S, van Duren LA, Martini E, Vandenbruwaene W, Callaghan DP, Balke T, Biermans G, Klaassen PC, van Steeg P, Dekker F, van de Koppel J, de Vries MB, Herman PMJ (2013) Organism traits determine the strength of scale-dependent bio-geomorphic feedbacks: a flume study on three intertidal plant species. Geomorphology 180–181:57–65. doi:10.1016/j.geomorph.2012.09.005
Brys R, Ysebaert T, Escaravage V, Van Damme S, Van Braeckel A, Vandevoorde B, Van Den Bergh E (2005) Afstemmen van referentiecondities en evaluatiesystemen in functie van de KRW : afleiden en beschrijven van typespecifieke referentieomstandigheden en/of MEP in elk Vlaams overgangswatertype vanuit de—overeenkomstig de KRW—ontwikkelde beoordelingssystemen voor biologische kwaliteitselementen. Report of the Research Institute for Nature and Forest (IN.O.2005.7).
Cahoon D, Reed D (1995) Relationships among marsh surface topography, hydroperiod, and soil accretion in a deteriorating Louisiana salt marsh. J Coast Res 11:357–369.
Carniello L, Defina A, D'Alpaos L (2009) Morphological evolution of the Venice lagoon: evidence from the past and trend for the future. J Geophys Res Earth Surf 114:1–10. doi:10.1029/2008JF001157
Christiansen T, Wiberg PLL, Milligan TGG (2000) Flow and sediment transport on a tidal salt marsh. Estuar Coast Shelf Sci 50:315–331. doi:10.1006/ecss.2000.0548
Coen I (2008) De eeuwige Schelde? Ontstaan en ontwikkeling van de Schelde. Report of Flanders Hydraulics Research. Borgerhout
Cox T, Maris T, De Vleeschauwer P, De Mulder T, Soetaert K, Meire P (2006) Flood control areas as an opportunity to restore estuarine habitat. Ecol Eng 28:55–63. doi:10.1016/j.ecoleng.2006.04.001
D'Alpaos A, Lanzoni S, Mudd SM, Fagherazzi S (2006) Modeling the influence of hydroperiod and vegetation on the cross-sectional formation of tidal channels. Estuar Coast Shelf Sci 69:311–324. doi:10.1016/j.ecss.2006.05.002
De Moor G (1983) Cryogenic structures in the Weichselian deposits of Northern Belgium and their significance. Polarforschung 53:79–86
De Smedt P (1969) Geomorfologie van slikken en schorren langsheen het Schelde-estuarium op Belgisch grondgebied. Acta Geogr Lovan 7:49–63
Du Laing G, Vandecasteele B, De Grauwe P, Moors W, Lesage E, Meers E, Tack FMG, Verloo MG (2007) Factors affecting metal concentrations in the upper sediment layer of intertidal reedbeds along the river Scheldt. J Environ Monit 9:449. doi:10.1039/b618772b
Fettweis M, Sas M, Monbaliu J (1998) Seasonal, neap-spring and tidal variation of cohesive sediment concentration in the Scheldt Estuary, Belgium. Estuar Coast Shelf Sci 47:21–36. doi:10.1006/ecss.1998.0338
Friedrichs CT, Perry JE (2001) Tidal salt marsh morphodynamics: a synthesis. J Coast Res SI 27:7–37
Gribsholt BTH, Boschker S, Struyf E, Andersson M, Tramper A, De Brabandere L, van Damme S, Brion N, Meire P, Dehairs F, Middelburg JJ, Heip CHR (2005) Nitrogen processing in a tidal freshwater marsh: a whole-ecosystem 15 N labeling study. Limnol Oceanogr 50:1945–1959. doi:10.4319/lo.2005.50.6.1945
Kearney MS, Grace RE, Stevenson JC (1988) Marsh loss in Nanticoke Estuary, Chesapeake Bay. Geogr Rev 78:205. doi:10.2307/214178
Kiden P (1991) The lateglacial and holocene evolution of the middle and lower River Scheldt, Belgium. In: Starkel L, Gregory KJ, Thornes JB (eds) Temperate palaeohydrology. Fluvial processes in the temperate zone during the Last 15,000 Years, pp 283–299
Kiden P (2006) Schelde in Belgie en Zuidwest-Nederland na de laatste ijstijd. Belgeo 3:279–294
Kirwan ML, Megonigal JP (2013) Tidal wetland stability in the face of human impacts and sea-level rise. Nature 504:53–60. doi:10.1038/nature12856
Leonard LA, Reed DJ (2002) Hydrodynamics and sediment transport through tidal marsh canopies. J Coast Res 36:459–469
Leonard LA, Croft AL (2006) The effect of standing biomass on flow velocity and turbulence in Spartina alterniflora canopies. Estuar Coast Shelf Sci 69:325–336. doi:10.1016/j.ecss.2006.05.004
Maris T, Cox T, Temmerman S, De Vleeschauwer P, Van Damme S, De Mulder T, Van den Bergh E, Meire P (2007) Tuning the tide: creating ecological conditions for tidal marsh development in a flood control area. Hydrobiologia 588:31–43. doi:10.1007/s10750-007-0650-5
Meire P, Ysebaert T, Van Damme S, Van Den Bergh E, Maris T, Struyf E (2005) The Scheldt estuary: a description of a changing ecosystem. Hydrobiologia 540:1–11. doi:10.1007/s10750-005-0896-8
Meire P, Dauwe W, Maris T, Peeters P, Coen L, Deschamps M, Rutten J, Temmerman S (2014) Estuaries in focus sigma plan proves efficiency: the recent "Saint-Nicholas" storm surge in the Scheldt estuary: the Sigma plan proves its efficiency! ECSA Bull 62:19–23
Möller I (2006) Quantifying saltmarsh vegetation and its effect on wave height dissipation: results from a UK East coast saltmarsh. Estuar Coast Shelf Sci 69:337–351. doi:10.1016/j.ecss.2006.05.003
Möller I, Spencer T (2002) Wave dissipation over macro-tidal saltmarshes: effects of marsh edge typology and vegetation change. J Coast Res 521:506–521. ISSN:0749-0208
Möller I, Kudella M, Rupprecht F, Spencer T, Paul M, van Wesenbeeck BK, Wolters G, Jensen K, Bouma TJ, Miranda-Lange M, Schimmels S (2014) Wave attenuation over coastal salt marshes under storm surge conditions. Geosci, Nat. doi:10.1038/ngeo2251
Moskalski SM, Sommerfield CK (2012) Suspended sediment deposition and trapping efficiency in a Delaware salt marsh. Geomorphology 139–140:195–204. doi:10.1016/j.geomorph.2011.10.018
Mudd SM, Fagherazzi S, Morris JT, Furbish DJ (2013) Flow, sedimentation, and biomass production on a vegetated salt marsh in South Carolina: toward a predictive model of marsh morphologic and ecologic evolution. In: The ecogeomorphology of tidal marshes. American Geophysical Union, pp 165–188. doi:10.1029/CE059p0165
Neumeier U, Ciavola P (2004) Flow resistance and associated sedimentary processes in a Spartina maritima salt-marsh. J Coast Res 202:435–447
Neumeier U, Amos CL (2006) The influence of vegetation on turbulence and flow velocities in European salt-marshes. Sedimentology 53:259–277. doi:10.1111/j.1365-3091.2006.00772.x
Penland S, Wayne L, Britsch D, Williams SJ, Beall AD, Butterworth V (2000) Geomorphic classification of coastal land loss between 1932 and 1990 in the Mississippi river delta plain, Southeastern Louisiana. USGS Open File Report 00-417
Reed DJ, Spencer T, Murray AL, French JR, Leonard L (1999) Marsh surface sediment deposition and the role of tidal creeks: implications for created and managed coastal marshes. J Coast Conserv 5:81–90. doi:10.1007/BF02802742
Schepers L, Kirwan M, Guntenspergen G, Temmerman S (2017) Spatio-temporal development of vegetation die-off in a submerging coastal marsh. Limnology and Oceanography 62:137–150. doi: 10.1002/lno.10381
Smolders S, Plancke Y, Ides S, Meire P, Temmerman S (2015) Role of intertidal wetlands for tidal and storm tide attenuation along a confined estuary: a model study. Nat Hazards Earth Syst Sci 15:1659–1675. doi:10.5194/nhess-15-1659-2015
Spencer KL, Harvey GL (2012) Understanding system disturbance and ecosystem services in restored saltmarshes: Integrating physical and biogeochemical processes. Estuar Coast Shelf Sci 106:23–32. doi: 10.1016/j.ecss.2012.04.020
Stoddart DR, Reed DJ, French JR (1989) Understanding salt-marsh accretion, Scolt Head Island, Norfolk. England. Estuaries 12:228. doi:10.2307/1351902

Struyf E, Temmerman S, Meire P (2007) Dynamics of biogenic Si in freshwater tidal marshes: Si regeneration and retention in marsh sediments (Scheldt estuary). Biogeochemistry 82:41–53. doi:10.1007/s10533-006-9051-5

Temmerman S, Govers G, Meire P, Wartel S (2003a) Modelling long-term tidal marsh growth under changing tidal conditions and suspended sediment concentrations, Scheldt estuary, Belgium. Mar Geol 193:151–169. doi:10.1016/S0025-3227(02)00642-4

Temmerman S, Govers G, Wartel S, Meire P (2003b) Spatial and temporal factors controlling short-term sedimentation in a salt and freshwater tidal marsh, scheldt estuary, Belgium, SW Netherlands. Earth Surf Process Landforms 28:739–755. doi:10.1002/esp.495

Temmerman S, Govers G, Meire P, Wartel S (2004a) Simulating the long-term development of levee-basin topography on tidal marshes. Geomorphology 63:39–55. doi:10.1016/j.geomorph.2004.03.004

Temmerman S, Govers G, Wartel S, Meire P (2004b) Modelling estuarine variations in tidal marsh sedimentation: response to changing sea level and suspended sediment concentrations. Mar Geol 212:1–19. doi:10.1016/j.margeo.2004.10.021

Temmerman S, Bouma TJ, Govers G, Lauwaet D (2005a) Flow paths of water and sediment in a tidal marsh: relations with marsh developmental stage and tidal inundation height. Estuaries 28:338–352. doi:10.1007/BF02693917

Temmerman S, Bouma TJ, Govers G, Wang ZB, De Vries MB, Herman PMJ (2005b) Impact of vegetation on flow routing and sedimentation patterns: three-dimensional modeling for a tidal marsh. J Geophys Res Earth Surf 110:1–18. doi:10.1029/2005JF000301

Temmerman S, Govers G, Bouma TJ, De Vries M, Wartel S, Meire P (2006) Opslibbing van schorren en overstromingsgebieden langs de Schelde: een onvermijdelijk natuurlijk proces. Water 26:1–9

Temmerman S, Bouma TJ, van de Koppel J, van der Wal D, De Vries MB, Herman PMJ (2007) Vegetation causes channel erosion in a tidal landscape. Geology 35:631–634. doi:10.1130/G23502A.1

Temmerman S, Moonen P, Schoelynck J, Govers G, Bouma TJ (2012) Impact of vegetation die-off on spatial flow patterns over a tidal marsh. Geophys Res Lett 39:1–5. doi:10.1029/2011GL050502

Temmerman S, Meire P, Bouma TJ, Herman PMJ, Ysebaert T, De Vriend HJ (2013) Ecosystem-based coastal defence in the face of global change. Nature 504:79–83. doi:10.1038/nature12859

Teuchies J, Vandenbruwaene W, Carpentier R, Bervoets L, Temmerman S, Wang C, Maris T, Cox TJS, Van Braeckel A, Meire P (2013) Estuaries as filters: The role of tidal marshes in trace metal removal. PLoS One 8:e70381. doi:10.1371/journal.pone.0070381

Van Braeckel A, Piesschaert F, Van den Bergh E (2006) Historische analyse van de Zeeschelde en haar getijgebonden zijrivieren. 19e eeuw tot heden. Report of the Research Institute for Nature and Forest (INBO.R.2006.29). Brussel

Van Braeckel A, Mikkelsen JH, Dillen J, Piesschaert F, Van den Bergh E, Coen L, De Mulder T, Ides S, Maximova T, Peeters P, Plancke Y, Mostaert F (2009) Inventarisatie en historische analyse van Zeescheldehabitats Vervolgstudie: resultaten van het tweede jaar. Report of the Research Institute for Nature and Forest (INBO.IR.2009.34)

Van Braeckel A, Elsen R, Van den Bergh E (2014) MONEOS—geomorfologie. Hoogteraaien van slik en schor in de Zeeschelde. Evolutie van toestand tot 2012. Report of the Research Institute for Nature and Forest (INBO.R.2014. 1860252). Brussel

Van Damme S, Frank D, Micky T, Olivier B, Eric S, Britta G, Oswald VC, Patrick M (2009) Tidal exchange between a freshwater tidal marsh and an impacted estuary: the Scheldt estuary, Belgium. Estuar Coast Shelf Sci 85:197–207. doi:10.1016/j.ecss.2009.08.005

van de Koppel J, Van der Wal D, Bakker JP, Herman PMJ (2005) Self-organization and vegetation collapse in salt marsh ecosystems. Am Nat 165:E1–E12. doi:10.1086/426602

van de Koppel J, Bouma TJ, Herman PMJ (2012) The influence of local- and landscape-scale processes on spatial self-organization in estuarine ecosystems. J Exp Biol 215:962–967. doi:10.1242/jeb.060467

Van den Bergh E, Huiskes A, Criel B, Hoffmann M, Meire P (2001) Biodiversiteit op de Scheldeschorren. Levende Nat 102:62–66

Van Den Bergh E, Ysebaert T, Meire P (2005) Water bird communities in the Lower Zeeschelde: long-term changes near an expanding harbour. Hydrobiologia 540:237–258. doi:10.1007/s10750-004-7147-2

van der Wal D, Pye K (2004) Patterns, rates and possible causes of saltmarsh erosion in the Greater Thames area (UK). Geomorphology 61:373–391. doi:10.1016/j.geomorph.2004.02.005

Van der Wal D, Wielemaker-Van den Dool A, Herman PMJ (2008) Spatial patterns, rates and mechanisms of saltmarsh cycles (Westerschelde, The Netherlands). Estuar Coast Shelf Sci 76:357–368. doi:10.1016/j.ecss.2007.07.017

Vandenbruwaene W, Maris T, Cox T, Cahoon DR, Meire P, Temmerman S (2011a) Sedimentation and response to sea-level rise of a restored marsh with reduced tidal exchange: comparison with a natural tidal marsh. Geomorphology 130:115–126. doi:10.1016/j.geomorph.2011.03.004

Vandenbruwaene W, Temmerman S, Bouma TJ, Klaassen PC, De Vries MB, Callaghan DP, Van Steeg P, Dekker F, Van Duren LA, Martini E, Balke T, Biermans G, Schoelynck J, Meire P (2011b) Flow interaction with dynamic vegetation patches: implications for biogeomorphic evolution of a tidal landscape. J Geophys Res Earth Surf 116:1–13. doi:10.1029/2010JF001788

Vandenbruwaene W, Meire P, Temmerman S (2012) Formation and evolution of a tidal channel network within a constructed tidal marsh. Geomorphology 151–152:114–125. doi:10.1016/j.geomorph.2012.01.022

Vandenbruwaene W, Bouma TJ, Meire P, Temmerman S (2013) Bio-geomorphic effects on tidal channel evolution: Impact of vegetation establishment and tidal prism change. Earth Surf Process Landforms 38:122–132. doi:10.1002/esp.3265

Vandenbruwaene W, Schwarz C, Bouma TJ, Meire P, Temmerman S (2015) Landscape-scale flow patterns over a vegetated tidal marsh and an unvegetated tidal flat: implications for the landform properties of the intertidal floodplain. Geomorphology 231:40–52. doi:10.1016/j.geomorph.2014.11.020

Verbeek H, Storm C (2001) Tidal wetland restoration in The Netherlands. J Coast Res SI 27:192–202

Vroon J, Storm C, Coosen J (1997) Westerschelde, stram of struis? Eindrapport van het Project Oostwest, een studie naar de beïnvloeding van fysische en verwante biologische patronen in een estuarium. Report RIKZ, 97.023.

Wang C, Temmerman S (2013) Does biogeomorphic feedback lead to abrupt shifts between alternative landscape states?: an empirical study on intertidal flats and marshes. J Geophys Res Earth Surf 118:229–240. doi:10.1029/2012JF002474

Weerman EJ, van de Koppel J, Eppinga MB, Montserrat F, Liu Q-X, Herman PMJ (2010) Spatial self-organization on intertidal mudflats through biophysical stress divergence. Am Nat 176:E15–E32. doi:10.1086/652991

Widdows J, Brinsley M (2002) Impact of biotic and abiotic processes on sediment dynamics and the consequences to the structure and functioning of the intertidal zone. J Sea Res 48:143–156. doi:10.1016/S1385-1101(02)00148-X

Ysebaert T, Meininger PL, Meire P, Devos K, Berrevoets CM, Strucker RCW, Kuijken E (2000) Waterbird communities along the estuarine salinity gradient of the Schelde estuary, NW-Europe. Biodivers Conserv 9:1275–1296. doi:10.1023/A:1008976306651

The Flemish Valley: Response of the Scheldt Drainage System to Climatic and Glacio-Eustatic Oscillations

18

Irénée Heyse and Alain Demoulin

Abstract

Beyond its focus on the climatic and eustatic controls of the Flemish Valley evolution, this chapter explores more broadly the Plio-Quaternary history of the Belgian part of the Scheldt basin, in which the Flemish Valley is essentially a buried incised-valley system that once drained the Scheldt area. The Scheldt is typically a lowland river flowing across a subdued hilly topography in its middle course and on the flat plain chiefly made of its own alluvial deposits in the lower course. We first describe the terraces located on the interfluves and valley sides of the Middle Scheldt region, providing also a tentative chronology of their development. Then, the morphology and the sedimentary infill of the Flemish Valley are described in relation to alternating erosional and aggradational episodes since the Saalian. The transition from stepped terraces to cut-and-fill processes at the Flemish Valley level is attributed to the waning of the tectonically driven uplift of the area during the Middle Pleistocene and the bedrock incision episode that lastly created the Flemish Valley is tentatively related to the base level fall induced by the catastrophic breaching of the Dover Strait dam during the Late Saalian. Finally, the Late Glacial and Holocene fate of the Flemish Valley is evoked, emphasizing the role of the Late Pleniglacial/Late Glacial coversand morphology in deflecting the river's former NNW-striking course to the east and over Antwerp and pointing to the first impact of human activities on the Scheldt behaviour in the last third of the Holocene.

Keywords

Flemish Valley • Lowlands' quaternary evolution • Scheldt basin • Glacio-eustatic control of river evolution • W Belgium

18.1 Introduction

The lowland landscape of western Belgium is essentially flat, with only subdued small-scale landforms that hardly animate the topography. However, the subsurface records an eventful Quaternary history mainly of the response of the Scheldt drainage system to changing climatic and base level controls during the Quaternary. The main feature is the so-called Flemish Valley (FV), a complex system of valleys incised in the Tertiary substratum and Quaternary deposits and now completely filled up with sediments attesting $\sim$200 ka of fluvial evolution in the Scheldt basin. Originally coined by Tavernier (1943) to describe the low plain north of Ghent, then meant as the erosional contact surface between the Quaternary filling of the Scheldt basin's buried paleo-valleys and their Tertiary substratum (De Moor 1963), the name "Flemish Valley" is now used to define this complex

I. Heyse (✉)
Department of Geography, University of Ghent, Krijgslaan 281-S8, Ghent, Belgium
e-mail: Irenee.Heyse@UGent.be

A. Demoulin
Department of Physical Geography and Quaternary, University of Liège, Sart Tilman, B11, 4000 Liège, Belgium
e-mail: ademoulin@ulg.ac.be

A. Demoulin (ed.), *Landscapes and Landforms of Belgium and Luxembourg*,
World Geomorphological Landscapes, DOI 10.1007/978-3-319-58239-9_18

incised-valley system as a whole, reaching depths of 25 m or more below the current sea level. As such, the FV extends along the lower and middle Scheldt and its main tributaries up to, e.g., Tournai on the Scheldt, Estaires on the Leie, and Aarschot on the Rupel-Demer axis (Fig. 18.1). The levelled modern topography resulting from the FV infill is further smoothed by the widespread deposition of Weichselian aeolian sediments. Consequently, the reconstruction of the FV's history is essentially based on data obtained from drillings, mining pits and outcrop profiles. Beyond the initial focus on the climatic and glacio-eustatic controls on the Scheldt valley evolution, more recent studies investigated the possible causes of the sudden change in the vertical behaviour of the drainage system that led more or less classical terrace staircases to give way to alluvial/marine-estuarine bodies nested in a vertically stabilized FV since the mid-Middle Pleistocene (Beerten 2010). However, despite the extraordinary wealth of lithological data, a persisting main concern relates to dating of the events, with very scarce, if any, reliable time constraints on the pre-Eemian history of the area. Here, we briefly review the available sedimentary evidence and describe the inferred landscape evolution from the Early Pleistocene Scheldt drainage system to the later FV, before ending with an overview of some landforms created atop the FV fill during the Late Glacial and the Holocene.

18.2 Geographical and Geological Settings

In addition to headwaters and uppermost catchment situated in northwestern France, the $\sim$22,000-km^2-wide Scheldt basin includes almost all lowland and low plateau areas of Low and Middle Belgium. While the catchment's highest elevation of 212 m is observed on top of the southern part of the Scheldt–Meuse divide near Anderlues, the part of the basin encompassing the FV most generally lies below 50 m, except for strings of >100-m-high lithologically controlled hills in the south (Fig. 18.1). With a source at $\sim$95 m TAW (*Tweede Algemene Waterpassing*), the Scheldt has a very low mean channel gradient of 0.27‰, which reduces to $\sim$0.12‰ downstream of Ghent. It is a typical lowland river characterized by low discharges, a wide floodplain, very gently sloping valley sides and subdued interfluves. Currently, the Scheldt is approximately 350 km long. In its downstream part the river becomes an estuary, up to several-km-wide at its outlet in the Southern Bight at Vlissingen. Tides propagate up to Ghent, 160 km upstream of Vlissingen, where the tidal range still amounts to 2 m (see this chapter).

Geologically, the FV area belongs to the Cenozoic Belgian Basin, where an extensive succession of Paleocene to Pliocene sediments outcrops from south to north. Resting on the underlying Caledonian Brabant Massif exposed only in the downcut upper valleys of some SE tributaries, this whole succession of sandy and clayey formations is tilted northward, attesting the long-lasting trend to uplift of the south-lying Artois axis and subsidence of the Belgian basin. Having possibly started as early as the Mid-Eocene (Van Vliet-Lanoë et al. 1998; King 2006; Ziegler and Dèzes 2007), the uplift of the Weald-Artois axis was repeatedly reactivated, the last time as recently as around 1.4 Ma (Meijer and Preece 1995; Van Vliet-Lanoë et al. 1998). The Quaternary behaviour of the Weald-Artois axis is especially important as a regional-scale contributor to the FV evolution through its role in the intermittent opening of the Dover Strait and connection between Channel and Southern Bight, which constitutes the Scheldt's base level.

The final emergence of the northern half of the Scheldt basin is fairly recent. In the Pliocene, the seas drowning the subsiding Campine were gradually more restricted to the northernmost margin of the current Scheldt basin east of Antwerp. A long-held belief stated that the oldest traits of the drainage system of the middle and lower Scheldt basin, corresponding to the systematic (N)NE orientation of all major rivers from the Leie in the west to the Gete in the east, had been established following the Late Tortonian sea regression as a pattern of parallel streams developing consequently to, and at the pace of, the Pliocene seas gradually retreating toward NE (e.g., Cornet 1904; Tavernier and De Moor 1974; Laurant 1976). However, Houthuys (2014) concluded from the revised age of the Diestian sands crowning the Flemish hills that a low-slope regression surface and its consequent drainage originated already during the Miocene due to a slow but persisting tectonic uplift in the Boulonnais—Artois area, by which the sea retreated to the NE.

Another striking feature of the Scheldt's drainage network is the major E–W trending Scheldt-Rupel-Demer axis, which strikes through the basin from Ghent to Hasselt and collects the waters of all parallel NE-flowing rivers. Like smaller river reaches of similar orientation, this axis is structurally controlled. When the drainage network began to incise the uniform regression surface on which it had developed in this area, differential erosion affected the NNE-dipping strata and the more resistant cohesive clay layers (e.g., the Boom and Asse clayey formations) formed cuestas whose S–facing escarpments, though of limited height, sufficed to create subsequent E–W drainage axes at their foot. Likewise, lines of iron-cemented deposits of the Diest Formation locally forced W- to WSW-trending drainage.

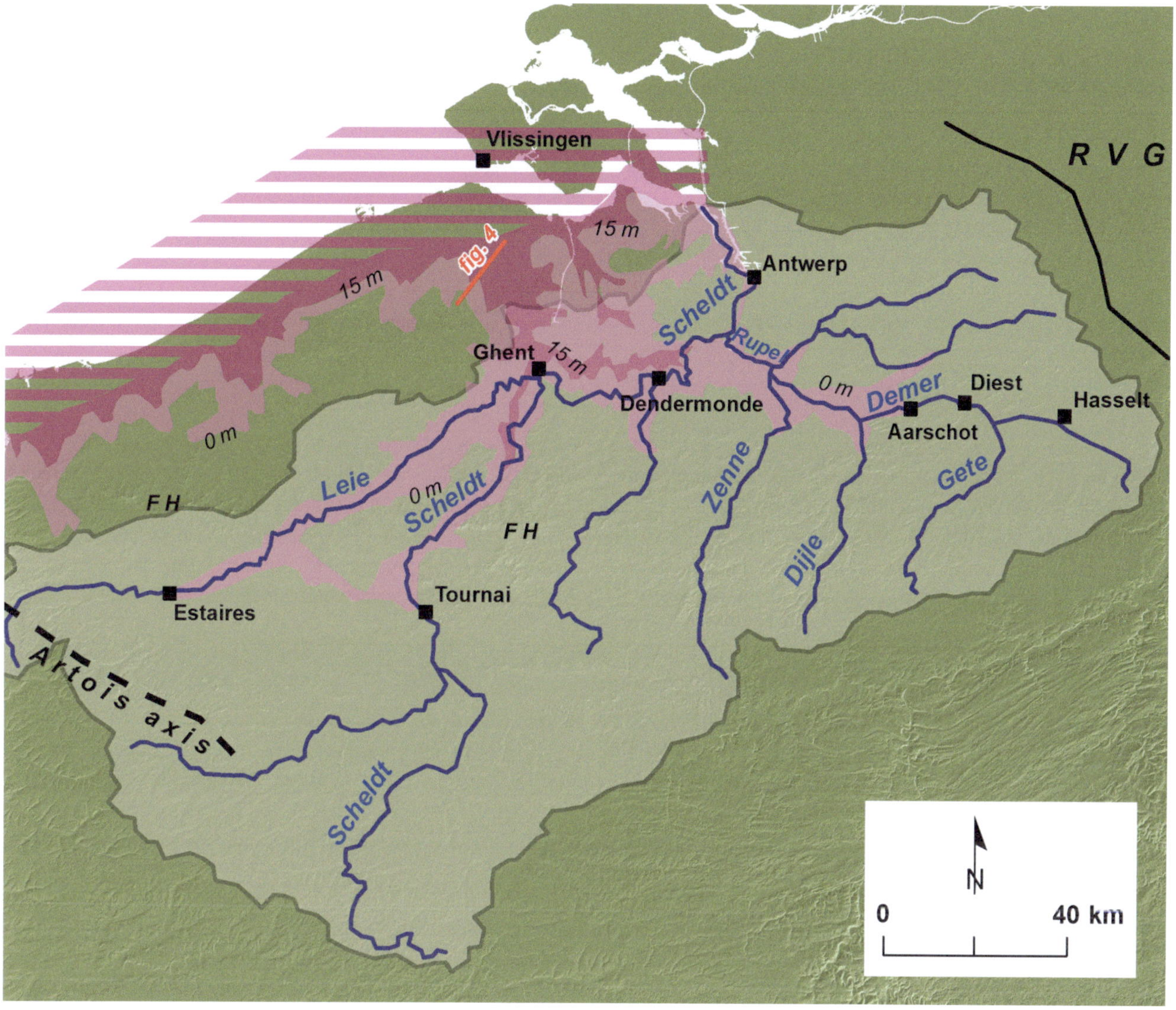

Fig. 18.1 The Flemish Valley in the Scheldt basin. The modern basin is shaded in *light olive green*; depths of the base of the Flemish Valley are indicated in light (0 to −15 m TAW) and dark (<−15 m TAW) *purple*. *RVG* Roer Valley Graben; *FH* Flemish hills

18.3 The Plio-Pleistocene Scheldt Drainage System

18.3.1 Higher Strath Versus Lower Cut-and-Fill Terraces

The latest synthesis by Tavernier and De Moor (1974) of fluvial terrace remnants preserved above the modern floodplains provides a qualitative overview of the pre-FV evolution of the Scheldt basin (see also De Moor and Heyse 1978). Although variable numbers of such levels have been identified in the Scheldt and its tributary valleys, only the terrace sequence of the Scheldt-Leie area is summarized hereafter (Fig. 18.2). The highest levels are not terraces sensu stricto but rather correspond to assumed alluvial gravel deposits scattered on planation surfaces parsimoniously preserved on top of the Flemish Hills (e.g., Mont Cassel) or benches along the flanks of these hills (e.g., Hotondberg) (Tavernier and De Moor 1974). Although the depositional nature (marine, coastal, fluviatile?) and the age of these sediments are highly questionable (de Heinzelin 1963; Pierre 2003; Houthuys 2014), their high degree of chemical weathering points to a probable pre-Pliocene age. Located between ~100 and ~20 m above present floodplains, the next four levels typically cover interfluves and, thus, cannot be assigned to a specific river valley. Only local remnants of the lower one (Meulebeke terrace) clearly pertain to the Leie valley. These strath terrace levels are characterized by mostly thin flint- and quartz-bearing gravel deposits occasionally interlayered with sands. Beyond indications of progressively colder environments and tentative correlations with glacial periods (De Moor 1963; Tavernier and De Moor

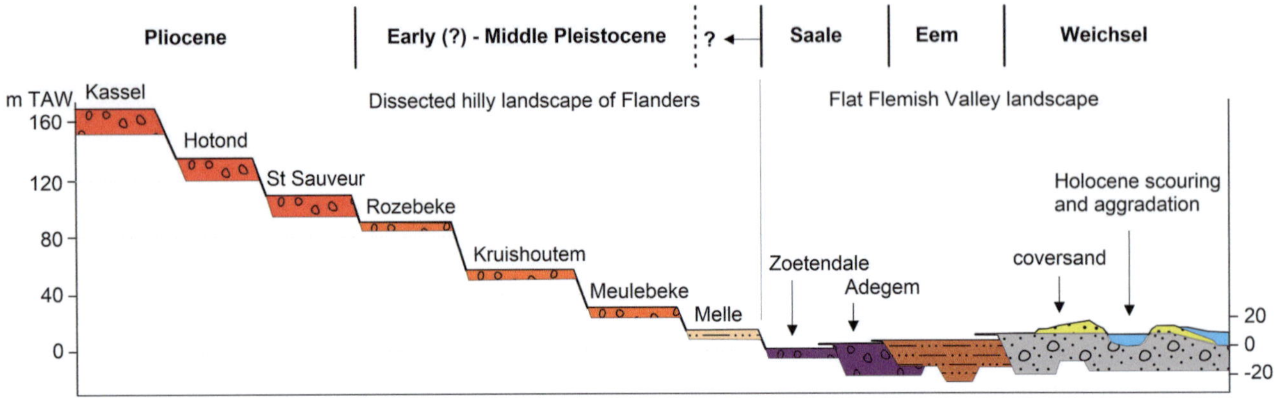

Fig. 18.2 Synthetic terrace staircase in the middle Scheldt-Leie area

1974), no reliable age data is so far available to establish any chronology of the Pliocene to Middle Pleistocene incisional history told by this terrace sequence in the Scheldt basin. Finally, the lowest pre-FV terrace of Melle, whose surface lies 5–10 m above the modern floodplain, is known only very locally, SE of Ghent. Although the presence of *Azolla filiculoides* in peat layers led De Groote (1977) to assign it to the Holstein interglacial, this view was contradicted by Vanhoorne (1992), who placed it in the Middle Saalian Wacken interstadial (MIS 7, according to the latest stratigraphy of Stephan 2014), leaving some uncertainty about the timing of the later FV development marked by cut-and-fill processes.

18.3.2 The Filling of the Flemish Valley

18.3.2.1 FV Topography

In the broadest sense, the FV is the most inland digitate segment of a major incised-valley system that was repeatedly scoured during sea-level glacio-eustatic lowstands to drain NW Belgium and its emerged continental shelf toward the axis of the Southern Bight and its Dover Strait outlet (Mathys 2009) (Fig. 18.3). The proper FV includes the buried mates of the middle Scheldt and Leie valleys, which an eastern branch encompassing the Scheldt-Rupel-Demer axis and the lower reaches of its tributaries joins at Ghent to form the downstream part of the FV, opening widely toward NNW. Based on the reflection seismic data, Mathys (2009) shows that, in the present North Sea W of Vlissingen, the system is divided into the WNW-flowing Dutch Channels and the chief WSW-trending Coastal Valley, which in turn goes into the Ostend Valley (Maréchal and Henriet 1983; Mostaert et al. 1989) flowing northwestward into the major Axial Channel of the Southern Bight. The observed maximum depth of the FV bottom is ~−25 m TAW NW of Ghent (Tavernier and De Moor 1974), consistent with the ~ −28 m recorded in the Ostend Valley (Mathys 2009). The 0 m contour line of the Quaternary base surface is a good indicator of the margins of the FV while the −15 m line highlights the incision depth far inland (Fig. 18.1).

18.3.2.2 FV Filling

Owing to the complex evolution of the FV and the spatial variability of the depositional environments, we focus here on a representative sedimentary sequence exposed along a SW–NE cross-section between Adegem and Sint Margriet (Fig. 18.4). This section synthesizes innumerable observations from mechanical drillings, hand augering, and outcrop profiles.

The oldest FV sediments are two stepped levels of flint- and quartz-bearing alluvial sands and gravels (the Zoetendale and Adegem terraces with bases at ~−8 and −10 to −15 m; Tavernier and De Moor 1974) buried beneath the younger fill on the flanks of the FV (Fig. 18.5a). Their Saalian I and II ages (hard to locate within the current MIS stratigraphy) are inferred from the assumed Eemian age of the overlying sediments and the presence of cryoturbations in the gravelly sands of Adegem. Though both levels have been heavily eroded, the relative elevation of their remnants suggests that the lower lying Adegem terrace originally covered the higher Zoetendale terrace (Tavernier and De Moor 1974), thus initiating a cut-and-fill process. The FV bottom itself is covered with a 5- to 20-m-thick sequence composed of fluvial, estuarine, and marine deposits, depending on the location within the system. Not only makes the current inland location of the tidal deposits clear that this ensemble belongs to an interglacial but molluscs, pollen, and diatom all point also to an Eemian age (Paepe and Vanhoorne 1967; Tavernier and De Moor 1974; De Groote 1977; Verbruggen 1999; Gullentops et al. 2001). The tidal deposits are lithostratigraphically named the Oostende

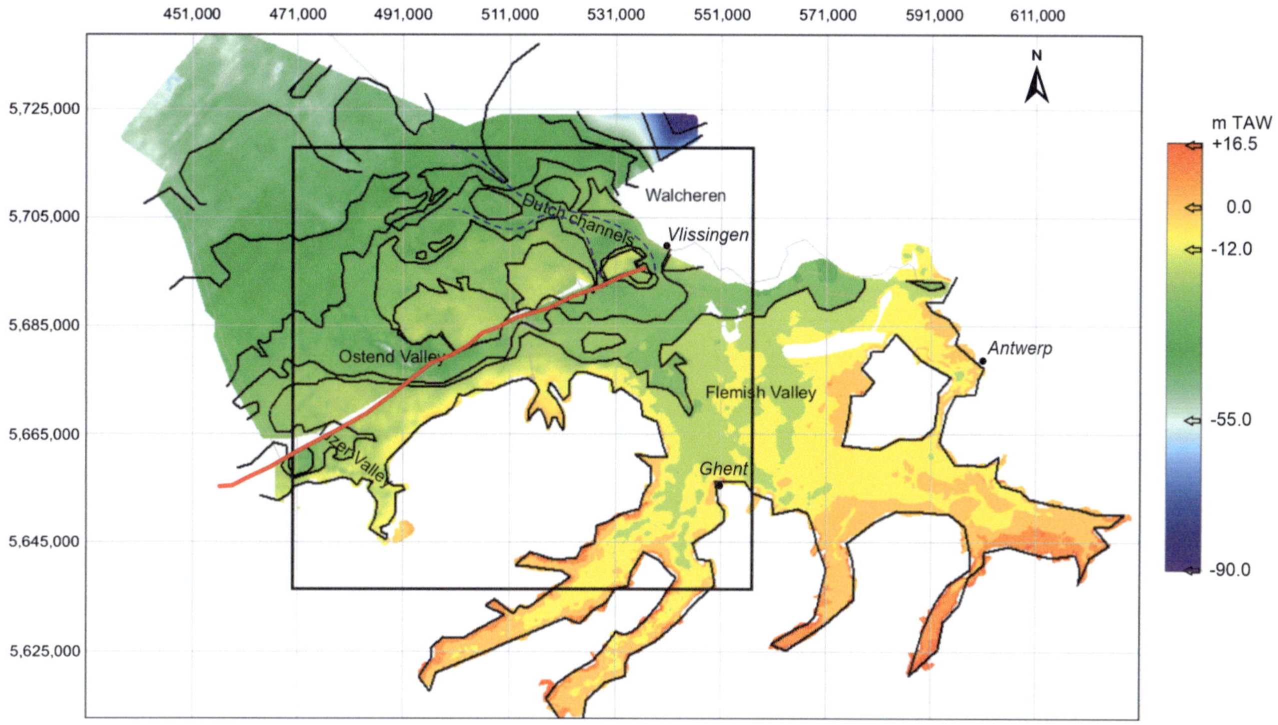

Fig. 18.3 **a** Elevation map of the Quaternary base surface (or, equivalently, top surface of the Paleogene) of western Belgium and the continental shelf showing the system of Quaternary valleys scoured in the paleogene, in which the FV is continued to the west by the Ostend Valley. The *red line* denotes the present coastline (modified after Mathys 2009)

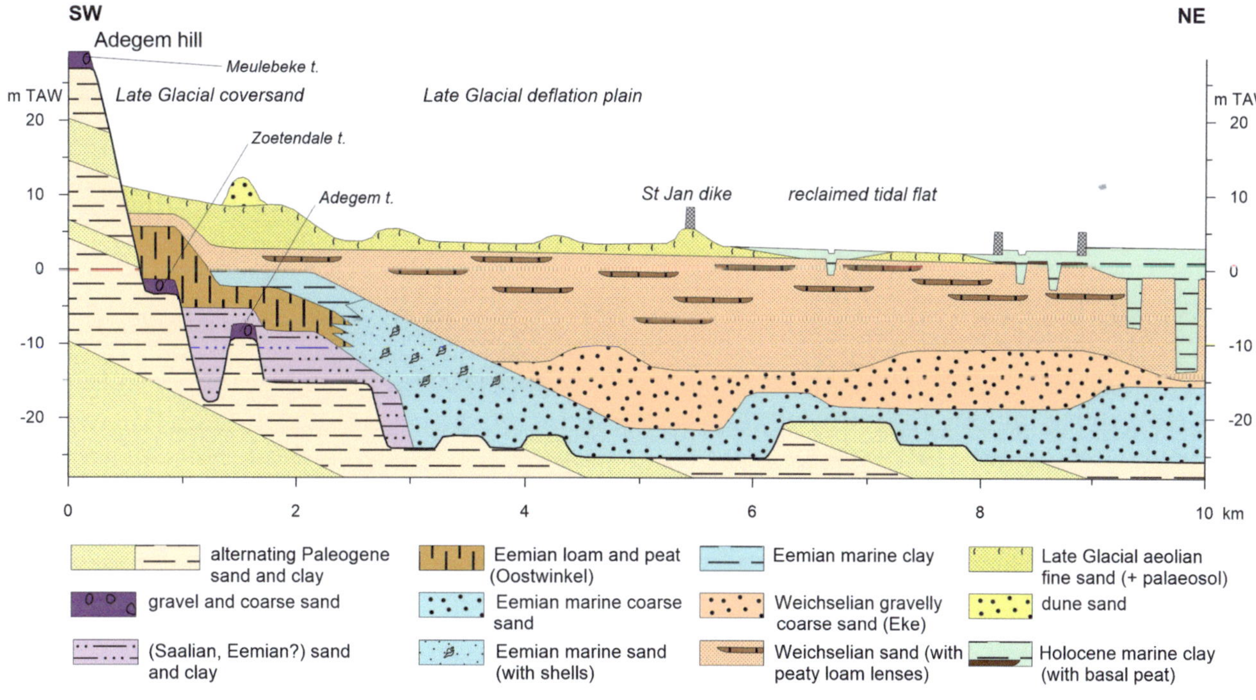

Fig. 18.4 Adegem—Sint Margriet cross-section of the FV filling (see location in Fig. 18.1)

Fig. 18.5 **a** Ledeberg, FV margin 1 km SE of Ghent. At +10 m TAW, contact surface between glauconitic sands of the Eocene substratum and thin Quaternary deposits made of a base gravel overlain by reworked aeolian sands. A presumably Saalian or Weichselian cryoturbation pocket contains mainly sandstone pieces of the Quaternary base gravel. **b** Uitbergen, FV margin 10 km E of Ghent. Bioturbated laminated silts and clays of the Oostwinkel formation (Eemian). **c** Beernem, western branch of the FV, 8 km SE of Brugge. Deeply cryoturbated peat (Eemian?) and Early Weichselian sands. **d** Nieuwdorp, northern FV, near Stekene. Relict ice wedge cast in Early Weichselian sands. **e** Eke, ~10 km SW of Ghent, left bank of the Scheldt. Braided channel laminated sandy deposits of late glacial age. The overlying lag gravel attests to aeolian deflation. **f** Wintam, eastern branch of the FV. Holocene channel scour (basal sands going upward into silty peat) into the sands of the late glacial valley

Formation whereas the fluvial deposits, mainly preserved along the SW border of the FV, are known as the Oostwinkel Formation (Fig. 18.5b).

The bulk of the FV infill is however made of fluvial deposits dating from the cold Weichselian period and collectively forming the Eeklo Formation (Gullentops et al. 2001). These deposits display a spatially highly irregular base surface and thicknesses varying from a few to more than 25 m (Tavernier and De Moor 1974). They begin with gravels and coarse sands of the Dendermonde Member embedding boulders, peat pebbles, large fossil mammal bones, tree trunks, and locally showing cryoturbations (De Moor 1974). Based on their stratigraphic position, an assumed Early Weichselian (MIS 4) age is assigned to the Dendermonde Member (De Moor and Heyse 1978). However, a 129-ka ESR age from mammoth bones in the eastern branch of the FV suggests it might have started slightly earlier (Germonpré et al. 1993). Then, one finds in vertical or lateral continuation extended sand bodies with intercalated silt, gravel, and peat layers and syngenetic cryoturbations and ice wedges evidencing braided river sedimentation under cold conditions (Eke Member) (Fig. 18.5c–e). Numerous radiocarbon dates of the Eke Member consistently place its deposition in MIS 3 and 2 (e.g., De Moor and Heyse 1978). Similar medium to fine sands, intensely cryoturbated and interlayered with peat, could also accumulate in the more protected environment of minor branches of the FV such as at Beernem. Finally, the braided channel facies gives way to horizontally laminated silts, often terminated by a loamy peat layer, and sandy silts of the Oostakker Member (Tavernier and De Moor 1969). This topzone of the main FV filling, which shows multiple levels of cryoturbation, formed as overbank deposits in the quiet environment of Pleniglacial braided-river floodplains (Miall 1977).

The filling of more upstream parts of the FV may be briefly illustrated by the sedimentary sequence described in the Dijle-Demer-Nete-Zenne confluence area by Vandenberghe and de Smedt (1979) and Bogemans (1993). According to Vandenberghe and de Smedt (1979), the lowest terraces above the modern floodplains would be Saalian in age, implying that the filled valleys have been incised in post-Saalian times. Their geoelectric data suggest that the incision, which locally reaches −3 to −7 m TAW, occurred through multistage regressive erosion of a network of narrow valleys and gullies. The bulk of the gully filling, aided in levelling the landscape by phases of cryopediment development, is made of Weichselian gravels. Bogemans (1993) however, notes that meandering river deposits, up to 5-m-thick, are very locally found in the deepest parts of the incised FV eastern branch in the Dijle-Zenne confluence area. Pollen and plant macrorests as well as the presence of *Corbicula fluminalis* place them in the Eemian, confirming that the Saalian last strath terrace was followed by erosion at the Saalian-Eemian transition. Following a limited erosion phase at the Eemian-Weichselian transition, braided rivers deposited sand and gravel at the beginning of the Weichselian, then sand during most of the Weichselian, only briefly interrupted by an episode during which meandering rivers were active. Pollen analyses assign this meandering episode to the Amersfoort and Brørup interstadials (Early Weichselian) (Bogemans 1993; Verbruggen 1999).

As for the Ostend Valley, downstream of the FV, reflection seismic data analysis and calibration with sediment core data evidence a filling made of three seismic units bounded by erosional unconformities (Mathys 2009). In the absence of datable material, Mathys (2009) proposes a tentative general chronology based on geometric relations with the sea-level curve and the FV infill but she concludes that no direct continuation may be found between the estuarine infilling of the Ostend and Flemish Valleys because an episode of rapid sea-level rise and marine erosion probably prevented keeping track of the migration of the estuarine facies from the Ostend to the Flemish Valley.

18.3.3 Plio-Pleistocene Evolution of the Scheldt System

While the basic characteristics and causes of the pre-Saalian history of the Scheldt drainage system are reasonably identified, its tentative timing remains fundamentally speculative beyond the approximate Holstein to Early Saale age of the last stage preceding the creation of the FV (terrace of Melle, De Groote 1977; Vanhoorne 1992). It is essentially a history of river incision and development of terraces in response to the long-lived tilt to NE of the area between the uplifting Weald-Artois axis and the subsiding Roer Valley Graben (Fig. 18.6). The consequent NE-oriented river system still prevailed in the Late Miocene when valleys now buried beneath the Diestian Formation were incised (Vandenberghe et al. 2014). As for the current average NNE direction of most rivers of West and Central Belgium, it probably results from modifications of the primitive drainage system by stream captures when the main regional tilt component turned to NNW in parallel with the more and more eastward location of maximum uplift in the Ardenne during the Plio-Pleistocene (Houthuys 2014). In this context of a long period of slow valley downcutting since the Oligocene, it is so far hardly possible to assign ages to the successive terrace levels of the Scheldt and its tributaries. While the lower levels certainly result from the combined effect of climate oscillations and accelerated uplift during the Middle Pleistocene (see Chap. 11), the upper ones refer to the fluvial response to slow Neogene tilt toward the Roer Valley Graben. However, unlike in the terrace staircase of the Ardennian valleys, there is here no clear hiatus in the vertical spacing of the successive terraces that would point to the remote effect of the Middle Pleistocene uplift pulse in the Ardenne. Considering the Scheldt response to regional-scale tectonic signals, one thus may only hypothesize that, if applicable in the Scheldt basin, this pulse, which occurred prior to the terrace of Melle, might have been responsible for the ~30-m incision leading from the Kruishoutem to the Meulebeke level (Fig. 18.2). Indeed, the latter terrace is the first for which some remnants are located inside the valleys rather than on top of interfluves, suggesting the Middle

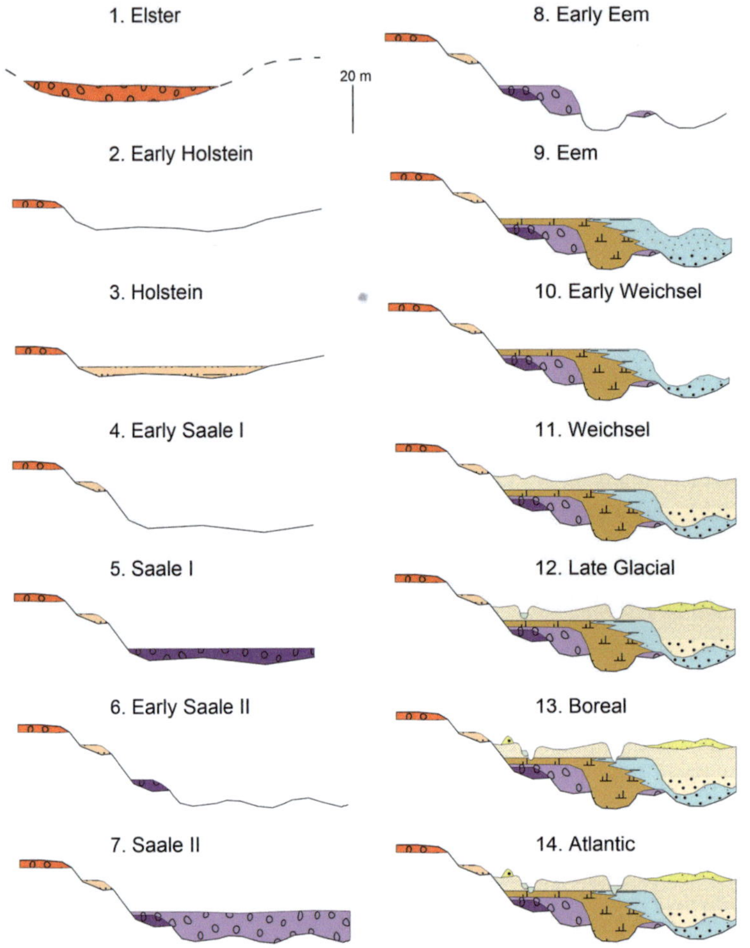

Fig. 18.6 Evolution of the Flemish Valley along the cross-section of Fig. 18.2 from the Elster (Meulebeke terrace) to the mid-holocene. See Figs. 18.2 and 18.4 for the legend

Pleistocene uplift/incision event fixed the position of valleys previously laterally mobile. The equally large incision above the Kruishoutem level might then be understood as a response to the Weald–Artois uplift that closed the Dover Strait around 1.4 Ma (Van Vliet-Lanoë et al. 1998).

By contrast, and although age issues are still debated, the more recent history of the FV is better documented (Fig. 18.6). After slowdown of downcutting in all valleys of the Scheldt basin from Elsterian to Saalian times (e.g., Tavernier and De Moor 1974; De Moor and Heyse 1978; Vandenberghe and de Smedt 1979) (Fig. 18.4), a phase of generalized renewed incision started during the Saalian and culminated with the Early Eemian transgression in the Ostend Valley and its continuation toward the FV (e.g., Mathys 2009). However, recent findings tend to confine this multi-stage incision to the Saalian (Bogemans et al. 2016). After this phase, a cut-and-fill evolution was initiated, which was no more interrupted by significant further deepening of the FV. While the post-Elsterian waning of river incision giving progressively way to FV cut-and-fill points to a phase of tectonic quiescence in the area, a specific cause has to be found for the Saalian incision. In the absence of tectonic evidence, it was very soon related to the opening of the Dover Strait (Paepe and Vanhoorne 1976; Vandenberghe and de Smedt 1979). This picture was later confirmed and refined when, based on the high-resolution bathymetry of the English Channel from sonar data, Gupta et al. (2007) showed the latter underwent two megaflood events. These events most likely occurred during the Elsterian (MIS 12) and the last glacial phase of the Saalian (MIS 6) (e.g., Gibbard 1995; Toucanne et al. 2009; Cohen et al. 2014), when most NW- and central-European rivers, including the Scheldt, drained into a large proglacial lake in the Southern Bight, dammed by the ice sheets in the north and the uplifted Weald–Artois axis in the SW (Fig. 18.7). Spilling of the lake waters over the lowest point of the Weald–Artois axis rapidly evolved in torrential flow and caused catastrophic breaching of the Dover Strait dam. As the base level of the resulting Channel River was located ~120 m below the present sea level west of the coasts of Ireland and Cornwall

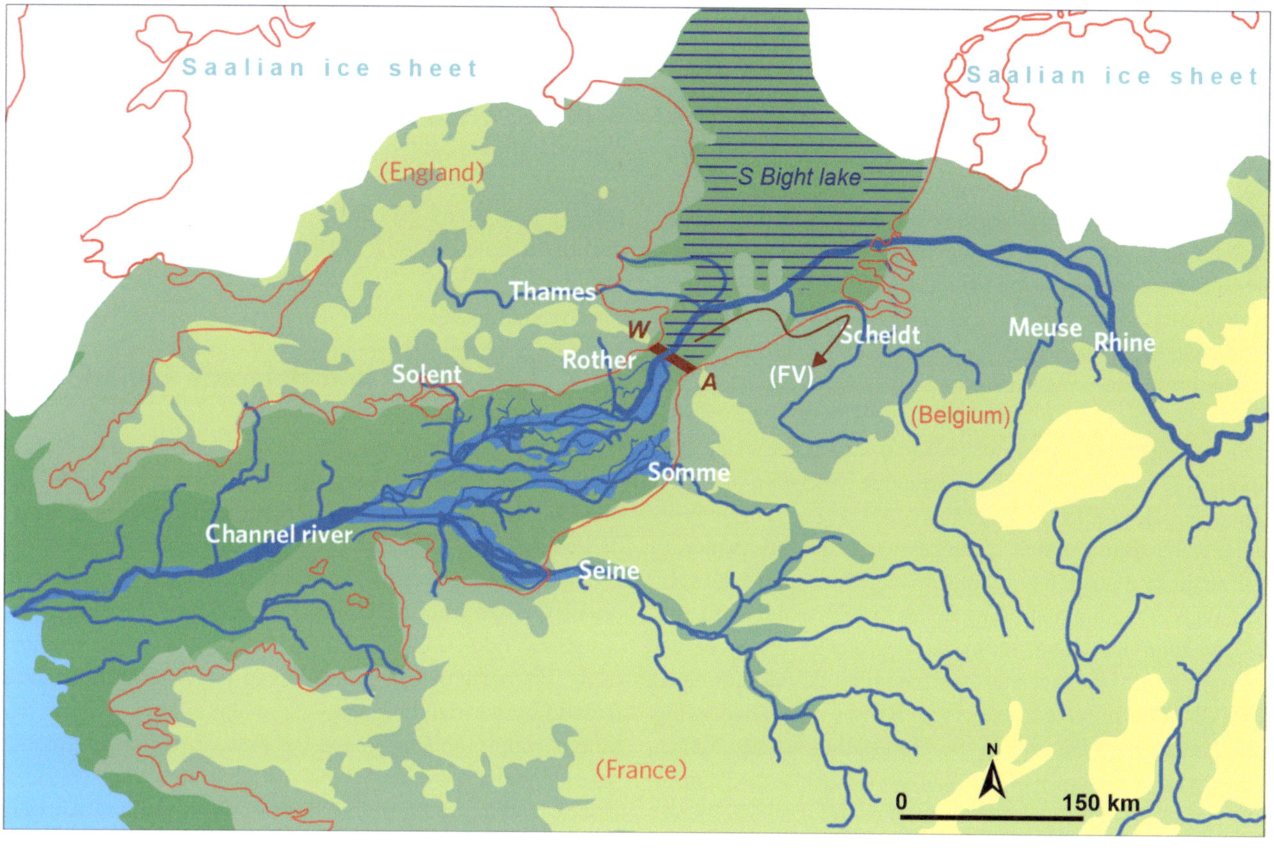

Fig. 18.7 Map of NW Europe illustrating how Saalian breaching of the Dover Strait dam (corresponding to the Weald (W)—Artois (A) uplifted axis, in *red brown*) induced partial emptying of the S Bight lake and regressive incision of the FV in the Scheldt basin (*red brown arrow*) (modified after Gibbard 2007)

and the Channel floor lay significantly lower than the dam top, each breaching event involved a wave of regressive erosion that propagated far into the Scheldt basin. While the Elsterian event did not leave any identified trace in the Scheldt basin, perhaps because it interfered with the ongoing tectonically driven incision of the valley, the Saalian one, inducing further 15- to 20-m deepening of the Channel River (Gupta et al. 2007), is held most probably responsible for the inland deep incision that created the FV (Beerten 2010; Beerten et al. 2013).

Once the FV was established, the incision and aggradation stages basically reflected the combined controls of glacio-eustatic changes of the FV base level, alternation of transgressive and regressive regimes, and climate-driven variations in hillslope sediment delivery. Noteworthily, while the oldest sediments preserved in the lower FV are Saalian in age, it seems that incision of, and early deposition in, its upper reaches, e.g., of the eastern branch occurred slightly later, during the Eemian (Bogemans 1993). The precise timing of the "Saalian" FV incision may at first glance seem unclear, not only because the ages of the Zoetendale and Adegem terraces proposed by Tavernier and De Moor (1974) are inferred rather than positively determined but also because the Saalian chronology is still debated, notably with respect to the inclusion of MIS 9 and 10 in the Saale complex (e.g., Busschers et al. 2008; Stephan 2014). However, based on the OSL dating of Rhine/Meuse sediments in the Netherlands, Busschers et al. (2008) have shown that the Drente maximum advance of the Scandinavian/British ice sheets over the North Sea occurred in MIS 6. Therefore, even if Beets et al. (2005) discuss the possibility of an MIS 8 ice advance in the North Sea basin, the catastrophic breaching of the Dover Strait dam and the subsequent FV incision most probably occurred during MIS 6. Moreover, if the Melle terrace was indeed formed during the Wacken interstadial (i.e., MIS 7, see Stephan 2014), rather than depending on glacio-eustatic oscillations within MIS 6, the two-stage (Zoetendale/Adegem) incision of the FV and certainly also the stepped flanks of the Ostend Valley (Maréchal and Henriet 1983) probably resulted from a similar multi-stage deepening of the breach across the Dover Strait dam by repeated lake overspill and emptying, which led to temporary interruption(s) and partial removal of the sediments before the fluvio-periglacial filling of the valley resumed.

Then, at the Saalian-Eemian transition, when the sea level was first still low, the strongly decreased supply of hillslope sediment to the rivers under warming conditions allowed the drainage system to re-incise in its Saalian filling (Fig. 18.6). Prolonged by the action of migrating tidal channels during the following transgression (Mathys 2009), this removed the largest part of the filling and locally further carved the Paleogene bedrock down to −25 m, and even a −28 m maximum in the downstream located Ostend Valley. At the same time, the eastern branch of the FV developed considerably upstream as a subsequent axis along the Asse and Boom clay cuestas in the Dendermonde-Mechelen area (De Moor and Heyse 1978; Bogemans 1993) (Fig. 18.1). With the Eemian transgression, a succession of fluvial, estuarine and shallow marine environments migrated landward, progressively invading and filling again the FV with sediments. Except south of Ghent or in more protected areas of the northern part of the river system, the initial fluvial sediments were largely removed from the FV by the succeeding high-energy tidal currents in the propagating estuary, which left only a basal layer of residual gravel below 5–20 m of tidal channel and tidal flat sediments (Heyse 1979). At maximum Eemian sea level a few metres above the current one, the marine influence was felt as far as 40 km inland, with Eemian marine deposits preserved up to +2.5 m TAW in the FV (De Moor and Heyse 1974).

The return to falling sea levels at the onset of the Early Weichselian (∼−40 m and ∼−50 m during MIS 5d and 5b, respectively; Waelbroeck et al. 2002) entailed renewed erosion in the FV and removal of much of its Eemian filling (Fig. 18.6). Based on pollen data, Verbruggen et al. (1991) showed that the climate was cool and very wet in Low and Middle Belgium during the Early Weichselian (∼115–70 ka), which, combined with the base level fall, allowed rivers to incise deeply, locally down to −17 m (Tavernier and De Moor 1974), in the still unfrozen fill. A similar incision is concurrently observed downstream in the reemerging Ostend Valley (Mathys 2009), whereas the upstream part of the FV eastern branch shows only hints of limited erosion by meandering channels (Bogemans 1993). The worsening climatic conditions of the Early Pleniglacial, colder than in the Early Weichselian but still wet according to Verbruggen et al. (1991), caused then a lot of hillslope material to be delivered by periglacial slope processes to the valleys, where braided rivers were most often incapable to cope with such a load. This led first to the accumulation of alluvial fans in the FV, going rapidly laterally and vertically into cryoturbated sandy channel fills during the Middle Pleniglacial, which in turn were overlain by overbank deposits testifying to quiet floodplain environments that still largely determine the modern flat topography of the FV. At the turn of the Late Pleniglacial, while most of the up to 20-m-thick Weichselian fill of the FV was accumulated, the regional climate not only was very cold but became also extremely dry, leading to the onset of a predominant aeolian activity that would give the final touch to the FV landscape during the Late Pleniglacial and Late the Glacial (Heyse and De Moor 1979).

18.4 Late Glacial and Holocene Evolution and Landforms

The aeolian activity that prevailed during the Late Pleniglacial and the Late Glacial in north and central Belgium had an unexpected major consequence on the layout of the FV. Indeed, coversand deposition developed a thick E–W oriented ridge from Maldegem to Stekene (Fig. 18.8). This coversand ridge obstructed the former NNW-oriented course of the FV north of Ghent and forced its eastward deflection, marked by hesitant drainage on the aggradation plain in that time, through the Durme and the present-day Dendermonde Scheldt and a preexisting gap in the Boom clay cuesta at Hoboken (Fig. 18.8), giving birth to the N-striking Antwerp Scheldt (De Moor 1981; Heyse 1983). In that time, this new course kept its northern flow direction beyond Antwerp and did not follow the modern estuary path (Kiden 2006). OSL dating of samples across the whole thickness of the ridge show that it started to form around 17.5–15.5 ka, i.e., at the time when the Late Pleniglacial Beuningen desert pavement, an important stratigraphic marker in northern Belgium (see also Chap. 14), was also formed (Derese et al. 2010). However, a large part of the sand was deposited later, during the Younger Dryas, suggesting higher probability of a Younger Dryas age for the completion of the Scheldt redirection toward Antwerp. This history is also documented by the evolution of the depressions at the foot of the damming ridge, for instance at Moervaart and Ede, which evolved into shallow lakes when the permafrost melted and the water table rose during the Bølling and the Allerød, as testified by their filling of alternating gyttja and marl. The level of the lakes substantially lowered and they came to an end at the Allerød-Younger Dryas transition (Heyse 1998; Crombé et al. 2013).

The N-striking downstream continuation of the Late Glacial Antwerp Scheldt across eastern Zeeland in the Netherlands was documented in the 1990s thanks to numerous boreholes, which revealed the presence of alluvial deposits buried beneath up to >20-m-thick tidal flat and marsh deposits (Kiden 1995; Vos and Van Heeringen 1997). Sand dunes extending on top of the Pleniglacial coversands

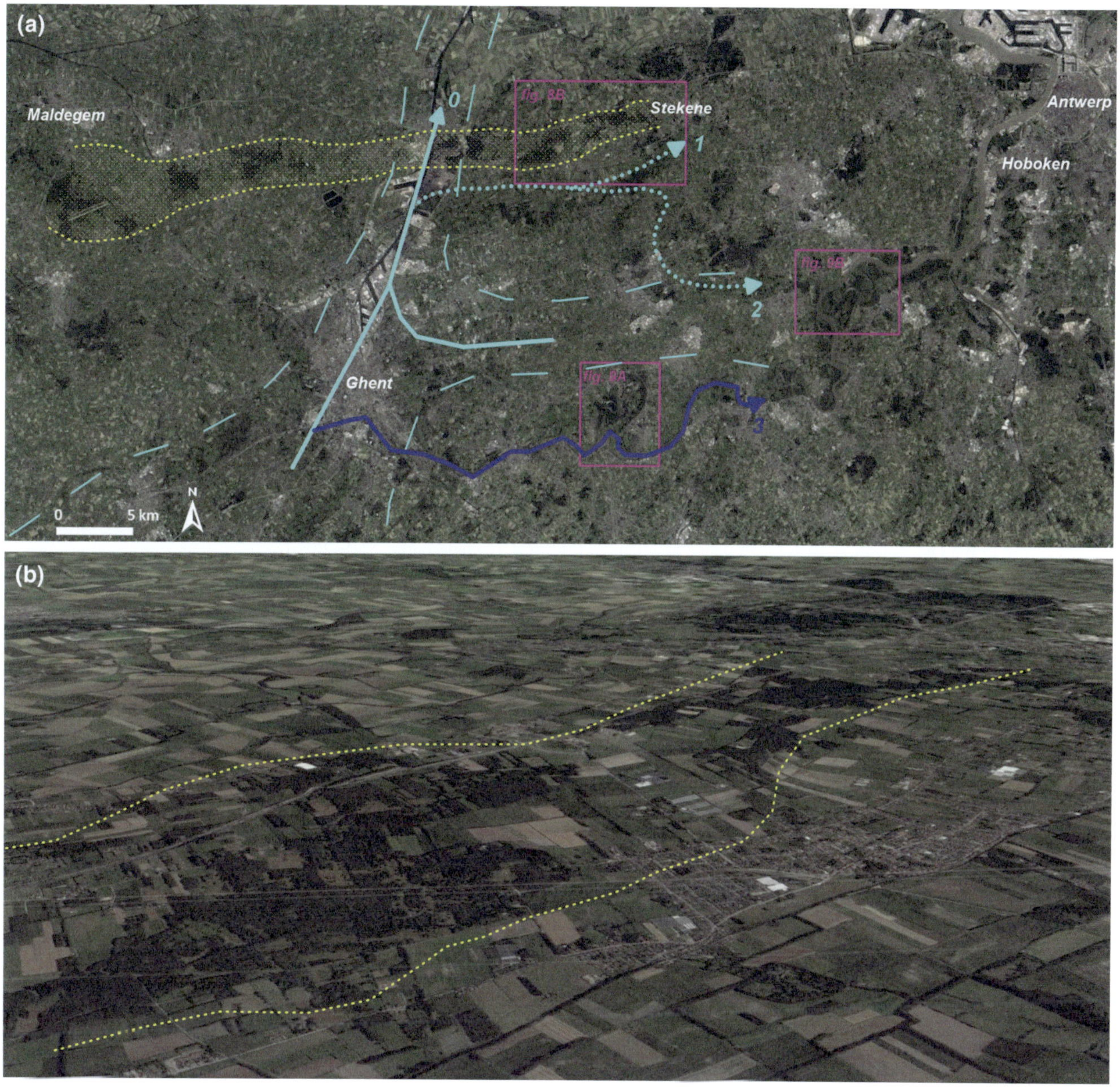

Fig. 18.8 **a** Google Earth view (25.08.2016) of the area N and E of Ghent, showing the eastward redirection of the FV caused by development of the Maldegem-Stekene coversand ridge (*fine yellow dots*). *Dashed light blue* lines delimit the FV. *0* Pre-coversand FV trajectory. *1, 2* Successive Holocene trajectories. *3* Present Scheldt course. **b** oblique zoom on the coversand ridge topography (in between *yellow dotted lines*) and, in the foreground, the lower lying Moervaart depression at Moerbeke (elevation factor 3)

along the leeward side of the Scheldt north of Antwerp and sourced from the exposed surface of the former braided course of the river were dated to the Younger Dryas (Vandenberghe et al. 2004). Dunes of the same age and likewise fed by the nearby alluvial plain are also locally observed along the Scheldt near Wichelen, west of Dendermonde (Bogemans and Vandenberghe 2011). Then, owing to Termination I's sea-level rise and drowning of the Zeeland area, deep tidal channels developed, one of which diverted the Scheldt through the Eastern Scheldt forming the northern limit of the former Walcheren Island (now a peninsula) sometime between 7.4 and 6.3 ka (Kiden 2006). Finally, local ruptures in the coastal barrier constructed in the time the sea level had been stabilized (~5.7 ka) reestablished a marine influence in the Lower Scheldt area and induced changes in the pattern of tidal channels. This gave birth to the present-day Western Scheldt estuary around 800–1000 AD, a waterway that became predominant from

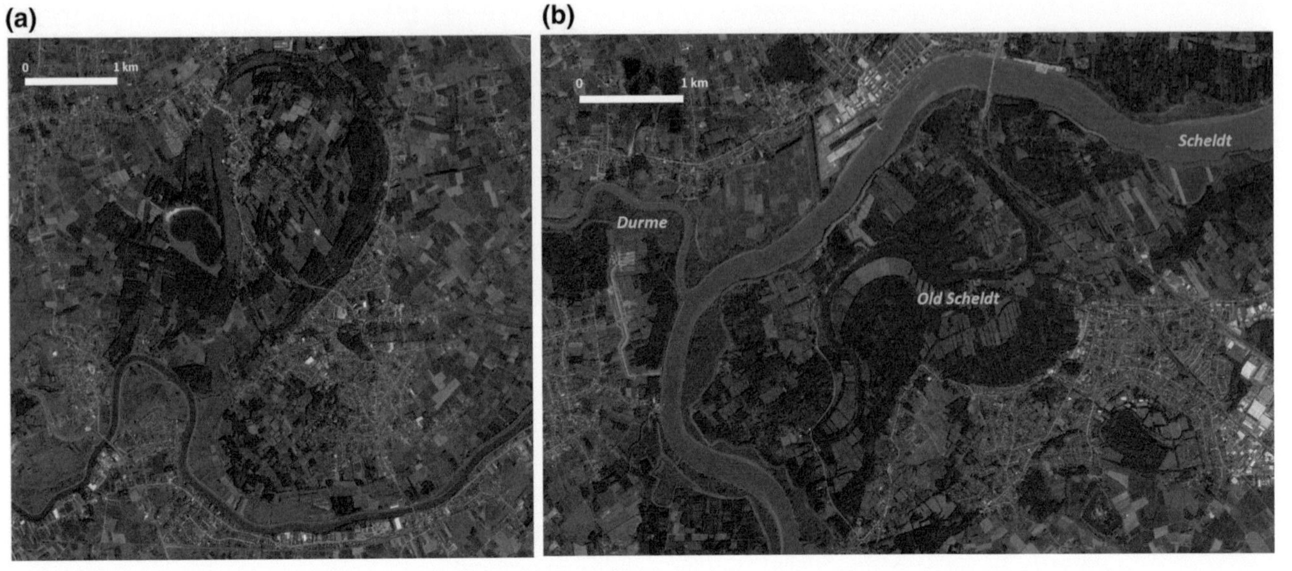

Fig. 18.9 **a** Satellite photograph of the abandoned large Late Glacial meander of the Scheldt at Overmere-Donk. **b** Satellite *photograph* of the underfit course of the Old Scheldt at Temse. See image locations in Fig. 18.8. (Google Earth, 17.08.2016)

~1500 AD and exclusive when the link to the Eastern Scheldt was definitively silted up in the nineteenth century (Kiden 2006).

Similar to what had happened at the Saalian–Eemian transition, channel pattern changed again from braided to meandering in the Scheldt basin (upstream of the time-transgressing tidal channel domain) with the onset of warming in the interstadials of the Late Glacial. Incision of the meandering system generally did not exceed a few metres, with a maximum of 10 m in the Ghent area (Tavernier and De Moor 1974) (Fig. 18.6). This is less than the corresponding Early Eemian incision depth in the same area because the distance from Ghent to the river outlet in the North Sea was now much longer through the Antwerp Scheldt and reduced proportionately the average gradient and the tide-driven erosional component of the Late Glacial meandering river. By contrast, the only ~1-kyear-old scouring of up to 25-m-deep tidal channels in the modern Western Scheldt estuary (Codde and De Keyser 1967) is much more comparable to the Eemian incision in the lower FV.

The time of Late Glacial incision is not yet accurately determined but in any case predated the Allerød, whose sediments are found at the base of the meander infill (Bogemans et al. 2012). Owing to the migration of large meanders, up to more than 2 km in amplitude, i.e., 3–5 times larger than their modern equivalents, active lateral erosion created a wide alluvial plain, in which important point bar deposits accumulated (Kiden 1991; Bogemans et al. 2012; Meylemans et al. 2013). Abandoned large meanders of this period, for instance at Kalken and Overmere-Donk, still make an important element of the Scheldt landscape especially between Ghent and Dendermonde (Fig. 18.9a). Aggradation and meander filling started at the Allerød and continued during the Holocene (Fig. 18.5f). From the second half of the Preboreal, peat accumulation progressively replaced gyttja and clastic deposition as channel fill, while the Scheldt evolved into an underfit river. From the Late Atlantic on, the additional effect of sea-level rise brought back brackish environments and weak tidal influence at the height of the Dutch–Belgian border (Vos and de Wolf 1997) and the rise of the groundwater table led to the extension of organic accumulation over the whole width of the Scheldt meander belt, where the almost complete absence of channel traces suggests extremely slow water flow across the swampy alder carr forest that covered the area in that time (Kiden 2006; Bogemans et al. 2012). A first major anthropogenic impact on landscape evolution is recorded in the Late Subboreal, when significant deforestation induced the return of clastic sedimentation in the alluvial plain around 4 ka (Verbruggen et al. 1991; Meylemans et al. 2013). This was followed by the evolution of the anabranching pattern of the underfit Scheldt into meandering channels, still underfit with respect to the Late Glacial system, sometime before the Roman period (Kiden 1991). This evolution was possibly related with the shift toward colder and wetter conditions in NW Europe at the Subboreal-Subatlantic transition (~800 cal BC, Van Geel et al. 1996). Finally, after partial filling of this channel around the second century AD (Meylemans et al. 2013), the opening and progressive deepening of the Western Scheldt estuary, partly supplemented by the effect of human activities, increased the tide-controlled processes in the Lower Scheldt. Much of the Early Subatlantic river course became erased, of which the 'Old Scheldt' reach remains a rare well-preserved witness south of Temse (Kiden 2006) (Fig. 18.9b). The result of the Late Pleistocene and Holocene

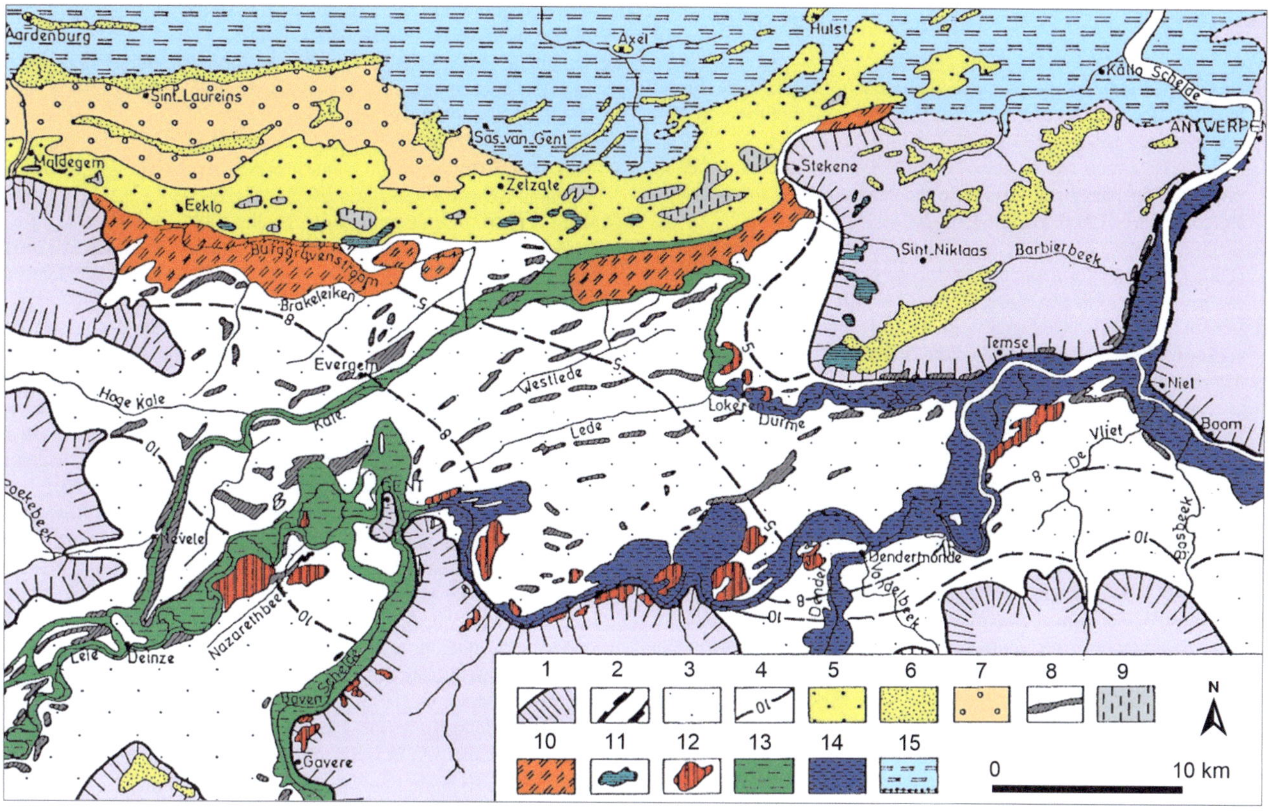

Fig. 18.10 Current morphological features in the Flemish Valley. *1* Hilly interfluves. *2* Gap of Hoboken. *3* Levelled top of the fluvioperiglacial fill of the FV. *4* Contour line of fill top. *5* Major coversand ridge of Maldegem-Stekene. *6* Low coversand ridge. *7* Deflation plain. *8* Linear fluvio-aeolian coversand embankment. *9* Shallow depression in the coversand. *10* Depression dammed by the Maldegem-Stekene ridge. *11* Local drift sand. *12* River dune, formed from reworked nearby fluvial sand. *13* High water inundation plain. *14* Late Glacial/Holocene valley floor. *15* Reclaimed Scheldt tidal flat

evolution of the FV is presented on Fig. 18.10, which shows the various morphological features constituting the present-day landscape as a transient topography waiting for the next major morphogenic event.

18.5 Conclusion

This brief overview of the Plio-Pleistocene evolution of the Scheldt basin and its landscapes, with a particular focus on the post-Saalian history of the Flemish Valley, the buried valley system that once formed the major drain of the basin, has emphasized how the tiniest details of a monotonous, subdued topography are meaningful for reconstructing the environments and processes that guided the Quaternary morphogenesis of a lowland plain. Nevertheless, the bulk of the information has been provided by careful, detailed analysis and synthesis of innumerable borehole and outcrop profiles. Beyond the uncovering of sedimentological and geomorphic consequences of the glacio-eustatically driven appearance and withdrawal of continental, estuarine, and marine environments in the Lower Scheldt area, also beyond the awareness of the climatic imprint left in the sedimentary and geomorphic records, a major lesson learned from reconstructing the Scheldt and FV Quaternary evolution is probably the determining effect of long-term tectonics and large-scale events on the grand geomorphic picture of the region. While waning of the regional NW tectonic tilt between the Artois–Weald axis and the Roer Valley Graben sometime in the mid-Middle Pleistocene caused the transition from a terrace staircase developed by a vertically incising river to vertically stabilized river and landscape, the catastrophic flooding of the English Channel during the Elsterian and, more effectively, the Saalian glacials was most probably responsible for the last bedrock incision event that created the Flemish Valley. Finally, in the last millennia, the later cut-and-fill evolution of the FV underwent the multiple effects of another kind of event, namely the advent and increasing importance of human interference with the natural geomorphic processes.

References

Beerten K (2010) Geomorphological evolution of the Nete basin: identification of past events to assess the future evolution. Extern rep belg nucl res cent 137:40

Beerten K, De Craen M, Wouters L (2013) Patterns and estimates of post-rupelian burial and erosion in the campine area, north-eastern Belgium. Phys Chem Earth 64:12–20

Beets D, Meijer T, Beets C, Cleveringa P, Laban C, van der Spek A (2005) Evidence for a middle pleistocene glaciation of MIS 8 age in the southern North Sea. Quatern Int 133:7–19

Bogemans F (1993) Quaternary geological mapping on basis of sedimentary properties in the eastern branch of the Flemish Valley. Mém serv géol de Belg 35:49

Bogemans F, Vandenberghe D (2011) OSL dating of an inland dune along the lower River Scheldt near Aardt (East Flanders, Belgium). Neth J Geosci 90:23–29

Bogemans F, Meylemans E, Jacops J, Perdaen Y, Storme A, Verdurmen I, Deforce K (2012) The evolution of the sedimentary environment in the lower River Scheldt valley (Belgium) during the last 13000 a BP. Geol Belg 15:102–112

Bogemans F, Roe H, Baeteman C (2016) Incised Pleistocene valleys in the Western Belgium coastal plain: age, origins and implications for the evolution of the southern North Sea basin. Palaeogeogr Palaeoclimatol Palaeoecol 456:46–59

Busschers F, Van Balen R, Cohen K, Kasse C, Weerts H, Wallinga J, Bunnik F (2008) Response of the Rhine-Meuse fluvial system to Saalian ice-sheet dynamics. Boreas 37:377–398

Codde R, De Keyser L (1967) Mer du Nord. Littoral/estuaire de l'Escaut. Escaut maritime. Atlas de Belgique, Comité national de Géographie, p 62

Cohen K, Gibbard P, Weerts H (2014) Northe Sea palaeogeographical reconstructions for the last 1 Ma. Neth J Geosci 93:7–29

Cornet J (1904) Etudes sur l'évolution des rivières belges. Ann Soc Géol Belg 31:M261–500

Crombé P, De Smedt P, Davies N, Gelorini V, Zwertvaegher A, Langohr R, Van Damme D, Demiddele H, Van Strydonck M, Antrop M, Bourgeois J, De Maeyer P, De Reu J, Finke P, Van Meirvenne M, Verniers J (2013) Hunter-gatherer responses to the changing environment of the Moervaart palaeolake (NW Belgium) during the late glacial and early holocene. Quatern Int 308–309:162–177

De Groote V (1977) Pollenanalytisch onderzoek van Midden-en Boven-Pleistocene afzettingen in Vlaanderen. Unpublished PhD Thesis, University of Ghent, p 98

de Heinzelin J (1963) Le réseau hydrographique de la région gallo-belge au Néogène. Essais de reconstitution. Bull Soc belg Géol, Paléontol, Hydrol 72:137–148

De Moor G (1963) Bijdrage tot de kennis van de fysische landschapsvorming in Binnen-Vlaanderen. Tijdschr Belgische Ver aardrijkskundige Stud 32:329–433

De Moor G (1974) De afzetting van Dendermonde en haar betekenis voor de jong-kwartaire evolutie van de Vlaamse Vallei. Natuurwetenschappelijk Tijdschr 56:44–75

De Moor G (1981) Periglacial deposits and sedimentary structures in the upper-Pleistocene infilling of the Flemish Valley (N.W. Belgium). Biul Peryglac 28:277–290

De Moor G, Heyse I (1974) Litostratigrafie van de kwartaire afzettingen in de overgangszone tussen de kustvlakte en de Vlaamse Vallei in Noordwest-België. Natuurwetenschappelijk Tijdschr 56:85–109

De Moor G, Heyse I (1978) De morfologische evolutie van de Vlaamse Vallei. De Aardrijkskunde 1978–4:343–375

Derese C, Vandenberghe D, Zwertvaegher A, Court-Picon M, Crombé P, Verniers J, Van den Haute P (2010) The timing of aeolian events near archaeological settlements around Heidebos (Moervaart area, N Belgium). Neth J Geosci 89:173–186

Germonpré M, Bogemans F, Van Neer W, Grün R (1993) The dating of two Pleistocene mammal assemblages from the Flemish Valley, Belgium. Contrib Tert Quat Geol 30:147–153

Gibbard P (1995) The formation of the Strait of Dover. In: Preece R (ed) Island Britain: a Quaternary perspective. Geological Society, London, Special Publication 96, pp 15–26

Gibbard P (2007) Europe cut adrift. Nature 448:259–260

Gullentops F, Bogemans F, De Moor G, Paulissen E, Pissart A (2001) Quaternary lithostratigraphic units (Belgium). Geol Belg 4: 153–164

Gupta S, Collier J, Palmer-Felgate A, Potter G (2007) Catastrophic flooding origin of shelf valley systems in the English channel. Nature 448:342–346

Heyse I (1979) Bijdrage tot de geomorfologische kennis van het noordwesten van Oost-Vlaanderen (België). *Verhandelingen van de Koninklijke Academie voor Wetenschappen, Letteren en Schone Kunsten van België*, Klasse der Wetenschappen 155, p 257

Heyse I (1983) Preliminary results of the study of a Vistulian Late Glacial drainage pattern in the Scheldt basin (Belgium—Flemish Valley—Moervaart depression). Quat Stud Pol 4:135–143

Heyse I (1998) An environmental model of valley damming due to aeolian action during the Latre Weichselian Glacial period, data based on a pipeline section study (Ede valley—Belgium). Natuurwetenschappelijk Tijdschr 78:149–159

Heyse I, De Moor G (1979) Morphology of Würm lateglacial and Holocene deposits in the Flemish Valley (North Belgium.) Acta Univ. Ouluensis, A: Sci. Rer. Nat. 82(3):121–131.

Houthuys R (2014) A reinterpretation of the Neogene emersion of central Belgium based on the sedimentary environment of the Diest Formation and the origin of the drainage pattern. Geol Belg 17: 211–235

Kiden P (1991) The Lateglacial and holocene evolution of the middle and lower River Scheldt, Belgium. In: Starkel L, Gregory K, Thornes J (eds) Temperate palaeohydrology. Wiley, Chichester, pp 283–299

Kiden P (1995) Holocene relative sea-level change and crustal movement in the southwestern Netherlands. Mar Geol 124:21–41

Kiden P (2006) De evolutie van de Beneden-Schelde in België en Zuidwest-Nederland na de laatste ijstijd. Belgeo 3(2006):279–294

King C (2006) Paleogene and neogene: uplift and a cooling climate. In: Brenchley P, Rawson P (eds) The geology of England and Wales, 2nd edn. Geol Soc, London, pp 395–428

Laurant A (1976) La formation du réseau hydrographique de la Belgique. In: Pissart A (ed) Géomorphologie de la Belgique. Université de Liège, Laboratoire de Géologie et Géographie physique, pp 29–50

Maréchal R, Henriet JP, (1983) Seismisch onderzoek op het Belgisch Continentaal Plat. Eerste fase. Ontginningszone 2. Unpublished report, Ministerie voor Economische Zaken, p 32

Mathys M, (2009) The Quaternary geological evolution of the Belgian continental shelf, southern North Sea. PhD Thesis, Ghent University, Belgium, p 382

Meijer T, Preece R, (1995) Malacological evidence relating to the insularity of the British Isles during the quaternary. In: Preece R (ed) Island Britain: a Quaternary perspective, Geological Society Special Publications 96, pp 89–110

Meylemans E, Bogemans F, Storme A, Perdaen Y, Verdurmen I, Deforce K (2013) Lateglacial and Holocene fluvial dynamics in the Lower Scheldt basin (N-Belgium) and their impact on the presence, detection and preservation potential of the archaeological record. Quat Int 308–309:148–161

Miall A (1977) A review of the braided-river depositional environment. Earth Sci Rev 13:1–62

Mostaert F, Auffret JP, De Batist M, Henriet JP, Moons A, Sevens E, Van den Broeke I, Verschuren M (1989) Quaternary shelf deposits and drainage patterns off the French and Belgian coasts. In: Henriet JP, De Moor G (eds) The quaternary and tertiary geology of the southern bight, North Sea. Belgian Geological Survey, Ghent, pp 111–118

Paepe R, Vanhoorne R (1967) The stratigraphy and palaeobotany of the late-pleistocene in Belgium. Toelichtende Verhandelingen geol mijnkaarten Belg 8:96

Paepe R, Vanhoorne R (1976) The Quaternary of Belgium in its relationship to the stratigraphical legend of the geological map. Mém serv l'explication Cartes géol min Belg 18:37

Pierre G (2003) L'origine du réseau hydrographique en Thiérache ardennaise ou l'histoire de la "Meuse lorraine". Cah Géog Phys 14 (2003):30–36

Stephan H (2014) Climato-stratigraphic subdivision of the Pleistocene in Schleswig-Holstein, Germany and adjoining areas. Quat Sci J (E&G) 63:3–18

Tavernier R (1943) De Kwartaire afzettingen van België. Natuurwetenschappelijk Tijdschr 25:121–137

Tavernier R (1946) L'évolution du Bas-Escaut au Pléistocène supérieur. Bull Soc Belg Géol, Paléontol, Hydrol 55:106–125

Tavernier R, De Moor G (1969) Paleomorfologisch kader en kwartaire afzettingen in Binnen-Vlaanderen. BELQUA excursie van 14 Jun 1969, Ghent University, p 8

Tavernier R, De Moor G (1974) L'évolution du bassin de l'Escaut. Société Géologique de Belgique, Colloque du Centenaire (Liège, 9–13 Sept 1974), L'évolution quaternaire des bassins fluviaux de la Mer du Nord méridionale, pp 159–231

Toucanne S, Zaragosi S, Bourillet JF, Gibbard P, Eynaud F, Giraudeau J, Turon JL, Cremer L, Cortijo E, Martinez P, Rossignol L (2009) A 1.2 Ma record of glaciation and fluvial discharge from the West European Atlantic margin. Quat Sci Rev 28:29 74–2981

Van Geel B, Buurman J, Waterbolk H (1996) Archaeological and palaeoecological indications of an abrupt climate change in the Netherlands, and evidence for climatological teleconnections around 2650 BP. J Quat Sci 11:451–460

Van Vliet-Lanoë B, Mansy JL, Margerel JP, Vidier JP, Lamarche J, Everaerts M (1998) Le Pas de Calais, un détroit cénozoïque à ouvertures multiples, Comptes-rendus de l'Académie des Sciences de Paris. Earth Planet Sci 326:729–736

Vandenberghe D, Kasse C, Hossain S, De Corte F, Van den Haute P, Fuchs M, Murray A (2004) Exploring the method of optical dating and comparison of optical and ^{14}C ages of Late Weichselian coversands in the southern Netherlands. J Quat Sci 19:73–86

Vandenberghe J, de Smedt P (1979) Palaeomorphology in the eastern Scheldt basin (Central Belgium). The Dijle-Demer-Grote Nete conflu area. Catena 6:73–105

Vandenberghe N, Harris W, Wampler J, Houthuys R, Louwye S, Adriaens R, Vos K, Lanckacker T, Matthijs J, Deckers J, Verhaegen J, Laga P, Westerhoff W, Munsterman D (2014) The implications of K–Ar glauconite dating of the Diest Formation on the paleogeography of the upper miocene in Belgium. Geol Belg 17:161–174

Vanhoorne R (1992) Azolla and Salvinia species (Azollaceae and Salviniaceae, Pteridophyta) from the caenozoic of Belgium. Bull K Belge Inst Natuurwetenschappen 62:229–255

Verbruggen C (1999) Quaternary palaeobotanical evolution of Northern Belgium. Geol Belg 2:99–110

Verbruggen C, Denys L, Kiden P (1991) Paleo-ecologische en geomorfologische evolutie van Laag- en Midden-België tijdens het Laat-Kwartair. De Aardrijkskunde 1991–3:357–376

Vos P, de Wolf H (1997) Palaeo-environmental diatom study of the Holocene deposits of the Province of Zeeland (SW Netherlands). Mededelingen Ned Ins Toegepaste Geowetenschappen TNO 59:111–141

Vos P, Van Heeringen R (1997) Holocene geology and occupation history of the province of Zeeland. Meded Ned Ins Toegepaste Geowetenschappen TNO 59:5–109

Waelbroeck C, Labeyrie L, Michel E, Duplessy JC, McManus J, Lambeck K, Balbon E, Labracherie M (2002) Sea-level and deep water temperature changes derived from benthic foraminifera isotopic records. Quat Sci Rev 21:295–305

Ziegler P, Dèzes P (2007) Cenozoic uplift of Variscan massifs in the Alpine foreland: timing and controlling mechanisms. Glob Planet Change 58:237–269

The Coastal Plain of Belgium, Joint Product of Natural Processes and Human Activities

19

Cecile Baeteman

Abstract

The coastal plain is a flat low-lying area with hardly any expression of relief at the surface. Its subsurface, however, contains a rich archive of the depositional history that started about 10,000 years ago, and that represents the history of infill of a major pre-Holocene palaeovalley. The infill of the palaeovalley was controlled by several factors, among which relative sea level rise was initially the main driver. In the course of time, the balance between sediment supply and accommodation space took over the control of infill. Toward the end of the infill, humans played a prominent role in the further evolution of the plain, causing changes that ultimately determined its present-day characteristics. The complex interaction of regional and local processes is explained in the geological setting of the area. The development of De Moeren and the Zwin region, two protected areas that experienced a specific evolution, different from the general history of infill, illustrates the effect of local influences in controlling spatial and temporal patterns of sedimentation in response to variations in coastal processes.

Keywords

Tidal environments • Sea level rise • Coastal processes • Land subsidence • Human interventions • Holocene

19.1 Introduction

The Belgian coastal plain is a unique landscape, shaped by the sea but modified by humans. The numerous ditches, canals, and sluices attest to the extensive human activities since the mediaeval period. They are essential for the sustainability of the area, whose high groundwater table is controlled by ingenious drainage systems. For the western part of the plain, the drainage systems join all together in Nieuwpoort in a complex of sluices, which offer a spectacular view when the gates are opened at low tide. It is from this site that a part of the plain was voluntarily inundated during the First World War.

The Holocene coastal plain belongs to the marine-dominated North Sea Lowlands which run from the Cretaceous marl outcrop at Blanc Nez, near Calais in northern France, to northern Denmark. In Belgium, the plain is a 15- to 20-km-wide embanked lowland at altitudes ranging from +2 to +5 m TAW. As the Belgian ordnance datum (TAW—second general levelling) refers to mean spring low water, this corresponds to 2–3 m below spring high tide (datums for The Netherlands—NAP—and France—NGF—are 2.33 and 2.29 m higher, respectively). The plain is crossed by a small river, the IJzer, which at present is canalized and debouches into the North Sea at Nieuwpoort and along which the plain extends further south (Fig. 19.1). The river, together with its tributaries, drains a small and relatively low-lying basin to the south and southwest of the coastal plain.

C. Baeteman (✉)
Quaternary Environments and Humans, Royal Belgian Institute of Natural Sciences, Jennerstraat 13, 1000 Brussels, Belgium
e-mail: cecile.baeteman@naturalsciences.be

A. Demoulin (ed.), *Landscapes and Landforms of Belgium and Luxembourg*, World Geomorphological Landscapes, DOI 10.1007/978-3-319-58239-9_19

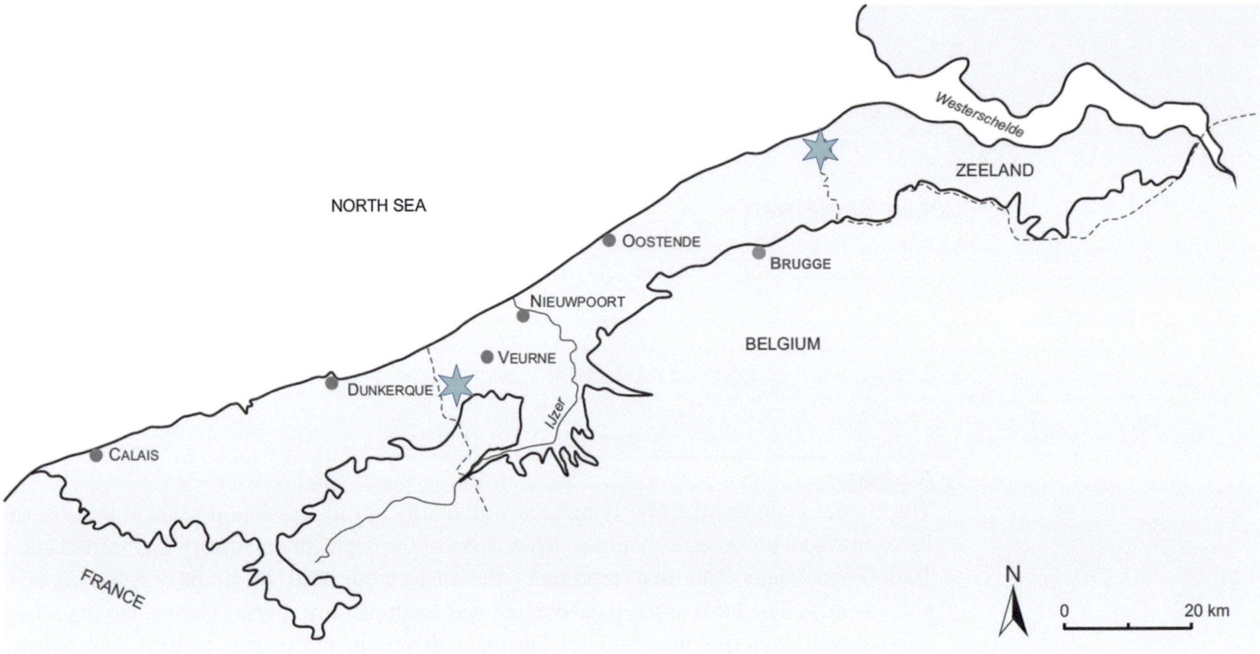

Fig. 19.1 Map of the coastal plains of northern France, Belgium, and Zeeland with *stars* indicating the location of two sites presented in detail: De Moeren in the west and the Zwin area in the east (reproduced with permission of Brepols)

The plain lies behind a straight coastline. Coastal dunes and a hard seawall in some places protect it from flooding during storm surges. The dunes, however, are strongly degraded or even destroyed in many places by housing projects for tourism since a few decades. Therefore, parts of the coastal stretch are often nicknamed 'Atlantic Wall'. A wide complex of splendid high dunes still exist only near De Panne in the west, close to the French border. The beaches are also being modified continuously for touristic purposes. In many places, beach nourishment is needed now and will become more frequent to cope with the predicted accelerated sea level rise, although human interventions are so far causing more harm than the rising sea level.

19.2 Geological Setting

Geologically speaking, the coastal plain is a young area, though with a very rich geological history. We thus address here its development during the Holocene (last 11.7 ka) because it explains all notable features of the plain.

The characteristics of the present landscapes are inherited from what happened before, in particular in terms of driving processes. The coastal plain landscape displays only the terminal geomorphic surface but not the record of the development stages. These stages are preserved in the subsurface. Because of the presence of a major pre-Holocene valley, the western part of the plain contains the most complete sequence of Holocene deposits, reaching a thickness of about 25–30 m at the coast and thinning out to the south, where Late Pleistocene deposits crop out.

19.2.1 Agents and Controls on Sedimentation: Sea Level and Tides

The coast of Belgium is tide-dominated (macro- to mesotidal), with a tidal range of 5 m in the west decreasing to 4.5 m in the east (Fig. 19.2). The shifting play of ebbs and flows has produced a series of sedimentary environments whose past deposits form the core of the coastal plain. These sedimentary environments are the coastal barrier, tidal channels, mud- and sand flats, salt and freshwater marshes.

Understanding the coastal plain development requires placing it in the context of dynamic sedimentary environments. An environmental interpretation of the sediment sequences is essential to document coastal changes because each tidal sedimentary environment has its specific relation to sea level. Besides, their relative positions also show typical spatial patterns controlled by the location of tidal channels (Fig. 19.3). These environments form a dynamic system with inland or seaward shifts and overlappings in response to the following factors: rate of relative sea level (RSL) rise, sediment supply, and accommodation space.

An extensive network of tidal channels acting as local sediment suppliers dissects the tidal flat while creeks form the drainage system in the supratidal zone (salt marsh). When the sediment supply exceeds the accommodation

Fig. 19.2 View of the IJzer estuary with associated mudflat and salt marsh (**a**) at the occasion of extreme water levels at a spring equinox (Nieuwpoort). (**b**) salt marshes and mudflat are completely inundated at extreme high water level

space created by RSL rise, the area rapidly silts up to high water level. Consequently, the frequency of tidal inundation decreases, a part of the drainage network of the area becomes redundant and silts up. Eventually the area becomes out of the reach of the daily tidal inundation and the salt marsh encroaches onto the mudflat, followed by peat accumulation if sufficient time is available for the formation of a freshwater lens in the shallow subsoil. However, at the same time, intertidal deposition continues in adjoining parts until they in turn experience the same evolution. If no sufficient time is available for a freshwater lens to form (which was the case in the period characterized by a rapid RSL rise), only vegetation horizons can develop. Consequently, vegetation horizons or peat beds are not synchronous and have no

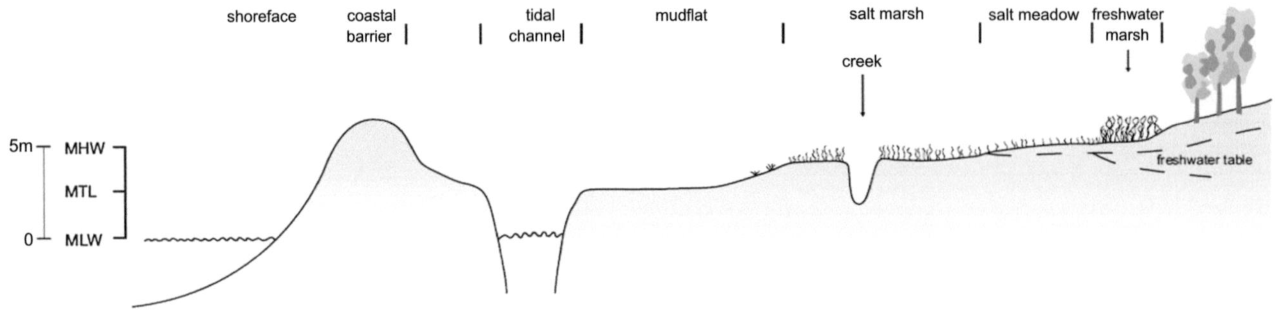

Fig. 19.3 Schematic representation of the different sedimentary environments in a tidal flat in relation to water levels. *MHW* mean high water spring; *MTL* mean tidal level; *MLW* mean low water spring (reproduced with permission of Brepols)

stratigraphical meaning, their occurrence depending on whether sufficient time and sediment were available to allow silting up of a particular part of the tidal flat. Moreover, being related to the position of the tidal channels, the evolution of one environment into another cannot be generalized for the entire plain.

The coastal barrier forms the boundary between the open sea and the tidal environments. It is a huge sand body extending below and above sea level, and including the shoreface, tidal deltas and inlets, washovers, beach and dunes. Coastal barriers are essentially transitory features in dynamic equilibrium with rising sea levels through the landward transfer of sand, eroded from the shoreface, to the tidal flat via tidal inlets and overwash processes. This process nowadays is called coastal erosion. Landward shifting of the barrier is fairly continuous while sea level is rising relatively rapidly but may become intermitted when the sea level rise slows down. In addition, the rate of inland shifting is reduced as the accommodation space for sediment in the tidal flat decreases, i.e. when the tidal flat silts up until high water level. The role of sediment supply then becomes dominant. With a positive sediment budget, inland shifting of the barrier is reversed and the barrier starts shifting seaward, a phenomenon called progradation (Roy et al. 1995). The dynamic character of the sedimentary environments implies that at any time throughout the Holocene all the sedimentary environments existed next to each other, even over short distances.

19.2.2 The Depositional History During the Holocene

Geological investigations, especially in the western part of the coastal plain, of the sediment characteristics and sedimentary environments together with age determination resulted in the understanding of the development and evolution of the plain. The major steps of this development will be recalled here, summarized from Baeteman (1999, 2004a, 2005a, 2013, 2008b), Baeteman and Van Strydonck (1989), Baeteman et al. (1999), and Baeteman and Declercq (2002).

The development of the coastal plain is controlled by the rate of RSL rise, the morphology of the flooded surface, the sediment budget, and the accommodation space (effected by sediment and peat compaction). During the infill of the area, initially caused by the RSL rise, the relative importance of the individual factors changed with time. In the last 2000 years, human activities played a prominent role.

Knowledge of the morphology of the pre-Holocene surface, i.e. the original topography at the beginning of the postglacial sea level rise, is crucial to the understanding of the sediment successions. In the west, the landscape prior to the marine flooding was characterized by four relatively small and shallow rivers joining in the central part of the western plain and forming a southeast–northwest depression (Fig. 19.4). This depression is the palaeovalley of an ancient river IJzer formed during the glacial sea level low stand and located several km to the west of the present-day river. It is not known when and why the river shifted to its modern position. The palaeovalley has a gentle slope with depths of −12 and −18 m TAW in the central and seaward part of the plain, respectively. This relief of the pre-Holocene surface in the western part is in contrast to the eastern part of the coastal plain, with consequences for its further development. The eastern part has a flat, gently sloping morphology and higher altitudes, ranging between +5 and −5 m TAW. Consequently, it was influenced much later by the Holocene transgression.

The RSL rise is recorded in the western plain as from ca 9500 cal BP. The rising sea invaded the area by way of the palaeovalley, which developed into an estuary with tidal channels and flats. As groundwater level rose with sea level, the flooded area was fringed by freshwater marshes in which peat accumulated. This peat is the basal peat, a transgressive unit shifting inland with the rising sea level. It forms the basis of the Holocene sediment succession where it has not been eroded by late Holocene tidal channels. Age and altitude of the basal peat were used to construct the relative

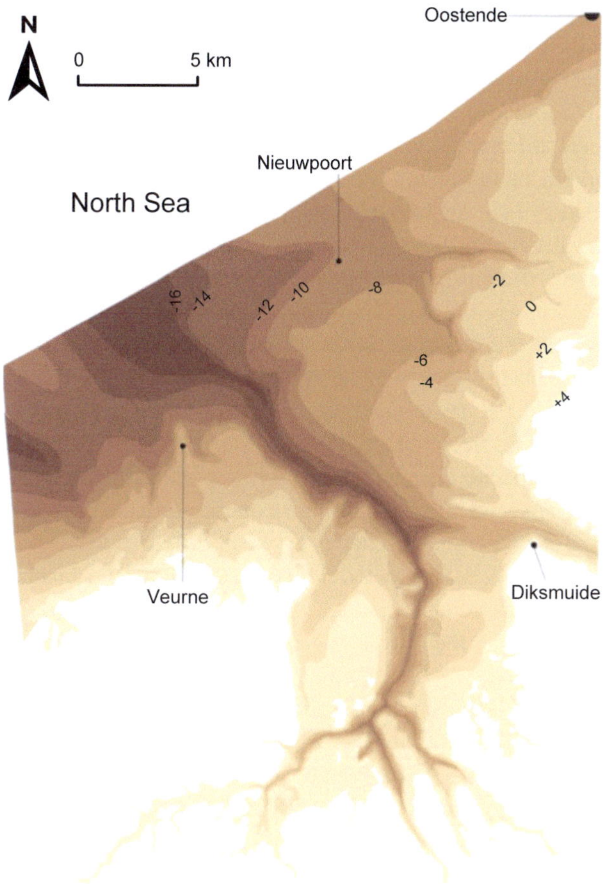

Fig. 19.4 Contour map of the pre-Holocene surface at a 2 m interval relative to TAW. Because of the vertical exaggeration, the map gives a false impression of valley depth. In reality, they are shallow with gentle slopes (adapted from Baeteman and Declercq 2002; reproduced with permission of Brepols)

sea-level curve of the area (Denys and Baeteman 1995) (Fig. 19.5).

The general trend of the relative sea-level curve shows that initially the RSL rise was very rapid at an average rate of 7 m/ka in the period before ca 7500 cal BP. This resulted in a rapid shift of the sedimentary environments across the continental shelf toward a position close to the present-day boundary of the coastal plain. This was associated with a significant vertical sediment accretion and the development of vegetation horizons. Wave erosion at the shoreface resulted in an inland shift of the coastal barrier. At ca 7500–7000 cal BP, the RSL curve shows a distinct retardation to an average of 2.5 m/ka. Consequently, the rapid landward shift of the environments stopped and the position of the coastal barrier became more or less stable. Sediment supply was now in balance with the accommodation space created by the RSL rise and parts of the tidal flat were infilled. Periods of emergence lasted much longer and freshwater conditions prevailed for short periods of about 200–300 years. Salt marsh vegetation evolved into reed growth resulting in peat accumulation. Initially these freshwater marshes were short-lived and developed only locally in the relatively higher silted-up areas that were out of the reach of the tidal flooding and, consequently, deprived of sediment for some time. However, by ca 6800 cal BP, peat growth developed on a more regional scale. As a result of the lower rate of RSL rise, the sequence deposited between ca 7800 and 5500 cal BP consists of successive peat beds alternating with tidal flat deposits (Fig. 19.5). The formation of this alternation was governed by the position of the tidal channels and thus by sediment supply, and not by sea-level fluctuations as generally assumed in former times (Ameryckx 1959; Maréchal 1992; Gullentops and Broothaers 1996). The depressions of the small palaeovalleys in the

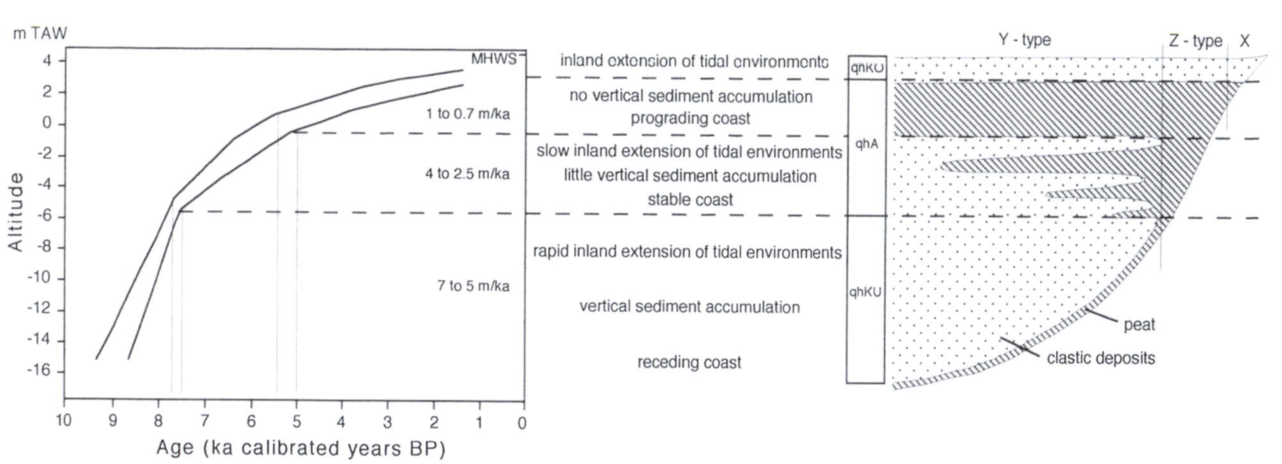

Fig. 19.5 Schematic representation of the depositional history and sediment succession of the Holocene coastal plain in relation to changes in the rate of the RSL rise. Age and altitude of the basal peat are used to construct the RSL curve (adapted from Baeteman 2005b; reproduced with permission of Brepols)

landward part of the plain are filled with a continuous peat accumulation that can attain thicknesses of 4 m.

The rate of RSL rise continued to decrease and, after ca 5500–5000 cal BP, it fell to an average of 0.70 m/ka. Sea level was close to its maximum and sediment supply had exceeded the creation of accommodation space. Inland migration of the sedimentary environments stopped completely, and the stabilization of the coastal barrier changed into barrier progradation, with the coastline prograding beyond the present-day one. Periods of peat growth lasted longer and the lateral extension of the freshwater marshes became more widespread. This period corresponds with the development of the (now) 1- to 3-m-thick uppermost intercalated peat bed known as surface peat (even though the latter started to accumulate as early as 6200 cal BP in certain areas). Between about 5500 and 4500 cal BP, almost the entire plain had changed into a freshwater marsh with peat accumulation. In the very western and seaward part of the plain, however, tidal sedimentation went on (see Sect. 19.3.1). This peat accumulation, which lasted 2000–3000 years, could keep pace with the slow RSL rise of the time. Traces of tidal influence in the thick peat bed of the landward area (Denys 1993) suggest that the major tidal channels remained open, although they were almost silted up, and served as drainage for the peat bog and the runoff from the Pleistocene higher grounds outside the plain. This also implies that a continuous and uninterrupted dune belt did not exist as generally assumed.

19.2.3 Return of the Tidal System in the Late Holocene

The late Holocene deposits of the coastal plain were formed under conditions of return of the tidal environment via tidal channels. Brackish clastic sediments replaced the peat accumulation. From the available data, it is known that the end of the peat growth was not synchronous. However, the end of the peat formation is extremely difficult to date accurately because of likely erosion of the peat surface during the initial stages of inundation in most cases. Moreover, the top of the peat may have been affected by some degree of intrusion, oxidation, or reworking making radiocarbon dates less reliable (Waller et al. 2006). In the western plain, only few locations show a gradual transition with the overlying sediments and lend thus trust in the dates. The all-youngest age for the top of the peat taken from the very landward part of the plain where no erosion or oxidation is assumed is about 1525 cal BP (ca 400 AD), other dates ranging from 3370 to 1525 cal BP, with a cluster between 2250 and 2000 cal BP.

19.2.3.1 Reestablishing and Incising a Tidal Channel Network

The renewed tidal extension entered the area via the mid-Holocene tidal channels that were meandering and landward branching across the peat bog until the landward margin of the plain. The channels were almost completely filled and the major ones served as drainage for the peat bog and the higher lying hinterland. It is suggested that a negative sediment balance (Beets and van der Spek 2000) and an excessive runoff from the hinterland (Baeteman 2005a, b) were at the origin of this tidal extension. Increased runoff was probably caused by an abrupt climate change around ca 2800 cal BP (van Geel et al. 1998). This involved an increase in precipitation, which eroded the upper part of the mid-Holocene channels. Excessive runoff was most probably enhanced by tree cutting in the hinterland during the Iron Age period, similarly to what Smyth and Jennings (1990) have documented in England. Once the channels were partly reincised, tidal waters could enter again. The extension of the tidal system, however, occurred in successive steps. At first, the tidal waters inundated the peat in the areas adjacent to the channels. This happened with modest erosion of the peat along the banks of small crevasse channels (Fig. 19.6). Formation of a crevasse channel with shallow erosion could be dated to 2140–1940 cal BP in Veurne and 2700–2440 cal BP in Raversijde, nearby Oostende.

In addition, in the areas adjacent to the channels, seawater stopped the growth of the peat, which consequently lost its potential of water retention, resulting in compaction of the bog (in the sense of volume reduction) and subsidence of the land adjacent to the channels. Peat extraction during Roman times had the same effect (Fig. 19.7) (Baeteman and Pieters 2015). The land subsidence provided accommodation space for the deposition of intertidal sediments. Interlaminated mud and sand with peat detritus, indicative of tidal deposition, was formed. Although erosion was modest during the first stage, it had far-reaching consequences, causing progressive drainage and de-watering of the peat bog with compaction and collapse in the vicinity of the channels. This resulted in a gradual increase of the tidal prism (the area the channels drain), to which the channels adapted by enlarging their cross section (depth–width). Because peat has a great resistance to erosion, the channels scoured deeply into the easily erodible material of the sand-filled early and mid-Holocene channels. Detailed mapping showed that the major late Holocene channels reoccupied the location of their mid-Holocene predecessors and sometimes eroded even into underlying Pleistocene deposits down to depths of 25 m (Baeteman 2004b). However, tidal currents in the channels detached large masses of peat that slumped and

Fig. 19.6 Erosion of the upper peat bed by a small late Holocene tidal channel with chaotic infill, which indicates that the fill happened shortly after the scouring. A vegetation horizon dated to 2375 cal BP (*arrow*) developed on the channel sediments and was in turn overlain by intertidal deposits at the *top* of which *S. plana* shells were dated to 2085 cal BP. The age of the vegetation horizon indicates that the breaching of the peat swamp must have happened shortly before 2375 cal BP, at least at this location nearby Veurne (from Baeteman et al. 1999)

Fig. 19.7 An example of peat extraction from the Roman period illustrating large excavated areas. Note the undisturbed mud and sandy clay overlying the peat, testifying that they were deposited after peat extraction (from Baeteman 2007b; reproduced with permission of Aksant)

eventually became buried in the channel fill. Radiocarbon dates of peat boulders in the channel fills indicate that they originate from the upper peat bed, including those in the lags at the base of the channels. During this phase, the channel network enlarged erosively with the continuously increasing tidal prism, affecting larger parts of the peat bog that collapsed because of de-watering. This ultimately contributed to the extension of an increasingly dense channel network in the plain (Fig. 19.8).

19.2.3.2 Controls on Sediment Distribution

In the areas where the upper peat bed has not been entirely eroded, the overlying deposits show various sediment characteristics. In the most distal reaches of the channels, little or no sediment differentiation is discernible vertically in the 1- to 2-m-thick clay. Surfaces indicative of breaks in the sedimentation are lacking, suggesting that the clay was deposited by one single process at the onset of the tidal inundation. In the proximity of the sand-filled channels,

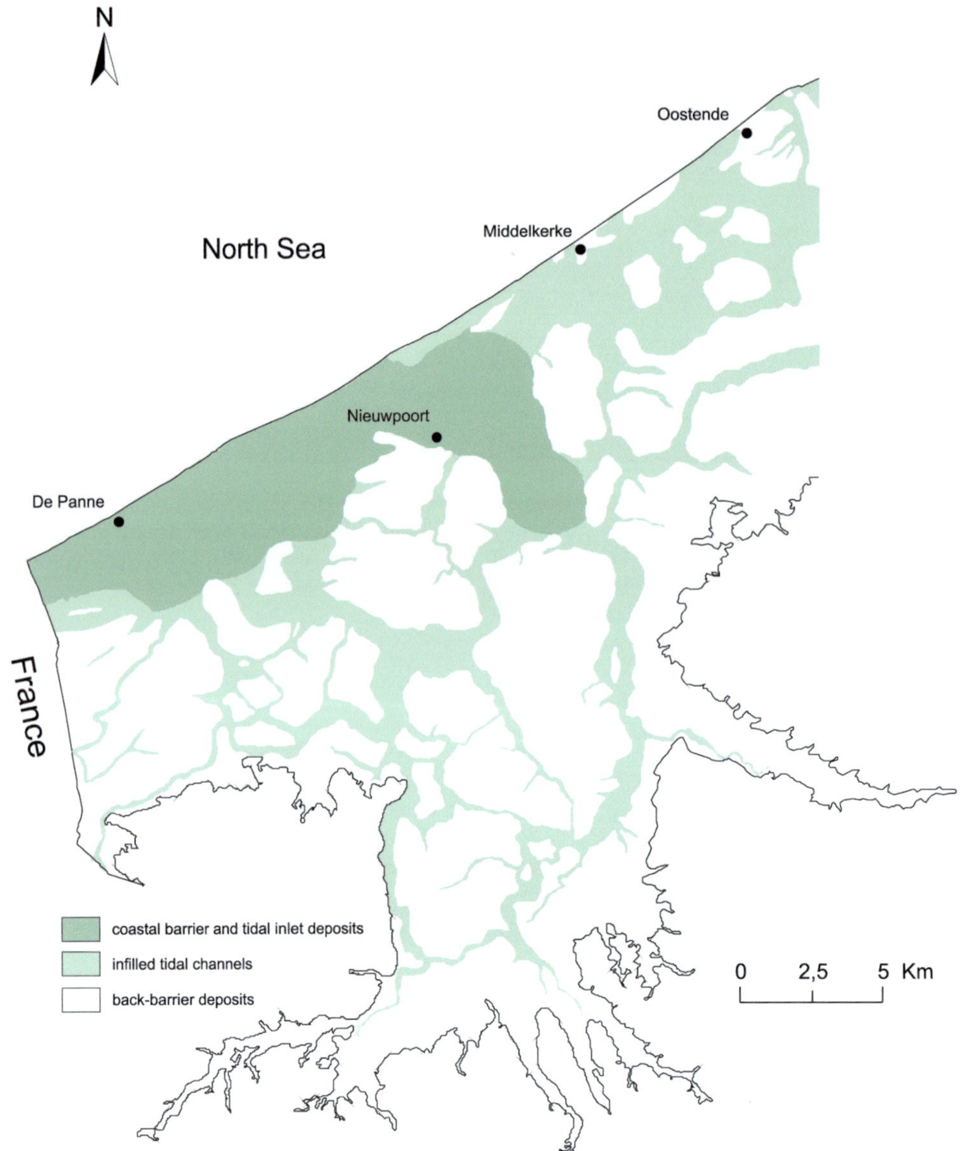

Fig. 19.8 Map of the western part of the Belgian coastal plain showing the location of the late Holocene sand-filled channels (adapted from Baeteman 2008a)

however, sediment variations are frequent. A sedimentological study together with radiocarbon dating of shallow outcrops in the proximity of sand-filled tidal channels allowed documenting the mechanisms and processes that caused such complex sediment variations and the evolution of the changing coastal landscape during the last 2000 years (Baeteman 2007a, 2008a; Baeteman et al. 2002).

Time and sediment gaps between the top of the peat and the overlying sediments suggest that some areas experienced a long period of little or no sediment deposition. Sites of peat collapse often found themselves in a subtidal position with minimum sediment deposition, assumedly as a result of negative sediment balance (Beets and van der Spek 2000) because all available sediments were used to fill the space produced by the deep scouring tidal channels. Moreover, during the 2–3 ka of uninterrupted peat growth, sea level has risen for about 2 m. Together with the collapse of the peat bog, this implies that a huge accommodation space had first to be filled before a state of dynamic equilibrium was reached between the channel cross-section, tidal prism, sediment supply, and the sea level. Sediment partly came from the in situ available early and mid-Holocene channel fills. However, owing to the large volume needed, the seaward area had to supply additional material, explaining the inland migration of the shoreline by erosion of the tidal deltas and shoreface.

Once dynamic equilibrium was attained between the controlling factors, the sediment surface reached an intertidal position. The youngest ages of the peat top indicate that peat growth came to an end and was replaced by a tidal flat in the entire plain between about 1600 and 1500 cal BP (ca 350–450 AD). While the major part of the plain changed into an intertidal flat and silted up into a supratidal flat, most of the channels eventually filled until they had reached an intertidal position. This happened in the period between ca 1400–1200 cal BP (ca 550–750 AD). Because of the silting up of the area and the very weak RSL rise, accommodation space was no longer created. Therefore, the channels started to migrate laterally due to meandering. This caused shallow erosion and reworking of the upper part of the channel fill and adjacent supratidal flats and explains the sediment variations. The available dates indicate that this reworking of the upper channel fill did not happen simultaneously across the entire area. Again, not every channel reacted in a similar way because of the interplay of local factors, explaining the variations in nature and timing of the sedimentation. This implies that sea level fluctuations were not responsible for the sediment variations recorded in the late Holocene deposits.

19.2.3.3 The Ultimate Shaping of the Plain

Finally, the channels and tidal inlets silted up to a large extent and the channel networks contracted. Archaeological and historical evidence document that, already from 550 AD, channels started to fill and salt marshes became suitable for human activity. From the sixth century onwards, the plain was suitable for pastoral activities and a free peasants' society existed. Only the major tidal channels remained open watercourses until their embankment in the first half of the twelfth century (Ervynck et al. 1999; Loveluck and Tys 2006; Tys 2013).

However, the major part of the salt marshes, originally under control of the Counts of Flanders, was progressively embanked well before the end of the twelfth century, leaving only the larger tidal channels in their natural state. The transformation of the early mediaeval salt marsh landscape into a polder required high hydraulic technological skills like building dikes, ditches, dams, sluices, and canals. This allowed the counts to develop a valuable power base for their new principality and to organize a landscape and society rather than protect the area from floods. The drainage channels were also used as commercial traffic routes. Through this, the counts succeeded in creating a centrally organized technical hydraulic landscape where the environment was used to secure and increase power and control over economy and society (Tys 2013). The hydraulic infrastructures were necessary for the drainage of rainfall in the polder characterized by water-saturated deposits in its subsurface and a high groundwater table. However, they also resulted in compaction of the surficial deposits and subsidence of the land, which explains why the 1000-year-old mediaeval polder now lies about 2 m below high tide level.

The sand-filled channels that are now concealed in the subsurface were much less sensitive to compaction. Therefore, their surface is at a slightly higher elevation due to differential compaction with their surroundings after embankment. On the other hand, the channels' waterlogged, loosely packed sand is very vulnerable to liquefaction. A shock applied to such sediments can change the packing and increase the pore fluid pressure to the extent that the sediment undergoes temporary liquefaction (Collinson and Thomson 1989). Digging ditches can also produce a failure in the hydraulic pressure resulting in a similar phenomenon (Fig. 19.9). The observation that the major late Holocene channels reoccupied the same position as their predecessors suggests that the sand-filled channels with easily erodible sediments will be similarly affected in the case of catastrophic flows due to storm surges and/or accelerated sea level rise, in particular if dikes and dunes along the shore are not sufficiently high or resistant.

19.3 Two Contrasting Landscapes

In the very west and very east of the coastal plain, two particular landscapes exist, which experienced contrasted evolutions, also different from the general history of infill. Moreover, De Moeren, in the west, is a reclaimed area characterized by a very low land surface while in the east, tides are penetrating higher lying salt marshes of the Zwin area.

19.3.1 De Moeren

De Moeren, situated in the west of the plain, is a very special landscape (Fig. 19.10), differing from the rest of the coastal plain in many aspects. It is a very flat area at altitudes between +0.20 and +1.20 m TAW, which is about 3 m lower than the average in the coastal plain and about 4–5 m below high water level. Another peculiarity is the complete absence of the uppermost thickest peat bed. The landscape is characterized by a remarkable regular pattern of ditches and

Fig. 19.9 Photograph of a sand-filled tidal channel that incised into former deposits. Note sand collapse due to liquefaction

roads resulting from land reclamation of the area, which used to be a pool and had to be drained at several attempts, the first one in 1626 (Fig. 19.11). The surrounding canals, called Ringslot and Kontergracht, together with two remaining windmills witness the history of the area (Fig. 19.12).

The area of De Moeren is enclosed by the Inland dunes (also called the 'Cabourg Duinen') in the north and by outcropping Pleistocene deposits forming the Izenberge Plateau in the south (Fig. 19.13). The Inland Dunes, located between Ghyvelde (in France) and Adinkerke, are a 6- to 8-m-high isolated ridge of ~6 km length and ~700 m width at 3 km inland from the present-day shoreline. This is the only occurrence of such an inland dune in the coastal plain. It is generally believed that this dune was part of a coastal barrier representing a former shoreline position at about 5000 years ago (Tavernier 1948; Sommé 1969).

The origin of the pool and hence the geological history of the area have been subject of debate for more than hundred years (see Baeteman 1985 for an overview). The absence of the upper peat bed was explained by peat excavation during the mediaeval period (Moormann 1951; Ameryckx 1978). It was generally believed that in this area a high-domed raised peat bog existed that grew so high that it was never flooded again. The peat bog would have developed until the fourth century AD when it was systematically and completely excavated, leaving sand deposits at the surface. Consequently, it was also believed that the entire subsoil of De Moeren consists of sand (De Moor and Pissart 1992). The idea of peat excavation arose from historical sources documenting accounts of extended peat digging organized by several abbeys in the Middle Ages. However, these accounts must be assigned to the adjacent area called 'De Buitenmoeren', where the typical mediaeval peat digging, leaving only a few centimetres of peat in situ, is obvious (Fig. 19.14). Investigation of the sediment succession by means of numerous undisturbed cores and temporary shallow outcrops together with radiocarbon dates shed new light on the questions why the peat is absent, why the area became a pool, and why it is situated at such a low altitude (Baeteman 1985, 2004a).

19.3.1.1 Sedimentary Environments and Age Constraints

A series of transects correlating facies of sedimentary environments between boreholes show that not only the peat bed is absent but the entire subsoil of De Moeren differs also from the sediment succession in the rest of the plain. Figures 19.15 and 19.16 are examples of such transects crossing the area. They document the main features characteristic of the subsoil in De Moeren. First, the slope of the pre-Holocene (Pleistocene) deposits is much steeper here than in the rest of the plain (Fig. 19.4). This slope might be at the origin of the different coastal development. The initial Holocene infill started with mud from the mudflat covering the basal peat. The deepest, and hence oldest age of the basal peat that was recovered at −15.25 m (9370–8650 cal BP in borehole Wo, Fig. 19.16) indicates that the infill of the area started about 9000 years ago. The age of 8700 cal BP in borehole B5B (Fig. 19.15) is from a *Cerastoderma edule* and has to be considered with care because the shell might have been transported. The mudflat environment prevailed

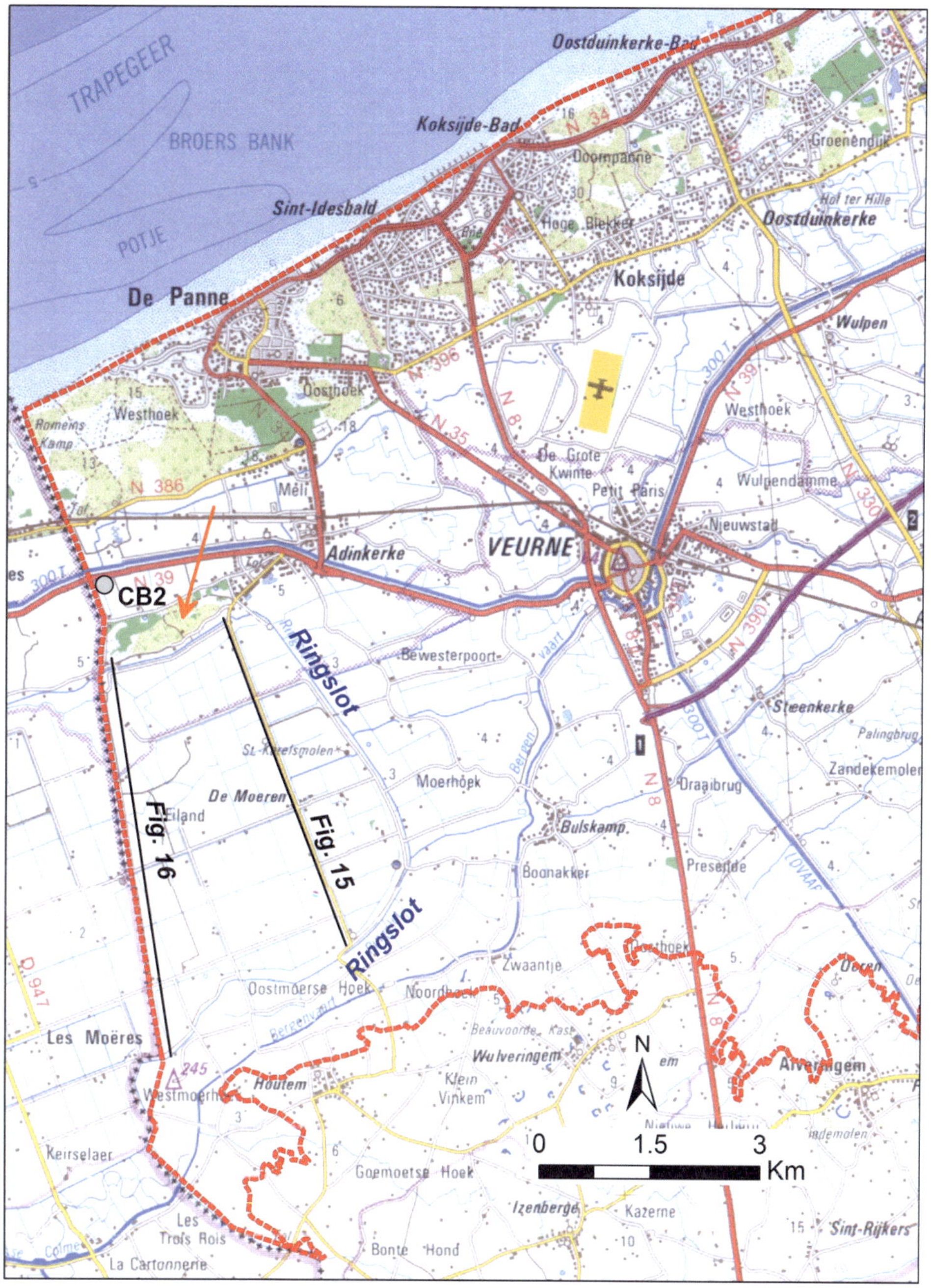

Fig. 19.10 Location map of De Moeren *encircled* by the Ringslot and located south of the Inner Dunes (*arrow*) at the French-Belgian border, illustrating the regular pattern of ditches (*in blue*) and roads. In the south, the *red dotted line* represents the boundary between the coastal plain and the Izenberge Plateau (excerpt of the NGI Topographic Map West-Vlaanderen 1/100,000)

for quite some time, resulting in a deposit that attains a thickness of up to 6 m in some places (Fig. 19.15). Vegetation horizons indicative of the temporary silting up into supratidal flats have been found in only few locations. This implies continuous tidal inundations with sufficient supply of sediment together with a high rate of RSL rise.

Then, a sudden and profound change occurred. The mudflat deposits are truncated by sand from most probably a sandflat and also tidal channels. Open-marine sand can be excluded, because of the presence of tidal flat shells and absence of typical sea shells. These sands extend far inland, suggesting deposition under high-energy conditions at a time that is poorly known because the major part of the mud has been eroded. The age of a *Scrobicularia plana*, a typical mudflat mollusc dated to 8380–8175 cal BP in B5B at −8 m (Fig. 19.15), seems rather old, as this would imply that 6 m of mud were deposited in about 500 years and, consequently, the rate of RSL rise would have been higher than the generally accepted average of ca 7 m/ka. However, considering the time–depth relation in the RSL curve (Fig. 19.5), there are chances that this occurred before the rate of RSL rise slowed down to an average of 4–2.5 m/ka at about 7500 years ago. This is the period when the first intercalated peat beds started to develop in the rest of the

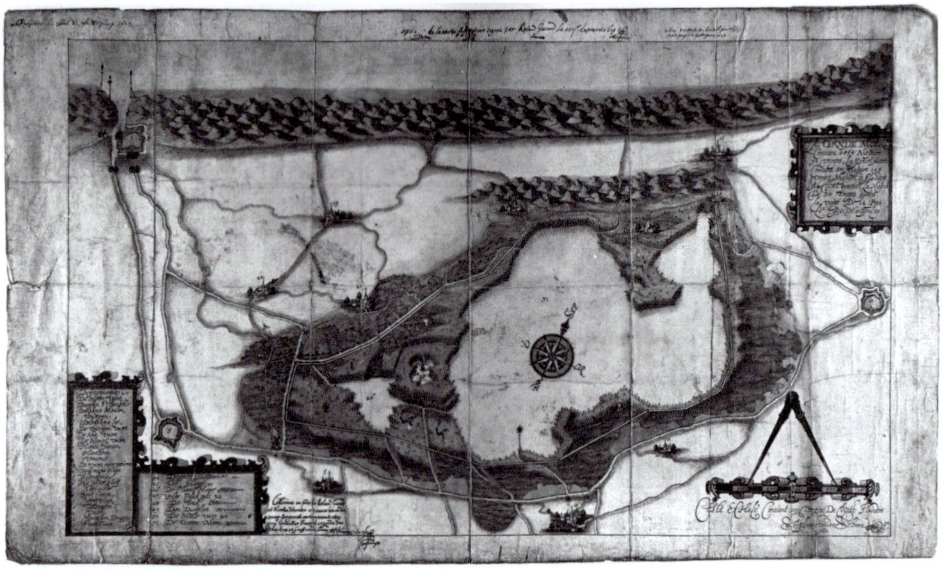

Fig. 19.11 Historical map of 1619 by R. Gerard showing the coastal dunes, Inner Dunes, and the pool of the French-Belgian Moeren (Algemeen Rijksarchief Brussel, n° 3024; from Bruneel 1979)

Fig. 19.12 View of De Moeren's landscape with the remains of a windmill on the surrounding dike (*photograph* by Laëtitia Dupin)

plain. Here, instead, a so far unexplained high-energy tidal flat was installed, similar to what has also been found across the border in northern France (Dewaide 2007). This might perhaps be related to the 8.2 ka event characterized by a sea-level jump attributed to the final drainage of proglacial Lake Agassiz (Hijma and Cohen 2010).

The sandflat environment prevailed for more than 1000 years before changing into a mixed flat at the altitude of −2 to 0 m. *S. plana* shells in borehole B5B date the transition to 6770–6520 cal BP. In borehole CB2 (Figs. 19.10 and 19.17) another *S. plana* at about 1 m above the transition was dated to 5275–4915 cal BP. In the central part of De Moeren, the mixed flat deposits are present up to the present-day surface (e.g. bh 223 in Fig. 19.15, bh 215 in Fig. 19.16). This is an atypical situation where silting up into mudflat and, eventually, salt marshes would have been expected. On the other hand, in the very northern part at the margin of the Inner Dunes, the mixed flat is capped by a peat bed forming the present-day surface (Fig. 19.17). The base of this peat bed, at +1.30 m, was dated to 3340–3070 cal BP in borehole CB2, and to 3580–3380 cal BP in a temporary outcrop in the Inner Dunes at an altitude of +2.35 m. This

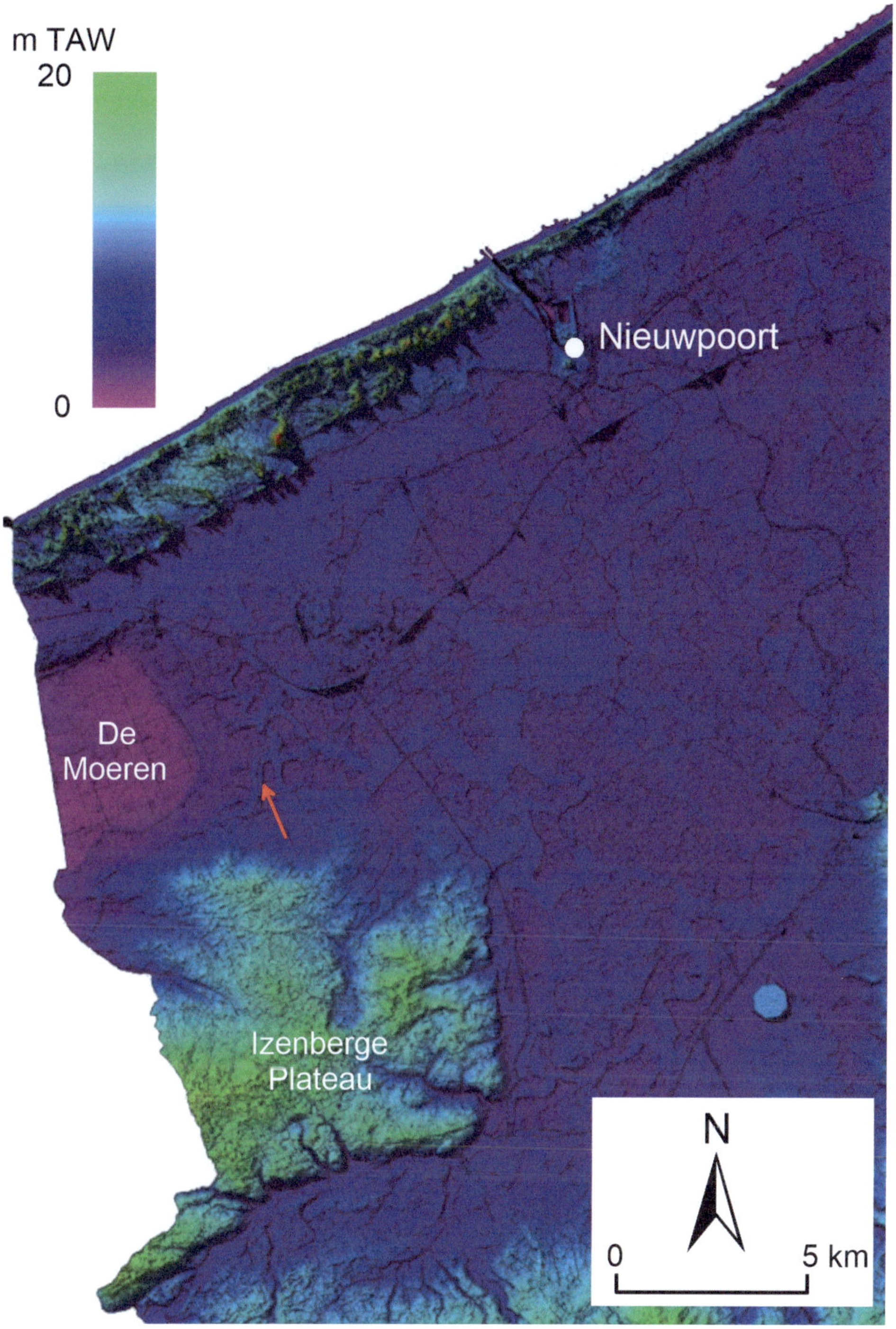

Fig. 19.13 Digital Terrain Model of the western part of the Belgian coastal plain clearly showing the very low altitude of De Moeren located north of the higher lying Izenberge Plateau. Note the well developed coastal dunes west of Nieuwpoort. The sand-filled tidal channels (*red arrow*) are also well expressed by their slightly higher position

peat bed is very much related to the Inner Dunes because it originated as a result of seepage. It documents that this area was out of the reach of tidal inundation as from about 3500 cal BP.

However, in the southern part of De Moeren, tidal channels incised deeply into the former deposits (e.g. bh 216, 224, 205). Here too, the sediment succession does not show the typical silting up into inter- and supratidal flats. The sand of the channels and their associated deposits crop out at the surface, suggesting that toward the end of the infill strongly reduced tidal inundations deprived the area of sediment.

19.3.1.2 Reconstruction of the Past Landscapes

Borehole and outcrop data were combined with about 40 ^{14}C dates to reconstruct the palaeogeographical evolution of De Moeren and the area 3 km east of it (Fig. 19.18). It should be noted that the spatial distribution of the sedimentary

Fig. 19.14 An example of a typical mediaeval peat extraction pit completely filled with sods covering a thin remnant of in situ peat. Note the great difference with the Roman peat extraction pits on Fig. 19.7 (from Baeteman 2007b; reproduced with permission of Aksant)

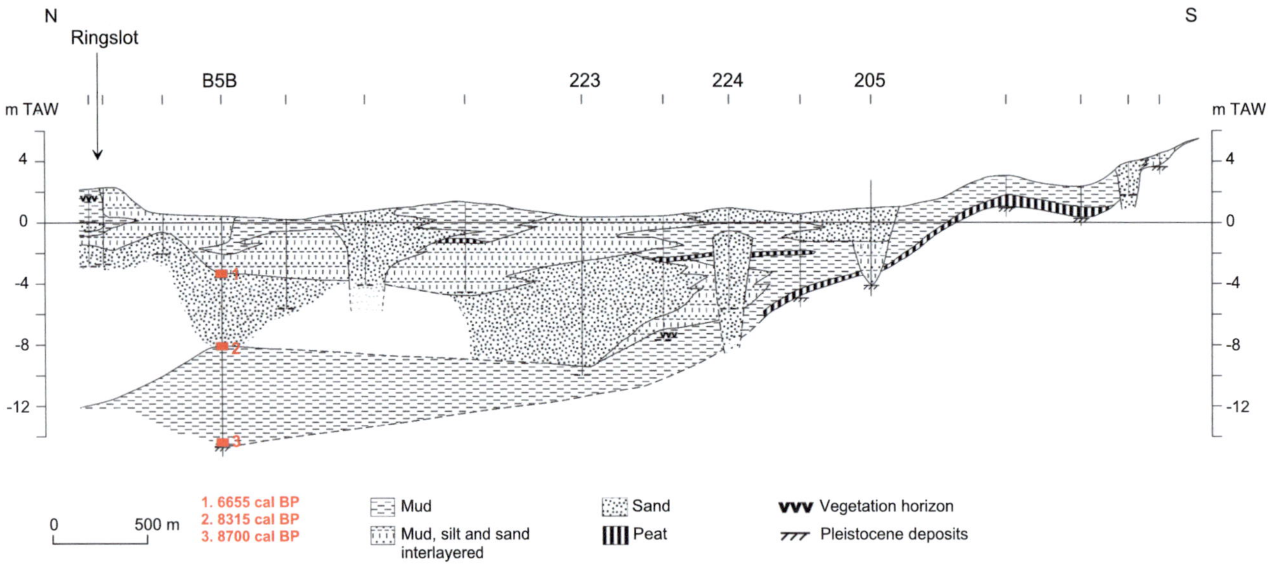

Fig. 19.15 Transect crossing De Moeren in the central part (from Baeteman 2004a)

environments is far from complete, and that time indications are approximate. The position of the former shorelines has been inferred from associated environments. The time slices are chosen according to the available ^{14}C ages and the major changes in the coastal development. Owing to low density of data, the reconstruction is more schematic especially for the early periods. The present-day coastline corresponds to the NW corner of the maps.

The area became inundated at about 9000 years ago and by 8700 cal BP a tidal embayment fringed by freshwater marshes was installed in the depressions of the Pleistocene headland. The shoreline was far beyond its present-day position. The high rate of RSL rise caused a significant lateral extension of the tidal embayment. While partial preservation of the basal peat as well as deposition of mud document tranquil water conditions of inundation, severe

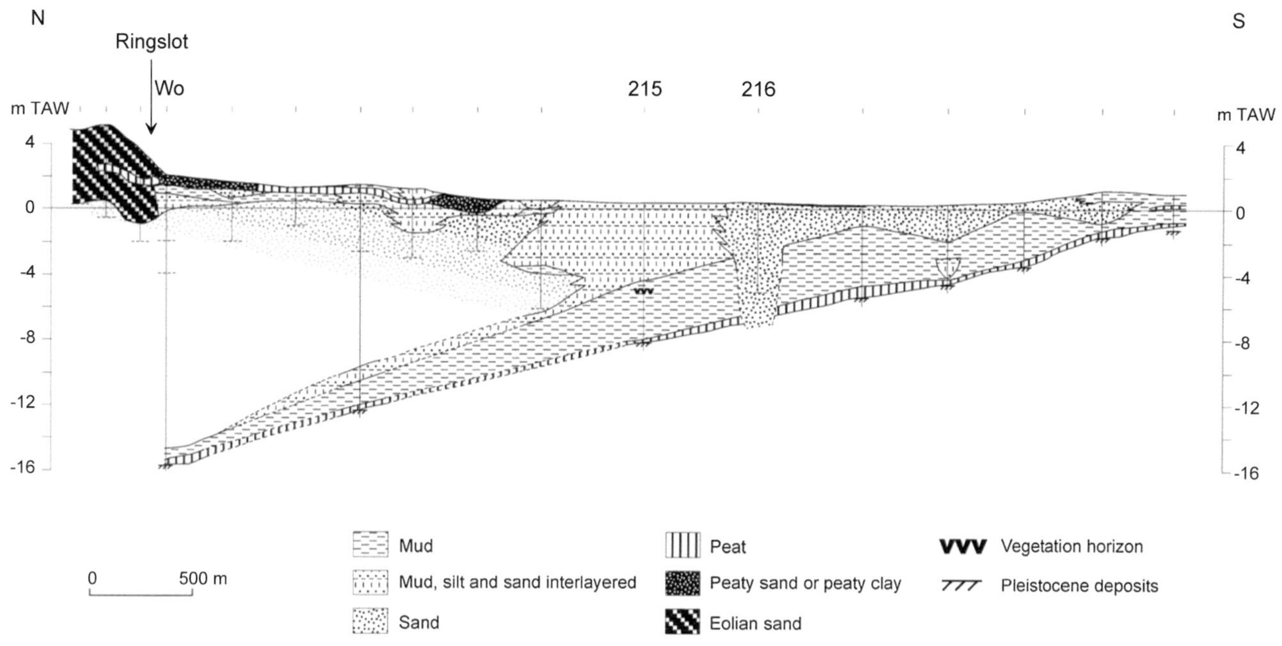

Fig. 19.16 Transect crossing De Moeren at the French-Belgian border (from Baeteman 2004a)

erosion took place by 8250 cal BP and a sand flat environment with tidal channels invaded the tidal embayment. However, despite rapid RSL rise, its landward margin did not shift significantly because of the steep slope of the pre-Holocene surface, which rather caused a large extension of the sand flat to the south, leaving little space for the development of a mud- and supratidal flat at its margin. The period between 8000 and 7000 cal BP is then characterized by continued extension of the sand flat, but this time with a significant shift of its landward margin. Also the shoreline transgressed inland.

This situation came to an end at about 7000 cal BP as a result of the slowdown of the RSL rise. In the southern part of the area, the sand flat started to silt up and a mixed flat and mudflat developed while the shoreline started to prograde. This tendency toward silting up continued until at least 2500 cal BP. Because of the silting up, the tidal channels fell out of use and larger areas became progressively out of the reach of the daily tidal inundation. In De Moeren, salt marshes locally evolved into vegetation horizons and peat accumulated in a few places free of tidal influence.

Another decrease in the rate of RSL rise at about 5000 cal BP caused a change of the major part of the area into a freshwater marsh, where peat started to accumulate on a larger scale. This was particularly the case in the east (De Buitenmoeren) and in the south. However, as said above, the absence of proper peat beds in De Moeren indicates that this area remained an intertidal flat with continued sediment supply. The sandflat environment is now very much restricted to the northern part. Along its southern margin, a surge drift line formed the core for eolian deposition, which eventually became the Inner Dunes. The shoreline stabilized slightly south of the present-day position, where a foredune probably developed. Extension of the peat bog, which is also well developed around the Inner Dunes, peaked around 3500 cal BP. De Moeren, however, remained an intertidal flat, reaching supratidal levels only locally in the central part.

19.3.1.3 Shaping the Present de Moeren Landscape

At about 2500 cal BP, the area was affected by the general return of the tides with the formation of tidal channels, which eroded deeply into the former deposits. In the east, a major tidal channel reached far inland, eroding parts of the peat bog. There, channel incision-dependent dewatering caused the muddy and peaty subsoil to subside by 1–2 m. This surface lowering created an important vertical accommodation space, which was eventually filled up to supratidal levels about 1500 cal BP. However, the major part of De Moeren changed again into a sand flat. In its northern part, continued peat accumulation documenting absence of tidal inundations suggests that the tidal channel(s) entered De Moeren via the southwest.

Extreme tides or surges interrupted occasionally the peat growth around the Inner Dunes. However, peat accumulation prevailed overall because a supratidal fronting sandflat and a dune protected it from the daily tidal inundation. Moreover, as the subsoil of this particular area consists entirely of sand, dewatering-induced compaction and subsidence are kept to a minimum, so that the area silted up

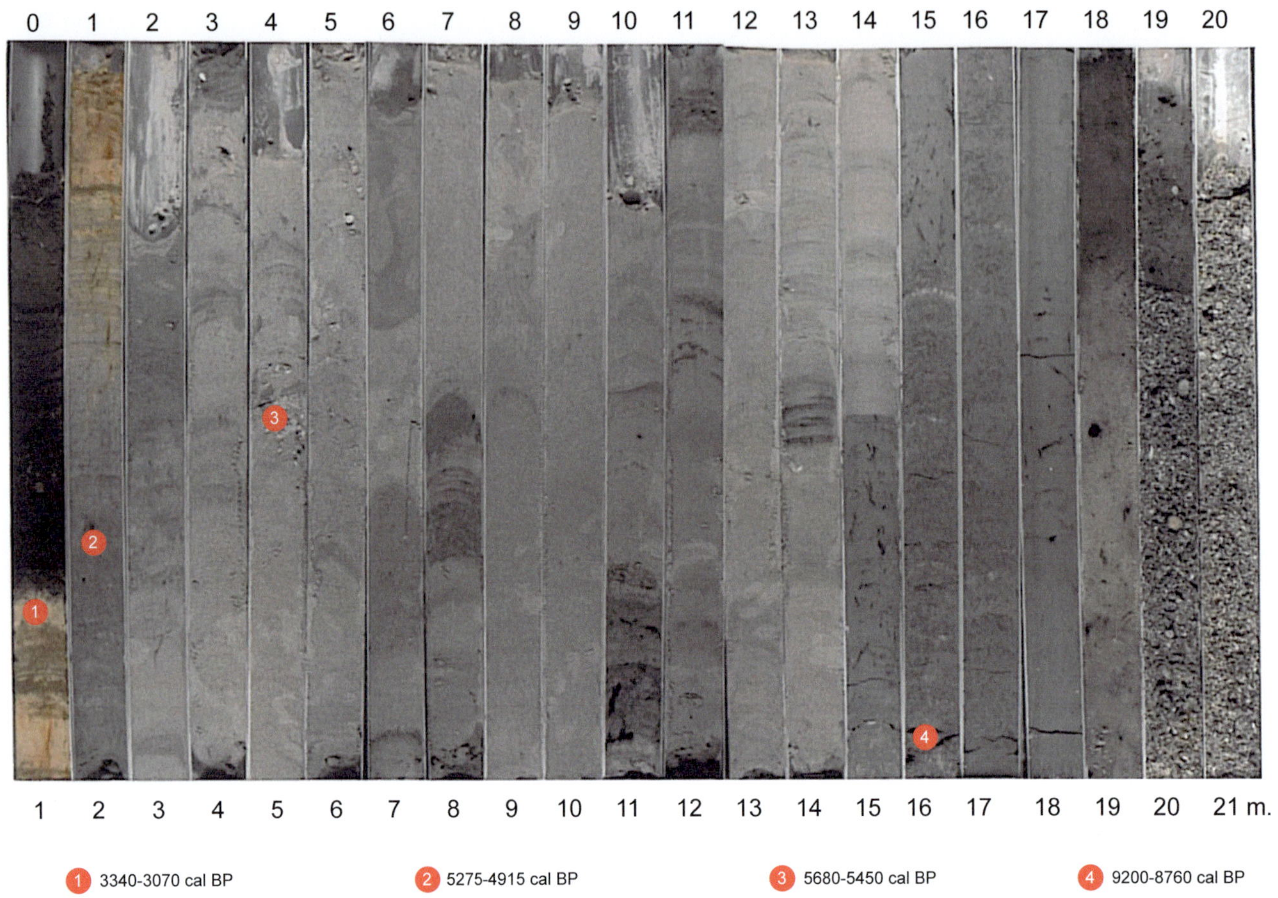

Fig. 19.17 Photograph of borehole CB2 clearly showing the erosional boundary between the mudflat and sandflat/channel deposits at a depth of 14.50 m. The organic content at the *top* of the Pleistocene deposits at 18 m originates from the development of the basal peat. Radiocarbon sample *1* is a thin peat bed; *2*, *3* and *4* are shells. See Fig. 19.10 for the location of the borehole (reproduced with permission of Mobility and Public Works, Governement of Flanders)

rapidly to high water level after each extreme event, and peat growth could resume.

The situation in the rest of De Moeren was quite different. The intertidal sand flat and subtidal channels did not silt up at all anymore. The tidal channel that fed De Moeren from the southwest most probably silted up in the area outside De Moeren, leading to tidal prism reduction and, ultimately, closure of the tidal inlet entry. This process was most likely enhanced by anthropogenic reduction of the tidal prism due to embankment of the supratidal areas. As a result, De Moeren was cut off from all sediment supply and remained at an intertidal level (i.e. ~2 m lower than the areas in supratidal position). Ultimately, it evolved into an isolated freshwater lake until it was drained and reclaimed. The drainage through ~2.5-m-deep ditches and a 7-m-deep canal, together with the pumping of groundwater, caused sediment compaction and at least 1 m land subsidence.

This palaeogeographical reconstruction shows that De Moeren was an intertidal flat throughout the major part of its Holocene history. Therefore, intercalated peat beds could not develop on a large scale as in the rest of the coastal plain. After the return of the tidal system had changed the area again into a sand flat, sediment deprivation left it in an intertidal position. Land subsidence due to reclamation works resulted in an even lower position. De Moeren together with the Inland Dunes are now classified as a protected area.

19.3.2 The Zwin Region

The Zwin region, the only extensive salt marsh environment in Belgium, is a unique area located in the very east of the coastal plain at the Belgian-Dutch border. It is a nature reserve, recognized as *international bird and habitat directive area*. The salt marsh is at the level of spring high water, tidal amplitude being there of ~4.5 m. The Zwin area consists of a tidal inlet with channel (the Zwin proper), intertidal sand flats with magnificent ripples, and salt marshes, which in late summer turn purple with the flowering

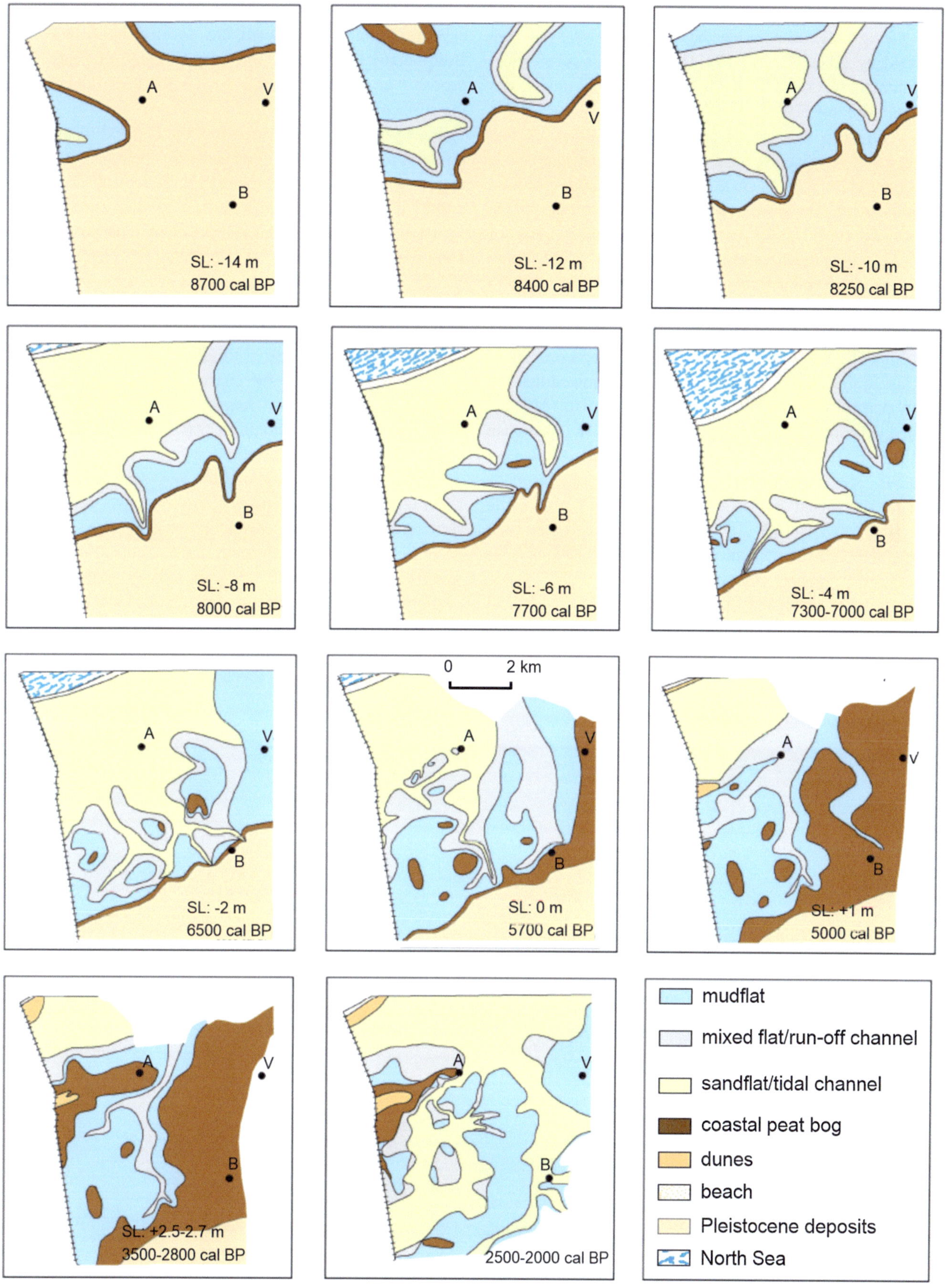

Fig. 19.18 Palaeogeographical reconstruction of the area of De Moeren and Buitenmoeren during the Holocene. *Sl* sea level of the time (from Baeteman 2004a)

sea lavender. The tidal environments are fronted by a narrow dune facing an even narrower beach. Landwards, they are bordered by dikes (Fig. 19.19). The area west of the tidal environment, separated from it by a low dike, lies at altitudes between 4.50 and 5.00 m TAW.

In contrast to the western part of the coastal plain, few geological investigations have been carried out here but a wealth of literature exists since the 1950s about the evolution of the Zwin region, based on historical sources and maps, and soil surveys. The historical data relate to the timing of embankments and settlements since about 1000 AD. However, the data were interpreted in the context of the outdated rigid chronology of successive transgressions since the Roman period (Verhulst 1964).

A limited geological investigation with deep cores in an area just to the west of the Zwin tidal zone showed that the major part of the Holocene sequence consists of tidal channel deposits down to a depth of-10 m TAW (Baeteman 2005c) (Fig. 19.20). Predominant sands include mud, peat remnants, shell fragments, and shell accumulation beds containing tidal flat and open-marine shells. Two *C. edule* were dated to 1550–1690 AD at −0.10 m, and to 1540–1680 AD at −0.97 m. The uppermost part of the deposits consists of sand most probably from a sandflat.

At altitudes of about +3 m TAW, the northern and southern part of the Zwin area show different situations (Fig. 19.21). In the south, salt marsh deposits overlying mudflat deposits crop out at the surface, one core displaying two fossil salt marsh surfaces. In the north, 2-m-thick beach deposits thinning southward are covered by eolian sand. Both deposits extend 1.2 km inland. A bed of peaty sand, probably originating through seepage from the dunes, forms the surface in the central part.

Based on this investigation and the detailed palaeogeography of nearby Zeeland (Vos and van Heeringen 1997), the major aspects of the Holocene evolution of the Zwin area appear strongly influenced by the morphology of the Pleistocene subsoil. In this eastern part of the coastal plain, the relatively high Pleistocene 'Headland of Walcheren and Vlaanderen' extended at least 20 km seawards. In the Zwin region, it has been almost entirely eroded by tidal scouring in the late Holocene, which prevents from retrieving when exactly the RSL rise started to affect the area. Comparison with the central part of the coastal plain and Zeeland suggests this might have happened between 7000 and 6000 years ago. The shoreline was then more than 5 km beyond its present-day position and tidal flat environments developed at its back, reaching the very seaward part of the Zwin region. Shoreline and tidal environments continued to shift further inland as a result of the RSL rise, though at decreasing rate. The tidal environments were fringed in the south by a freshwater marsh with accumulation of (basal) peat above the Pleistocene subsoil. At ca 5500 cal BP, further slowdown of the RSL rise combined with a sufficient sediment supply caused shoreline progradation, and tidal flat silting up to supratidal level. In parallel, the freshwater

Fig. 19.19 Aerial photograph of the Zwin region in 2005 showing the tidal channel crossing the salt marshes (© 2015 Aerodata International Surveys NASA)

Fig. 19.20 Photograph of a core showing the typical tidal channel deposits in the Zwin area (reproduced with permission of Mobility and Public Works, Governement of Flanders)

marsh continued to extend slowly further inland in relation with the very slow rate of RSL rise, allowing uninterrupted peat accumulation in the Zwin area for about 3000 years.

This situation came to an end with the return of the tidal system (see Sect. 19.2.3). The shoreface erosion and inland shifting of the shoreline caused erosion of the tidal environments that developed at the back of the barrier then situated seaward of the present-day shoreline. Newly formed tidal channels scoured into the tidal flat deposits. However, erosion was unequal throughout the area, sparing islands such as Wulpen, an embanked salt marsh with four villages well known from historical maps. These islands, often erroneously taken for barrier islands, ultimately vanished because of the continuous inland shifting of the shoreline associated with shoreline erosion.

The return of the tidal system and formation of tidal channels is not dated in the Zwin region. According to the palaeogeography of Zeeland, this must have happened after 300 AD. The Zwin region was then characterized by tidal channels penetrating the peat bog. Channel bank erosion and peat collapse lowered the surface, resulting in a progressive enlargement of the tidal prism with a positive feedback on channel evolution. The core analysis evidenced a lateral shift of the tidal channels from west to east that historical sources suggest to relate to the effects of several severe storm surges that affected the area. Despite continued coastline inland displacement, the channels in the eastern and southern parts silted up and evolved into intertidal sand flats out of the reach of strong tidal currents. The *Cerastoderma* ages document that this happened shortly after 1550–1700 AD. Further silting up then changed the area into mudflats and, eventually, supratidal salt marshes in the southern part. The inland shifting of the shoreline finally reached the Zwin region and beach deposits transgressed over the salt marshes.

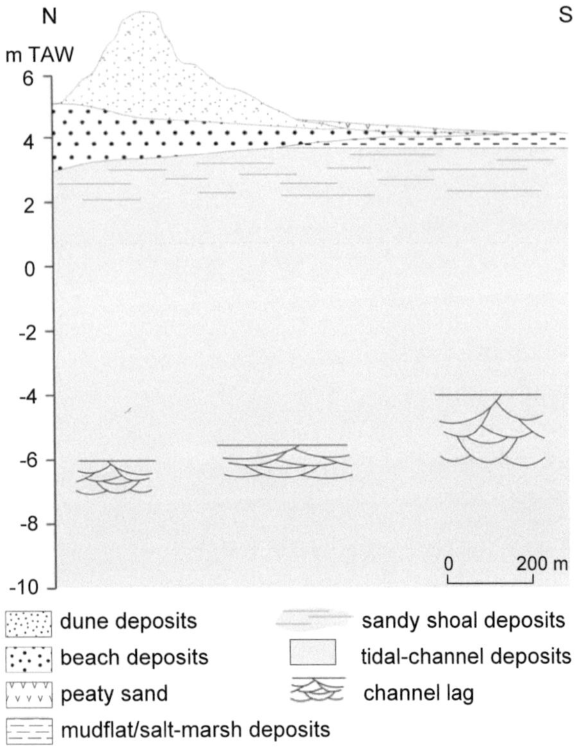

Fig. 19.21 Schematic representation of the sediment succession in the Zwin region

A thin bed of beach deposits in the landward part of the area indicates that the beach was situated inland for a short period. Finally, filling up of the area led to shoreline progradation up to its present-day position, where it stabilized, and a coastal dune started to form. The tidal channels silted up completely and evolved into inter- and supratidal environments, except the present-day Zwin. However, this is not a natural situation. The Zwin actually is a degenerating inlet and channel in closure stage. As a "dead" sediment sink, it shoals progressively, leading to tidal prism decrease and shrinking of the sand transport capacity (Bowman 1993). Only frequent dredging since many decades prevents the channel from a complete silting up. From 2016, a breakdown of the International Dike at the southern margin of the Zwin area is also planned in order to increase the sand transport capacity by deepening and broadening the channel, thus enlarge the tidal environment, and transform former embanked areas into a natural landscape.

The channel network and associated tidal environments that developed with the return of the tidal system reached several kilometres inland, up to the coastal plain current boundary. One of the channels reached as far as the city of Brugge (Fig. 19.1) and joined a small river, the Reie. This favourable situation changed this originally small settlement into a booming economic centre in the mid-twelfth and thirteenth centuries AD. By the fourteenth and fifteenth centuries, this centre had evolved into a flourishing harbour with international contacts. The architecture of the fourteenth century houses of Brugge is still original in some places (e.g. along the Spiegelrei in the centre of the city). However, severe restauration works around 1900 modified the original architecture of many facades into a late-Gothic style (D. Tys, personal communication). Another settlement, now called Damme, existing already at the end of the twelfth century along the channel that reached Brugge, also evolved into an important centre that served as transit harbour. Seagoing ships reshipped their cargo into small boats that could navigate the shallow waters until Brugge.

The area progressively silted up and evolved into salt marshes while the tidal channels and their tributaries filled and fell out of use. The area became then suitable for human occupation. The large amount of archaeological remains dating from the period between the seventh and tenth century AD documents that this happened quite early in this region (B. Hillewaert, personal communication). Ultimately, numerous embankments since 1000 AD changed the salt marshes into polders.

19.4 Final Considerations

The Belgian coastal plain and its bounding shoreline form a key area where the interaction of natural processes and human activities is obvious. Human interventions at the shoreline unfortunately changed the environment to such an extent in some places that there is no space available anymore to respond to natural processes without causing damages. These areas are consequently at great risk from the impact of future major storm surges. Because dunes form the most secure protection for the entire coastal area, it is essential to take great care of those that are left over.

By contrast, the plain, with its remarkable extensiveness, offers a peaceful impression. Some places have hardly changed since the Middle Ages and strolling around the green meadows crossed by the numerous ditches gives the illusion that time has had no impact since then. Therefore, some areas, like for instance the meadows surrounding the picturesque village of Lampernisse in the centre of the plain, have been classified as protected areas. However, because of the geological history hidden beneath the surface and human activities that have enhanced the lowering of the land surface, the coastal plain is a vulnerable area prone to a series of disasters, of which sudden and extensive inundation stands on top.

References

Ameryckx JB (1959) De ontstaansgeschiedenis van de zeepolders. Biekorf 60(11):1–26

Ameryckx JB (1978) De Moeren: historisch geografische schets. Het ingenieursblad 47:221–222

Baeteman C (1985) The origin of De Moeren. In: Van Molle M. (ed) Recent trends in physical geography in Belgium. Liber Amicorum Prof Dr. L. Peeters, Study Series of the Vrije Universiteit Brussel, pp 31–43

Baeteman C (1999) The Holocene depositional history of the IJzer palaeovalley (western Belgian coastal plain) with reference to factors controlling peat beds. Geologica Belgica 2(3–4):39–72

Baeteman C (2004a) The Holocene development of a tide-dominated coastal lowland. Western coastal plain of Belgium. Field Guide. The QRA third international postgraduate symposium Fieldtrip, 17 Sept 2004. Belgian Geological Survey. Belgium, Brussels

Baeteman C (2004b) Geologische kaart van België 1/25.000. Profieltypenkaart van de Holocene Kustafzettingen. Belgische Geologische Dienst, Brussel

Baeteman C (2005a) How subsoil morphology and erodibility influence the origin and pattern of late Holocene tidal channels: case studies from the Belgian coastal lowlands. Quatern Sci Rev 24:2146–2162

Baeteman C (2005b) The streif classification system: a tribute to an alternative system for organising and mapping Holocene coastal deposits. Quatern Int 133–134:141–149

Baeteman C (2005c) De laat Holocene evolutie van de Zwinduinen en -polders. Unpublished report for the Integrale gebiedsvisie voor het Vlaams natuurreservaat "De Zwinduinen en -polders" te Knokke - Heist, met aandacht voor het recreatief medegebruik. Aminal - Afdeling Natuur - Cel Kustzonebeheer

Baeteman C (2007a) De laat holocene evolutie van de Belgische kustvlakte: Sedimentatieprocessen versus zeespiegelschommelingen en Duinkerke transgressies. Geoarchaeological and Bioarchaeological Studies 8:1–17

Baeteman C (2007b) Roman peat-extraction pits as possible evidence for the timing of coastal changes. An example from the Belgian coastal plain. In: Beenakker JJ, Horsten FH, De Kraker AMJ, Renes H (eds) Landschap in ruimte en tijd, Aksant Amsterdam, pp 16–25

Baeteman C (2008a) Radiocarbon-dated sequences from the Belgian coastal plain: testing the hypothesis of fluctuating or smooth late-Holocene relative sea-level rise. Holocene 18(8):1219–1228

Baeteman C (2008b) De Holocene geologie van de Belgische Kustvlakte. Geological survey of Belgium Professional Paper 2008/2 - N. 304

Baeteman C (2013) History of research and state of the art of the Holocene depositional history of the Belgian coastal plain. In: Thoen E, Borger GJ, de Kraker A, Soens T, Tys D, Vervaet L, Weerts H (eds) Landscapes or seascapes? The history of coastal environment in the North Sea area reconsidered. CORN Publication Series 13. Brepols Publishers, Turnhout, pp 11–30

Baeteman C, Declercq P-Y (2002) A synthesis of early and middle Holocene coastal changes in the western Belgian lowlands. Belgeo 2:77–107

Baeteman C, Pieters M (2015) Hoe en waarom het landschap veranderde tijdens de Romeinse periode te Raversijde (Oostende, Belgische Kustvlakte). West-Vlaamse Archeologica 25:1–22

Baeteman C, Van Strydonck M (1989) Radiocarbon dates on peat from the Holocene coastal deposits in West-Belgium. Prof Pap Geol Sur Belgium 241(6):59–91

Baeteman C, Beets DJ, Van Strydonck M (1999) Tidal crevasse splays as the cause of rapid changes in the rate of aggradation in the Holocene tidal deposits of the Belgian Coastal Plain. Quatern Int 56:3–13

Baeteman C, Scott DB, Van Strydonck M (2002) Changes in coastal zone processes at a high sea-level stand: a late Holocene example from Belgium. J Quat Sci 17:547–559

Beets DJ, van der Spek AJF (2000) The Holocene evolution of the barrier and the back-barrier basins of Belgium and the Netherlands as a function of late Weichselian morphology, relative sea-level rise and sediment supply. Netherlands Journal of Geosciences-Geologie en Mijnbouw 79:3–16

Bowman D (1993) Morphodynamics of the stagnating Zwin inlet, The Netherlands. Sed Geol 84:219–239

Bruneel D (1979) Bijdrage tot de kennis van de historische geografie van De Moeren. MSc thesis UGent, 159 pp

Collinson JD, Thompson DB (1989) Sedimentary structures. Unwin Hyman Ltd., London

De Moor G, Pissart A (1992) Het reliëf. Geografie van België. Nationaal Comité voor Geografie, Gemeentekrediet, Brussel, pp 129–215

Denys L (1993) Palaeoecologisch diatomeeënonderzoek van de Holocene afzettingen in de westelijke Belgische kustvlakte, Unpublished PhD thesis, Universitaire Instelling Antwerpen

Denys L, Baeteman C (1995) Holocene evolution of relative sea-level and local mean high water spring tides in Belgium—a first assessment. Mar Geol 124:1–19

Dewaide L (2007) Evolution des environnements sédimentaires holocènes de la région des Moëres. Msc thesis, Université de Liège

Ervynck A, Baeteman C, Demiddele H, Hollevoet Y, Pieters M, Schelvis J, Tys D, Van Strydonck M, Verhaeghe F (1999) Human occupation because of a regression, or the cause of a transgression? A critical review of the interaction between geological events and human occupation in the Belgian coastal plain during the first millennium AD. Probleme der Küstenforschung im südlichen Nordseegebiet 26:97–121

Gullentops F, Broothaers L (1996) Overzicht van de geologie van Vlaanderen. In: Gullentops F, Wouters L (eds) Delfstoffen in Vlaanderen. Antilope, Lier, pp 1–27

Hijma MP, Cohen KM (2010) Timing and magnitude of the sea-level jump preluding the 8200 yr event. Geology 38:275–278

Loveluck C, Tys D (2006) Coastal societies, exchange and identity along the Channel and southern North Sea shores of Europe, AD 600–1000 AD. J Marit Archaeol 1:1–30

Maréchal R (1992) Kwartairgeologie, Lithologie van de oppervlakkige lagen. Tweede Atlas van België, Blad II, 3:25 p

Moormann FR (1951) De bodemgesteldheid van het Oudland van Veurne Ambacht. Natuurwetenschappelijk Tijdschrift 33:3–124

Roy PS, Cowell PJ, Ferland MA, Thom BG (1995) Wave-dominated coasts. In: Carter RWG, Woodroffe CD (eds) Coastal evolution: late quaternary shoreline morphodynamics. University Press, Cambridge, pp 121–186

Smyth C, Jennings S (1990) Late Bronze Age-Iron Age valley sedimentation in east Sussex, southern England. In: Boardman J (ed) Soil erosion on agricultural land. Wiley, Chichester, pp 273–284

Sommé J (1969) La plaine maritime. Ann Soc Géol Nord 89(1):117–126

Tavernier R (1948) De jongste geologische geschiedenis der Vlaamse Kustvlakte. Handelingen der Maatschappij voor Geschiedenis en Oudheidkunde te Gent 3:107–115

Tys D (2013) The medieval embankment of coastal Flanders in context. In: Thoen E, Borger GJ, de Kraker A, Soens T, Tys D, Vervaet L, Weerts H (eds) Landscapes or seascapes? The history of coastal environment in the North Sea area reconsidered. CORN Publication Series 13. Brepols Publishers, Turnhout, pp 199–239

van Geel B, Raspopov O, van der Plicht J, Renssen, H (1998) Solar forcing of abrupt climate change around 850 calendar years BC. In: Peiser BJ, Palmer T, Bailey ME (eds) Natural catastrophes during

Bronze Age civilisations. BAR International Series 728, pp 162–168
Verhulst A (1964) Het landschap in Vlaanderen. Willemsfonds 202. 128 pp
Vos PC, van Heeringen RM (1997) Holocene geology and occupation history of the Province of Zeeland. In: Fischer MM (ed) Holocene evolution of Zeeland (SW Netherlands). Mededelingen Nederlands Instituut voor Toegepaste Geowetenschappen TNO 59:5–109
Waller MP, Long AJ, Schofield JE (2006) Interpretation of radiocarbon dates from the upper surface of late-Holocene peat layers in coastal lowlands. Holocene 16:51–61

Landslides in Belgium—Two Case Studies in the Flemish Ardennes and the Pays de Herve

20

Olivier Dewitte, Miet Van Den Eeckhaut, Jean Poesen and Alain Demoulin

Abstract

Most landslides in Belgium, and especially the largest features, do not occur in the Ardenne, where the relief energy and the climate conditions seem most favourable. They appear in regions located mainly north of them where the lithology consists primarily of unconsolidated material. They develop on slopes that are relatively smooth, and their magnitude is pretty large with regard to that context. An inventory of more than 300 pre-Holocene to recent landslides has been mapped. Twenty-seven percent of all inventoried landslides are shallow complex landslides that show signs of recent activity. The remaining landslides are deep-seated features and rotational earth slides dominate ($n > 200$). For such landslides, the average area is 3.9 ha, but affected areas vary from 0.2 to 40.4 ha. The exact age of the deep-seated landslides is unknown, but it is certain that during the last century no such landslides were initiated. Both climatic and seismic conditions during the Quaternary may have triggered landslides. The produced landslide inventory is a historical inventory containing landslides of different ages and triggering events. Currently, only new shallow landslides or reactivations within existing deep-seated landslides occur. The focus on the Hekkebrugstraat landslide in the Flemish Ardennes allows us to understand the recent dynamics of a large reactivated landslide. It shows the complexity of the interactions between natural and human-induced processes. The focus on the Pays the Herve allows for a deeper understanding of landslide mechanisms and the cause of their origin in natural environmental conditions. These two examples are among the best-studied landslides and their analysis allows us to highlight the main processes at play and to better unravel their interactions.

Keywords

Mass movement • Flemish Ardennes • Pays de Herve • Landslide reactivation • Landslide cause • Triggering factors

O. Dewitte (✉)
Department of Earth Sciences, Royal Museum for Central Africa, Leuvensesteenweg 13, 3080 Tervuren, Belgium
e-mail: olivier.dewitte@africamuseum.be

M. Van Den Eeckhaut
Arcadis Belgium, Koningsstraat 80, 1000 Brussels, Belgium
e-mail: miet.vandeneeckhaut@arcadis.com

J. Poesen
Department of Earth and Environmental Sciences, KU Leuven, Celestijnenlaan 200E, 3001 Leuven-Heverlee, Belgium
e-mail: jean.poesen@kuleuven.be

A. Demoulin
Department of Physical Geography and Quaternary, University of Liège, Sart Tilman, B11, 4000 Liège, Belgium
e-mail: ademoulin@ulg.ac.be

A. Demoulin (ed.), *Landscapes and Landforms of Belgium and Luxembourg*, World Geomorphological Landscapes, DOI 10.1007/978-3-319-58239-9_20

20.1 Introduction

At the European level, Belgium does not appear as a country prone to landsliding (Günther et al. 2014). Topographic conditions such as those found in mountain environments are not present and processes capable of triggering landslides, whether seismic or climatic, occur very rarely. Therefore, large slope instabilities, such as those found in the Alps for instance (Crosta et al. 2013), are not observed. However, an interesting point about landslide processes in Belgium is that most instabilities, and especially the largest ones, do not occur in the Ardenne where the relief energy and the climate conditions are the most favourable. They appear in regions located mainly north of them, on slopes that are relatively smooth, and their magnitude is pretty large with regard to that context (Fig. 20.1). The objective of this chapter is to provide a general picture of the landslides in these regions and then to focus on two representative cases. These two examples are among the best-studied landslides in Belgium and their understanding allows for highlighting the main processes at play and better unravelling their interactions.

20.2 Landslides in Middle Belgium

20.2.1 Geographical and Geological Settings

The Flemish hills located in Southern Flanders are clearly marked by landslide processes (Fig. 20.2). The 4850 km^2 area of Southern Flanders contains five hilly regions where slope angles are often steeper than 3° (Fig. 20.2a). From west to east these are the West Flemish hills (up to 155 m asl), the hills of the Flemish Ardennes (up to 156 m asl), the Pajottenland hills (up to 112 m asl), the Hageland hills (up to 105 m asl) and the Haspengouw hills (up to 128 m asl). These hilly regions are separated by wide valleys of the Leie, Scheldt, Dender, Dijle and Gete rivers (Fig. 20.2). The lithology consists primarily of unconsolidated Late Miocene to Early Palaeocene sediments, with some west–east

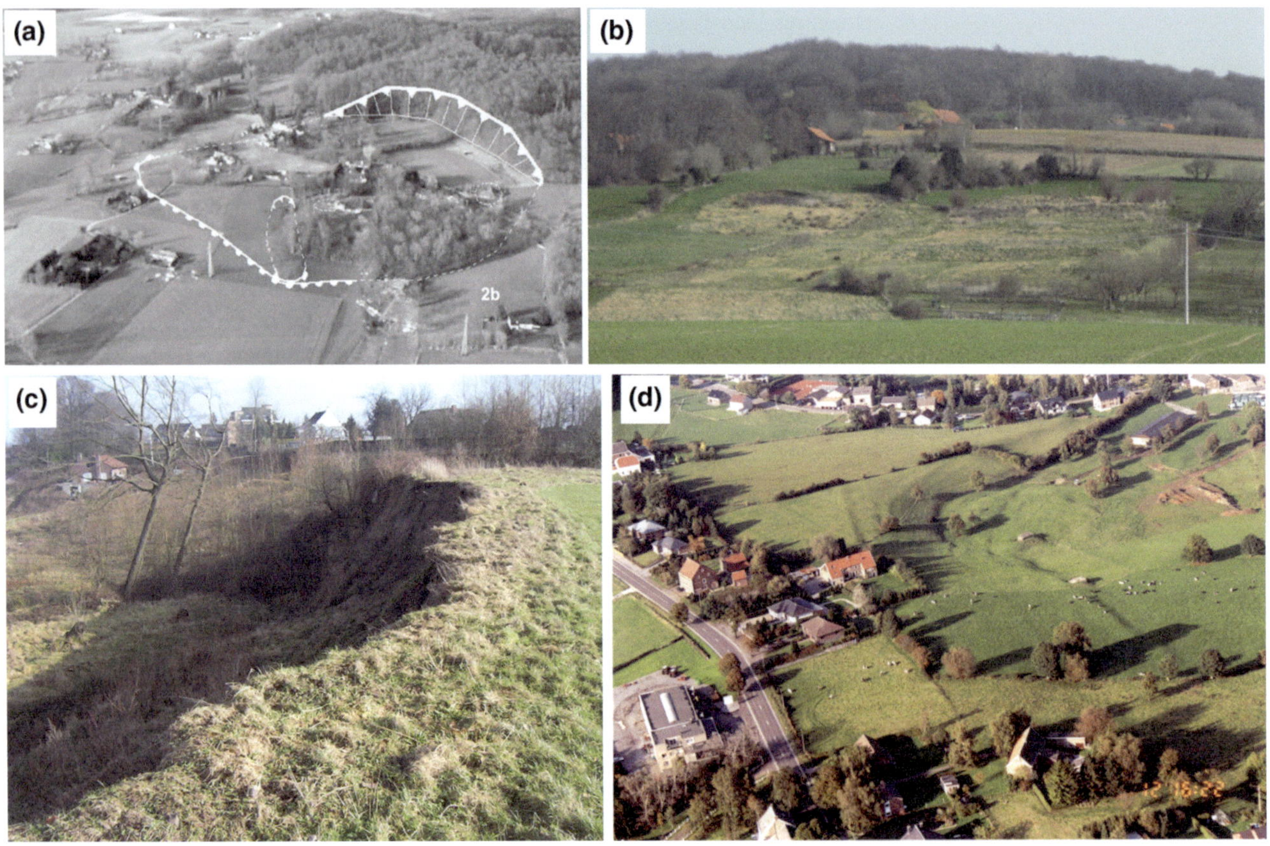

Fig. 20.1 Landslide examples in Middle Belgium and Pays de Herve. **a** Typical large deep-seated landslide in the Flemish Ardennes (Wittentak, March 2004, photo by J. Poesen in Van Den Eeckhaut et al. 2005). The *white lines* represent the limits of the landslide: main scarp (*right*) and toe (*left*). See Fig. 20.2f for location. **b** Frontal view of two adjacent shallow complex slides located in the West Flemish hills (Dranouter landslide, March 2009, photo by M. Van Den Eeckhaut in Van Den Eeckhaut et al. 2011). See Fig. 20.2d for location. **c** Main scarp of an old deep-seated landslide reactivated in March 2001 (Maarkedal, February 2003, photo by O. Dewitte in Dewitte 2006). See Fig. 20.2f for location. **d** Middle-sized deep-seated landslide in the Pays de Herve (Chaineux, 1999, photo by A. Pissart). See Fig. 20.10 or location

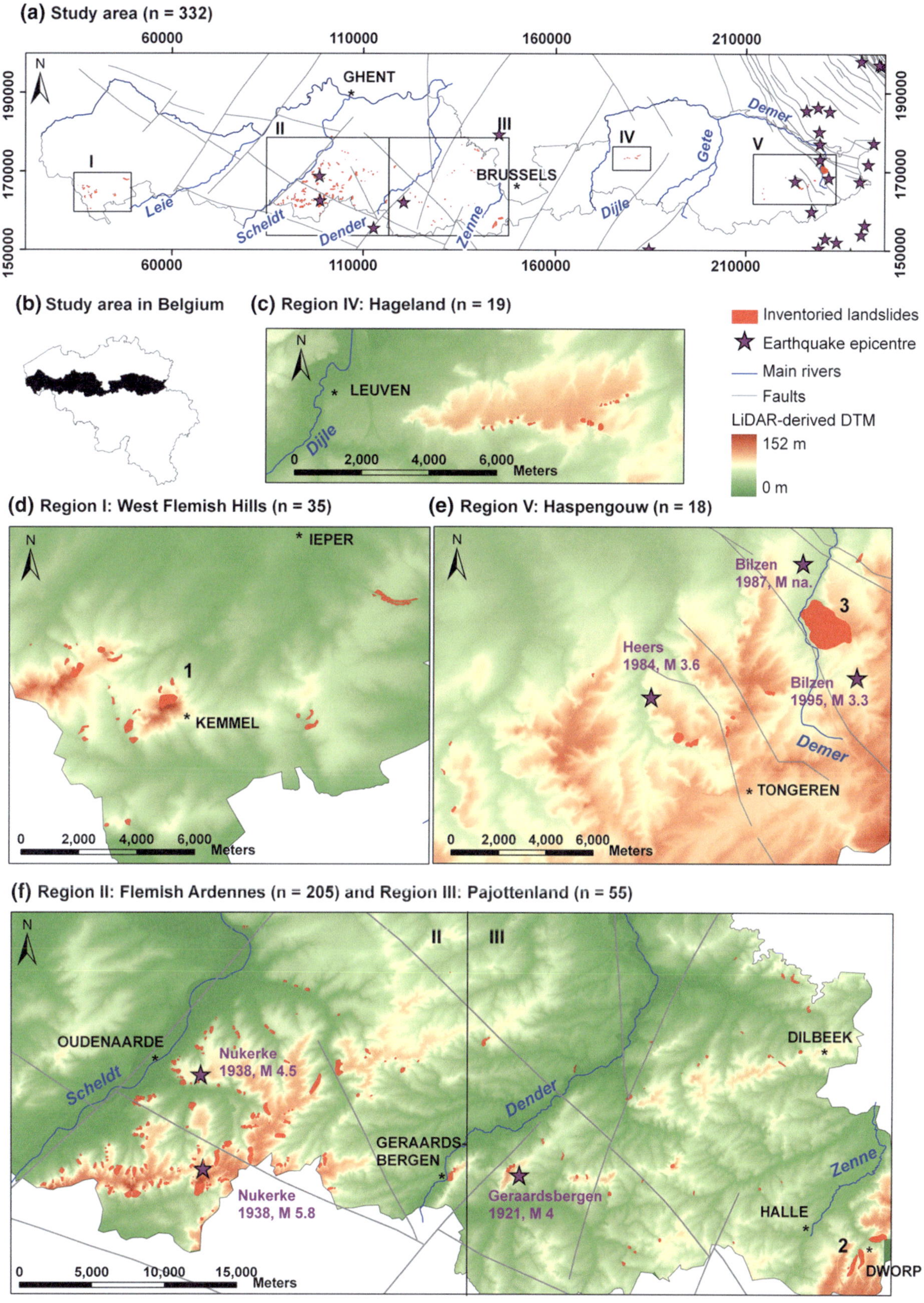

Fig. 20.2 **a** Regions in Southern Flanders (Middle Belgium) where landslides have been mapped (Belgian Lambert 72 coordinates in m). **b** Location of the represented areas within Belgium. **c** Landslides in Hageland. **d** Landslides in the West Flemish hills. **e** Landslides in Haspengouw. **f** Landslides in the Flemish Ardennes and Pajottenland. Excerpts from the Flemish LiDAR DTM (from Van Den Eeckhaut et al. 2011)

variability in the distribution of the lithostratigraphic units. These Cenozoic formations consist of subhorizontal layers of alternating sand, clay and clay-silt dipping at 0.2° towards the northeast (Jacobs et al. 1999).

Analysis of LiDAR-derived imagery (shaded relief, slope, surface roughness and contour maps) combined with intense field work led to the discovery of more than 330 old and recent large landslides in the region (Van Den Eeckhaut et al. 2007b, 2011) (Figs. 20.2 and 20.3). The Flemish Ardennes (Fig. 20.2f) is the region where a majority of the landslides (more than 200) was observed. In the field, all landslides were classified according to Cruden and Varnes (1996). Based on field estimate of the depth of the surface of rupture, landslides with a surface of rupture deeper than 3 m, but generally deeper than 10 m, and landslides with this surface at a depth of less than 3 m were classified as deep-seated and shallow, respectively. Deep-seated rotational earth slides dominate, accounting for 65% (n = 217) of the mapped landslides (Van Den Eeckhaut et al. 2011). Eight percent of the deep-seated movements are classified as complex. Twenty-seven percent (n = 90) of all inventoried landslides are shallow complex landslides (Fig. 20.1b). For about 50 of them, detailed information on the initiation or reactivation date was found (Van Den Eeckhaut et al. 2011). With the exception of four landslide events reported between AD 1828 and AD 1926, all dated historical landslide events occurred after AD 1960. Moreover, many of the remaining undated shallow landslides show signs of recent activity (Van Den Eeckhaut et al. 2011).

There is a high degree of variability in the metrics of individual slides in each landslide type (Van Den Eeckhaut et al. 2011). The smallest shallow landslide identified has a landslide-affected area of 0.5–0.75 ha. The mean area of deep-seated rotational earth slide is 3.9 ha, but area varies from 0.2 to 40.4 ha, similar to the range shown by deep-seated complex slides. Most landslides have mean slope gradients in the order of 10–11°. Susceptibility analyses for deep-seated landslides in the Flemish Ardennes show that slope gradient and lithology are the controlling factors with the highest predictive power (Van Den Eeckhaut et al. 2006, 2009b).

To complete the picture, it is worth mentioning that two landslides in Haspengouw (Fig. 20.2e) were classified as deep-seated gravitational slope deformations (DSGSDs), i.e. mass movement defined to affect an entire slope and be generated by mechanical processes such as deep-seated block sliding and lateral spreading (Van Den Eeckhaut et al. 2011). The Alden Biesen DSGSD (nr. 3 in Fig. 20.2e) is located in the eastern part of Flanders near the town of Bilzen. With an area of 266.5 ha, it is the largest landslide in Southern Flanders, being significantly larger than the second largest form in the region (i.e. the 42.0 ha Muziekberg landslide in the Flemish Ardennes). With the exception of the Kolmont landslide (7.0 ha) located less than 10 km southwest of Alden Biesen, no other DSGSD was found in Southern Flanders (Van Den Eeckhaut et al. 2011).

The exact age of the deep-seated landslides is unknown, but it is certain that during the last century no such landslides

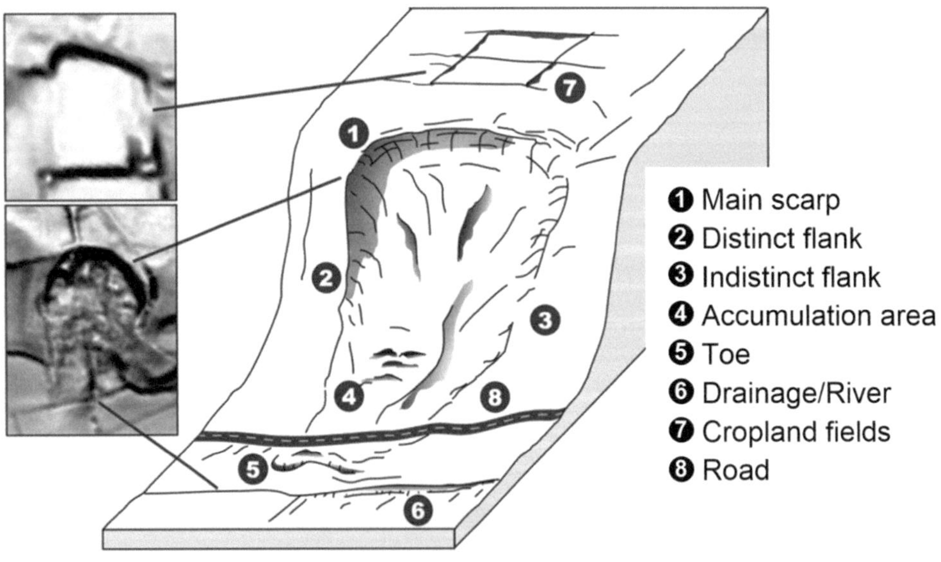

Fig. 20.3 Sketch of a typical landslide in its typical physical environment in the Flemish hills, with excerpts of the LiDAR-derived shaded relief map illustrating two characteristic topographic features of the study area, namely a cropland field with earth banks, or lynchets (*top left*) and the main scarp of a deep-seated landslide (*bottom left*). This conceptualization is also valid for the Pays the Herve where landslide features are very similar (see Fig. 20.11). Landslide features are defined according to IAEG Commission on Landslides (1990) (from Van Den Eeckhaut et al. 2012)

were initiated (Ost et al. 2003; Van Den Eeckhaut et al. 2005, 2011). Only one study in the Flemish Ardennes attempted to unravel the age and history of an old landslide and obtained an AMS ^{14}C age of 8700–8440 cal year BP that dates the Collinabos landslide to the beginning of the Boreal (Van Den Eeckhaut et al. 2007c). This corresponds to a relatively wet period coinciding with permafrost melting (Van Den Eeckhaut et al. 2011) and it is likely that other landslides were initiated in Flanders during this period. However, the subdued topography of some landslides, especially the DSGSDs, suggests they may be pre-Holocene (Van Den Eeckhaut et al. 2011).

Apart from a climatic trigger, the possibility of a seismic origin of the landslides has been evoked as the highest landslide densities often coincide with the presence of Quaternary active faults (Fig. 20.2a; Ost et al. 2003; Van Den Eeckhaut et al. 2011). Indeed, the many landslides of the Flemish Ardennes concentrate in an area where two earthquakes occurred during the twentieth century, the Geraardsbergen 1921 ($M_L = 4$) and Nukerke 1938 ($M_L = 5.8$) earthquakes (Fig. 20.2f), and where Camelbeeck (1993) locates the seismogenic South Brabant Fault Zone. Although Somville (1939) reported that the 1938 earthquake had an important impact on the groundwater flow, forming new springs and relocating existing ones, no historical document linking landslide activity to earthquakes has been found.

Both climatic and seismic conditions during the Quaternary may have caused landslides in Southern Flanders. The landslide inventory is a historical inventory containing landslides of different ages and triggering events. Currently, only new shallow landslides or reactivations within existing deep-seated landslides occur in the study area (Ost et al. 2003; Dewitte et al. 2008, 2009; Van Den Eeckhaut et al. 2009a, 2011). Despite their relatively small displacement distances, these movements are a continuously operating process causing considerable economic damage (Vranken et al. 2013). In the Flemish Ardennes alone, the annual landslide risk is estimated to vary between 51 and 86 million € (2011 values; Vranken et al. 2014). Note that the inventory of Southern Flanders, as other historical inventories, is assumed to be incomplete as morphological characteristics of old shallow landslides can be obliterated by human activities (e.g. land levelling) and soil erosion (see Chap. 15) or hidden by dense vegetation (Van Den Eeckhaut et al. 2007a).

20.2.2 Sixty-Five Years of Slope Movements Documented in the Flemish Ardennes: The Hekkebrugstraat Landslide

As stated above, large deep-seated landslides in the Flemish Ardennes occurred in an undefined epoch far before present time and due to specific environmental conditions. In contrast, observation of their current activity strongly suggests that the most probable mass movement hazard in these regions is actually not the occurrence of new such landslides but reactivation of the pre-existing ones (Dewitte and Demoulin 2005; Dewitte et al. 2006, 2008, 2009, 2010).

Located on the Leupegem hill, Flemish Ardennes, next to the town of Oudenaarde (Landslide 1, Fig. 20.4), the Hekkebrugstraat landslide (HL) is a representative example of these deep-seated rotational earth slides. It suffered a deep reactivation in 1995 that caused ground displacements of several metres. Eyewitness reports of its evolution since the 1950s as well as of various human activities within and near the landslide have been documented by Van Den Eeckhaut et al. (2007a). In addition, quantitative estimates of the displacements have been obtained for the period 1952–2002 (Dewitte et al. 2008, 2009). This well-documented landslide thus allows for unravelling the processes that drive its present activity.

The HL dimensions are typical of those of other landslides in the Flemish hills. It extends over an area of ca. 9 ha, has a total length of 370 m, and a maximum width of 260 m. The maximum height of its main scarp is 9 m. The HL cuts across the Aalbeke Member of the Kortrijk Formation, which consists of 10-m-thick homogeneous blue massive clays highly sensitive to landsliding (Van Den Eeckhaut et al. 2006, 2009b, 2011).

20.2.2.1 Insights from Archives and Eyewitness Reports

Although the emphasis is put on the HL history since the 1950s, a short overview of the information on both the landslide and land use prior to 1945 is relevant. For this purpose, two historical maps, the Ferraris map of 1771 (Fig. 20.5a) and a map from 'The atlas of the rural roads' (1841) were analysed (Van Den Eeckhaut et al. 2007a). There is no indication of a landslide on the Ferraris map (Fig. 20.5a). At the end of the eighteenth century, the area currently affected by the landslide was forested, while the upslope contributing area was under cropland or pasture. The fact that the Hekkebrug road is indicated by a thick brown line on the Ferraris map suggests that the road was a sunken lane, and that the hillslope was well drained, preventing topsoil saturation. Despite the absence of the HL on the Ferraris map, there are four reasons to believe that the hillslope had already failed at that time. The first is that none of the other large landslides mapped in the Flemish Ardennes are shown on this map. Second, there is a lack of historical documents describing the initiation of the landslide after the eighteenth century. As it is located only 2 km from the city of Oudenaarde (Fig. 20.4), and as there were already six houses within less than 500 m of the present-day landslide in 1771 (Fig. 20.5a), it seems very unlikely that the initiation would have post-dated 1771 without having been

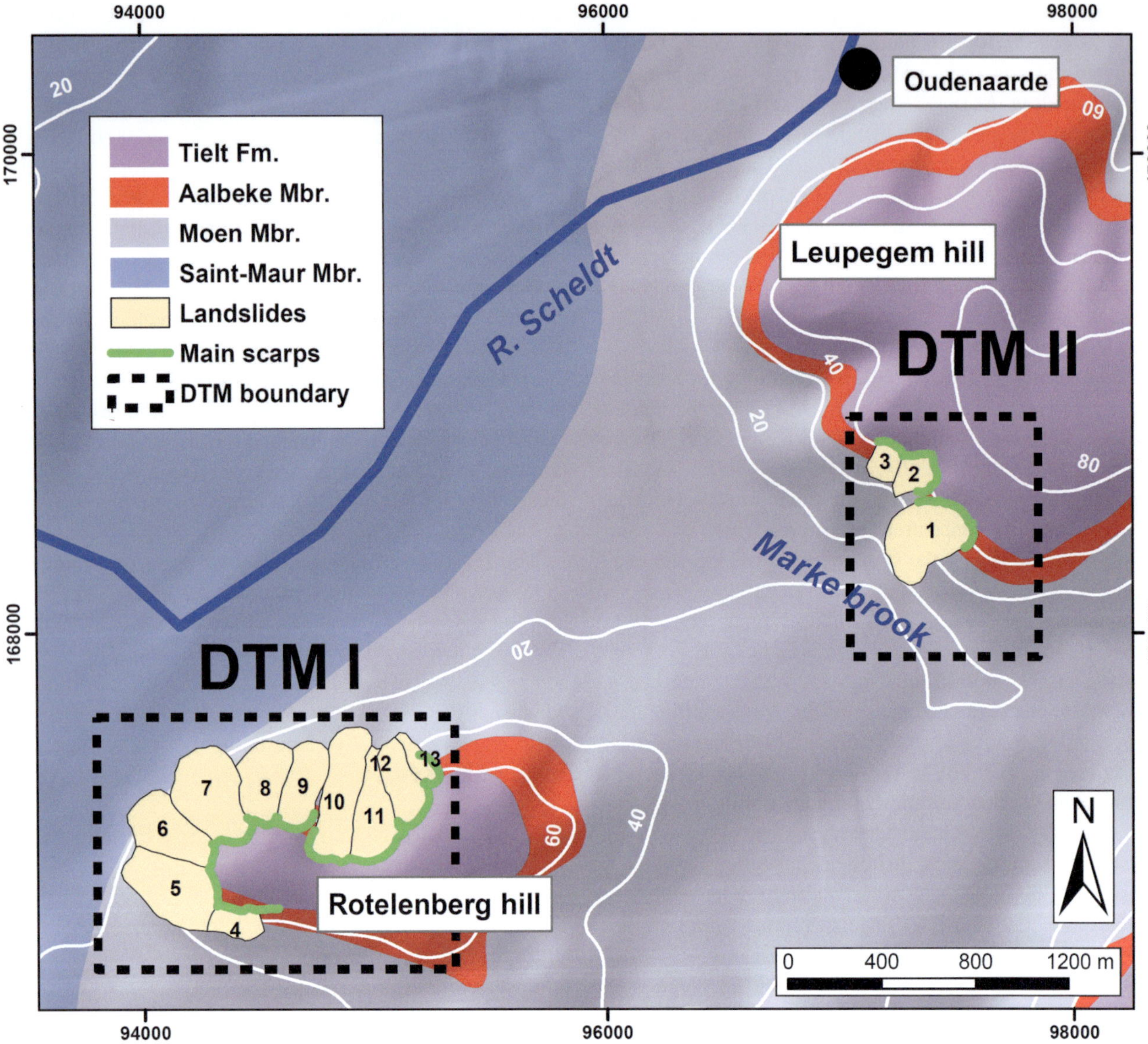

Fig. 20.4 Location map of the Hekkebrugstraat landslide (landslide 1 in the figure) and the surrounding landslides with the lithological and topographic context. The landslides are located next and south of the town of Oudenaarde (see Fig. 20.2f for general location). The scars and the main scarps of the landslides are superposed on the regional geology (Eocene). The quadrangles DTM I and DTM II locate the Digital Terrain Models used for ground displacement measurements within the 13 landslides (see Dewitte et al. 2008, 2009 for additional information). In this chapter, only ground displacements for landslide 1 are shown (Figs. 20.7 and 20.8). The altitudes (*white contour lines*) and coordinates (Belgian Lambert 72) are in metres (from Dewitte et al. 2010)

reported. The third reason is that, in contrast to most surrounding hillslope sections, which were used as either cropland or pasture, the HL site itself was forested. This strongly suggests that the microtopographical and/or hydrological conditions of the site were not suitable for agricultural activities. Finally, the curved shape of the path leading to the farm, superposed to the present position of the landslide scarp, was most probably imposed by an existing slope discontinuity. The map from 'The atlas of the rural roads' (1841) only shows roads and field plots. The curved shape of both the plots downslope and the footpath upslope of the present-day main scarp location again indicate the presence of a break in the slope pointing to a pre-existing slope failure. The slipped area was at least partly cultivated in the mid-nineteenth century, as the depletion zone was divided into small plots.

From 1950, more detailed information is available from aerial photographs and interviews with local inhabitants (Van Den Eeckhaut et al. 2007a). Undoubtedly, there was a landslide, and the affected area was partly cultivated (Fig. 20.5b). From 1952 (Fig. 20.5b) to 1996 (Fig. 20.5c), land-use changes are obvious. They are explained by several

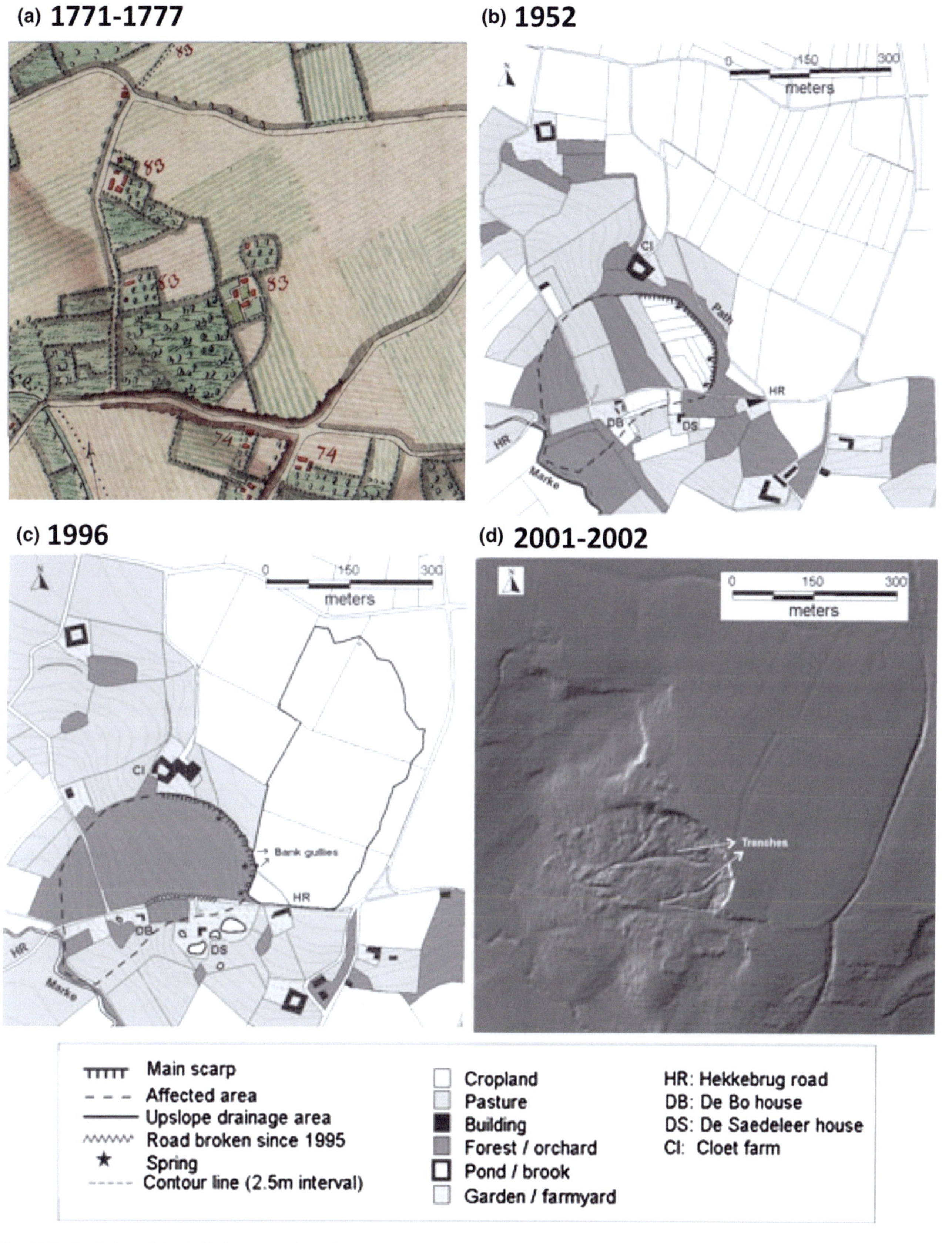

Fig. 20.5 Evolution of the Hekkebrugstraat landslide site as revealed by old maps. **a** Excerpt of the Ferraris map (1771–1777). **b** Situation around 1952 based on aerial photograph interpretation. **c** Situation around 1996 based on aerial photograph interpretation. **d** Topography around 2001–2002 shown on a LIDAR-derived hillshade map (DEM of Flanders 2007) (from Van Den Eeckhaut et al. 2007a)

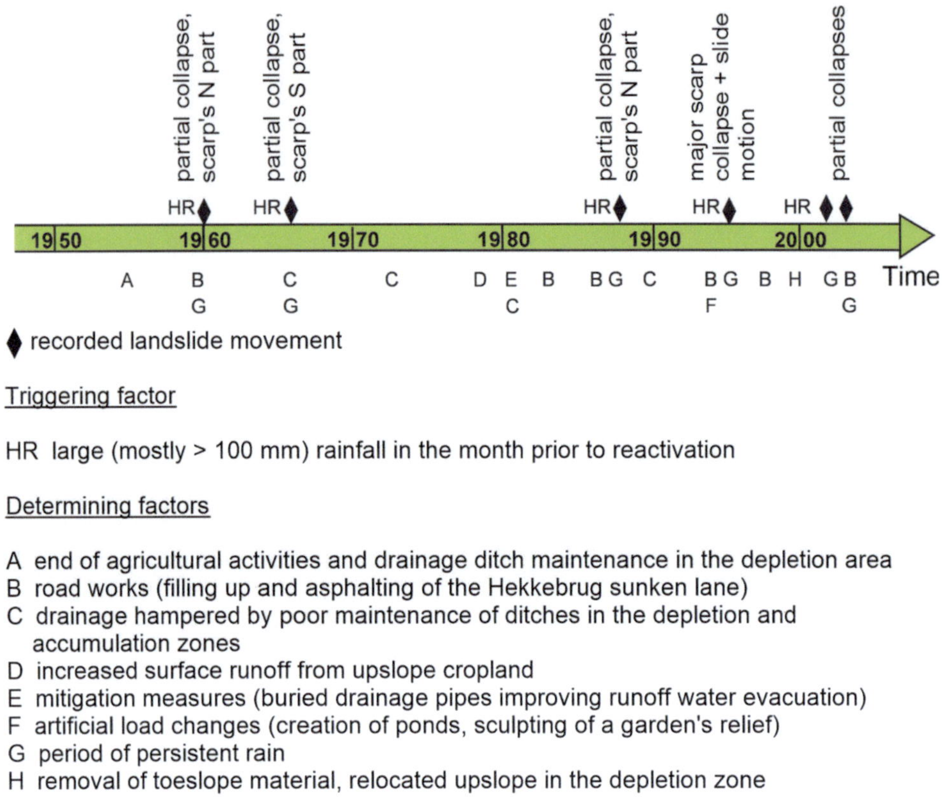

Fig. 20.6 Inventory of the determining and triggering factors responsible for reactivations of the Hekkebrugstraat landslide from the 1950s to 2005 (from Van Den Eeckhaut et al. 2007a)

natural and anthropogenic factors reported by Van Den Eeckhaut et al. (2007a) and summarized in Fig. 20.6. Since 1950, movements of HL have occurred at least six times. The reactivation of 1995 was the most catastrophic. Most displacements occurred after monthly rainfall depths above 100 mm during the previous months. Long periods of high rainfall, and the presence of relatively steep slopes, impermeable clays and springs are important natural causal factors. Equally important, however, are anthropogenic determining factors (Van Den Eeckhaut et al. 2007a). The information collected by Van Den Eeckhaut et al. (2007a) is of primary importance for the validation of the landslide displacements analysed below.

20.2.2.2 Vertical Displacements

The understanding of landslide mechanisms is greatly facilitated when accurate quantitative information on their horizontal and vertical displacements is available. To obtain such data, aerial stereophotogrammetry was applied to build 2-m resolution digital terrain models (DTMs) of HL at three epochs (1996, 1973, and 1952) (Dewitte et al. 2008). An additional DTM was interpolated from airborne LIDAR data acquired in 2002 by the Flemish Government (Fig. 20.5d) (Dewitte 2006; Dewitte et al. 2008). Simple DTM subtraction yielded direct estimates of the vertical displacement of each landslide pixel during the considered time interval with a confidence interval of ± 0.75 m (Fig. 20.7).

Between 1952 and 1973 (Fig. 20.7a), the topographical changes were very limited. The main scarp remained unchanged. Small collapses observed at the head of the slide might have resulted from tension crack formation and destabilization in response to high groundwater levels at the contact with the Aalbeke clays, similar to what is currently observed in other landslides of the Flemish Ardennes. The role of runoff is poorly known. No field data are available and one even ignores whether runoff induced a net volume loss within the landslide, or rather transported sediment from above the main scarp. However, it was however probably of little importance, due to the presence of a forested zone upslope of the landslide head during this entire period (Dewitte 2006). The elongated zone of uplift observed along the Hekkebrug road north of house 2 actually represents improvement works that consisted in adding layers of stone bricks and asphalt to increase road accessibility (Van Den Eeckhaut et al. 2007a).

By contrast, significant movements occurred during the 1973–1996 period (Fig. 20.7b), revealing a typical rotational pattern, with a subsiding zone at the head, an intermediate area without vertical displacement, and an uplifted zone of accumulation towards the foot of the landslide. Immediately downslope of the main scarp, large values of apparent collapse, corresponding to the height of the scarp ($\sim$7 m), actually indicate a scarp retreat. The horizontal retreat can be approximated by the width of the band with collapsed terrain

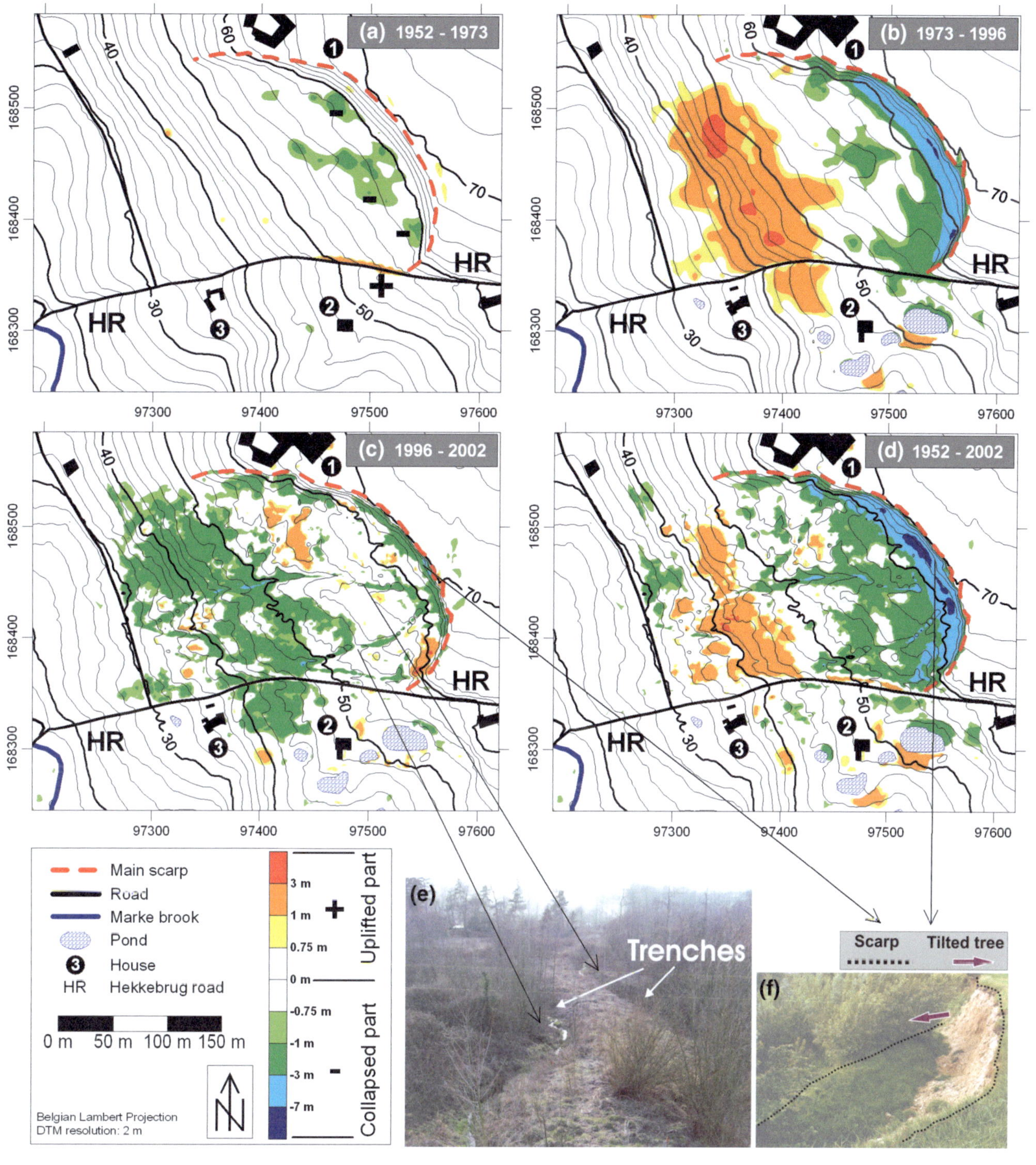

Fig. 20.7 Maps of vertical ground displacements within the Hekkebrugstraat landslide (landslide 1 in Fig. 20.4) inferred from DTM comparisons. **a** From 1952 to 1973; **b** from 1973 to 1996; **c** from 1996 to 2002; **d** from 1952 to 2002. In view **a** and **b**, the underlying contour lines describe the 1973 and 1996 topographies, respectively, whereas the 2002 topography is shown in views (**c**) and (**d**). Only vertical movements >1σ (0.75 m) uncertainty are represented. **e** West-looking view from the *top* of the main scarp of two 2-m-deep drainage ditches dug in 2000 (picture taken in February 2003). Young trees growing within the landslide are mainly poplar trees planted in place of their tall predecessors cut in September 1995 after the main reactivation. **f** Main scarp reactivation occurred in March 2002 showing clearly the listric geometry of the slip surface (picture taken in May 2002) (from Dewitte et al. 2008)

Fig. 20.8 Horizontal displacements measured within the Hekkebrugstraat landslide (landslide 1 in Fig. 20.4) between 1973 and 1996. **a** General view. **b** Focused view on wall displacements close to house 2. The areas collapsed and uplifted between 1973 and 1996 (see Fig. 20.7) are also represented (*grey* tones). The three photographs attest to traces of horizontal displacements (April 2004). (*A*) Picture taken from the Hekkebrug road upslope house 2. The backward toppling of the trees towards the landslide head is due to lateral displacement of the terrain. (*B*) Picture taken from the Hekkebrug road in front of house 2. The main entry, which originally faced the house, has moved several metres downslope since the February 1995 reactivation. (*C*) Picture taken on the road crossing the toe of the landslide. The backward toppling of the trees towards the landslide head is also due to lateral displacement of the terrain (from Dewitte et al. 2008) ▶

and reaches locally 15 m. Beyond artificial soil surface elevation changes around the ponds near house 2 (Fig. 20.7b and c) (Van Den Eeckhaut et al. 2007a), no area outside of the landslide shows significant height difference between the two models, thus validating the comparison.

Van Den Eeckhaut et al. (2007a) pointed out that ground movements during the 1973–1996 period were mainly due to the reactivation event of February 1995. This reactivation was triggered by heavy rainfall during the winter 1994–1995. In addition, the rain effect was further enhanced by increased surface runoff produced on bare cultivated areas upstream of the main scarp during that winter (Fig. 20.5c), resulting in an increase of the water volume flowing directly into the landslide zone. Moreover, the artificial loads brought by the improvement of the Hekkebrug road and the digging of several ponds close to house 2, and the absence of well-maintained drainage ditches after the cessation of the agricultural activities within the landslide zone in the 1950s contributed also to reactivations (Van den Eeckhaut et al. 2007a).

A decrease in landslide activity is observed after 1996 (Fig. 20.7c). This period is characterized by the subsidence of large areas within the landslide, mainly due to compaction upon drainage in response notably to the digging of two 2-m-deep ditches across the landslide area in 2000 (Fig. 20.5d, 20.7c, e). Starting from the base of the main scarp, these ~250-m-long ditches drain the spring water exfiltrating from the aquifer perched above the Aalbeke clays, and the runoff water (Dewitte et al. 2008). As a consequence, little displacement has been observed in the zone of accumulation since 2001 (Van Den Eeckhaut et al. 2007a). The main scarp, however, was still active during this period, as shown in Fig. 20.7f. A 25-m-long, 4-m-wide soil slice collapsed in March 2002. Soil material was also artificially removed and displaced within the landslide, explaining, e.g. the anomalous uplifted zone just downslope of the main scarp close to the Hekkebrug road. Figure 20.7d sums up the vertical soil surface displacements induced by the landslide reactivation and the main human interventions, mainly since February 1995.

The volume change of the landslide can be estimated from the comparison of the DTMs (Dewitte et al. 2008). According to this comparison, and after correction for the impact of road improvements, a volume loss of ca. 5000 ± 1000 m^3 was recorded between 1952 and 1973. Between 1973 and 1996, ca. 37,000 ± 5000 m^3 were lost in the upper part of the landslide, approximately balanced by the ca. 38,000 ± 6000 m^3 deposited downslope. From 1996 to 2002, a volume loss of ca. 35,000 ± 6000 m^3 was computed. Unless the LiDAR data are systematically biased, this volume loss, much larger than that calculated for the period 1952–1973, is likely to result mainly from natural and artificial compaction of the recently displaced soil and, to some extent, removal of material induced by the creation of the two longitudinal drains (Fig. 20.7c, e).

20.2.2.3 Horizontal Displacements

Horizontal ground displacements were measured both at the main scarp and within the landslide (Fig. 20.8) (Dewitte et al. 2008). As the main scarp is located at the border of a plateau, its retreat may be compared to a horizontal displacement, locally reaching 20 m between February 1995 and April 2006 (Fig. 20.8a). Obviously, this retreat did not result from a single reactivation affecting the whole scarp, but rather from successive collapses like that observed in 2002 (Fig. 20.7f).

The position of walls located between houses 2 and 3 was also measured in 1973 and 1996 (Fig. 20.8). Whereas the brick walls next to house 3 were not affected by the reactivation, the displacements of the walls in front of house 2 clearly testify to the downslope movement of this area (Fig. 20.8b). The displacements were in general larger along the road (Fig. 20.8b), i.e. closer to the centre of the moving mass, where the shear strength is smaller. Road displacements within the transfer area also bear witness to lateral motion (Fig. 20.8), but no control point is available to allow an accurate estimate of movement.

Topographic profiles are very illustrative of the combination of vertical and horizontal displacements (Fig. 20.9). Figure 20.9a shows that the main scarp retreat, amounting to 14 m in this profile, is clearly associated with the zone of the highest subsidence at the landslide head. According to the conceptual model developed by Casson et al. (2005) for rotational landslides, the distribution of the topographic variations between 1973 and 1996 corresponds to a large rotation angle dividing the landslide into three areas

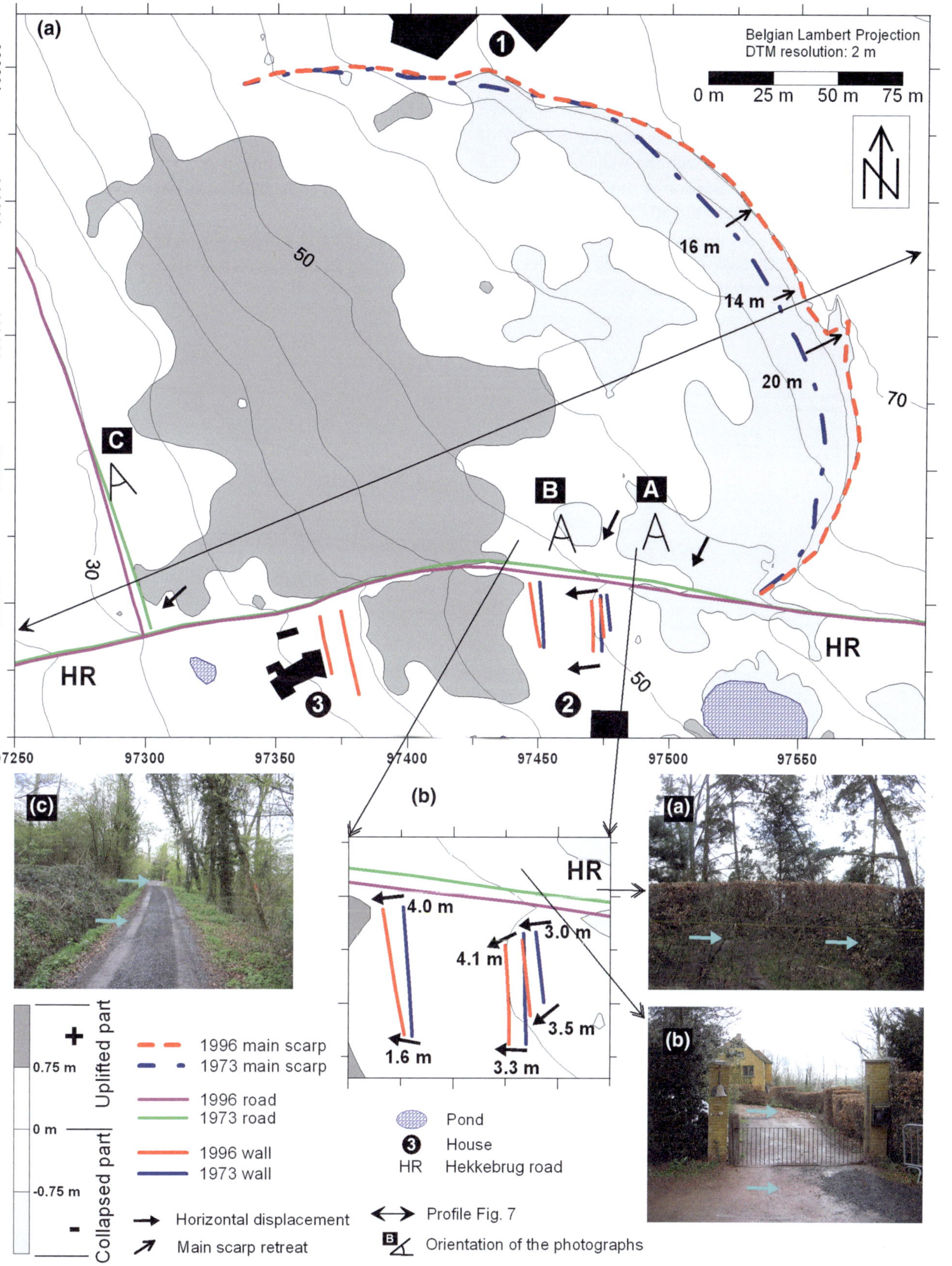

(a)
Belgian Lambert Projection
DTM resolution: 2 m
0 m 25 m 50 m 75 m
16 m
14 m
20 m
50
70
30
HR
A
B
C
168550
168500
168450
168400
168350
97250
97300
97350
97400
97450
97500
97550
(b)
HR
4.0 m
3.0 m
4.1 m
3.5 m
1.6 m
3.3 m
(c)
(a)
(b)
0.75 m
0 m
-0.75 m
Uplifted part
Collapsed part
1996 main scarp
1973 main scarp
1996 road
1973 road
1996 wall
1973 wall
Horizontal displacement
Main scarp retreat
Pond
House
HR Hekkebrug road
Profile Fig. 7
Orientation of the photographs

(depletion–transfer–accumulation). The distribution of elevation changes within the profile 1973–1996, the shape of the visible part of the surface of rupture at the landslide head, and the alternation of subhorizontal clayey sand and clay layers suggest that the ~220-m-long surface of rupture of the February 1995 reactivation is situated at a depth of ~20 m (Fig. 20.9a), in contact with the Aalbeke clays. The uplift in the zone of accumulation is not a true vertical surface movement, but mainly the effect of the downslope lateral displacement of the slipped mass along the curved rupture surface (Fig. 20.9a). Likewise, the transfer area where no vertical displacement of the surface is observed between the collapsed and uplifted parts of the landslide corresponds to a zone of mainly horizontal motion. This model agrees with the values of lateral displacement measured at walls (Fig. 20.8b). The small uplifts measured in the landslide toe, downslope of the intersection between the lower part of the surface of rupture and the original ground surface (Fig. 20.9a), correspond with several small earthflows subsequent to the main reactivation (Van Den Eeckhaut et al. 2007a).

Another issue is whether this deep reactivation occurred along the pre-existing rupture surface of the original landslide or developed into the landslide debris. Electric

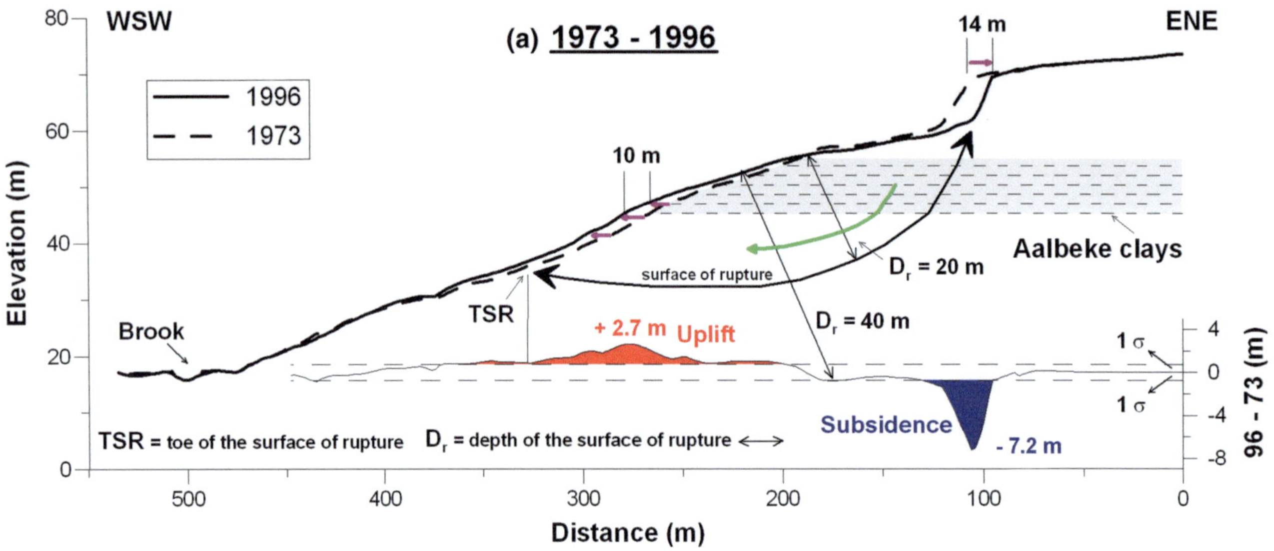

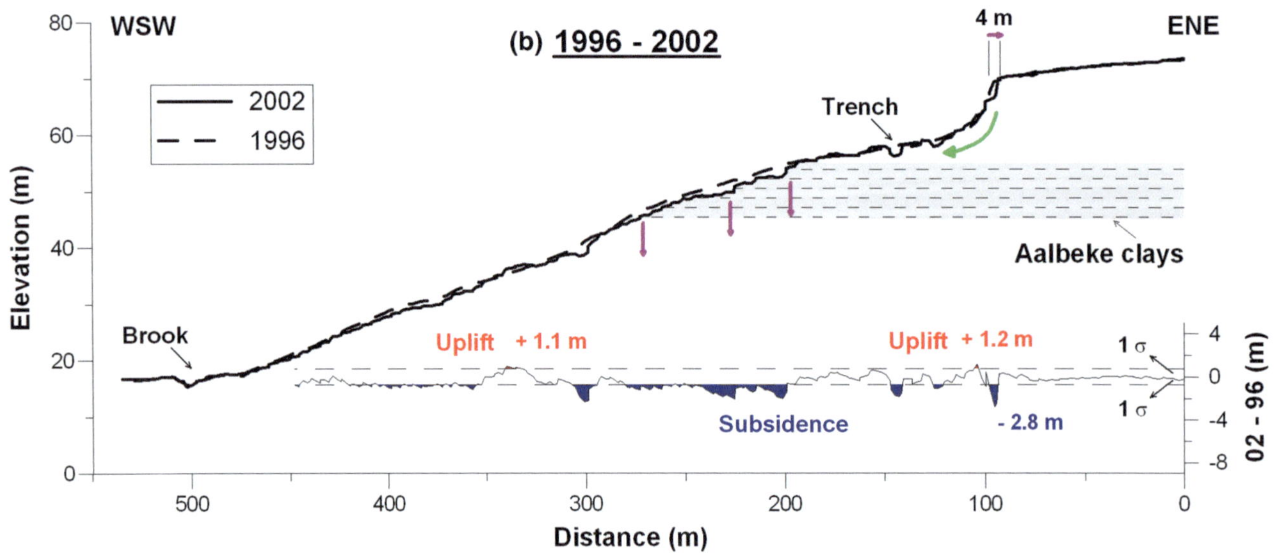

Fig. 20.9 DTM cross-sections of the Hekkebrugstraat landslide (landslide 1 in Fig. 20.4) compared with vertical displacement profiles (see location in Fig. 20.8). **a** Ground displacements between 1973 and 1996. **b** Ground displacements between 1996 and 2002. Lateral and vertical motions of the main scarps and ridges are represented by *horizontal* and *vertical purple arrows*, rotational motions by *curved green arrows*. In *blue* and *red*, associated profiles of the >1σ (0.75 m) vertical displacements obtained by DTM subtraction. Delimiting the collapsed and uplifted areas. Maximum and minimum vertical displacements are also provided (from Dewitte et al. 2008)

resistivity profile measurements carried out across a similar but dormant rotational earth slide revealed a surface of rupture at 15 m depth (Van Den Eeckhaut et al. 2007c), pretty similar to what is found for the HL reactivation, and also in contact with the sensitive clays of the Aalbeke Member (Fig. 20.9a). The Hekkebrugstraat landslide, as many others in the Flemish Ardennes and, in particular, the one studied by Van Den Eeckhaut et al. (2007c), is probably a multiple rotational slide, the original rupture surface of which was partly reused by the February 1995 reactivation. Assuming a 20 m depth of the surface of rupture, Dewitte et al. (2008) computed a landslide volume of $\sim 625{,}000\ m^3$.

The 1996–2002 comparison of topographic profiles shows a succession of positive and negative vertical variations of topography (Fig. 20.9b). Until 2000 and the construction of the two drainage ditches, several flow-like reactivations affected the zone of accumulation (Van Den Eeckhaut et al. 2007a). Starting from the top, the first subsidence/uplift succession is due to the reactivation of the main scarp in March 2002 (e.g. Fig. 20.7f). According to the model of Casson et al. (2005), this profile corresponds to a small rotation angle, suggesting that the corresponding reactivation did affect the landslide only shallowly. Consequently, the zone of accumulation mainly underwent compaction and, locally, the effects of anthropogenic material displacements.

20.2.2.4 Reactivations and Other Displacements

The HL example aimed at providing a better understanding of the mechanisms and controlling factors of ground movements within old landslides. Dewitte et al. (2009) carried out such an analysis for the other 12 landslides located in Fig. 20.4. They showed that the setting and the magnitude of the displacements vary spatially and temporally. Within the most active landslides, the movement pattern is typically rotational deep-seated failure with subsidence in the zone of depletion and uplift within the zone of accumulation. However, a detailed analysis of the vertical and horizontal movements allowed them to recognize various slope processes and the related controlling factors. A distinction can be made between reactivation and non-reactivation features. Large reactivations affect the landslides in depth by reusing pre-existing surfaces of rupture. They are associated with the largest amounts of subsidence and uplift. Smaller reactivations are more confined and affect only the landslide head. These findings suggest an overall retrogressive activity of multiple rotational earth slides. Other measured displacements were produced by shallow planar slide-earthflows occurring in the zone of accumulation, sometimes as a consequence of a large reactivation, and small shallow failures occurring independently of the presence of the old landslides.

Most slope movements are triggered by intense rainfall. Among the factors that control their spatial and temporal distribution, vegetation clearly reduces the potential for shallow failures in the forested parts, due to an increase in shear strength via root cohesion and soil moisture abstraction during evapotranspiration, and anthropogenic activities alter the probability of movement occurrence.

These results support the hypothesis that slope movements affect the old landslides more frequently than suggested by the archive analysis. They stress the need for considering old landslides in land management and mitigation strategies in hilly areas, even if they seem stabilized at first glance.

On a final note, it should be mentioned that zones affected by landslides in the Flemish Ardennes are also often affected by soil piping and soil pipe collapses (Verachtert et al. 2010), which induce mean soil loss rates of 2.4–4.6 ton/ha/year (Verachtert et al. 2011). It is hypothesized that poorly drained landslides may create favourable conditions for pipe development (Verachtert et al. 2012). No field evidence was found that pipe development enhanced landsliding, probably because the subsurface drainage discharge generated upslope of the landslide and flowing through the pipes is too small. Even when pipes become blocked, it is more likely that new pipes develop and new pipe collapses occur than that they trigger or reactivate landslides. A conceptual model, summarizing all elements that influence piping erosion in the Flemish Ardennes, including the role of landslides, is presented by Verachtert et al. (2012).

20.3 Other Landslides in Belgium

Beyond the LiDAR-data-based mapping of landslides in the Flemish part of Middle Belgium (Van den Eeckhaut et al. 2011), several studies identified ancient and modern landslide features in a few places of the Meso-Cenozoic cover of Wallonia and the Grand-Duchy of Luxemburg (Ozer et al. 1998; Demoulin et al. 2003; Demoulin and Chung 2007; see also Chap. 23) and the recent LiDAR DTM of the Walloon region confirms and extends unpublished observations by A. Ozer that landsliding also affected the Permian fill of the graben of Malmédy in northern Ardenne (Mreyen et al. 2016). While modern landslide activity or reactivation in Middle Belgium has been addressed in the previous section, we will focus here on the origin of ancient deep-seated landslides in the Pays de Herve, eastern Belgium. Indeed, the limited modern landsliding phenomena caused by combined climatic and anthropogenic triggers in the most landslide-prone areas of Belgium (Flemish Ardennes, Pays de Herve) are in sharp contrast with the much larger size of the mapped old landslides in these areas, raising the question of a possible different origin of the latter.

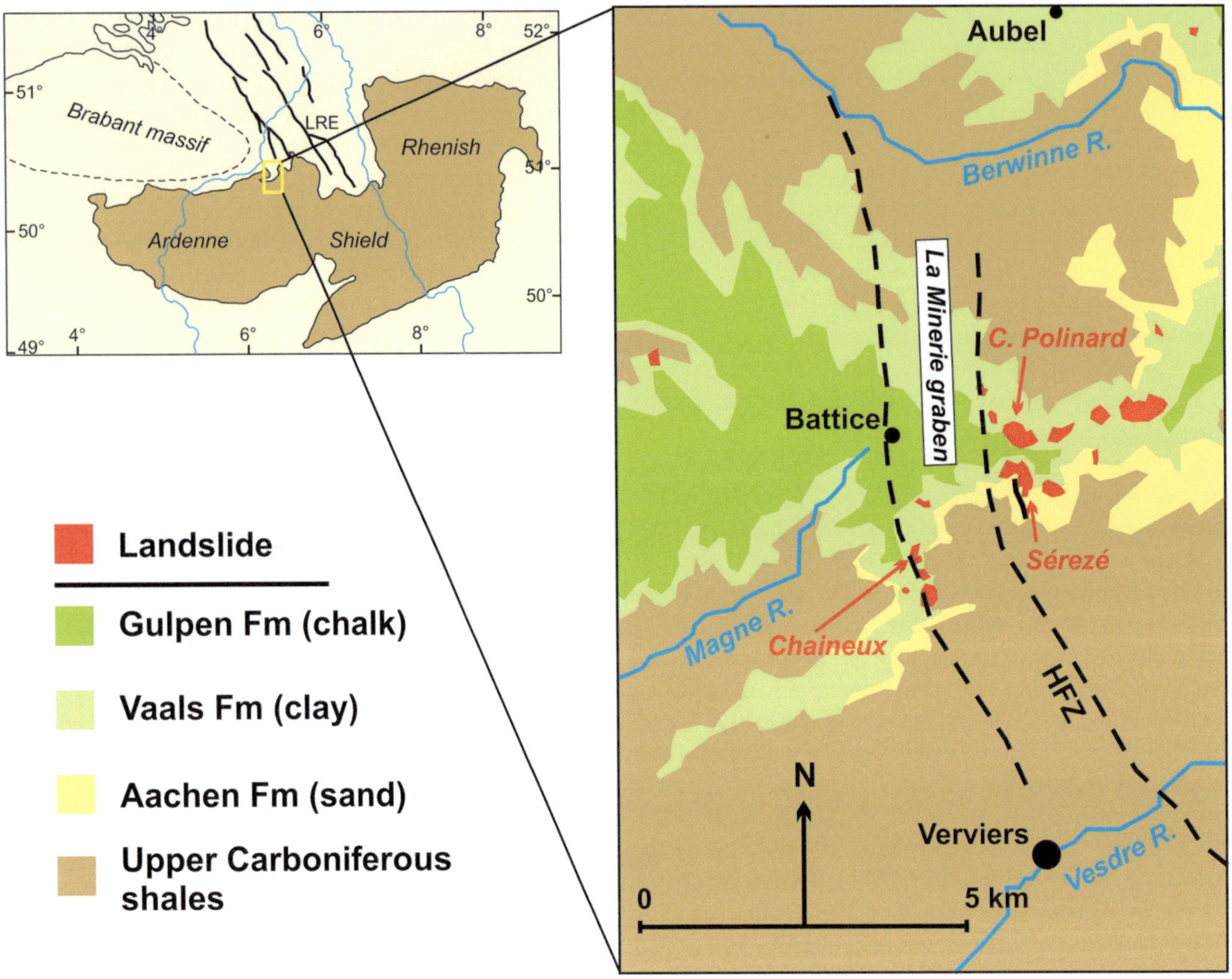

Fig. 20.10 Location map and simplified geology of the Pays de Herve in Belgium, showing also the landslide distribution within the study area, with the location of the Sérezé, Croix Polinard, and Chaineux landslides, described or mentioned in the text. *HFZ* Hockai fault zone

20.3.1 Geographical and Geological Settings

Geographically, the Pays de Herve is situated east of Liège and the Meuse valley, where it forms the easternmost part of the northern Ardennian foreland at altitudes between 200 and 350 m (Fig. 20.10). However, contrasting with the low plateaus of Middle Belgium, which are separated from the Ardenne itself by the intervening Condroz plateau and the Meuse valley, it is adjacent to the Ardennian margin and represents a kind of intermediate setting between Condroz, with which it shares its largely outcropping Paleozoic basement, and Hesbaye (or Haspengouw, in Flemish), similarly preserving an extended Cretaceous cover.

The geological and topographic contexts of landsliding in the Pays de Herve are fairly similar to those encountered in the Flemish Ardennes. The landslides affect poorly consolidated sediments of the Cretaceous cover that makes up the bulk of the ridges running across the Herve Plateau. These subhorizontal cover deposits rest on the Upper Carboniferous shales of the folded Paleozoic basement. The basal Aachen Formation is mainly comprised of fine sands interlayered with coarse sands and clays. Although the area where landslides are mapped corresponds to the westernmost extension of the formation, the sands are between 2–3 and 10 m thick. The overlying Vaals Formation displays glauconiferous clays and marls. Its maximum thickness is 25 m in the area affected by landslides. To the east, it gradually becomes sandier and contains indurated layers. Resting on the Vaals clays, the chalks of the Gulpen Formation are well preserved on top of most ridges of the Pays de Herve but are almost absent in the landslide zone. In their upper metres, they are generally weathered to clay-with-flints, frequently displaying solution pockets filled with remnants of Tertiary marine sands and Quaternary loess. The landslides are essentially developed within the clays of the Campanian Vaals Formation but in close connection with the presence of the underlying Aachen sands.

While the WSW-striking Variscan thrust faults of the basement have been inactive for a long time, several NNW-striking Variscan faults were episodically reactivated as transtensional or pure normal faults during the Cenozoic. In particular, such active faults, and especially the Ostende fault, which represents the local segment of the seismogenic Hockai fault zone (see Chap. 13), define the small graben of La Minerie. Parallel to the nearby bounding faults of the Roer Valley Graben, the major Hockai fault zone runs across the main landslide cluster (Fig. 20.10).

Topographically, the Pays de Herve is a moderately dissected tableland with a relief of 80–100 m and slope angles rarely exceeding 12°. The landslides occur in the upper part of the slopes, most of them along the main E-W divide separating the Berwinne and Vesdre drainage basins, and a few others along a secondary ridge extending S and then WSW within the Vesdre basin (Fig. 20.10).

With respect to hydrogeology, the main aquifer of the Pays de Herve lies within the Gulpen chalks, but numerous small aquifers also exist within the Aachen sands, where they are trapped between the clays resulting from the Mesozoic weathering of the Palaeozoic basement, clay layers inside the Aachen Formation and the Vaals clays. Another aquifer is also present in the underlying fissured Palaeozoic shales. As for the climatic context, present-day yearly rainfall is 900 mm, and the centennial daily rainfall is of 99 mm at the Thimister station, right in the centre of the landslide zone.

20.3.2 Anatomy of the Sérezé Landslide

Extending over 12 ha, the Sérezé landslide is one of the largest and best representative of a set of 15 large landslides concentrated in an area of ~5 × 3 km^2 in the south-central Pays de Herve (Fig. 20.10). Located on the southern slope of the main E–W ridge and facing to the southwest, it has a width of 550 m and extends on 250 m downslope of a 10 to 15-m-high head scarp. Though not recently activated, this double-arcuate scarp displays slopes between 17 and 22° that abruptly interrupt the gently sloping (by 8–9°) ridge flank in its upper part (Fig. 20.11). At its foot are several elongate tilted and rotated blocks with marked counterslopes that determine intervening depressions. As for the main part of the landslide body, it features a hummocky topography, especially in its lower half, and is terminated by a steep ~2-m-high toe scarp. The great length of this body in comparison with that of the upslope rotated blocks suggests that a significant translational slide component is also present, making the Sérezé landslide a compound feature in the sense of Hutchinson (1988).

An electrical resistivity tomography (ERT) of the landslide, complemented by a series of electrical and seismic refraction soundings, was carried out in order to image its overall internal structure (Demoulin et al. 2003). Aided by the local lithostratigraphy known from borehole data, the tomographic analysis of the rock column upslope of the slide shows the superposition of a 5 to 10-m-thick moderately resistive upper layer corresponding to the slightly weathered chalks of the Gulpen Formation, then a ~20-m-thick layer of very low resistivity, indistinctly including the Vaals clays and the underlying Aachen sands, and a basal layer of increasing resistivity representing the Paleozoic basement and featuring a possible steeply dipping fault (Fig. 20.12). Inside the landslide, deformation of the Vaals clays evidences at least two ~25-m-thick rotated blocks whose bottom surface attests the listric shape of the basal shear in the upper half of the disturbed volume. By contrast, the downslope half of the landslide appears as a 10-m-thick slope-parallel body extending over 100 m and slightly thinning in its distal extremity. This landslide structure fully confirms the multiple rotational plus translational characters also inferred from the surface topography of the landslide.

A 20-m-long trench dug in the landslide toe, parallel to the slide direction, provided a detailed view of how the displaced material moved to accumulate and form the toe scarp. A second trench in another nearby landslide, shown in Fig. 20.13 with the ^{14}C ages obtained for three charcoal pieces, revealed a similar structure. At Sérezé, the original ground surface is identified by a podzolic soil developed on colluviated silts of mainly eolian origin. Superposing a grey humic horizon, a bleached eluviated horizon and a variegated illuvial horizon, this soil is ~0.6-m-thick under the distal part of the toe scarp, then displays local bulging and thickening to >1 m before wedging and rapidly disappearing inward. The silts rest on 1 to 2-m-thick colluviated green glauconiferous Vaals clays that shallow auger borings showed lying over a thin (~0.3 m) bed of water-saturated purple sands of the Aachen Formation and Paleozoic schists weathered to black clay. Covering the soil and the colluviated material, the slipped masses are mainly made of light grey-brown clays with few flints and a subordinate silt component, a mixture of all sediments cropping out upslope. In contrast with the somewhat blurred contact surface between the undisturbed soil and the overlying slipped clays, an undulating sharp shear surface separates the latter in two distinct bodies, from which the lower one shows lumps of soil especially in its upper part. This exposed section suggests that, before leaving the soil on which it slid intact at the far end of its toe, the landslide, which had started several tens of metres upslope, partly scraped it, stacking broken soil

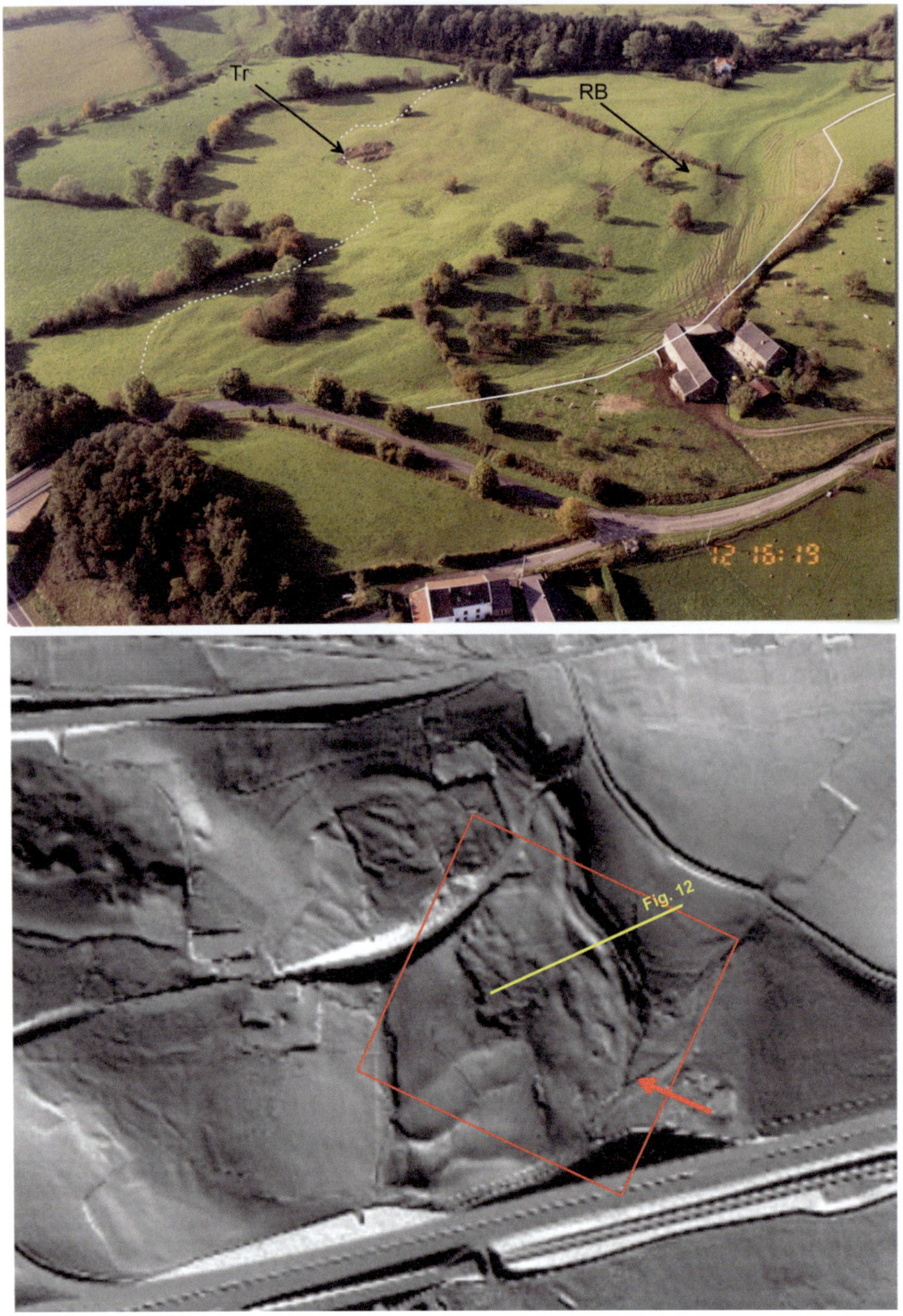

Fig. 20.11 The Sérezé landslide on the southern side of the main E–W ridge of the Pays de Herve (see location in Fig. 20.10). *Top* oblique aerial photograph of the southern part of the landslide. The *white line* denotes the *top* of the headscarp, the *dashed white line* follows the landslide toe. *RB* rotated block; *Tr* trench opened in the toe of the landslide. *Bottom* excerpt of the LiDAR DTM of Wallonia showing the general bilobate shape of the Sérezé landslide with two coalesced shear surfaces. The *red arrow* indicates the direction in which the photograph is looking, the *red rectangle* delimits the area covered by the photograph

pieces in a final bulge (Fig. 20.13). However, no positive field observation can resolve whether the two slipped bodies were emplaced simultaneously or during different slip episodes, except maybe that the lower one is, as a whole, strikingly more rubefied than the other.

Fortunately, numerous charcoal pieces present not only in the buried original soil but also in the soil patches at the top of the lower body allowed for ^{14}C dating. A similar situation was encountered in another trench cut across a second landslide, north of the ridge (Fig. 20.13). Multiple ^{14}C ages obtained in the two trenches have brought consistent insight into a common oldest time of landslide probable initiation between 50 and 250 AD, followed by several reactivation episodes in one or the other place around 400, 700, and 1250 AD (Fig. 20.14). In particular, these dates demonstrate that the upper body in the Sérezé landslide toe moved onto the lower one during a reactivation event in the thirteenth century.

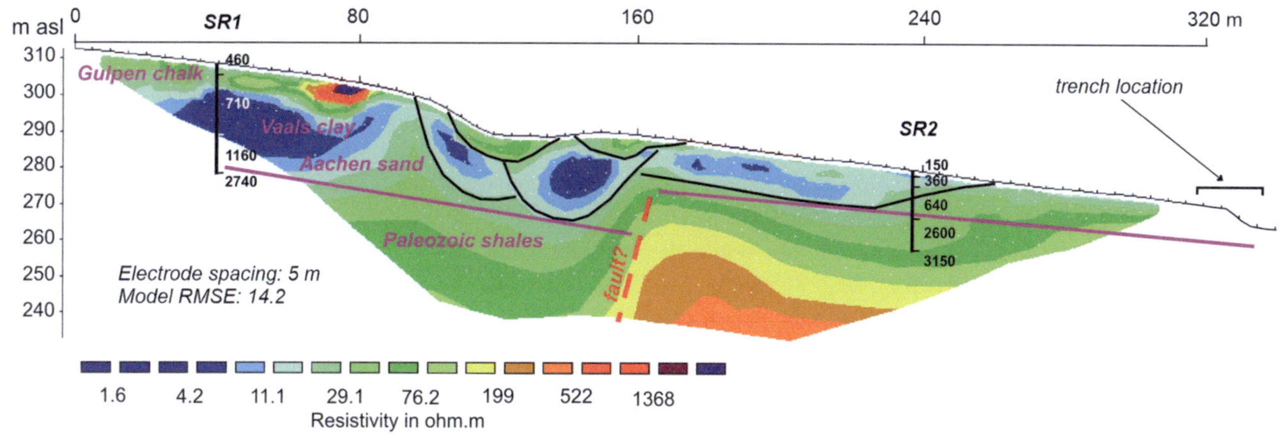

Fig. 20.12 Electrical resistivity tomography across the Sérezé landslide, with interpretation of the subsurface structures (see location on Fig. 20.11). *Underlined in black*, several rotated blocks are identified in the *upper part* of the landslide, whereas its *lower part* corresponds to a single translational unit. SR1 and SR2 are seismic refraction soundings with seismic velocity given in m/s. The *magenta lines* correspond approximately to the base of the Aachen Formation as it can be extrapolated upslope from its position in the trench, based on the observed resistivity and seismic velocity contrasts

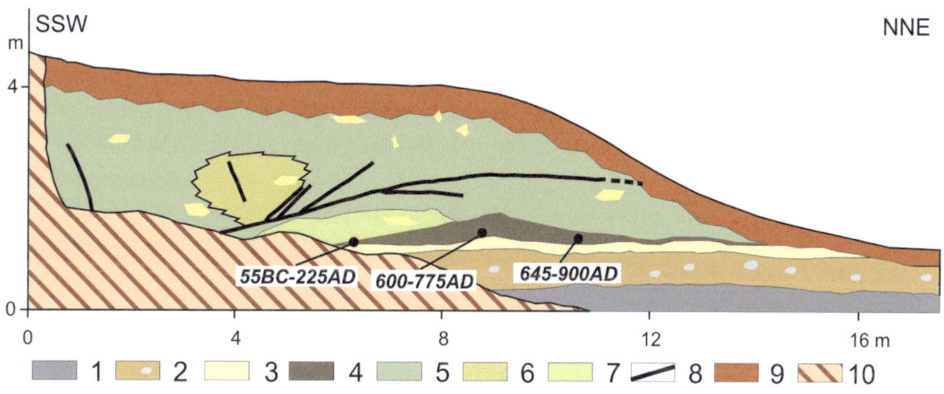

Fig. 20.13 Section across the toe of the Croix Polinard landslide on the north-looking side of the Battice ridge (see location on Fig. 20.10). *1* Paleozoic shales weathered to clay; *2* clay-with flint pre-landslide colluvium; *3*, *4* eluvial and humic horizons, respectively, of a podzolic soil developed in a silt of mainly eolian origin; note the bulging of the humic horizon in the centre of the section, with location of the ^{14}C-dated samples (calibrated 2σ ranges); *5*, *6*, *7* slipped masses, chiefly made of reworked Vaals clays with sparse sandstone blocks; *8* major shear surfaces; *9* present-day soil; *10* waste hiding the base of the section

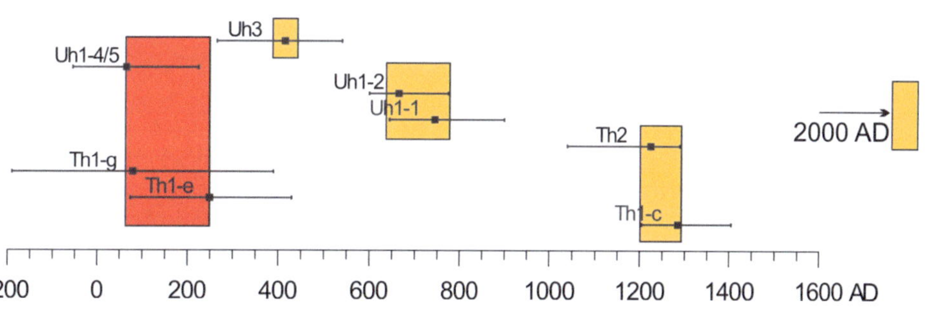

Fig. 20.14 ^{14}C date distribution (calibrated ages with 2σ range) of landslide motion in the Sérezé (Th) and Croix Polinard (Uh) landslides, and movements observed (1998 event) and monitored (1999–2005) in recent years. In *red*, assumed time period of landslide initiation; in *orange*, reactivation phases in either landslide

20.3.3 Mechanism and Causes of Landsliding in the Pays de Herve

20.3.3.1 Mechanism

The landslide surface topography, its ERT-revealed overall structure and the toe internal structure consistently stress the compound character of the Sérezé landslide and the translational movement that prevailed in its lower half. While all landslides affect mainly the Vaals clays, shallow geophysics and field data also show that the lower planar part of the basal shear is always located in the Aachen sands or at their interface with the overlying clays. Moreover, no landslide (except perhaps metre-scale shallow slides) is observed in the clays where the underlying Aachen sands are absent. Demoulin et al. (2003) identified therefore sand liquefaction as the primary process initiating the large deep-seated landslides of the Pays de Herve. Indeed, it has been long since laboratory measurements and field observation have shown that these sands are highly liquefiable (e.g. Graulich 1969). Sand liquefaction having initiated translational downslope motion of the Vaals clays, this in turn destabilized the upslope part of the hillside, causing successive rotational movements in a retrogressive way. A unique confirmation of this motion sequencing is provided by one landslide where only the initial mid-slope translation occurred, leaving a graben-like depression upslope of the displaced mass. This complex process can hardly be envisioned other than catastrophic, made of a sequence of fast movements caused by a sudden overwhelming trigger.

Slope stability analyses reached a similar conclusion (Demoulin et al. 2003). While the sand-clay interface dips ~0.8° NNW (the majority of landslides having occurred accordingly on N- or NW-facing slopes), they showed that, in static conditions, the safety factor (F_s, ratio of shear strength over stress along the shear surface) remains well in the stability domain for translational surfaces sloping by 4° even when the sediments are water-saturated. Likewise, in case of rotational motion along a listric slip surface, F_s comes close to instability only for highly improbable conditions of shear surface geometry and geotechnical properties of the sands, therefore suggesting that a very unusual trigger is required for impulsing large-scale landsliding.

20.3.3.2 Causes

Since no observation supports footslope sapping by river erosion or human interference as possible causes of initiation of the deep-seated landslides of the Pays de Herve around 50–250 AD, triggers of large-scale landsliding have been searched in either climatic circumstances or a seismic event (Demoulin et al. 2003). Key to the discussion was the contrast observed between the major event during which all large landslides occurred, moving tens to hundreds of metres, and the several reactivation episodes that seem in general to have affected only a part of the existing landslides and caused metre-scale displacements.

The best documented reactivation occurred in the autumn of 1998, when a more than centennial daily rainfall of 126 mm created a new 100-m-long, 1.5-m-high head scarp and caused 1 m downslope displacement of part of an existing landslide that underwent anthropogenic perturbation since three decades at Manaihan (Demoulin and Glade 2004) (Fig. 20.10). Beyond this, this event barely provoked a few small ground cracks within the scarps of some of the other old landslides in the area. Continued monitoring of the destabilized Manaihan landslide showed that the persisting human-induced high pore-water pressures within the landslide body involved annual displacements in the order of 0.15–0.25 m $year^{-1}$ taking place essentially at the end of the winter season, while extremely wet summer months (e.g. 288 mm in July 2000) were largely unable to cause ground motion (Demoulin 2006). In other words, such a rainfall event was unable even to significantly reactivate existing landslides in natural conditions. Likewise, two of the ^{14}C-dated reactivation episodes occurred around 700 and 1250 AD, two periods that various proxies (palynology, lake levels, testate amoebae) identify as especially wet, with cool wet winters, in NW Europe (Hendon et al. 2001; Holzhauser et al. 2005; Magny 2013), and they were nevertheless barely capable of inducing metre-scale displacements in some pre-existing landslides.

This strongly suggests that, although initiation of the large landslides occurred during another comparatively wet period, at least in west-central Europe (Holzhauser et al. 2005; Magny 2013), a period that in any case slightly postdates the wettest, highly landslide-prone, beginning of the Subatlantic (~2.8–2.0 ka cal BP: van Geel et al. 1996; Dikau and Schrott 1999; Chambers et al. 2010), a climatic trigger is certainly insufficient to explain such a major landsliding event. While this negative evidence is the first hint of a seismic origin of the landslides of the Pays de Herve, several positive arguments that together satisfy most criteria defined by Crozier (1992) to view landslides as earthquake-triggered convincingly support this interpretation too. These arguments are:

- All available data point to the dated landslides having been initiated simultaneously. Though not decisive and needing confirmation by dating further landslides, this is a necessary condition in order to consider a seismic trigger.
- The spatial distribution of the landslides cannot be explained on the basis of geological (rock type) or geomorphological (slope) conditions. By contrast, it is tightly linked to the Ostende fault, a segment of the

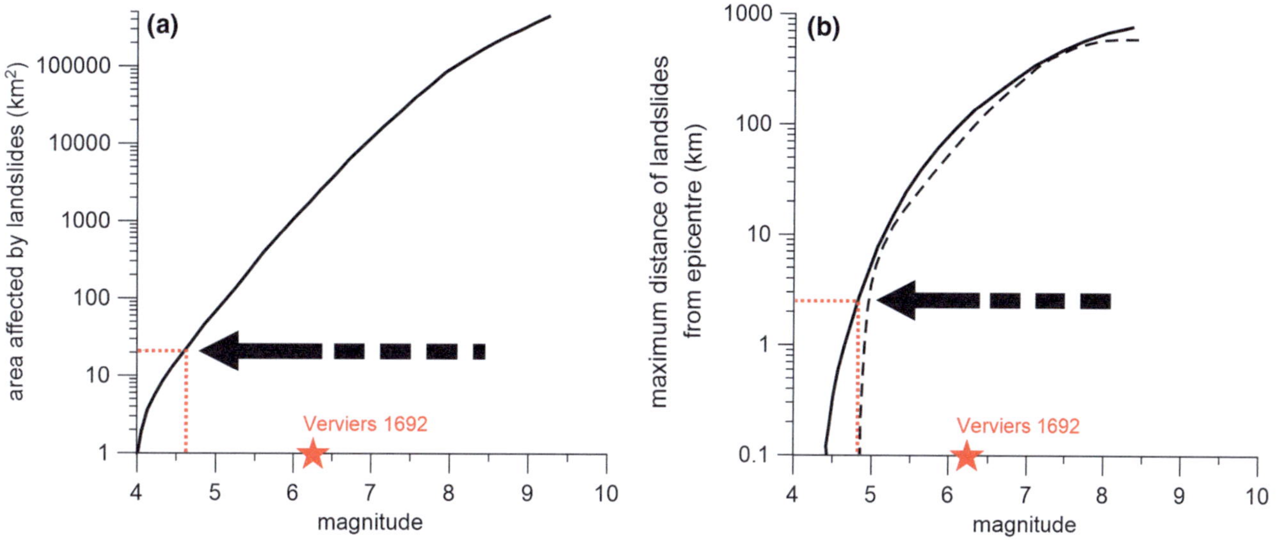

Fig. 20.15 Position of the Pays de Herve landsliding event within the empirical relationships between magnitude of the triggering earthquake and characteristics of the landslide distribution (**a** total area affected by landslides; **b** maximum distance of landslides from epicentre) in case of a seismic origin of the landslides (Keefer 1984). In **b**, the *solid curve* refers to 'coherent slides' and the *dashed curve* to 'lateral spreads' (and liquefaction-induced mass movements). Both relations point consistently to a possible M_L 4.5–5.0 earthquake (*red dotted lines*). The *red stars* locate the Verviers 1692 earthquake, which, based on the available data, seems to have been unable to cause landsliding in the study area

active Hockai fault zone in an area whose ongoing seismic activity is demonstrated (see Chap. 13) (Fig. 20.15).

- Stability analyses showed there is very little probability that the slopes be destabilized and cause extended landsliding in static conditions, even in the case of saturated sediments. A dynamic trigger was most likely, either seismic or possibly related to a catastrophic intense rainfall event at the hour-to-day scale.
- Liquefaction, which was the primary cause for setting the ground in motion, is in most cases associated with the dynamic oscillations of saturated sediments at the passage of seismic waves.
- The size characteristics of the landslides and their position in the upper part of gently sloping hillsides are typical of earthquake-triggered landslides.

Finally, if one admits the seismic origin of extended landsliding in the Pays de Herve in the period 50–250 AD, a concluding remark is that no exceptional earthquake was required to produce the observed set of deep-seated landslides. According to the empirical relationships between earthquake magnitude and different characteristics of the landslide distribution (maximum distance from fault rupture zone, total area affected by landsliding) established by Keefer (1984), and assuming that the seismic source was indeed the Ostende fault, an M_S 4.7–5.0 earthquake would have sufficed to cause their observed distribution. The same author also noted that the landslide type present in the Pays de Herve, i.e. coherent soil block slide, has been described in relation with earthquakes of magnitude M_S as low as 4.5. Such estimates are consistent with seismic hazard assessments for the area, which yield a return period of 447 years for an M_S 5.0 earthquake (Leynaud et al. 2000).

20.4 Conclusions

Through two representative examples, we provided insight into the landslide processes that are the most specific for Belgium. Even though the literature that focuses on these case studies is fairly abundant, we also showed that the understanding of these landslides is still suffering from several gaps, in particular when it comes to the timing of initiation and the factors at play at that moment. In the two cases, all available data suggest that major landsliding was caused by extreme events linked to former climatic and/or seismic conditions.

Although the landslides had an origin in natural environmental conditions, their recent dynamics, dominated by reactivation of ancient features, highlights the interactions with anthropogenic activities. This is demonstrated by using detailed qualitative analysis of information on landslide history and land-use changes provided by local inhabitants, aerial photographs and newspaper articles and internal reports, coupled with quantitative analysis.

In the regions west of Brussels (Fig. 20.2), we focused on landslides in Flanders. However, the geological and topographic settings typical of the Flemish Hills extend also in W Wallonia, and similar landslides are found there as well (Ozer et al. 1998). Likewise, despite an apparently much contrasted general landscape, the same underlying landslide-prone conditions also characterize the landsliding area of the Pays de Herve in eastern Belgium.

The information presented here is not meant to be static. The wide range of LiDAR derivative products has recently become a powerful tool in landslide research, particularly for landslide identification and landslide inventory mapping (Jaboyedoff et al. 2012; Guzzetti et al. 2012). Recently, the first analysis of new high-resolution LiDAR data for the hilly regions south of the Flemish hills revealed new old-landslide features (Akkermans 2014). Similarly, the LiDAR DTM of Wallonia, available since 2015, already allowed for the identification of previously unknown landslides in the Malmédy area (Mreyen et al. 2016), while even the Ardennian slopes cut in Lower Devonian shales are not devoid of local landsliding (R. Maquil, oral communic.).

Finally, landslides themselves are dynamic features responding rapidly to all kinds of change. The reader visiting Belgian landslides with this book in hand will have to remember that most results presented here are from studies carried out in the early 2000s and much may have changed since then.

References

Akkermans B (2014) Grondverschuivingen in de Henegauwse en Waals-Brabantse Leemstreek. Unpublished thesis, Master of Science, Geography, KU Leuven—VU Brussel

Camelbeeck T (1993) Introduction à l'étude des tremblements de terre. La séismicité en Belgique et dans le nord de la France. Annales de la Société Géologique du Nord, T. 2 (2ème série) 5–22

Casson B, Delacourt C, Allemand P (2005) Contribution of multi-temporal remote sensing images to characterize landslide slip surface—application to the La Clapiere landslide (France). Nat Hazards Earth Syst Sci 5:425–437

Chambers F, Daniell J, ACCROTELM Members (2010). Peatland archives of late-Holocene climate change in northern Europe. PAGES News 18:4–6

Crosta GB, Frattini P, Agliardi F (2013) Deep seated gravitational slope deformations in the European Alps. Tectonophysics 605:13–33. doi:10.1016/j.tecto.2013.04.028

Crozier M (1992) Determination of paleoseismicity from landslides. In: Bell D (ed) Landslides (Glissements de terrain), proceedings 6th international symposium, vol 2, pp 1173–1180

Cruden DM, Varnes DJ (1996) Landslide types and processes. In: Turner AK, Schuster RL (eds) Landslides: investigation and mitigation, pp 36–75. Special report 247. Transportation Research Board National Research Council, National Academy Press, Washington, DC

DEM of Flanders (2007) Digital elevation model of Flanders (obtained by laser scanning). http://www.gisvlaanderen.be/gis/diensten/geo-vlaanderen/?catid=82007

Demoulin A (2006) Monitoring and mapping landslide displacements: a combined DGPS-stereophotogrammetric approach for detailed short- and long-term rate estimates. Terra Nova 18:290–298

Demoulin A, Chung C-J (2007) Mapping landslide susceptibility from small datasets: a case study in the Pays de Herve (E Belgium). Geomorphology 89:391–404

Demoulin A, Glade T (2004) Recent landslide activity in Manaihan, East Belgium. Landslides 1:305–310. doi:10.1007/s10346-004-0035-z

Demoulin A, Pissart A, Schroeder C (2003) On the origin of late Quaternary palaeolandslides in the Liege (E Belgium) area. Int J Earth Sci 92:795–805

Dewitte O (2006) Kinematics of landslides in the Oudenaarde area and prediction of their reactivation: a probabilistic approach. Unpublished PhD thesis, University of Liege

Dewitte O, Demoulin A (2005) Morphometry and kinematics of landslides inferred from precise DTMs in West Belgium. Nat Hazards Earth Syst Sci 5:259–265

Dewitte O, Chung C-J, Demoulin A (2006) Reactivation hazard mapping for ancient landslides in West Belgium. Nat Hazards Earth Syst Sci 6:653–662

Dewitte O, Jasselette JC, Cornet Y, Van Den Eeckhaut M, Collignon A, Poesen J, Demoulin A (2008) Tracking landslide displacements by multi-temporal DTMs: a combined aerial stereophotogrammetric and LIDAR approach in western Belgium. Eng Geol 99:11–22

Dewitte O, Van Den Eeckhaut M, Poesen J, Demoulin A (2009) Decadal-scale analysis of ground movements in old landslides in western Belgium. Zeitschrift für Geomorphol 53:23–45

Dewitte O, Chung CJ, Cornet Y, Daoudi M, Demoulin A (2010) Combining spatial data in landslide reactivation susceptibility mapping: a likelihood ratio-based approach in W Belgium. Geomorphology 122:153–166

Dikau R, Schrott L (1999) The temporal stability and activity of landslides in Europe with respect to climatic change (TESLEC): main objectives and results. Geomorphology 30:1–12

Graulich JM (1969) La géologie de l'autoroute Liège-Aachen entre Herve et La Saute (Clermont). Professional paper service Géologique de Belgique 5, 61 p

Günther A, Van Den Eeckhaut M, Malet J-P, Reichenbach P, Hervás J (2014) Climate-physiographically differentiated Pan-European landslide susceptibility assessment using spatial multi-criteria evaluation and transnational landslide information. Geomorphology 224:69–85. doi:10.1016/j.geomorph.2014.07.011

Guzzetti F, Mondini AC, Cardinali M, Fiorucci F, Santangelo M, Chang KT (2012) Landslide inventory maps: new tools for an old problem. Earth Sci Rev 112:42–66. doi:10.1016/j.earscirev.2012.02.001

Hendon D, Charman D, Kent M (2001) Palaeohydrological records derived from testate amoebae analysis from peatlands in northern England: within-site variability, between-site comparability and palaeoclimatic implications. The Holocene 11:127–148

Holzhauser HP, Magny M, Zumbühl H (2005) Glacier and lake-level variations in west-central Europe over the last 3500 years. The Holocene 15:789–801

Hutchinson J (1988) Morphological and geotechnical parameters of landslides in relation to geology and hydrology, General Report. In: Bonnard C (ed) Landslides, Proceedings 5th international symposium on landslides, vol 1, pp 3–35

IAEG Commission on Landslides (1990) Suggested nomenclature for landslides. Bull Int Assoc Eng Geol 41:13–16

Jaboyedoff M, Oppikofer T, Abellán A, Derron M, Loye A, Metzger R, Pedrazzini A (2012) Use of LIDAR in landslide investigations: a review. Nat Hazards 61:5–28

Jacobs P, De Ceukelaire M, De Breuck W, De Moor G (1999) Text clarifying the Belgian geological Map, Flemish Region, Map Shet 29 Kortrijk, Map Scale 1/50,000 (in Dutch). Ministerie van Economische Zaken and Ministerie van de Vlaamse Gemeenschap

Keefer D (1984) Landslides caused by earthquakes. Geol Soc Am Bull 95:406–421

Leynaud D, Jongmans D, Teerlynck H, Camelbeeck T (2000) Seismic hazard assessment in Belgium. Geologica Belgica 3:67–86

Magny M (2013) Orbital, ice-sheet, and possible solar forcing of Holocene lake-level fluctuations in west-central Europe: a comment on Bleicher. The Holocene 23:1202–1212

Mreyen A, Havenith H, Fernandez-Steeger T (2016) Geophysical investigation of landslides and fault scarps in the Hockai Fault Zone, Belgium. Geophys Res Abstr 18:2350

Ost L, Van Den Eeckhaut M, Poesen J, Vanmaercke-Gottigny MC (2003) Characteristics and spatial distribution of large landslides in the Flemish Ardennes (Belgium). Zeitschrift für Geomorphol 47:329–350

Ozer A, Bonino E, Closson D, Ozer P, Petit F, Pissart A (1998) Etude des contraintes physiques et géotechniques du Mont-de-l'Enclus. Convention DGATLP-ULG. Laboratoire de Géomorphologie et de Télédétection. Rapport final, 87 p

Somville O (1939) Le tremblement de terre belge du 11 juin 1938. Annales de l'Observatoire Royal de Belgique, 3e série 2(5):273–287

Van Den Eeckhaut M, Poesen J, Verstraeten G, Vanacker V, Moeyersons J, Nyssen J, van Beek LPH (2005) The effectiveness of hillshade maps and expert knowledge in mapping old deep-seated landslides. Geomorphology 67:351–363

Van Den Eeckhaut M, Vanwalleghem T, Poesen J, Govers G, Verstraeten G, Vandekerckhove L (2006) Prediction of landslide susceptibility using rare events logistic regression: a case-study in the Flemish Ardennes (Belgium). Geomorphology 76:392–410. doi:10.1016/j.geomorph.2005.12.003

Van Den Eeckhaut M, Poesen J, Dewitte O, Demoulin A, De Bo H, Vanmaercke-Gottigny MC (2007a) Reactivation of old landslides: Lessons learned from a case-study in the Flemish Ardennes (Belgium). Soil Use Manag 23:200–211. doi:10.1111/j.1475-2743.2006.00079.x

Van Den Eeckhaut M, Poesen J, Verstraeten G, Vanacker V, Nyssen J, Moeyersons J, van Beek LPH, Vandekerckhove L (2007b) Use of LIDAR-derived images for mapping old landslides under forest. Earth Surf Process Land 32:754–769

Van Den Eeckhaut M, Verstraeten G, Poesen J (2007c) Morphology and internal structure of a dormant landslide in a hilly area: the Collinabos landslide (Belgium). Geomorphology 89:258–273

Van Den Eeckhaut M, Muys B, Van Loy K, Poesen J, Beeckman H (2009a) Evidence for repeated re-activation of old landslides under forest. Earth Surf Process Land 34:352–365. doi:10.1002/esp.1727

Van Den Eeckhaut M, Reichenbach P, Guzzetti F, Rossi M, Poesen J (2009b) Combined landslide inventory and susceptibility assessment based on different mapping units: an example from the Flemish Ardennes, Belgium. Nat Hazards Earth Syst Sci 9:507–521. doi:10.5194/nhess-9-507-2009

Van Den Eeckhaut M, Poesen J, Gullentops F, Vandekerckhove L, Hervás J (2011) Regional mapping and characterisation of old landslides in hilly regions using LiDAR-based imagery in Southern Flanders. Quat Res 75:721–733. doi:10.1016/j.yqres.2011.02.006

Van Den Eeckhaut M, Kerle N, Poesen J, Hervás J (2012) Geomorphology object-oriented identification of forested landslides with derivatives of single pulse LiDAR data. Geomorphology 173–174:30–42. doi:10.1016/j.geomorph.2012.05.024

Van Geel B, Buurman J, Waterbolk H (1996) Archaeological and paleoecological indications of an abrupt climate change in The Netherlands, and evidence for climatological teleconnections around—2650 BP. J Quat Sci 11:451–460

Verachtert E, Van Den Eeckhaut M, Poesen J, Deckers J (2010) Factors controlling the spatial distribution of soil piping erosion on loess-derived soils: a case study from central Belgium. Geomorphology 118:339–348

Verachtert E, Maetens W, Van Den Eeckhaut M, Poesen J, Deckers J (2011) Soil loss due to piping erosion. Earth Surf Proc Land 36:1715–1725

Verachtert E, Van Den Eeckhaut M, Poesen J, Deckers J (2012) Spatial interaction between collapsed pipes and landslides in hilly regions with loess-derived soils. Earth Surf Proc Land 38:826–835

Vranken L, Van Turnhout P, Van Den Eeckhaut M, Vandekerckhove L, Poesen J (2013) Economic valuation of landslide damage in hilly regions: a case study from. Sci Total Environ 447:323–336

Vranken L, Vantilt G, Van Den Eeckhaut M, Vandekerckhove L, Poesen J (2014) Landslide risk assessment in a densely populated hilly area. Landslides 12:787–798. doi:10.1007/s10346-014-0506-9

Spy and Scladina Caves: A Neandertal's Story

21

Stéphane Pirson, Michel Toussaint, Dominique Bonjean and Kévin Di Modica

Abstract

Following a brief presentation of some geologic and geomorphologic specificities of Belgium in connection with Neandertals (e.g. caves, flint), we briefly focus on the changing environment in Neandertal times. Some historiographic aspects are then developed, recalling that in the nineteenth century, several Belgian sites played a key role in the emerging sciences of prehistory and palaeoanthropology. The first ever found hominin anatomically distinct from modern man was unearthed in 1829–1830 in Engis Cave, near Liège, while the world famous discovery of Spy in 1886 helped to definitively demonstrate the existence of Neandertal man. General information about Neandertals introduces the presentation of the caves of Spy and Scladina. The recent complete reassessment of the archaeological, anthropological, and faunal collections at Spy exemplifies the interest of re-examining ancient collections. As for Scladina, it illustrates the potential of new excavations where modern investigation techniques are applied directly in the field. In these two sites, modern interdisciplinary researches led to prominent results on topics such as ancient DNA, dental development, age at death, palaeodiet, or archaeology, complemented in the case of Scladina by site formation processes and detailed palaeoenvironmental and chronostratigraphic reconstructions. Finally, a few aspects related to territorial exploitation by Neandertals are exposed, focusing on lithic resource management.

Keywords

Caves • Neandertal • Palaeoanthropology • Middle palaeolithic • Natural resources • Palaeoenvironment • Spy Cave • Scladina Cave

21.1 Introduction

Amongst the wide variety of landscapes and landforms dealt with in this book, caves offer a peculiar atmosphere, in addition to highly scenic views (see Chaps. 9 and 10). In Belgium, the presence of caves is essentially related to outcropping Palaeozoic limestones. Despite the relatively small size of the country, many hundreds of caves have been recognized so far, among which more than 300 have yielded prehistoric remains from Lower Palaeolithic to Neolithic periods (Toussaint and Pirson 2007). The nature of these remains is variable: abandoned tools (e.g. knapped flints,

S. Pirson (✉)
Service public de Wallonie, rue des Brigades d'Irlande 1, 5100 Jambes, Belgium
e-mail: stephane.pirson@spw.wallonie.be

M. Toussaint
rue de Coenhez 1, 4590 Ouffet, Belgium

D. Bonjean · K. Di Modica
Scladina Cave Archaeological Centre, rue Fond des Vaux 339d, 5300 Sclayn-Andenne, Belgium

A. Demoulin (ed.), *Landscapes and Landforms of Belgium and Luxembourg*, World Geomorphological Landscapes, DOI 10.1007/978-3-319-58239-9_21

bone retouchers), portable art, remnants of consumed fauna, burned material (wood and bone) or even, in some cases, human bones. The very good preservation of bones in calcareous environments makes caves highly interesting places to study ancient Prehistory, by comparison to open-air sites where osseous material, either human or animal, is most often lacking.

From a historiographic perspective, key discoveries from several Belgian prehistoric caves, along with those from a few other karstic regions in the southwest of France, in Croatia and in Rhineland, have played a fundamental role in the genesis and development of two disciplines during the nineteenth century: prehistory and palaeoanthropology (e.g. Toussaint 2001; Toussaint and Pirson 2006). The first ever found fossil hominin distinct from anatomically modern human (AMH) is Neandertal man, and the first Neandertal bone ever discovered was found in Belgium in 1829–1830, in the second cave of Engis, near Liège (Fig. 21.1). However, it was not recognized as different from AMH at that time. The very recognition of the existence of a man anatomically distinct from modern humans was connected with two other Belgian caves, representing two additional milestones in the history of palaeoanthropology: La Naulette in 1866 and Spy in 1886.

Belgium is still a key area to study Neandertals and their Middle Palaeolithic cultures. It is the area with the highest concentration of Neandertal human bones and teeth in Northwestern Europe, and one of the richest in Europe. So far, eight caves yielded Neandertal bones or teeth (Fig. 21.1). In addition, more than 440 open-air and karstic locations representing several sedimentary environments and landforms, but also diverse chronological and climatic contexts, have yielded Middle Palaeolithic artefacts (Di Modica 2010). This situation makes Belgium a favourable place to improve our understanding of the relationships between Neandertals and their environment, especially with respect to the exploitation of various natural resources (e.g. Di Modica et al. 2016). The high quality of organic matter preservation in caves from Belgium further enhances their interest, with exceptional DNA results (Orlando et al. 2006) as well as Neandertal diet studies focusing on N/C isotopic studies (Bocherens et al. 2001; Bocherens et al., in Rougier and Semal 2013; Wissing et al. 2016).

All these aspects are addressed in this chapter focusing on two iconic prehistoric cave sites associated with Neandertal studies: the Spy and Scladina Caves. While Spy was excavated by pioneers in the nineteenth century, Scladina exemplifies an ongoing modern excavation. Accessible to the public, they both have been recently the subject of interdisciplinary scientific monographs, showing the complementarity of the reassessment of ancient collections and careful examination of new field data (Rougier and Semal 2013; Toussaint and Bonjean 2014).

21.2 Setting

21.2.1 Geographical and Geological Setting

Despite its small size (30,528 km^2), Belgium presents highly contrasted geomorphological and geological settings (see Chaps. 1 and 2). Landforms may radically change over kilometres, while varied lithologies represent a large part of the geological timescale, from Early Palaeozoic to Holocene.

Broadly speaking, Belgium lies at the border of two worlds. The northern half of the country, encompassing Lower and Central Belgium, is mainly made of unconsolidated Cenozoic sediments and is dominated by plains and plateaus related to the European Plain. In contrast, the southern half (Upper Belgium) presents a completely different situation, with deep valleys incised in the uplifted basement of the Ardenne-Rhenish Massif mainly made of Devonian and Carboniferous rocks.

The Palaeozoïc substratum of the northern part of Upper Belgium includes Middle Devonian and Lower Carboniferous limestone formations in which many caves developed (Ek 1995; see also Chaps. 8 and 9) (Fig. 21.2). Caves are also known in Lower Carboniferous formations outcropping in the southern part of Central Belgium. The presence of caves, used as natural shelters, is important for Palaeolithic archaeology because they acted as sedimentary traps, allowing the preservation of Palaeolithic items as well as animal and human bones. They were also frequently used as burial places during Mesolithic and Neolithic times (Toussaint 2013).

In Central Belgium, a thick loess cover is preserved, whose detailed loess/palaeosol succession (Haesaerts 1984; Haesaerts et al. 2016) allows for accurate chronostratigraphic positioning of Palaeolithic artefacts (Pirson and Di Modica 2011; Di Modica et al. 2016). The presence of loess in the surroundings of caves also helps establish correlations between the two sedimentary environments, in particular benefitting chronostratigraphy and palaeoenvironmental reconstructions of cave entrance sedimentary sequences, and thus of the associated human occupations (Pirson et al. 2006; Pirson 2007; Pirson, in Pirson et al. 2011). Central Belgium also exposes Cretaceous chalk outcropping in the Mons Basin and the Liège-Tongres area (Fig. 21.2), from which good quality flint can be recovered.

Tephras originating in the Quaternary volcanic fields of nearby Eifel are also valuable as accurate chronostratigraphic markers in Palaeolithic sequences. So far, four tephras have been recognized in Belgium, all dating from the Late Pleistocene (Juvigné 1993; Pirson and Juvigné, in Pirson et al. 2011). Finally, we refer to Chap. 10 for a presentation of the evolution of the hydrographic network. A few fluvial terrace staircases are known, some of which

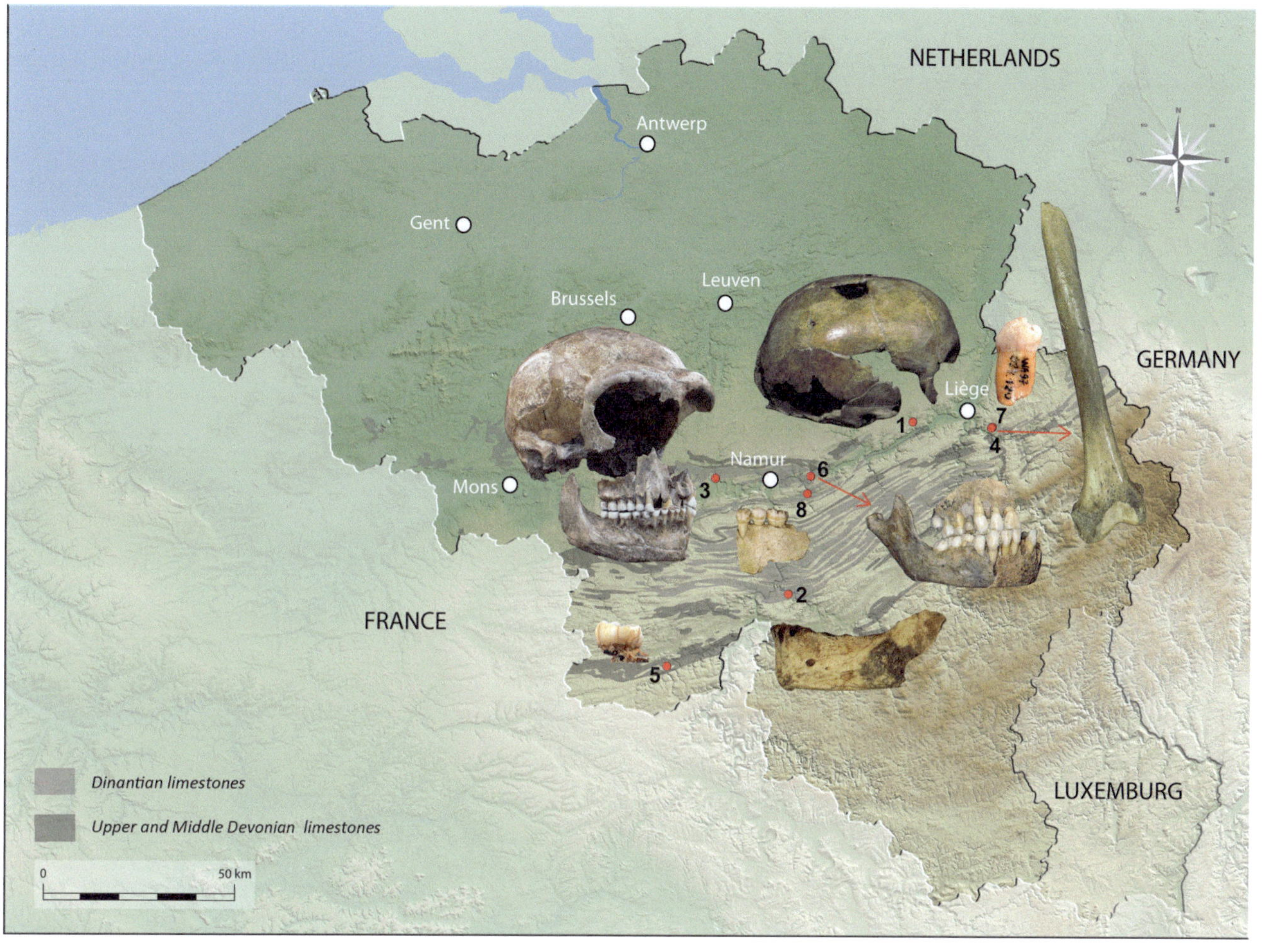

Fig. 21.1 Location of the caves which have yielded Neandertal remains in Belgium. *1* Engis (1829–1830); *2* La Naulette (1866); *3* Spy (1886); *4* Fonds-de-Forêt (1895); *5* Couvin—*Trou de l'Abîme* (1984); *6* Scladina (1993–2006 [1990]); *7* Walou (1999 [1997]); *8* Goyet (2004 [±1870]). [] = date of the discovery of the first fossil from the site but not recognized either as Neandertal (Goyet) or as a human remain (Scladina and Walou)

yielded more or less developed connection with Palaeolithic archaeology (e.g. Pirson et al. 2009; Rixhon et al. 2014).

21.2.2 Palaeoenvironmental Setting

Quaternary is characterized by major climate changes, which form the background of the story of humankind. Oxygen isotope data define 52 cycles over the 2.6 Ma of the Quaternary, each cycle composed of a cold and a more temperate period (Cohen and Gibbard 2011) (Fig. 21.3a). These cycles also show a great internal climatic instability, as illustrated by the last climatic cycle: during the last 130 ka, over 25 millennial climatic oscillations made of a cold stadial and a milder interstadial were recorded in the Greenland ice cores (Rasmussen et al. 2014) (Fig. 21.3b).

Neandertals and their Pre-Neandertal ancestors occupied NW Europe from ca. 500 to 40 ka. In the frame of the high climatic instability described above, these European populations underwent five temperate and five cold phases, and dozens of short-termed climatic oscillations over these 500 ka. While Neandertals at times had to face particularly cold phases, at other times they lived in temperate forests, enjoying climatic conditions similar to ours. Their development in Europe during this long, highly contrasted period underlines their remarkable adaptability to changing environments (Pirson and Toussaint 2011).

During maximum advances of the Scandinavian ice sheet, northern Europe was covered by ice down to the southern half of the Netherlands and was completely uninhabited (van Gijssel 2006; Ehlers et al. 2007a, b). A several 100-km-wide tract of land at the southern margin of the ice cap (thus including Belgium) was characterized by permafrost (Haesaerts and Van Vliet-Lanoë 1981; Haesaerts 1984). The mass of water frozen in the ice caps lowered the sea level by ~120 m and vast coastal areas emerged. During these

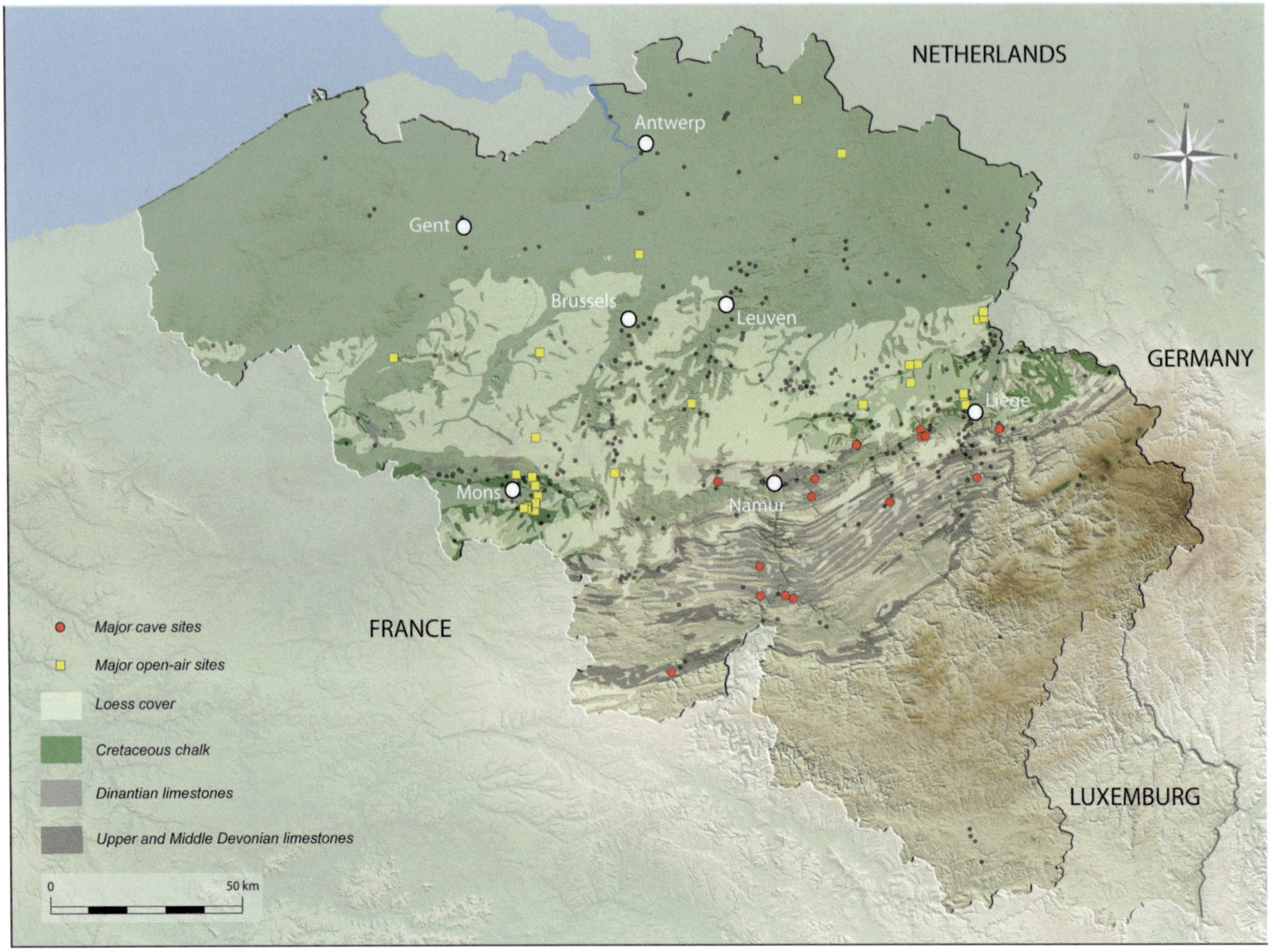

Fig. 21.2 Distribution of the Lower and Middle Palaeolithic find-sites in Belgium, superposed to a map of the Quaternary loess cover, the Cretaceous outcrops, and the Palaeozoic limestone outcrops (data after de Bethune 1954). *Black dots* denote the 442 find-sites (Di Modica 2010; Di Modica et al. 2016)

glacial phases, Neandertals shared an open steppe-like environment with animals such as the mammoth and woolly rhinoceros, reindeer, lion and hyena, and bear. Forest cover was scarce to inexistent, with trees surviving in protected valleys and spreading out only during short interstadials. In the coldest phases (e.g. between 70 and 60 ka), Neandertal populations were not present in our regions (Pirson and Di Modica 2011; Di Modica et al. 2016). With the return of interglacial conditions, the ice sheet melted and receded, allowing Neandertals to occupy a much wider territory. They then lived in temperate forests (e.g. Antoine et al. 2006; Gaudzinski-Windheuser et al. 2014) with fallow and red deers, boars and aurochs, along with lions and bears.

21.3 Milestones in Prehistoric Research in Belgium

Until the end of the eighteenth century, and even beyond, history of the Earth and history of Man had to comply with the Biblical account. According to the chronology established by Archbishop Ussher in the seventeenth century, the Earth had been created 4004 years BC, the human species itself was believed to be very young, and the dogma of the fixity of species still largely prevailed. Progress in the fields of palaeontology and stratigraphy, and the emerging notion of transformism would slowly modify these paradigms from the turn of the nineteenth century.

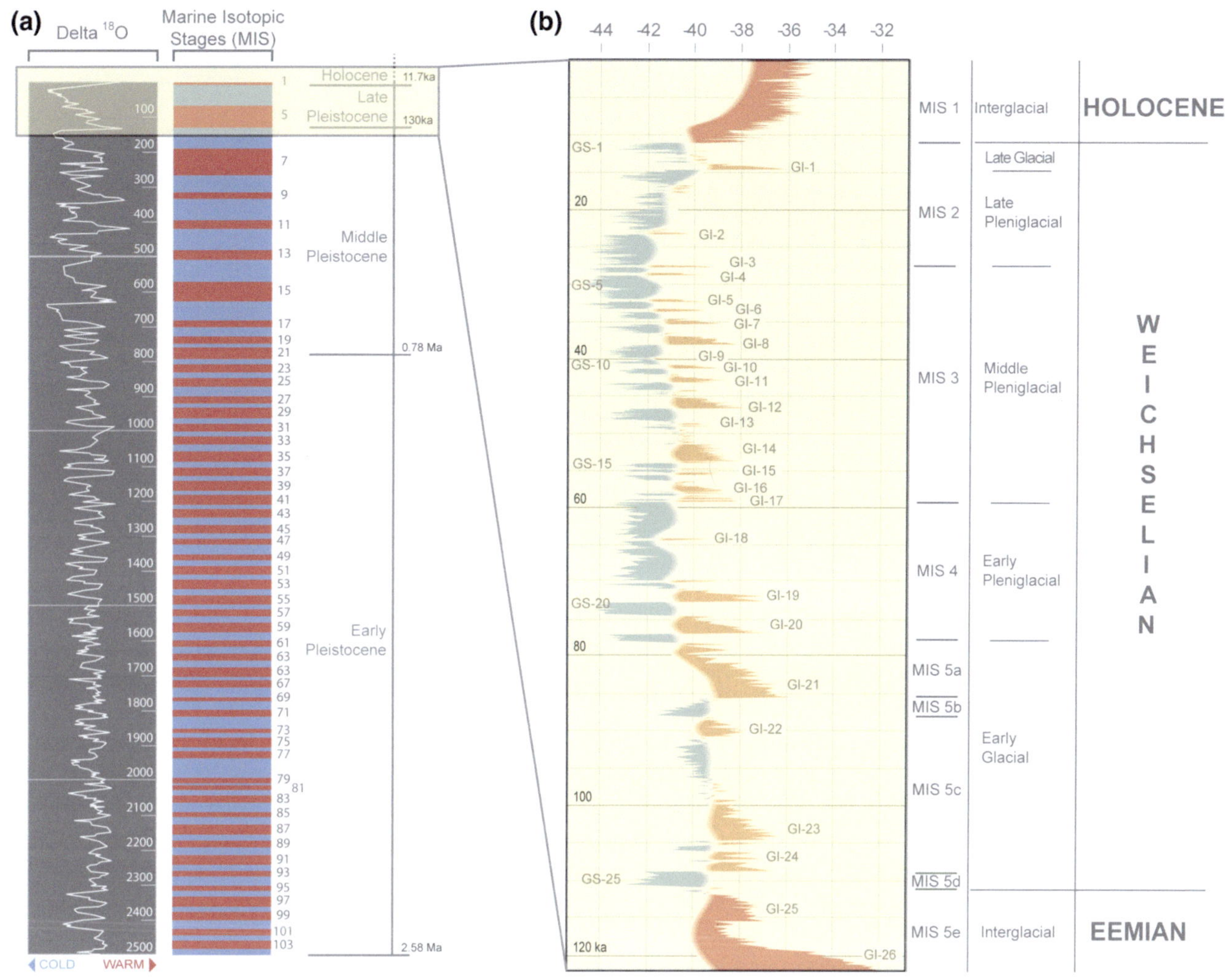

Fig. 21.3 **a** Oxygen isotope data obtained from tests of fossil benthic foraminifera in the deep-ocean core DSDP Site 607 in the Pacific (modified after Wilson et al. 2000), paralleled by the 52 reconstructed cold/warm climate cycles of the Quaternary (modified after Toussaint and Pirson 2013). **b** Zoom on the millennial climate changes of the Last Glacial period recorded through oxygen isotope data from Greenland ice cores. *GI* Greenland interstadials; *GS* Greenland stadials (modified after Toussaint and Pirson 2013; data from NorthGRIP Members 2004 and Andersen et al. 2007; © SPW)

In the beginning of the nineteenth century, Cuvier's ideas were dominant (e.g. Trinkaus and Shipman 1992). No convincing argument was yet put forward to demonstrate the antiquity of man. In this context, in the winter of 1829–30, Philippe-Charles Schmerling, a physician and nonconformist naturalist, discovered two skullcaps associated with woolly rhinoceros and mammoth bones in a cave near Engis, not far from the city of Liège (Fig. 21.4). These finds, as well as other investigations in regional caves, convinced him of the contemporaneousness of man and large extinct prehistoric mammals (Schmerling 1833–1834) and helped him realize that flint objects found in association with human and animal bones were actually man-made. Some of his contemporaries came to similar, though less well-argued conclusions, notably in France (e.g. Picard, Tournal) and in England (e.g. Frere, MacEnery). It however took the scientific community more than a quarter of a century to accept at last Schmerling's theories (Toussaint 1992, 2001). As two of his nineteenth-century's biographers noted, Schmerling "*was wrong because he was right too early*" (Cels and Jacques 1898). Several key events eventually led to the acceptance of Schmerling's ideas, namely (1) prehistoric archaeology had become more firmly established, thanks to Jacques Boucher de Perthes' dedication (e.g. Boucher de Perthes 1849); (2) Charles Darwin's *The Origin of Species* had introduced the concept of species evolution (Darwin 1859); and (3) the

Fig. 21.4 The skullcap of the Engis Neandertal child (after Schmerling 1833–1834) (*left*). Philippe-Charles Schmerling (1790–1836) (*right*)

famous Scottish geologist and father of the uniformitarianism Charles Lyell, who had not been convinced by Schmerling's theories when he visited him in 1833, had finally come round to Schmerling's ideas in his book "*The geological evidences of the antiquity of man*" (Lyell 1863).

Once the antiquity of man was accepted, the next step was demonstrating the existence of a man anatomically distinct from modern humans. This idea was only largely admitted between the 1860s and the 1880s. The first "fossil man" to be unearthed was a Neandertal man. And the first Neandertal fossil ever found was one of the two skullcaps found by Schmerling in Engis (Toussaint 1992, 2001). However, due to his juvenile characteristics, it took a century to demonstrate it was different from anatomically modern humans (Fraipont 1936). Meanwhile, in 1856, the Neandertal eponymous discovery was made in Feldhofer Cave near Düsseldorf, in the Düssel valley[1] (Germany), generating a passionate controversy. These fossils were considered by some to differ from modern humans (Fuhlrott 1859) but the idea was far from being unanimously acknowledged. They were considered as pathological by eminent anthropologists like Rudolph Virchow (1872), or even as a Cossack War victim (Mayer 1864). It is also to be noted that, discovered by workers during stone quarrying, these fossils were out of context.

Ten years later, a key discovery in another Belgian cave helped to move forward. In 1866, Édouard Dupont, who would become Director of the Natural History Museum in Brussels, unearthed a human mandible in La Naulette Cave, in the Lesse valley (Toussaint 1992, 2001) (Fig. 21.5). Contrary to the discovery in the Feldhofer Cave, the fossil's antiquity was confirmed by a precise stratigraphic context and its association with extinct large mammals (Dupont 1866). That is why, according to Paul Broca, the famous anthropologist from Paris, the mandible from La Naulette constitutes: "(…) *the first fact providing Darwinists with anatomical evidence*, (…) *the first link in the chain which, according to them, extends from man to the apes*" (Pruner-Bey et al. 1866). In the meantime, William King had proposed that the Feldhofer remains represented a distinct species, *Homo neanderthalensis* (King 1864), in other words, the Neandertals.

Two decades after the La Naulette discovery, a more receptive scientific community willingly accepted the "Man of Spy" skeletal remains unearthed in June and July 1886 in front of the entrance of the Betche aux Rotches Cave (Rougier and Semal 2013). Thanks to multidisciplinary researches conducted by geologist Maximin Lohest, archaeologist Marcel De Puydt and palaeontologist Julien Fraipont, this discovery led to the definitive acceptance of Neandertal man as a human type (Trinkaus and Shipman 1992; Toussaint 1992, 2001). More importantly, it provided the first definitive evidence for the existence of prehistoric fossil humans morphologically slightly different from anatomically modern humans (De Puydt and Lohest 1887; Fraipont and Lohest 1887). For the first time, a lot of bones were found pertaining to two Neandertal adult skeletons in a well-controlled stratigraphic context (following the time's standards), associated with "Mammoth age fauna" and Mousterian artefacts, the latter being absent in La Naulette.

[1]The valley was later renamed Neander valley (*tal* = valley in German) in honour of the seventeenth-century music composer and poet Joachim Neuman. The name "Neuman" ("new man" in English) was translated into Greek as Neander following a family custom largely spread at the time (Trinkaus and Shipman 1992).

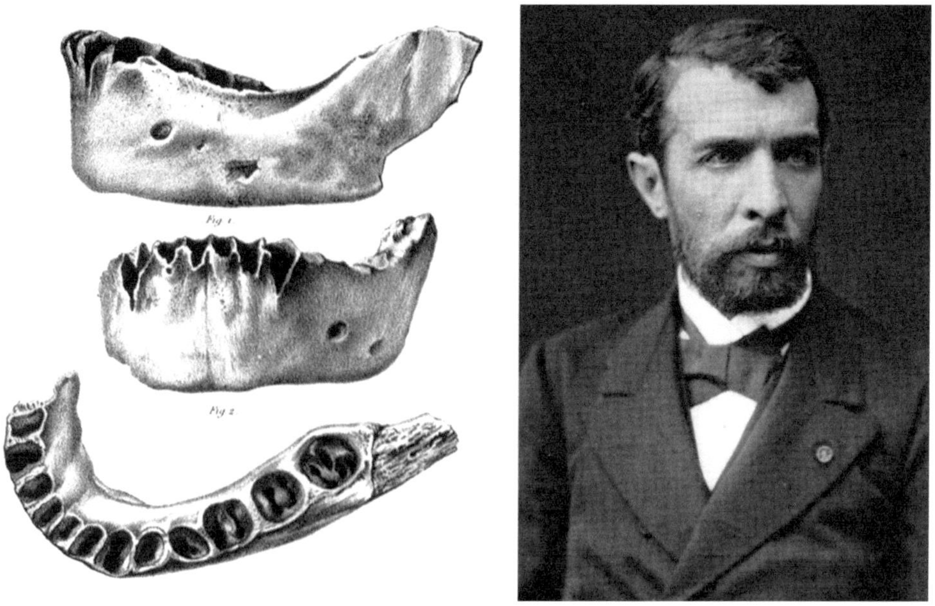

Fig. 21.5 The mandible found at La Naulette (after Dupont 1866) (*left*). Edouard Dupont (1841–1911) (*right*)

21.4 Who Were the Neandertals?

Neandertals are the typical European hominins. They descend from the first human groups that spread from Africa throughout Europe about 2 Ma ago, evolving gradually over hundreds of millennia. First Pre-Neandertals such as those from the famous sites of Tautavel, Eastern Pyrenees (France), and Swanscombe, Kent (England), appeared around 500 ka with emerging specific traits. Following a progressive anatomical specialization in mosaic pattern, these forerunners developed into recent Pre-Neandertals and then classic Neandertals, whose anatomical traits reached their utmost expression about 120,000 years ago. It is also in Europe that Neandertals disappeared some 40,000 years ago, while at the same time AMHs began to populate the continent (Trinkaus et al. 2003; Higham et al. 2014; Nigst et al. 2014; Hublin 2015).

Though very close from the anatomical point of view, Neandertal shows several characteristics that differentiate him from Modern Man (Vandermeersch and Maureille 2009; Pirson and Toussaint 2011) (Fig. 21.6). Some are archaic features shared with other ancient populations (plesiomorphic trait), such as a receding symphyseal region on the mandible. Others are specialized, i.e. supposedly typical Neandertal (apomorphic trait), such as a retromolar space or a mandibular condylar process projecting partly outside of the branch (Fig. 21.6). Neandertal's average cranial capacity is around 1450 cc, with maximum values over 1800 cc. It is similar, though slightly superior, to Modern man (average 1350 cc).

Neandertal's status has been very controversial since the eponymous discovery in 1856, either considered as a subspecies of Modern Man (*Homo sapiens neanderthalensis*) or as a distinct species (*Homo neanderthalensis*). Recent advances linked with DNA sequencing have renewed the debate, indicating that Neandertals and modern men did interbreed several times (Green et al. 2010; Fu et al. 2015). As shown recently, the proportion of Neandertal DNA in Anatomically Modern Humans decreased from 3 to 6% during the Upper Palaeolithic to ~2% in present-day Eurasian populations (Fu et al. 2016).

Neandertals were nomadic hunters just like all Palaeolithic people. According to isotopic studies, they were highly carnivorous (e.g. Bocherens et al. 2001). Depending on the advance or retreat of the ice sheets, they occupied practically all Europe and migrated as well towards the Near East and Siberia (Vandermeersch and Maureille 2009; Pirson and Toussaint 2011). From a cultural point of view, while the first Pre-Neandertals lived at the end of the Lower Palaeolithic, most of the lineage is linked with the Middle Palaeolithic, from ~300 to ~40 ka (Fig. 21.7).

Far from the cliché of "Cave Man", prehistoric hunters–gatherers, and thus Neandertals, did not live only in cave entrances. They settled in small groups both in cave entrances or rock shelters and in open-air campsites (Vandermeersch and Maureille 2009; Pirson and Toussaint 2011;

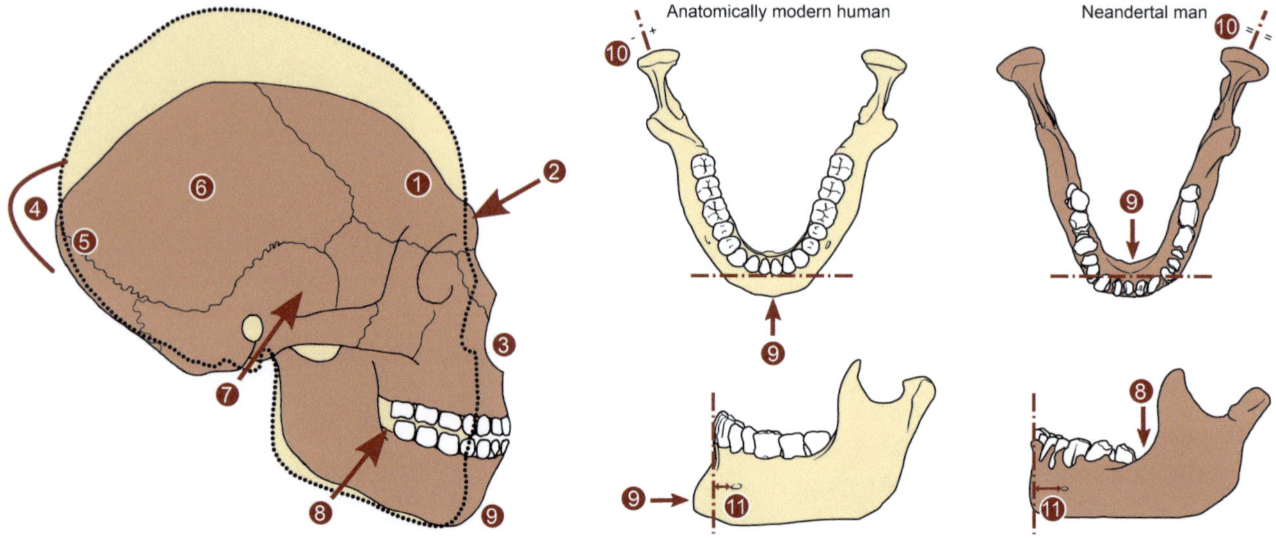

Fig. 21.6 Comparison of the skull and the lower jaw of the anatomically modern human (in *yellow*) and the Neandertal man (in *brown*) (modified after Pirson and Toussaint 2011). The most characteristic features of Neandertals are highlighted as follows: *1* Receding forehead; *2* Supraorbital torus; *3* Large nasal cavity; *4* Acute angle between *planum nuchale* and *planum occipitale*; *5* Occipital bun and suprainiac fossa; *6* Rectangular parietal bone; *7* Low squamous portion of the temporal; *8* Retromolar space (gap behind M3); *9* Little or no real chin, and *planum alveolare* at the posterior face of the symphysis; *10* Mandibular condylar process projecting partly outside of the branch (in modern human, the projection is inside); *11* Posterior position of mental foramen generally below M1 (but below P2 in modern human). *Computer graphics* S. Lambermont (AWEM) and J.-F. Lemaire (SPW); © SPW

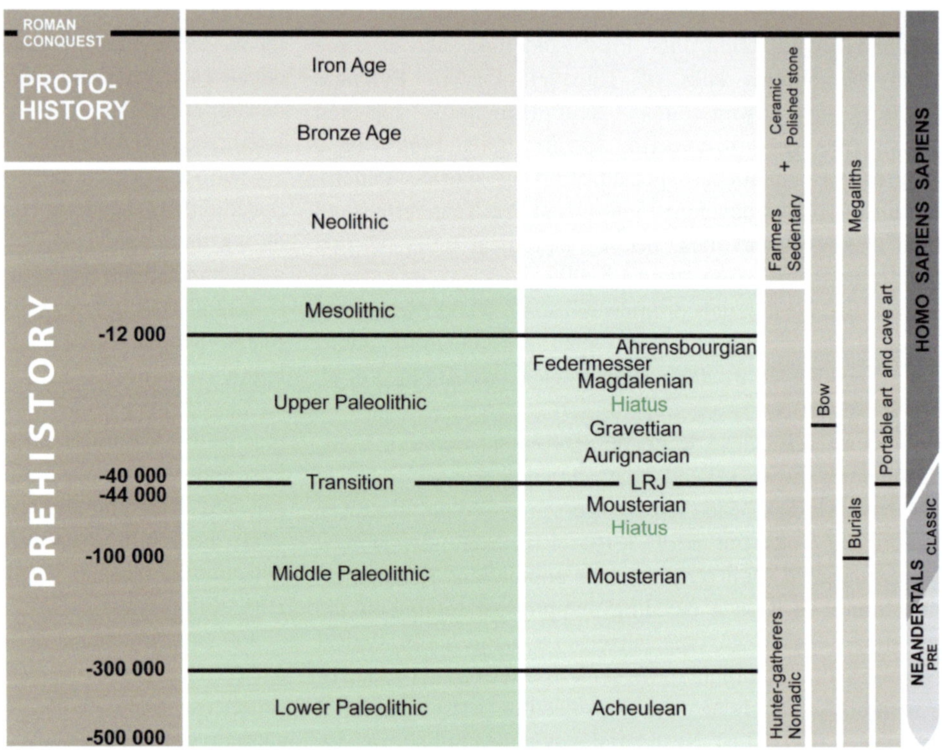

Fig. 21.7 Main divisions of prehistory in Belgium (from Pirson and Toussaint 2011; © SPW)

Toussaint and Pirson 2013). A lot of settlements are known from areas deprived of caves, such as Northern France. Even in regions like Belgium where caves are numerous, most finding spots are in open-air situations. Out of the 442 locations where artefacts related to Neandertals or their direct ancestors have been recovered, only 47 are related to caves (Di Modica 2010) (Fig. 21.2). Among these 442 locations, 48, from which 17 are caves, are major sites because they include human remains (8 caves) or a large quantity of artefacts, or yielded important historic contributions or contextual data linked with modern excavations.

Neandertals are the first known hominin to bury their dead, with oldest burials dated ~110 to 90 ka in the Near East. The existence of burials testifies that Neandertal possessed some kind of spirituality. Some forty Neandertal tombs have been found so far (Vandermeersch and Maureille 2009), all in caves or rock shelters. Women and men as well as children of various ages were buried, including foetuses. In Belgium, various clues suggest that at least one of the two adult Neandertal skeletons found in the Spy Cave was buried in a tomb. Nothing however allows for similar conclusions in the other Belgian sites where Neandertal remains have been found (Engis, La Naulette, Goyet, Fonds-de-Forêt, Couvin, Walou, and Scladina) (Toussaint et al. 2001). A recent reassessment of the faunal collections from the Goyet Cave led to identify Neandertal bones showing possible traces of cannibalism (Rougier et al. 2016), a practice long suspected at other Neandertal sites in Europe such as Krapina (Croatia) and, more recently, Moula-Guercy (Ardèche, France).

21.5 Spy Cave: A Key Historical Discovery

In 1886, the discovery of two partial skeletons of Neandertals in a well-documented stratigraphic context at Spy largely contributed to the definitive acceptance by the scientific community of the existence of fossil humans with a more archaic morphology than modern populations. These finds, coupled with the remarkable monograph published in 1887 by the discoverers, made Spy a reference site for European palaeoanthropology. It is also a major site for Palaeolithic archaeology. Tens of thousands of lithic artefacts and animal skeletal fragments have been unearthed at the site during several archaeological campaigns. Between 2004 and 2007, a careful re-examination of the huge Spy collections led to the reassessment of the Spy anthropological and archaeological material, while historical and geoarchaeological contextualization of previous researches complemented the revision of the material. This further highlighted the major importance of the site at the European scale. A first monograph presenting these results has been published, dealing with historiography, stratigraphy, archaeology, and faunal material, as well as biogeochemistry and ^{14}C dating (Rougier and Semal 2013). A second volume focusing on palaeoanthropology should soon follow.

The highly scenic site of Spy Cave can be visited today. It is set in the middle of a nature reserve. A new Interpretation Centre opened in December 2011, the EHoS (*Espace de l'Homme de Spy*; http://www.hommedespy.be/), where the anatomy, culture, and environment of Neandertals are presented and discussed.

21.5.1 Site Location and Description

Spy Cave (Fig. 21.8) is located in the village of Spy, ~12 km west of Namur (Figs. 21.1 and 21.2). This famous site, also known as the cave of Betche aux Rotches, is situated on the left bank of the Orneau River, a tributary of the Sambre (Fig. 21.8). Cut in a Visean limestone massif, its main entrance opens to the southwest, some 18 m above the present level of the Orneau, below an extended plateau (Pirson et al. 2013a). The cave is relatively small. The cave porch opens onto the principal chamber, which leads towards short narrow galleries. Secondary entrances are known on the slope.

The Orneau River flows in southern Central Belgium from Gembloux in the north to Jemeppe-sur-Sambre in the south, across the southern flank of the Lower Palaeozoic Brabant Massif and its Upper Palaeozoic cover. The valley exposes one of the most complete outcrops in the Palaeozoic series of the area, the cave itself being developed in the top of the massive and brecciated part of the Lives Formation. In its downstream course, the Orneau crosses the Upper Carboniferous Coal Measures. Lutetian marine sands of the Bruxelles Formation still cover the plateau in the vicinity of the cave. These sands are locally cemented in glazed sandstones ("Fayat sandstones") that were used by Neolithic people in the nearby area of Velaine-sur-Sambre as polishing stones and as a standing stone (Rutot 1887; Pirson et al. 2003).

The Quaternary evolution of the Orneau valley is poorly known (Pirson et al. 2013a). In front of Spy Cave, the oversized valley of the modern stream is interpreted as a cut-off meander of the Sambre now partly used by the Orneau (Fichefet 1933). The timing of this meander cut-off is so far unknown. Fichefet (1933) reports the presence of pebbles pertaining to several alluvial terraces, presumably of the Sambre, in the lower course of the present Orneau Valley. One such terrace lies at 118 m asl, i.e. ~20 m above the valley bottom, and is found on both sides of the valley, and notably near the cave. According to this author, before the meander was cut off, the Orneau-Sambre confluence was situated somewhere near the Spy Cave.

Fig. 21.8 **a** View of Spy Cave's entrance (*Photo* G. Focant; © SPW). **b** Map of Spy Cave with position of ancient excavations and approximate location of the two adult skeletons (Spy1 and Spy2) discovered in 1886 (from Toussaint 2001, modified after de Loë and Rahir 1911; © SPW). **c** Location of Spy Cave in the Orneau valley (Lidar-derived image; © SPW). The position of the cut-off meander of the Sambre (*dashed blue line*) is from Fichefet (1933)

21.5.2 History of Excavations

The site of Spy has been explored at multiple occasions (see Semal et al., in Rougier and Semal 2013, for a recent overview of the excavation history), first by amateurs. Maximin Lohest, geologist, and Marcel De Puydt, archaeologist, began the first scientific excavation campaign in August 1885 (Fig. 21.9). In June and July 1886, their researches were "*crowned with undreamed success*" (De Puydt and Lohest 1887: 208), human remains having been unearthed in their stratigraphic context and in association with extinct animal remains and knapped flints (De Puydt and Lohest 1886). Julien Fraipont, palaeontologist, was entrusted with the study of the animal and human bones

Fig. 21.9 The three discoverers of the "Men of Spy", from *left* to *right*: Max Lohest (1857–1926), Marcel De Puydt (1855–1940), and Julien Fraipont (1857–1910) (*Drawing* B. Clarys; © SPW)

(Fraipont and Lohest 1887). Given the importance of the 1885–1886 discoveries, several excavation campaigns were subsequently undertaken until the early 1980s in the cave, on the terraced ground in front of it, and on the slope deposits down to, and including, the alluvial plain (Semal et al., in Rougier and Semal 2013). They were conducted by Baron Alfred de Loë and Edmond Rahir between 1903 and 1909, Joseph Hamal-Nandrin from 1927 to 1933, François Twiesselmann from 1948 to 1954, and finally under the direction of Michel Dewez from 1979 to 1981. Many amateur collectors also explored the site, mainly between 1955 and 1985.

21.5.3 Stratigraphy

Relevant lithostratigraphic information is available only from the De Puydt and Lohest and de Loë and Rahir excavations (Pirson et al. 2013b). Although general information about the main lithostratigraphic units can be found in the stratigraphic descriptions they published respectively, a detailed correlation between their sequences is impossible. Descriptions of the stratigraphy by the same team may even vary in successive publications, including discrepancies in the distribution of the archaeological material in the sedimentary units (Pirson et al. 2013b; Di Modica et al., in Rougier and Semal 2013). This is notably the case with the work of De Puydt and Lohest, during which most of the in situ Neandertal remains were recovered. Their most complete stratigraphic description lists six layers (A–F) grouped into three "fauna-bearing levels" (FBLs). No true section drawing exists for these 1885–1886 excavations. The only available stratigraphic documents are several versions of a sketch portraying the same generalized section of the terrace deposits made after the discovery of the Neandertal remains (Fig. 21.10). Unravelling the stratigraphic context of the archaeological and anthropological remains in these conditions is rather elusive.

21.5.4 Context of the Fossil Discovery

De Puydt and Lohest's (1886) excavation was the most interesting for palaeoanthropology. Dozens of human skeletal bones were discovered in the "third fauna-bearing level" before the cave mouth. Later, Twiesselmann found some more Neandertal bone fragments and teeth in the De Puydt and Lohest's excavation backfill (Semal et al., in Rougier and Semal 2013). An isolated vertebra was discovered by an amateur on the slope leading from the cave to the Orneau. Finally, the recent reassessment of the Spy collections led to the discovery of many unpublished human remains among the unsorted fauna (Rougier et al. 2004; Semal et al. 2009; Rougier and Semal 2013).

Information concerning De Puydt and Lohest's excavation conditions is scarce and sometimes contradictory (Semal et al., in Rougier and Semal 2013). A trench and timbered galleries were dug in 1885, and both excavation techniques were probably repeated in 1886. As Lohest would later admit, these methods were far from exemplary

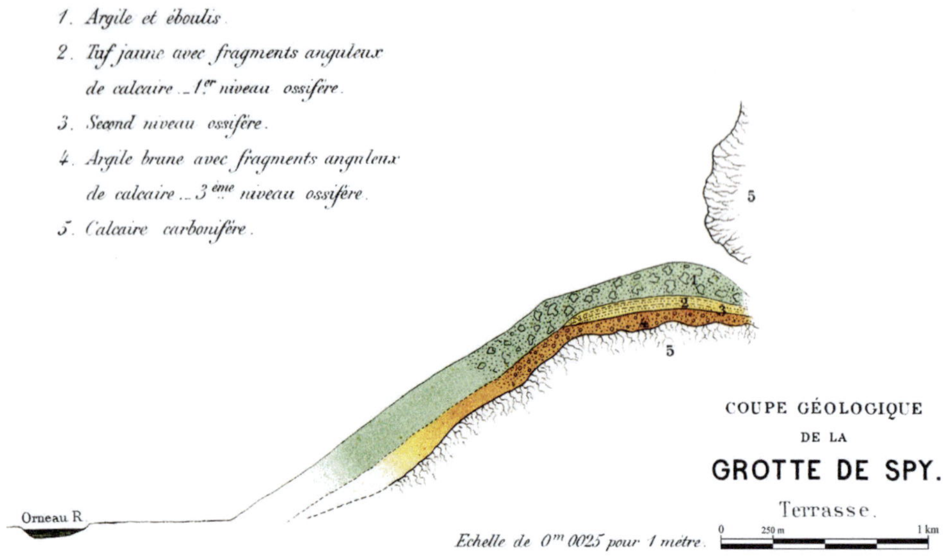

Fig. 21.10 Sketch of the Spy Cave terrace profile recorded during the 1885–1886 excavations (from De Puydt and Lohest 1887, planche I)

even for the period, especially the timbered galleries (Lohest et al. 1925). They resulted in the lack of distribution plans of the Neandertal remains and in the mixing of various archaeological occupations demonstrated by later typological studies (Ulrix-Closset 1975; Otte 1979; Pirson et al. 2013b).

21.5.5 The Neandertal Fossils

The monograph published the year after the discovery of the Spy fossils by Fraipont and Lohest (1887) is the first-ever comprehensive analysis on Neandertal skeletons. In this study, the Spy hominin skeletons were thought to belong to two adults, known as Spy I and Spy II. Following the 2004–2007 reassessments of the nineteenth- and twentieth-century collections, the number of human remains has increased from 89 numbered bones and teeth to more than 1800 pieces representing all skeletal elements (Semal et al. 2009). These include at least 24 new Neandertal remains belonging either to the two already known specimens (Spy I and II) or to one additional immature individual recognized through two mandibular fragments and four teeth (Spy VI; Crevecoeur et al. 2010). However, most of the new skeletal remains come from a Neolithic collective burial.

Spy I and Spy II are each represented by a skullcap, maxillar, and mandible elements, and by a series of often incomplete postcranial remains belonging equally to upper and lower members. One of the skulls ($\sim$1300 cc) belonged to a woman, the other ($\sim$1500 cc) to a rather young man. Next to many plesiomorphic characters (e.g. low elongated cranial vault, receding mandibular symphysis), the Spy remains show derived traits specific to Neandertals (e.g. retromolar space). The postcranial bones, like those of other Neandertals, have robust epiphyses compared to modern man.

The diet of Spy I and II was investigated through C and N isotopic analyses of a scapula sample (Bocherens et al. 2001). The isotopic composition of the sample suggests a diet high in protein, provided by large herbivores. This highly carnivorous Neandertal diet included also plant foods (nenuphar), as shown by microfossils extracted from dental calculus (Henry et al. 2011).

To all appearances, at least one adult skeleton was the object of intentional burial, even though the early excavators denied it. Spy I was found "lying on its side, one hand laid against its lower jaw" (Fraipont and Lohest 1887), which strongly resembles what has since often been observed in Neandertal burials (Vandermeersch and Maureille 2009).

21.5.6 Archaeology

While the archaeological material from the different layers was initially ascribed to a single culture, later typological studies confirmed mixed material in all the layers defined in the field, in all the collections from the site. The most striking examples of such admixtures are the presence of pottery in the second FBL (e.g. De Puydt and Lohest 1887), and new radiocarbon dates obtained on human material demonstrating that some bones initially attributed to the Spy I and Spy II Neandertals are in fact Neolithic (Semal et al. 2009). Multiple factors probably contributed to the mixing of material from different periods, including the excavation technique, the thinness of the second fauna-bearing level, natural disturbances inherent to such cave entrance sequences, and post-excavation mixing (Pirson et al. 2013b). As a result, the different cultures and/or

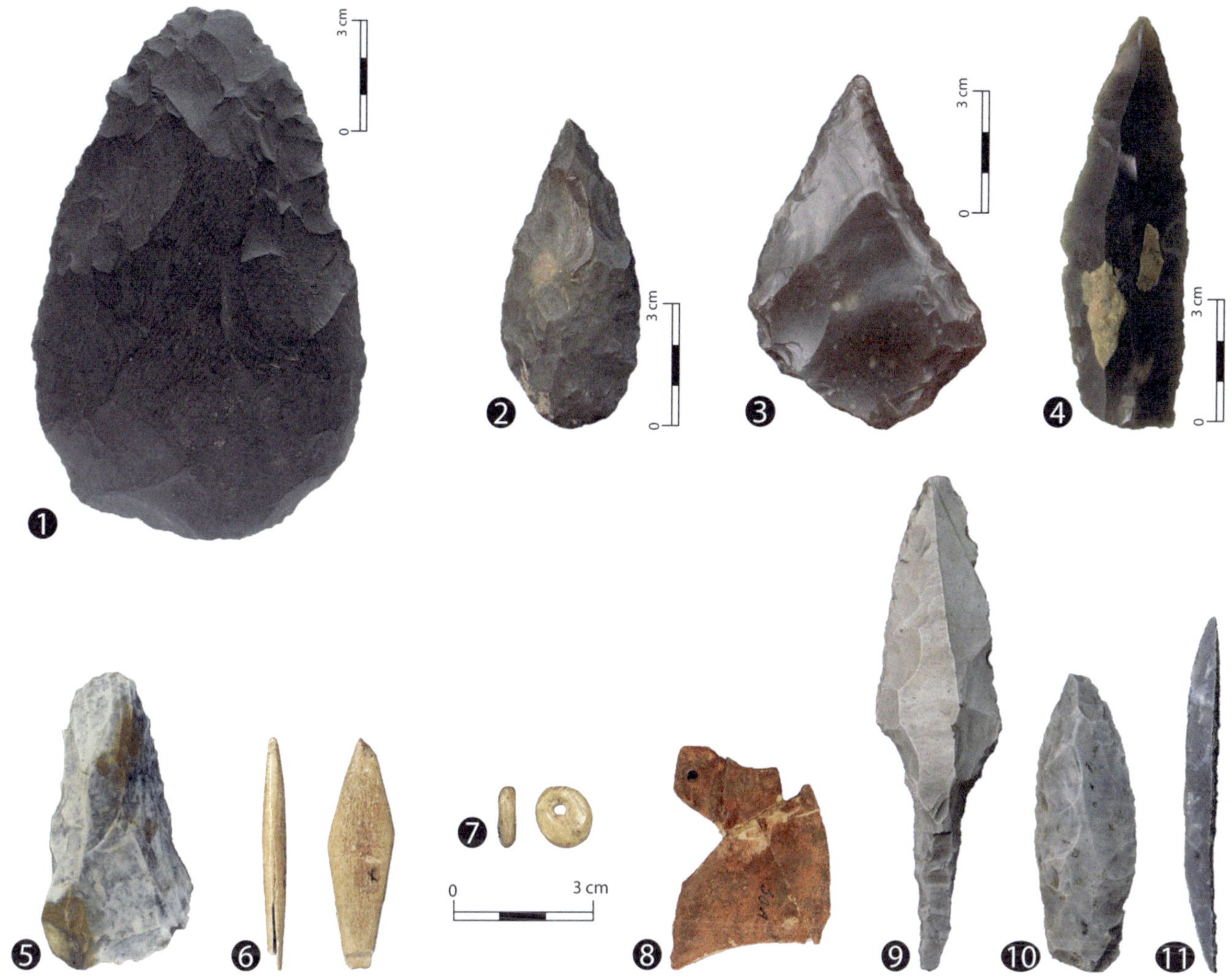

Fig. 21.11 Characteristic artefacts from Spy Cave, from Middle Palaeolithic (*1–3*), Lincombian-Ranisian-Jerzmanowician (*4*), Aurignacian (*5–8*), and Gravettian (*9–11*) (modified after Semal et al. 2011). *1* MTA biface (phtanite); *2* Foliate piece of Keilmesser type (flint); *3* Mousterian point (flint); *4* Jerzmanowice point (flint); *5* Carinated end scraper (flint); *6*. Split-based point (antler); *7* Bead (ivory); *8* Pendant with an ear-shape stained with ochre (ivory); *9* Font-Robert point (flint); *10* Maisières point (flint); *11* Gravette point (flint)

techno-complexes currently identified at Spy are evaluated independently of the three FBLs and any lithostratigraphic context.

The Spy archaeological sequence appears very rich (Rougier and Semal 2013). At least 14 distinct archaeological phases have been identified, including 10 Palaeolithic assemblages: (i) a minimum of three Middle Palaeolithic facies, (ii) a transitional Middle to Upper Palaeolithic industry, called Lincombian-Ranisian-Jerzmanowician, (iii) several Early Upper Palaeolithic assemblages including at least two Aurignacian and three Gravettian facies, and (iv) Late Upper Palaeolithic material (Pirson et al. 2013b, and references therein) (Fig. 21.11). The Aurignacian assemblage is one of the richest in Northern Europe for both the many lithic artefacts and the osseous artefacts and pendants (Flas et al., in Rougier and Semal 2013). Dozens of beads made of mammoth ivory have been found, displaying all manufacturing stages from preforms to ready-made objects (Khlopachev, in Rougier and Semal 2013). Beyond the Palaeolithic, material from the Mesolithic, Neolithic, Protohistoric, and historical periods is also acknowledged. In brief, together with Goyet Caves less than 20 km apart, Spy is the most important Palaeolithic site in Belgium and one of the major sites in Europe.

21.5.7 Palaeoenvironmental and Chronostratigraphic Data

Owing to the excavation conditions, palaeoenvironmental data are almost inexistent. Though large mammals have been recovered during the excavations (Germonpré et al., in

Rougier and Semal 2013), mixing problems prevent any reliable interpretation. Other data sets classically used in palaeoenvironmental reconstructions of cave entrance sequences, such as small mammals, malacology, palynology, anthracology, or pedology, are completely lacking here.

Likewise, very limited chronostratigraphic data is available for the Spy sequence. Although they help provide a general chronostratigraphy (Rougier and Semal 2013), the available radiocarbon dates must be considered with caution because of the ill-defined stratigraphic location of the dated material, which moreover mostly did not come from culturally attributable pieces, such as bone points. The best chronostratigraphic tool is therefore an archaeostratigraphy built from archaeological sequences dated at other sites (Pirson et al. 2013b).

The age of the Neandertal fossils is much better bracketed. The human remains of the 2004–2007 reassessment that fit to the original collection are of particular interest since, unlike the original consolidated and varnished specimens, they were recovered from unwashed fauna, which makes them usable for isotopic analyses. A number of AMS ^{14}C dates have been obtained from the two Neandertal adults. However, several of them are undoubtedly biased towards excessively young ages due to contamination and diagenetic influence. Three samples treated by two distinct laboratories using both traditional and ultrafiltration pretreatment protocols yielded reliable dates indicating that the two individuals lived around 41 ka cal BP (Semal et al. 2009; Higham et al. 2014). The Neandertals from Spy thus represent one of the latest Neandertal populations in Europe, which implicates them in the fierce debate about the Middle to Upper Palaeolithic transition and the replacement of Neandertals by Anatomically Modern Humans throughout Eurasia (Roebroeks 2008; Higham et al. 2014; Nigst et al. 2014; Hublin 2015). Unfortunately, the poor excavation conditions prevent any firm association with either the Mousterian culture or the Lincombian-Ranisian-Jerzmanowician (Flas 2011).

21.6 Scladina Cave: A Modern Excavation

Since 1978, Scladina Cave has been the object of several multidisciplinary researches dealing mainly with archaeology, archaeozoology, large and small mammals, palynology, stratigraphy, and sedimentology (e.g. Otte 1992; Otte et al. 1998). Over the first 15 years of field research, the interest in Scladina was essentially archaeological, linked to the discovery of Middle Palaeolithic artefacts. More than 17,500 lithic artefacts have been collected in the cave, corresponding to four main Mousterian occupations. Thousands of faunal remains have also been found throughout the sequence. However, the most important find from Scladina is undoubtedly the nearly complete juvenile mandible of a Neandertal child along with a fragment of right maxilla, unearthed from 1990 onwards (Toussaint and Bonjean 2014) (Fig. 21.12). It constitutes the main palaeoanthropological discovery of the twentieth century in Belgium, joining the exclusive club of the famous Pleistocene fossils found in Spy, Engis, or La Naulette (Toussaint and Pirson 2006). In the last 10 years, major interdisciplinary studies focusing on stratigraphy, sedimentary dynamics, palaeoenvironmental reconstructions, and chronostratigraphy were conducted in Scladina. The sedimentary infill appears as one of the most complete cave entrance sequences in Belgium (Pirson 2007; Pirson and Pirson et al., in Toussaint and Bonjean 2014; Pirson et al. 2008) and makes Scladina a reference site for cave entrance sedimentology in NW Europe.

Scladina Cave is still currently excavated. Beside the scientific project, the site is also equipped to receive visitors and offers the possibility of visiting an ongoing Palaeolithic excavation (http://www.scladina.be/). It has become a field school where students from Belgium and abroad improve their investigation skills in prehistorical research in cave sites.

21.6.1 Site Location and Description

Scladina Cave is located in the village of Sclayn, near Andenne (Bonjean et al. 2011) (Figs. 21.1, 21.2 and 21.13a). It is situated on the west side of the small Fond des Vaux Valley drained by the Ri de Pontine, a zero-order tributary of the Meuse (Fig. 21.13c). The cave opens towards the east in a Visean limestone massif at ~138 m altitude, approximately 7 m below the levelled surface of the ~0.4-km-wide interfluve separating the Fond des Vaux from the Meuse valley, and about 30 m above the local valley bottom. Quartz and quartzite river pebbles on the interfluve and two benches interrupting its western side have been reported, the three surfaces being interpreted as Meuse terrace deposits at respectively 115, 130, and 150 m (Gullentops and Deblaere, in Otte 1992). Gullentops and Deblaere (in Otte 1992) correlated the 115 m terrace with the so-called main terrace of the Meuse (see Chap. 10).

The cavity is a long cylindrical karst structure extending along the axial plane of an anticline developed in the Haut-le-Wastia Member of the Lives Formation (Pirson 2007). It consists in a 6-m-high, 6–20-m-wide gallery. After 35 years of almost permanent fieldwork, excavations have reached 40 m beyond the entrance (Fig. 21.13b). Speleological prospecting indicates that the cave extends for at least another 12 m. It is connected with other caves, namely the Saint-Paul Cave, which intersects Scladina 20 m to the

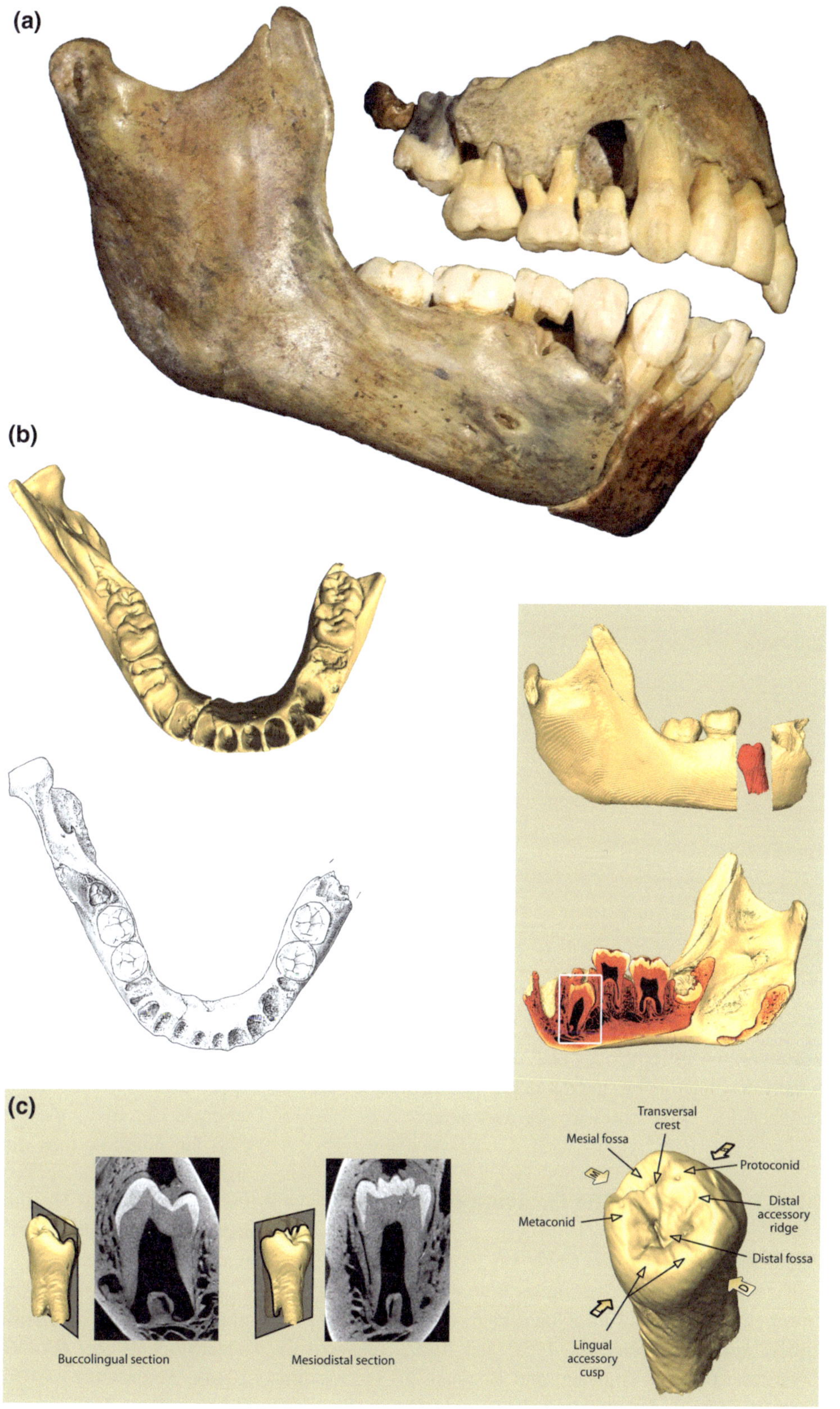

Fig. 21.12 The Scladina child. **a** The Scladina mandible and maxillar fragment with the isolated teeth refitted (*Photograph* Joël Eloy, AWEM; © Archéologie Andennaise and SPW); **b** Scan and drawing of the Scladina mandible in superior view (from Toussaint, in Toussaint and Bonjean 2014); **c** Virtual extraction and 3D reconstruction of the unerupted right second premolar from the mandible through micro-CT (computerized tomography) (from Toussaint, in Toussaint and Bonjean 2014)

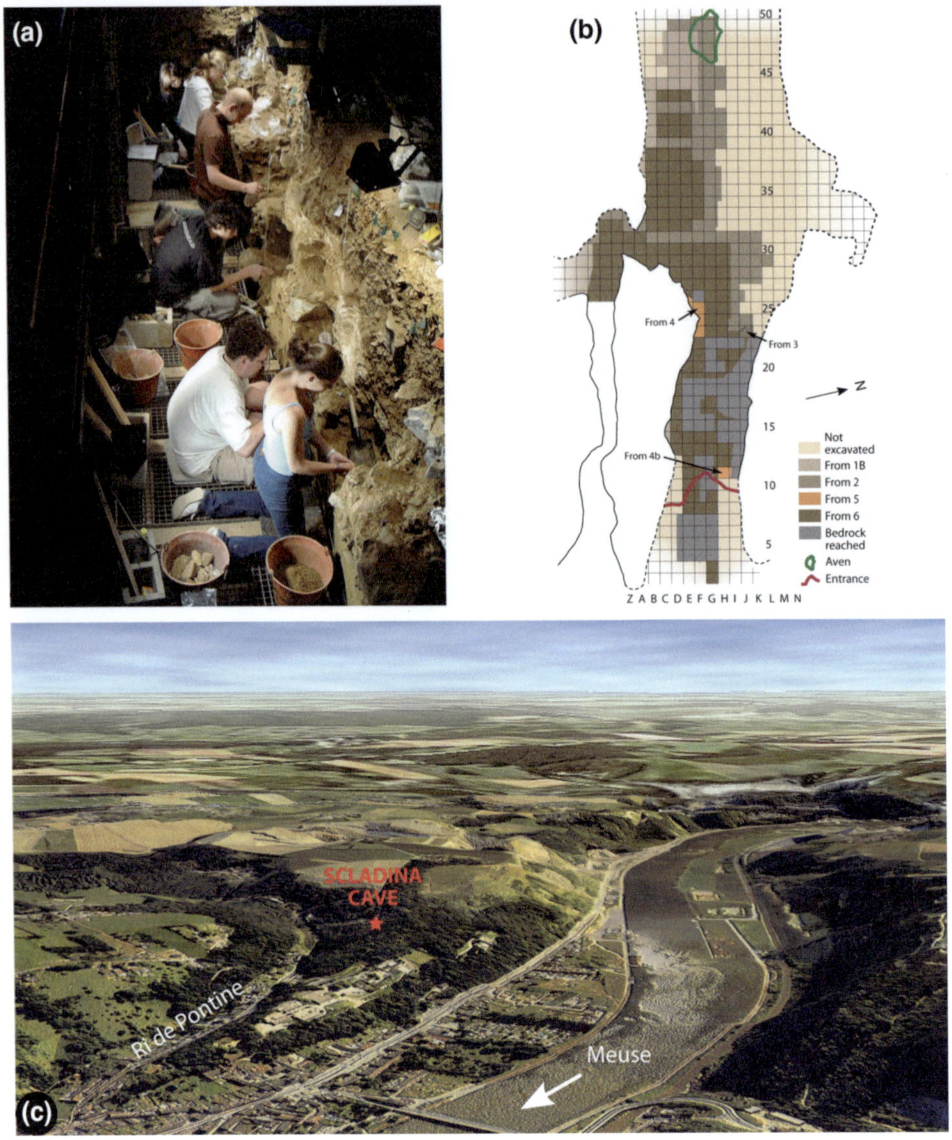

Fig. 21.13 Scladina Cave. **a** Excavators at work in the cave (*Photo* and © Archéologie Andennaise). **b** Map of Scladina Cave (from Bonjean et al. in Toussaint and Bonjean 2014). Archaeological research began on the terrace, in front of the entrance, and has extended 39 m inside the cave, from 11 to 50 m. **c** 3D representation of the Meuse Valley and its small tributary, the Ri de Pontine, with location of the cave (modified after Toussaint in Toussaint and Bonjean 2014; *Computer graphic* J. Eloy, AWEM)

northwest of its entrance, and the Sous-Saint-Paul Cave, which develops a few metres below the two others and is connected to Saint-Paul Cave by a chimney.

21.6.2 History of the Excavations

The area around Scladina Cave was prospected by speleologists and amateur archaeologists from the beginning of the twentieth century onwards. Saint-Paul and Sous-Saint-Paul caves were discovered in 1951 and 1953, respectively (Bonjean et al., in Toussaint and Bonjean 2014). Scladina Cave itself was also discovered by speleologists in 1971. Until 1977, they worked on the first 15 m, removing the upper 2 m of sediments under the ceiling and discovering artefacts. In 1977, they contacted professional archaeologists, an initiative that would save the cave. A rich faunal and archaeological collection ranging from the Middle Palaeolithic to the Neolithic was exhumed from Scladina, including an important Neolithic burial, unfortunately without any context.

Professional excavation of the site thus started in 1978, under the scientific authority of the University of Liège, and never stopped since (Bonjean et al., in Toussaint and Bonjean 2014). At present, the excavations are carried out by the Scladina Cave Archaeological Centre in conjunction with the

University of Liège, with support from the City of Andenne and the Service public de Wallonie. In 1996, the cave was classified as an archaeological site of exceptional interest (*site exceptionnel de Wallonie*, protected by regional laws).

21.6.3 Stratigraphy

Contrary to Spy Cave, Scladina Cave meets the quality criteria of a modern excavation. Dozens of sedimentary profiles were studied in detail throughout the whole site (e.g. Otte 1992; Otte et al. 1998; Pirson 2007; Pirson, in Toussaint and Bonjean 2014), illustrating the complexity of the stratigraphy. Important lateral sedimentary variations were highlighted as well as complex geometries and varied lithologies, eventually leading to the definition of the actual 15-m-thick Scladina stratigraphic sequence encompassing ca. 120 layers grouped in 30 "units" (Pirson 2007) (Fig. 21.14). These field studies revealed the diversity of the sedimentary sources and a large variety of sedimentary dynamics dominated by slope processes, including debris flow, run-off, rock fall, solifluction, settling, torrential flow, and speleothem formation (Pirson 2007; Pirson, in Toussaint and Bonjean 2014). Complex diagenetic processes have also been recognized.

21.6.4 Context of the Fossil Discovery

The first Neandertal remains have been identified in July 1993 in Scladina Cave, when a right hemimandible was found in situ in the sedimentary complex 4A (Bonjean et al., in Toussaint and Bonjean 2014) (Fig. 21.14a, b). Several teeth and a small maxillary fragment unearthed from 1990 onwards were later recognized as belonging to the same individual. The last remain was discovered in 2006, but future excavation is planned in the sedimentary complex 4A.

Although several Neandertal teeth were not identified in the field, accurate excavation methods allowed for their precise stratigraphic positioning and interdisciplinary studies enabled reconstructing the major steps of their complex history. Using all pieces of evidence (spatial distribution, projections onto the nearest sedimentary profiles, detailed anatomical descriptions, taphonomy, analysis of drawings, field notes and pictures) (Toussaint and Bonjean 2014), all the Neandertal remains were recently relocated in the new stratigraphic record of the cave sequence, showing that they are all in secondary spatial position in sedimentary units 4A-CHE and 4A-POC (Pirson et al., in Toussaint and Bonjean 2014) (Fig. 21.14b). The first redistribution phase is linked with the development of the large gully of Unit 4A-CHE, while some fossils were reworked from the top of Unit 4A-CHE during the deposition of Unit 4A-POC (Fig. 21.14b, c). At least two spatial redistribution phases are therefore attested.

21.6.5 The Neandertal Fossils

Nineteen human remains were found so far at Scladina, between 1990 and 2006. They consist in two fragments forming a nearly complete mandible, a maxillary fragment and 16 isolated teeth belonging to the same juvenile Neandertal. Histological studies suggest child's death occurred at the approximate age of 8 years (Smith et al. 2007), confirmed by the 8.5-year estimate obtained from a model for dental maturation (Toussaint et al., in Toussaint and Bonjean 2014). It is tempting to suggest that the child may be female (Toussaint, in Toussaint and Bonjean 2014), even if a sex determination based on limited juvenile cranial fragments is not very reliable.

This individual has been directly dated to 127 + 46/−32 ka by gamma spectrometry (Toussaint et al. 1998). However, combination of bone taphonomy, sedimentogenesis, palaeoenvironmental reconstructions, and chronostratigraphic markers of the whole sequence rather points to an approximate age of either 87 (GS 22 of Rasmussen et al. 2014; MIS 5b) or 80 ka (GI 21.1d of Rasmussen et al. 2014; end of MIS 5a), in a cold episode of the Weichselian Early Glacial (Pirson et al., in Toussaint and Bonjean 2014).

Morphologically, the Scladina mandible combines plesiomorphic traits, derived Neandertal features (e.g. a highly excavated submandibular fossa and a deeply excavated pterygoid fossa), and other features not unanimously accepted as derived though frequent among Neandertals (e.g. a laterally expanded condyle). Characterization of a Neandertal mandible lies in fact in the unique combination of a set of traits that, taken individually, are not typical of the taxon. In this regard, the Scladina mandible matches the general pattern of Neandertals (Toussaint, in Toussaint and Bonjean 2014). The maxilla also displays features typical of the Neandertal group (Verna and Toussaint, in Toussaint and Bonjean 2014). As for the Scladina teeth, they do not exhibit any single morphological character strictly specific to Neandertal but they possess the combination of features that is the signature of Neandertals against AMHs (De Vries et al., in Toussaint and Bonjean 2014; Toussaint, in Toussaint and Bonjean 2014).

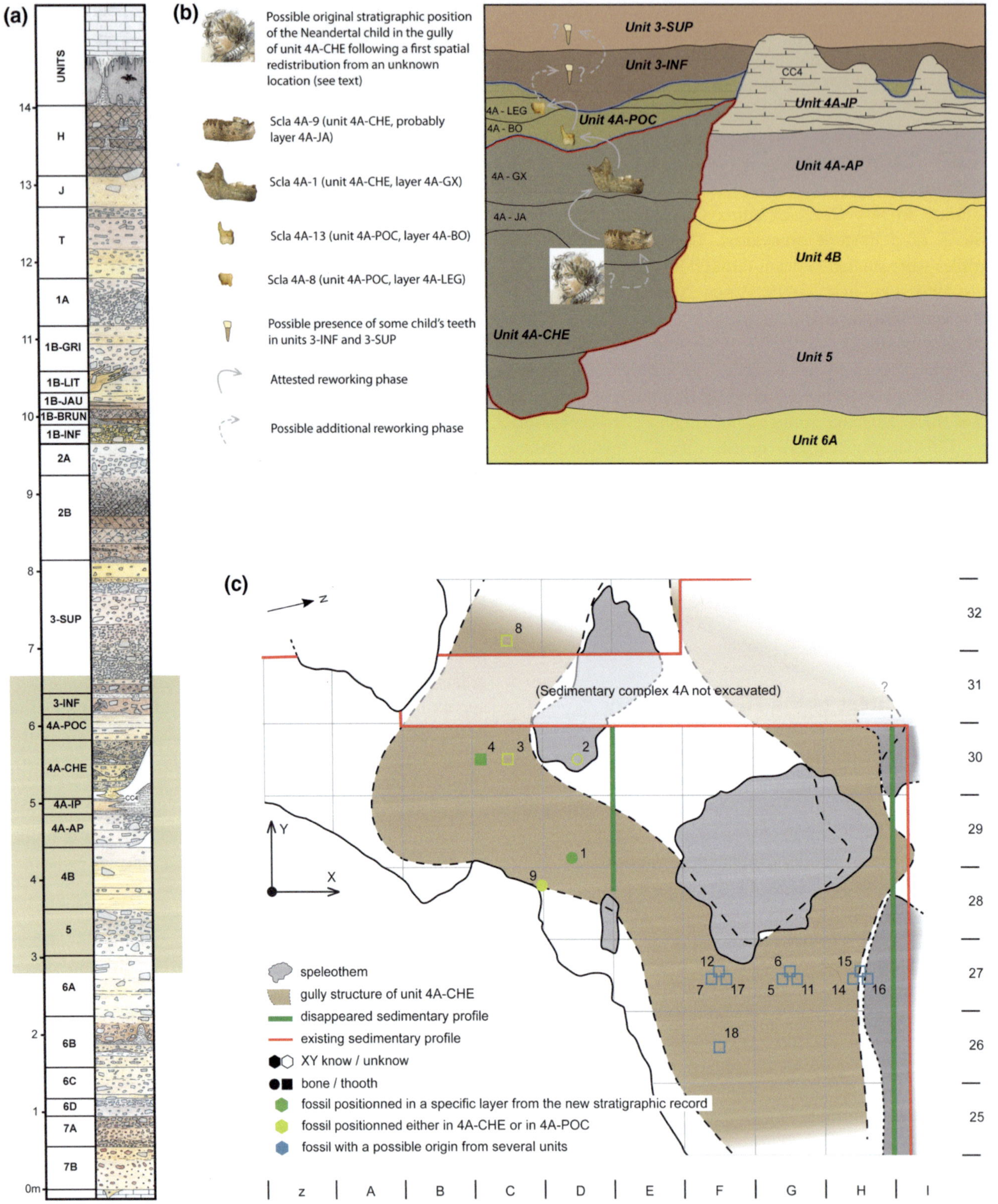

Fig. 21.14 Stratigraphic position of the Neandertal child (after Pirson et al. in Toussaint and Bonjean 2014). **a** Scladina Cave synthetic lithostratigraphic sequence (the *grey rectangle* highlights the units illustrated in b). **b** Sketch describing the stratigraphic relationships between the units of Sedimentary Complex 4A and summarizing the complex depositional history of the Neandertal child remains. **c** Spatial distribution (plan view) of the 19 juvenile Neandertal remains, and their relationship with the course and extent of the gully structure of Unit 4A-CHE

Isotopic biogeochemistry (C/N, Bocherens, in Toussaint and Bonjean 2014) and dental microwear texture (El Zaatari et al., in Toussaint and Bonjean 2014) provided insights about the diet of the child, indicating that he was omnivorous but mainly ate meat. Biogeochemistry points to consumption of open-environment herbivores, while microwear data also reveal a diet that most likely included small amounts of abrasive plant foods. Furthermore, micro-spatial analysis of Ba/Ca ratios in the upper first right maxillary molar (Austin et al. 2013) shed light onto the weaning process, showing that the child exhibits a pattern of exclusive breastfeeding during seven months, followed by seven further months of supplementation before a probable abrupt cessation of breastfeeding at the age of 1.2 years.

Finally, mtDNA analysis on a deciduous molar of the Scladina child provided interesting results, demonstrating that this >80-ka-old mtDNA sequence is more distantly related to those of modern humans than sequences of more recent Neandertals are. These results suggest that the Neandertal population of W Europe experienced a significant demographic bottleneck in the last 30–50 ka preceding their extinction (Orlando et al. 2006). In any case, the Scladina sequence has revealed unexpected levels of genetic diversity among Neandertals (Orlando and Hänni, in Toussaint and Bonjean 2014).

21.6.6 Archaeology

Scladina is one of the major Palaeolithic archaeological sites in Belgium. Two numerically important Middle Palaeolithic artefact series have been recovered (Fig. 21.15), namely the Unit 5 ($\sim$9000 artefacts) and Unit 1A ($\sim$4500) assemblages. These assemblages have been extensively studied, especially with respect to typology, technology, petrography, spatial distribution, and statistics (see Bonjean et al., in Toussaint and Bonjean 2014).

The Unit 5 assemblage dates from a cold phase of the Weichselian Early Glacial, some 100 ka ago. Flint nodules of distant provenance were used together with local stones such as quartz and quartzite river pebbles, limestone, and chert. Several stone hammers were recovered (Di Modica 2010), as well as 26 bone retouchers (Fig. 21.15), in some cases made of cave bear bones (Abrams et al. 2014). Faunal analysis showed that six whole chamois had been brought to the site and processed in order to obtain skins, meat, tendons, and marrow (Patou-Mathis, in Otte et al. 1998). The predation of small game was also illustrated by the presence of a hare coxal bone bearing 18 butchery marks (Bonjean et al. 2011) (Fig. 21.15). More than 1000 burned bone fragments have been collected, evidencing the existence of one or several hearths (Bonjean et al., in Toussaint and Bonjean 2014).

The Unit 1A assemblage was produced in a time period ranging from $\sim$44,600 to $\sim$42,200 cal BP, as suggested by the combination of radiocarbon dates (Bonjean et al. 2013) with the general stratigraphic framework, taphonomic processes on bone material, and the sedimentary processes that controlled the introduction of the archaeological material in the cave (Di Modica et al. 2016). Due to transport and deposition of the artefacts through debris flow and run-off (Pirson 2007), this assemblage is in a poorer state of preservation than that from Unit 5 (Bonjean et al., in Toussaint and Bonjean 2014). Local flint and quartzite river pebbles were used (Di Modica 2010). Although bad preservation of the bones prevented detailed archaeozoological study, the presence of $\sim$200 fragments of burned bones suggests they were used as fuel (Abrams et al. 2010). Detailed analysis of a black mineral pigment associated with the archaeological material unequivocally identified its sources in the area of Franquenies, $\sim$40 km to the northeast of Scladina Cave, implying transportation to the cave (Bonjean et al. 2015).

Aside from these two major archaeological assemblages, several small, mostly Middle Palaeolithic lithic assemblages, have also been recovered, each sedimentary unit having actually provided at least some artefacts. Scattered Upper Palaeolithic and Mesolithic pieces probably reworked from the plateau were extracted from bioturbated settings (Bonjean et al., in Toussaint and Bonjean 2014).

21.6.7 Palaeoenvironmental and Chronostratigraphic Data

Scladina yielded one of the best interdisciplinary-studied cave entrance sequences in Belgium, together with Walou Cave, east of Liege (Draily et al. 2011; Pirson et al. 2011). The palaeoenvironments have been documented through large and small mammals, amphibians, reptiles, palynological, anthracological, sedimentological, magnetic susceptibility, and pedological studies (e.g. Otte 1992; Otte et al. 1998; Ellwood et al. 2004; Pirson 2007; Pirson et al. 2008; Pirson et al., in Toussaint and Bonjean 2014; Blain et al. 2014). More than twenty alternating climatic phases have been documented, including cold stadials, short interstadials,

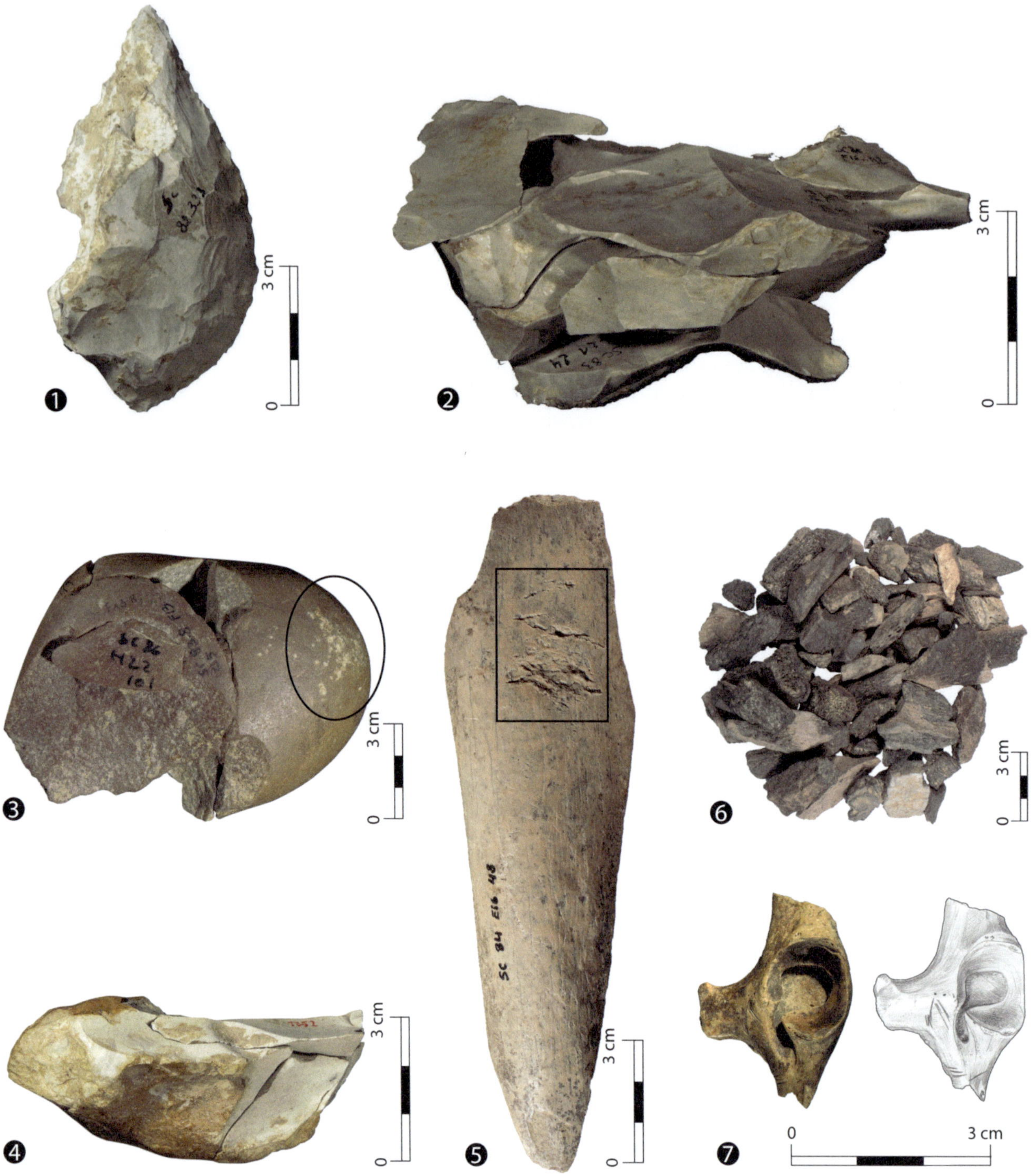

Fig. 21.15 Characteristic artefacts from Scladina Cave (Middle Palaeolithic) (*Photos* and ©: Archéologie Andennaise). *1* Simple convex side scraper from Unit 5 (flint); *2* Refitted core from Unit 5 (flint); *3* Refitting of several quartzite flakes from Unit 5 illustrating the reduction process of a quartzite river pebble. After reduction, the nucleus was used as a stone hammer, as shown by the traces of percussion on the *right* (within *black ellipse*); *4* Refitting on a small alluvial flint pebble from Unit 1A; *5* Bone retoucher from Unit 5 made from a cave bear right femur. The *black rectangle* locates the used area, with visible knapping-marks in which residual flint inclusions have been identified by SEM (from Abrams et al. 2014); 6. Burned bone fragments suggesting the use of bone as fuel. Such burned bones have been recovered from Units 5 and 1A; 7. Hare coxal fragment from Unit 5 exhibiting 18 cut marks, attesting the hunting of small game (*Drawing* S. Lambermont, AWEM)

and longer temperate phases, either interglacial or early glacial. The most recent chronostratigraphic interpretation of this complex sequence is based on climatostratigraphic, biostratigraphic, and mineralogical data, numerical dates, and comparison with the reference loess sequence from Central Belgium (Fig. 21.16). With 70 available dates (TL, ESR, U/Th, and radiocarbon), Scladina is the most densely dated prehistoric site in Belgium.

21.7 Territorial Exploitation by Neandertals

In Neandertal times, important and rapid climatic changes occurring in contrasted geomorphological and geological settings greatly impacted the availability of useful resources, such as hunted games or gathered plants. The diversity of the geological substratum, together with the vegetation cover and the river channel form (braided versus meandering) also determined lithic raw material availability (Di Modica 2010; Di Modica et al. 2016).

Evidence for mobility patterns of, and landscape exploitation by, Neandertal populations is rather scarce. However, the combination of lithic resources management and archaeozoology data allowed for establishing a scheme of resource exploitation and inter-regional variability (Di Modica 2010; Di Modica et al. 2016). Owing to the very poor preservation of bone material at open-air sites and the bad conditions of excavation in most early studies, lithic raw material in Palaeolithic sites constitutes indeed the most valuable source of information about population mobility and land-use strategies.

Cretaceous flint was the most exploited raw material in Belgium during the whole Lower and Middle Palaeolithic. All 442 known finding spots yielded flint artefacts. On the contrary, most of the other raw materials were used only locally, when flint sources were lacking in the vicinity of the site. The distribution map of Lower and Middle Palaeolithic sites in Belgium (Fig. 21.2) suggests a preference for areas characterized by good quality flint sources related with Cretaceous outcrops, even though parameters linked to site preservation and history of research should not be neglected (Di Modica et al. 2016).

The comparative study of lithic raw material assemblages from several sites highlighted differences related to their geomorphological and geological background (Fig. 21.17). Numerous open-air sites are concentrated in Central Belgium, on plateaus and valleys where Cretaceous chalks outcrop. In such contexts of favourable flint supply, tool manufacturing occurred largely or completely at the find site. Average nucleus size and techno-morphological characteristics of the flakes point to a highly standardized lithic tool production based on specific concepts, and abandonment of the nucleus before complete exhaustion. The same situation is observed for some cave sites located close to good quality flint sources, particularly in the Mehaigne and Vesdre Basins. This suggests that, when flint nodules were easily accessible, tool quality prevailed over raw material saving.

There are also cave and open-air sites located far from Cretaceous outcrops. Lithic assemblage studies from these sites indicate that good quality flint was brought to the site, sometimes from up to 50 km, pointing to a planned mobility of populations that transported the lithic resources required for their activities from one region to another where they knew flint was lacking. Once at the site, the reduction processes were adapted to the situation, aiming at maximizing the tool production at the expense of standardization. Next to imported flint, locally available raw materials were also exploited for stone tool production. In some cases, small flint pebbles available in the site surroundings were used preferentially as an additional supply in good quality flint. In their absence, other locally available raw materials such as phtanite, chert and limestone, sandstone, quartz, and quartzite compensated for the lack of large good quality flint nodules.

The black mineral pigment included in the archaeological assemblage 1A from Scladina Cave provides further valuable insight into Neandertal territorial exploitation. Raman spectroscopy analysis allowed for distinguishing between the different potential geological sources, pointing to the Franquenies area in the Dijle valley, ~40 km apart from Scladina Cave (Bonjean et al. 2015). This confirms the long-distance displacements evidenced by the flint exploitation strategies.

Settling away from flint sources was not primarily determined by the possibility of stone tool production from other raw materials (Di Modica et al. 2016). More plausible reasons are the shelter offered by caves, the attractiveness of deep narrow valleys in Upper Belgium, the exploitation of specific mineral resources (e.g. pigments), an easier access to particular organic resources, or even cultural reasons. These hypotheses are not exclusive, multiple motivations potentially explaining human settlements distant from rich flint sources. In summary, all the above elements suggest a complementary occupation of contrasted environments implying displacements from one region to another according to the needs in either lithic resources or plant and fauna resources.

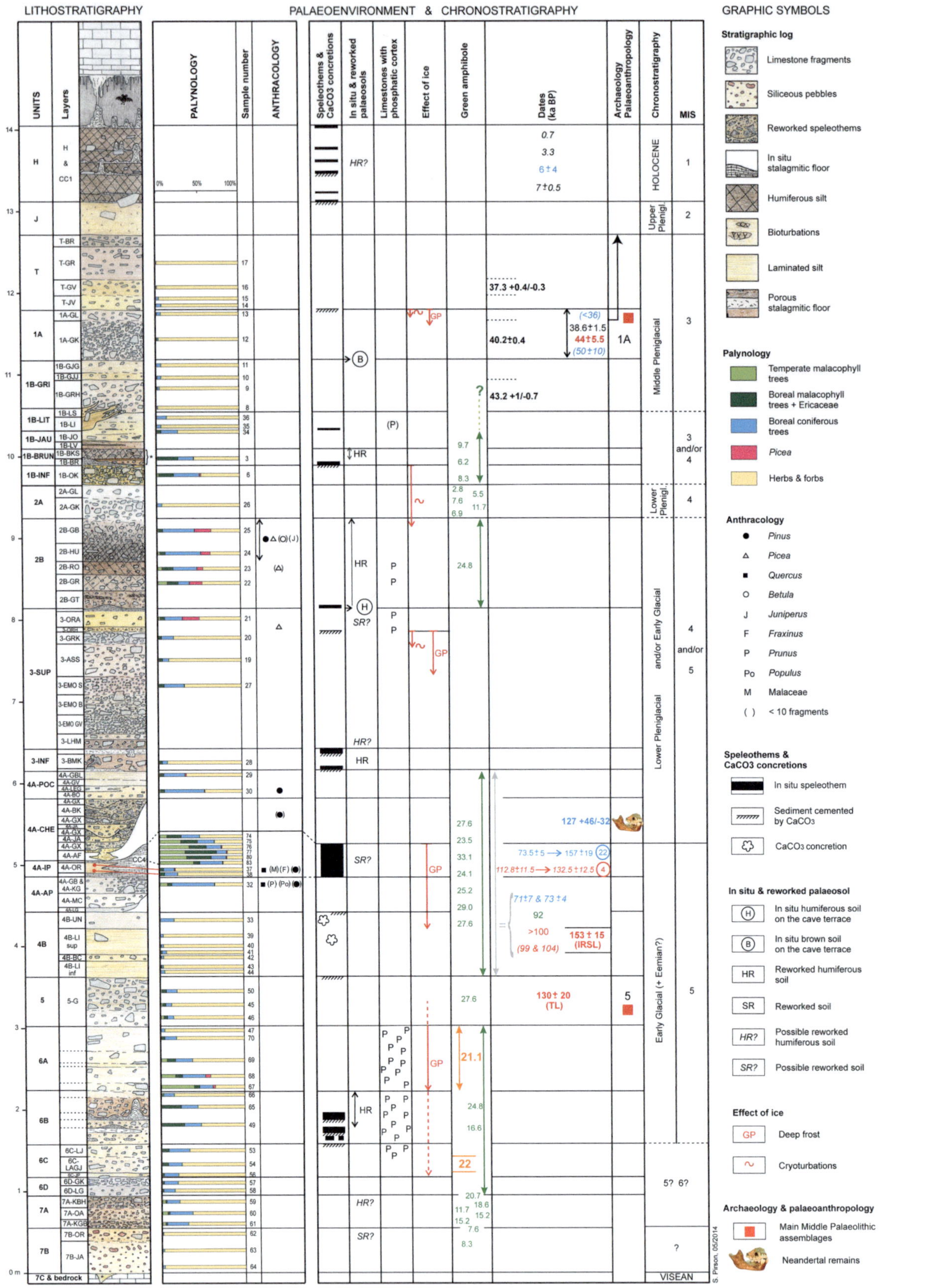
LITHOSTRATIGRAPHY
PALAEOENVIRONMENT & CHRONOSTRATIGRAPHY
UNITS
Layers
PALYNOLOGY
Sample number
ANTHRACOLOGY
Speleothems & CaCO3 concretions
In situ & reworked palaeosols
Limestones with phosphatic cortex
Effect of ice
Green amphibole
Dates (ka BP)
Archaeology Palaeoanthropology
Chronostratigraphy
MIS
HOLOCENE
Upper Plenigl.
Middle Pleniglacial
Lower Plenigl.
Lower Pleniglacial and/or Early Glacial
Early Glacial (+ Eemian?)
VISEAN
GRAPHIC SYMBOLS
Stratigraphic log
Limestone fragments
Siliceous pebbles
Reworked speleothems
In situ stalagmitic floor
Humiferous silt
Bioturbations
Laminated silt
Porous stalagmitic floor
Palynology
Temperate malacophyll trees
Boreal malacophyll trees + Ericaceae
Boreal coniferous trees
Picea
Herbs & forbs
Anthracology
Pinus
Picea
Quercus
Betula
Juniperus
Fraxinus
Prunus
Populus
Malaceae
< 10 fragments
Speleothems & CaCO3 concretions
In situ speleothem
Sediment cemented by CaCO3
CaCO3 concretion
In situ & reworked palaeosol
In situ humiferous soil on the cave terrace
In situ brown soil on the cave terrace
Reworked humiferous soil
Reworked soil
Possible reworked humiferous soil
Possible reworked soil
Effect of ice
Deep frost
Cryoturbations
Archaeology & palaeoanthropology
Main Middle Palaeolithic assemblages
Neandertal remains
S. Pirson, 05/2014

◀ **Fig. 21.16** Scladina Cave stratigraphic log including a synthesis of the palaeoenvironmental and chronostratigraphic data (from Pirson et al., in Toussaint and Bonjean 2014). **Green amphibole**: percentages of green amphibole in the silt fraction, in *green* calculated from data of Gullentops and Deblaere (1992), in *orange* data from Pirson (2007). **Dates**: synthesis of all reliable dates available for the Scladina sequence. In *black* published ^{14}C date on bone; *black italic* ^{14}C dates on calcite (in situ and reworked speleothems); *black bold* ^{14}C dates on teeth accurately positioned in the new stratigraphic system; *blue* U/Th date on in situ speleothem; *blue italic* U/Th dates on reworked fragments of speleothem; *blue bold* U/Th dates (gamma spectrometry) obtained on the Neandertal mandible; *red* thermoluminescence date on sediment; *red italic* thermoluminescence dates on in situ speleothem; *red bold* luminescence dates on burnt flint (TL) and on sediment (IRSL); *green* coupled U/Th–ESR dates on bone. *Circled numbers* number of dates for CC4 interval. Problematic dates in *brackets*. **Chronostratigraphy**: interpretation of the Scladina sequence based upon all the available analyses. **MIS**: suggested correlation of cave sequence with the marine oxygen isotopic stages

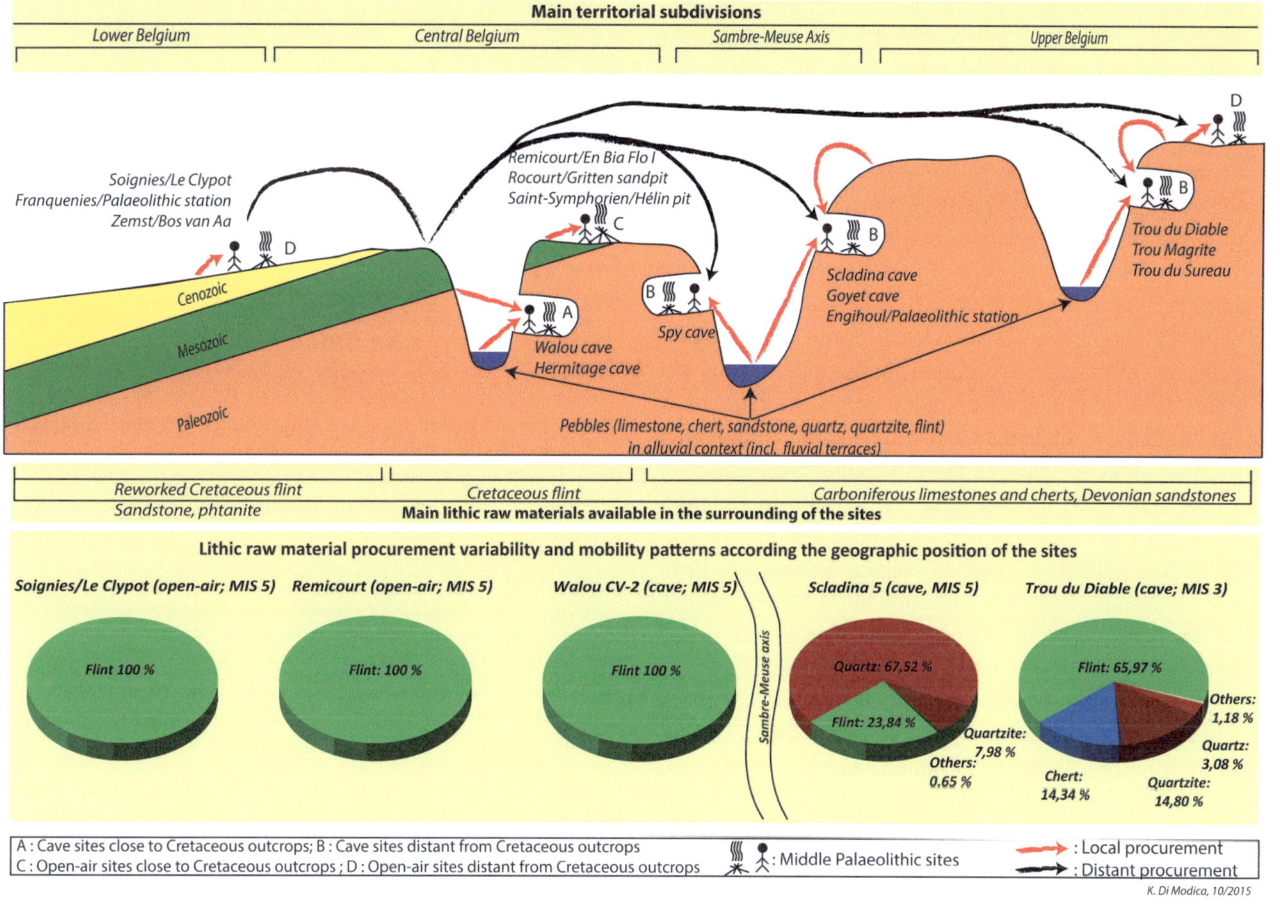

Fig. 21.17 Raw lithic material acquisition and transportation patterns for the Middle Palaeolithic in Belgium (from Di Modica et al. 2016)

21.8 Conclusion

In the nineteenth century, the tenacity of forerunners such as Schmerling, Dupont, Lohest, De Puydt, and Fraipont highlighted Belgium as a key area for early studies in prehistory and palaeoanthropology. Excavations in the caves of Engis, La Naulette, and Spy documented the contemporaneousness of Man and extinct large mammals, and allowed for definitive demonstration of the existence of a fossil hominin more "archaic" than anatomically modern humans, i.e. Neandertal Man.

In the last decades, while major anthropological and/or archaeological discoveries have been made in Eurasian Middle Palaeolithic sites such as Atapuerca in Spain, Denisova in Siberia, Caours in France, Neumark-Nord 2 in Germany, new sites have also been unearthed in Belgium and several classic sites have been re-examined. Belgian caves having yielded Neandertal remains now amount to eight, making the Meuse Basin by far the richest anthropological area in northwestern Europe. Scladina and Walou Caves as well as the open-air site of Remicourt are three major places of recent excavations, and the reassessment of the Spy and Goyet collections is another recent achievement that benefitted from a modern interdisciplinary approach. They all brought significant contributions in various research fields such as the genetic characterization of Neandertals, their diet, anatomy, culture, symbolic behaviour, and chronology. These advances went hand in hand with recent technological improvements related to DNA extraction, isotopic biogeochemistry, 3D-reconstruction from computed tomography scans, or AMS radiocarbon dating. Likewise, progress in the reconstruction of the changing paleoclimates and paleoenvironments of the Pleistocene allowed for a better understanding of the relations between Neandertals and their environments. Advances in sedimentology and pedology combined with experiments in active periglacial environments also led to better understanding of the site formation processes and assessment of site integrity.

In brief, the richness of archaeological and anthropological remains, the diversity of the geological substratum, the country's position in the heart of NW Europe where influences and groups from western and central Europe met and intermingled, and the simultaneous presence of loess and caves favouring detailed palaeoenvironmental and chronostratigraphic reconstructions all together make Belgium a very favourable place for the study of the human response, and in particular that of Neandertal societies, to environmental changes.

Popularization has not been left behind. In Belgium, apart from the EHoS in Spy, several museums, such as the Royal Belgian Institute of Natural Sciences in Brussels, the Gallo-Romeins Museum in Tongres, the Prehistomuseum in Ramioul, showcase Neandertal culture and environments. The virtual visit of Scladina Cave will soon be possible, based on the 3D reconstruction of the archaeological site supported by the European Commission (*Horizon 2020*). The cultural value of the Belgian Neandertal sites, all located in Wallonia, is acknowledged by the Walloon Region, which, together with other concerned European regions, plans to apply for a UNESCO World Heritage serial property including the major Neandertal sites.

Acknowledgements The authors gratefully acknowledge Ken Dethier and Jean-François Lemaire (Service public de Wallonie), Joël Eloy and Sylviane Lambermont (Association wallonne d'Etudes mégalithiques) and Ivan Jadin (Royal Belgian Institute of Natural Sciences) for their technical assistance. We are also very grateful to Paul Haesaerts, Freddy Damblon, Mona Court-Picon, Patrick Semal and Paul Spagna (Royal Belgian Institute for Natural Sciences), Dominique Bosquet, Christelle Draily and Pierre van der Sloot (Service public de Wallonie), Grégory Abrams (Archeologie Andennaise), and Marie Wéra (Prehistomuseum, Ramioul) for many fruitful discussions. Last but not least, we would like to thank Alain Demoulin for having invited us to submit a paper dealing with Neandertal caves from Belgium, which was slightly off-topic from the main theme of this volume, and for helpful comments, which significantly improved the manuscript.

References

Abrams G, Bonjean D, Di Modica K, Pirson S, Otte M, Patou-Mathis M (2010) Les os brûlés de l'ensemble sédimentaire 1A de Scladina (Andenne, Belgique): apports naturels ou restes de foyer(s) néandertalien(s)? Notae Praehistoricae 30:5–13

Abrams G, Bello SM, Di Modica K, Pirson S, Bonjean D (2014) When Neanderthals used cave bear (Ursus spelaeus) remains: bone retouchers from unit 5 of Scladina Cave (Belgium). Quatern Int 326–327:274–287

Andersen KK, Bigler M, Clausen HB, Dahl-Jensen D, Johnsen SJ, Rasmussen SO, Seierstad I, Steffensen JP, Svensson A, Vinther BM, Davies SM, Muscheler R, Parrenin F, Röthlisberger R (2007) A 60 000 year Greenland stratigraphic ice core chronology. Clim Past Discuss 3:1235–1260

Antoine P, Limondin-Lozouet N, Auguste P, Locht J-L, Galheb B, Reyss J-L, Escudé E, Carbonel P, Mercier N, Bahain J-J, Falguères C, Voinchet P (2006) Le tuf de Caours (Somme, France): mise en évidence d'une séquence eemienne et d'un site paléolithique associé. Quaternaire. Revue de l'Association française pour l'étude du Quaternaire 17(4):281–320

Austin C, Smith TM, Bradman A, Hinde K, Joannes-Boyau R, Bishop D, Hare DJ, Doble P, Eskenazi B, Arora M (2013) Barium distributions in teeth reveal early-life dietary transitions in primates. Nature 498:216–219

Blain H-A, López-García JM, Cordy J-M, Pirson S, Abrams G, Di Modica K, Bonjean D (2014) Middle to Late Pleistocene herpetofauna from Scladina and Sous-Saint-Paul Caves (Namur, Belgium). C.R. Palevol 13:681–690

Bocherens H, Billiou D, Mariotti A, Toussaint M, Patou-Mathis M, Bonjean D, Otte M (2001) New isotopic evidence for dietary habits of Neandertals from Belgium. J Hum Evol 40:497–505

Bonjean D, Di Modica K, Abrams G, Pirson S, Otte M (2011) La grotte Scladina: bilan 1971–2011. In: Toussaint M, Di Modica K, Pirson S (eds) Le Paléolithique moyen en Belgique. Mélanges Marguerite Ulrix-Closset, pp 323–334. Bulletin de la Société belge d'études Géologiques et Archeologiques «Les Chercheurs de la Wallonie», hors-série n° 4 et Etudes et Recherches archeologiques de l'Université de Liège, 128

Bonjean D, Abrams G, Di Modica K, Pirson S (2013) Andenne/Sclayn: la grotte Scladina, nouvelles datations au radiocarbone. Chronique de l'Archéologie wallonne 20:241–244

Bonjean D, Vanbrabant Y, Abrams G, Pirson S, Burlet C, Di Modica K, Otte M, Vander Auwera J, Golitko M, McMillan R, Goemaere E (2015) A new Cambrian black pigment used during the late Middle Palaeolithic discovered at Scladina Cave (Andenne, Belgium). J Archaeol Sci 55:253–265

Boucher de Perthes J (1849) Antiquités celtiques et antédiluviennes: Mémoire sur l'industrie primitive et leurs arts à leur origine. Tome premier. Treuttel et Wurtz, Derache, Dumoulin et Didron, Paris, 628 p. et LXXX pl.p

Cels A, Jacques V (1898) Schmerling. Contribution à l'histoire du préhistorique en Belgique. Bulletin de la Société d'Anthropologie de Bruxelles XV:86–101

Cohen KM, Gibbard PL (2011) Global chronostratigraphical correlation table for the last 2.7 million years (v. 2010). Subcommission on

quaternary stratigraphy (International Commission on Stratigraphy). http://www.quaternary.stratigraphy.org.uk/. Cambridge, England
Crevecoeur I, Bayle P, Rougier H, Maureille B, Higham T, Van der Plicht J, De Clerck N, Semal P (2010) (online version). The Spy VI child: a newly discovered Neandertal infant. J Human Evol 59(6): 1–16
Darwin C (1859) On the origin of species by means of natural selection, or the preservation of favoured races in the struggle for life. Murray, London
de Bethune P (1954) Carte géologique de la Belgique au 500.000e
de Loë A, Rahir E (1911) Nouvelles fouilles à Spy, grotte de la Betche-aux-Rotches. Bulletin de la Société d'Anthropologie de Bruxelles 30:40–58
De Puydt M, Lohest M (1886) Exploration de la grotte de Spy. Annales de la Société géologique de Belgique 13:34–39
De Puydt M, Lohest M (1887) L'homme contemporain du Mammouth à Spy (Namur). Annales de la Fédération archéologique et historique de Belgique, vol 2. Compte rendu des travaux du Congrès tenu à Namur les 17–19 août 1886, pp 205–240
Di Modica K (2010) Les productions lithiques du Paléolithique moyen de Belgique: variabilité des systèmes d'acquisition et des technologies en réponse à une mosaïque d'environnements contrastésDépartement des Sciences historiques, Faculté de Philosophie et Lettres, Université de Liège et Ecole doctorale "Sciences de la Nature et de l'Homme" (ED 227), Museum National d'Histoire Naturelle, p 787
Di Modica K, Toussaint M, Abrams G, Pirson S (2016) The Middle Palaeolithic from Belgium: chronostratigraphy, territorial management and culture on a mosaic of contrasted environments. Quatern Int 411A:77–106
Draily C, Pirson S, Toussaint M (2011) La grotte Walou à Trooz (Belgique). Fouilles de 1996 à 2004, vol 2. Les sciences de la vie et les datations. Namur, Service public de Wallonie (Etudes et Documents, Archéologie, 21), 241 p
Dupont E (1866) Etude sur les fouilles scientifiques exécutées pendant l'hiver de 1865–1866 dans les cavernes des bords de la Lesse. Bulletins de l'Académie royale des Sciences, des Lettres et des Beaux-Arts de Belgique 2e, XXII:31–54
Ehlers J, Astakhov VI, Gibbard PL, Mangerud J, Svendsen JI (2007a) Glaciations—mid-quaternary in Eurasia. In: Elias SA (ed) Encyclopedia of quaternary science, vol 2. Elsevier, Netherlands, pp 1036–1043
Ehlers J, Astakhov VI, Gibbard PL, Mangerud J, Svendsen JI (2007b) Glaciations—late pleistocene glaciations in Europe. In: Elias SA (ed) Encyclopedia of quaternary science, vol 2. Elsevier, Netherlands, pp 1085–1095
Ek C (1995) Grottes et rivières des régions calcaires. In: Demoulin A (ed) L'Ardenne. Essai de géographie physique. Hommage au Professeur Albert Pissart, Département de Géographie physique et Quaternaire, Université de Liège, Liège, pp 178–193
Ellwood BB, Harrold FB, Benoist SL, Thacker PT, Otte M, Bonjean D, Long GJ, Shahin AM, Hermann RP, Grandjean F (2004) Magnetic susceptibility applied as an age-depth-climate relative dating technique using sediments from Scladina Cave, a Late Pleistocene cave site in Belgium. J Archaeol Sci 31:283–293
Fichefet J (1933) Observations sur trois méandres recoupés de la Sambre. Bulletin du cercle des géographes liégeois 5, fascicule 2, 15–22
Flas D (2011) The Middle to Upper Paleolithic transition in Northern Europe: the Lincombian-Ranisian-Jerzmanowician and the issue of acculturation of the last Neanderthals. World Archaeol 43(4): 605–627
Fraipont C (1936) Les hommes fossiles d'Engis. Archives de l'Institut de Paléontologie humaine, mémoire 16. Masson, Paris, p 53, 4 pl
Fraipont J, Lohest M (1887) La race humaine de Néanderthal ou de Canstadt en Belgique. Recherches ethnographiques sur des ossements humains découverts dans les dépôts quaternaires d'une grotte à Spy et détermination de leur âge géologique. Archives de Biologie (Gand) 7:587–757
Fu Q, Hajdinjak M, Moldovan OT, Constantin S, Mallick S, Skoglund P, Patterson N, Rohland N, Lazaridis I, Nickel B, Viola B, Prüfer K, Meyer M, Kelso J, Reich D, Pääbo S (2015) An early modern human from Romania with a recent Neanderthal ancestor. Nature 524:216–219
Fu Q et al (2016) The genetic history of Ice Age Europe. Nature 534:200–205
Fuhlrott C (1859) Menschliche Ueberreste aus einer Felsengrotte des Dusselthals. Ein Beitrag zur Frage uber die Existenz fossiler Menschen. Verhandlungen des Naturhistorischen Vereins der Preußischen Rheinlande und Westphalens 16:131–153
Gaudzinski-Windheuser S, Kindler L, Pop E, Roebroeks W, Smith G (2014) The Eemian Interglacial lake-landscape at Neumark-Nord (Germany) and its potential for our knowledge of hominin subsistence strategies. Quatern Int 331:31–38
Green RE et al (2010) A draft sequence of the neandertal genome. Science 328:710–722
Gullentops F, Deblaere C (1992) Erosion et remplissage de la grotte ScladinaRecherches aux grottes de Sclayn. Volume 1: Le contexte. Etudes et Recherches Archéologiques de l'Université de Liège, 27, Liège, p 9–31
Haesaerts P (1984) Aspects de l'évolution du paysage et de l'environnement en Belgique au Quaternaire. In: Cahen D, Haesaerts P (eds) Peuples chasseurs de la Belgique préhistorique dans leur cadre naturel, Bruxelles, pp 27–39
Haesaerts P, Van Vliet-Lanoë B (1981) Phénomènes périglaciaires et sols fossiles observés à Maisières-Canal, à Harmignies et à Rocourt. Biul Peryglac 28:291–324
Haesaerts P, Damblon F, Gerasimenko N, Spagna P, Pirson S (2016) The late pleistocene loess-palaeosol sequence of Middle Belgium. Quatern Int 411:25–43
Henry AG, Brooks AS, Piperno DR (2011) Microfossils in calculus demonstrate consumption of plants and cooked foods in Neanderthal diets (Shanidar III, Iraq; Spy I and II, Belgium). Proc Natl Acad Sci USA 108:486–491
Higham T et al (2014) The timing and spatio-temporal patterning of Neanderthal disappearance. Nature 512:306–309
Hublin J-J (2015) The modern human colonization of western Eurasia: when and where? Quatern Sci Rev 118:194–210
Juvigné E (1993) Contribution à la Téphrostratigraphie du Quaternaire et son application à la Géomorphologie. Service géologique de Belgique, Mémoires pour servir à l'Explication des Cartes Géologiques et Minières de la Belgique, Bruxelles, no 36, 66 p
King W (1864) The reputed fossil man of the Neanderthal. Q J Sci 1:88–97
Lohest M, Julin C, Rutot A (1925) Notice sur Julien Fraipont, membre de l'Académie. Annuaire de l'Académie royale des Sciences, des Lettres et des Beaux-Arts 91:131–197
Lyell C (1863) Geological evidences of the antiquity of man: with remarks on theories of the origin of species by variation. J. Murray, London
Mayer F (1864) Ueber die fossilen Ueberreste eines menschlichen Schadels und Skelettes in einer Felsenhohle des Dussel- oder Neander-Thales. Archiv für Anatomie, Physiologie und wissenschaftliche Medicin 1–26
Nigst PR, Haesaerts P, Damblon F, Frank-Fellner C, Mallol C, Viola B, Götzinger M, Niven L, Trnka G, Hublin JJ (2014) Early modern human settlement of Europe north of the Alps occurred 43,500 years ago in a cold steppe-type environment. Proc Nat Acad Sci (PNAS) Early Edition 1–6

NorthGRIP-Members (2004) High-resolution record of Northern Hemisphere climate extending into the last interglacial period. Nature 431:147–151

Orlando L, Darlu P, Toussaint M, Bonjean D, Otte M, Hänni C (2006) Revisiting Neandertal diversity with a 100,000 year old mtDNA sequence. Curr Biol 16(11):R400–R402

Otte M (1979) Le paléolithique supérieur ancien en Belgique. In: Otte M (ed) Musées royaux d'Art et d'Histoire, vol 5. Monographies d'Archéologie Nationale, Bruxelles, p 684

Otte M (1992) (dir.) Recherches aux grottes de Sclayn. Volume 1: Le contexte, vol 27. Etudes et Recherches Archéologiques de l'Université de Liège, Liègep, p 178

Otte M, Patou-Mathis M, Bonjean D (dir.) (1998) Recherches aux grottes de Sclayn. Volume 2: L'Archéologie, vol 79. Etudes et Recherches Archéologiques de l'Université de Liège, Liège, 437 p

Pirson S (2007) Contribution à l'étude des dépôts d'entrée de grotte en Belgique au Pléistocène supérieur. Stratigraphie, sédimentogenèse et paléoenvironnement. Faculté des Sciences, Université de Liège, Liège, 435 p

Pirson S, Di Modica K (2011) Position chronostratigraphique des productions lithiques du Paléolithique ancien en Belgique: un état de la question. In: Toussaint M, Di Modica K, Pirson S (eds) Le Paléolithique moyen en Belgique. Mélanges Marguerite Ulrix-Closset, vol 128. Bulletin de la Société belge d'études Géologiques et Archeologiques «Les Chercheurs de la Wallonie», hors-série n° 4 et Etudes et Recherches archéologiques de l'Université de Liege, pp 105–148

Pirson S, Toussaint M (2011) Neandertal, l'Européen, Service public de Wallonie, Namur, p 127

Pirson S, Toussaint M, Frébutte C (2003) Les matières premières des mégalithes de Belgique: état de la question Notae Praehistoricae 23:147–172

Pirson S, Haesaerts P, Court-Picon M, Damblon F, Toussaint M, Debenham NC, Draily C (2006) Belgian cave entrance and rock-shelter sequences as palaeoenvironmental data recorders: the example of Walou cave. Geologica Belgica 9(3–4):275–286

Pirson S, Court-Picon M, Haesaerts P, Bonjean D, Damblon F (2008) New data on geology, anthracology and palynology from the Scladina Cave pleistocene sequence: preliminary results. Mem Geol Surv Belgium 55:71–93

Pirson S, Haesaerts P, Di Modica K (2009) Cadre chronostratigraphique des principaux gisements du Paléolithique moyen du bassin de la Haine: un état de la question. In: Di Modica K, Jungels C (eds) Paléolithique moyen en Wallonie. La collection Louis Eloy, Collections du Patrimoine culturel de la Communauté française, n° 2, pp 58–77

Pirson S, Draily C, Toussaint M (dir.) (2011) La grotte Walou à Trooz (Belgique). Fouilles de 1996 à 2004, vol 1. Les sciences de la terre. Namur, Service public de Wallonie (Etudes et Documents, Archéologie, 20), 208 p

Pirson S, Delcambre B, Goemaere E (2013a) Chapter V. Geological context. In: H. Rougier, P. Semal (eds) Spy Cave. 125 years of multidisciplinary research at the Betche aux Rotches (Jemeppe-sur-Sambre, Province of Namur, Belgium), vol 1. Anthropologica et Praehistorica, 123 (2012), Bruxelles, pp 73–90

Pirson S, Di Modica K, Jungels C, Flas D, Hauzeur A, Toussaint M, Semal P (2013b) Chapter VI. The stratigraphy of Spy cave. A review of the available lithostratigraphic and archaeostratigraphic information. In: Rougier H, Semal P (eds) Spy Cave. 125 years of multidisciplinary research at the Betche aux Rotches (Jemeppe-sur-Sambre, Province of Namur, Belgium), vol 1. Anthropologica et Praehistorica, 123 (2012), Bruxelles, pp 91–131

Pruner-Bey F, Broca P, de Mortillet G (1866) Discussion sur la mâchoire humaine de La Naulette (Belgique). Bulletin de la Société d'Anthropologie de Paris, 2ème série 1:584–603

Rasmussen SO, Bigler M, Blockley SPE, Blunier T, Buchardt SL, Clausen HB, Cvijanovic I, Dahl-Jensen D, Johnsen SJ, Fischer H, Gkinis V, Guillevic M, Hoek WZ, Lowe JJ, Pedro JB, Popp T, Seierstad IK, Steffensen JP, Svensson AM, Vallelonga P, Vinther BM, Walker MJC, Wheatley JJ, Winstrup M (2014) A stratigraphic framework for abrupt climatic changes during the Last Glacial period based on three synchronized Greenland ice-core records: refining and extending the INTIMATE event stratigraphy. Quat Sci Rev 106:14–28

Rixhon G, Bourlès D, Braucher R, Siame L, Cordy J-M, Demoulin A (2014) 10Be dating of the Main Terrace level in the Amblève valley (Ardennes, Belgium): new age constraint on the archaeological and palaeontological filling of the Belle-Roche palaeokarst. Boreas 43:528–542

Roebroeks W (2008) Time for the Middle to Upper Paleolithic transition in Europe. J Hum Evol 55(5):918–926

Rougier H, Semal P (dir.) (2013) Spy Cave. 125 years of multidisciplinary research at the Betche aux Rotches (Jemeppe-sur-Sambre, Province of Namur, Belgium), vol 1. Anthropologica et Praehistorica, 123 (2012), Bruxelles, p 380

Rougier H, Crèvecoeur I, Fiers E, Hauzeur A, Germonpré M, Maureille B, Semal P (2004) Collections de la Grotte de Spy: (re) découvertes et inventaire anthropologique Notae Praehistoricae 24:181–190

Rougier H, Crevecoeur I, Beauval C, Posth C, Flas D, Wißing C, Furtwängler A, Germonpré M, Gómez-Olivencia A, Semal P, van der Plicht J, Bocherens H, Krause J (2016) Neandertal cannibalism and Neandertal bones used as tools in Northern Europe. Nature (Scientific Reports 6): 29005:1–11. doi:10.1038/srep29005

Rutot A (1887) Sur l'âge du grès de Fayat. Bulletin de la Société belge de Géologie de Paléontologie et d'Hydrologie 1:42–48

Schmerling PC (1833–1834) Recherches sur les ossemens fossiles découverts dans les cavernes de la province de Liège, vol 2. P. J. Collardin, Liège, 167 and 195 p

Semal P, Rougier H, Crèvecoeur I, Jungels C, Flas D, Hauzeur A, Maureille B, Germonpré M, Bocherens H, Pirson S, Cammaert L, De Clerck N, Hambucken A, Higham T, Toussaint M, van der Plicht J (2009) New data on the late neandertals: direct dating of the belgian spy fossils. Am J Phys Anthropol 138:421–428

Semal P, Jungels C, Di Modica K, Flas D, Hauzeur A, Toussaint M, Pirson S, Khlopachev G, Pesesse D, Tartar E, Crèvecoeur I, Rougier H, Maureille B (2011) La grotte de Spy (Jemeppe-sur-Sambre; prov. Namur). In: Toussaint M, Di Modica K, Pirson S (eds) Le Paléolithique moyen en Belgique. Mélanges Marguerite Ulrix-Closset, vol 128. Bulletin de la Société belge d'études Géologiques et Archeologiques «Les Chercheurs de la Wallonie», hors-série n° 4 et Etudes et Recherches archeologiques de l'Université de Liege, pp 305–321

Smith TM, Toussaint M, Reid DJ, Olejniczak AJ, Hublin J-J (2007) Rapid dental development in a Middle Paleolithic Belgian Neanderthal. PNAS 104(51):20220–20225

Toussaint M (1992) The role of Wallonia in the history of paleoanthropology. In: Toussaint M (ed) 5 millions d'années. L'aventure humaine, vol 56. Etude et Recherche Archéologique de l'Université de Liège, Liège, pp 27–41

Toussaint M (2001) Les hommes fossiles en Wallonie. De Philippe-Charles Schmerling à Julien Fraipont, l'émergence de la paléoanthropologie. Ministère de la Région wallonne. Direction générale de l'Aménagement du Territoire, du Logement et du Patrimoine, division du Patrimoine, Namur, 60 p

Toussaint M (2013) Transitions, ruptures et continuité dans les pratiques sépulcrales préhistoriques du karst mosan belge et de ses abords. In: Jaubert J, Fourment N, Depaepe P (eds) Transitions, ruptures et continuité en Préhistoire. Actes du XXVIIe Congrès préhistorique de France, Bordeaux/Les Eyzies, 31 mai-5 juin 2010.

Vol. 1: Évolution des techniques - Comportements funéraires - Néolithique ancien, pp 183–200
Toussaint M, Pirson S (2006) Neandertal studies in Belgium: 2000–2005. Periodicum Biologorum 108(3):373–387
Toussaint M, Pirson S (2007) Aperçu historique des recherches concernant l'homme préhistorique dans le karst belge aux XIXe et XXe siècles: archéologie, géologie, paléoanthropologie, paléontologie, datations. In: Evin J (ed) Un siècle de construction du discours scientifique en préhistoire, vol 2. Actes du XXVIe Congrès préhistorique de France. Avignon, 21–25 septembre 2004. Société préhistorique française, pp 117–142
Toussaint M, Pirson S (2013) L'archéologie en Wallonie. Les chasseurs-cueilleurs du Paléolithique et du MésolithiqueCarnets du Patrimoine, vol 109. Service public de Wallonie, Namur, p 68
Toussaint M, Bonjean D (dir.) (2014) The Scladina I-4A Juvenile Neandertal (Andenne, Belgium). Palaeoanthropology and Context, vol 134. Etudes et Recherches archéologiques de l'Université de Liège, Liège, 464 p
Toussaint M, Otte M, Bonjean D, Bocherens H, Falguères C, Yokoyama Y (1998) Les restes humains néandertaliens immatures de la couche 4A de la grotte Scladina (Andenne, Belgique). Comptes rendus de l'Académie des Sciences de Paris 326: 737–742
Toussaint M, Pirson S, Bocherens H (2001) Neandertals from Belgium. Anthropologie et Préhistoire 112:21–38
Trinkaus E, Shipman P (1992) The Neandertals. Changing the image of mankind, Seuil, Paris
Trinkaus E, Moldovan O, Milota S, Bîlgar A, Sarcina L, Athreya S, Bailey SE, Rodrigo R, Mircea G, Higham T, Bronk Ramsey C, Van der Plicht J (2003) An early modern human from the Pestera cu Oase, Romania. PNAS 100(20):11231–11236
Ulrix-Closset M (1975) Le Paléolithique moyen dans le bassin mosan en Belgique. Universa, Wetteren, 221 p
van Gijssel K (2006) A continent-wide framework for local and regional stratigraphies: application of genetic sequence and event stratigraphy to the Middle Pleistocene terrestrial succession of Northwest and Central EuropeFaculteit der Wiskunde en Natuurwetenschappen en die der Geneeskunde Ph.D. thesis, University of Leiden, Leiden, p 117
Vandermeersch B, Maureille B (2009) Les Néandertaliens. Biologie et cultures, vol 23. Editions du Comité des travaux historiques et scientifiques. Documents préhistoriques, Paris, p 342
Virchow R (1872) Untersuchung des Neanderthal-Schadels. Verhandlungen der Berliner Gesellschaft für Anthropologie, Ethnologie und Urgeschichte 4:157–165
Wilson RCL, Drury SA, Chapman JL (2000) The great ice age. Climate change and life. Routledge, The Open University, London, p 267
Wissing C, Rougier H, Crevecoeur I, Germonpré M, Naito YI, Semal P, Bocherens H (2016) Isotopic evidence for dietary ecology of Late Neandertals in North-Western Europe. Quatern Int 411A:327–345

The Semois Valley in Southern Ardenne: Short-Wavelength, Large-Amplitude Meanders Incised into a Slaty Basement

22

François Petit, Eric Hallot and Geoffrey Houbrechts

Abstract

The incised meanders of the Semois Valley, in Southern Ardenne, are remarkably elongated, forming a nearly 100-km-long meander belt. Their elongation is conditioned by the relative orientation of the meander loops and the regional slaty cleavage, as theorized by Strahler. These meanders already existed in the Early Quaternary, some cut-off meanders lying more than 100 m above the present floodplain. Although the Quaternary incision rates of the Semois were low (≤ 0.05 mm/year), several meanders look more entrenched than ingrown, probably because lateral erosion was fairly slow (~ 0.3–0.5 mm/year), owing to the large quantity of material to evacuate by a rather small river. The combined aesthetic and scientific interests of this particularly well-developed meander belt speak in favour of the preservation of the nicest meanders as geomorphosites.

Keywords

Entrenched meanders • Strahler theory • Slaty cleavage • Elongated meanders • River erosion

22.1 Introduction

The Ardennian Semois Valley is among the most picturesque valleys of Belgium and features remarkable geomorphological landscapes (Paulissen et al. 1979). The river has its source in the Jurassic cover south of the Ardenne and develops barely incised broad meanders in the $\sim$100-km-long first half of its course in a 5- to 6-km-wide depression cut into Liassic marls at the foot of the Ardennian slopes. Then the Semois definitively enters the resistant Paleozoic basement of the Ardenne 3 km downstream of Chassepierre, where it cuts a meandering $\sim$100-km-long course down to its confluence with the Meuse at Monthermé, France. Its catchment area is 1350 km^2. The succession of an upstream course in soft rocks and a downstream course in very resistant rocks sharply contrasts with the usual watershed organization. According to Pissart (1961), a Late Miocene capture of the Semois by the present Ardennian Meuse could have induced a wave of erosion that allowed the river to capture former tributaries of the Chiers flowing onto the Mesozoic cover at the foot of the massif, thus creating this surprising pattern (see Fig. 10.3 in Chap. 10, and Fig. 23.10 in Chap. 23). However, beyond doubts that can be raised about these captures (see Chaps. 10 and 23), river reaches in the Jurassic are observed already downstream of the hypothesized capture points. As for the elongated shape of the Semois catchment, which mainly results from the very short development of its southern tributaries, it probably also originates in the ancient geomorphic history of the area, when a superimposed Semois continued to flow WNW, outside the erosion corridor that developed towards the Moselle trough in the Oligocene (Demoulin 1995) (Fig. 22.1). Moreover,

F. Petit (✉) · G. Houbrechts
Department of Physical Geography and Quaternary, University of Liège, Sart Tilman, B11, 4000 Liège, Belgium
e-mail: Francois.Petit@ulg.ac.be

G. Houbrechts
e-mail: G.Houbrechts@ulg.ac.be

E. Hallot
ISSeP, rue du Chéra 200, 4000 Liège, Belgium
e-mail: e.hallot@issep.be

A. Demoulin (ed.), *Landscapes and Landforms of Belgium and Luxembourg*, World Geomorphological Landscapes, DOI 10.1007/978-3-319-58239-9_22

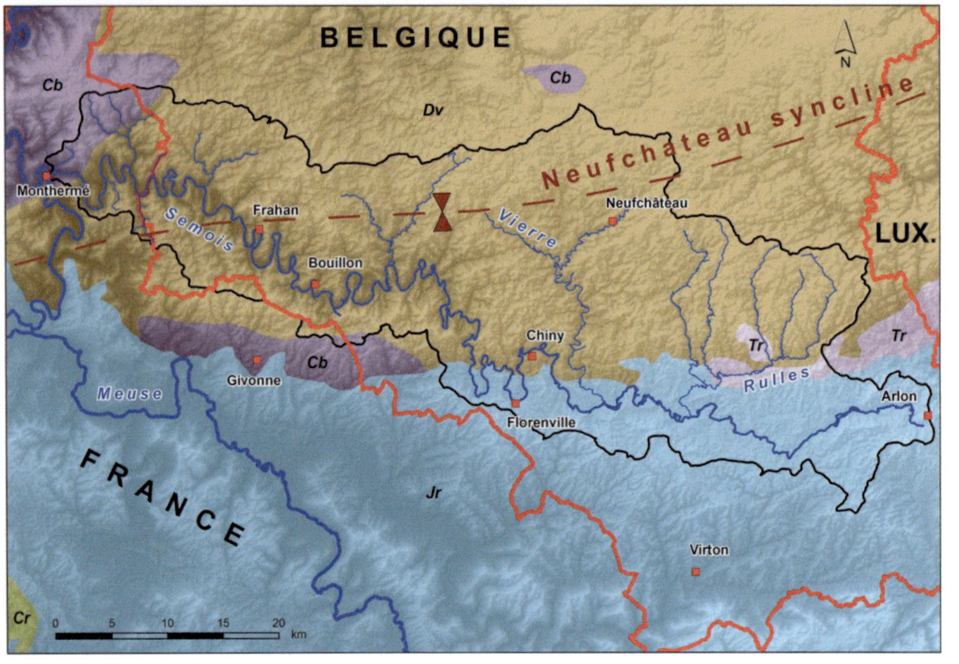

Fig. 22.1 Location map of the Semois watershed (*black outline*), with the geology draped over the SRTM 1″ DEM. *Cb.* Cambrian; *Dv.* Devonian; *Tr.* Triassic; *Jr.* Jurassic; *Cr.* Cretaceous

the northern tributaries coming from the Ardenne also drain limited catchments, with the relative exceptions of the Rulles and the Vierre, with catchments of 222 and 259 km^2, respectively.

The physiognomy of the valley changes dramatically when the Semois enters the Ardenne. The river has incised a ~150-m-deep narrow valley in the plateau. Moreover, the course of the river becomes much more sinuous, following a succession of elongated incised meanders which flow around extremely narrow necks over more than 60 km (Fig. 22.2).

22.2 Geological and Geographical Settings

The Paleozoic basement of the southern Ardenne is mainly composed of alternating Lower Devonian shales and sandstones outcropping in the E-W trending, slightly E-plunging syncline of Neufchateau which extends from Charleville, France, in the west to Wiltz, Luxembourg, and beyond in the east. In the west, the Cambrian Rocroi and Gedinne massifs outcrop in the axis of the Ardenne and Givonne anticlines, respectively. While shales and sandstones, which occur randomly throughout the area, are of similar resistance to erosion, the Cambrian quartzites and their rim of Lochkovian conglomerates are much more resistant (Pissart 1962; see also Chap. 2). To the south, the Triassic/Jurassic cover lies unconformably on the Paleozoic of the Ardenne. As the Triassic and Liassic seas drowned the massif diachronically from east to west, the Mesozoic rocks resting on the Paleozoic become younger to the west. In the study area, they are weak Lower Liassic marls.

Morphologically, the southern margin of the Ardenne massif exhibits a continuous and regular slope, leading upward to a subhorizontal erosion surface that truncates the massif at elevations around 430 m. While this surface, which widens westward, was formed in the Selandian (see Chap. 5), the 3- to 4-km-wide marginal slope itself represents another, much older and tilted topography, the so-called post-Hercynian surface, on which the Triassic–Liassic seas proceeded and which was later exhumed as an erosional scarp when a younger surface was cut into the Selandian surface in Oligo–Miocene times (Demoulin 1995).

Finally, crucial for the geomorphic phenomenon presented in this chapter is the presence of a regional slaty cleavage in the Neufchateau syncline, oriented east–west, parallel to the fold axes and the stratification (Price and Cosgrove 1990). Note that the course of the Ardennian Semois also closely follows, although at a slightly oblique angle, the axis of the syncline.

22.3 Meander Shape and Structure

Figure 22.3 shows a reminder of what defines a meander, i.e. a succession of two alternate half loops separated by an inflection point. A meander is described by two parameters, namely its wavelength and amplitude. The main zone of vertical and lateral erosion is located in the outer concave portions of the meander banks, tending to gradually narrow the meander neck. Eventually, lateral erosion in adjacent loops of same orientation puts their outer banks into contact, creating a flow path through the neck that shortcuts, or cuts

Fig. 22.2 View of Frahan from Rochehaut (see location in Fig. 22.12), showing the gentle slope on the convex side of the meander, with terraces on which the village of Frahan developed

off the meander. Flow energy is minimal on the inner convex meander banks, which thus do not experience erosion and tend instead to accumulate sediment (point bar deposits) (Fig. 22.3).

Two types of meanders occur: free meanders and entrenched meanders. The latter are incised into a plateau and the meandering channel is located within a generally narrow floodplain, in frequent contact with the valley slopes, which constrain its lateral migration. In contrast, free meanders, such as those featured by the Upper Semois in Lorraine, develop within a broad floodplain without lateral constraints imposed by the hillslopes.

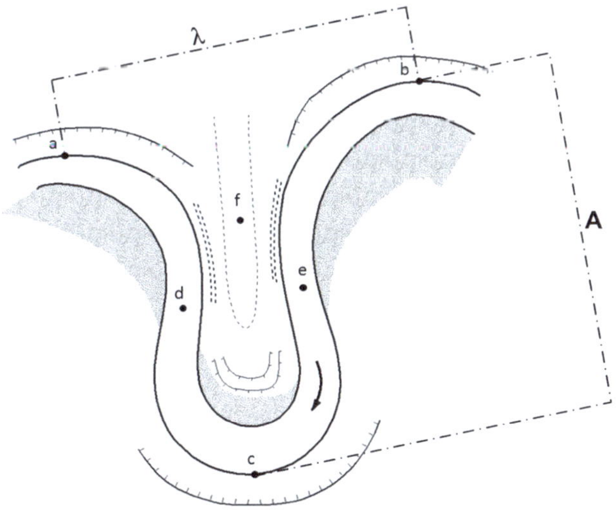

Fig. 22.3 Sketch illustrating the components of an entrenched meander and the parameters used to describe it. *a*, *b*, *c* Meander loop apexes; *d*, *e* Crossing points, or inflections; *f* Meander neck; *A* Meander amplitude; *λ* Meander wavelength. Stippled areas denote point bar deposits. Terrace risers are featured between neck f and the corresponding point bar

As localized lateral erosion causes a meander train to shift downstream, meanders tend to have an asymmetric transverse profile. In incised meanders, the concave bank and the upstream part of the convex bank have steep slopes, with frequent rock outcrops (Fig. 22.4). Owing to continuous removal by the river of the material reaching their foot, these ~30°-steep equilibrium slopes, similar to Richter slopes, are essentially cut in bedrock and display only a thin veneer of transiting debris. By contrast, downstream of the meander apex, sediment deposition forms a gently sloping surface in the convexity, the slip-off slope. If episodes of vertical erosion interfere with this process of lateral migration, stepped terraces will develop instead of a uniform slope. However, the terrace deposits are often covered by slope deposits that may re-create a smooth slope, hiding more or less the terrace morphology.

Once a meander has been cut off and, consequently, disconnected from the incising river, its bottom is progressively filled with hillslope deposits. The height of each cut-off meander above the modern floodplain and, to some extent, the thickness of its colluvial fill increase with time, thus defining their relative chronology. In the case of incised meanders, the cut-off process is obviously very long, considering the quantity of debris to evacuate and the bedrock resistance to erosion. Figure 22.5 shows that the entrenched meanders of the Ardennian Semois are characterized by a strong N–S elongation, perpendicular to the direction of the regional slaty cleavage. A vast majority of the meanders

Fig. 22.4 NW-looking view of the Semois at Mortehan (village in the foreground). The river flows from the *right* to the *left* of the photograph. Downstream of the bridge, the steep slope of the concave side (left bank) extends farther downstream than the crossing point of the meander. Stepped terraces, on which the village of Cugnon and several campsites are located, interrupt the slope of the convex side (*Photograph* D. Kemp)

between Chiny and Monthermé, including many cut-off meanders, conform to this orientation.

Meander elongation is readily described by their amplitude/wavelength ratio A/λ (Fig. 22.3). The relationship between these two variables is illustrated in Fig. 22.6 for the elongated, live and cut off, meanders of the Ardennian Semois and, for comparison, for cleavage-independent meanders of other Ardennian rivers. The A/λ ratio, and thus the elongation, of the meanders of the Ardennian Semois is more than twice that of 'normal' meanders in the Ardenne (regression slope of 1.13 vs. 0.48) but, more interestingly, the ratio of cut-off meanders is also much

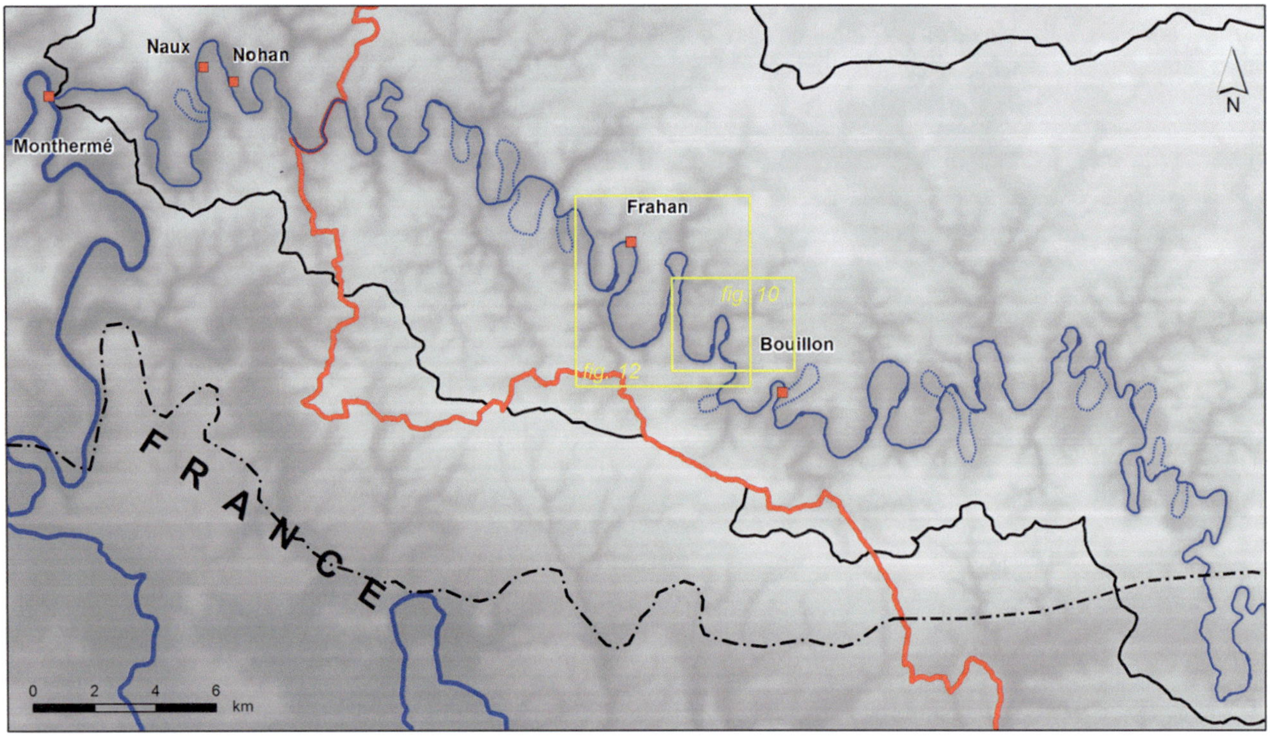

Fig. 22.5 The course of the Ardennian Semois between Chiny (lying ~10 km upstream of the river's entry in the map frame) and Monthermé, with its quasi-continuous train of elongated meanders, including 13 cut-off meanders. The *dashed-dotted line* represents the limit Paleozoic–Mesozoic (modified after Pissart 1961; Gob et al. 2005)

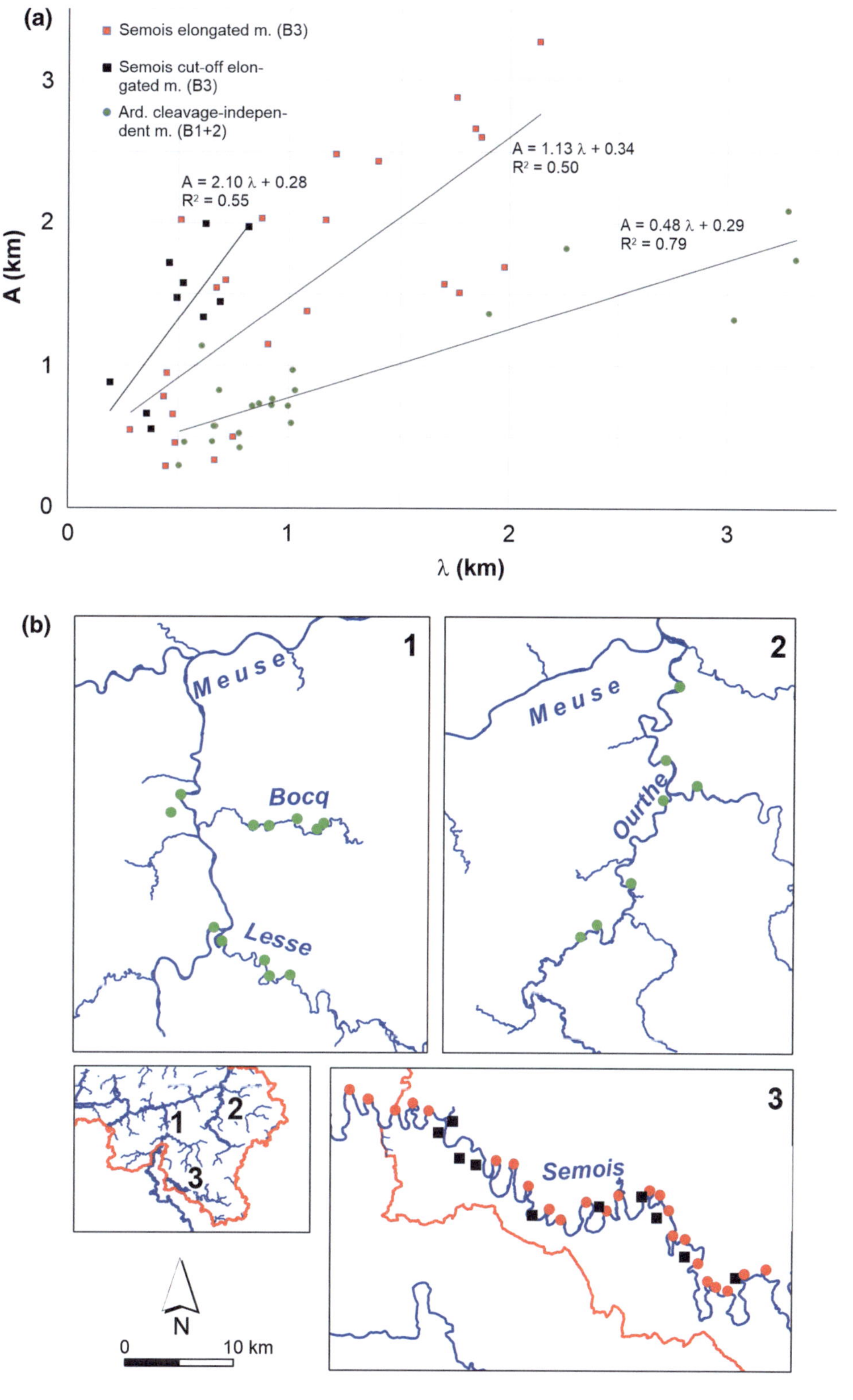

Fig. 22.6 **a** Relationship between wavelength and amplitude of the meanders. *Red squares* Active cleavage-controlled elongate meanders of the Ardennian Semois. *Black squares* Cut-off elongate meanders of the Semois. *Green circles* Cleavage-independent incised meanders in various rivers of the Ardenne. **b** Location map of the meanders used to establish the relations in *A*

larger than that of active meanders (regression slope of 2.10 vs. 1.13), which suggests that, whatever the meander size, there is a threshold elongation around 2.1 beyond which the narrowing of the neck makes cut-off highly likely (Fig. 22.7). More surprisingly, we note also that the set of cut-off meanders of the Semois does not include meanders with $A > 2$ km, whereas the amplitude of active meanders may reach 3 and more kilometres.

Fig. 22.7 South-looking view of the meander of Poupehan. Despite a wavelength λ of 2.2 km, this extremely elongated meander (A = 3.3 km) preserves only a narrow neck. The flat horizon line corresponds to the Selandian erosion surface of the southern Ardenne (see Chap. 5). Note also the narrowness of the floodplain lacking a road network in the largest part of the meander (*Photograph* D. Kemp)

22.4 Mechanisms of Meander Formation and Evolution

The relation between meander elongation and the presence of a slaty cleavage in the bedrock was highlighted by Strahler (1946) in the USA. Subsequently, a similar link has been proposed in Belgium for meanders of the Ourthe (Alexandre 1956) and the Semois (Pissart 1962).

In bedrock with slaty cleavage, meander lobes do not develop at the same speed in every direction. In addition to the general downstream shift of any meander train, individual meanders tend to grow faster and lengthen in the direction perpendicular to the cleavage, i.e. the part of the meander bank that is parallel to the cleavage direction retreats faster, owing to more effective lateral erosion. Struck by water where flow energy is maximum, rock sheets and plates slide past each other along the cleavage planes and are easily eroded by the river (Fig. 22.8). In contrast, in the part of the meander perpendicular to the cleavage, sheet and plate edges do not give an easy hold to lateral erosion, neighbouring plates, moreover, supporting each other mutually, which limits the lateral erosion efficiency in extracting rock plates from the bank. Thus, when the general river course is parallel to the slaty cleavage and the meanders develop in the perpendicular direction, they tend to become especially elongated. In contrast, in the case of a river course perpendicular to the regional cleavage, the original smoothly meandering channel may be modified by the addition of apexes and meanders developing perpendicular to the cleavage, i.e. lengthening parallel to the general river course (Fig. 22.9).

Based on the systematic observation that a bedrock platform is developed at the foot of the concave bank close to the low water level, Seret (1979) built on Strahler's explanation and suggested that erosion of the bank parallel to cleavage is mainly governed by frost shattering during winter low water levels, the river's role being then reduced to the removal of the debris. Although controlled by a different erosion process, the relation between cleavage orientation and meander elongation nevertheless remains because frost shattering is also much more effective in walls parallel to cleavage. However, it should be noted that, though potentially effective in periglacial environments, this process has been observed in rivers of the Famenne region, where

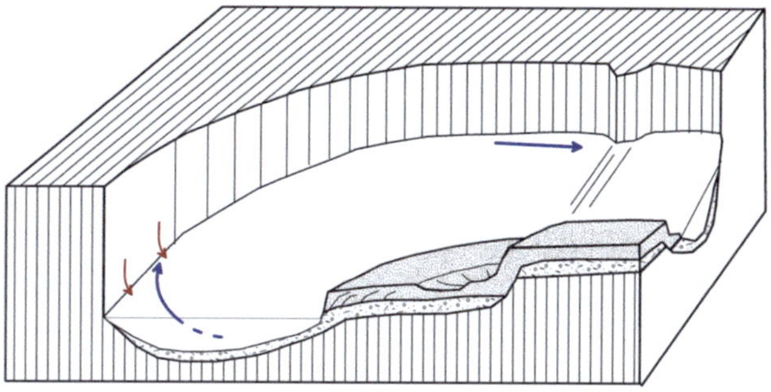

Fig. 22.8 Diagram showing the relationship between stream flow and slaty cleavage of the bedrock (*lining*) in the different parts of a meander. Stippled areas are terraces on the inner convex side of the meander (modified after Strahler 1946)

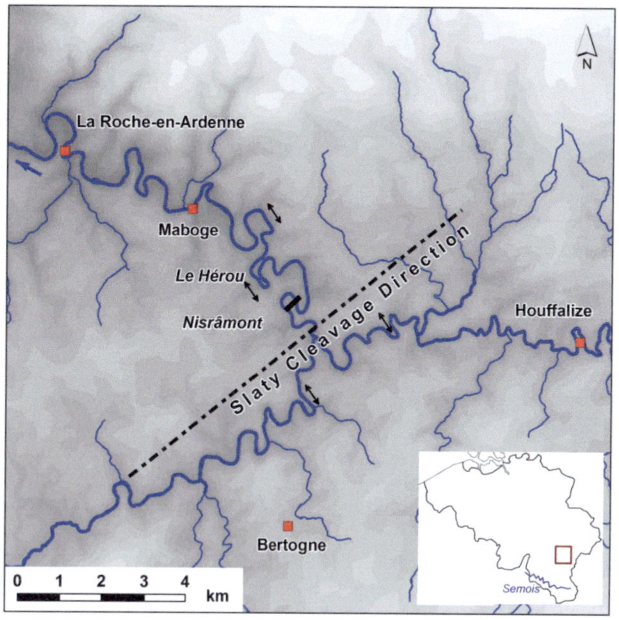

Fig. 22.9 Meanders of the Ourthe River in the confluence area of its W and E upstream courses, illustrating the variable meander morphology controlled by the angle between flow and cleavage directions in rivers flowing across a slaty bedrock (Alexandre 1956). The W and E upstream branches of the Ourthe flow parallel to the regional cleavage and form meanders that typically become elongated perpendicularly to the general stream course. In contrast, downstream of the confluence, the course of the river is perpendicular to the cleavage so that, if meanders are to become elongated, this must occur in the same direction as the general river course, thus producing complex meander forms visible, e.g. near Le Hérou. *Double arrows* denote the direction of meander elongation

the outcropping Upper Devonian shales are much less resistant to erosion than the Lower Devonian shales and slates of the Semois catchment. Pissart (1961), who believed that the Ardennian Semois course was as old as Late Miocene, concluded that its large meanders initiated and developed in parallel with the river incision into the Palaeozoic bedrock. Indeed, the stepped terraces preserved on the convex bank of some meanders indicate that at least these meanders developed during incision, contributing to the increase of the river sinuosity during the Quaternary. However, Pissart (1961) also noted that some of the 14 cut-off meanders he mapped are located high, up to 137 m at Mouzaive, above the present floodplain, suggesting that they were abandoned during the Early Pleistocene. This implies that these meanders appeared early in the incision history of the river and adapted quickly to the structure.

22.5 Dating the Meanders

Pissart (1961) studied the terraces of the Semois and the Meuse in their Ardennian courses. He identified about sixty terrace remnants in the Semois valley, which he grouped into ten levels, connecting some of them to those of the Meuse valley. Thus, the three lower terraces of the Semois, located 6, 9, and 14 m, respectively, above the present floodplain, seem to connect to the three levels of the Meuse grouped together as the Main Terrace, whose middle level has been dated north of Liege to 725 ± 120 ka by Rixhon et al. (2011), while the lower level is accordingly dated at ~625 ka (see Chap. 10). Whereas the Meuse has incised ~60 m in the Liège-Visé area, north of the Ardenne, since formation of this youngest Main Terrace level, downcutting of the Meuse and Semois valleys did not exceed ~15 m in the same time at Monthermé, south of the massif, which yields an average Middle Pleistocene to Holocene incision rate ≤25 m/Ma in the Semois catchment. However, Pissart (1961) also notes that the cutting off of numerous meanders caused incision rates to be locally higher in the Ardennian Semois.

The stepped terraces aligned on the convex bank of several meanders reflect their growing elongation and shift of apex with time and, if dated, would allow an estimation of the rate of lateral erosion of the concave bank, assuming constant valley floor width. Observed shifts range from 500 m (e.g. at Nohan) to 1500 m (at Naux), starting however from distinct terrace levels (shift-related incision has been 60 and 95 m at Nohan and Naux, respectively). As the only (weak) time anchor available in the Semois valley is the Younger Main Terrace (625 ka), the ~250 m shift of the apex of the Naux meander that occurred since abandonment of this terrace suggests a local lateral erosion rate of ~0.4 mm/year. However, several mostly south-looking meanders display no terrace remnants but instead show a steep convex valley side without noticeable shift of the meander apex. For instance, a steep 90-m-high inner slope rises directly from the present floodplain in the meander east of Laforêt, which even seems to have undergone retrogradation in the course of incision. Likewise, a 120-m-high inner valley side with a uniform ~17° slope is observed for the meander west of

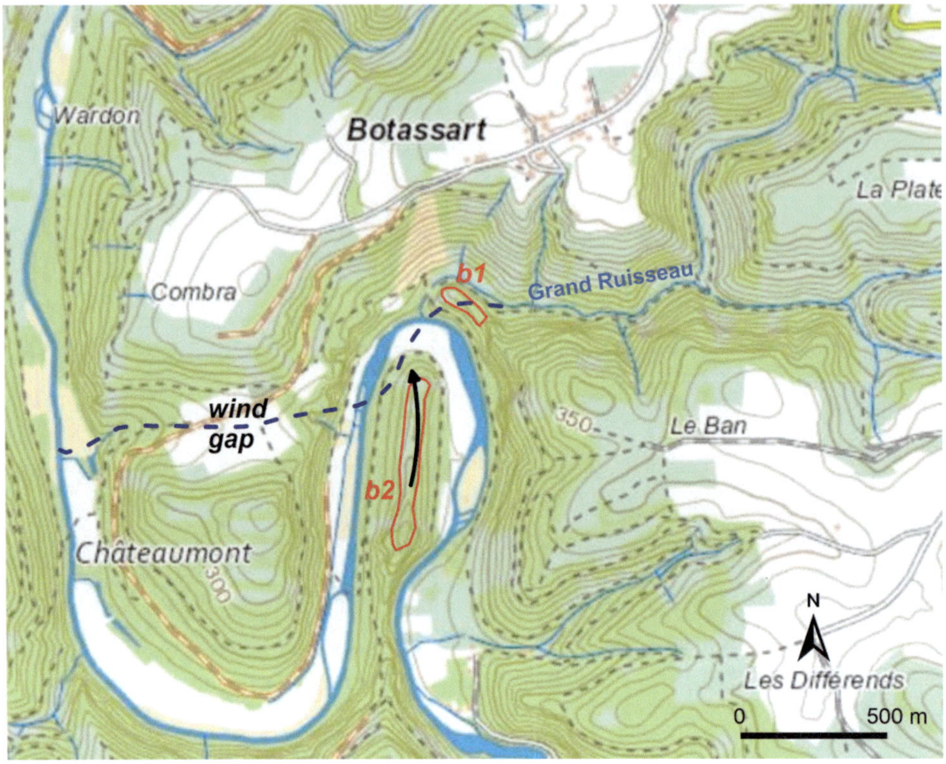

Fig. 22.10 The Sainte-Austreberthe-type capture of the Grand Ruisseau caused by the northward elongation of the Tombeau du Géant meander (see location in Fig. 22.5). The beheaded valley is now left as a wide wind gap across the neck of the next meander. *Dashed-blue line* Pre-capture course of the Grand Ruisseau. The *black arrow* figures the progressive elongation of the meander. The long bevelled neck *b2* of the meander lies at altitudes just above that of the capture (suggested by the presence of bench *b1*), highlighting how an episode of active lateral erosion and meander elongation caused the capture, after which quasi-zero lateral erosion occurred. Cartographic background from http://www.ngi.be/topomapviewer/public

Rochehaut, indicating that almost no meander elongation (i.e. significantly less than 0.1 mm/year) occurred there since a very long time. While these overall very low lateral erosion rates partly result from the volume of fairly resistant material the incising river had to remove, the contrasting behaviour of different meanders reveals a complex evolution of the river sinuosity with time, resulting in the juxtaposition of ingrown and entrenched meanders (ingrown meanders result from low downcutting rates allowing lateral erosion to develop an asymmetric transverse valley profile, while entrenched meanders are produced by high incision rates accompanied by insignificant lateral erosion, resulting in a symmetric transverse profile of the meanders). This suggests also that the Ardennian Semois had already formed a part of its well-developed meanders before it started to incise its valley into the Paleozoic bedrock from the dawn of the Quaternary onwards.

Progress of lateral erosion during meander incision is also confirmed by a nice case of Sainte-Austreberthe-type stream capture in the 'Tombeau du Géant' meander, south of Botassart (Fig. 22.10). Lateral erosion and northward shift of the meander made it tangent to the Grand Ruisseau valley, resulting in the diversion of the Grand Ruisseau stream, whose beheaded reach now appears as a wind gap across the meander neck immediately downstream. Although this probably much colluviated windgap lies 70 m above the present floodplain of the Semois, a bench located 40 m above the floodplain along the Grand Ruisseau valley where the latter currently joins the Semois suggests the capture occurred at this relative altitude. As the wind gap, though colluviated, attests significant pre-capture incision into the plateau, the capture event confirms that this meander grew and lengthened during incision. In this respect, it is worth noting that an extended bench crowns the Tombeau du Géant neck at 245–255 m asl, i.e. just above the altitude of the capture. This bench probably bears witness to an episode of active meander lengthening that would have ended in the Grand Ruisseau capture. However, the reconstructed path of the Grand Ruisseau inside the capturing meander allows estimating the post-capture elongation to no more than ~ 120 m, which, based on a rough capture age of 1.2–1.5 Ma, translates

Fig. 22.11 View of the town of Bouillon with its medieval castle perched on *top* of the meander neck. The town was established in the floodplain widened at the height of a cut-off meander (*Photograph* D. Kemp)

into a very low rate of lateral erosion of 0.08–0.10 mm/year in the last 1–1.5 million years.

22.6 Geomorphic Control on Human Settlements and Communication Modes in the Ardennian Valley of the Semois

The sinuous course of the Ardennian Semois also influenced human settlements. Relatively few villages are present in the bottom of the valley because the floodplain is generally fairly narrow and the elongated meanders make communications difficult. Exceptions are observed on the convex banks where extended stepped terraces allowed settlement outside of flooding areas and offer arable soils (Figs. 22.2 and 22.4), and in abandoned meanders (e.g. Dohan, Laforet). The site of the small town of Bouillon, with its medieval castle perched on the meander neck, combines the advantages of both an enlarged floodplain and an abandoned meander (Figs. 22.5 and 22.11).

Although roads at places link villages across a meander neck, the main road network is located on the plateau with dead-end access to the villages in the valley bottom, which sometimes makes meanders devoid of village, such as the Poupehan and Tombeau du Géant meanders. They are almost inaccessible and may require very long travel to link villages close to each other as the crow flies (Fig. 22.12). It is poorly known, however, how the modern road network reflects the travel conditions when the villages and hamlets were first settled. Another sign of a deficient road network in the valley was that wood floating was practiced on the

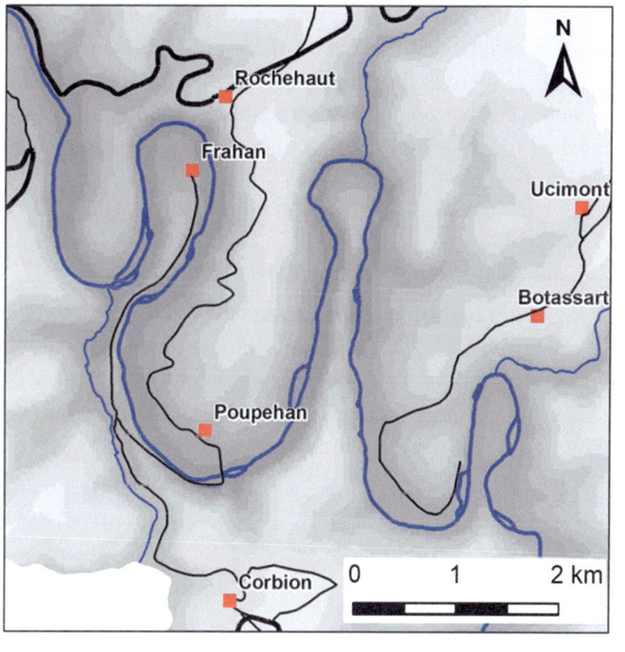

Fig. 22.12 An example of the impracticality of the road network caused by the morphology of the meanders in the Ardennian Semois valley. In this example (see location on Fig. 22.5), the two villages of Rochehaut and Frahan, which are only 1 km apart as the crow flies, are linked by a single 9-km-long road

Semois until the Second World War. Indeed, this expensive technique was used in the Ardenne only where poor accessibility ruled out every other way to transport the logs. Marquet (1973) provides several examples on the Ourthe, particularly between Nisramont and Maboge, where the river also forms tight entrenched meanders (Fig. 22.9). Though wood floating was conducted along the entire course of the Semois, several places were especially known for this practice, such as Cugnon, Bouillon, Bohan, and Poupehan. Near each village, dams were built to collect floating firewood and brushwood. Only timber intended for mine galleries was floated down the Semois and the Meuse. Such poor accessibility likely contributed to the good preservation of most natural sites along the Semois, especially also because the region is not densely inhabited and no major urban area exists in the immediate vicinity.

22.7 Conclusion

We have presented and discussed the elongated meanders of the Ardennian Semois as a remarkable case of structural influence on the evolution of meandering rivers. Present along the whole stretch of the river across the Paleozoic basement, these meanders were partly existent in the Early Quaternary, when the Semois began to incise its valley into the plateau. Despite very vague dating, field evidence shows that, while many meanders grew in length during incision, some others seem instead to be more of the entrenched type. The overall Quaternary rates of lateral erosion were, and still are, low, probably owing to the resistance to erosion of the rocks and the volume of material to be removed by the river.

The rarity of such a 100-km-long belt of elongate entrenched meanders makes the aesthetically appealing landscapes of the Ardennian Semois also geomorphologically highly valuable and, beyond the creation of geomorphosites in the best-preserved meanders, should encourage the implementation of a protective status for the entire Ardennian Semois valley. Of course, this will require willingness, firmness, and vigilance by the authorities if the Semois meanders are to be preserved in a state as natural as possible. As an example, it took more than fifty years before the Frahan meander was classified as an exceptional site of the Walloon Region in 1997 and campsites were removed from the meander.

References

Alexandre J (1956) Les méandres de l'Ourthe supérieure. Ann Soc Géol Belg 80:75–90

Demoulin A (1995) L'Ardenne des plateaux, héritage des temps anciens: surfaces d'érosion en Ardenne. In Demoulin A (ed) L'Ardenne: Essai de Géographie Physique, Hommage au Professeur A. Pissart. Université de Liège, Liège, pp 68–93

Gob F, Houbrechts G, Hiver J-M, Petit F (2005) River dredging, channel dynamics and bedload transport in an incised meandering river (the River Semois, Belgium). River Res Appl 21(7):791–804

Marquet L (1973) Le flottage du bois. Enquêtes du musée de la Vie Wallonne 13:163–204

Paulissen E, Vlassenbroeck, Antrop M (1979) Acht Belgische Landschappen in Stereovisie: De Ardense Semois. De Aardrijkskunde, 225–238

Pissart A (1961) Les terrasses de la Meuse et de la Semois. La capture de la Meuse lorraine par la Meuse de Dinant. Annales de la Société Géologique de Belgique 84:1–108

Pissart A (1962) Les aplanissements tertiaires et les surfaces d'érosion anciennes de l'Ardenne du Sud-Ouest. Annales de la Société Géologique de Belgique 85:71–150

Price NJ, Cosgrove JW (1990) Analysis of geological structures. Cambridge University Press, Cambridge, p 502

Rixhon G, Braucher R, Bourlès D, Siame L, Bovy B, Demoulin A (2011) Quaternary river incision in NE Ardennes (Belgium)-Insights from $^{10}Be/^{26}Al$ dating of the terraces. Quat Geochronol 6:273–284

Seret G (1979) L'étirement des méandres encaissés perpendiculairement à la direction du clivage schisteux: un processus périglaciaire. Bulletin Société belge de Géologie 88:129–135

Strahler AN (1946) Elongate entrenched meanders of Conodoguinet creek Pa. Am J Sci 244:31–40

Cuestas in Gutland (S Luxembourg) and Belgian Lorraine: Evolution of a Structurally Controlled Landscape

23

François Petit, Robert Maquil, Birgit Kausch and Eric Hallot

Abstract

The cuesta landscape of Belgian Lorraine and Luxembourg's Gutland belongs to the north-eastern rim of the Paris Basin, characterized by the wide syncline of the Gulf of Luxembourg. Highlighting the underlying geological structure and lithology, the cuestas were progressively sculpted in close relation with the contrasted evolutions of the Meuse and Rhine river systems. The landscape was established during the Cenozoic in Triassic and Jurassic rocks. These are formed of alternating hard pervious and soft, generally impervious units, the latter with a limited resistance to superficial weathering under the climatic conditions of the Neogene and Quaternary. The hard units frequently appear as thick homogenous layers, not only making the core of the cuestas but featuring also many escarpments. Height and lateral extent of the cuestas are controlled by the thickness of the hard layers and lateral changes in facies, as exemplified in the Middle Liassic sandstone and Triassic rocks. Remnants of an extended Neogene erosional surface cut the cuesta summits at ~400–420 m asl. Later fluvial erosion incised this surface, allowing for the development of lower planation surfaces down to 300 m, including a prominent level at ~350 m. Lower structural surfaces also exist on top of hard layers. Finally, landsliding is shown to participate actively in sculpting the cuesta landscape.

Keywords

Belgian Lorraine • Gutland • Cuesta • Structural geomorphology • Silcrete • Ferricrete

23.1 Introduction

Coming down from the deeply incised Ardennes and Eislek (or Oesling) Paleozoic plateaus (e.g. driving from Bastogne to Arlon, see Fig. 23.1), one discovers to the south the widely open undulating landscape of the Belgian Lorraine and Luxemburgian Gutland, which appears as an extended depression limited by a continuous ridge at the horizon (Fig. 23.2). Developed in the Mesozoic cover of the Paris Basin, this depression is in fact subdivided by a series of more or less parallel asymmetric ridges of variable height and highest altitudes of ~400–420 m. These ridges systematically display steep N-facing escarpments and gentle slopes on their southern side (Fig. 23.1). With the intervening troughs, they form the typical cuesta landscape of the eastern Paris Basin. In the Belgian Lorraine, three ridges run

F. Petit (✉)
Department of Physical Geography and Quaternary, University of Liège, Sart Tilman, B11, 4000 Liège, Belgium
e-mail: Francois.Petit@ulg.ac.be

R. Maquil
91, rue Clairefontaine, 9220 Diekirch, Luxembourg
e-mail: robert.maquil@pt.lu

B. Kausch
Service Géologique du Luxembourg/Geological Survey of Luxembourg, 23, rue du Chemin de Fer, 8057 Bertrange, Luxembourg
e-mail: b.kausch@web.de

E. Hallot
ISSeP, rue du Chéra 200, 4000 Liège, Belgium
e-mail: e.hallot@issep.be

A. Demoulin (ed.), *Landscapes and Landforms of Belgium and Luxembourg*, World Geomorphological Landscapes, DOI 10.1007/978-3-319-58239-9_23

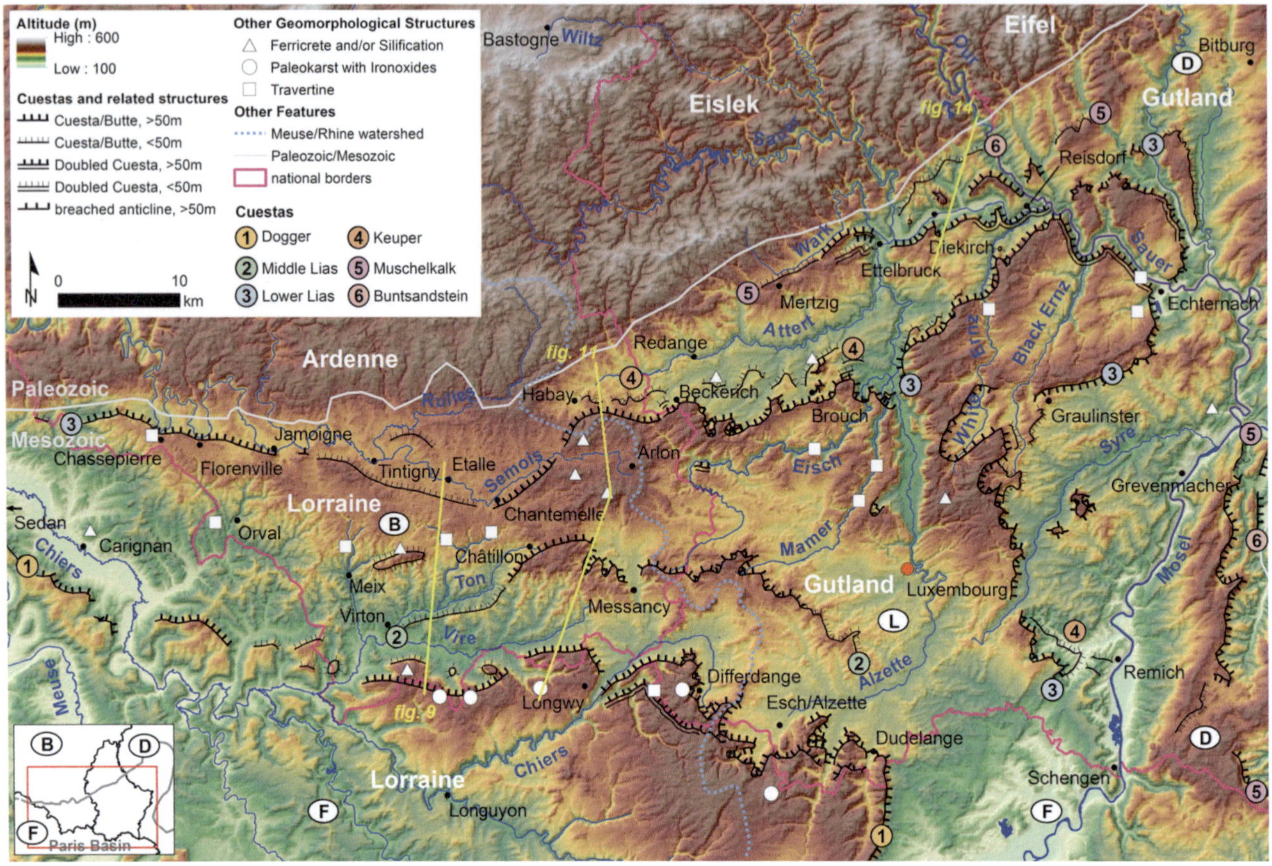

Fig. 23.1 Geomorphological map of the cuesta landscape on a background Digital Elevation Model (DEM) of southern Belgium and Luxembourg. It is admitted that cuestas do exist also in Germany, to the northeast of our study area. *Database* Jarvis et al. (2008)

roughly from west to east. The depression widens progressively eastward, where six separate ridges are distinguished in the Luxembourg Gutland. The southernmost, highest ridge, which is also the most continuous across the study area, may be followed as an arcuate morphology from Sedan, France, in the west, through Belgian Lorraine and Gutland, to Nancy, France, in the south.

23.2 Geographical and Geological Settings

Luxembourgian Gutland and Belgian Lorraine together make the northeasternmost part of the Paris Basin. Structurally, the Belgian Lorraine is characterized by E–W oriented, gently S-sloping formations of mainly Early and early Middle Jurassic age successively covering the southern slope of the Ardennes Paleozoic massif from east to west. Eastwards, the Mesozoic sequence starts with more and more important Triassic deposits bordering Ardennes and Eislek in the Gutland (see Fig. 2.1 of Chap. 2) and becoming almost exclusive to the east in Germany. In the Gutland, the simple homoclinal structure of the Mesozoic strata evolves in a more complex NE-trending syncline, the Gulf of Luxembourg, displaying secondary undulations more or less parallel to the syncline axis and several regional-scale faults. Details about the Mesozoic cover will be provided in the following sections in relation with the landforms corresponding to specific formations.

The water divide between the drainage basins of the rivers Meuse and Rhine runs in a N–S direction across the study area, approximately separating the Belgian Lorraine from the Gutland and cutting all cuestas at right angle (Fig. 23.1). The rivers of Belgian Lorraine belong to the Meuse catchment, those in Gutland mostly to the Rhine catchment. In Belgian Lorraine, the drainage network tends to conform to the underlying geological structure. By contrast, in Gutland, most rivers flow north- or northeastward as superimposed courses independent of the geological structure (see, e.g. the obsequent course of the Alzette cutting across hard Hettangian sandstones north of Luxembourg City, Fig. 23.1; see also Chap. 4).

It is common belief that the cuestas were produced by Late Cenozoic differential erosion of a gently undulating erosion surface preserved on their top at elevations of 400–420 m. Sparse remnants of older surfaces are observed locally at higher altitudes, especially near the water divide.

Fig. 23.2 View from the Ardennes plateau to the south showing the peripheral subsequent depression, the wooded front of the Lower Lias cuesta in the middle ground (with the city of Arlon and its church on *top*) and the Dogger cuesta in the far left background. In the *right* background, the Hirtzenberg, capped by ferricretes

However, contrasted points of view have been proposed about the formation and evolution of the cuestas, which will be discussed hereafter, after a detailed description of the individual cuestas.

23.3 Cuesta Morphology: The Basics

Generally, a cuesta is characterized by an asymmetric ridge opposing a steep front-slope cut in hard rocks and a structural slope corresponding to the gently dipping bedding planes on top of the hard strata (Fig. 23.3). This dip-slope progressively goes into the surface cutting through softer rocks at lower altitudes. The resulting cuesta landscape thus consists of alternating ridges and troughs. Steep escarpments underline the outcrop area of resistant strata in the landscape, their height and lateral extent being controlled by the thickness and lateral facies variations of the strata. In Lorraine and Gutland, the flat top of the cuestas shows that an ancient erosion surface truncates the upper part of the dip-slope in the hard layers. Minor structural surfaces may also occur at slightly lower altitudes in the rear of a cuesta when a layer of moderately resistant rocks subsist above the hard rocks. The width of the troughs, also called subsequent depressions, is largely controlled by the thickness of the weak formation and, still more, the dip of the strata. Changes in dip direction are strongly related to the geometry of the underlying geological structure.

Two cases of cuesta splitting into two close, parallel front-slopes may occur. A cuesta is said to be double when the two front-slopes correspond to two superposed hard layers separated by a (generally thin) soft unit (Figs. 23.3 and 23.4), thus featuring a complex escarpment made of a steep upper slope going down into a concave slope on the intervening soft layer and, finally, a bottom cliff. Where a river flows almost parallel to the cuesta and cuts across the front-slope at an acute angle, a single hard layer may display two close parallel escarpments, which together form a doubled cuesta (Figs. 23.3 and 23.5). Isolated hills capped by remnants of the hard layer in front of the cuesta are called buttes (Figs. 23.3 and 23.6). Finally, where the Mesozoic cover is slightly folded in Gutland, breached anticlines rimmed by mutually facing scarp slopes, such as those observed in the Alzettte and White Ernz valleys, north of

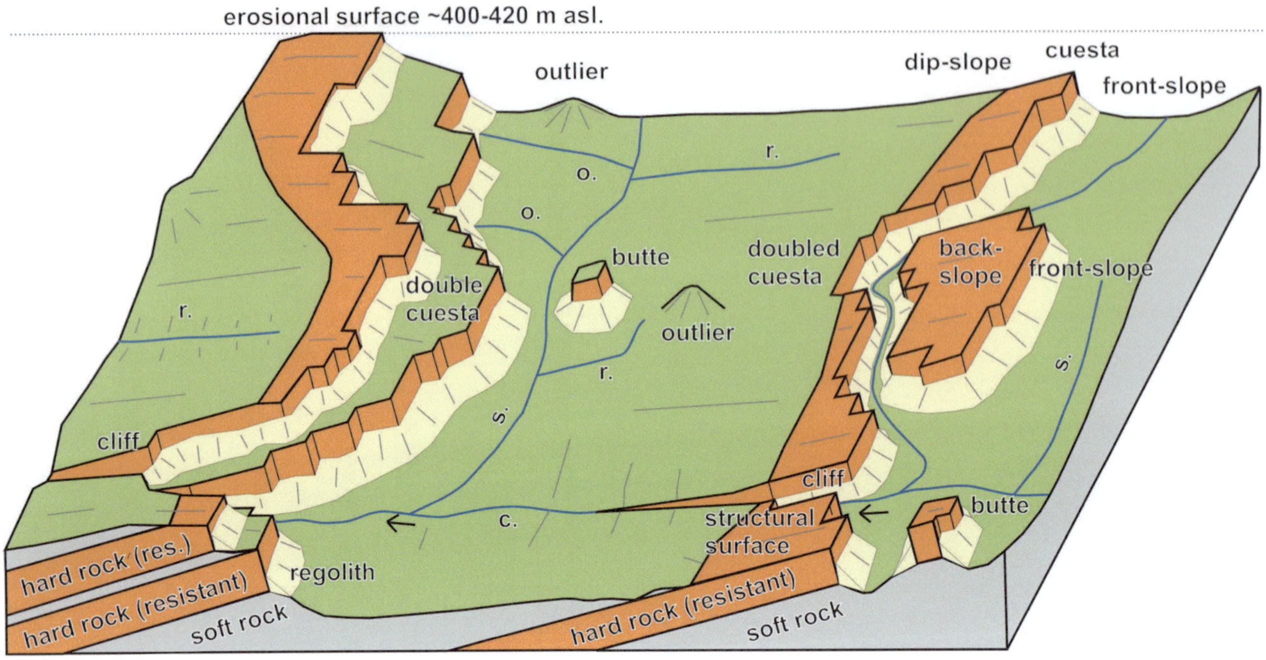

Fig. 23.3 Block diagram showing the typical elements of a cuesta landscape and its hydrographical network. *c*, *s*, *o*, *r* Consequent, subsequent, obsequent, and resequent river courses

Luxembourg City, result from exhumation and erosion of the folded layers.

Adaptation of the drainage network to a cuesta landscape results, according to Davis (1899), in a fairly regular pattern (Fig. 23.3). Consequent rivers are major streams flowing in the direction of the dip of the strata and often cut deep water gaps through several cuestas. Subsequent (orthoclinal) rivers flow in the soft-rock troughs, parallel to the front of the cuestas. Minor streams are either resequent (cataclinal) when flowing down the dip-slope, in accordance with the dip of the strata, or obsequent (anaclinal) when coming down from the scarp slope, thus flowing against the dip of the strata. In Gutland, obsequent rivers of some importance cutting across several cuestas exist (e.g. the Alzette, see Fig. 23.1), possibly pointing to superimposition.

23.4 The Cuestas of Belgian Lorraine and Luxembourg's Gutland

Lithology and facies distribution are chiefly controlled by the position of the coast line during deposition. In the study area, it moved from east to west along the southern edge of the Ardennes relief during the Triassic and the Early Jurassic. In the east, Lower Triassic sandstones (Buntsandstein) rest unconformably on the pre-Triassic topography cut in the Devonian basement (Lucius 1948), followed by Middle Triassic (Muschelkalk) and Upper Triassic (Keuper) sediments while, in the west, Lower Jurassic (Liassic) and, farther west, in France, lower Middle Jurassic (Dogger) sediments directly lie on the post-Hercynian erosion surface, which had then levelled the Ardennian basement (Demoulin 1995; see also Chap. 5). Along a WNW transect from Luxembourg City, the thickness of the Mesozoic cover reduces from 1000 m in the syncline axis to ~300 m in its margin in Belgian Lorraine (e.g. Schintgen and Förster 2013) (Fig. 23.7). Triassic sediment facies vary from detritic in the west to marly and dolomitic platform type deposits in the east. Then, the Lower Lias sand sedimentation is remarkably diachronic from east to west and south to north (Bintz et al. 1970) while, at any time, the transition from sandy detritic to marly calcitic facies occurs over short distances of a few kilometres. Overall, the complexity of these changes both in time and space makes local correlations sometimes difficult and, geomorphologically, causes (spatially discontinuous) cuestas pass from one to another stratigraphic formation along its strike. Owing to the obliquely transgressing character of the Mesozoic seas over the Ardennian margin, a well-developed marginal trough occur between the Ardennes plateau and the first cuesta only where the first Mesozoic units resting on the basement are made of soft rocks, for instance between Tintigny and Florenville where it is carved in Hettangian Jamoigne marls in the east and Sinemurian Warcq marls in the west (Figs. 23.1 and 23.7).

The next sections describe all cuestas individually, beginning with the southernmost one, which corresponds to the youngest resistant formation, because it is the most

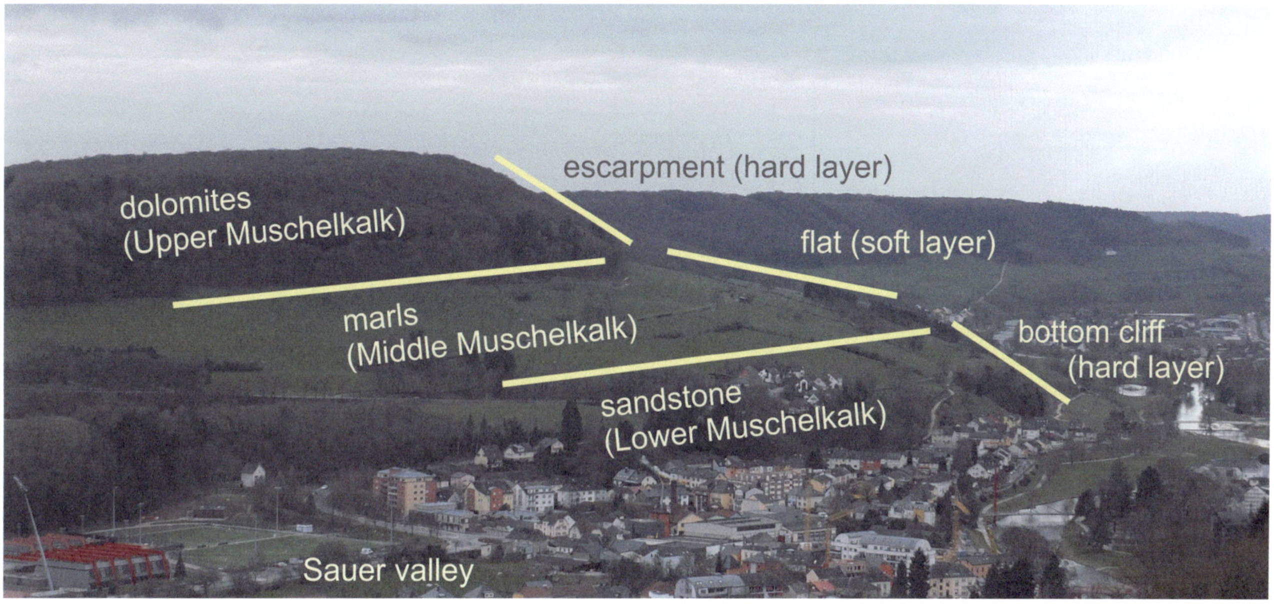

Fig. 23.4 Double cuesta in the Upper Muschelkalk (*top*) and Lower Muschelkalk (*bottom*) of Gutland. View from the Härebierg (394 m asl), north of Diekirch, to the southwest. A similar slope profile may be observed all along the Sauer valley

Fig. 23.5 View to the (south)east of a doubled cuesta in the Muschelkalk, with distinct front-slopes on both sides of the Sauer valley east of Diekirch. In the background, the Lower Lias cuesta

Fig. 23.6 Buttes between Differdange and Esch/Alzette in front of the Dogger cuesta (visible in the *left* background)

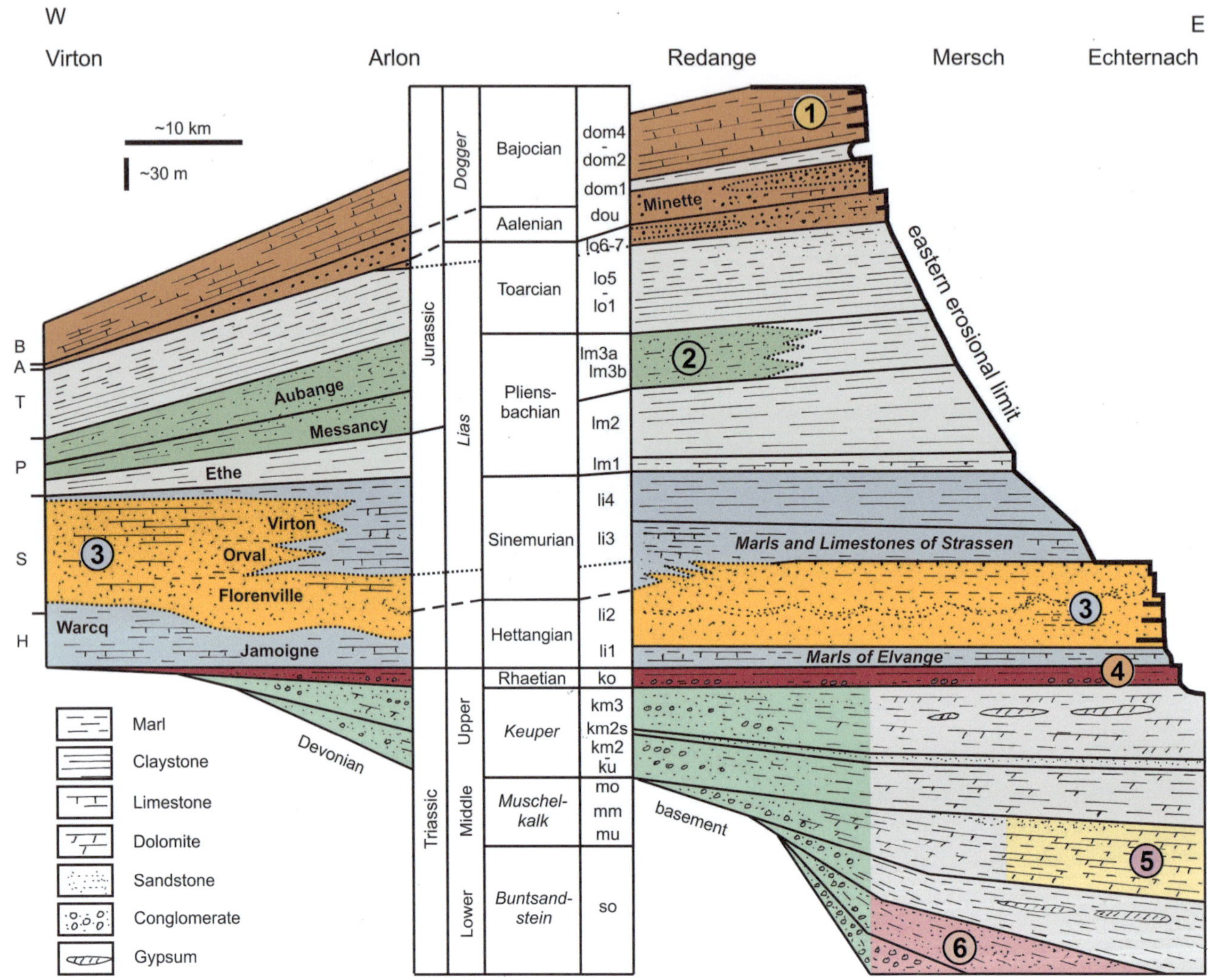

Fig. 23.7 Lithological log of the Mesozoic formations in Belgian Lorraine and Luxembourgian Gutland from Virton to Echternach (after Monteyne 1965; Boulvain et al. 2001; Bouezmarni 2011; Poty et al., Goemare and Belanger for the Belgian Lorraine). The numbers refer to the cuestas mapped in Fig. 23.1

continuous and easy to follow throughout the study area, thus providing a good anchor point for the mapping of the other cuestas. All cuestas are named after their common stratigraphical position.

23.4.1 The Dogger Cuesta

The Dogger cuesta (numbered 1 in Fig. 23.1) is named after the regional name of the Middle Jurassic series. Bajocian hard limestones form the upper part of the escarpment and Aalenian ironstones, going eastward into the Toarcian/Aalenian *Minette* iron ore, build its lower part. The thickness of this famous iron ore, exploited by an important local steel industry in the twentieth century, increases from ~30 to 60 m to the east. The strike of the cuesta is west-east along the Belgian-French border. Owing to the presence of the Gulf of Luxembourg syncline, it turns to southeast in Luxembourg and to south of the French border, forming the Moselle cuesta in France. To the south of Esch/Alzette, in Gutland, a layer of marls separating the Bajocian limestones from the *Minette* allows for the formation of a double cuesta, whose two steps become separated into two distinct cuestas by an up to 2-km-wide bench east of Esch/Alzette (Fig. 23.1). In the area of Longwy, the Dogger cuesta is doubled by incision of the Reierbaach, a small tributary of the Chiers (Fig. 23.1).

The cuesta top is truncated by an erosion surface at altitudes culminating around 430 m (Désiré-Marchand 1985). The scarp height reaches 150 m in Belgian Lorraine. West of the Belgian-French border, although it is strongly dissected by rivers and decreases in altitude, its height remains fairly similar due to the important incision of the Chiers (Fig. 23.1). The broad subsequent depression at the foot of the cuesta is carved into soft Toarcian marls and claystones. Several rivers of the Meuse catchment (Vire, upper Chiers) run subsequently in this depression, where a few buttes, such as the residual hill of Montquintin near Virton (see Chap. 24) and the buttes along the water divide between Differdange and Esch/Alzette, are preserved (Fig. 23.6).

23.4.2 The Pliensbachian (Middle-Lias) Cuesta

The 30-km-long Pliensbachian cuesta (Fig. 23.1, nr 2) is restricted to the eastern Belgian Lorraine and western Gutland. It is formed by siltites and calcareous, often ferruginous sandstones (the so-called Aubange *macignos*). In Gutland, the Middle Lias sandstones reach a thickness of about 50 m, thinning out progressively and passing to marls to the east (Fig. 23.7). At the foot of the cuesta, the depression is carved in homogeneous Pliensbachian marls in the east, passing to the shales of the Ethe Formation in the west (Fig. 23.7). The cuesta reaches maximum elevation (~410 m) and height (~110 m) in its central part, where the Messancy creek cuts it consequently. It progressively disappears westward because of decreased thickness of the Pliensbachian hard rocks and eastward because of facies variations in the upper Pliensbachian (Lucius 1948).

23.4.3 The Lower Lias Cuesta

The Lower Lias cuesta (Fig. 23.1, nr 3) best delineates the underlying geological structure. Being in direct contact with the Paleozoic basement ~25 km west of Chassepierre, its straight trace highlights the homoclinal structure in Belgian Lorraine before describing the arcuate shape of the Gulf of Luxembourg syncline in the east and ending with a disappearing E-facing scarp near the Moselle valley at Remich. Prominent buttes in front of the cuesta near Brouch and east of Luxembourg City point to secondary undulations or faults within the syncline.

The stratigraphic position of the cuesta is complex, in relation with the diachronism of sedimentation of the sandstone unit that makes the bulk of the ridge. While it is of Sinemurian age in the west (Fig. 23.7), large parts were already deposited during the Hettangian in the south and east, where the sandstone is first embedded as a thin lenticular unit in an alternation of marls and limestones (Hanzo 2012) before thickening northward to an 80- to 100-m-thick homogeneous unit. There, the overlying Marls and Limestones of Strassen and the underlying Marls of Elvange have reduced thicknesses (see Chap. 4). In Belgium, the Sinemurian sandstones and sandy limestones of Virton, Orval and Florenville are locally separated by interstratified marls (e.g. the Marls of Strassen, Fig. 23.7). Despite these variations in the lithology and age of the hard rock body, the cuesta shows a remarkable continuity in the landscape throughout Lorraine and Gutland.

While mean elevation of the escarpment top is ~400 m in Belgian Lorraine, it rises eastward to ~420 m near Arlon and locally to 420–430 m in Gutland (Fig. 23.1). Between Chassepierre and Florenville, the well-marked escarpment is nearly 100 m high, partly because of active erosion of the subsequent Semois (Fig. 23.8). By contrast, it becomes progressively smaller east of Florenville and disappears almost completely at Etalle (Fig. 23.9), in relation with decalcification and loosening of the sandstones (Monteyne 1965) where joint density is higher or interlayered marls determine the presence of perched aquifers (Lebaut 2001). On top of the Stockem hill, west of Arlon, the sandstone has loosened to completely bleached sand probably when chemical weathering led to the formation of the silcretes that, together with ferricretes, currently protect the hill against erosion (Hufty 1959). Intense decalcification is also observed in Gutland, for instance between Beckerich and Brouch or in the Graulinster area (Figs. 23.1 and 23.10).

The sequence of erosion surfaces is somewhat complex in the area of the Lower Lias cuesta. Beyond stepped structural benches locally developed at the foot of the main sandstone escarpment, Souchez-Lemmens (1968, 1971) distinguished two main such surfaces in eastern Belgian Lorraine. Dotted with ferricrete remnants, the higher surface extends near Arlon at and above 410–415 m, dominated by a few 20- to 50-m-high hills such as the Hirtzenberg (465 m), whose shaly substrate is protected against erosion by a ferricrete cap (Fig. 23.11).

As for the lower surface, it is especially well developed in the north-central Belgian Lorraine, notably forming the so-called Fratin glacis at ~350 m elevation in the area between Tintigny and Etalle (Figs. 23.1 and 23.9). According to Souchez (1963a), residual hills rising a few dozen metres above the surface (e.g. Bois du Bonlieu, 382 m) bear witness to an older erosional, rather than structural, level.

23.4.4 Transitional Lias-Triassic Cuestas

Small scarps up to 20 m in height are developed locally in the north and east at the foot of the Lower Lias cuesta, due to the presence of harder units within the Hettangian marls and of Rhaetian sandstones, sometimes including m-thick conglomerates (Souchez-Lemmens 1971; Alexandre 1976; Liedtke et al. 2010) (Fig. 23.1, nr 4). These small scarps are often obscured by landsliding in the Rhaetian claystones, then forming an irregular footslope of the Lower Lias cuesta, or by a thick mantle of regolith. Otherwise, they locally emerge from the subsequent marginal depression extending

Fig. 23.8 Viewed from its own *top*, the Lower Lias cuesta at Chassepierre, Belgium, and the marginal depression lying at its foot and leading to the foot of the post-Hercynian surface forming the Ardennian margin in the *left* background

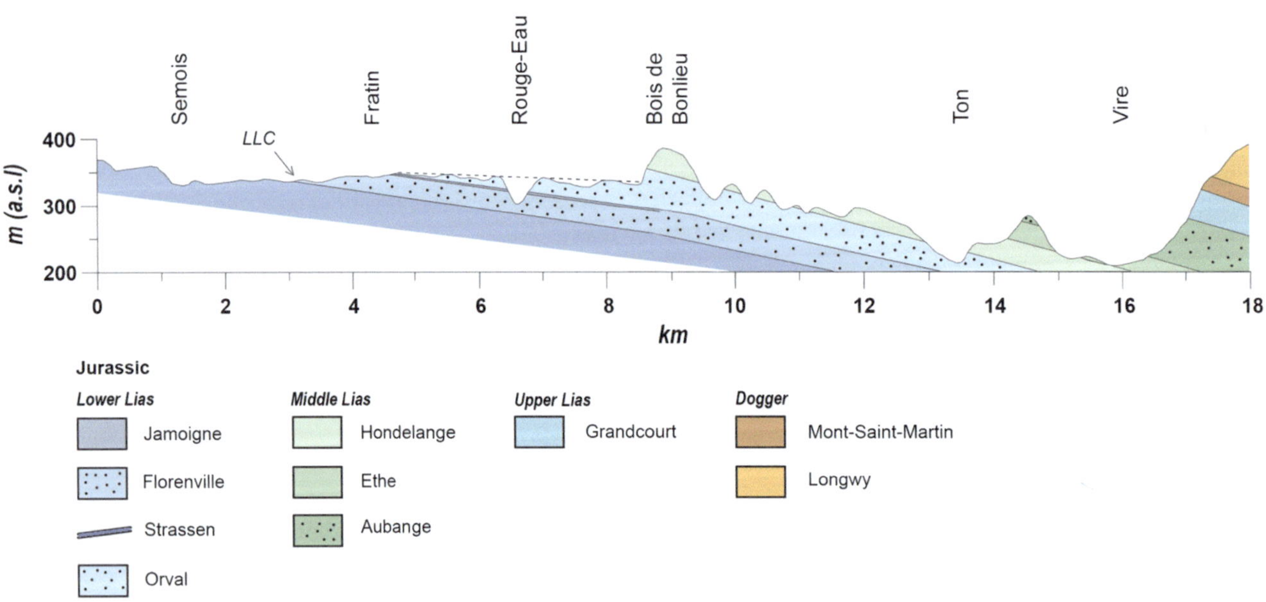

Fig. 23.9 Cross section between Etalle and the Vire valley highlighting the quasi-inexistent front-slope of the Lower Lias cuesta (LLC) in this area where a (local?) erosion surface, the Fratin glacis (*thin dashed line*), cuts indifferently through all rocks. The Pliensbachian and Dogger cuestas are visible in the right part of the section (see location in Fig. 23.1)

Fig. 23.10 Decalcification and bleaching of the Luxembourg Sandstone in a small sand quarry cut in the front-slope of the Lower Lias cuesta at Beckerich, Luxembourg

between the cuesta and the Ardennian margin in the north as, e.g. NE of Tintigny. Carved in Upper Triassic (Rhaetian and Keuper) rocks, this softly undulating depression widens from 2 km at Chassepierre in the west to ~10 km south of Mertzig in the east. East of the Belgian-Luxembourgian border, Middle and Lower Triassic deposits interpose progressively between the two reliefs and the depression, disconnected from the Ardennes, narrows from Diekirch to Reisdorf (Fig. 23.1). Along its long subsequent course in the western part of the depression, the laterally unconstrained W-flowing Semois has developed a wide alluvial plain where it draws large free meanders (Fig. 23.12). In the east, pertaining to the Rhine catchment, the more deeply incised Attert is also subsequent whereas the Sauer-White Ernz-Black Ernz system was probably inherited by superimposition. On the other limb of the syncline, in SE Gutland, the corresponding but much broader Upper Triassic depression at the foot of the Lower Lias cuesta is subsequently crossed by the Moselle and its tributary, the Syre, which even cut into Muschelkalk strata of secondary anticlinal undulations (Fig. 23.1).

23.4.5 Triassic Cuestas

In NE Gutland, two more cuestas, the Upper Muschelkalk and Buntsandstein cuestas are developed in hard Triassic strata (Figs. 23.1 and 23.13). Their counterpart on the eastern limb of the Gulf of Luxembourg syncline lies in Germany, east of the Moselle, beyond the wide depression carved in very thick Upper Triassic marls (Fig. 23.1). Important facies variations are observed, from reddish, mostly sandy and conglomeratic Triassic sediments in the west to Keuper marls overlying a dolomitic Upper Muschelkalk, a marly gypsiferous Middle Muschelkalk, and resistant sandstones predominating in the Lower Muschelkalk and the Buntsandstein in the east and southeast.

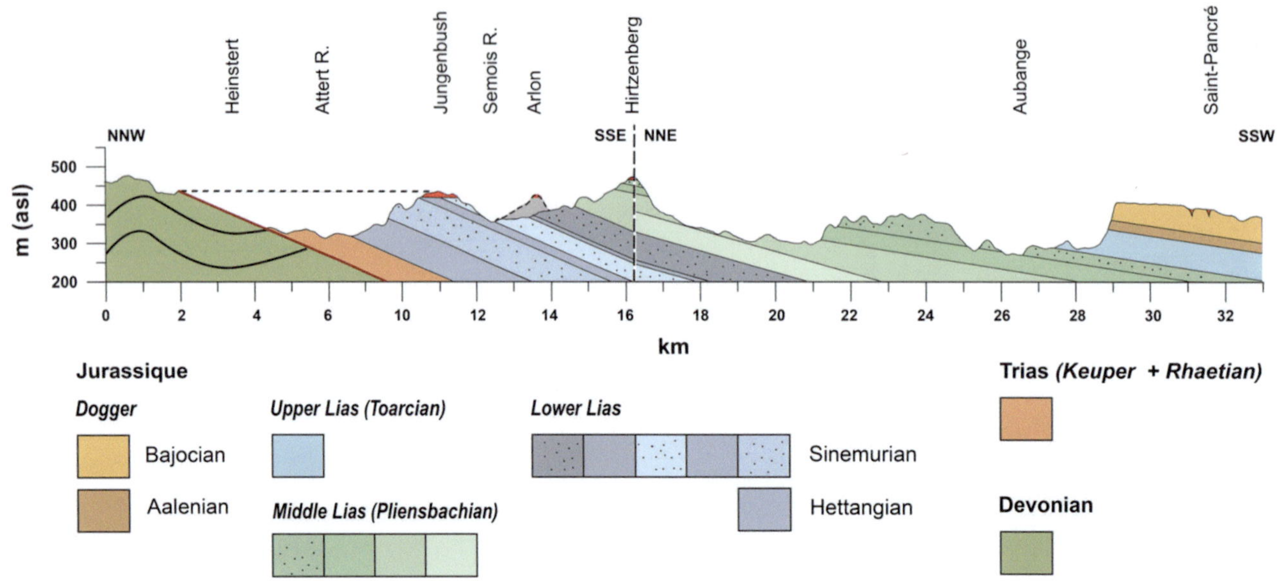

Fig. 23.11 Cross section close to the Luxembourgian-Belgian border highlighting the cuesta relief in the Lower Lias Florenville and Virton sandstones near Arlon (see location in Fig. 23.1). Note that the back slope of the cuesta is not purely structural here and some residual hills (Hirtzenberg, Jungenbusch) do not witness the cuesta itself but instead an erosion surface underlined by duricrusted sandstones (*in red*). The *bold brown line* denotes the base surface of the Mesozoic cover, whose exhumed part corresponds to the post-Hercynian surface (modified after Alexandre, 1976)

Fig. 23.12 Oblique aerial view of the Semois large meanders near Tintigny in the marginal trough between the Lower Lias cuesta and the wooded Ardennian margin in the background (*Photograph* D. Kemp)

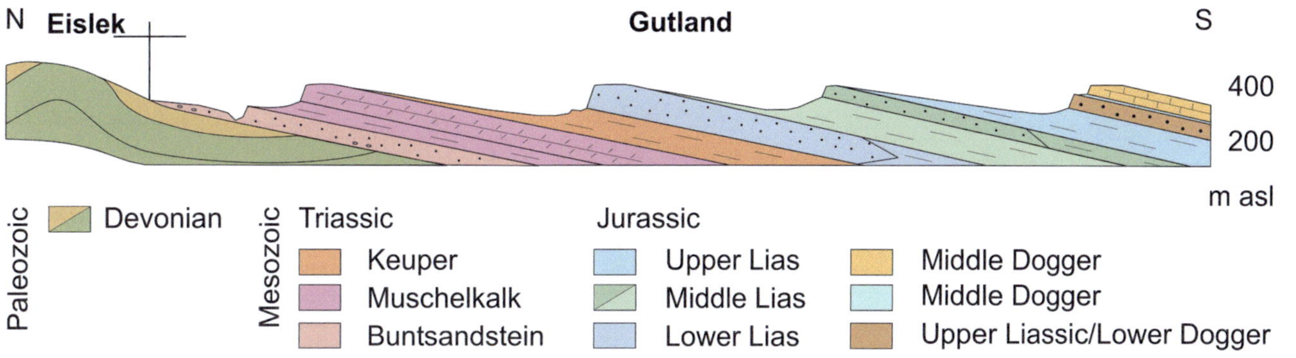

Fig. 23.13 General geological cross-section of Luxembourg's Gutland. Note that the first cuesta south of the Ardenne, absent in Belgian Lorraine, is determined here by the Upper Muschelkalk dolomites. N–S variations from detritic to marly facies are schematically figured for the Middle Lias and the Lower Lias sandstones

Following the Eislek margin in NE Gutland, one first meets a single cuesta formed by undifferentiated Triassic rocks and underlined by the subsequent reach of the Wark flowing at its foot (Fig. 23.1, nr 5). Due to facies variations, it soon evolves into a double cuesta near Mertzig, consisting of a dolomitic and sandy Upper Muschelkalk escarpment and a bottom cliff developed in Lower Muschelkalk/Buntsandstein sandstones, separated by thick Middle Muschelkalk marls (Fig. 23.14). Farther northeast, the Lower Triassic sandy and conglomeratic strata become individualized and form the independent minor Buntsandstein cuesta at lower elevations (~350 m; Fig. 23.1, nr 6). By contrast, with an up to 160-m-high escarpment near Ettelbrück, the Muschelkalk cuesta is very distinct in the landscape, especially in the Diekirch-Reisdorf area where it is doubled by the incision of the Sauer valley (Figs. 23.1 and 23.14). Its top corresponds to an erosion surface at elevations of almost 400 m to the north of the Sauer and ~350 m south of the valley (Verhoef 1966).

While the Triassic cuestas of the eastern limb of the Gulf of Luxembourg syncline are located in Germany, the Moselle has cut a deep valley into the dip-slope of the Muschelkalk cuesta, with the development of important dolomitic cliffs. There, water circulation in the Keuper dolomites, locally concentrated along faults, favours destabilization of the landslide-prone more marly Keuper strata on top of the cliffs. In the area of Grevenmacher for instance, several large landslides have been described, among which the historical landslide of the Longkaul has recorded activity over several centuries. Initial sliding starts in the Keuper strata outcropping in the upper part of the slope, accumulating material that loads the regolith and the Middle Muschelkalk gypsiferous marls of the lower slope at the foot of the dolomitic cliff and induces further downslope movements towards the valley bottom (Cordier et al. 2012) (Figs. 23.1 and 23.15).

23.4.6 Weathering and Duricrusts

Weathering features and duricrusts constitute interesting accessory elements of the cuesta landscape of Belgian Lorraine and Luxembourgian Gutland. A first type of weathering products is related to the carbonate content of many formations in the Mesozoic cover. Dissolution of, especially, the Middle Jurassic (Dogger) limestones developed karstic features in particular on top of the Dogger cuesta, where Fizaine (2005, 2012) describes mega-lapiés, up to ~15 m in depth and width (e.g. at Saint-Pancré, just south of the French-Belgian border), which he interprets as a cryptokarst developed most probably in parallel with its covering Lower Cretaceous deep weathering mantle (see below).

Although they also develop from springs in the Dogger limestones, carbonate deposition and travertine formation are more typically related to the Lower Lias sandstones, e.g. north of Meix (Couteaux 1969; Debaut 1987), where they originate from springs at marly levels in the carbonate sandstones, the calcitic matrix of the sandstone and the marls and limestones of Strassen equally providing dissolved carbonate. Souchez (1963b) suggested that travertine develops more easily in the Lower Lias carbonate sandstones than in the Bajocian limestones because the sandstone open porosity allows the water to come closer to saturation. Travertine deposits may be found on the cuesta front-slope but groundwater flow following the dip of the strata more often favours their occurrence on the dip-slope, where deposits up to 20 m in height may be found (Figs. 23.16 and 23.17). Pollen data and ^{14}C dating show that travertine formation began with the onset of Holocene warming in the Preboreal and Boreal (Couteaux 1969; Souchez and Paepe 1972). At Kopstal, in the Mamer valley, north of Luxembourg City, a 7-m-thick deposit accumulated during most of the Holocene (Meyrick 2000; Dabkowski et al. 2015). Travertine deposition is still active in many places.

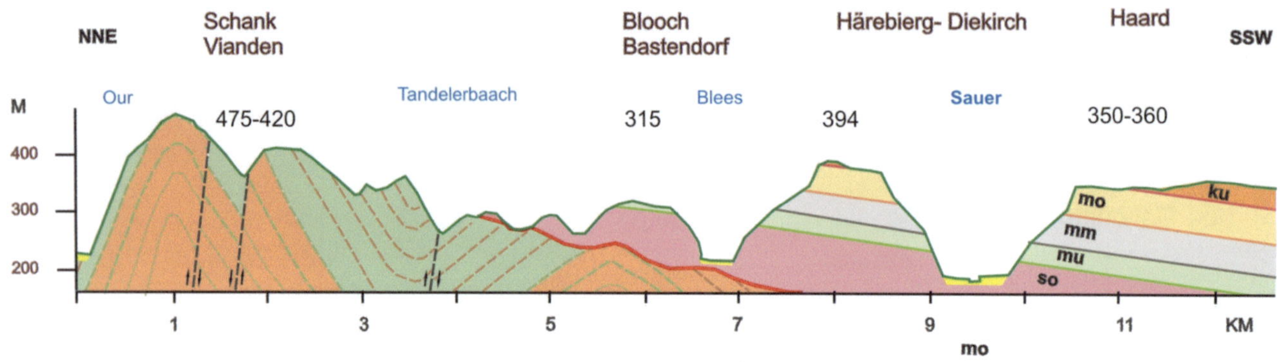

Fig. 23.14 NNE–SSW geomorphological-geological profile Härebierg showing the Mesozoic cover unconformably resting on the Paleozoic basement and displaying the double Muschelkalk (mo + mu)-Buntsandstein (so) cuesta, doubled by the incised valley of the Sauer (see location in Fig. 23.1). The sandy facies of the Lower Muschelkalk (mu) forms the bottom cliff with the sandy conglomeratic Buntsandstein (so). Both cuestas are truncated by the same structurally controlled, tilted erosion surface at 350–360 and ~400 m asl, respectively. Where it is exposed on the margin of the Eislek plateau, the Paleozoic/Mesozoic unconformity (*bold red line*) corresponds to the pre-Triassic topography. *Numbers* are elevation data. ku. Lower Keuper. mo, mm, mu. Upper, Middle, and Lower Muschelkalk, respectively. so. Upper Buntsandstein

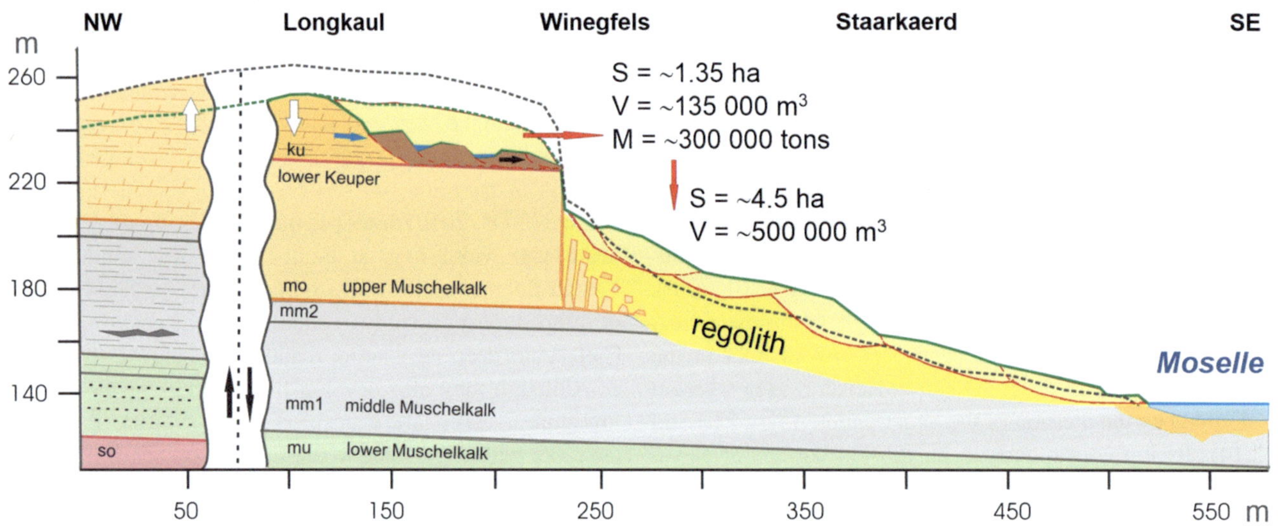

Fig. 23.15 NW–SE geomorphological-geological profile Longkaul juxtaposing two parallel profiles on the two sides of a fault oriented obliquely to them. The dashed black line describes the topography of the uplifted block with a Muschelkalk escarpment dominating a characteristic slope in the landslide-prone gypsiferous Muschelkalk. The green topographic profile passes through the Longkaul earthslide, which is periodically activated by the hydraulic pressures created in the Lower Keuper dolomites by water infiltration during heavy rainfall. The sliding material in turn loads and may destabilize the regolith on the lower slope

Beyond the presence of the Aalenian *Minette* iron ore of sedimentary origin, Belgian Lorraine and Luxembourgian Gutland are also characterized by scattered ferricrete occurrences (Lucius 1948; Kaboth 1969; Fizaine 2005). These ferricretes range from true in situ lateritic crust of Lower Cretaceous age trapped in the paleokarst of the Dogger cuesta (e.g. the Borne de Fer at Ottange, south of the French-Luxembourgian border; Fizaine 2005, 2012; Théveniaut et al. 2007), through elements of such old crusts reworked in the paleokarst (e.g. Saint-Pancré, in nearby French Lorraine; Fizaine 2012), in ancient deposits on the Cenozoic erosion surfaces (e.g. near Arlon; Souchez-Lemmens 1971) or in fluvial terraces (Lucius 1948), to groundwater ferricretes either associated with the 400-m erosion surface (Hufty 1959; Souchez-Lemmens 1968) or of probable Plio-Quaternary age at lower levels (Souchez-Lemmens 1971). They are frequently observed cementing the Sinemurian sandstones on top of the Lower Lias cuesta and are responsible for the preservation of residual hills in weak but indurated rocks.

Silcrete remnants are also known in the study area (Baeckeroot 1929; Alexandre 1976; Lucius 1948). They appear as either silicified blocks up to 1–2 m in size dispersed on the summit surfaces and called 'Pierre de Stonne' after the eponymous village in N France or fragments of such blocks. The silicified material is generally thought to be

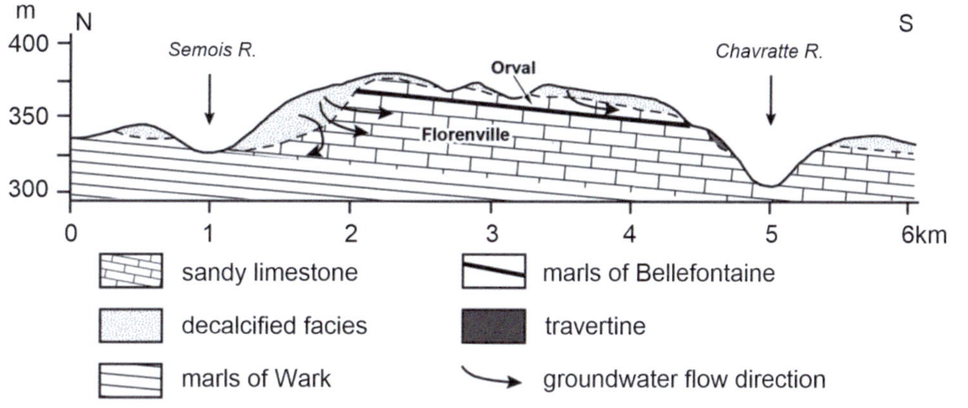

Fig. 23.16 N–S cross section of the interfluve between Semois and Chiers (not represented in the figure) at the height of Tintigny, showing the carbonate-loaded groundwater flow guided by the dip of the strata and causing travertine formation 3 km farther south where the dip-slope is cut by the Chavratte valley. The Chavratte is a S-flowing subtributary of the Chiers (after Monteyne, in Lebaut, 2001)

Fig. 23.17 Active travertine deposition at mid-height of the valley side of the Rouge Eau creek

Paleocene or Oligocene sand (Tricart 1949; Demoulin 1995; Quesnel 2003). It is mostly cemented by authigenic quartz overgrowth on the sand grains and microquartz precipitation in the voids, resulting in an essentially quartzitic structure. While Voisin (1988) proposed that similar blocks to the west of the study area indicate groundwater silicification, which he tentatively linked with silica transfer from deep weathering mantles developing on the nearby Ardennes during the Thanetian (see also Chap. 5), Quesnel (2003) observed pedogenic features (amorphous silica cutans and caps on intraformational gravels) in many silicified blocks, which she consequently associated with the development of an extended Paleogene erosion surface over the NE Paris Basin corresponding to the ~400–420 m surface of Belgian Lorraine and Luxembourgian Gutland. While pedogenic silcretes indeed require a warm climate with contrasted wet and dry seasons to form and can thus be associated with an erosion surface-forming stage, groundwater silcretes are no reliable climatic indicator. Moreover, as silcrete remnants are found only in reworked position, they are of no use in the reconstruction of the long-term evolution of Lorraine and Gutland.

23.5 Age and Formation of the Cuestas

23.5.1 Main Lines of the Cenozoic Regional Evolution

Although continental sediments are scarce and, in any case, not dated in Belgian Lorraine and Luxembourgian Gutland, the main stages of landscape evolution may be identified from the general morphology and correlations with neighbouring areas. The highest, oldest landscape elements correspond to residual hills rising a few dozen metres above the envelope surface of the region, such as the Hirtzenberg (465 m) south of Arlon. Another hill in similar topographic position is the Borne de Fer (450 m) dominating the Dogger cuesta by ~40 m in nearby French Lorraine. Paleomagnetic dating of a lateritic crust at the Borne de Fer makes it belong to a Lower Cretaceous topography (Théveniaut et al. 2007), from which Hirtzenberg is therefore probably another remnant in the study area. Liedkte et al. (2010) instead suggest they bear witness to a 450–470-m surface which they date from the beginning of the Paleogene. However, the most remarkable ancient landscape stage still surviving in the present topography is the envelope surface that truncates all cuestas at elevations progressively increasing eastward, from ~370 m near Florenville to ~390 m around Etalle, and 400–420 m near Arlon and across Gutland. Integration of this surface in the pattern of stepped surfaces described for the Ardennes-Eifel area tends to suggest that it was formed in Oligo-Miocene times (Demoulin 1995; see also Chap. 5). Though they are poor climatic indicators, the groundwater ferricretes and silcretes often associated with the surface are compatible with its Neogene age. As for the lower surface mentioned by Souchez-Lemmens (1971) in the Fratin area around 350 m asl., i.e. at elevations lower than the top of the nearby cuestas, it remains open whether it corresponds to a subdued element of the higher surface on locally weaker or more fractured rocks, thus also Oligo-Miocene in age, or it was carved after the cuestas had already begun to become individualized in the landscape, and is therefore of younger Neogene age. A surface of similar altitude in the Diekirch area (Fig. 23.14) is interpreted as a younger erosion level developed when the fluvial system was already fixed (Verhoef 1966). In any case, recent (Plio-Quaternary) uplift of the Ardennes—Eifel—NE Paris basin region (see Chap. 10) induced incision of the drainage network in Lorraine and Gutland. From that time, the river systems developed on the erosion surface, possibly originally born on an Eocene or Oligocene cover, began to adapt progressively to, and thus enhance, the underlying geological structure by sculpting the cuestas. Only some rivers of Gutland (e.g. in the Sauer area) kept a superimposed character. It is generally agreed that the front-slope of the cuestas hardly retreated since they started to emerge in the eroding landscape (Tricart 1949; Busche et al. 2005).

23.5.2 Early Drainage Network Evolution

The ancestor of the present drainage network of Belgian Lorraine and Luxembourgian Gutland appeared at the latest when the region emerged after regression of a last Paleogene sea of Eocene (Quesnel 2003) or, more probably, Oligocene age (Tricart 1949; Demoulin 1995). At that moment, the regional landscape was characterized by a low-lying topography developed around a ~100-m-high relief encompassing the east-central Ardenne and central and north Eifel, where the Danian erosional landscape had been preserved by the Paleogene erosion and the development of Paleocene and Oligocene surfaces including a corridor corresponding to the present-day Rhine and Moselle troughs at ~400 m asl (Quitzow 1982; Demoulin 1995; see also Chap. 5). The rivers draining Lorraine and Gutland towards the northern seas of the time had thus to bypass this relief either by the west, as did probably the ancestral Meuse system (see Chap. 10) or, following the Moselle-Rhine trough, by the east. In the beginning, they flowed in broad valleys over either a cover of Paleogene sediments overlying the ~400–420 m erosion surface or the weathering mantle associated with that surface. This was probably the time when many groundwater ferricretes formed in these valleys. Due to onset of fast subsidence in the Lower Rhine Embayment in the Late Oligocene, the proto-Rhine system, to which the Sauer and its tributaries in Gutland already belonged, underwent a first episode of incision into the ~400 m surface, superimposing its original drainage pattern into the underlying geological structure. While some rivers in Gutland, like the Sauer and the Alzette, have preserved a typically superimposed course through later incision, others show NE-striking courses inherited from the original drainage towards the Moselle trough and appearing now either as subsequent when they flow on weak strata (Attert, Syre) or obsequent when their original course headed against the plunge of the Gulf of Luxembourg syncline axis (e.g. Mamer, Eisch).

On the other side of the divide between Rhine and Meuse basins, incision in the Semois and Chiers systems responded to the evolution of their base level in the Meuse. Based on the presence of wind gaps on top of the Lower Lias cuesta and scarce alluvial gravels containing Ardennian pebbles preserved to the south of this cuesta, Pissart (1961) suggested that S-flowing Ardennian streams, in particular the Vierre and the Rulles, were once continuing their southern course beyond the Lower Lias cuesta and were tributaries of the Chiers, via the Ton River (Fig. 23.18). As the wind gaps lie approximately at the level of the ~350 m surface near Fratin, he also thought that these ancient courses had already begun to superimpose in the underlying structure when they were captured by the Semois. This was consistent with the reconstruction he proposed for the drainage history of the Meuse south of the Ardennes, where Late Miocene headward erosion of a N-flowing river across the Ardennes first captured a Paleo-Semois that formerly flowed westward, before capturing also the Meuse (Pissart 1961; see also Chap. 10). According to him, this lent a temporary advantage to the Semois against the Chiers, whose post-capture headward erosion was delayed until the Meuse was captured in turn. However, the Semois and Meuse captures are far from demonstrated, as Pissart (1974) himself acknowledged, and has many contradictors (e.g. Bustamante and Voisin 1975; Etlicher 2005). Moreover, beyond arguments about the Meuse evolution exposed in Chap. 10, local evidence

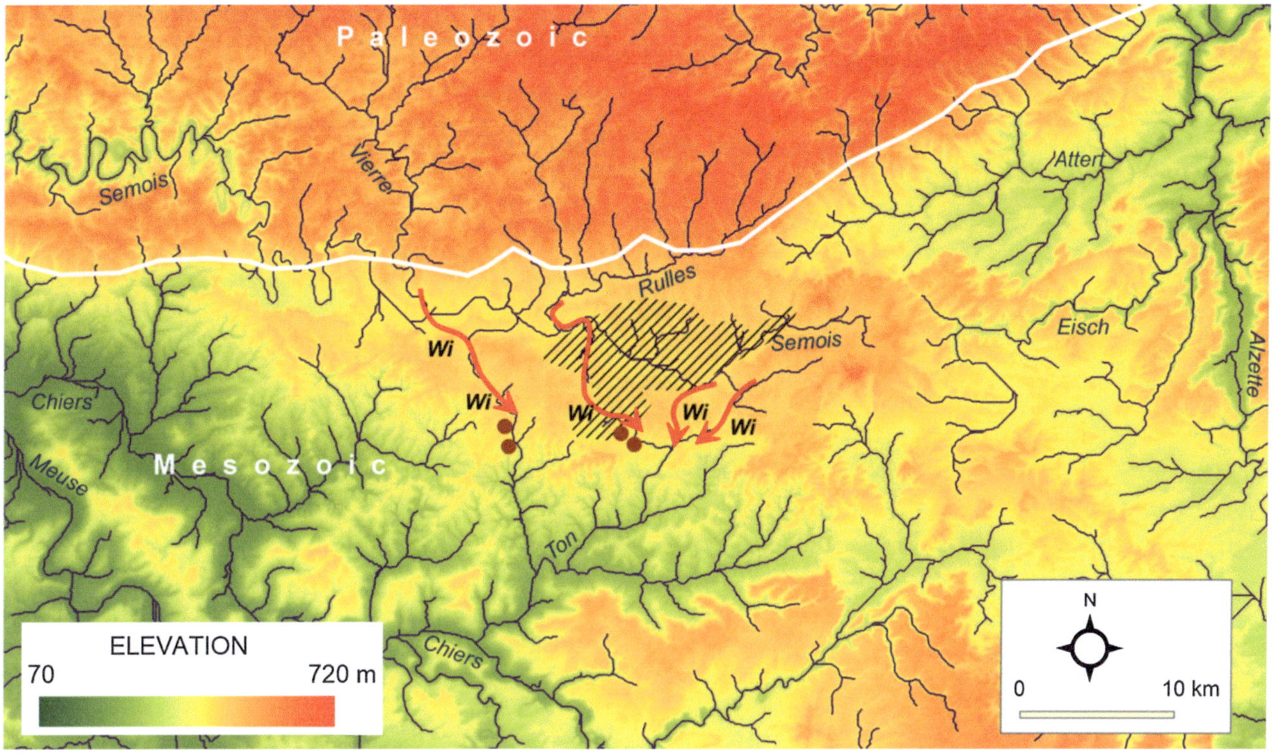

Fig. 23.18 Capture of former tributaries of the Chiers River (especially the Vierre and the Rulles) by the Semois as hypothesized by Pissart (1961) on the basis of geomorphic (*Wi* wind gaps) and sedimentary (*brown dots* fluvial gravel of Ardennian origin) evidence. The hatched area corresponds to the Fratin glacis. This hypothesis is barely credible because the Ton/Chiers base level in the south is much lower (~ 100 m) than the Semois base level in the west

seems also to contradict the Vierre and Rulles captures, which the very limited amount of Semois incision in Lorraine since the supposed events, the erosion rate of the Chiers through Mesozoic rocks much higher than that of the Semois in the Paleozoic, and the nearby base level of the Ton and Chiers lying ~ 100 m lower than the capturing Semois make highly improbable. Even the development of the Fratin lower surface and the quasi-absence of front-slope showing a very weak lithological control on erosion in the capture area are not favourable to the stream piracy hypothesis. The conundrum posed especially by the presence of Ardennian pebbles south of the Lower Lias cuesta ridge thus requires further observation to be solved within an improved understanding of the regional long-term drainage evolution.

23.6 Conclusion

This chapter highlighted how geological structure and long-term climate and tectonic evolution interacted to create the cuesta landscape of Belgian Lorraine and Luxembourgian Gutland. Local-scale features, from alternation of weak and resistant strata to facies variations and local-scale tectonic deformation, intermingle with the regional or larger scale landscape evolution characteristics. Enhancing the geological structure in the topography has obviously strong consequences for the drainage patterns and their changes through time, which this study case showed to be generally difficult to unravel, especially when going deeper in time in erosional landscapes that preserve only scarce witnesses of their past. In any case, the cuesta landscape of Belgian Lorraine and Luxembourgian Gutland, together with its continuation in NE France, is a long-studied but still worth future research archetype of structural landscapes.

References

Alexandre J (1976) Les surfaces de transgression exhumées et les surfaces d'aplanissement. Géomorphologie de la Belgique, Hommage au Professeur P. Macar, Université de Liège, pp 75–92

Baeckeroot G (1929) Sur l'existence de la Pierre de Stonne entre la Meuse et la Moselle. Annales de la Société géologique du Nord 54:87–94

Bintz J, Guerin-Franiatte S, Mouterde R, Muller A, Feuth-Siedeck I (1970) Contributions à la connaissance du Lias inférieur du Grand-Duché de Luxembourg. Publ. du Serv. Géol. du Lux, Luxembourg

Boulvain F, Belanger I, Delsate D, Dosquet D, Ghysel P, Godefroit P, Laloux M, Roche M, Teerlynck H, Thorez J (2001) New lithostratigraphical, sedimentological, mineralogical and

palaeontological data on the Mesozoic of Belgian Lorraine: a progress report. Geologica Belgica 3(1–2):3–33
Bouezmarni M. 2011 (in revision). Carte hydrogéologique de Wallonie, 68/3–4, Nobressart—Attert, Service Public de Wallonie
Busche D, Kempf J, Stengel I (2005) Landschaftsformen der Erde. Bildatlas der Geomorphologie, Wissenschaftliche Buchgesellschaft, Darmstadt
Bustamante L, Voisin L (1975) La capture de la Meuse Lorraine par la Meuse de Dinant. Remise en question partielle basée sur l'analyse des minéraux lourds, Service géologique de Belgique, 17 p
Coûteaux M (1969) Formation et chronologie palynologique des tufs calcaires du Luxembourg Belgo-Grand Ducal. Bulletin de l'Association française pour l'étude du Quaternaire. 6(3):179–206
Cordier S, Brou L, Colbach R, Harmand D, Kausch B, Löhnertz W, Losson B, Maquil R, Naton H-G (2012) New insights on the Quaternary evolution of the Moselle River and its tributaries (Luxembourg, France, Germany). FLAG Biennial meeting 2012, Field excursion guide, Fluvial archives from past to present—Introduction, 126 pp
Dabkowski J, Brou L, Naton H-G (2015) New stratigraphical and geochemical data on the Holocene environment and climate from a tufa deposit at Direndall (Mamer Valley, Luxembourg). The Holocene 25(7):1153–1164
Davis WM (1899) The geographical cycle. Geogr J 14:481–504
Debaut V (1987) Etude de l'aquifère sinémurien du Sud-Luxembourg. Actes du Congrès Scientifique «Les eaux souterraines en Wallonie: bilan et perspectives» , Faculté Polytechnique de Mons, pp 77–88
Demoulin A (1995) Les surfaces d'érosion méso-cénozoïques en Ardenne-Eifel. Bulletin Société géologique France 166(5):573–585
Désiré-Marchand J (1985) Notice de la carte géomorphologique du Grand-Duché de Luxembourg. Publications du Service Géologique du Luxembourg, Bulletin no 13/1985, 47 p
Etlicher B (2005) French and Belgian uplands. In: Physical The (ed) Koster E. Oxford University Press, Geography of western Europe, pp 231–250
Fizaine J-P (2005) Contribution à l'étude des paléokarsts et des ferricrètes dans les régions frontalières du Pays-Haut, de la Lorraine belge et du Gutland. Mémoire de DEA, Université de Nancy 2, 137 pp
Fizaine J-P (2012) Les paléokarsts et les formations ferrugineuses associées dans le Bassin parisien et ses bordures: caractérisations et évolutions géomorphologiques. Unpublished PhD thesis, Université de Lorraine, Nancy, 658 p
Goemare E, Belanger I. Planchette 68/3–4, Nobressart—Attert, Carte géologique de Wallonie, non publiée
Hanzo M (ed) (2012) Stratotype Hettangien. Collection Patrimoine géologique, Biotope and Museum national d'Histoire naturelle, 332 p
Hufty A (1959) Ferruginisations en Lorraine belge et leur utilisation en Géomorphologie. Ann Soc Géologique de Belgique 83:71–86
Jarvis A, Reuter H, Nelson A, Guevara E (2008) Hole-filled SRTM for the globe, Version 4 -http://srtm.csi.cgiar.org
Kaboth D (1969) Zur Genese der Bohnerze und Raseneisenerze Luxemburgs. Unpublished PhD thesis, University of Münster
Lebaut S (2001) L'apport de l'analyse et de la modélisation hydrologiques de bassins versants dans la connaissance du fonctionnement d'un aquifère: les grès d'Ardenne-Luxembourg. Mosella, 26(1–2), 271 pp
Liedtke H, Deshaies M, Gamez P, Harmand D, Preusser H (2010) Die Oberflächenformen in der Grenzregion Saarland-Lothringen-Luxemburg. Les formes du relief dans la région frontalière Sarre-Lorraine-Luxembourg. Institut für Landeskunde im Saarland (Band 48)/Deutsche Akademie für Landeskunde, Saarbrücken/-Leipzig, 367 p
Lucius M (1948) Das Gutland. Erläuterungen zu der geologischen Spezialkarte Luxemburgs, Geologie Luxemburgs. Publ Serv Géol Lux. V, Luxembourg, 405 p
Monteyne R (1965) Calcaire sableux d'Orval et calcaire sableux de Florenville dans la région de Virton. Bull Soc Belge de Géologie, Paléontologie et Hydrologie 74:60–79
Meyrick RA (2000) Holocene molluscan faunal history and environmental change from tufa at Direndall, Luxembourg. Bulletin de la Société Préhistorique Luxembourgeoise 22:55–75
Poty E, Belanger I, Boulvain F, Laloux M, Monteyne R. Planchette 68/7–8 et 69/5, Habay-la-Neuve—Arlon—Sterpenich, Carte géologique de Wallonie, non publiée
Pissart A (1961) Les terrasses de la Meuse et de la Semois. La capture de la Meuse lorraine par la Meuse de Dinant. Ann Soc Géologique de Belgique 84:1–108
Pissart A (1974) La Meuse en France et en Belgique. Formation du bassin hydrographique. Les terrasses et leurs enseignements. Centenaire de la Société Géologique de Belgique - L'Evolution Quaternaire des Bassins Fluviaux de La Mer du Nord Méridionale, Liège, pp 105–131
Quesnel F (2003) Paleoweathering and paleosurfaces from northern and eastern France to Belgium and Luxembourg: geometry, dating and geodynamic implications. Géol Fr 2003(1):95–104
Quitzow H (1982) Die Hochflächenlandschaft der zentralen Eifel und der angrenzenden Teile des Rhein-Troges und Neuwieder Beckens. Mainzer geowiss. Mitteilungen 11:173–208
Schintgen T, Förster A (2013) Geology and basin structure of the Trier-Luxembourg Basin—implications for the existence of a buried Rotliegend graben. Zeitschrift der Deutschen Gesellschaft für Geowissenschaften 164:615–637
Souchez R (1963a) A propos des reliefs résiduels gréseux au nord de Virton. Revue belge de Géographie 87(2):171–176
Souchez R (1963b) Corrosion chimique comparée du grès calcaire sinémurien et du calcaire bajocien de Lorraine belge. Erdkunde 17:228–232
Souchez R, Paepe R (1972) Radiocarbon ages and valley floor evolution in South Belgium. Bull Soc Belge Paléont Hydrol 81(3–4):221–225
Souchez-Lemmens M (1968) Accumulations ferrugineuses absolues en Lorraine belge septentrionale. Ann Soc Géologique de Belgique 91:145–158
Souchez-Lemmens M (1971) Les indurations ferrugineuses et l'évolution géomorphologique de la Lorraine belge nord-orientale. Revue belge de Géographie 95(1–2):143 p
Théveniaut H, Quesnel F, Wyns R, Hugues G (2007) Paleomagnetic dating of the «Borne de Fer» ferricrete (NE France): Lower Cretaceous continental weathering. Palaeogeogr, Palaeoclimatol, Palaoecol 253:271–279
Tricart J (1949) La partie orientale du Bassin de Paris. Etude morphologique, SEDES, Paris, p 474
Verhoef P (1966) Geomorphological and pedological investigations in the Redange-sur-Attert area (Grand-Duchy of Luxembourg). PhD thesis, University of Amsterdam, Publikaties van het Fysisch-Geografisch Laboratorium van de Universiteit van Amsterdam, 8, 531 p
Voisin L (1988) Introduction à l'étude de la Pierre de Stonne et des formations siliceuses associées au sud-ouest de l'Ardenne. Mém Soc Hist Nat Ard, Charleville-Mézières, 44 p

Geomorphosites: Function and Geoheritage Preservation in Belgium

24

André Ozer and Michiel Dusar

Abstract

This chapter opens with stressing the need for geomorphosite recognition, and situating the concept of geomorphosite in the broader context of geosites and the relations between geodiversity and biodiversity. Then, after mentioning the measures taken at the national level, notably with the creation in 1835 of the Royal Commission for Monuments, whose responsibilities were extended to sites in the first half of the twentieth century, it reviews the separate legislative evolution for site protection in Wallonia and Flanders after the competence had been transferred to the regional level in the 1980s. In parallel, the sites with a major geomorphic component listed as protected exceptional heritage in Wallonia are briefly reviewed and other sites so far unlisted but worth of interest are also pointed. A similar list of protected geosites is presented for Flanders, where the situation is however very different. Flat landscapes and densely populated areas make geosites far less spectacular there. The number of protected sites with a geologic or geomorphic component is nevertheless similar in Flanders and Wallonia, in each case about 20. Flanders promotes initiatives taken to arouse public interest for geosites by, e.g. integrating them in hiking or cycling trails. Recently, local groups of mixed lay people and geoscientists are also highly active in attempts to create geoparks, which in some cases gather all conditions required by UNESCO for a nomination as Global Geopark.

Keywords

Geomorphosite • Geodiversity • Site conservation • Wallonia site protection legislation • Flanders site protection legislation

24.1 Introduction

In countries like Belgium, with continuous human occupation of the land and subsequent landscape reshaping for at least two millennia, the environment in which we live seems utterly cultural and almost devoid of natural interest, except for the other living creatures which share our environment and struggle to survive within their own tiny part of a common habitat. The Great Acceleration witnessed during the last 150 years has allowed current generations to dominate their environment to the point that humanity has become the driving force for evolution, at considerable cost to the natural ecosystems. The current knowledge economy seems to be detached from the geological substrate. However, the critical minerals issue and legitimate fears of dependency on countries monopolising the exploitation of geological resources—or more correctly, practising long-term geological exploration and maintaining an extractive industry—have awakened sobering thoughts at the European level that we need the blessings of the Earth just as we need to know how to deal with its misfortunes.

A. Ozer (✉)
University of Liège, Liège, Belgium
e-mail: aozer@ulg.ac.be

M. Dusar
Geological Survey of Belgium, Royal Belgian Institute of Natural Sciences, Jennerstraat 13, 1000 Brussels, Belgium
e-mail: michiel.dusar@naturalsciences.be

A. Demoulin (ed.), *Landscapes and Landforms of Belgium and Luxembourg*,
World Geomorphological Landscapes, DOI 10.1007/978-3-319-58239-9_24

Geo-information is regaining importance, but making it visible and meaningful to the public and decision makers is still a challenge (Dusar 2012; Dusar and Dreesen 2012). "*Raising public awareness is the key to sustainable geo-conservation success*" (Gonggrijp 2000). Geoheritage ("*the legacy of the geological past, in rocks, soils and landforms*") conservation and management has to play a vital role, especially in countries where daily life is no longer connected to the resources on which our well-being is based.

There is no fundamental difference between geoheritage and cultural heritage values to be conserved or promoted, being tangible or intangible. Whereas the intangible cultural heritage is widely acknowledged, this aspect is of great importance for the geoheritage as well. Geoheritage derives its value from its interpretation and connection to former and current exploitation and use. Geodiversity, biodiversity and cultural heritage are all legacies of the past, moments in the evolution of the earth and humanity, adapted to the physical conditions of the environment, modified in mutual interaction, following unique pathways to shape the diversity of physical and cultural landscapes. Conservation is then about ways of preserving the morphology, structure and function of the heritages, adapted to an evolving society, reserving or restoring their setting and ensuring their integrity. This requires interdisciplinary understanding of the different heritages by checking interventions (e.g. new construction, change of groundwater level, change in vegetation) on potential adverse effects (Hardin 1998).

Geodiversity contributes fundamentally to the landscape, wildlife, economy and scenic beauty, and is, therefore, not only a scientific resource, but also a part of our economic, cultural and ecological inheritance. In analogy with biodiversity, geodiversity can be described as "*the variety of non-living or abiotic nature*". Geodiversity corresponds to "*the natural range of geological (rocks, minerals, fossils), geo-morphological (landforms, active exogenous processes) and soil features*" (Gray 2004).

Geodiversity protection and management pertains to geosites of scientific quality, rarity, aesthetic appeal, educational value, and will be also relevant to sites of archaeological, ecological, historical and cultural interest. Geosite conservation refers to a practical, site-specific conservation management strategy that encompasses documenting the importance of the site, planning and implementing conservation and protection measures, site monitoring and enhancement (Ellis et al. 1996). Protection and maintenance of geoheritage and associated ecosystems in a healthy, sustainable state is most likely when they become geotouristic assets. Geosite conservation principles depend on site typology. Exposure sites are parts of a rock that is extensive and also continuing below the ground surface, such as old quarry faces; maintenance of exposure is thus generally sufficient. Conservation of active landforms, such as dunes, is complex and requires good insight in dynamic processes. Integrity sites contain finite deposits or (relict) landforms that are irreplaceable if destroyed, such as caves; a strictly protective approach is required for their preservation.

Representing a particular type of geosite, the fairly recent concept of geomorphosite was introduced in the scientific literature by Panizza (2001). "*Geomorphosites are geo-morphological landforms that have acquired a scientific, cultural/historical, aesthetic and/or social/economic value due to human perception or exploitation*" (Reynard and Panizza 2005). The International Association of Geomorphologists (IAG) also launched in 2001 a working group aimed at defining, evaluating, mapping and preserving geomorphosites. Beyond the fact that most geomorphosites pertain to the category of dynamic geosites (Fort et al. 2015), subgroup boundaries cannot be strictly defined because of pervasive interaction between their constituting factors (lithology, tectonics, climate, hydrology, biosphere, human intervention and time).

In contrast to neighbouring countries, where pressures by the scientific community led land-planning authorities to vote protective measures for geomorphosites (e.g., Serrano and Gonzales Trueba 2011), most scientists, not to speak of public administrations, are reluctant to use this very word in Belgium, except in the informal mode. Two attempts to rally geoscientists on supporting geoheritage and geodiversity cases and to put this on the agenda of the regional governments have been raised to no avail, first by Patric Jacobs in 1991 on behalf of the Belgian—Luxembourg Union of Geologists (Jacobs 1991), and later by Roland Dreesen in 2012 on behalf of the IUGS Contact Forum Geoheritage, Geoconservation and Geotourism (Dreesen 2012). However, despite this lack of focused attention to sites of geomorphological interest, current practice distinguishes geomorphological properties among the natural characteristics of a protected site (Gelaudé 2012).

By contrast, historically, the idea of site protection emerged very soon after the independence of Belgium in 1831. Aimed to maintain and protect the architectural heritage, the Royal Commission for Monuments was created by a Royal Decree of 1835. However, the first law on the conservation of the 'beauty' of landscapes was voted only in 1911, with specific purposes such as limiting the impact of mining and quarrying activities on landscapes by tree planting (see Chap. 12) or, conversely, avoiding afforestation of heaths and dunes. This law induced the creation in 1912 of the Site section in the Commission and, from 1931, a new law ruled the conservation of monuments and sites.

An overview of open landscapes, worth of conservation based on their botanical and geological characteristics and vulnerable to loss of value or even utter destruction, was provided by Massart (1912) for the whole of Belgium. Ironically, the first legally protected site was the limestone cliff of Marche-les-Dames, 8 km east of Namur, where King Albert I died after a solo climb on 17 February, 1934, less than 2 months after having signed the classification decree on 30 December, 1933.

24.2 Geomorphosites in Wallonia

In 1980, a constitutional amendment transferred all matters of monument and site heritage to the Communities, the competence being further transferred to the Flemish, Walloon and Brussels Regions in 1989. As an exception, the heritage of the German-speaking part of Wallonia has been managed by the German-speaking Community of Belgium since 1994.

In the Walloon Region, the heritage protection is regulated since 1991 by the Walloon code for land planning, urbanism and heritage, familiarly known as CWATUP. A list of the exceptional immovable heritage of Wallonia was established in 1993, to be revised every third year. In 1999, an innovating decree included the notion of 'landscape interest' in the definition of the heritage. Based on the UNESCO operational guidelines, criteria for a natural site to be listed as a major site of the Walloon Region, also adopted by the Royal Commission for Monuments, Sites, and Archaeological Excavations, include its ecological wealth or scenic quality as a work of nature or joint work of man and nature. Further details are provided in an internal report of 2012: "… a natural site of exceptional scientific, conservation, or scenic value … a natural geological, geomorphological or biological monument, or combining several of these origins, which has an exceptional aesthetic or scientific value …", where the notion of geomorphological origin appears for the first time.

Recently, private initiatives attempted to create a specific label 'Grand site de Wallonie' similar to what exists in France since 1976 to describe sites as, e.g. the Bay of Somme. So far, such a label has no legal existence in Wallonia.

At the global scale, UNESCO ratified the 'Global Geopark' label on 17 November, 2015 that may be applied to "*single, unified geographical areas where sites and landscapes of international geological significance are managed with a holistic concept of protection, education and sustainable development*". Applications have been submitted for two areas of Wallonia. While the Famenne-Ardenne aspiring Geopark, encompassing the limestone area of the Calestienne and centred on the cave system of Han-sur-Lesse (see Chap. 9), is ready to obtain the label, the process started more recently for the Mons basin Geopark, which notably includes the exceptional sites of the Neolithic flint mines of Spiennes and the Malogne phosphate mines (Mortier et al. 2013). We briefly review hereafter the sites with geomorphic interest listed as exceptional heritage, mentioning also other valuable geomorphosites that would deserve inclusion into this list. For each site, municipality and date of classification are indicated in brackets. All cited sites are located in Fig. 24.1.

24.2.1 Underground Geomorphosites

Some underground geomorphosites are man-made mines, among which the phosphate mines of La Malogne (Mons 19-11-1990), the underground marl mines at Folx-les-Caves (Orp-Jauche 08-07-1993), the flint mines of Spiennes (Mons 07-11-1991), also listed as an archaeological site and, since December 2009, as UNESCO World Heritage (Fig. 24.2), and finally the Montagne St-Pierre (Visé 12-10-1981), a natural site listed for its great biodiversity but also including an extended network of underground galleries attesting former exploitation of the chalk (see Chap. 11).

Only three natural caves are recognised as exceptional, namely the Remouchamps cave (Aywaille 31-12-1941 and 11-06-1986) (see Chap. 8), the Lyell and Rosée caves (Engis 08-07-1977) and the Schmerling cave (Flémalle 14-05-1938), which is however listed as an archaeological site, owing to the discovery of two Neandertal skullcaps by Schmerling in 1829 (see Chap. 21). The Remouchamps cave is famous for the 600-m-long navigable section of its underground stream, the Rubicon, collecting the waters of the 'Vallons des Chantoirs' (Sinkhole Dale) (see Chap. 8), and for its many speleothems. Despite remarkable features, such as the Grandchamps sinkhole (see Chap. 8, Fig. 8.2), and renewed proposals since Rahir (1932) suggested the creation of a National Park including a large part of the Amblève valley, the Remouchamps cave and the Sinkhole Dale, the latter is not classified. The Lyell and Rosée caves, with no public access for conservation purposes, contain exceptional speleothems, such as ~2-m-long, ~5-mm-wide tubular stalactites (soda straws) and helictites (Fig. 24.3). In attendance of its classification by the Royal Commission for Monuments and Sites, access to the recently discovered Noû Bleû cave, which also possesses exceptional speleothems (see Chap. 8), is strictly limited, which is facilitated by the location of its entrances on private property. Finally, it is surprising that the Han caves and their geomorphological surroundings have not been listed as an exceptional site so far (see Chap. 9).

24.2.2 Natural Landscapes and Landforms as Geomorphosites

Exceptional natural and scenic sites are mainly rock scarps, rock pinnacles and fluvial landforms.

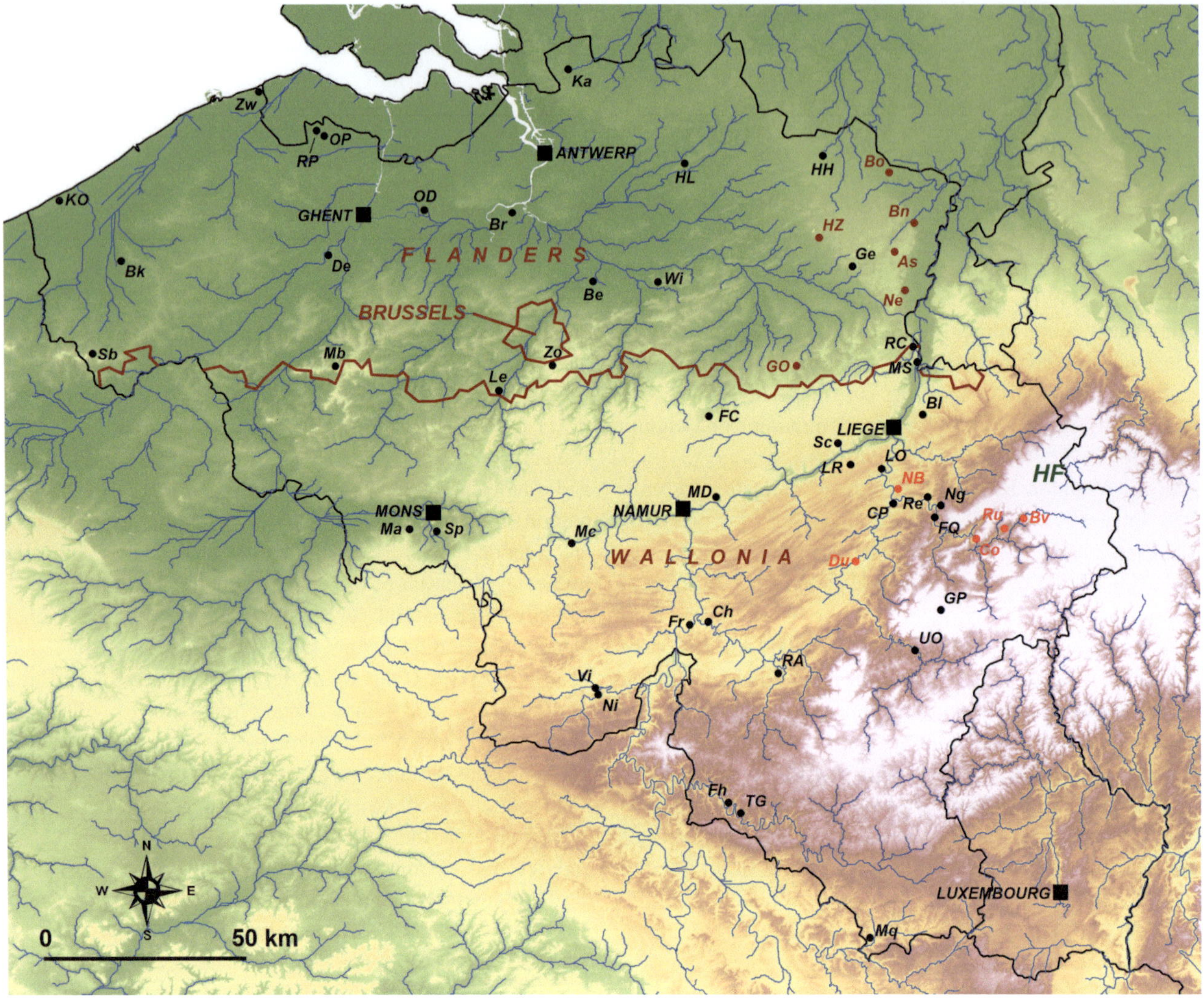

Fig. 24.1 Location map showing all sites mentioned in the text. *As* As railway station. *Be* Berg. *Bk* Blankaart. *Bl* Blégny. *Bn* Bergerven. *Bo* Bocholt plain. *Br* Bornem. *Bv* Bévercé. *Ch* Chaleux. *Co* Coo. *CP* Comblain-au-Pont. *De* Deurle. *Du* Durbuy. *FC* Folx-les-Caves. *Fh* Frahan. *FQ* Fonds de Quarreux. *Fr* Freÿr. *Ge* Genk (De Maten). *GO* Gelinden-Overbroek. *GP* Grand passage. *HF* Hautes Fagnes. *HH* Hoeverbergen-Holven. *HL* Herenthals-Lichtaart. *HZ* Lillo Fault scarp between Houthalen and Zolder. *Ka* Kalmthout. *KO* Koksijde-Oostduinkerke. *Le* Lembeek. *LO* Lower Ourthe. *LR* Lyell and Rosée caves. *Ma* La Malogne. *Mb* Muziekberg. *Mc* Marcinelle. *MD* Marche-les-Dames. *Mq* Montquintin. *MS* Montagne Saint-Pierre/Sint-Pietersberg. *NB* Noû Bleû. *Ne* Neerharerheide. *Ng* Ninglinspo. *Ni* Nismes (Fondry des Chiens). *OD* Overmere-Donk. *OP* Oostpolder. *RA* Ry d'Ave. *RC* Roosburg-Caster. *Re* Remouchamps. *RP* Roeselaerepolder. *Ru* Rustave valley. *Sb* Scherpenberg. *Sc* Schmerling cave. *Sp* Spiennes. *TG* Tombeau du Géant. UO. Upper Ourthe. Vi. Viroinval. Wi. Wijngaardberg. Zo. Zoniënwoud/Forêt de Soignes. *Zw* Zwin. In *black*, sites with a significant geomorphic component listed as exceptional; in *red*, unclassed sites in Wallonia; in *dark green*, unclassed nature reserve in Wallonia; in *brown*, other sites mentioned in Flanders. *Bold red-brown* lines denote regional boundaries

24.2.2.1 Rock Escarpments

Remarkable rock escarpments and pinnacles are found along the incised Ardennian valleys. Listed sites are present in the Meuse valley at Marche-les-Dames (Namur 30-12-1933), with a dolomitic limestone cliff cut in separate walls by joints, and Freÿr (Dinant 08-08-1944 and 03-08-1956), where largely dolomitized Waulsortian reefs display up to 120-m-high vertical rock walls; in the Lesse valley at Chaleux (Houyet 04-04-1939 and 29-10-1981), where limestone dissolution has left rock pinnacles (Fig. 24.4a); in the Ourthe valley near Comblain-au-Pont (five decrees from 18-06-1946 to 26-07-1977), where thick subvertical limestone beds enhanced by differential erosion emerge in the landscape as 'Tartines' (literally, bread loaves) (Macar 1946) (Fig. 24.4b); and in the Viroin valley at the Eau Noire—Eau Blanche confluence (Viroinval 20-10-1947), with the impressive hill formed by a massive Frasnian bioherm of geological, geomorphological and biological interest.

While all the listed rock scarps are thus cut into limestones, it is surprising that no similarly impressive rock walls

Fig. 24.2 Europe's largest and earliest Neolithic mines (4300–2200 BC), the flint mines of Spiennes remarkably illustrate the transition between opencast and underground mining for flint nodules (see location in Fig. 24.1). They are listed as an archaeological site of the Walloon Region and, since December 2000, as a UNESCO World Heritage site (*Photo* Guy Focant © SPW-Patrimoine)

Fig. 24.3 Exceptional up to 2-m-long tubular stalactites in the Rosée Cave near Engis, in the Meuse valley (see location in Fig. 24.1). They justified classifying the cave as exceptional site of Wallonia in 1977 (*Photo* Guy Focant © SPW-Patrimoine)

Fig. 24.4 **a** Rock pinnacles at Chaleux, Houyet, Lesse valley, due to differential dissolution of the limestone (*Photo* Guy Focant © SPW-Patrimoine). **b** Emerging from the background valleyside, subvertical limestone sheets enhanced by differential erosion near Comblain-au-Pont, Ourthe valley (see location of the two sites in Fig. 24.1)

featuring other rock types are included so far, such as the Bévercé kegels (tower karst) in the Permian conglomerates of the Malmédy graben (Ozer 1967, 1969) or numerous cliffs made of Lower Devonian rocks in the Ardenne, most often the product of differential erosion between resistant quartzites, arkoses and conglomerates, and weaker slates and shales (Prick and Ozer 1995; Dejonghe and Jumeau 2007).

The 'Fondry des Chiens' (Nismes/Viroinval 04-09-2002), another exceptional site in the Middle Devonian limestones outcropping in the Nismes area, west of the Meuse, exposes a spectacular, up to 20-m-deep, emptied paleo-cryptokarst (Fig. 24.5). Humid tropical climates of the Miocene triggered active dissolution of the limestones and shaped mega-lapiés under the cover of Oligocene sands. Contemporaneous weathering of these glauconitic sands resulted in iron accumulation within the solution pockets. Later, regional uplift in the Plio-Quaternary and exploitation of the limonite ore by man since prehistoric times exhumed the paleokarst, which now forms a site of high geomorphological and biological (calcareous grassland) value (Ek 1995; Quinif 1993).

24.2.2.2 Structural Landforms

A few structural landforms have also been acknowledged as of exceptional scenic (and, implicitly, geological and geomorphological) value, such as the Ry d'Ave, or Wavreille, anticline (Han-sur-Lesse/Rochefort 26-05-1982), remarkably exposed in a water gap of the superimposed Ry d'Ave (see Chap. 9), and the Montquintin butte (Dampicourt/Rouvroy 22-01-1973), located in front of the Dogger cuesta cut in hard Bajocian limestones in Belgian Lorraine (see Chap. 23). However, and though crowned by the ruins of a mediaeval castle that adds to its scenic attractiveness, the latter site owes its national-scale 'exceptional' character only to its uniqueness as a butte in Belgium. By contrast, other sites of similar value, as the rock wall cut across the Durbuy anticline, are not at all protected.

24.2.2.3 Fluvial Landforms

Other geomorphosites listed as exceptional heritage of Wallonia are mainly related to fluvial erosion, being especially located within incised meandering rivers in the Ardenne. They include elongate incised meanders of the Semois at Frahan (Bouillon 24-04-97) and Botassart, known as the 'Tombeau du Géant' (Bouillon 06-12-1976 and 24-05-1977) (see Chap. 22) (Fig. 24.6) and the Hérou and Cheslé meanders of the Upper Ourthe, cut in Lower Devonian rocks near Nisramont (Houffalize/Laroche 15-10-1937 and 12-11-2003). The Cheslé meander retains traces of a Celtic oppidum.

Developed at the foot of the limestone spur of 'Roche aux Faucons', which houses many karstic features, the remarkable Fèchereux meander of the Lower Ourthe, downstream of Esneux, and its polygenic river terraces make another listed site (Esneux/Neupré, four decrees from 16-03-1936 to 22-05-2015).

Two further sites lie in the Amblève valley of NE Ardenne. The first one corresponds to the boulder-bed sector of the Amblève channel located at Fonds de Quarreux, a site emphatically described by Rahir (1932), where the river cuts across resistant Revinian (Upper Cambrian) quartzites that deliver huge blocks to the valley bottom (Aywaille/Stoumont 06-02-1970 and 23-10-1978) (see Chap. 6). A few kilometres downstream of Quarreux, the

Fig. 24.5 The "Fondry des Chiens" at Nismes, Viroin valley, an emptied up to 20-m-deep mega-lapié developed in the Middle Devonian limestones under the cover of Oligocene sand probably in the Miocene and exhumed as a consequence of regional uplift in the Plio-Pleistocene and limonite ore exploitation (see location in Fig. 24.1). The site was classified as exceptional natural heritage of Wallonia in 2002 (*Photo* Guy Focant © SPW-Patrimoine)

second one encompasses the deeply incised valley of the Ninglinspo, a small tributary of the Amblève whose geomorphic and aesthetic interest lies in the presence of several giant potholes and small cascades distributed along its very steep and narrow channel (Aywaille 08-04-1949). Surprisingly, the Coo waterfall, resulting from a meander cutoff in the Middle Amblève valley, is not classed even at the regional level (see Chap. 6). Finally, enclosed in the already mentioned exceptional site of Freÿr, the deep cut carved by the tiny Colébi creek that enlarged a fracture in the limestones of the steep Meuse valley flank at Falmignoul is also noteworthy.

24.2.2.4 Varia

Located in the Tailles Plateau, the peat bog of the 'Fange du Grand Passage' (Houffalize/Vielsalm 16-03-1979) is not only a listed geomorphosite but also a small national nature reserve. Its geomorphological interest lies in the presence of periglacial ramparted depressions attesting the development of lithalsas during the Younger Dryas (see Chap. 7). However, it is by far not the best site and most representative area for these relics of periglacial landforms, which abound in large tracts of the Hautes Fagnes Plateau, NE Ardenne, with forms up to ~80 m in diameter (see Chap. 7). The Hautes Fagnes Plateau, large parts of which are also nature reserves, is a region of global-scale major geomorphological interest that would deserve recognition as a World Natural Heritage of the UNESCO, not to mention a status of exceptional geomorphosite of the German-speaking Community of Belgium, on which territory it lies.

Finally, owing to its rich coal mining and industrial past, Wallonia also possesses anthropogenic landforms listed as sites of exceptional value, such as large slag heaps at Marcinelle, where the sadly famous Bois du Cazier mining disaster occurred in 1956 (Charleroi 28-05-1990 and 22-08-2011) and Blégny-Mine (Blégny 22-08-2011). Together with Grand-Hornu and Bois-du-Luc, two other well-preserved coal mining sites in the west of the Walloon coal basin, Bois-du-Cazier and Blégny-Mine constitute a UNESCO World Heritage Property listed since 2012. The Royal Commission for Monuments and Sites is also considering whether the Rustave valley, displaying a dense

Fig. 24.6 An elongate incised meander of the Semois at Botassart, near Bouillon, known as the "Tombeau du Géant" (Giant's Grave) and classified as exceptional natural site of Wallonia in 1977) (see location in Fig. 24.1) (*Photo* Guy Focant © SPW-Patrimoine)

network of smaller scale anthropogenic landforms such as field balks and lynchets, which underline ancient field parcel limits near Stavelot, should be classified (Ozer 1967, 1969).

24.3 Geomorphosites in Flanders

Protection of nature and maintenance of biodiversity are the driving forces behind the creation of nature reserves or nature parks. In this sense, with reference to the biome concept, ecosystem diversification would be very limited in Flanders, resulting in rich but uniform ecosystems. However, taking account of geoheritage, i.e. geodiversity and its exploitation by man, results in a great variety of half-natural landscapes. In flat-lying areas like Flanders, long-abandoned traditional land uses and past extraction of mineral resources (peat, clay, sand) have created diverse landscapes, which are at the origin of nature reserves.

The presentation of the most emblematic but also threatened sites (sensu landscapes) by Massart (1912) has maintained its value until today. It included (Fig. 24.1):

- the coastal dunes between Koksijde and Oostduinkerke, currently supplemented by the Westhoek nature reserve (see Chap. 19);
- the Zwin sea breach, actually being restored to its early nineteenth century situation, as was already urged by Massart (see Chap. 19);
- the Blankaart drowned salt marshes at the landward end of the polders;
- the creeks of the Meetjesland, especially the Oostpolderkreek and the Roeselaerekreek;
- the ancient course of the Scheldt near Bornem, the 'wielen' (former dyke breaches) and the tidal-fluvial deposits at the Scheldt—Rupel confluence, currently being restored (see Chaps. 17 and 18);
- the paleomeander of Overmere-Donk, forming a nature reserve and leisure area (Daels 1991);
- river dunes along the Leie meanders at Deurle, built up today, except for a golf course;
- the (former) botanical hotspot in the wet valley of Berg at the transition from the Brabant loam plateau to the Flemish Valley (currently Modeltuinwijk Ter Bronnen

and nature reserve Torfbroek, Kampenhout, close to the urban sprawl and airport of Brussels);
- Wijngaardberg as the most typical ferruginous sandstone hill of the Hageland, currently partly protected with holistic geo-bio-cultural conservation strategy (Abts 2015; see also Chap. 14);
- the remnant hills of Scherpenberg and Muziekberg, also capped by ferruginous sandstones and located in the Flemish Hills and Flemish Ardennes, respectively;
- the Zoniënwoud, or Soignes Forest, near Brussels, preserving natural and traditional anthropogenic landforms (see Chap. 15);
- the drift sand dunes of the Kalmthout nature reserve, being currently incorporated in the aspiring transboundary Delta Geopark, managed by the Brabantse Wal Foundation;
- the Herentals—Lichtaart sand ridge and the parallel Nete depression (Meesters 1992);
- the parabolic sand dunes of Hoeverbergen and Holven in Lommel (Gullentops 1957; Van Uytven 2013; see also Chap. 12);
- different heathlands and the Stiemer valley with sand dunes in Genk (Van Uytven and Dreesen 2014), from which only the Schemmersberg heath and the 'De Maten' nature reserve (Berten 1990) currently remain (Fig. 24.7);
- the (obliterated) quartzite quarry of Rodenem and the diorite quarry of Lembeek in the Senne valley near Halle;
- Roosburg and Caster hill sites in the Mergelland, with their calcareous grasslands and underground quarries (see Chap. 11).

A fundamental requirement for the preservation of valued landscapes is the inventory of preserved traditional landscape elements. In 1995, the Flemish landscape atlas started, mapping relics of traditional landscapes, i.e. areas where historic land use is recognisable today through high density of fairly intact and coherent natural and cultural heritage. This landscape atlas became operational in 2001 (Hofkens and Roosens 2001). Relic zones are widespread, covering structures such as alluvial valleys, dunes, old forests, or rural areas in general. Next to the relic zones, anchor places are identified as intact, representative and/or unique complexes of natural and cultural heritage. Anchor places are considered the most valuable and appealing landscapes in Flanders. Initially, 381 anchor places were designated, covering 16.5% of the Flemish surface (Hofkens and Roosens 2001). Heritage values must be described according to a framework based on natural, historic, sociocultural, aesthetic and spatial criteria. Abiotic factors, such as geology, topography, hydrography are generally listed as background values, not only of scientific value but also contributing to the historical evolution and current aesthetic appreciation of the site (Gelaudé 2012).

The conversion of anchor places into protected sites, named heritage landscapes, has been made possible by a

Fig. 24.7 Ecotope sequence from permanent ponds to drift sands on dunes along the Stiemer River, a small stream draining the Campine Plateau towards SW (Diepenbeek pediment). De Maten nature reserve, Genk (see location in Fig. 24.1) (© 2016 Google)

Fig. 24.8 Surface rupture along the passively activated Lillo Fault in the mining subsidence area between collieries of Houthalen and Zolder, due to differential coal extraction and ground subsidence on both sides of the fault (see location in Fig. 24.1)

Fig. 24.9 Experiencing the Feldbiss Fault in the otherwise flat topography of the Campine, climbing De Warre fault scarp close to the Bergerven nature reserve (Janssen and Dreesen 2010) (see location in Fig. 24.1)

regional decree dated 13-02-2004, completing the regional decree of 16-04-1996 on landscape policy. In this way, the national law on the conservation of monuments and sites (including landscapes) of 07-08-1931 has been adapted to the realities of regionalization. The conservation policy is regulated by the regional decree of 22-06-2012, following the restructuring of the Flemish Agency on the immovable heritage by decree of 10-06-2011. With few exceptions such as the marl quarry of Gelinden—Overbroek (Fig. 24.1), no particular attention is given to geosites or geomorphosites, despite the requirement to describe the physiographic features of protected sites. This is in striking contrast to the attention to archaeological objects. Of over 83,000 heritage objects (constructions and sites) protected so far, 424 have listed geomorphological heritage properties. These do not qualify as geomorphosites but represent a wide variety of mixed natural, cultural and historical objects, including village centres having preserved their geomorphic and hydrographic context, historically stable land use adapted to the local soil type, parklands around castles and abbeys, old forests, heathlands, relatively unspoilt stretches of dune belts and alluvial valleys, abandoned meanders, springs, incised valleys and steep slopes, wetlands such as fens, marshes, peat extraction pits and estuarine floodlands (schorren), remnant hills with a ferruginous sandstone cap (see Chap. 14), few old quarry faces, some closed depressions (see Chap. 15), and even a golf course established in a river dune complex at Sint-Martens-Latem.

The comprehensive inventories of anchor sites and heritage objects provide a fairly complete overview of valuable geomorphic features in Flanders and can be considered as an update of the inventory by Massart (1912), although lacking a uniform assessment of the intrinsic values of geosites and geomorphosites. A thematic geosite inventory according to geological processes as proposed by Kiden et al. (1991) has not been performed. Dreesen et al. (2001) listed 23 geomorphological objects of scientific interest in the province of Belgian Limburg, irrespective of their conservation status. The most important category was formed by aeolian deposits on the Campine Plateau, followed by objects of anthropogenic origin mainly related to coal mining (see Chap. 12) (Fig. 24.8), then broader landscapes deriving their characteristics from natural processes and neotectonic scarps (Janssen and Dreesen 2012; see also Chap. 13) (Fig. 24.9), and more seldom occurrences such as a thermal spring and a calcareous tufa deposit.

Protection of landscapes and sites is mostly inspired by threats to their integrity or even existence and aims to obtain means for restoration and maintenance, which represents a never ending battle (e.g. Daels 1991). Regretfully, few

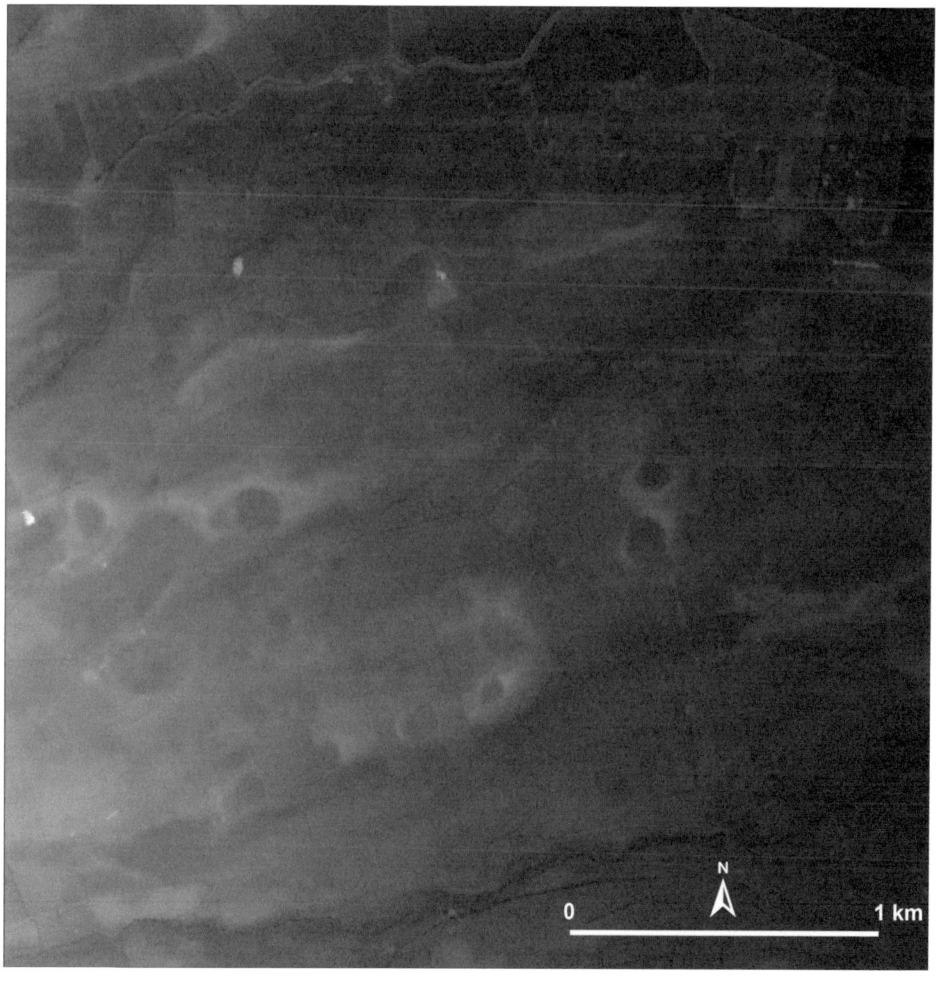

Fig. 24.10 Flat blowouts (1 m average height difference) originating in areas with low sand supply and high water table, preserved only in areas where no tillage was ever practiced in the Bocholt Plain northeast of Bree (see location in Fig. 24.1) (from Dusar et al. 2008)

valuable geo(morpho)sites are preserved in urban settings or in densely populated areas marked by drastic changes of the land surface (Gelaudé 2012). Moreover, ignoring the geological substrate and the dynamic processes involving interaction between the geosphere, the hydrosphere and the biosphere leads to problems, aggravated by climate change. This means that the rare and endangered species for which the nature reserves are created in priority may be highly vulnerable and, as a result, biodiversity gradually loses its most unique representatives. This problematic fact that protecting the living nature is insufficient in sustaining what is rare and exceptional is observed worldwide and it is now realised that it may be due in a significant way to a conceptual flaw, the separation of geo- and biodiversity. Remedial action now appears top down at the international level. The International Union for Conservation of Nature (IUCN) adopted at its World Conservation Congress, held in Barcelona, 5–14 October 2008, a motion emphasising the link between geoheritage and biodiversity and the need to incorporate the conservation of geodiversity and geoheritage in its actions for nature conservation (IUCN 2008). Sadly, the notion that 'geo' is too complex to understand or even that it is detrimental to 'bio' seems to be still in the minds of some conservationists. A striking example is the first protected geosite in Flanders, near the railway station in As, which is now incorporated in a gateway to the National Park Hoge Kempen. When management of this geosite was taken over by the National Park, the explanatory panel on the gravel deposits which are essential to the morphology and functioning of the Campine Plateau disappeared and has been replaced by general information on the social benefits of the national park. New threats also arise from the vigorous restoration of supposedly natural ecosystems, which sometimes leads to sacrifice vulnerable landforms. For instance,

Fig. 24.11 Participants to a geologic/ecological walking trail in the humid heath of Neerharerheide at the southeastern foothill of the Campine Plateau, where the water table shows at land surface due to underlying impermeable Oligocene clays (Van Uytven and Dreesen 2014) (see location in Fig. 24.1) (*Photo* R. Houthuys)

blowouts uniquely preserved in the Bocholt plain (NE Limburg) have been levelled to create new wetlands (Dusar et al. 2008) (Fig. 24.10).

Popular support for protecting geomorphosites seems more readily obtained in already green areas and is linked to the growing success of geotourism. As understood by geoscientists, geotourism is tourism focused on geology and landscapes. However, tourists also appreciate open spaces, greenery, cultural authenticity for outdoor activities, thereby creating new economic opportunities and providing a strong motivation for landscape and site protection, especially along walking or cycling trails or in leisure areas (Vanneste et al. 2016). The geological cycling trails issued by the Province of Limburg with the assistance of Geolim, a working group of amateur geologists, are fine examples of citizen science succeeding in raising interest for geologic and geomorphic features in the landscape, adopting a holistic approach (e.g. Janssen and Dreesen 2010; Van Uytven and Dreesen 2014) (Fig. 24.11).

24.4 Conclusion

The concept of heritage is a cultural construction, a value attributed to natural sites or areas and/or to their historical use. Similarly to biodiversity and cultural heritage, geoheritage must be recognised and promoted for scientific, educational, aesthetic and/or recreational purposes. The necessary public consent to, and support by the authorities for, protection and conservation of geoheritage require popularising geology and geomorphology and making them understandable for lay people. This is particularly true in Flanders, where the geological component of many heritage sites often remains unnoticed. One also needs acknowledge that geodiversity and active geological processes are often vital for maintaining the biodiversity and health of an ecosystem. Sustainable conservation of geoheritage can only succeed when this cooperation with living nature conservationists is functional (Dusar 2012; Dusar and Dreesen 2012).

There is a marked difference in the visibility and acknowledgement of geomorphosites between Flanders and Wallonia, the latter particularly south of the Sambre-et-Meuse axis where the vertical dimension of landscapes becomes more striking. Wallonia is rich in geomorphosites, which diversely combine scientific, educational, aesthetic, scenic and historical values. Only a selection of them have been classified so far as protected exceptional sites by the Royal Commission for Monuments, Sites, and Archaeological Excavations, which is aware of the necessity of extending this protection to other sites of major geomorphic interest.

Flemish geomorphosites tend to be less spectacular, more vulnerable and their geomorphic interest thus less obvious to the public. However, they are as much intricately connected to the evolution of the landscapes they characterise and representative for geological and geomorphological processes. An effective way to enhance their visibility and to guarantee their preservation is their integration in the well-established networks of hiking and cycling trails.

References

Abts H (ed) (2015) De Wijngaardberg in Wezemaal, een erfgoed- en natuurmonument van Europees belang. Regionale Vereniging Natuurpunt Oost-Brabant, Dossier, 24 p

Berten R (1990) Natuur en Flora in Limburg. Uitgave R. Berten, Halveweg 45 - 3600 Genk, 235 p

Daels L (1991) Landschapsbescherming in Vlaanderen. Bulletin de la Société belge de Géologie 100(3/4):307–311

Dejonghe L, Jumeau F (2007) Les plus beaux rochers de Wallonie. Géologie et petite histoire. Service géologique de Belgique, 358 p

Dreesen R (2012) Objective. Definitions. In: Dreesen R (ed) Contactforum – Forum de Contact Geoheritage, Geoconservation and Geotourism, Koninklijke Vlaamse Academie van België voor Wetenschappen en Kunsten, Brussels, Nov 15, 2012. Royal Belgian Institute of Natural Sciences, Geological Survey of Belgium, pp 1–6

Dreesen R, Dusar M, Beerten K (2001) De onderkant van Limburg, verwaarloosd erfgoed. LIKONA, Limburgse Koepel voor Natuurstudie, Jaarboek 2000, pp 19–31

Dusar M (2012) Geoheritage in Belgium: unknown, unloved? In: Dreesen R (ed) Contactforum – Forum de Contact Geoheritage, Geoconservation and Geotourism, Koninklijke Vlaamse Academie van België voor Wetenschappen en Kunsten, Brussels, Nov 15, 2012. Royal Belgian Institute of Natural Sciences, Geological Survey of Belgium, pp 70–83

Dusar M, Dreesen R (2012) Challenges to geoheritage conservation and sustainable development in Belgium. Eur Geol 34:8–11

Dusar M, Matthijs J, Dreesen R, Poesen J (2008) Cirkelvormige depressies in de Vlakte van Bocholt. LIKONA Jaarboek 2007 (17):20–35

Ek C (1995) Grottes et rivières des régions calcaires. In: Demoulin A (ed) L'Ardenne. Essai de géographie physique. Département de Géographie physique, University of Liège, pp 178–193

Ellis N, Bowen D, Campbell S, Knill J, McKirdy A, Prosser C, Vincent M, Wilson R (1996) An introduction to the geological conservation review. Joint Nature Conservation Committee, Peterborough, 131 p

Fort M, Bétard F, Arnaud-Fassetta G (2015) Géomorphologie dynamique et environnement. Collection U., Armand Colin, Paris, 336 p

Gelaudé F (2012) The geology in the natural value of a landscape. In: Dreesen R (ed) Contactforum – Forum de Contact Geoheritage, Geoconservation and Geotourism, Koninklijke Vlaamse Academie van België voor Wetenschappen en Kunsten, Brussels, Nov 15, 2012. Royal Belgian Institute of Natural Sciences, Geological Survey of Belgium, pp 54–61

Gonggrijp GP (2000) Planning and management for geoconservation. In: Barretino D, Wimbledon W, Gallego E (eds) Geological heritage: its conservation and management. Instituto Tecnologico Geominero de Espana, Madrid, pp 29–45

Gray M (2004) Geodiversity: valuing and conserving abiotic nature. Wiley, Chichester 434 p

Gullentops F (1957) Quelques phénomènes géomorphologiques depuis le Pléni-Wurmien. Bulletin de la Société belge de Géologie 66:86–95

Hardin G (1998) Extensions of "The Tragedy of the Commons". Science 280:682–683

Hofkens E, Roosens I (eds) (2001) Nieuwe impulsen voor de landschapszorg. De landschapsatlas, baken voor een verruimd beleid, Ministerie van de Vlaamse Gemeenschap, Afdeling Monumenten en Landschappen, Brussel, 191 p

IUCN (2008) Motion CGR4.MOT055—IUCN on conservation of geodiversity and geological heritage. World Conservation Congress, Barcelona, 5–14 Oct 2008

Jacobs P (1991) Geoconservation. Doel van het symposium - but du symposium. Bulletin de la Société belge de Géologie 100(3/4): 247–248

Janssen J, Dreesen R (2010) Geologische fietsroute. Tussen Kempen and Maas. Provincie Limburg, Provinciaal Natuurcentrum, 75 p

Kiden P, Denys L, Verbruggen C, Paulissen E, Langohr R (1991) Earth science conservation and the Quaternary record in northern and middle Belgium. Bulletin de la Société belge de Géologie 100 (3/4):313–320

Macar P (1946) Principes de géomorphologie normale. Vaillant-Carmanne, Liège, 304 p

Massart J (1912) Pour la Protection de la Nature en Belgique. Société Royale de Botanique de Belgique, Volume Jubilaire Cinquantenaire, 308 p

Meesters L (1992) De heuvelrug tussen Herentals en Kasterlee. Monumenten and Landschappen 11(2):25–40

Mortier T, Baele JM, Blaivie G, Collet H, Cornet C, De Bast E, Dejax J, Delaby S, Dupuis C, Dupont N, Folie A, Godefroit P, Goemaere E, Hennebert M, Kaufmann O, Licour L, Martin T, Papier S, Pirson S, Prestianni C, Quinif Y, Robaszynski F, Smith T, Spagna P, Yans J (2013) Itinéraires pédagogiques du géoparc du Bassin de Mons. ASBL La Malogne, 6 Volumes

Ozer A (1967) Contribution à l'étude géomorphologique des régions où affleure le Poudingue de Malmédy. Unpublished M.Sc. thesis, University of Liège, 189 p

Ozer A (1969) Microrelief et dessin parcellaire. Rideaux et «Ackerberghe» dans la région de Stavelot. Bulletin de la Société géographique de Liège 5:111–120

Panizza M (2001) Geomorphosites: concepts, methods and example of geomorphological survey. Chin Sci Bull 46:4–6

Prick A, Ozer A (1995) Les paysages physiques de l'Ardenne. In: Demoulin A (ed) L'Ardenne. Essai de géographie physique. Département de Géographie physique, University of Liège, pp 1–52

Quinif Y (1993) Le karst d'avant les grottes actuelles: les Abannets. Lapiaz, Numéro spécial 'phénomènes karstiques de la région de Couvin-Nismes' 8:22–23

Rahir F (1932) Pourquoi le Pays de l'Amblève doit-il devenir Parc National. Les amis de l'Amblève, Bruxelles, 71 p

Reynard E, Panizza M (2005) Géomorphosites: définition, évaluation et cartographie. Géomorphologie 11/3:177–180

Serrano E, Gonzales Trueba JJ (2011) Environmental education and landscape leisure. Geotourist map and geomorphosites in the Picos de Europa National Park. GeoJ Tourism Geosites IV/2(8):295–308

Van Uytven D (2013) Geomorfologie van Bosland. LIKONA Jaarboek 2012 (22):16–21

Van Uytven D, Dreesen R (2014) Geologische fietsroute Hoge Kempen. Provincie Limburg, Provinciaal Natuurcentrum, 144 p

Vanneste D, Vandeputte C, Poesen J (2016) Geotourism and the underestimated potential of 'ordinary' landscapes. The Belgian case. Proceedings of TCL2016 Conference, INFOTA 2016, pp 587–598

MIX
Papier aus verantwortungsvollen Quellen
Paper from responsible sources
FSC® C105338

If you have any concerns about our products,
you can contact us on
ProductSafety@springernature.com

In case Publisher is established outside the EU,
the EU authorized representative is:
Springer Nature Customer Service Center GmbH
Europaplatz 3, 69115 Heidelberg, Germany

Printed by Libri Plureos GmbH
in Hamburg, Germany